Scientific and Common Names

of 7,000 Vascular Plants in the United States

Lois Brako ❖ Amy Y. Rossman ❖ David F. Farr

APS PRESS
The American Phytopathological Society
St. Paul, Minnesota USA

Authors:

Lois Brako
Missouri Botanical Gardens
and University of Missouri, St. Louis

Amy Y. Rossman
David F. Farr
Systematic Botany and Mycology Laboratory
Beltsville Agricultural Research Center
USDA-Agriculture Research Service
Beltsville, Maryland

This publication is No. 7 in the series *Contributions from the U.S. National Fungus Collections*

This book has been reproduced directly from computer-generated copy submitted in final form to APS Press by the authors. No editing or proofreading has been done by the Press.

Reference in this publication to a trademark, proprietary product, or company name by personnel of the U.S. Department of Agriculture or anyone else is intended for explicit description only and does not imply approval or recommendation to the exclusion of others that may be suitable.

Library of Congress Catalog Card Number: 94-79381
International Standard Book Number: 0-89054-171-X

Published 1995 by The American Phytopathological Society

Printed in the United States of America on acid-free paper

The American Phytopathological Society
3340 Pilot Knob Road
St. Paul, Minnesota 55121-2097, USA

Contents

Scientific and Common Names

of 7,000 Vascular Plants in the United States

INTRODUCTION

Scientific and Common Names of 7,000 Vascular Plants in the United States is a reference to the scientific and common names of vascular plants in the United States. Common or vernacular names are the names often used in conversation by both lay-people and scientists. Unlike the scientific or Latin names of plants that are governed by the *International Code of Botanical Nomenclature* (Greuter, W. et al. 1988. Koeltz Scientific Books, Königstein, Germany), no rules exist for recognizing one accepted common name. The common name of a plant often varies over time and by region within the United States. Each of the plants listed in this reference has at least one and in some cases up to thirteen common names. Common names are not considered reliable for use in scientific publications in which research results must be accurately communicated, yet common names of plants are frequently encountered by scientists in their work. Thus, there is a need to relate the scientific names of vascular plants with their common names.

This book consists of four sections. The first section is a list of scientific names in alphabetical order by genus. As in the following example, generic names are listed with the author followed by the plant family in parentheses. Common names pertaining to that genus, if any, are listed on the next line. Species names are presented with the author and one or more common names in alphabetical order.

> *Abies* Mill. (Poaceae)
>   fir
>   *alba* Mill. - silver fir
>   *amabilis* Douglas ex Forbes - Cascade fir,
>   Pacific silver fir, white fir

The second section is a list of common names followed by the scientific name. Each common name is indexed on every word of that name. In the example cited above, the scientific name *Abies amabilis* is indexed under "fir" listing "Cascade fir", "Pacific silver fir", and "white fir" as well as under the words "Cascade," "Pacific," "silver," and "white." Thus, with the exception of a few words such as "not" and "like", any word in a common name can be used to locate the scientific name(s) associated with that common name. The third section is a list of important synonyms of the scientific name in regular typeface cross referenced to the accepted scientific name in italics. The fourth section is a list of the genera of included vascular plants arranged by plant family.

Scientific and Common Names of 7,000 Vascular Plants in the United States includes a majority of the plants of interest to plant pathologists and other agricultural scientists. It is based on the vascular plants on which fungi were reported in Farr et al. (1989. *Fungi on Plants and Plant Products in the United States*. APS Press, St. Paul, Minnesota). Additional names were obtained from selected references as follows:

1. Elias, T. 1987. *The Complete Trees of North America. Field Guide and Natural History.* Gramercy, New York, New York.
2. Dennis, L. J. 1980. *Gilkey's Weeds of the Pacific Northwest.* Oregon State University Press, Corvallis, Oregon.
3. Gleason, H. A. and A. Cronquist 1991. *Manual of Vascular Plants of Northeastern United States and Adjacent Canada*, Second edition. The New York Botanical Garden, Bronx, New York.
4. Muenscher, W. C. 1980. *Weeds.* Second edition. Cornell University, Ithaca, New York.
5. Patterson, D. T., ed. 1989. *Composite List of Weeds.* Weed Science Society of America, Champaigne, Illinois.
6. Peterson, R. T. and M. McKenny 1986. *A Field Guide to Wildflowers. Northeastern and Northcentral North America.* Houghton Mifflin, Boston, Massachusetts.

All scientific names of vascular plants were verified for accuracy against one or more recent authoritative references. The scientific names were reviewed by comparing the names electronically with three vascular plant databases:

1. Terrell E. E., S. R. Hill, J. H. Wiersema, and W. E. Rice. 1986. *A Checklist of Names for 3,000 Vascular Plants of Economic Importance.* USDA-Agric. Handbook 505.
2. Germplasm Resources Information Network (GRIN). USDA, Agriculture Research Service, Beltsville, Maryland.
3. Wiersema, J. H., J. H. Kirkbride, and C. R. Gunn. 1990. *Legume (Fabaceae) Nomenclature in the USDA Germplasm System.* USDA Tech. Bull. 1757.

Recent monographic and floristic treatments and reference resources were consulted to resolve inconsistencies between names in these databases and to verify names that were not in any of the databases.

These are listed below:

1. Anonymous 1991. *PLANTS: Plant List of Accepted Nomenclature, Taxonomy, and Symbols. Alphabetical Listing Report.* U.S. Department of Agriculture, Soil Conservation Service.
2. Bailey, L. H. and E. Z. Bailey 1976. *Hortus Third. A Concise Dictionary of Plants Cultivated in the United States and Canada.* Macmillan Publishing Co., New York, New York.
3. Gleason, H. A. and A. Cronquist 1991. *Manual of Vascular Plants of Northeastern United States and Adjacent Canada*, Second edition. New York Botanical Garden, Bronx, New York.
4. Hickman, J. D. (ed.) 1993. *The Jepson Manual. Higher Plants of California.* University of California, Berkeley, California.
5. Huxley, A., M. Griffiths, and M. Levy 1992. *The New Royal Horticultural Society Dictionary of Gardening.* Macmillan, London. 4 vols.
6. Kartesz, J.T. 1994. *A Synonymized Checklist of the Vascular Flora of the United States, Canada and Greenland.* Second edition. vol. 1. Checklist. vol. 2 Thesaurus. Timber Press, Portland, Oregon.
7. Lellinger, D. B. 1985. *A Field Manual of the Ferns and Fern Allies of the United States and Canada.* Washington, DC.
8. Royal Botanic Garden, Kew. 1993. *Index Kewensis on Compact Disk.* Oxford University Press, Oxford.

Genera are recognized and listed according to Gunn, et al. (1992. *Families and Genera of Spermatophytes Recognized by the Agricultural Research Service.* USDA Tech. Bull. 1796). Author names were standardized according to Brummitt and Powell (1992. *Authors of Plant Names.* Royal Botanic Gardens, Kew).

Spelling, capitalization, and punctuation of common names were standardized using the following guidelines. The use of the word "common" was considered redundant and eliminated from the list. Names were checked in *The Plant Book. A Portable Dictionary of the Higher Plants.* (Mabberly, D. J. 1987. Cambridge University Press, Cambridge) for consensus on the application of common names to specific genera, such as "apple" for the genus *Malus.* Any other application using the common name "apple" is hyphenated, such as "wood-apple" for *Limonia acidissima.* Exceptions to this hyphenation rule include geographical or habitat names such as "Mexican apple" and "meadow anemone," and proper names that are not fanciful such as "Apache pine." Also, the term "false" or "wild" before a common name is not hyphenated. All fanciful names such as "kiss-me-over-the-garden-gate" are hyphenated. For suffixes such as -berry, -flower, -grass, -leaf, -root, -weed, -wood, -wort, and numerous others, the *American Heritage Standard Dictionary of the English Language* (William Morris, ed., 1981, Houghton Mifflin Company, Boston) was consulted. If the name was used in the dictionary as one word such as "bluegrass", the spelling was accepted for this word. If the name was two separate words or not present in the dictionary, a hyphen was used. Exceptions to this format are common names using the word "grass" which were not hyphenated unless they were members of families other than the grass family (Poaceae). Capitalization follows the *American Heritage Standard Dictionary.* For common names referring to animals, we followed guidelines set forth in Kartesz and Thieret (1991. Common names for vascular plants: guidelines for use and application. *Sida* 14:421-434) and Rickett (1965. The English names of plants. *Bull. Torrey Botanical Club* 92:137-139).

Vascular Plant Names

Abelia R. Br. (Caprifoliaceae)
×grandiflora (André) Rehder - glossy abelia
Note = *A. chinensis* × *A. uniflora.*

Abelmoschus Medik. (Malvaceae)
esculentus (L.) Moench - gumbo, lady's-finger, okra

Abies Mill. (Pinaceae)
fir
alba Mill. - silver fir
amabilis Douglas ex Forbes - Cascade fir, Pacific silver fir, silver fir, white fir
balsamea (L.) Mill. - balsam fir, fir-balsam
bracteata (D. Don) Nutt. - bristle-cone fir, Santa Lucia fir
cephalonica Loudon - Greek fir
concolor (Gordon & Glend.) Lindl. - white fir
fraseri (Pursh) Poir. - Fraser's balsam fir, Fraser's fir, she-balsam
grandis (Douglas) Lindl. - balsam fir, grand fir, lowland fir, lowland white fir, silver fir
lasiocarpa (Hook.) Nutt. - alpine fir, subalpine fir
lasiocarpa (Hook.) Nutt. var. *arizonica* (Merriam) Lemmon - cork fir, cork-bark fir
magnifica A. Murray bis - California red fir, red fir
procera Rehder - noble fir

Abronia Juss. (Nyctaginaceae)
sand-verbena
latifolia Eschsch. - yellow sand-verbena

Abrus Adans. (Fabaceae)
precatorius L. - coral-bead-plant, crab's-eye, Indian licorice, jequirity bean, licorice-vine, love pea, prayer-beads, precatory bean, red-bead-plant, rosary pea, weather-plant, weather-vine, wild licorice

Abutilon Mill. (Malvaceae)
flowering-maple, Indian mallow, parlor-maple
incanum (Link) Sweet - Indian mallow
pictum (Gillies ex Hook. & Arn.) Walp. - flowering-maple
theophrasti Medik. - butter-print, butter-weed, China jute, cotton-weed, Indian hemp, Indian mallow, piemacker, velvet-leaf, velvet-weed

Acacia Mill. (Fabaceae)
angustissima (Mill.) Kuntze - prairie acacia
berlandieri Benth. - Berlandier's acacia, guajillo
constricta Benth. - mescat acacia, white-thorn acacia
cyclopis A. Cunn. - coastal wattle, cultivated acacia
farnesiana (L.) Willd. - cassie, huisache, popinac, sponge-tree, sweet acacia, West Indian black-thorn
greggii A. Gray - cat-claw acacia, Gregg's cat's-claw, long-flowered cat's-claw, Texas mimosa
koa A. Gray - koa
melanoxylon R. Br. - Australian black-wood, black-wood, black-wood acacia
rigidula Benth. - black-brush acacia
tortuosa (L.) Willd. - cat's-claw, twisted acacia

Acalypha L. (Euphorbiaceae)
gracilens A. Gray - short-stalk copper-leaf, slender copper-leaf
ostryifolia Riddell - hop-hornbeam copper-leaf, rough-pod copper-leaf
rhomboidea Raf. - rhombic copper-leaf
virginica L. - mercury-weed, three-seeded-mercury, Virginia copper-leaf, wax-balls
wilkesiana Müll. Arg. - beefsteak-plant, fire-dragon, Jacob's-coat, painted copper-leaf

Acanthospermum Schrank (Asteraceae)
australe (Loefl.) Kuntze - Paraguay bur, Paraguay star-bur
hispidum DC. - bristly star-bur

Acca O. Berg (Myrtaceae)
sellowiana (O. Berg) Burret - feijoa, pineapple-guava

Acer L. (Aceraceae)
maple
barbatum Michx. - Florida maple, southern sugar maple, sugar-tree
campestre L. - field maple, hedge maple
circinatum Pursh - vine maple
ginnala Maxim. - Amur maple
glabrum Torr. - dwarf maple, Rocky Mountain maple
glabrum Torr. subsp. *douglasii* (Hook.) Wesm. - Douglas's maple, mountain maple
grandidentatum Nutt. - big-tooth maple, sugar maple
leucoderme Small - chalk maple, white-bark maple
macrophyllum Pursh - big-leaf maple, canyon maple, Oregon maple
mono Maxim. - mono maple
negundo L. - ash-leaf maple, box elder
negundo L. subsp. *californicum* (Torr. & A. Gray) Wesm. - California box elder
nigrum F. Michx. - black maple, black sugar maple, rock maple
palmatum Thunb. - Japanese maple
pensylvanicum L. - moosewood, Pennsylvania maple, striped maple, whistle-wood
platanoides L. - Norway maple
pseudoplatanus L. - mock-plane, sycamore maple
rubrum L. - red maple, scarlet maple, soft maple, swamp maple
rubrum L. var. *trilobum* Torr. & A. Gray ex K. Koch - trident maple
saccharinum L. - river maple, silver maple, soft maple, white maple
saccharum Marsh. - hard maple, rock maple, sugar maple
spicatum Lam. - moose maple, mountain maple
truncatum Bunge - Shantung maple

Achillea L. (Asteraceae)
yarrow
millefolium L. - bloodwort, milfoil, nose-bleed, sanguinary, thousand-leaf, thousand-seal, western yarrow, yarrow
ptarmica L. - sneezeweed, sneezeweed yarrow, sneezewort, sneezewort yarrow

Achlys DC. (Berberidaceae)
triphylla (Sm.) DC. - deer's-foot, vanilla-leaf

Acinos Mill. (Lamiaceae)
arvensis (L.) Moench - alpine calamint, mother-of-thyme

Acoelorrhaphe H. Wendl. (Arecaceae)
wrightii (Griseb. & H. Wendl. ex Griseb.) H. Wendl. ex Becc. - Everglades palm, Paurotis's palm, saw cabbage palm, silver saw palm

Acokanthera G. Don (Apocynaceae)
bushman's-poison, poison-bush, poison-tree
oppositifolia (Lam.) Codd - bushman's-poison

Aconitum L. (Ranunculaceae)
aconite, monk's-hood, wolf-bane
carmichaelii Debeaux - azure monk's-hood
napellus L. - aconite, bear's-foot, friar's-cap, garden monk's-hood, garden wolf-bane, helmut-flower, soldier's-cap, Turk's-cap
reclinatum A. Gray - trailing wolf-bane
uncinatum L. - southern monk's-hood, wild monk's-hood

Acorus L. (Acoraceae)
calamus L. - calamus, flag-root, sweet-flag

Acroptilon Cass. (Asteraceae)
repens (L.) DC. - Russian knapweed

Acrostichum L. (Pteridophyta)
swamp fern
aureum L. - leather fern

Actaea L. (Ranunculaceae)
baneberry, cohosh, necklace-weed
alba (L.) Mill. - doll's-eyes, white baneberry, white cohosh
rubra (Aiton) Willd. - baneberry, red baneberry, snakeberry

Actinidia Lindl. (Actinidiaceae)
deliciosa (A. Chev.) C. S. Liang & A. R. Ferguson - Chinese gooseberry, kiwi, kiwi-fruit

Adenium Roem. & Schult. (Apocynaceae)
obesum (Forssk.) Roem. & Schult. - desert-rose

Adenocaulon Hook. (Asteraceae)
bicolor Hook. - pathfinder, trail-plant

Adenostoma Hook. & Arn. (Rosaceae)
fasciculatum Hook. & Arn. - chamise, greasewood

Adiantum L. (Pteridophyta)
maidenhair fern
capillaris-veneris L. - Venus's-hair fern
pedatum L. - American maidenhair fern, five-finger fern, northern maidenhair fern
raddianum C. Presl cv.'decorum' - delta maidenhair fern
tenerum Sw. - brittle maidenhair fern, fan maidenhair fern

Adlumia Raf. ex DC. (Papaveraceae)
fungosa (Aiton) Britton, et al. - Allegheny-vine, climbing fumitory

Adonis L. (Ranunculaceae)
annua L. - bird's-eye, fall adonis, pheasant's-eye, pheasant-eye adonis
vernalis L. - spring adonis

Adoxa L. (Adoxaceae)
moschatellina L. - moschatel, musk-root, townhall's-clock

Aechmea Ruiz & Pav. (Bromeliaceae)
air-pine, living-vase
fasciata (Lindl.) Baker - urn-plant
orlandiana L. B. Sm. - finger-of-God

Aegilops L. (Poaceae)
goat grass
cylindrica Host - jointed goat grass
triuncialis L. - barbed goat grass

Aegopodium L. (Apiaceae)
podagraria L. - bishop's gout-weed, gout-weed

Aeschynanthus Jack (Gesneriaceae)
pulcher (Blume) G. Don - climbing-beauty, lipstick-plant, lipstick-vine, pipe-plant, red bugle-vine, royal-red-bugler, scarlet basket-vine

Aeschynomene L. (Fabaceae)
americana L. - American joint-vetch, joint-vetch
indica L. - Indian joint-vetch, kat-sola
virginica (L.) Britton, et al. - curly-indigo, northern joint-vetch

Aesculus L. (Hippocastanaceae)
buckeye, horse-chestnut
californica (Spach) Nutt. - California buckeye, California horse-chestnut
×*carnea* Hayne - red horse-chestnut
Note = *A. hippocastanum* × *A. pavia*.
flava Sol. - sweet buckeye, yellow buckeye
glabra Willd. - fetid buckeye , Ohio buckeye
glabra Willd. var. *arguta* (Buckley) B. L. Rob. - Texas buckeye
hippocastanum L. - European horse-chestnut, horse-chestnut
parviflora Walter - bottlebrush buckeye, dwarf horse-chestnut
pavia L. - red buckeye
sylvatica W. Bartram - dwarf buckeye, Georgia buckeye, painted buckeye
turbinata Blume - Japanese horse-chestnut

Aethusa L. (Apiaceae)
cynapium L. - fool's-parsley

Agalinis Raf. (Scrophulariaceae)
purpurea (L.) Pennell - gerardia

Agapanthus L'Hér. (Liliaceae)
africanus (L.) Hoffmanns. - African lily, blue agapanthus, lily-of-the-Nile

Agastache Clayton ex Gronov. (Lamiaceae)
giant hyssop
foeniculum (Pursh) Kuntze - anise-hyssop, blue giant hyssop, fennel giant hyssop, fragrant giant hyssop, lavender giant hyssop
nepetoides (L.) Kuntze - catnip, catnip giant hyssop, giant hyssop, yellow giant hyssop
scrophulariaefolia (Willd.) Kuntze - purple giant hyssop
urticifolia (Benth.) Kuntze - nettle-leaf giant hyssop

Agave L. (Agavaceae)
americana L. - American century-plant, century-plant, maguey
parryi Engelm. - mescal

sisalana Perrine - hemp-plant, sisal, sisal-hemp

Ageratina Spach (Asteraceae)
adenophora (Spreng.) R. M. King & H. Rob. - crofton-weed
altissima (L.) R. M. King & H. Rob. - deer-wort, Indian sanicle, richweed, squaw-weed, white sanicle, white snakeroot

Ageratum L. (Asteraceae)
floss-flower, pussy's-foot
conyzoides L. - tropic ageratum, white-weed
houstonianum Mill. - ageratum

Aglaomorpha Schott (Pteridophyta)
meyeniana Schott - bear's-paw fern

Aglaonema Schott (Araceae)
modestum Schott ex Engl. - Chinese evergreen
simplex Blume - Chinese evergreen

Agrimonia L. (Rosaceae)
agrimony, cocklebur, harvest-lice
eupatoria L. - agrimony, medicinal agrimony
gryposepala Wallr. - agrimony, beggar-ticks, feverfew, stickweed
striata Michx. - roadside agrimony

Agropyron Gaertn. (Poaceae)
wheat grass
cristatum (L.) Gaertn. - crested wheat grass, fairway crested wheat grass
desertorum (Fisch. ex Link) Schult. - desert crested wheat grass, standard crested wheat grass
fragile (Roth) Candargy subsp. *sibiricum* Willd. - Siberian wheat grass

Agrostemma L. (Caryophyllaceae)
githago L. - corn cockle, corn-campion, corn-mullein, corn-rose, crown-of-the-field, old-maid's-pink

Agrostis L. (Poaceae)
bent grass
canina L. - brown bent grass, velvet bent grass
capillaris L. - bent grass, brown-top, colonial bent grass, Rhode Island bent grass
exarata Trin. - spike redtop
gigantea Roth - black bent grass, redtop
hyemalis (Walter) Britton, et al. - hair grass, tickle grass, winter bent grass
idahoensis Nash - Idaho redtop
oregonensis Vasey - Oregon bent grass
pallens Trin. - dune bent grass, thin grass
perennans (Walter) Tuck. - autumn bent grass, brown bent grass, upland bent grass
rossiae Vasey - Ross's redtop
scabra Willd. - fly-away grass, hair grass, rough bent grass, tickle grass
stolonifera L. - creeping bent grass
thurberiana Hitchc. - Thurber's redtop

Ailanthus Desf. (Simaroubaceae)
altissima (Mill.) Swingle - ailanthus, copa-tree, tree-of-heaven, varnish-tree

Aira L. (Poaceae)
caryophyllea L. - silver hair grass

Ajuga L. (Lamiaceae)
chamaepitys (L.) Schreb. - yellow bugleweed
genevensis L. - erect bugle, Geneva bugle, standing bugle
reptans L. - bugle, carpet bugle, carpet bugleweed, creeping bugleweed

Akebia Decne. (Lardizabalaceae)
quinata Houtt. - chocolate-vine, five-leaf akebia

Albizia Durazz. (Fabaceae)
julibrissin Durazz. - mimosa, mimosa-tree, silk-tree, silk-tree albizia
lebbeck (L.) Benth. - fry-wood, lebbeck, lebbek-tree, siris-tree
saman (Jacq.) F. Muell. - rain-tree, saman

Alcea L. (Malvaceae)
ficifolia (L.) Cav. - Antwerp hollyhock
rosea L. - hollyhock

Alchemilla L. (Rosaceae)
lady's-mantle
alpina L. - alpine lady's-mantle
arvensis (L.) Scop. - parsley-piert
microcarpa Boiss. & Reut. - parsley-piert, slender parsley-piert
xanthochlora Rothm. - lady's-mantle

Aletris L. (Liliaceae)
colic-root, star grass
aurea Walter - yellow colic-root
farinosa L. - ague-root, crow-corn, unicorn-root

Aleurites J. R. Forst. & G. Forst. (Euphorbiaceae)
moluccana (L.) Willd. - candleberry, candleberry-tree, candlenut, candlenut-tree, country-walnut, Indian walnut, Otaheite-walnut, varnish-tree

Alhagi Gagnebin (Fabaceae)
maurorum Medik. - camel-thorn
pseudalhagi (M. Bieb.) Desv. - camel-thorn

Alisma L. (Alismataceae)
water-plantain
gramineum Lej. - narrow-leaf water-plantain
lanceolatum With. - lance-leaf water-plantain
plantago-aquatica L. - large-flowered water-plantain, water-plantain

Allamanda L. (Apocynaceae)
cathartica L. - allamanda, golden-trumpet
schottii Pohl - bush allamanda

Alliaria Heist. ex Fabr. (Brassicaceae)
petiolata (M. Bieb.) Cavara & Grande - garlic mustard

Allionia L. (Nyctaginaceae)
incarnata L. - trailing four-o'clock, windmills

Allium L. (Liliaceae)
onion
ampeloprasum L. - elephant garlic, great-headed garlic
amplectens Torr. - wild onion
ascalonicum L. - shallot
canadense L. - meadow garlic, meadow leek, rose leek, wild garlic, wild onion
cepa L. - Egyptian onion, eschalot, multiplier onion, onion, potato onion, shallot, tree onion
cepa L. var. *viviparum* (Metzg.) Alef. - top onion

cernuum Roth - lady's leek, nodding onion, nodding wild onion, wild onion

drummondii Regel - wild onion

fistulosum L. - bunching onion, ciboule, Japanese bunching onion, Spanish onion, spring onion, stone leek, two-bladed onion, Welsh onion

porrum L. - leek, purret

sativum L. - garlic

schoenoprasum L. - chives, ezo-negi, schnittlauch

stellatum Ker Gawl. - prairie onion, wild onion

textile A. Nelson & J. F. Macbr. - wild onion

tricoccum Aiton - ramp, wild leek

tuberosum Rottler ex Spreng. - Chinese chives, kui ts'ai, nira, Oriental garlic

validum S. Watson - swamp onion

vineale L. - crow garlic, field garlic, stag's garlic, wild garlic, wild onion

Alnus Mill. (Betulaceae)
alder

glutinosa (L.) Gaertn. - black alder, European alder

incana (L.) Moench - gray alder, speckled alder, white alder

japonica (Thunb.) Steud. - Japanese alder

maritima Muhl. ex Nutt. - brook alder, seaside alder

oblongifolia Torr. - Arizona alder, New Mexico alder

rhombifolia Nutt. - sierra alder, white alder

rubra Bong. - Oregon alder, red alder

rugosa (Du Roi) Spreng. - hazel alder, smooth alder, speckled alder

serrulata (Aiton) Willd. - hazel alder, smooth alder

sinuata (Regel) Rydb. - green alder, Sitka alder

tenuifolia Nutt. - mountain alder, thin-leaf alder

viridis (Chaix) DC. - European green alder

viridis (Chaix) DC. subsp. *crispa* (Aiton) Turrill - American green alder, green alder, mountain alder

Alocasia (Schott) G. Don (Araceae)
elephant-ear-plant

cucullata (Lour.) G. Don - Chinese taro

Aloe L. (Aloaceae)

variegata L. - kanniedood aloe, partridge-breast, pheasant's-wings, tiger aloe

vera (L.) Burm. f. - Barbados aloe, bitter aloe, Curaçao aloe, medicinal aloe, true aloe, Unguentine-cactus

Alopecurus L. (Poaceae)

aequalis Sobol. - short-awn foxtail

alpinus Sm. - alpine foxtail

arundinaceus Poir. - creeping foxtail, reed foxtail

carolinianus Walter - Carolina foxtail

geniculatus L. - bent foxtail, floating foxtail, marsh foxtail, water foxtail, water timothy

myosuroides Huds. - black grass, mouse foxtail, slim-spike foxtail

pratensis L. - meadow foxtail

rendlei Eig - Rendle's foxtail

Aloysia Juss. (Verbenaceae)

gratissima (Gillies & Hook.) Tronc. - Texas white-brush, white-brush

Alpinia Roxb. (Zingiberaceae)
ginger-lily

purpurata (Vieill.) K. Schum. - red-ginger

zerumbet (Pers.) B. L. Burtt & R. M. Sm. - shell-ginger, shellflower

Alstonia R. Br. (Apocynaceae)

macrophylla Wall. ex G. Don - batino

Alternanthera Forssk. (Amaranthaceae)

bettzichiana (Regel) G. Nicholson - calico-plant

caracasana Kunth - khaki-weed, mat chaff-flower

ficoidea (L.) R. Br. var. *amoena* (Lem.) L. B. Sm. & Downs - joy-weed, parrot-leaf, shoofly

philoxeroides (Mart.) Griseb. - alligator-weed, periquito-saracura

sessilis (L.) R. Br. - sessile joy-weed

Althaea L. (Malvaceae)

cannabina L. - hemp-leaved mallow

officinalis L. - marsh mallow

Alysicarpus Neck. ex Desv. (Fabaceae)

vaginalis (L.) DC. - Alyce clover, one-leaf-clover

Alyssum L. (Brassicaceae)
madwort

alyssoides (L.) L. - yellow alyssum

desertorum Stapf - dwarf alyssum

minus (L.) Rothm. - field alyssum

saxatile L. - basket-of-gold, gold-dust, golden-tuft alyssum, golden-tuft madwort, rock madwort

Amaranthus L. (Amaranthaceae)
amaranth

albus L. - prostrate amaranth, tumble pigweed, tumbleweed, tumbling pigweed, white pigweed

arenicola I. M. Johnst. - sand-hills amaranth

australis (A. Gray) J. D. Sauer - giant amaranth

bigelovii Uline & Bray - Torrey's amaranth

blitoides S. Watson - prostrate amaranth, prostrate pigweed

caudatus L. - Inca wheat, love-lies-bleeding, quilete, tassel-flower

cruentus L. - purple amaranth

fimbriatus (Torr.) Benth. - fringed pigweed

graecizans L. - mat amaranth, prostrate amaranth, prostrate pigweed, spreading pigweed, tumble pigweed, tumbleweed

hybridus L. - amaranth pigweed, green amaranth, pigweed, red amaranthus, rough pigweed, smooth pigweed, spleen amaranthus, wild beet

lividus Willd. - livid amaranth

palmeri S. Watson - Palmer's amaranth

powellii S. Watson - Powell's amaranth

retroflexus L. - amaranth pigweed, careless-weed, green amaranth, pigweed, red-root, red-root pigweed, rough pigweed, wild beet

rudis Sauer - water-hemp

spinosus L. - soldier-weed, spiny amaranth, spiny pigweed, thorny amaranth

tricolor L. - Chinese amaranth, Ganges amaranth, Joseph's-coat, tampala

tuberculatus (Moq.) Sauer - tall water-hemp

viridis (L.) Britton, et al. - green amaranth, pakai, slender amaranth

Amaryllis L. (Liliaceae)

belladonna L. - belladonna-lily, Cape belladonna, naked-lady-lily

Ambrosia L. (Asteraceae)
ragweed
acanthicarpa Hook. - annual bur-sage, annual burweed, black-weed, hay-fever-weed
artemisiifolia L. - bitter-weed, hogweed, mayweed, ragweed, Roman wormwood, wild tansy
bidentata Michx. - lance-leaf ragweed
confertiflora DC. - poverty-weed, slim-leaf bur-sage
grayi (A. Nelson) Shinners - woolly-leaf bur-sage, woolly-leaf poverty-weed
psilostachya DC. - naked-spiked ragweed, perennial ragweed, western ragweed
psilostachya DC. var. *coronopifolia* (Torr. & A. Gray) Farw. - perennial ragweed, western ragweed
tenuifolia Spreng. - false ragweed
tomentosa Nutt. - bur ragweed, silver-leaf poverty-weed, skeleton-leaf bur, skeleton-leaf bur-sage, white-leaf franseria, white-weed
trifida L. - bitter-weed, buffalo-weed, crown-weed, giant ragweed, great ragweed, horse-weed, king's-head, tall ambrosia, wild hemp
trifida L. var. *texana* Scheele - blood ragweed

Amelanchier Medik. (Rosaceae)
Juneberry, serviceberry, shad, shadbush, sugarplum
alnifolia (Nutt.) Nutt. - Saskatoon serviceberry, western serviceberry
alnifolia (Nutt.) Nutt. var. *cusickii* (Fernald) C. L. Hitchc. - Cusick's serviceberry
alnifolia (Nutt.) Nutt. var. *pumila* (Nutt.) Nelson - cluster serviceberry
arborea (F. Michx.) Fernald - downy serviceberry
bartramiana (Tausch) Roem. - mountain Juneberry, mountain serviceberry
canadensis (L.) Medik. - downy serviceberry, eastern serviceberry, shadbush
interior Nielsen - inland serviceberry
laevis Wiegand - Allegheny serviceberry, smooth serviceberry
sanguinea (Pursh) DC. - New England serviceberry, round-leaf serviceberry
utahensis Koehne - Utah serviceberry

Ammannia L. (Lythraceae)
auriculata Willd. - red-stem
coccinea Rottb. - purple ammannia
latifolia L. - pink ammannia

Ammi L. (Apiaceae)
majus L. - greater ammi
visnaga L. - bisnaga, toothpick ammi

Ammophila Host (Poaceae)
arenaria (L.) Link - European beach grass, marram grass
breviligulata Fernald - American beach grass

Amorpha L. (Fabaceae)
false indigo
canescens Pursh - lead-plant
fruticosa L. - bastard-indigo, false indigo, false indigo-bush, indigo-bush
nana Nutt. - dwarf wild indigo, fragrant false indigo, smooth lead-plant

Amorphophallus Blume ex Decne. (Araceae)
devil's-tongue, snake-palm
rivieri Durieu - devil's-tongue, leopard palm, snake-palm, umbrella arum

Ampelopsis Michx. (Vitaceae)
aconitifolia Bunge - monk's-hood-vine
arborea (L.) Koehne - pepper-vine
cordata Michx. - heart-leaf ampelopsis, racoon-grape

Amphiachyris (A. DC.) Nutt. (Asteraceae)
dracunculoides (DC.) Nutt. - annual broom-weed, broom-weed

Amphicarpaea Elliott ex Nutt. (Fabaceae)
bracteata (L.) Fernald - hog-peanut

Amsinckia Lehm. (Boraginaceae)
douglasiana A. DC. - Douglas's fiddlehead
intermedia Fisch. & C. A. Mey. - buckthorn-weed, finger-weed, tarweed
lycopsoides Lehm. - coast fiddle-neck, fiddle-necks, tarweed fiddle-neck, yellow burn-weed, yellow forget-me-not
menziesii (Lehm.) A. Nelson & J. F. Macbr. - rancher's fire-weed, small-flowered fiddle-neck
retrorsa Suksd. - palouse tarweed, rigid fiddle-neck
tessellata A. Gray - devil's-lettuce, western fiddle-neck

Amsonia Walter (Apocynaceae)
blue-star
tabernaemontana Walter - blue dogbane, blue-star, willow amsonia

Anacardium L. (Anacardiaceae)
occidentale L. - caju, cashew, marañon

Anagallis L. (Primulaceae)
arvensis L. - eye-bright, poison-chickweed, poorman's-weatherglass, red-chickweed, scarlet pimpernel, shepherd's-clock

Ananas Mill. (Bromeliaceae)
comosus (L.) Merr. - pineapple

Anaphalis DC. (Asteraceae)
everlasting, life-everlasting
margaritacea (L.) Benth. & Hook. - cotton weed, moonshine, pearly everlasting, poverty-weed, silver button, silver-leaf

Anchusa L. (Boraginaceae)
arvensis (L.) M. Bieb. - small bugloss
azurea Mill. - Italian bugloss
officinalis L. - alkanet, bugloss

Andromeda L. (Ericaceae)
bog-rosemary
glaucophylla Link - bog-rosemary

Andropogon L. (Poaceae)
beard grass, blue-stem
elliottii Chapm. - Elliott's beard grass, Elliott's blue-stem
gerardii Vitman - beard grass, big blue-stem, turkey-foot
glomeratus (Walter) Britton, et al. - blue-stem, bushy beard grass, bushy beard-stem, bushy blue-stem
hallii Hack. - Hall's blue-stem, sand blue-stem, turkey-foot
virginicus L. - beard grass, broom sedge blue-stem, broom-sedge, sedge grass, Virginia beard grass

Androsace L. (Primulaceae)
rock-jasmine
filiformis Retz. - slender-stem androsace

occidentalis Pursh - western androsace

Anemone L. (Ranunculaceae)
anemone, lily-of-the-field, wind-flower
canadensis L. - Canadian anemone, meadow anemone
coronaria L. - poppy anemone
cylindrica A. Gray - candle anemone, long-headed anemone, long-headed thimbleweed, thimbleweed
hupehensis (Lemoine) Lemoine var. *japonica* (Thunb.) Bowles & Stearn - Japanese anemone
nemorosa L. - European wood anemone
patens L. - American pasqueflower, eastern pasqueflower, Hartshorn's-plant, lion's-beard, pasqueflower, prairie-smoke, wild crocus
quinquefolia L. - wood anemone
virginiana L. - tall anemone, thimbleweed

Anemopsis Hook. & Arn. (Saururaceae)
californica (Nutt.) Hook. & Arn. - yerba-mansa

Anethum L. (Apiaceae)
graveolens L. - dill

Angelica L. (Apiaceae)
archangelica L. - garden angelica
atropurpurea L. - alexanders, American angelica, great angelica, masterwort, purple-stem angelica
triquinata Michx. - filmy angelica, mountain angelica

Anisacanthus Nees (Acanthaceae)
thurberi (Torr.) A. Gray - chuparosa, desert-honeysuckle

Annona L. (Annonaceae)
cherimola Mill. - cherimalla, cherimola, cherimoya, chirimoya, custard-apple
glabra L. - pond-apple
reticulata L. - bullock's-heart, corazón, custard-apple, soursop
squamosa L. - custard-apple, sugar-apple, sweetsop

Anoda Cav. (Malvaceae)
cristata (L.) Schltdl. - spurred anoda, violettas

Antennaria Gaertn. (Asteraceae)
everlasting, lady's-tobacco, pussy-toes
neglecta Greene - field pussy-toes
neglecta Greene var. *canadensis* (Greene) Cronquist - Canadian pussy-toes
plantaginifolia (L.) Richardson - early everlasting, lady's-tobacco, mouse's-ears, plantain-leaf everlasting, plantain-leaf pussy-toes, pussy-toes, white-plantain
solitaria Rydb. - solitary pussy-toes

Anthemis L. (Asteraceae)
chamomile, dog-fennel
arvensis L. - corn-chamomile, field-chamomile
cotula L. - dill-weed, dog-chamomile, dog-fennel, fetid-chamomile, mayweed, mayweed-chamomile, stinking-chamomile, stinking-daisy, stinkweed
tinctoria L. - golden marguerite, yellow-chamomile

Anthoxanthum L. (Poaceae)
odoratum L. - sweet vernal grass

Anthriscus Pers. (Apiaceae)
caucalis M. Bieb. - bur chervil, bur-beak chervil
cerefolium (L.) Hoffm. - chervil
sylvestris (L.) Hoffm. - wild chervil

Anthurium Schott (Araceae)
andraeanum André - flamingo-flower, flamingo-lily, oil-cloth-flower

Anthyllis L. (Fabaceae)
vulneraria L. - kidney-vetch, lady's-fingers, woundwort

Antidesma L. (Euphorbiaceae)
bunius (L.) Spreng. - bignay, Chinese laurel, salamander-tree

Antirrhinum L. (Scrophulariaceae)
snapdragon
majus L. - garden snapdragon, snapdragon
orontium L. - lesser snapdragon

Aphanes L. (Rosaceae)
occidentalis (Nutt.) Rydb. - western lady's-mantle

Aphelandra R. Br. (Acanthaceae)
squarrosa Nees - saffron-spike, zebra-plant

Apios Fabr. (Fabaceae)
americana Medik. - American potato bean, groundnut, wild bean

Apium L. (Apiaceae)
graveolens L. - celery
graveolens L. var. *dulce* (Mill.) Pers. - celery
graveolens L. var. *rapaceum* (Mill.) Gaudin - celeriac, knob celery, turnip-rooted celery

Aplectrum Torr. (Orchidaceae)
hyemale (Muhl. ex Willd.) Torr. - Adam-and-Eve, puttyroot

Apocynum L. (Apocynaceae)
dogbane
androsaemifolium L. - bitter dogbane, dogbane, honey-bloom, milk ipecac, spreading dogbane, wandering milkweed
cannabinum L. - American hemp, bowman's-root, Choctaw-root, hemp dogbane, Indian hemp, Indian physic, rheumatism-weed
sibiricum Jacq. - clasping dogbane, prairie dogbane

Aquilegia L. (Ranunculaceae)
columbine
canadensis L. - Canadian columbine, columbine, honeysuckle, meeting-houses, wild columbine
vulgaris L. - European columbine, European crowfoot, garden columbine, garden crowfoot

Arabidopsis Heynh. (Brassicaceae)
thaliana (L.) Heynh. - mouse-ear-cress, thale-cress

Arabis L. (Brassicaceae)
rock-cress
canadensis L. - sickle-pod
glabra (L.) Bernh. - tower mustard
hirsuta (L.) Scop. - hairy rock-cress
laevigata (Muhl. ex Willd.) Poir. - smooth rock-cress
missouriensis Greene - green rock-cress

Arachis L. (Fabaceae)
hypogaea L. - earth-nut, goober, grass-nut, groundnut, manî, monkey-nut, peanut, pindar

Aralia L. (Araliaceae)
californica S. Watson - elk-clover
cordata Thunb. - oudo, spikenard, udo
hispida Vent. - bristly sarsaparilla, dwarf-elder

nudicaulis L. - wild sarsaparilla
racemosa L. - American spikenard, life-of-man, petty-morel, spikenard
spinosa L. - angelica-tree, devil's-walking-stick, Hercules's-club, prickly-ash

Araucaria Juss. (Araucariaceae)
angustifolia (Bertol.) Kuntze - Brazilian pine, Paraná pine
araucana K. Koch - Chilean pine, monkey-puzzle-tree
columnaris (G. Forst.) Hook. - New Caledonia pine
heterophylla (Salisb.) Franco - house-pine, Norfolk Island pine

Arbutus L. (Ericaceae)
menziesii Pursh - madrone, Pacific madrone
unedo L. - cane-apples, strawberry-tree

Arceuthobium M. Bieb. (Viscaceae)
pusillum Peck - dwarf mistletoe, eastern dwarf mistletoe

Archontophoenix H. Wendl. & Drude (Arecaceae)
alexandrae (F. Muell.) H. Wendl. & Drude - Alexandra's palm, king's palm, northern Bangalow palm

Arctagrostis Griseb. (Poaceae)
latifolia (R. Br.) Griseb. - polar grass

Arctium L. (Asteraceae)
beggar's-button, burdock, clotbur
lappa L. - beggar's-button, cockle-button, cuckold, edible burdock, gobo, great burdock, harlock
minus (Hill) Bernh. - burdock, clotbur, cockle-button, cuckoo-button, lesser burdock, smaller burdock
tomentosum Mill. - cotton burdock, woolly burdock

Arctostaphylos Adans. (Ericaceae)
bearberry, manzanita
alpina (L.) Spreng. - alpine bearberry, black bearberry
columbiana Piper - hairy manzanita
glandulosa Eastw. - Eastwood manzanita
hookeri G. Don - Monterey manzanita
manzanita Parry - Parry's manzanita
nevadensis A. Gray - pine-mat manzanita
nummularia A. Gray - Bragg's manzanita
patula Greene - green-leaf manzanita
pumila Nutt. - dune manzanita, sand-mat manzanita
pungens Kunth - Mexican manzanita, point-leaf manzanita
tomentosa (Pursh) Lindl. - woolly manzanita
uva-ursi (L.) Spreng. - bear-grape, bearberry, creashak, hog cranberry, kinnikinick, meal-berry, mountain-box, sand-berry
viscida Parry - white-leaf manzanita

Arctotis L. (Asteraceae)
African daisy
stoechadifolia Bergius - blue-eyed African daisy

Ardisia Sw. (Myrsinaceae)
crenata Sims - coral-berry, spiceberry
crispa (Thunb.) A. DC. - ardisia
escallonioides Schltdl. & Cham. - marlberry

Arenaria L. (Caryophyllaceae)
sandwort
caroliniana Walter - long-root, pine-barren sandwort
groenlandica (Retz.) Spreng. - mountain daisy, mountain sandwort
serpyllifolia L. - thyme-leaf sandwort
stricta Michx. - rock sandwort

Arenga Labill. (Arecaceae)
pinnata (Wurmb) Merr. - areng palm, black fiber palm, Gomuti palm, sugar palm

Argemone L. (Papaveraceae)
argemony, prickle poppy
albiflora Hornem. - blue-stem prickle poppy, white prickle poppy
mexicana L. - Mexican poppy, Mexican prickly poppy, prickly poppy
platyceras Link & Otto - crested prickle poppy
polyanthemos (Fedde) Ownbey - annual prickly poppy

Argyreia Lour. (Convolvulaceae)
nervosa (Burm. f.) Bojer - elephant-climber, woolly morning-glory

Arisaema Mart. (Araceae)
dracontium (L.) Schott - dragon-arum, dragonroot, green-dragon
triphyllum (L.) Torr. - dragon-root, Indian turnip, Jack-in-the-pulpit, small Jack-in-the-pulpit, swamp Jack-in-the-pulpit

Aristida L. (Poaceae)
three-awn
adscensionis L. - prairie three-awn, six-weeks three-awn
arizonica Vasey - Arizona three-awn grass
dichotoma Michx. - church-mouse three-awn, poverty grass, wiregrass
divaricata Humb. & Bonpl. ex Willd. - poverty three-awn
longespica Poir. - cana-brava, slim-spike three-awn
orcuttiana Vasey - beggar-tick grass
pansa Wooton & Standl. - Wooton's three-awn
purpurascens Poir. - arrow-feather, arrow-feather three-awn
purpurea Nutt. - purple three-awn, red three-awn, Reverchon's three-awn
stricta Michx. - pine-land three-awn
ternipes Cav. - spider grass

Aristolochia L. (Aristolochiaceae)
birthwort
macrophyllum Lam. - Dutchman's-pipe, pipe-vine
serpentaria L. - Virginia snakeroot

Armeria (DC.) Willd. (Plumbaginaceae)
sea-pink, thrift
maritima (Mill.) Willd. subsp. *sibirica* (Turcz.) Lawr. - fox-flower

Armoracia P. Gaertn., B. Mey. & Scherb. (Brassicaceae)
rusticana P. Gaertn., B. Mey. & Scherb. - horseradish, red-cole

Arnoseris Gaertn. (Asteraceae)
minima (L.) Schweigg. & Körte - dwarf nipplewort, lamb-succory

Aronia Medik. (Rosaceae)
chokeberry
arbutifolia (L.) Elliott - red chokeberry
melanocarpa (Michx.) Elliott - black chokeberry
prunifolia (Marsh.) Rehder - Florida chokeberry, purple chokeberry

Arrhenatherum P. Beauv. (Poaceae)
elatius (L.) J. Presl & C. Presl - tall oat grass, tuber oat grass

Artemisia L. (Asteraceae)
mugwort, sagebrush, wormwood
abrotanum L. - old-man, southern wormwood, southernwood
absinthium L. - absinth wormwood, absinthe, absinthium, wormwood
annua L. - annual wormwood, sweet Annie, sweet wormwood
arbuscula Nutt. - low sagebrush
biennis Willd. - biennial wormwood, bitter-weed, false tansy
californica Less. - California sagebrush
campestris L. - sage-wort
campestris L. subsp. *caudata* (Michx.) H. Hall & Clem. - field sage-wort
cana Pursh - silver sagebrush
douglasiana Besser - California mugwort, mugwort
dracunculus L. - estragon, tarragon
filifolia Torr. - sand sagebrush
frigida Willd. - estafiata, fringed sagebrush, prairie sage-wort
ludoviciana Nutt. - cudweed, Louisiana sagebrush, Louisiana wormwood, silver wormwood, western mugwort, white sage
pontica L. - Roman wormwood
stelleriana Besser - beach wormwood
tridentata Nutt. - basin sagebrush, big sagebrush, sagebrush
vulgaris L. - felon-herb, mugwort, wormwood

Arthraxon P. Beauv. (Poaceae)
hispidus (Thunb.) Makino - joint-head arthraxon

Artocarpus J. R. Forst. & G. Forst. (Moraceae)
altilis (Parkinson) Fosberg - breadfruit
heterophyllus Lam. - jackfruit

Aruncus L. (Rosaceae)
dioicus (Walter) Fernald - goat's-beard
sylvester Kostel. - goat's-beard

Arundinaria Michx. (Poaceae)
bamboo, cane
gigantea (Walter) Muhl. - cane-break bamboo, giant cane, large cane, southern cane
gigantea (Walter) Muhl. subsp. *tecta* (Walter) Maclure - giant cane, switch-cane

Arundo L. (Poaceae)
donax L. - cana-brava, carrizo, giant reed

Asarum L. (Aristolochiaceae)
asarabacca, wild ginger
canadense L. - snakeroot, wild ginger

Asclepias L. (Asclepiadaceae)
milkweed
amplexicaulis Sm. - blunt-leaf milkweed, clasping milkweed
curassavica L. - blood-flower milkweed, false ipecac, red-head
eriocarpa Benth. - woolly-pod milkweed
exaltata L. - poke milkweed, tall milkweed
fascicularis Decne. - Mexican whorled milkweed, narrow-leaf milkweed
incarnata L. - swamp milkweed
labriformis Jones - labriform milkweed
lanceolata Walter - few-flowered milkweed, lanceolate milkweed
latifolia (Torr.) Raf. - broad-leaf milkweed

purpurascens L. - purple milkweed
quadrifolia Jacq. - four-leaved milkweed
rubra L. - red milkweed
speciosa Torr. - Greek milkweed, showy milkweed
subverticillata (A. Gray) Vail - poison milkweed, western whorled milkweed, whorled milkweed
sullivantii Engelm. - smooth milkweed, Sullivant's milkweed
syriaca L. - cotton weed, milkweed, silkweed
tuberosa L. - butterfly milkweed, butterfly-weed, chigger-flower, Indian paintbrush, orange milkweed, orange-root, pleurisy-root, tuber-root, white-root
variegata L. - white milkweed
verticillata L. - eastern whorled milkweed, horsetail milkweed, whorled milkweed
viridiflora Raf. - green milkweed

Asimina Adans. (Annonaceae)
incana (Bartram) Exell - flag pawpaw
parviflora (Michx.) Dunal - dwarf pawpaw
reticulata Shuttlew. ex Chapm. - net-leaf pawpaw
triloba (L.) Dunal - American pawpaw, pawpaw

Asparagus L. (Liliaceae)
asparagoides (L.) Druce - smilax asparagus
densiflorus (Kunth) Jessop - asparagus-fern, Sprenger's asparagus
officinalis L. - asparagus, garden asparagus
setaceus (Kunth) Jessop - asparagus-fern, fern asparagus, lace-fern

Asperugo L. (Boraginaceae)
madwort
procumbens L. - catch-weed

Asphodelus L. (Liliaceae)
fistulosus L. - onion-weed

Aspidistra Ker Gawl. (Liliaceae)
elatior Blume - barroom-plant, cast-iron-plant

Asplenium L. (Pteridophyta)
spleenwort
nidus L. - bird's-nest fern
rhizophyllum L. - walking fern
trichomanes L. - maidenhair spleenwort

Aster L. (Asteraceae)
aster, frost-flower, starwort
acuminatus Michx. - whorled aster, whorled wood aster
alpinus L. - rock aster
cordifolius L. - blue heart-leaf aster, blue-wood aster, heart-leaf aster
divaricatus L. - white heart-leaf aster, white-wood aster
dumosus L. - bushy aster, long-stalked aster
ericoides L. - heath aster, squarrose white aster
hemisphericus Alexander - southern aster
laevis L. - smooth aster
lanceolatus Willd. - frost-flower, panicled aster, white field aster
lateriflorus (L.) Britton - calico aster, goblet aster, starved aster
laterifolius (L.) Britton - small white aster
linariifolius L. - savory-leaf aster, stiff aster
lowrieanus Porter - Lowrie's aster
macrophyllus L. - big-leaf aster, large-leaf aster
nemoralis Aiton - bog aster, leafy bog aster

novae-angliae L. - New England aster
novi-belgii L. - New York aster
oolentangiensis Riddell - azure aster
patens Aiton - clasping aster, large purple aster
pilosus Willd. - awl aster, white heath aster
puniceus L. - bristly aster, purple-stem aster
sagittifolius Willd. - arrow-leaf aster
schreberi Nees - Schreber's aster
shortii Lindl. - midwestern blue heart-leaf aster, Short's aster
subulatus Michx. - annual salt marsh aster
tenuifolius L. - perennial salt marsh aster
tradescantii L. - shore aster, Tradescant's aster
umbellatus Mill. - flat-top white aster, tall flat-top white aster
undulatus L. - clasping heart-leaf aster, wavy-leaf aster

Astilbe Buch.-Ham. ex D. Don (Saxifragaceae)
perennial spiraea
biternata Vent. & Britton - false goat's-beard

Astragalus L. (Fabaceae)
locoweed, milk-vetch, poison-vetch
aboriginorum Richardson - Indian milk-vetch
adsurgens Pall. - standing milk-vetch
agrestis G. Don - field milk-vetch, purple loco
allochrous A. Gray - half-moon loco
alpinus L. - alpine milk-vetch
bisulcatus (Hook.) A. Gray - two-grooved milk-vetch, two-grooved poison-vetch
cicer L. - cicer milk-vetch
earlei Greene ex Rydb. - Big Bend loco
falcatus Lam. - Russian sickle milk-vetch
flexuosus (Hook.) G. Don - flexible milk-vetch, pliant milk-vetch
glycyphyllos L. - fit's-root, licorice milk-vetch
lentiginosus Hook. - fleckled milk-vetch, rattle-weed, spotted loco, spotted locoweed
lotiflorus Hook. - lotus milk-vetch
miser Douglas - timber milk-vetch, weedy milk-vetch
miser Douglas var. *hylophilus* (Rydb.) Barneby - Yellowstone milk-vetch
miser Douglas var. *oblongifolius* (Rydb.) Cronquist - Wasatch milk-vetch
miser Douglas var. *serotinus* (A.Gray) Barneby - Columbia milk-vetch, late milk-vetch
missouriensis Nutt. - Missouri milk-vetch
mollissimus Torr. - woolly loco, woolly locoweed
nuttallianus A. DC. - Nuttall's milk-vetch, small-flowered milk-vetch
nuttallii (Torr. & A. Gray) J. T. Howell - Nuttall's milk-vetch
pectinatus Douglas ex Hook. - narrow-leaf milk-vetch, tine-leaf milk-vetch
tenellus Pursh - loose-flowered milk-vetch, purse milk-vetch
wootonii E. Sheld. - Wooton's loco

Astranthium Nutt. (Asteraceae)
integrifolium (Michx.) Nutt. - western daisy

Astrophytum Lem. (Cactaceae)
star cactus
myriostigma (Salm-Dyck) Lem. - bishop's-cap, bishop's-hood, monk's-hood
ornatum (DC.) Britton & Rose - bishop's-cap, ornamental monk's-hood, star cactus

Athyrium Roth (Pteridophyta)
felix-femina (L.) Roth - lady fern
felix-femina (L.) Roth var. *michauxii* Mett. - northern lady fern
thelypteroides (Michx.) Desv. - silvery spleenwort

Atriplex L. (Chenopodiaceae)
orache, saltbush
arenaria Nutt. - sea beach orache
argentea Nutt. - silver-scale, silver-scale saltbush
canescens (Pursh) Nutt. - ceniza, chamiza, four-wing saltbush
confertifolia (Torr. & Frém.) S. Watson - shad-scale, sheep's-fat, spiny saltbush
elegans (Moq.) D. Dietr. - wheel-scale saltbush
elegans (Moq.) D. Dietr. var. *fasciculata* (S. Watson) M. E. Jones - Salton's saltbush
hortensis L. - garden orache
patula L. - spear oracle, spreading orache
patula L. subsp. *hastata* (L.) H. Hall & Clem. - halberd-leaf orache
polycarpa (Torr.) S. Watson - all-scale, desert saltbush
rosea L. - red orache, red-scale, tumbling atriplex, tumbling oracle
semibaccata R. Br. - Australian saltbush, berry saltbush

Atropa L. (Solanaceae)
belladonna L. - belladonna, deadly nightshade

Aucuba Thunb. (Cornaceae)
japonica Thunb. - Japanese aucuba, Japanese laurel

Aureolaria Raf. (Scrophulariaceae)
pedicularia (L.) Raf. - false foxglove
virginica (L.) Pennell - downy false foxglove

Avena L. (Poaceae)
oats
barbata Brot. - slender oat, slender wild oat
fatua L. - drake, flaver, potato oats, red oats, Tartarian oat
ludoviciana Durand - winter wild oat
nuda L. - naked oat
sativa L. - cultivated oat, oat
sterilis L. - animated oat, sterile oat
strigosa Schreb. - bristle oat, sand oat, small oat

Averrhoa L. (Oxalidaceae)
carambola L. - blimbing, caramba, carambola, country gooseberry, star-fruit

Avicennia L. (Verbenaceae)
germinans (L.) L. - black mangrove, black-wood

Axonopus P. Beauv. (Poaceae)
affinis Chase - carpet grass
compressus (Sw.) P. Beauv. - savanna grass, tropical carpet grass
furcatus (Flüggé) Hitchc. - big carpet grass

Axyris L. (Chenopodiaceae)
amaranthoides L. - Russian pigweed

Azara Ruiz & Pav. (Flacourtiaceae)
microphylla Hook. - aromo, chin-chin

Azolla Lam. (Azollaceae)
caroliniana Willd. - azolla, Carolina mosquito-fern, water-fern
pinnata R. Br. - pinnate mosquito-fern

Baccaurea Lour. (Euphorbiaceae)
racemosa (Reinw.) Müll. Arg. - kapundung, menteng

Baccharis L. (Asteraceae)
halimifolia L. - consumption-weed, eastern baccharis, groundsel-bush, groundsel-tree, sea-myrtle, silverling
pilularis DC. - chaparral-broom, coyote-brush, dwarf baccharis, kidney-wort
pilularis DC. var. *consanguinea* (DC.) Kuntze - kidney-wort baccharis
salicifolia (Ruiz & Pav.) Pers. - seep-willow, sticky baccharis, water-motie, water-willow
salicina Torr. & A. Gray - willow baccharis
sarothroides A. Gray - broom baccharus, desert-broom, rosin-brush
viminea DC. - mule's-fat

Bacopa Aubl. (Scrophulariaceae)
caroliniana (Walter) B. L. Rob. - Carolina water-hyssop
eisenii (Kellogg) Pennell - Eisen's water-hyssop
innominata (Gomez Maza) Alain - nameless water-hyssop
monnieri (L.) Wettst. - water-hyssop
rotundifolia (Michx.) Wettst. - disc water-hyssop

Bactris Jacq. ex Scop. (Arecaceae)
spiny-club palm
gasipaes Kunth - chonta, peach palm, pejibeye, pejivalle, pupuna

Baileya Harv. & A. Gray ex A. Gray (Asteraceae)
multiradiacata Harv. & A. Gray ex A. Gray - desert-marigold

Ballota L. (Lamiaceae)
fetid horehound
nigra L. - black horehound

Balsamorhiza Nutt. (Asteraceae)
balsam-root, sunflower
sagittata (Pursh) Nutt. - Oregon sunflower

Bambusa Schreb. (Poaceae)
bamboo
multiplex (Lour.) Raeusch. - hedge bamboo
vulgaris Schrad. ex H. Wendl. - bamboo

Baptisia Vent. (Fabaceae)
false indigo
australis (L.) R. Br. - blue false indigo, blue wild indigo

Barbarea R. Br. (Brassicaceae)
orthoceras Ledeb. - winter-cress
verna (Mill.) Asch. - early winter-cress, scurvy-grass
vulgaris R. Br. - bitter-cress, pot-herb, rocket-cress, St. Barbara's cress, upland-cress, water mustard, winter-cress, yellow-rocket, yellow-weed

Barleria L. (Acanthaceae)
cristata L. - Philippine violet

Bartonia Muhl. (Gentianaceae)
paniculata (Michx.) Muhl. - screw-stem
virginica Britton, et al. - five-hook bassia

Bartsia L. (Scrophulariaceae)
alpina L. - alpine bartsia, velvet-bells

Basella L. (Basellaceae)
Malabar nightshade
alba L. - country-spinach, Indian spinach, Malabar spinach, vine-spinach

Bassia All. (Chenopodiaceae)
hyssopifolia (Pall.) Kuntze - five-hook bassia, hyssop bassia, thorn orache
scoparia (L.) A. J. Scott - belvedere, burning-bush, kochia, summer-cypress

Bauhinia L. (Fabaceae)
galpinii N. E. Br. - nasturtium bauhinia, red bauhinia
hookeri F. Muell. - mountain-ebony
variegata L. - mountain-ebony, orchid-tree

Beckmannia Host (Poaceae)
syzigachne (Steud.) Fernald - American slough grass

Begonia L. (Begoniaceae)
×*rex-cultorum* L. H. Bailey - beefsteak-geranium, rex begonia
Note = A group of interrelated cultivars of complex hybrid origin derived mostly from *B. rex* and related rhizomatous species.
×*tuberhybrida* Voss - hybrid tuberous begonia
Note = A group of cultivars of complex hybrid origin derived through hybridization and selection from several Andean species.

Belamcanda Adans. (Iridaceae)
chinensis (L.) DC. - blackberry-lily, leopard-flower

Bellis L. (Asteraceae)
daisy
perennis L. - English daisy, European daisy

Benincasa Savi (Cucurbitaceae)
hispida (Thunb.) Cogn. - ash gourd, Chinese fuzzy gourd, Chinese preserving melon, Chinese watermelon, Chinese winter melon, tunka, wax gourd, white gourd, white pumpkin, zit-kwa

Berberis L. (Berberidaceae)
barberry
aquifolium Pursh - blue barberry, holly barberry, holly mahonia, mountain-grape, Oregon grape
canadensis Mill. - Allegheny barberry, American barberry
fendleri A. Gray - Colorado barberry
fremontii Torr. - Fremont's mahonia
haematocarpa Wooton - red barberry
nervosa Pursh - Cascades mahonia
pinnata Lag. - cluster mahonia
repens Lindl. - creeping barberry
thunbergii DC. - Japanese barberry
vulgaris L. - barberry, European barberry, jaundice-berry, piprage

Berchemia Neck. ex DC. (Rhamnaceae)
scandens (Hill) K. Koch - Alabama supple-jack, rattan-vine, supple-jack

Berteroa DC. (Brassicaceae)
incana (L.) DC. - hoary-alyssum

Besseya Rydb. (Scrophulariaceae)
rubra (Douglas) Rydb. - kitten's-tail

Beta L. (Chenopodiaceae)
  beet
  vulgaris L. - beet, beetroot, garden beet, mangel, mangold, sea beet, sugar beet
  vulgaris L. subsp. *cicla* (L.) W. Koch - chard, spinach beet, Swiss-chard

Betula L. (Betulaceae)
  birch
  alleghaniensis Britton - gray birch, yellow birch
  ×*caerulea* Blanch. - blue birch
  glandulosa Michx. - bog birch, dwarf birch, resin birch
  lenta L. - black birch, cherry birch, mahogany birch, mountain-mahogany, sweet birch
  minor (Tuck.) Fernald - dwarf white birch
  nigra L. - black birch, red birch, river birch
  occidentalis Hook. - mountain birch, water birch
  papyrifera Marsh. - canoe birch, paper birch, white birch
  papyrifera Marsh. var. *kenaica* (W. H. Evans) A. Henry - Kenai birch
  pendula Roth - European white birch
  populifolia Marsh. - fire birch, gray birch, old-field birch, white birch
  pumila L. - gray birch, low birch, swamp birch, yellow birch
  ×*sandbergii* Britton - Sandberg's birch
  uber (Ashe) Fernald - Ashe's birch

Bidens L. (Asteraceae)
  beggar-ticks, bur-marigold, pitchforks, Spanish-needles, stick-tights, tickseed, water-marigold
  aristosa (Michx.) Britton - bearded beggar-ticks, midwestern tickseed-sunflower, tickseed-sunflower
  bigelovii A. Gray - Bigelow's beggar-ticks
  bipinnata L. - beggar-ticks, cuckold, Spanish-needles
  cernua L. - nodding beggar-ticks, nodding bur-marigold, pitchforks, smaller bur-marigold, stick-tights
  connata Muhl. - connate beggar-ticks, purple-stem beggar-ticks
  coronata (L.) Britton - northern tickseed, tickseed-sunflower
  frondosa L. - devil's beggar-ticks, sticktight
  hyperborea Greene - estuary beggar-ticks, northern estuarine beggar-ticks
  mitis (Michx.) Sherff - coastal plain tickseed, marsh beggar-ticks
  pilosa L. - beggar-ticks, bur-marigold, hairy beggar-ticks, Spanish-needles
  polylepis Blake - coreopsis beggar-ticks, Ozark tickseed
  tripartita L. - trifid beggar-ticks
  vulgata Greene - tall beggar-ticks, western beggar-ticks

Bignonia L. (Bignoniaceae)
  capreolata L. - cross-vine, quarter-vine

Bischofia Blume (Euphorbiaceae)
  javanica Blume - toog

Bixa L. (Bixaceae)
  orellana L. - achiote, annatto, lipstick-tree

Blechnum L. (Pteridophyta)
  spicant (L.) Roth - deer fern, hard fern

Blepharoneuron Nash (Poaceae)
  tricholepis (Torr.) Nash - hairy-drop-seed

Blephilia Raf. (Lamiaceae)
  ciliata (L.) Benth. - downy woodmint
  hirsuta (Pursh) Benth. - hairy wood mint, wood mint

Blighia K. D. Koenig (Sapindaceae)
  sapida K. D. Koenig - akee

Boehmeria Jacq. (Urticaceae)
  cylindrica (L.) Sw. - bog-hemp, false nettle
  nivea (L.) Gaudich. - China grass, Chinese silk-plant, ramie

Boerhavia L. (Nyctaginaceae)
  spiderling, wine-flower
  coulteri (Hook. f.) S. Watson - Coulter's spiderling
  diffusa L. - red spiderling
  erecta L. - erect spiderling

Boisduvalia Spach (Onagraceae)
  densiflora (Lindl.) S. Watson - spike-primrose

Bombax L. (Bombacaceae)
  ceiba L. - red silk cotton-tree

Borago L. (Boraginaceae)
  borage
  officinalis L. - borage, cool-tankard, tale-wort

Bothriochloa Kuntze (Poaceae)
  barbinodis (Lag.) Herter - cane beard grass, cane blue-stem
  ischaemum (L.) Keng - King Ranch blue-stem, plains blue-stem, Turkestan blue-stem, yellow blue-stem
  ischaemum (L.) Keng var. *songarica* (Fisch. & C. A. Mey.) Celarier & Harlan - King Ranch blue-stem
  pertusa (L.) A. Camus - camagueyana, comagueyana, hurricane grass, pitted beard grass, pitted blue-stem
  saccharoides (Sw.) Rydb. - silver beard grass, silver blue-stem

Botrychium Sw. (Pteridophyta)
  grape fern, moonwort
  lanceolatum (C. Gmel.) Ångstr. - lance-leaf grape fern
  lunaria (L.) Sw. - moonwort
  simplex E. Hitchc. - little grape fern
  virginianum (L.) Sw. - rattlesnake fern

Bougainvillea Comm. & Juss. (Nyctaginaceae)
  glabra Choisy ex DC. - paper-flower
  umbellifera (J. R. Forst. & G. Forst.) Seem. - kepau, papala

Bouteloua Lag. (Poaceae)
  grama, mesquite grass
  aristidoides (Kunth) Griseb. - needle grama, six-weeks needle grama
  barbata Lag. - six-weeks grama
  curtipendula (Michx.) A. Gray - side-oats grama
  eriopoda (Torr.) Torr. - black grama
  gracilis (Kunth) Lag. ex Steud. - blue grama
  hirsuta Lag. - hairy grama
  radicosa (E. Fourn.) D. A. Griffiths - purple grama
  rothrockii Vasey - Rothrock's grama
  simplex Lag. - mat grama

Boykinia Nutt. (Saxifragaceae)
  aconitifolia Nutt. - brook saxifrage

Brachiaria (Trin.) Griseb. (Poaceae)
  ciliatissima (Buckley) Chase - fringed signal grass

fasciculata (Sw.) Parodi - brown-top millet, brown-top panicum
mutica (Forssk.) Stapf - Pará grass
piligera (F. Muell.) Hughes - narrow-leaf signal grass
plantaginea (Link) Hitchc. - Alexander grass
platyphylla (Griseb.) Nash - broad-leaf signal grass
ramosa (L.) Stapf - brown-top millet
subquadripara (Trin.) Hitchc. - small-flowered Alexander grass

Brachychiton Schott & Endl. (Sterculiaceae)
acerifolius A. Cunn. ex F. Muell. - Australian flame-tree, flame bottle-tree

Brachypodium P. Beauv. (Poaceae)
distachyon (L.) P. Beauv. - false brome
pinnatum (L.) P. Beauv. - Japanese false brome grass
sylvaticum (Huds.) P. Beauv. - slender false brome grass

Brasenia Schreb. (Cabombaceae)
schreberi J. F. Gmel. - purple wan-dock, water-shield

Brassica L. (Brassicaceae)
chinensis L. - pak-choi
juncea (L.) Czern. - brown mustard, Chinese mustard, gai-choi, Indian mustard, kai-tsoi, karashina, leaf mustard, mustard cabbage, mustard-greens, ostrich-plume, southern cole, Swatow mustard
napus L. - colza, rape, rapeseed, rutabaga, swede rape, turnip
napus L. var. *napobrassica* (L.) Rchb. - rutabaga, swede, Swedish turnip
nigra (L.) W. Koch - black mustard, brown mustard, cadlock, scurvy, senvil, warlock
oleracea L. - cabbage, kale, kohlrabi, wild cabbage
oleracea L. var. *acephala* DC. - bore cole, braschette, cole, colewort, collards, flowering cabbage, kale
oleracea L. var. *botrytis* L. - broccoli, cauliflower
oleracea L. var. *capitata* L. - cabbage, savoy, savoy cabbage
oleracea L. var. *gemmifera* DC. - Brussels sprouts
oleracea L. var. *gongylodes* L. - kohlrabi
oleracea L. var. *italica* Plenck - asparagus broccoli, Italian broccoli, sprouting broccoli
pekinensis (Lour.) Rupr. - celery cabbage, Chinese cabbage, pe-tsai, Shantung cabbage
rapa L. - bird's-rape, bird's-rape mustard, field mustard, turnip
tournefortii Gouan - African mustard

Brickellia Elliott (Asteraceae)
brickel-bush
eupatorioides (L.) Shinners - false boneset
grandiflora (Hook.) Nutt. - tassel-flower

Briza L. (Poaceae)
media L. - perennial quaking grass, quaking grass
minor L. - little quaking grass

Brodiaea Sm. (Liliaceae)
coronaria (Salisb.) Engl. - harvest brodiaea

Bromus L. (Poaceae)
brome grass, chess
anomalus Rupr. ex E. Fourn. - nodding brome
arenarius Labill. - Australian chess, sand brome
arizonicus (Shear) Stebbins - Arizona brome
arvensis L. - field brome

briziformis Fisch. & C. A. Mey - quakegrass, rattlesnake brome, rattlesnake chess
carinatus Hook. & Arn. - California brome, mountain brome
catharticus Vahl - prairie grass, rescue grass, Schrader's brome
ciliatus L. - fringed brome
commutatus Schrad. - hairy chess, smooth brome grass
diandrus Roth - rip-gut brome
erectus Huds. - meadow brome, upright brome
hordeaceus L. - soft brome, soft chess
inermis Leyss. - awnless brome grass, Hungarian brome, smooth brome
japonicus Thunb. ex Js. Murray - Japanese brome, Japanese chess
madritensis L. - compact brome, foxtail chess, Madrid brome, Spanish brome, Spanish brome grass
purgans L. - Canadian brome
rigidus Roth - rip-gut brome, rip-gut brome grass, rip-gut grass
rubens L. - foxtail brome, foxtail brome grass, foxtail chess, red brome
secalinus L. - cheat, cheat grass, chess, cock grass, rye brome, wheat-thief
sterilis L. - barren brome, poverty brome, sterile brome
tectorum L. - cheat grass, downy brome, downy brome grass, downy chess, drooping brome, early chess, slender chess
trinii Desv. - Chilean chess

Broussonetia L'Hér. ex Vent. (Moraceae)
papyrifera (L.) Vent. - paper-mulberry, tapa-cloth-tree

Brunfelsia L. (Solanaceae)
americana L. - lady-of-the-night

Brunnichia Banks ex Gaertn. (Polygonaceae)
ovata (Walter) Shinners - buckwheat-vine, lady's-eardrops, red-vine

Bryonia L. (Cucurbitaceae)
alba L. - white bryony

Buchloe Engelm. (Poaceae)
dactyloides (Nutt.) Engelm. - buffalo grass

Buchnera L. (Scrophulariaceae)
blue-hearts
rubra (Douglas) Rydb. - kitten's-tail

Bucida L. (Combretaceae)
buceras L. - black-olive, ox-horn bucida

Buddleja L. (Loganiaceae)
butterfly-bush
davidii Franch. - butterfly-bush, orange-eyed buddleja, summer-lilac

Bunias L. (Brassicaceae)
erucago L. - crested bunias
orientalis L. - hill-mustard

Bupleurum L. (Apiaceae)
thorough-wax
rotundi-folium L. - thorough-wax

Bursera Jacq. ex L. (Burseraceae)
simaruba (L.) Sarg. - gum-elemi, gumbo, gumbo-limbo, West Indian birch

Butia (Becc.) Becc. (Arecaceae)
capitata (Mart.) Becc. - South American jelly palm

Butomus L. (Butomaceae)
umbellatus L. - flowering-rush, grassy-rush, water-gladiolus

Buxus L. (Buxaceae)
boxwood
microphylla Siebold & Zucc. - little-leaf boxwood
microphylla Siebold & Zucc. var. *japonica* (Müll.Arg. ex Miq.) Rehder & E.H.Wilson - Japanese boxwood
microphylla Siebold & Zucc. var. *koreana* Nakai ex Rehder - Korean boxwood
sempervirens L. - boxwood, Turkish boxwood

Byrsonima Rich. ex Kunth (Malpighiaceae)
crassifolia (L.) Kunth - nance

Cabomba Aubl. (Cabombaceae)
caroliniana A. Gray - purple fanwort
caroliniana A. Gray cv.'multipartita' - green fanwort

Cacalia L. (Asteraceae)
atriplicifolia L. - pale Indian plantain
muhlenbergii (Sch. Bip.) Fernald - great Indian plantain
suaveolens L. - hastate Indian plantain, sweet-scented Indian plantain

Caesalpinia L. (Fabaceae)
bonduc (L.) Roxb. - gray nickers, yellow nickers
gilliesii (Hook.) Benth. - bird-of-paradise, paradise poinciana
pulcherrima (L.) Sw. - dwarf poinciana, flower-fence poinciana, flowering-fence, paradise-flower, pride-of-Barbados

Cajanus DC. (Fabaceae)
pigeon pea
cajan (L.) Millsp. - Angola pea, cat-jang pea, Congo pea, dahl, pigeon pea

Cakile Mill. (Brassicaceae)
edentula (Bigelow) Hook. - American sea-rocket
maritima Scop. - sea-rocket

Caladium Vent. (Araceae)
bicolor (Aiton) Vent. - elephant's-ear, fancy-leaf caladium, heart-of-Jesus
×hortulanum Birdsey - fancy-leaf caladium

Calamagrostis Adans. (Poaceae)
reed grass
canadensis (Michx.) P. Beauv. - blue-joint
montanensis (Scribn.) Scribn. - plains reed grass
neglecta (Ehrh.) Gaertn., B. Mey. & Scherb. - narrow reed grass, northern reed grass
nutkaensis (J. Presl) Steud. - Pacific reed grass
purpurascens R. Br. - purple reed grass
rubescens Buckley - pine grass
scribneri Beal - Scribner's reed grass

Calamintha R. Br. (Lamiaceae)
nepeta (L.) Savi - calamint

Calamovilfa (A. Gray) Hack. ex Scribn. (Poaceae)
longifolia (Hook.) Scribn. - prairie sand-reed, sand-reed

Calandrinia Kunth (Portulacaceae)
rock-purslane
ciliata (Ruiz & Pav.) DC. - red-maids

Calceolaria L. (Scrophulariaceae)
mexicana Benth. - pocketbook-flower, pouch-flower, slipper-flower, slipperwort

Calendula L. (Asteraceae)
officinalis L. - pot-marigold

Calla L. (Araceae)
palustris L. - water-arum, water-dragon, wild calla

Calliandra Benth. (Fabaceae)
false mesquite, powder-puff
eriophylla Benth. - fairy-duster, mesquitilla, mock-mesquite
haematocephala Hassk. - red powder-puff

Callicarpa L. (Verbenaceae)
americana L. - American beauty-berry, French mulberry
dichotoma (Lour.) K. Koch - Chinese beauty-berry, purple beauty-berry

Callirhoe Nutt. (Malvaceae)
poppy mallow
alcaeoides (Michx.) A. Gray - pale poppy mallow, plains poppy mallow
involucrata (Torr. & A. Gray) A. Gray - purple poppy mallow

Callistemon R. Br. (Myrtaceae)
citrinus (Curtis) Skeels - crimson bottlebrush
viminalis (Sol. ex Gaertn.) Cheel - weeping bottlebrush

Callistephus Cass. (Asteraceae)
chinensis (L.) Nees - annual aster, China aster

Callitriche L. (Callitrichaceae)
water-chickweed, water-starwort
stagnalis Scop. - European water-starwort
verna L. - water-starwort

Calluna Salisb. (Ericaceae)
vulgaris (L.) Hull - heather, Scotch heather

Calocedrus Kurz (Cupressaceae)
decurrens (Torr.) Florin - California incense-cedar, incense-cedar

Calochortus Pursh (Liliaceae)
butterfly-tulip, globe-tulip, mariposa-lily, sago-lily, star-tulip
kennedyi Porter - desert mariposa
macrocarpus Douglas - green-banded mariposa
nudus S. Watson - sierra star-tulip
nuttallii Torr. & A. Gray - sego-lily
tolmiei Hook. & Arn. - pussy's-ears

Calophyllum L. (Clusiaceae)
inophyllum L. - Alexandrian-laurel, ati, Indian laurel, laurel-wood, tamanu, tamono

Calopogonium Desv. (Fabaceae)
swamp-pink
mucunoides Desv. - calopo, fisolilla

Calothamnus Labill. (Myrtaceae)
sanguineus Labill. - blood-red netbush, netbush, one-sided bottlebrush

Calotropis R. Br. (Asclepiadaceae)
gigantea (L.) Dryand. ex W. T. Aiton - bowstring-hemp, crown-flower, crown-plant, giant milkweed, madar, mudar

procera (Aiton) Dryand. ex W. T. Aiton - auricula-tree, dumb-cotton, madar, small crown-flower, small mudar

Caltha L. (Ranunculaceae)
marsh-marigold
palustris L. - cowslip, king's-cup, marsh-marigold, May-blob, meadow-bright

Calycanthus K. Schum. (Calycanthaceae)
sweet-shrub
floridus L. - Carolina allspice, pineapple-shrub, strawberry-shrub
occidentalis Hook. & Arn. - California sweet-shrub, spicebush, sweet-shrub

Calypso Salisb. (Orchidaceae)
bulbosa (L.) Oakes - fairy-slipper

Calyptocarpus Less. (Asteraceae)
vialis Less. - sprawling horseweed

Calyptranthes Sw. (Myrtaceae)
pallens (Poir.) Griseb. - pale lid-flower

Calyptridium Nutt. ex Torr. & A. Gray (Portulacaceae)
umbellatum (Torr.) Greene - pussy-paws

Calystegia R. Br. (Convolvulaceae)
bindweed
hederacea Wall. - Japanese bindweed
sepium (L.) R. Br. - bracted bindweed, devil's-vine, great bindweed, hedge bindweed, wild morning-glory
spithamaea (L.) Pursh - low bindweed

Camassia Lindl. (Liliaceae)
quamash (Pursh) Greene - camass, camosh, quamash
scilloides (Raf.) Cory - eastern camass, indigo-squill, meadow-hyacinth, wild hyacinth

Camelina Crantz (Brassicaceae)
false flax
microcarpa Andrz. - Dutch flax, false flax, flat-seeded false flax, Siberian oil-seed, small-seeded false flax, western flax
sativa (L.) Crantz - large-seeded false flax

Camellia L. (Theaceae)
japonica L. - camellia
sasanqua Thunb. - Sasanqua camellia
sinensis (L.) Kuntze - tea, tea-plant

Camissonia Link (Onagraceae)
bistorta (Nutt. ex Torr. & A. Gray) Raven - California sun-cup, sun-cup
subacaulis (Pursh) Raven - sun-cup

Campanula L. (Campanulaceae)
bellflower
americana L. - American bellflower, tall bellflower
aparinoides Pursh - bedstraw bellflower, marsh bellflower
carpatica Jacq. - tussock bellflower
divaricata Michx. - Appalachian bellflower, southern harebell
glomerata L. - clustered bellflower
medium L. - Canterbury bells
persicifolia L. - peach-bells, willow bellflower
rapunculoides L. - bellflower, creeping bellflower, purple bell, rover bellflower
rotundifolia L. - bluebell, harebell
trachelium L. - nettle-leaf bellflower, throat-wort

Campsis Lour. (Bignoniaceae)
radicans (L.) Seem. ex Bureau - cow-itch, trumpet-creeper, trumpet-honeysuckle, trumpet-vine

Canavalia DC. (Fabaceae)
ensiformis (L.) DC. - giant stock bean, horse bean, jack bean, sword bean, wonder bean
gladiata (Jacq.) DC. - sword bean

Canella P. Browne (Canellaceae)
winteriana (L.) Gaertn. - canella, cinnamon-bark, wild cinnamon

Canna L. (Cannaceae)
×*generalis* L. H. Bailey - garden canna
indica L. - achira, edible canna, gruya, Indian-shot, Queensland arrowroot, Spanish arrowroot, tous-les-mois

Cannabis L. (Cannabaceae)
hemp
sativa L. - gallow-grass, marijuana, neck-weed, red-root, soft hemp, true hemp

Canotia Torr. (Celastraceae)
holocantha Torr. - canotia

Caperonia A. St.-Hil. (Euphorbiaceae)
castaniifolia (L.) A. St.-Hil. - Mexican-weed
palustris (L.) A. St.-Hil. - Texas-weed

Capparis L. (Capparaceae)
caper-bush
cynophallophora L. - black-willow, Jamaican caper, Jamaican caper-tree

Capsella Medik. (Brassicaceae)
bursa-pastoris (L.) Medik. - shepherd's-purse

Capsicum L. (Solanaceae)
annuum L. - bell pepper, Cayenne pepper, chili pepper, garden pepper, green pepper, mango pepper, paprika pepper, pimento, pimiento
frutescens L. - bird pepper, Cayenne pepper, chili pepper, tabasco pepper
frutescens L. cv.'grossum' - bell pepper, sweet pepper

Caragana Fabr. (Fabaceae)
pea-tree
arborescens Lam. - Siberian pea-tree

Cardamine L. (Brassicaceae)
bitter-cress
bulbosa (Schreb. ex Muhl.) Britton, et al. - spring-cress
concatenata (Michx.) O. Schwarz - cut-leaf toothwort
diphylla (Michx.) A. Wood - crinkle-root, pepperwort, toothwort
douglasii (Torr.) Britton - purple-cress
hirsuta L. - hairy bitter-cress
oligosperma Nutt. - little bitter-cress
parviflora L. - dry-land bitter-cress, small-flowered bitter-cress
pensylvanica Muhl. ex Willd. - Pennsylvania bitter-cress
pratensis L. - cuckoo bitter-cress, cuckoo-flower, lady's-smock
rotundifolia Michx. - mountain winter-cress, trailing bitter-cress

Cardaria Desv. (Brassicaceae)
 chalepensis (L.) Hand.-Mazz. - lens-podded white-top
 draba (L.) Desv. - heart-padded hoary-cress, hoary
   pepperwort, hoary-cress, perennial peppergrass, white-top
 pubescens (C. A. Mey.) Jarm. - globe-podded hoary-cress,
   hairy white-top, Siberian mustard, white-top

Cardiospermum L. (Sapindaceae)
 halicababum L. - balloon-vine, heart-seed

Carduus L. (Asteraceae)
 acanthoides L. - plumeless thistle
 crispus L. - welted thistle
 nutans L. - musk thistle, nodding thistle, plumeless thistle
 pycnocephalus L. - Italian thistle
 tenuiflorus Curtis - slender-flower thistle

Carex L. (Cyperaceae)
   sedge
 adusta Boott - browned sedge
 aestivalis M. A. Curtis - summer sedge
 albolutescens Schwein. - yellow-white sedge
 albursina E. Sheld. - white bear sedge
 alopecoidea Tuck. - foxtail sedge
 angustata Boott - wide-fruit sedge
 aquatilis Wahlenb. - water sedge
 arcta Boott - northern clustered sedge
 arctata Boott - drooping wood sedge
 atherodes Spreng. - sugar-grass sedge
 atratiformis Britton - black sedge
 aurea Nutt. - golden-fruited sedge
 brunnescens (Pers.) Poir. - brownish sedge
 canescens L. - silvery sedge
 cephaloidea Dewey - thin-leaf sedge
 cephalophora Muhl. - oval-headed sedge
 cherokeensis Schwein. - wolf-tail sedge
 chordorrhiza L.f. - creeping sedge
 comosa Boott - bristly sedge
 concinnoides Mack. - low northern sedge
 conoidea Schkuhr - field sedge
 crinita Lam. - fringe sedge
 cristatella Britton - crested sedge
 deflexa Hornem. - northern sedge
 filifolia Nutt. - thread-leaf sedge
 foenea Willd. - hay sedge
 frankii Kunth - Frank's sedge
 geyeri Boott - elk sedge
 glaucescens Elliott - waxy sedge
 gracillima Schwein. - graceful sedge
 gravida L. H. Bailey - heavy sedge
 hystricina Muhl. - bottlebrush sedge, porcupine sedge
 interior L. H. Bailey - inland sedge
 intumescens Rudge - bladder sedge
 lacustris Willd. - rip-gut sedge
 lanuginosa Michx. - woolly sedge
 lasiocarpa Ehrh. - slender sedge, wool-fruit sedge
 laxiflora Lam. - loose-flowered sedge
 lenticularis Michx. - lenticular sedge
 leptalea Wahlenb. - bristle-stalk sedge, flaccid sedge
 limosa L. - mud sedge
 louisianica Bailey - Louisiana sedge
 lupuliformis Sartwell ex Dewey - hop-like sedge
 lupulina Muhl. ex Willd. - hop sedge
 lurida Wahlenb. - sallow sedge

 muricata L. - lesser prickly sedge
 nebrascensis Dewey - Nebraska sedge
 nigromarginata Schwein. var. *elliptica* (Boott) Gleason -
   black-edged sedge
 oligosperma Michx. - few-seeded sedge
 pallescens L. - pale sedge
 pauciflora Lightf. - few-flowered sedge
 pedunculata Muhl. - long-stalked sedge
 plantaginea Lam. - plantain-leaved sedge
 platyphylla J. Carey - thicket sedge
 podocarpa R. Br. - long-awn arctic sedge
 rariflora (Wahlenb.) Sm. - loose-flowered alpine sedge
 retrorsa Schwein. - retrorse sedge
 riparia Curtis - river bank sedge
 rosea Schkuhr - stellate sedge
 rostrata Stokes - beaked sedge
 saxatilis L. - russet sedge
 scabrata Schwein. - rough sedge
 scoparia Schkuhr - pointed broom-sedge
 senta Boott - rough sedge, rough senta
 siccata Dewey - dry-spiked sedge, hillside sedge
 sterilis Willd. - little prickly sedge
 stipata Muhl. - awn-fruited sedge
 straminea Willd. - straw sedge
 stricta Lam. - tussock sedge
 supina Willd. - weak arctic sedge
 tenera Dewey - marsh straw sedge
 tetanica Schkuhr - wood sedge
 torta Boott - twisted sedge
 tribuloides Wahlenb. - blunt-broom sedge
 trisperma Dewey - three-fruited sedge
 verrucosa Muhl. - nerved waxy sedge
 vestita Willd. - velvet sedge
 viridula Michx. - green sedge
 vulpinoidea Michx. - fox sedge

Carica L. (Caricaceae)
   papaya
 papaya L. - melon-tree, papaya, pawpaw

Carissa L. (Apocynaceae)
 carandas L. - karanda, perunkila
 macrocarpa (Eckl.) A. DC. - amatungulu, big num-num,
   Natal-plum

Carnegiea Britton & Rose (Cactaceae)
 gigantea (Engelm.) Britton & Rose - Arizona-giant, giant
   cactus, giant saguaro, saguaro, sahuaro

Carpinus L. (Betulaceae)
   hornbeam, ironwood
 betulus L. - European hornbeam
 caroliniana Walter - American hornbeam, blue-beech,
   ironwood, muscle-wood, water-beech

Carpobrotus N. E. Br. (Aizoaceae)
 chilensis (Molina) N. E. Br. - sea-fig
 edulis (L.) N. E. Br. - hottentot-fig

Carthamus L. (Asteraceae)
 lanatus L. - distaff thistle, woolly distaff thistle
 oxyacantha M. Bieb. - carthamus
 tinctorius L. - bastard-saffron, false saffron, safflower,
   strawberry-clover

Carum L. (Apiaceae)
carvi L. - caraway

Carya Nutt. (Juglandaceae)
   hickory
aquatica (F. Michx.) Nutt. - bitter hickory, bitter pecan, water hickory
cordiformis (Wangenh.) K. Koch - bitter-nut hickory, pig-nut, swamp hickory
floridana Sarg. - Florida hickory, scrub hickory
glabra (Mill.) Sweet - broom hickory, pig-nut hickory, small-fruited hickory
illinoensis (Wangenh.) K. Koch - pecan
laciniosa (F. Michx.) Loudon - king's-nut, shellbark hickory
myristiciformis (F. Michx.) Nutt. - nutmeg hickory
ovalis (Wangenh.) Sarg. - false shagbark, sweet pig-nut
ovata (Mill.) K. Koch - shagbark hickory, shellbark hickory
pallida (Ashe) Engl. & Graebn. - pale hickory, sand hickory
texana Buckley - black hickory, Buckley's hickory, Ozark hickory
tomentosa (Poir.) Nutt. - mocker-nut hickory, square-nut, white hickory, white-heart hickory

Caryota L. (Arecaceae)
   fishtail palm
mitis Lour. - Burmese fishtail palm, clustered fishtail palm, tufted fishtail palm
urens L. - fishtail palm, jaggery palm, kittul-tree, sago palm, toddy palm, wine palm

Casimiroa La Llave & Lex. (Rutaceae)
edulis La Llave & Lex. - Mexican apple, white sapote, zapote-blanco

Cassia L. (Fabaceae)
   senna, shower-tree
artemisioides Gaudich. - feathery cassia, shower-tree, wormwood cassia, wormwood senna
fistula L. - golden-rain, golden-shower, Indian laburnum, pudding-pipe-tree, purging cassia, purging fistula
javanica L. var. *indochinensis* Gagnep. - joint-wood, pink-and-white-shower

Castanea Mill. (Fagaceae)
   chestnut
alnifolia Nutt. - Florida chinkapin
crenata Siebold & Zucc. - Japanese chestnut
dentata (Marsh.) Borkh. - American chestnut, chestnut
mollissima Blume - Chinese chestnut
pumila (L.) Mill. - Allegheny chinquapin
sativa Mill. - Eurasian chestnut, European chestnut, Spanish chestnut

Castanopsis (D. Don) Spach (Fagaceae)
   chinquapin
chrysophylla (Douglas ex Hook.) A. DC. - giant chinquapin, golden chinquapin
sempervirens (Kellogg) T. R. Dudley - bush chinquapin, California chinquapin, sierra chinquapin

Castanospermum A. Cunn. ex Hook. (Fabaceae)
australe A. Cunn. & C. Fraser - Australian chestnut, black bean, Moreton Bay chestnut

Castilleja Mutis ex L.f. (Scrophulariaceae)
   Indian paintbrush, painted-cup

affinis Hook. & Arn. - Indian paintbrush
coccinea (L.) Spreng. - Indian paintbrush, painted-cup, scarlet paintbrush
foliolosa Hook. & Arn. - woolly Indian paintbrush, woolly painted-cup
latifolia Hook. & Arn. - Monterey Indian paintbrush, seaside painted-cup
sessiliflora Pursh - downy paintbrush

Casuarina L. (Casuarinaceae)
   Australian pine, beefwood-tree, she-oak
cunninghamiana Miq. - beefwood, Cunningham's beefwood, river she-oak
equisetifolia L. - Australian pine, beefwood-tree, horsetail casuarina, horsetail-tree, mile-tree, South Sea ironwood
glauca Sieber - Brazilian oak, scaly-bark beefwood
stricta Aiton - drooping she-oak

Catabrosa P. Beauv. (Poaceae)
   brook grass
aquatica (L.) P. Beauv. - water-hairbrush

Catalpa Scop. (Bignoniaceae)
bignonioides Walter - catalpa, catawba, Indian bean catalpa, southern catalpa
ovata G. Don - Chinese catalpa
speciosa (Warder ex Barney) Warder ex Engelm. - catawba-tree, cigar-tree, Indian bean, northern catalpa, western catalpa

Catha Forssk. ex Scop. (Celastraceae)
edulis (Vahl) Forssk. ex Endl. - Abyssinian tea, Arabian tea, cafta, chat, kat, khat, qat, Somali tea

Catharanthus G. Don (Apocynaceae)
roseus (L.) G. Don - bright-eyes, Madagascar periwinkle, rose periwinkle

Caulophyllum Michx. (Berberidaceae)
thalictroides (L.) Michx. - blue cohosh, papoose-root

Ceanothus L. (Rhamnaceae)
   red-root
americanus L. - Jersey tea ceanothus, mountain-sweet, New Jersey tea, red-root, wild snowball
arboreus Greene - Catalina ceanothus, Catalina mountain-lilac, felt-leaf ceanothus
cordulatus Kellogg - mountain whitethorn, snowbush
cuneatus (Hook.) Nutt. - buck-brush
cuneatus (Hook.) Nutt. var. *rigidus* (Nutt.) Hoover - Monterey ceanothus
gloriosus J. T. Howell - Point Reyes ceanothus, Point Reyes creeper
griseus (Trel.) McMinn - Carmel ceanothus
griseus (Trel.) McMinn var. *horizontalis* McMinn - Carmel creeper, Yankee Point ceanothus
integerrimus Hook. & Arn. - deer-brush, deer-bush
sanguineus Pursh - red-stem ceanothus, wild lilac
spinosus Nutt. - green-bark ceanothus, spiny ceanothus
thyrsiflorus Eschsch. - blue-blossum, blue-brush

Cedrus Trew (Pinaceae)
   cedar
atlantica (Endl.) G. Manetti ex Carrière - Atlas cedar
deodara (D. Don) G. Don - deodar, deodara cedar

Ceiba Mill. (Bombacaceae)
  pentandra (L.) Gaertn. - ceiba, kapok, silk cotton-tree, white silk cotton-tree

Celastrus L. (Celastraceae)
  bittersweet, shrubby bittersweet
  orbiculatus Thunb. - Oriental bittersweet
  scandens L. - American bittersweet, climbing bittersweet, false bittersweet, shrubby bittersweet, staff-vine, wax-work

Celosia L. (Amaranthaceae)
  argentea L. - celosia, veludo-branco, wool-flower
  argentea L. var. *cristata* (L.) Kuntze - cock's-comb, crista-de-galo

Celtis L. (Ulmaceae)
  hackberry, nettle-tree, sugarberry
  laevigata Willd. - Mississippi hackberry, southern hackberry, sugarberry
  occidentalis L. - hackberry, nettle-tree, northern hackberry, sugarberry
  pallida Torr. - desert hackberry, granjeno, spiny hackberry
  reticulata Torr. - net-leaf hackberry

Cenchrus L. (Poaceae)
  sandbur
  echinatus L. - hedgehog grass, southern sandbur
  incertus M. A. Curtis - bur grass, coast sandbur, field sandbur
  longispinus (Hack.) Fernald - long-spine sandbur, mat sandpur, sandbur
  tribuloides L. - dune sandbur

Centaurea L. (Asteraceae)
  knapweed
  americana Nutt. - American knapweed, basket-flower, cardo-del-valle, thornless thistle
  calcitrapa L. - purple star-thistle
  cineraria L. - dusty-miller
  cyanus L. - bachelor's-button, bluebottle, cornflower
  diffusa Lam. - diffuse knapweed, spreading knapweed, tumble knapweed
  iberica Trevir. ex Spreng. - Iberian star-thistle
  jacea L. - brown knapweed
  macrocephala Muss. Puschk. - big-head knapweed
  maculosa Lam. - spotted knapweed
  melitensis L. - Malta star-thistle, tocalote
  montana L. - mountain bluet
  nigra L. - black knapweed
  nigrescens Willd. - Vochin knapweed
  pratensis Thuill. - meadow knapweed
  repens L. - creeping knapweed, Russian knapweed, Turkestan thistle
  solstitialis L. - yellow star-thistle
  squarrosa Willd. - squarrose knapweed
  trichocephala M. Bieb. ex Willd. - feather-head knapweed

Centaurium Hill (Gentianaceae)
  erythraea Raf. - centaury

Centella L. (Apiaceae)
  asiatica (L.) Urb. - Asiatic pennywort

Centranthus Neck. ex Lam. & A. DC. (Valerianaceae)
  ruber (L.) DC. - fox-brush, Jupiter's-beard, red valerian

Centrosema (DC.) Benth. (Fabaceae)
  butterfly pea, cochita
  pubescens Benth. - butterfly pea
  virginianum (L.) Benth. - spurred butterfly pea

Cephalanthus L. (Rubiaceae)
  occidentalis L. - buttonbush

Cephalotaxus Siebold & Zucc. ex Endl. (Cephalotaxaceae)
  plum-yew
  harringtonia (Forbes) K. Koch - Harrington's plum-yew
  harringtonia (Forbes) K. Koch var. *drupacea* (Siebold & Zucc.) Koidz. - cow's-tail-pine, Japanese plum-yew, plum-fruited yew

Cerastium L. (Caryophyllaceae)
  mouse-ear chickweed
  arvense L. - field chickweed, meadow chickweed, starry glasswort
  glomeratum Thuill. - mouse-ear chickweed, sticky chickweed
  nutans Raf. - nodding chickweed
  tomentosum L. - snow-in-summer
  vulgatum L. - mouse-ear chickweed

Ceratonia L. (Fabaceae)
  siliqua L. - algarroba bean, carob, locust bean, St. John's-bread

Ceratophyllum L. (Ceratophyllaceae)
  demersum L. - coon's-tail

Ceratopteris Brongn. (Pteridophyta)
  pteridoides (Hook.) Hieron. - floating fern
  thalictroides (L.) Brongn. - floating water fern, water fern, water-sprite

Cercidiphyllum Siebold & Zucc. (Cercidiphyllaceae)
  japonicum Siebold & Zucc. - katsura-tree

Cercis L. (Fabaceae)
  canadensis L. - eastern redbud
  chinensis Bunge - Chinese redbud
  occidentalis Torr. - California redbud, western redbud

Cercocarpus Kunth (Rosaceae)
  betuloides Torr. & A. Gray - birch-lead mountain-mahogany
  montanus Raf. - mountain-mahogany

Cereus Mill. (Cactaceae)
  uruguayanus Kiesling - Peruvian apple cactus

Cestrum L. (Solanaceae)
  diurnum L. - day jessamine, day-blooming cestrum
  nocturnum L. - night jessamine

Chaenomeles Lindl. (Rosaceae)
  japonica (Thunb.) Spach - flowering quince, Maule's quince
  speciosa (Sweet) Nakai - Japanese quince

Chaenorhinum (DC. ex Duby) Rchb. (Scrophulariaceae)
  dwarf snapdragon
  minus (L.) Lange - dwarf snapdragon, lesser toadflax, small snapdragon

Chaerophyllum L. (Apiaceae)
  tainturieri Hook. - hairy-fruit chervil, southern chervil

Chamaebatia Benth. (Rosaceae)
  foliolosa Benth. - mountain-misery

Chamaebatiaria (Porter) Maxim. (Rosaceae)
  millefolium (Torrey) Maxim. - fern-bush

Chamaecrista Moench (Fabaceae)
fasciculata (Michx.) Greene - partridge pea
nictitans (L.) Moench - sensitive partridge pea, wild sensitive-plant

Chamaecyparis Spach (Cupressaceae)
false cypress
lawsoniana (A. Murray bis) Parl. - Lawson's cypress, Port Orford cedar
nootkatensis (D. Don) Spach - Alaska cedar, Alaska yellow-cedar, Nootka yellow-cedar
obtusa (Siebold & Zucc.) Siebold & Zucc. ex Endl. - Hinoki cypress, Hinoki false cypress, Japanese false cypress
pisifera (Siebold & Zucc.) Endl. - Sawara cypress
thyoides (L.) Britton, et al. - Atlantic white-cedar, southern white-cedar, swamp white-cedar

Chamaedaphne Moench (Ericaceae)
calyculata (L.) Moench - cassandra, leather-leaf

Chamaedorea Willd. (Arecaceae)
elegans Mart. - good-luck palm, pacaya, parlor palm
erumpens H. E. Moore - bamboo palm

Chamaemelum Mill. (Asteraceae)
chamomile
nobile (L.) All. - garden chamomile, Roman chamomile

Chamaesaracha (A. Gray) Benth. (Solanaceae)
coronopus (Dunal) A. Gray - dwarf ground-chervil

Chasmanthium Link (Poaceae)
latifolium (Michx.) Yates - wild oats

Cheilanthes Sw. (Pteridophyta)
lip fern
feei Moore - Fee's lip fern, slender lip fern
gracillima D. C. Eaton - lace lip fern
lanosa (Michx.) D. C. Eaton - lanate lip fern, woolly lip fern

Chelidonium L. (Papaveraceae)
majus L. - celandine, greater celandine, swallow-wort, wart-weed

Chelone L. (Scrophulariaceae)
snakehead, turtle-head
glabra L. - balmony, snakehead, turtle-head, white turtle-head
lyonii Pursh - red turtle-head

Chenopodium L. (Chenopodiaceae)
goosefoot, pigweed
album L. - bacon-weed, fat-hen, frost-blite, goosefoot, lamb's-quarters, meal-weed, pigweed, white goosefoot
ambrosioides L. - American worm-seed, Jerusalem tea, Mexican tea, Spanish-tea, strong-scented pigweed, worm-seed
berlandieri Moq. - net-seed lamb's-quarters, pit-seed goosefoot
bonus-henricus L. - all-good, fat-hen, good-King-Henry, mercury, perennial goosefoot, wild spinach
botrys L. - feather-geranium, Jerusalem oak goosefoot, turnpike goosefoot
capitatum (L.) Asch. - blite goosefoot, blite-mulberry, Indian-paint, strawberry pigweed, strawberry-blite, strawberry-spinach
desiccatum A. Nels. - narrow-leaf lamb's-quarters
ficifolium Sm. - fig-leaved goosefoot

glaucum L. - oak-leaf goosefoot
hybridum L. - maple-leaf goosefoot, sow-bane
incanum (S. Watson) A. Heller - mealy goosefoot
leptophyllum (Moq.) S. Watson - slim-leaf lamb's-quarters
missouriense Aellen - Missouri goosefoot
murale L. - nettle-leaf goosefoot, sow-bane, swine-bane
polyspermum L. - many-seeded goosefoot
quinoa Willd. - quinoa, quinua
rubrum L. - coast-blite, red goosefoot
simplex (Torr.) Raf. - maple-leaf goosefoot
strictum Roth var. *glaucophyllum* (Aellen) Wahl - late-flowering goosefoot
urbicum L. - city goosefoot
vulvaria L. - stinking goosefoot

Chilopsis D. Don (Bignoniaceae)
linearis (Cav.) Sweet - desert-willow, flowering-willow

Chimaphila Pursh (Ericaceae)
wax-flower, wintergreen
maculata (L.) Pursh - spotted wintergreen
umbellata (L.) Barton - pipsissewa, prince's-pipe
umbellata (L.) Barton var. *cisatlantica* S. F. Blake - pipsissewa, prince's-pipe

Chimonanthus Lindl. (Calycanthaceae)
praecox (L.) Link - winter-sweet

Chiococca P. Browne (Rubiaceae)
alba (L.) Hitchc. - snowberry

Chionanthus L. (Oleaceae)
virginicus L. - fringe-tree, old-man's-beard

Chloranthus Sw. (Chloranthaceae)
spinosa (Benth.) G. L. Nesom - spiny aster

Chloris Sw. (Poaceae)
finger grass, windmill grass
ciliata Sw. - fringed chloris
gayana Kunth - Rhodes's grass
petraea Sw. - rock finger grass
radiata (L.) Sw. - radiate finger grass
verticillata Nutt. - tumble windmill grass, windmill grass
virgata Sw. - feather finger grass

Chlorogalum (Lindl.) Kunth (Liliaceae)
pomeridianum (DC.) Kunth - amole, soap-plant, wild potato

Chlorophytum Ker Gawl. (Liliaceae)
comosum (Thunb.) Jacques - ribbon-plant, spider-ivy, spider-plant, walking anthericum

Chondrilla L. (Asteraceae)
juncea L. - gum succory, rush skeleton-weed, skeleton-weed

Chorisia Kunth (Bombacaceae)
speciosa A. St.-Hil. - floss-silk-tree

Chorispora R. Br. ex DC. (Brassicaceae)
tenella R. Br. ex DC. - bead-podded mustard, blue mustard

Chromolaena DC. (Asteraceae)
odorata (L.) R. M. King & H. Rob. - bitter-bush

Chrysalidocarpus H. Wendl. (Arecaceae)
lutescens H. Wendl. - areca palm, butterfly palm, cane palm, golden cane palm, golden feather palm, Madagascar palm, yellow butterfly palm, yellow palm

Chrysanthemum L. (Asteraceae)
balsamita L. - costmary, costmary chrysanthemum, mint-geranium
cinerariifolium (Trevir.) Vis. - Dalmatian insect-flower, Dalmatian pyrethrum, pyrethrum
coccineum Willd. - painted daisy, Persian insect-flower, pyrethrum
coronarium L. - crown daisy, garland chrysanthemum
frutescens L. - marguerite, Paris daisy, white daisy, white marguerite, white-weed
×*morifolium* Ramat. - florist's chrysanthemum, mum
segetum L. - corn chrysanthemum, corn-marigold

Chrysobalanus L. (Chrysobalanaceae)
icaco L. - coco-plum, icaco

Chrysopogon Trin. (Poaceae)
aciculatus (Retz.) Trin. - pilipiliula

Chrysopsis (Nutt.) Elliott (Asteraceae)
golden aster
camporum Green - prairie golden aster
falcata (Pursh) Elliott - falcate golden aster, sickle-leaf golden aster
graminifolia (Michx.) Elliott - grass-leaf golden aster, silk-grass
mariana (L.) Elliott - Maryland golden aster, shaggy golden aster

Chrysothamnus Nutt. (Asteraceae)
rabbit-brush
nauseosus (Pall. ex Pursh) Britton - gray rabbit-brush, rubber rabbit-brush
nauseosus (Pall. ex Pursh) Britton subsp. *graveolens* (Nutt.) Piper - green-plume rabbit-brush
paniculatus (Gray) Hall - desert rabbit-brush
parryi (A. Gray) Greene - Parry's rabbit-brush
pulchellus (A. Gray) Greene - southwestern rabbit-brush
viscidiflorus (Hook.) Nutt. - Douglas's rabbit-brush, yellow rabbit-brush

Cibotium Kaulf. (Pteridophyta)
chamissoi Kaulf. - hapuu-ii, Hawaiian tree-fern
splendens (Gaudich.) Krajina - blond tree-fern, hapuu, Hawaiian tree-fern, man tree-fern

Cicer L. (Fabaceae)
arietinum L. - Bengal gram, chickpea, Egyptian pea, garbanzo, yellow gram

Cichorium L. (Asteraceae)
chicory
endivia L. - endive, escarole
intybus L. - barbe-de-capuchin, blue daisy, blue-sailors, bunk, chicory, coffee-weed, succory, witloof

Ciclospermum Lag. & Segura (Apiaceae)
leptophyllum (Pers.) Britton & E. H. Wilson - wild celery

Cicuta L. (Apiaceae)
water-hemlock
bulbifera L. - bulb-bearing water-hemlock
douglasii (DC.) J. M. Coult. & Rose - western water-hemlock
maculata L. - beaver-poison, children's-bane, muskrat-weed, musquash-root, spotted cowbane, spotted water-hemlock, water-hemlock

Cimicifuga Wernisch. (Ranunculaceae)
bugbane, rattle-top
americana Michx. - American bugbane, mountain bugbane, summer cohosh
racemosa (L.) Nutt. - black cohosh, black snakeroot, bugbane

Cinchona L. (Rubiaceae)
officinalis L. - quinine

Cinna L. (Poaceae)
arundinacea L. - stout wood-reed, wood-reed
latifolia (Trevir. ex Göpp.) Griseb. - drooping wood-reed

Cinnamomum Schaeff. (Lauraceae)
camphora (L.) J. Presl - camphor-tree
verum J. Presl - Ceylon cinnamon, cinnamon

Circaea L. (Onagraceae)
enchanter's-nightshade
alpina L. - small enchanter's-nightshade
quadrisulcata (Maxim.) Franch. & Sav. - enchanter's-nightshade

Cirsium Mill. (Asteraceae)
plume thistle, thistle
altissimum (L.) Spreng. - tall thistle
arvense (L.) Scop. - Canadian thistle, creeping thistle, green thistle, perennial thistle, small-flowered thistle
brevifolium Nutt. - palouse thistle
discolor (Muhl.) Spreng. - field thistle
edule Nutt. - Indian thistle
flodmanii (Rydb.) Arthur - Flodman's thistle, prairie thistle
foliosum (Hook.) DC. - leafy thistle
horridulum Michx. - yellow thistle
muticum Michx. - swamp thistle
occidentale (Nutt.) Jeps. - western thistle
ochrocentrum A. Gray - yellow-spine thistle
palustre (L.) Scop. - marsh thistle
pumilim (Nutt.) Spreng. - bull thistle, fragrant thistle, pasture thistle
pumilum (Nutt.) Spreng. - pasture thistle
undulatum (Nutt.) Spreng. - wavy-leaf thistle
virginianum Hook. - plume thistle, spear thistle, Virginia thistle
vulgare (Savi) Ten. - bull thistle

Cissus L. (Vitaceae)
grape ivy, ivy, tree-bine
incisa (Nutt. ex Torr. & A. Gray) Des Moul. - marine ivy, marine-vine, possum-grape
rhombifolia Vahl - grape ivy, Venezuelan tree-bine
sicyoides L. - princess-vine
verticillata (L.) Nicolson & Jarvis - possum grape

Cistus L. (Cistaceae)
rock-rose
palhinhai Ingram - St. Vincent's cistus

Citrofortunella J. W. Ingram & H. E. Moore (Rutaceae)
Note = *Citrus* × *Fortunella*.
×*mitis* (Blanco) J. W. Ingram & H. E. Moore - calamondin, Panama orange
Note = *Citrus reticulata* × *Fortunella* sp.

Citrullus Schrad. ex Eckl. & Zeyh. (Cucurbitaceae)
colocynthis (L.) Schrad. - bitter-apple, vine-of-Sodom
lanatus (Thunb.) Matsum. & Nakai - watermelon

lanatus (Thunb.) Matsum. & Nakai var. *citroides* (Bailey) Mansf. - citron, citron-melon, preserving melon, stock melon

Citrus L. (Rutaceae)
aurantiifolia (L.) Swingle - key lime, lime, Mexican lime, West Indian lime
aurantium L. - bigarade, bitter orange, Seville orange, sour orange
jambhiri Lush. - rough lemon
latifolia (Yu. Tanaka) Tanaka - Tahitian lime
limon (L.) Burm. f. - lemon
×*limonia* Osbeck - lemandarin, Mandarin lime, Rangpur lime
Note = *C. limon* × *C. reticulata.*
maxima (Burm.) Merr. - pummelo, shaddock
medica L. - citron
×*paradisi* Macfad. - grapefruit, pamplemousse, pomelo
Note = *C. maxima* × *C. sinensis.*
reticulata Blanco - king orange, Mandarin orange, Satsuma orange, tangerine
sinensis (L.) Osbeck - orange, sweet orange
×*tangelo* J. W. Ingram & H. E. Moore - tangelo
Note = *C. paradisi* × *C. reticulata.*

Cladium P. Browne (Cyperaceae)
jamaicense Crantz - sawgrass

Cladrastis Raf. (Fabaceae)
lutea (F. Michx.) K. Koch - virgilia, yellow-wood

Clarkia Pursh (Onagraceae)
farewell-to-spring, godetia
amoena (Lehm.) A. Nelson - farewell-to-spring, satinflower

Clausena Burm. f. (Rutaceae)
lansium (Lour.) Skeels - wampi

Claytonia L. (Portulacaceae)
spring-beauty
caroliniana Michx. - broad-leaf spring-beauty, Carolina spring-beauty
megarrhiza (A. Gray) Parry ex S. Watson - spring-beauty
perfoliata Donn ex Willd. - miner's-lettuce, winter-purslane
sibirica L. - Siberian purslane
virginica L. - narrow-leaf spring-beauty, spring-beauty

Clematis L. (Ranunculaceae)
clematis, leather-flower, vase-vine, virgin's-bower
crispa L. - blue-jasmine
glaucophylla Small - leather-flower
ochroleuca Aiton - curly-heads
versicolor Small - leather-flower
viorna L. - leather-flower, vase-vine
virginiana L. - devil's-darning-needle, leather-flower, virgin's-bower, woodbine

Cleome L. (Capparaceae)
spider-plant
hassleriana Chodat - spider-flower
lutea Hook. - golden cleome, yellow bee-plant, yellow cleome
serrulata Pursh - Rocky Mountain bee-plant, stinking-clover
spinosa Jacq. - spiny spider-flower

Clerodendrum L. (Verbenaceae)
glory-bower, Kashmir boquet, tube-flower
thompsoniae Balf. - bag-flower, bleeding-glory-bower, glory-tree, tropical bleeding-heart

Clethra L. (Clethraceae)
white-alder
acuminata Michx. - mountain white-alder, sweet pepper-bush
alnifolia L. - alder-leaf pepper-bush, coast white-alder, summer-sweet, sweet pepper-bush

Clidemia D. Don (Melastomataceae)
hirta (L.) D. Don - camasey, soap-bush

Cliftonia Banks ex C. F. Gaertn. (Cyrillaceae)
monophylla (Lam.) Britton ex Sarg. - black titi, buckwheat-bush, buckwheat-tree, ironwood, titi-tree

Clinopodium L. (Lamiaceae)
vulgare L. - wild basil

Clintonia Raf. (Liliaceae)
borealis (Aiton) Raf. - blue-bead-lily, corn-lily, yellow clintonia
uniflora (Schult.) Kunth - bride's-bonnet, queen's-cup

Clitoria L. (Fabaceae)
mariana L. - butterfly pea
ternatea L. - aparajita, Asian pigeon-wings, blue pea, butterfly pea, conchitas, gokorna, zapatica-de-la-reina

Clusia L. (Clusiaceae)
rosea Jacq. - balsam-apple, copey, cupey, Scotch-attorney

Cnicus L. (Asteraceae)
benedictus L. - blessed thistle

Cnidoscolus Pohl (Euphorbiaceae)
stimulosus (Michx.) Engelm. & A. Gray - bull-nettle, spurge-nettle, tread-softly
texanus (Müll. Arg.) Small - Texas bull-nettle, tread-softly

Coccoloba P. Browne (Polygonaceae)
diversifolia Jacq. - dove-plum, pigeon-plum, snail-seed
uvifera (L.) L. - kino, platter-leaf, sea-grape

Coccothrinax Sarg. (Arecaceae)
fan palm
argentata (Jacq.) L. H. Bailey - Florida silver palm

Cocculus DC. (Menispermaceae)
carolinus (L.) DC. - Carolina moonseed, coral-beads, red moonseed, red-berry moonseed, snail-seed

Cochlearia L. (Brassicaceae)
officinalis L. - scurvy-grass

Cocos L. (Arecaceae)
nucifera L. - coconut, coconut palm

Codiaeum A. Juss. (Euphorbiaceae)
croton, variegated-laurel
variegatum (L.) Blume - garden-croton

Coffea L. (Rubiaceae)
arabica L. - arabica coffee, coffee
canephora Pierre ex A. Froehner - robusta coffee, wild robusta coffee

Coix L. (Poaceae)
lacryma-jobi L. - adlay, adlay millet, Job's-tears

Colchicum L. (Liliaceae)
autumnale L. - autumn crocus, meadow-saffron, mysteria, wonder-bulb

Collinsia Nutt. (Scrophulariaceae)
grandiflora Lindl. - blue-lips
verna Nutt. - blue-eyed-Mary, eastern blue-eyed-Mary, innocence

Collinsonia L. (Lamiaceae)
horse-balm, horseweed
canadensis L. - citronella, horse-balm, northern horse-balm, richweed, stone-root

Colocasia Schott (Araceae)
esculenta (L.) Schott - coco-yam, dasheen, eddo, elephant-ear-plant, kalo, taro, wild taro

Colubrina Rich. ex Brongn. (Rhamnaceae)
arborescens (Mill.) Sarg. - coffee colubrina, wild coffee
elliptica (Sw.) Brizicky & Stern - soldier-weed

Colutea L. (Fabaceae)
arborescens L. - bladder-senna

Comandra Nutt. (Santalaceae)
bastard-toadflax
umbellata (L.) Nutt. - bastard-toadflax

Commelina L. (Commelinaceae)
benghalensis L. - tropical spiderwort
communis L. - Asiatic dayflower, dayflower
diffusa Burm. f. - creeping dayflower, spreading dayflower
virginica L. - Virginia dayflower

Comptonia L'Hér. ex Aiton (Myricaceae)
asplenifolia L. - sweet-fern
peregrina (L.) J. M. Coult. - fern-gale, meadow-fern, shrubby-fern, spleenwort-bush, sweet-fern

Condalia Cav. (Rhamnaceae)
hookeri M. C. Johnst. - brasil
mexicana Schltdl. - Mexican blue-wood, Mexican condalia
spathulata A. Gray - squaw-bush

Conioselinum Hoffm. (Apiaceae)
pacificum (S. Watson) J. M. Coult. & Rose - hemlock-parsley

Conium L. (Apiaceae)
maculatum L. - California fern, deadly-hemlock, Nebraska fern, poison stinkweed, poison-hemlock, poison-parsley, snake-weed, spotted-hemlock, winter-fern, wode-whistle

Conocarpus L. (Combretaceae)
erectus L. - button mangrove, buttonwood
erectus L. var. *sericeus* DC. - silver button mangrove, silver buttonwood, silver-tree

Conoclinium DC. (Asteraceae)
coelestinum (L.) DC. - blue boneset, hardy ageratum, mistflower

Conringia Heist. ex Fabr. (Brassicaceae)
orientalis (L.) Dumort. - hare-ear mustard

Consolida S. F. Gray (Ranunculaceae)
ajacis (L.) Schur - rocket larkspur

Convallaria L. (Liliaceae)
majalis L. - lily-of-the-valley

Convolvulus L. (Convolvulaceae)
arvensis L. - bear-bind, bindweed, corn-bind, creeping-Jennie, field bindweed, green-vine, orchard morning-glory, small bindweed, small-flowered morning-glory, wild morning-glory
cneorum L. - silver-bush

Conyza Less. (Asteraceae)
bonariensis (L.) Cronquist - hairy fleabane
canadensis (L.) Cronquist - bitter-weed, blood-stanch, butterweed, fleabane, hogweed, horseweed, mare's-tail
floribunda Kunth - tall fleabane
ramosissima Cronquist - dwarf fleabane

Coprosma CJ. R. Forst. & G. Forst. (Rubiaceae)
repens A. Rich. - looking-glass-plant, mirror-plant

Coptis Salisb. (Ranunculaceae)
gold-thread
groenlandica (Oeder) Fernald - canker-root, gold-thread
trifolia (L.) Salisb. - canker-root

Corallorrhiza Gagnebin (Orchidaceae)
coralroot
maculata Raf. - large coralroot, spotted coralroot
odontorhiza (Willd.) Nutt. - autumn coralroot, late coralroot
trifida Chatelain - early coralroot, northern coralroot, pale coralroot

Corchorus L. (Tiliaceae)
aestuans L. - East Indian Jew's mallow

Cordia L. (Boraginaceae)
boissieri A. DC. - anacahuita

Cordyline Comm. ex Juss. (Agavaceae)
cabbage-tree
terminalis (L.) Kunth - good-luck-plant, ti, ti-palm, tree-of-kings

Coreopsis L. (Asteraceae)
tickseed
lanceolata L. - garden coreopsis, lance-leaf coreopsis, tickseed
palmata Nutt. - finger tickseed, stiff coreopsis
tinctoria Nutt. - calliopsis, plains coreopsis
tripteris L. - tall coreopsis, tall tickseed
verticillata L. - thread-leaf tickseed, whorled coreopsis

Coriandrum L. (Apiaceae)
sativum L. - Chinese parsley, cilantro, coriander, parsley

Corispermum L. (Chenopodiaceae)
hyssopifolium L. - hyssop-leaf tickseed

Cornus L. (Cornaceae)
cornel, dogwood
alternifolia L.f. - alternate-leaf dogwood, green osier, pagoda dogwood
amomum Mill. - knob-styled dogwood, red-willow, silky dogwood
asperifolia Michx. - small rough-leaf cornel
canadensis L. - bunchberry, cracker-berry, dwarf cornel, pudding-berry
drummondii C. A. Mey. - rough-leaf dogwood
florida L. - eastern flowering dogwood, flowering dogwood, white dogwood
glabrata Benth. - brown dogwood, western cornel
nuttallii Audubon - mountain dogwood, Pacific dogwood
obliqua Raf. - silky dogwood
occidentalis Coville - western osier

racemosa Lam. - gray dogwood, northern swamp dogwood, panicled dogwood
rugosa Lam. - round-leaf dogwood
sanguinea L. - blood-twig dogwood, dogberry, peg-wood
sericea L. - American dogwood, red osier dogwood
stricta Lam. - southern swamp dogwood, stiff dogwood, stiff-cornel dogwood

Coronilla L. (Fabaceae)
varia L. - ax-seed, crown-vetch, trailing crown-vetch

Coronopus Zinn (Brassicaceae)
 wart-cress
didymus (L.) Sm. - swine-cress

Cortaderia Stapf (Poaceae)
selloana (Schult. & Schult.f.) Asch. & Graebn. - pampas grass

Corydalis Vent. (Papaveraceae)
aurea Willd. - golden corydalis
flavula (Raf.) DC. - yellow corydalis, yellow fume-wort, yellow harlequin
micrantha (Engelm.) A. Gray - slender corydalis, slender fume-wort
sempervirens (L.) Pers. - pale corydalis, rock harlequin, Roman wormwood

Corylus L. (Betulaceae)
 filbert, hazel, hazelnut
americana Marsh. - American filbert, American hazelnut
avellana L. - European filbert, European hazelnut
cornuta Marsh. - beaked filbert, beaked hazelnut
cornuta Marsh. var. *californica* (A. DC.) W. M. Sharp - California hazelnut, western hazel
maxima Mill. - giant filbert

Cosmos Cav. (Asteraceae)
bipinnatus Cav. - cosmos, garden cosmos
sulphureus Cav. - orange cosmos, yellow cosmos

Costus L. (Costaceae)
malortieanus H. Wendl. - spiral-flag, spiral-ginger, stepladder-plant

Cotinus Mill. (Anacardiaceae)
coggygria Scop. - smoke-bush, smoke-tree, Venetian sumac, wig-tree
obovatus Raf. - American smoke-tree, chittamwood

Cotula L. (Asteraceae)
australis (Sieber ex Spreng.) Hook. f. - southern brass-buttons
coronopifolia L. - brass-buttons

Cowania D. Don (Rosaceae)
mexicana D. Don - cliff-rose

Crambe L. (Brassicaceae)
 sea kale
abyssinica Hochst. ex R. E. Fr. - colewort, crambe
maritima L. - scurvy-grass, sea kale

Crassula L. (Crassulaceae)
arborescens Willd. - Chinese jade-plant, silver jade-plant, silver-dollar
ovata (Mill.) Druce - cauliflower-ears, Chinese rubber-plant, dollar-plant, dwarf rubber-plant, jade-tree, Japanese rubber-plant

Crataegus L. (Rosaceae)
 hawthorn, red haw, thorn-apple
aestivalis (Walter) Torr. & A. Gray - hawthorn, May hawthorn
brainerdii Sarg. - Brainerd's hawthorn
calpodendron (Ehrh.) Medik. - black-thorn, pear hawthorn, pear-thorn
chrysocarpa Ashe - fireberry hawthorn, round-leaf hawthorn
coccinea L. - scarlet hawthorn
coccinioides Ashe - Kansas hawthorn
columbiana Howell - Columbia hawthorn
crus-galli L. - cockspur, cockspur hawthorn, cockspur-thorn
douglasii Lindl. - black hawthorn
erythropoda Ashe - cerro hawthorn
flabellata (Bosc) K. Koch - fan-leaf hawthorn
flava Aiton - summer-haw, yellow hawthorn, yellow-fruited-thorn
intricata Lange - Biltmore hawthorn, entangled hawthorn
laevigata (Poir.) DC. - English hawthorn, pear-thorn, quick-set-thorn
marshallii Eggl. - parsley hawthorn
mollis (Torr. & A. Gray) Scheele - downy hawthorn
monogyna Jacq. - English hawthorn, one-seeded hawthorn
phaenopyrum (L.f.) Medik. - Washington's hawthorn, Washington's-thorn
pruinosa (H. Wendl.) K. Koch - frosted hawthorn
punctata Jacq. - dotted hawthorn
rivularis Nutt. ex Torr. & A. Gray - river hawthorn
spathulata Michx. - little-hip hawthorn, pasture hawthorn
succulenta Schrad. - fleshy hawthorn
viridis L. - green hawthorn

Crepis L. (Asteraceae)
 hawk's-beard
biennis L. - rough hawk's-beard
capillaris (L.) Wallr. - smooth hawk's-beard
occidentalis Nutt. - western hawk's-beard
setosa Hallier f. - bristly hawk's-beard
tectorum L. - narrow-leaf hawk's-beard
vesicaria L. subsp. *haenseleri* (Boiss. ex DC.) P. D. Sell - European hawk's-beard

Crescentia L. (Bignoniaceae)
cujete L. - calabash, calabash-tree

Cressa L. (Convolvulaceae)
truxillensis Kunth - alkali-weed

Crinum L. (Liliaceae)
 crinum-lily, spider-lily
americanum L. - southern swamp crinum

Crossandra Salisb. (Acanthaceae)
infundibuliformis (L.) Nees - firecracker-flower

Crossopetalum P. Browne (Celastraceae)
ilicifolia (Poir.) Kuntze - Christmas-berry

Crotalaria L. (Fabaceae)
 rattle-box
brevidens Benth. - slender-leaf crotalaria
juncea L. - Indian hemp, Madras hemp, sunn hemp
lanceolata E. Mey. - lance-leaf crotalaria
pallida Aiton - smooth crotalaria, striped crotalaria
retusa L. - water-leaf rattle-box, wedge-leaf rattle-box
rotundifolia J. F. Gmel. - low rattle-box, rabbit-bells

sagittalis L. - rattle-box, rattle-weed, weedy rattle-box, wild pea

spectabilis Roth - showy crotalaria, showy rattle-box

Croton L. (Euphorbiaceae)
croton
capitatus Michx. - hog-wort, woolly croton
glandulosa L. - tooth-leaved croton
glandulosus L. var. *septentrionalis* Müll.Arg. - tropical croton
lindheimerianus Scheele - three-seeded croton
monanthogynus Michx. - one-seeded croton, prairie-tea, prairie-tea croton
texensis (Klotzsch) Müll. Arg. - skunk-weed, Texas croton

Crupina (Pers.) DC. (Asteraceae)
vulgaris Cass. - bearded-creeper, crupina

Cryptantha Lehm. ex G. Don (Boraginaceae)
intermedia (A. Gray) Greene - white forget-me-not

Cryptanthus Otto & A. Dietr. (Bromeliaceae)
earth-star
acaulis (Lindl.) Beer - starfish-plant
bromelioides Otto & A. Dietr. - pink cryptanthus
bromelioides Otto & A. Dietr. var. *tricolor* M. B. Foster - rainbow-star

Cryptogramma R. Br. (Pteridophyta)
parsley fern, rock brake
crispa (L.) R. Br. - European parsley fern, mountain parsley fern, parsley fern, rock brake
stelleri (S. G. Gmel.) Prantl - fragile cliff brake, slender cliff brake, Steller's rock brake

Cryptomeria D. Don (Taxodiaceae)
japonica (L.f.) D. Don - Japanese cedar

Cryptotaenia DC. (Apiaceae)
canadensis (L.) DC. - honewort, white chervil, wild chervil

Ctenium Panz. (Poaceae)
aromaticum (Walter) A. Wood - toothache grass

Cucumis L. (Cucurbitaceae)
anguria L. - bur gherkin, goar-berry gourd, gooseberry gourd, West Indian gherkin
dipsaceus Ehrenb. ex Spach - hedgehog gourd, teasel gourd, wild gourd
melo L. - melon
melo L. var. *cantalupensis* Naudin - cantaloupe
melo L. var. *inodorus* Naudin - casaba melon, honeydew melon, winter melon
melo L. var. *reticulatus* Naudin - muskmelon, netted melon, nutmeg melon, Persian melon
sativus L. - cucumber

Cucurbita L. (Cucurbitaceae)
gourd, pumpkin, squash
digitata A. Gray - finger-leaf gourd
foetidissima Kunth - buffalo gourd, calabazilla, fetid wild pumpkin, Missouri gourd, wild gourd, wild pumpkin
maxima Duchesne - autumn squash, marrow, pumpkin, squash, winter squash
moschata (Duchesne) Duchesne ex Poir. - Canadian pumpkin, crook-neck squash, pumpkin, winter squash
pepo L. - marrow, pumpkin, spaghetti squash, summer squash, warron, winter squash, zucchini

pepo L. var. *melopepo* (L.) Alef. - bush pumpkin, bush squash
pepo L. var. *ovifera* (L.) Alef. - yellow-flowered gourd
texana A. Gray - Texas gourd, wild marrow

Cuminum L. (Apiaceae)
cyminum L. - cumin

Cunila D. Royen ex L. (Lamiaceae)
origanoides (L.) Britton - American dittany, dittany, stone mint, sweet horsemint

Cunninghamia R. Br. (Taxodiaceae)
lanceolata (Lamb.) Hook. f. - China fir

Cuphea P. Browne (Lythraceae)
carthagenensis (Jacq.) J. F. Macbr. - tarweed cuphea
hyssopifolia Kunth - elfin-herb, false heather
ignea A. DC. - cigar-flower, firecracker-plant, red-white-and-blue-flower
viscosissima Jacq. - blue wax-weed, clammy cuphea, tarweed

Cupressus L. (Cupressaceae)
cypress
arizonica Greene - Arizona cypress, rough-bark Arizona cypress
bakeri Jeps. - Baker's cypress, modoc cypress
guadalupensis S. Watson - Guadalupe cypress, tecate cypress
lusitanica Mill. - cedar-of-Goa, Mexican cypress, Portuguese cypress
macnabiana A. Murr. bis - McNab's cypress
macrocarpa Hartw. - Monterey cypress
pygmaea (Lemmon) Sarg. - Mendocino cypress
sargentii Jeps. - Sargent's cypress
sempervirens L. - Italian cypress

Curcuma L. (Zingiberaceae)
longa L. - turmeric
petiolata Roxb. - queen-lily

Cuscuta L. (Cuscutaceae)
approximata Bab. - small-seeded alfalfa dodder
campestris Yunck. - field dodder, western flax dodder
coryli Engelm. - hazel dodder
epilinum Weihe - devil's-hair, flax dodder, hair-weed
epithymum (L.) L. - clover dodder, thyme dodder
gronovii Willd. ex Roem. & Schult. - dodder, gold-thread-vine, onion dodder, swamp dodder
indecora Choisy - collared dodder, large-seeded dodder
obtusiflora Kunth - Australian dodder, southern dodder
pentagona Engelm. - field dodder, large-seeded dodder, lespedeza dodder, love-vine
planiflora Tenore - small-seeded dodder
polygonorum Engelm. - polygonum dodder, smartweed dodder
sandwichiana Choisy - Sandwich's dodder
suaveolens Ser. - alfalfa dodder
umbellata Kunth - umbrella dodder
umbrosa Beyr. ex Hook. - large-fruit dodder

Cussonia Thunb. (Araliaceae)
spicata Thunb. - cabbage-tree

Cyamopsis DC. (Fabaceae)
tetragonoloba (L.) Taub. - cluster bean, guar

Cyanotis D. Don (Commelinaceae)
kewensis C. B. Clarke - teddy-bear-plant, teddy-bear-vine

Cyathea Sm. (Pteridophyta)
  tree-fern
  arborea (L.) Sm. - West Indian tree-fern
  australis (R. Br.) Domin - Australian tree-fern, rough tree-fern
  cooperi (F. Muell.) Domin - Australian tree-fern

Cycas L. (Cycadaceae)
  bread-palm, conehead, funeral-palm, sago conehead
  circinalis L. - fern-palm, queen sago, sago-palm
  revoluta Thunb. - Japanese fern-palm, Japanese sago-palm, sago-palm

Cyclamen L. (Primulaceae)
  persicum Mill. - florist's cyclamen

Cycloloma Moq. (Chenopodiaceae)
  atriplicifolium (Spreng.) J. M. Coult. - tumbleweed, winged pigweed

Cydista Miers (Bignoniaceae)
  aequinoctialis (L.) Miers - garlic-vine

Cydonia Mill. (Rosaceae)
  oblonga Mill. - quince

Cymbalaria Hill (Scrophulariaceae)
  muralis P. Gaertn., B. Mey. & Scherb. - coliseum-ivy, Kenilworth ivy, pennywort

Cymbopogon Spreng. (Poaceae)
  citratus (DC. ex Nees) Stapf - fever grass, lemongrass, West Indian lemongrass
  flexuosus (Nees ex Steud.) J. F. Watson - East Indian lemongrass, Malabar grass

Cymophyllus Mack. ex Britton & A. Br. (Cyperaceae)
  fraseri (Andr.) Mack. - Fraser's sedge

Cymopterus Raf. (Apiaceae)
  ibapensis M. E. Jones - ibapah spring parsley, spring parsley

Cynanchum L. (Asclepiadaceae)
  laeve (Michx.) Pers. - angle-pod, blue-vine, honey-vine, sand-vine
  louiseae Kartesz & Gandhi - black swallow-wort, climbing milkweed
  scoparium Nutt. - leafless cynanchum

Cynara L. (Asteraceae)
  cardunculus L. - cardoon
  scolymus L. - globe artichoke

Cynodon Rich. (Poaceae)
  dactylon (L.) Pers. - Bermuda grass, devil grass, dog-tooth grass, scutch grass, wiregrass
  transvaalensis Burtt-Davy - African Bermuda grass

Cynoglossum L. (Boraginaceae)
  beggar's-lice, hound's-tongue
  amabile Stapf & J. R. Drumm. - Chinese forget-me-not
  boreale Fernald - northern wild comfrey
  officinale L. - hound's-tongue
  virginianum L. - wild comfrey

Cynosurus L. (Poaceae)
  cristatus L. - crested dog's-tail, crested dog-tail grass
  echinatus L. - hedgehog dog-tail grass, rough dog's-tail

Cyperus L. (Cyperaceae)
  flat sedge, galingale, umbrella sedge

  alternifolius L. - umbrella sedge, umbrella-palm, umbrella-plant
  articulatus L. - jointed flat sedge
  brevifolius (Rottb.) Hassk. - green kyllinga
  compressus L. - annual sedge
  cylindricus Chapm. - pine-barren cyperus
  diandrus Torr. - sedge galingale
  difformis L. - small-flowered umbrella sedge
  eragrostis Lam. - tall umbrella-plant
  erythrorhizos Muhl. - red-root flat sedge, red-rooted cyperus
  esculentus L. - chufa, coco, coco sedge, earth-almond, edible galingale, northern nut-grass, nut sedge, rush-nut, tiger-nut, yellow nut sedge, yellow nut-grass
  filiculmis Vahl - slender cyperus
  globulosus Aubl. - globe sedge
  iria L. - rice flat sedge
  lecontei Torr. - Leconte's sedge
  odoratus L. - flat sedge
  ovularis (Michx.) Torr. - globose cyperus
  pseudovegetus Steud. - knob sedge, marsh cyperus
  retrorsus Chapm. - cylindric sedge
  rivularis Kunth - shining cyperus
  rotundus L. - coco sedge, coco-grass, nut sedge, nut-grass, purple nut sedge
  squarrosus L. - awned cyperus
  strigosus L. - false nut sedge, straw-colored cyperus
  surinamensis Rottb. - Surinam sedge
  virens Michx. - green sedge

Cyphomandra Mart. ex Sendtn. (Solanaceae)
  betacea (Cav.) Sendtn. - tomato-tree, tree-tomato

Cypripedium L. (Orchidaceae)
  lady's-slipper, moccasin-flower
  acaule Aiton - moccasin-flower, nerve-root, pink lady's-slipper, two-leaf lady's-slipper
  calceolus L. var. *pubescens* (Willd.) Correll - American valerian, golden-slipper, large lady's-slipper, nerve-root, Noah's-ark, umbil-root, Venus's-shoe, whip-poor-will-shoe, yellow Indian-shoe
  candidum Muhl. ex Willd. - small white lady's-slipper, white lady's-slipper
  reginae Walter - showy lady's-slipper

Cyrilla Garden ex L. (Cyrillaceae)
  racemiflora L. - black titi, he-huckleberry, ironwood, leatherwood, myrtle, red titi, southern leatherwood

Cyrtanthus Aiton (Liliaceae)
  angusitfolius (L.f.) Aiton - fire-lily
  elatus (Jacq.) Traub - George's-lily, Scarborough-lily

Cyrtomium C. Presl (Pteridophyta)
  falcata (L.f.) C. Presl - house holly fern, Japanese holly fern

Cystopteris Bernh. (Pteridophyta)
  bulbifera (L.) Bernh. - berry bladder fern, bulblet bladder fern
  fragilis (L.) Bernh. - brittle bladder fern, brittle fern, fragile fern
  montana (Lam.) Desv. - mountain bladder fern

Cytisus L. (Fabaceae)
  broom
  scoparius (L.) Link - broom, European broom, Scotch broom

Dactylis L. (Poaceae)
glomerata L. - cock's-foot, orchard grass

Dactyloctenium Willd. (Poaceae)
aegyptium (L.) Willd. - crowfoot grass, Egyptian grass

Dalbergia L.f. (Fabaceae)
sissoo Roxb. ex DC. - Indian rosewood, sissoo, sisu

Dalea L. (Fabaceae)
indigo-bush
candida Willd. - white prairie-clover
formosa Torr. - feather-plume
frutescens A. Gray - black dalea
leporina (Aiton) Bullock - foxtail dalea, hare-foot dalea
purpurea Vent. - purple prairie-clover
villosa (Nutt.) Spreng. - downy prairie-clover, silky prairie-clover

Dalibarda L. (Rosaceae)
repens L. - dewdrop, robin-run-away

Danthonia DC. (Poaceae)
oat grass
californica Bol. - California oat grass
intermedia Vasey - intermediate oat grass, timber oat grass
parryi Scribn. - Parry's oat grass
sericea Nutt. - downy oat grass
spicata (L.) P. Beauv. ex Roem. & Schult. - bonnet grass, June grass, old-fog, poverty grass, poverty oat grass, white oat grass, white-horse, wiregrass
unispicata (Thurb.) Munro ex Macoun - one-spike oat grass

Daphne L. (Thymelaeaceae)
cneorum L. - garland-flower, rose daphne
mezereum L. - February daphne, mezereum
odora Thunb. - winter daphne

Darlingtonia Torr. (Sarraceniaceae)
californica Torr. - California pitcher-plant, cobra-lily

Dasistoma Raf. (Scrophulariaceae)
macrophylla (Nutt.) Raf. - mullein-foxglove

Dasylirion Zucc. (Agavaceae)
bear-grass, sotol
wheeleri S. Watson - spoonflower

Datura L. (Solanaceae)
thorn-apple
discolor Bernh. - small datura
innoxia Mill. - angel's-trumpet, downy thorn-apple, Indian apple, sacred datura, tolguacha
metel L. - downy thorn-apple, Hindu datura, horn-of-plenty, metel
stramonium L. - Jamestown-weed, jimson-weed, mad-apple, stink-wort, stramonium thorn-apple, thorn-apple

Daucus L. (Apiaceae)
carota L. - bird's-nest, devil's-plague, Queen Anne's-lace, queen's-lace, wild carrot
carota L. subsp. *sativus* (Hoffm.) Arcang. - carrot
pusillus Michx. - rattlesnake-weed, southwestern carrot

Davallia Sm. (Pteridophyta)
trichomanoides Blume - ball fern, squirrel-foot fern

Decodon J. F. Gmel. (Lythraceae)
verticillatus (L.) Elliott - swamp loosestrife, water-oleander, water-willow

Decumaria L. (Hydrangeaceae)
barbara L. - climbing hydrangea, wood-vamp

Delonix Raf. (Fabaceae)
regia (Bojer ex Hook.) Raf. - flamboyant-tree, flame-tree, peacock-flower, royal poinciana

Delphinium L. (Ranunculaceae)
larkspur
andersonii A. Gray - desert larkspur
barbeyi (Huth) Huth - tall larkspur
cardinale Hook. - cardinal larkspur, scarlet larkspur
consolida L. - field larkspur, forking larkspur
elatum L. - candle larkspur
exaltatum Aiton - tall larkspur
geyeri Greene - Geyer's larkspur
glaucum S. Watson - mountain larkspur, pale larkspur
grandiflorum L. - bouquet larkspur
menziesii DC. - cow-poison, field larkspur, low larkspur, peco, poison-weed, stagger-weed, staves-acre
nuttallianum Walp. - dwarf larkspur, low larkspur, meadow larkspur
occidentale (S. Watson) S. Watson - dunce-cap larkspur
tricorne Michx. - dwarf larkspur, spring larkspur
trolliifolium A. Gray - cow-poison, tall larkspur, wood larkspur

Dendromecon Benth. (Papaveraceae)
rigida Benth. - bush poppy, tree poppy

Dennstaedtia Bernh. (Pteridophyta)
cup fern
punctilobula (Michx.) T. Moore - boulder fern, hairy dicksonia, hay-scented fern

Deschampsia P. Beauv. (Poaceae)
hair grass
atropurpurea (Wahlenb.) Scheele - mountain hair grass
caespitosa (L.) P. Beauv. - tufted hair grass
danthonioides (Trin.) Munro ex Benth. - annual hair grass
elongata (Hook.) Munro ex Benth. - slender hair grass
flexuosa (L.) Trin. - crinkled hair grass

Descurainia Webb & Berthel. (Brassicaceae)
incisa (A. Gray) Britton - western tansy mustard
pinnata (Walter) Britton - pinnate tansy mustard, tansy mustard
pinnata (Walter) Britton var. *brachycarpa* (Richardson) Fernald - green tansy mustard
pinnata (Walter) Britton subsp. *menziesii* (DC.) Detling - Menzies tansy mustard
richardsonii (Sweet) O. E. Schulz - Richardson's tansy mustard
sophia (L.) Webb ex Prantl - flix-weed, herb-Sophia, tansy mustard

Desmanthus Willd. (Fabaceae)
bundle-flower
illinoensis (Michx.) MacMill. ex B. L. Rob. & Fernald - bundle-flower, Illinois bundle-flower, prairie-mimosa, prickleweed
velutinus Scheele - velvet bundle-flower

Desmodium Desv. (Fabaceae)
  beggar-ticks, tick-trefoil
  canadense (L.) DC. - beggar-weed, Canadian tick-trefoil, sainfoil, showy tick-trefoil, tick-trefoil
  canescens (L.) DC. - hoary tick-clover, hoary tick-trefoil
  ciliare (Muhl.) DC. - little-leaf tick-trefoil, small-leaf tick-trefoil
  cuspidatum (Muhl.) Loudon - big tick-trefoil, large-bracted tick-trefoil
  glutinosum (Muhl.) Schindl. - cluster-leaf tick-trefoil, pointed-leaf tick-trefoil
  incanum DC. - creeping beggar-weed
  nudiflorum (L.) DC. - naked tick-trefoil, naked-flowered tick-trefoil
  paniculatum (L.) DC. - panicled tick-trefoil
  rotundifolium DC. - prostrate tick-trefoil, round-leaf tick-trefoil
  sessilifolium (Torr.) Torr. & A. Gray - sessile tick-clover, sessile tick-trefoil, sessile-leaf tick-trefoil
  strictum (Pursh) DC. - pine-barren tick-trefoil, stiff tick-trefoil
  tortuosum (Sw.) DC. - beggar-weed, Florida beggar-weed
  triflorum (L.) DC. - three-flowered beggar-weed
  uncinatum (Jacq.) DC. - silver-leaf desmodium, Spanish tick-clover
  viridiflorum (L.) DC. - velvety tick-trefoil

Dianthus L. (Caryophyllaceae)
  pink
  armeria L. - Deptford pink, grass pink
  barbatus L. - sweet-William
  caryophyllus L. - carnation, clove pink, divine-flower
  chinensis L. - rainbow pink
  deltoides L. - maiden pink
  plumarius L. - cottage pink, garden pink, grass pink
  prolifer L. - childing pink

Dicentra Bernh. (Papaveraceae)
  canadensis (Goldie) Walp. - squirrel-corn
  cucullaria (L.) Bernh. - Dutchman's-breeches
  formosa (Haw.) Walp. - western bleeding-heart
  spectabilis (L.) Lem. - bleeding-heart

Dichanthium Willemet (Poaceae)
  annulatum (Forssk.) Stapf - Brahman grass, Diaz's blue-stem, Kleberg's blue-stem, Kleberg's grass, ringed beard grass

Dichelostemma Kunth (Liliaceae)
  pulchellum (Salisb.) A. Heller - blue-dicks, wild hyacinth

Dichondra J. R. Forst. & F. Forst. (Convolvulaceae)
  carolinensis Michx. - Carolina dichondra, lawn-leaf

Dicranopteris Bernh. (Pteridophyta)
  linearis (Burm.) Underw. - savannah fern

Dictamnus L. (Rutaceae)
  albus L. - burning-bush, dittany, fraxinella, gas-plant

Didiplis Raf. (Lythraceae)
  diandra (Nutt.) Wood - water-purslane

Dieffenbachia Schott (Araceae)
  dumb-cane, mother-in-law's-tongue-plant, tuft-root
  maculata (Lodd.) G. Don - spotted dumb-cane
  seguine (Jacq.) Schott - dumb-cane, mother-in-law-plant

Diervilla Mill. (Caprifoliaceae)
  lonicera Mill. - bush-honeysuckle, northern bush-honeysuckle

Dietes Salisb. ex Klatt (Iridaceae)
  iridioides (L.) Klatt. - African iris

Digitalis L. (Scrophulariaceae)
  crabgrass, finger grass
  lanata Ehrh. - digitalis, Grecian foxglove
  lutea L. - straw foxglove
  purpurea L. - foxglove

Digitaria Haller (Poaceae)
  bicornis (Lam.) Roem. & Schult. - tropical crabgrass
  californica (Benth.) Henry - Arizona cotton-top
  ciliaris (Retz.) Koeler - southern crabgrass
  eriantha Steud. subsp. *pentzii* (Stent) Kok - pangola grass, woolly finger grass
  exilis (Kippist) Stapf - fonio, fundi, hungry-rice
  filiformis (L.) Koeler - slender crabgrass
  insularis (L.) Mez ex Ekman - sour grass
  ischaemum (Schreb.) Schreb. ex Muhl. - finger grass, small crabgrass, smooth crabgrass
  longiflora (Retz.) Pers. - Indian crabgrass
  sanguinalis (L.) Scop. - crowfoot grass, hairy crabgrass, large crabgrass, pigeon grass, Polish millet, purple crabgrass
  scalarum (Schweinf.) Chiov. - blue-couch
  velutina (Forssk.) Beauvais - velvet finger grass
  violascens Link - violet crabgrass

Dimorphotheca Moench (Asteraceae)
  Cape-marigold
  sinuata DC. - blue-eyed Cape-marigold

Diodia L. (Rubiaceae)
  teres Walter - button-weed, poor-Joe, rough button-weed
  virginiana L. - Virginia button-weed

Dionaea J. Ellis (Droseraceae)
  muscipula J. Ellis - Venus's-flytrap

Dioscorea L. (Dioscoreaceae)
  yam
  batatas Decne. - Chinese yam, cinnamon-vine
  bulbifera L. - aerial yam, air-potato, bulbil-bearing yam, potato yam
  villosa L. - colic-root, wild yam

Diospyros L. (Ebenaceae)
  persimmon
  kaki L.f. - date-plum, Japanese persimmon, kaki, kaki persimmon, keg-fig
  texana Scheele - black persimmon, chapote, Mexican persimmon, Texas persimmon
  virginiana L. - American persimmon, date-plum, persimmon, possum-apple, possum-wood

Diplotaxis DC. (Brassicaceae)
  muralis (L.) DC. - sand rocket
  tenuifolia (L.) DC. - large sand rocket, wall-rocket

Dipsacus L. (Dipsacaceae)
  fullonum L. - teasel, wild teasel
  laciniatus L. - cut-leaf teasel
  sativus (L.) Honck. - Fuller's teasel

Dirca L. (Thymelaeaceae)
  palustris L. - leatherwood, moosewood, rope-bark, wicopy

Disporum Salisb. ex D. Don (Liliaceae)
hookeri (Torr.) Nichols - fairy-bells
lanuginosum (Michx.) G. Nicholson - yellow mandarin
maculatum (Buckley) Britton - nodding mandarin

Distichlis Raf. (Poaceae)
salt grass
spicata (L.) Greene - salt grass, seashore salt grass
stricta (Torr.) Rydb. - desert salt grass

Dodecatheon L. (Primulaceae)
American cowslip, shooting-star
alpinum (A. Gray) Greene - alpine shooting-star
conjugens Greene - American cowslip, shooting-star
hendersonii A. Gray - mosquito-bills, sailor-caps
meadia L. - American cowslip, eastern shooting-star,
shooting-star

Dodonaea Mill. (Sapindaceae)
viscosa (L.) Jacq. - Florida hop-bush

Dopatrium Buch.-Ham. ex Benth. (Scrophulariaceae)
junceum (Roxb.) Buch.-Ham. - dopatrium

Doronicum L. (Asteraceae)
plantagineum L. - leopard-bane

Dovyalis E. Mey. ex Arn. (Flacourtiaceae)
caffra (Hook. f. & Harv.) Warb. - kai-apple, kau-apple, kei-
apple, umkokolo

Draba L. (Brassicaceae)
nemorosa L. - wood whitlow-grass
reptans (Lam.) Fernald - Carolina whitlow-grass
verna L. - spring whitlow-grass, whitlow-grass

Dracaena Vand. ex L. (Agavaceae)
draco (L.) L. - dragon-tree
fragrans (L.) Ker Gawl. cv.'massangeana' - corn-plant
sanderana Sander ex Mast. - Belgian evergreen
surculosa Lindl. - gold-dust dracaena, spotted dracaena

Dracocephalum L. (Lamiaceae)
parviflorum Nutt. - American dragonhead, dragonhead

Dracopis Cass. (Asteraceae)
amplexicaulis (Vahl) Cass. - coneflower

Drosera L. (Droseraceae)
capillaris Poir. - pink sundew
filiformis Raf. - dew-thread
rotundifolia L. - round-leaf sundew

Dryas L. (Rosaceae)
octopetala L. - mountain-avens
octopetala L. subsp. *alaskensis* (Porsild) Hultén - mountain-
avens

Drymaria Willd. ex Schult. (Caryophyllaceae)
arenarioides Kunth - alfombrilla
cordata (L.) Willd. ex Roem. & Schult. - heart-leaf drymary

Dryopteris Adans. (Pteridophyta)
shield fern, wood fern
arguta (Kaulf.) Watt - coastal shield fern, coastal wood fern
campyloptera Clarkson - mountain wood fern, spreading
wood fern
carthusiana (Vill.) H. P. Fuchs - fancy fern, florist's fern,
spinulose wood fern, toothed wood fern

marginalis (L.) A. Gray - leather wood fern, marginal shield
fern, marginal wood fern
pedata (L.) Fée - hand fern

Duchesnea Sm. (Rosaceae)
indica (Andr.) Focke - false strawberry, Indian mock-
strawberry, Indian strawberry, mock-strawberry

Duranta L. (Verbenaceae)
erecta L. - Brazilian sky-flower, garden dewdrop, golden
dewdrop, pigeon-berry, sky-flower

Dyssodia Cav. (Asteraceae)
dog-weed, fetid-marigold
papposa (Vent.) Hitchc. - false mayweed, fetid-marigold,
stinking marigold, stinkweed
tenuiloba (DC.) B. L. Rob. - Dahlberg's daisy, golden-fleece

Echinacea Moench (Asteraceae)
purple coneflower
pallida (Nutt.) Nutt. - pale echinacea, pale purple coneflower,
prairie-coneflower
purpurea (L.) Moench - purple coneflower

Echinocereus Engelm. (Cactaceae)
viridiflorus Engelm. - hedgehog cactus, pitaya

Echinochloa P. Beauv. (Poaceae)
cup grass
colona (L.) Link - jungle rice grass, jungle-rice
crus-galli (L.) P. Beauv. - barn grass, barnyard grass, cock-
foot panicum, cockspur grass, jungle rice grass, panic grass,
water grass
crus-pavonis (Kunth) Schult. - gulf cockspur
frumentacea Link - billion-dollar grass, Japanese millet,
Sanwa millet
oryzoides (Ard.) Fritsch - early water grass
phyllopogon (Stapf) Koss - late water grass

Echinocystis Torr. & A. Gray (Cucurbitaceae)
lobata (Michx.) Torr. & A. Gray - balsam-apple, mock
cucumber, mock-apple, prickly cucumber, wild balsam-
apple, wild cucumber

Echinodorus Rich. ex Engelm. (Alismataceae)
cordifolius (L.) Griseb. - bur-head, creeping bur-head, Texas
mud-baby

Echinops L. (Asteraceae)
globe thistle
sphaerocephalus L. - great globe thistle

Echinopsis Zucc. (Cactaceae)
spachiana (Lem.) Friedrich & G. D. Rowley - golden-column,
torch cactus, white torch cactus

Echium L. (Boraginaceae)
vulgare L. - blue-devil, blue-thistle, blueweed, snake-flower,
viper-bugloss

Egeria Planch. (Hydrocharitaceae)
densa Planch. - Brazilian elodea, Brazilian waterweed, egeria,
South American waterweed

Eichhornia Kunth (Pontederiaceae)
azurea (Sw.) Kunth - anchored water-hyacinth
crassipes (Mart.) Solms - water-hyacinth

Elaeagnus L. (Elaeagnaceae)
angustifolia L. - oleaster, Russian olive, silverberry, wild olive
commutata Bernh. ex Rydb. - silverberry
pungens Thunb. - thorny elaeagnus
umbellata Thunb. - autumn elaeagnus, autumn-olive

Elatine L. (Elatinaceae)
minima (Nutt.) Fisch. & C. A. Mey - small water-wort
triandra Schkuhr - water-wort

Eleocharis R. Br. (Cyperaceae)
spike rush
acicularis (L.) Roem. & Schult. - needle spike rush, slender spike rush
baldwinii (Torr.) Chapm. - slender spike rush
cellulosa Torr. - Gulf Coast spike rush
ovata (Roth) Roem. & Schult. - blunt spike rush, ovoid spike rush
palustris (L.) Roem. & Schult. - creeping spike rush
parvula (Roem. & Schult.) Link - dwarf spike rush, small spike rush
quadrangulata (Michx.) Roem. & Schult. - square-stem spike rush
rostellata (Torr.) Torr. - beaked spike rush
vivipara Link - sprouting spike rush

Elephantopus L. (Asteraceae)
elatus Bertol. - Florida elephant's-foot
nudatus A. Gray - purple elephant's-foot
tomentosus L. - devil's-grandmother, tobacco-weed

Elettaria Maton (Zingiberaceae)
cardamomum (L.) Maton - cardamon, Ceylon cardamom, Malabar cardamom

Eleusine Gaertn. (Poaceae)
coracana (L.) Gaertn. - African millet, finger millet, korakan, ragi
indica (L.) Gaertn. - crowfoot grass, goose grass, wiregrass, yard grass

Ellisia L. (Hydrophyllaceae)
nyctelea (L.) L. - ellisia, water-pod

Elodea Michx. (Hydrocharitaceae)
canadensis Michx. - elodea, waterweed
longivaginata H. St. John - long-sheath elodea
nuttallii (Planch.) H. St. John - free-flowered waterweed, Nuttall's waterweed, western elodea

Elymus L. (Poaceae)
Lyme grass, wild rye
canadensis L. - Canadian wild rye
elymoides (Raf.) Swezey - squirrel's-tail
glaucus Buckley - blue wild rye
hystrix L. - bottlebrush
junceus Fisch. - Russian wild rye
lanceolatus (Scribn. & J. G. Sm.) Gould - thick-spike wheat grass
sibiricus L. - Siberian wild rye
smithii (Rydb.) Gould - western wheat grass
trachycaulus (Link) Shinners - bearded couch, slender wheat grass
villosus Muhl. ex Willd. - downy wild rye, hairy wild rye
virginicus L. - Virginia wild rye

Elytrigia Desv. (Poaceae)
elongata (Host) Nevski - tall wheat grass
intermedia (Host) Nevski - intermediate wheat grass, pubescent wheat grass
repens (L.) Nevski - couchgrass, devil's grass, knotgrass, quackgrass, quick grass, quitch grass, scutch grass, Shelly's grass, wheat grass, witch grass

Emex Campd. (Polygonaceae)
australis Steinh. - three-cornered-jack
spinosa (L.) Campd. - spiny emex

Emilia Cass. (Asteraceae)
fosbergii Nicolson - Cupid's-shaving-brush
javanica (Burm.) Merr. - Flora's paintbrush, tassel-flower
sonchifolia (L.) DC. ex Wight - red tassel-flower

Empetrum L. (Empetraceae)
atropurpureum Fernald & Wiegand - purple crowberry
eamesii Fernald & Wiegand - rockberry
nigrum L. - black crowberry, crakeberry, crowberry, curlew-berry, monox

Engelmannia A. Gray ex Nutt. (Asteraceae)
pinnatifida Torr. & A. Gray ex Nutt. - Engelmann's daisy

Ephedra L. (Ephedraceae)
joint-fir
nevadensis S. Watson - gray ephedra, Nevada ephedra
sinica Stapf - Chinese ephedra
trifurca Torr. - long-leaf ephedra
viridis Coville - green ephedra, Mormon tea

Epifagus Nutt. (Orobanchaceae)
virginiana (L.) Barton - beech-drops, cancer-root

Epigaea L. (Ericaceae)
repens L. - mayflower, trailing arbutus

Epilobium L. (Onagraceae)
fireweed, willow-herb
angustifolium L. - burnt-weed, fireweed, flowering-willow, great willow-herb, Indian wickup, wickup
canum (Greene) Raven - California fuchsia, fire-chalice
coloratum Biehler - eastern willow-herb, purple-leaf willow-herb
davuricum Fisch. - arctic willow-weed
hirsutum L. - hairy willow-herb, hairy willow-weed, willow-herb
latifolium L. - river-beauty

Epipremnum Schott (Araceae)
aureum (Lindl. & André) Bunting - devil's-ivy, golden Ceylon creeper, golden pothos, hunter's-robe, ivy-arum, pothos, Solomon Island ivy, taro-vine, variegated philodendron

Equisetum L. (Equisetaceae)
horsetail, scouring-rush
arvense L. - bottlebrush, field horsetail, foxtail-rush, horse-pipes, horsetail, horsetail-fern, meadow-pine, pine-grass, scouring-rush, snake-grass
fluviatile L. - water horsetail
hyemale L. - scouring-rush
laevigatum A. Braun - smooth scouring-rush
palustre L. - marsh horsetail, shade horsetail
sylvaticum L. - Sylvan's horsetail, woodland horsetail

telmateia Ehrh. - giant horsetail
variegatum Schleich. ex F. Weber & D. Mohr - variegated horsetail, variegated scouring-rush

Eragrostis Wolf (Poaceae)
love grass
bahiensis Schrad. ex Schult. - Bahia love grass
barrelieri Daveau - Mediterranean love grass
capillaris (L.) Nees - lace grass
cilianensis (All.) Janch. - stink grass
ciliaris (L.) R. Br. - gopher-tail love grass
curvula (Schrad.) Nees - Boer's love grass, weeping love grass
elliottii S. Watson - Elliott's love grass
hirsuta (Michx.) Nees - big-top love grass
intermedia Hitchc. - plains love grass
lehmanniana Nees - Lehmann's love grass
mexicana (Hornem.) Link - Mexican love grass
pectinacea (Michx.) Nees - tufted love grass, tufted spear grass
pilosa (L.) P. Beauv. - Indian love grass
poaeoides P. Beauv. ex Roem. & Schult. - little love grass
secundiflora J. Prcsl & C. Presl - red love grass
spectabilis (Pursh) Steud. - purple love grass
tenella (L.) P. Beauv. ex Roem. & Schult. - feather tumble grass, love grass petticoat-climber
trichodes (Nutt.) A. Wood - sand love grass

Eranthemum L. (Acanthaceae)
pulchellum Andr. - blue-sage

Erechtites Raf. (Asteraceae)
glomerata (Poir.) DC. - New Zealand fireweed
hieracifolia (L.) Raf. ex DC. - American burn-weed, burn-weed, fireweed, pilewort
minima (Poir.) DC. - Australian burn-weed

Eremocarpus Benth. (Euphorbiaceae)
setigerus (Hook.) Benth. - dove-weed, turkey-mullein, woolly white drought-weed

Eremochloa Buse (Poaceae)
ophiuroides (Munro) Hack. - centipede grass, lazy-man's grass

Eremurus M. Bieb. (Liliaceae)
robustus Regel - desert-candle, king's-spear

Erica L. (Ericaceae)
heath
carnea L. - snow-heather, spring heath, winter heath
cinerea L. - Scotch heath, twisted heath
tetralix L. - cross-leaf heath
vagans L. - Cornish heath

Ericameria Nutt. (Asteraceae)
austrotexana M. C. Johnst. - false broom-weed

Erigenia Nutt. (Apiaceae)
bulbosa (Michx.) Nutt. - harbinger-of-spring, pepper-and-salt

Erigeron L. (Asteraceae)
fleabane
annuus (L.) Pers. - annual fleabane, daisy fleabane, sweet-scabious, white-top
philadelphicus L. - daisy fleabane, fleabane, Philadelphia fleabane, skevish

pulchellus Michx. - blue spring daisy, poor-Robin's fleabane, rose petty
speciosus (Lindl.) DC. - Oregon fleabane
strigosus Muhl. ex Willd. - daisy fleabane, rough daisy fleabane, rough fleabane, white-top

Eriobotrya Lindl. (Rosaceae)
japonica (Thunb.) Lindl. - Japanese medlar, Japanese plum, loquat

Eriochloa Kunth (Poaceae)
contracta Hitchc. - prairie cup grass
polystachya Kunth - carib grass, malojilla
villosa (Thunb.) Kunth - Chinese cup grass, woolly cup grass

Eriodictyon Benth. (Hydrophyllaceae)
californicum (Hook. & Arn.) Torr. - yerba-santa

Eriogonum Michx. (Polygonaceae)
umbrella-plant, wild buckwheat
fasciculatum Benth. - California buckwheat
giganteum S. Watson - St. Catharine's-lace
longifolium Nutt. - long-leaf wild buckwheat
umbellatum Torr. - sulphur-flower

Eriophorum L. (Cyperaceae)
cotton grass
russeolum Fr. - russet cotton grass
vaginatum L. - sheathed cotton grass, tussock cotton grass
virginicum L. - Virginia cotton grass

Erodium L'Hér. (Geraniaceae)
heron's-bill, stork's-bill
botrys (Cav.) Bertol. - broad-leaf filaree
cicutarium (L.) L'Hér. - alfilaria, pin grass, pin-clover, red-stem filaree, wild musk
moschatum (L.) L'Hér. ex Aiton - musk-clover, white-stem filaree
texanum A. Gray - Texas filaree

Eruca Mill. (Brassicaceae)
vesicaria (L.) Cav. subsp. *sativa* (Mill.) Thell. - garden rocket, salad rocket

Erucastrum C. Presl (Brassicaceae)
gallicum (Willd.) O. E. Schulz - dog-mustard

Eryngium L. (Apiaceae)
aquaticum L. - button snakeroot, eryngo, marsh eryngo, rattlesnake-master
yuccifolium Michx. - button snakeroot, rattlesnake-master

Erysimum L. (Brassicaceae)
blister-cress, treacle mustard, wallflower
asperum (Nutt.) DC. - prairie rocket, western wallflower
capitatum (Douglas) Greene - coast wallflower, western wallflower
cheiranthoides L. - treacle mustard, wallflower mustard, worm-seed mustard
cheiri (L.) Crantz - English wallflower, wallflower
hieraciifolium L. - tall worm-seed mustard, tall worm-seed wallflower
repandum L. - bushy wallflower, spreading mustard, treacle mustard

Erythrina L. (Fabaceae)
coral-tree
crista-galli L. - cockspur coral-tree, cry-baby-tree

flabelliformis Kearney - chilicote, southwestern coral bean, western coral bean

herbacea L. - cardinal-spear, Cherokee bean, coral bean, eastern coral bean, red cardinal

Erythronium L. (Liliaceae)
adder's-tongue, dog-tooth-violet, fawn-lily, trout-lily

albidum Nutt. - blonde-Lilian, white dog-tooth-violet

americanum Ker Gawl. - amber-bell, trout-lily, yellow adder's-tongue

grandiflorum Pursh - avalanche-lily, glacier-lily

Eschscholzia Cham. (Papaveraceae)
californica Cham. - California poppy

Espostoa Britton & Rose (Cactaceae)
lanata (Kunth) Britton & Rose - cotton-ball, new old-man cactus, Peruvian old-man cactus, Peruvian snowball cactus

Eucalyptus L'Hér. (Myrtaceae)
Australian gum, eucalypt, ironbark, stringy-bark

amplifolia Naudin - cabbage gum

calophylla Lindl. - marri, red gum

calycogona Turcz. - gooseberry mallee

camaldulensis Dehnh. - Murray's red gum, red gum, river red gum

cinerea F. Muell. ex Benth. - mealy stringy-bark, silver-dollar-tree, spiral eucalyptus

citriodora Hook. - lemon-scented gum

globulus Labill. - blue gum, Tasmanian blue gum

grandis A. W. Hill ex Maiden - rose gum

lehmannii (Schau) L. Preiss ex Benth. - bushy yate, Lehmann's gum

marginata Sm. - jarrah

niphophila Maiden & Blakeley - snow gum

obliqua L'Hér. - messmate, messmate stringy-bark

pauciflora Sieber ex Spreng. - cabbage gum, ghost gum, snow gum

pilularis Sm. - black-butt

platypus Hook. - round-leaf moort

polyanthemos Schauer - red box, round-leaf eucalypt, silver-dollar gum

populnea F. Muell. - bimble-box

pulverulenta Sims - money-tree, silver-leaf gum, silver-leaf mountain gum

punctata DC. - gray gum

robusta Sm. - swamp-mahogany

saligna Sm. - Sydney blue gum

sideroxylon A. Cunn. ex Maiden - red ironbark

sieber L. A. S. Johnson - black mountain-ash, silvertop-ash

tereticornis Sm. - forest red gum

tetragona (R. Br.) F. Muell. - tallerack, white-leaf marlock

viminalis Labill. - manna gum

Eucharis Planch. & Linden (Liliaceae)
grandiflora Planch. & Linden - Amazon lily, Eucharist-lily, lily-of-the-Amazon, Madonna-lily

Euclidium R. Br. (Brassicaceae)
syriacum (L.) R. Br. - Syrian mustard

Eugenia L. (Myrtaceae)
foetida Pers. - box-leaf eugenia

uniflora L. - Brazilian cherry, pitanga, Surinam cherry

Euonymus L. (Celastraceae)
burning-bush, spindle-tree

alatus (Thunb.) Siebold - winged burning-bush, winged euonymus, winged spindle-tree

americanus L. - bursting-heart, strawberry-bush

atropurpureus Jacq. - burning-bush, eastern wahoo, wahoo

europaeus L. - European euonymus, European spindle-tree

fortunei (Turcz.) Hand.-Mazz. - climbing euonymus

japonicus Thunb. - evergreen euonymus, Japanese spindle-tree

obovatus Nutt. - running strawberry-bush

occidentalis Nutt. ex Torr. - western burning-bush, western wahoo

Eupatorium L. (Asteraceae)
boneset, thoroughwort

album L. - white thoroughwort, white-bracted eupatorium

capillifolium (Lam.) Small - dog-fennel

compositifolium Walter - Yankee-weed

fistulosum Barratt - hollow Joe-Pye-weed, hollow-stemmed Joe-Pye-weed, trumpet-weed

hyssopifolium L. - hyssop-leaf boneset, hyssop-leaf thoroughwort

maculatum L. - Joe-Pye-weed, purple boneset, smoke-weed, spotted Joe-Pye-weed, tall boneset, trumpet-weed

perfoliatum L. - ague-weed, boneset, feverweed, purple boneset, sweating-plant, thoroughwort

purpureum L. - green-stemmed Joe-Pye-weed, Joe-Pye-weed, purple-node Joe-Pye-weed, sweet Joe-Pye-weed

rotundifolium L. - round-leaf eupatorium, round-leaf thoroughwort

serotinum Michx. - late boneset, late eupatorium, late-flowering thoroughwort

sessilifolium L. - upland boneset

Euphorbia L. (Euphorbiaceae)
spurge

albomarginata Torr. & A. Gray - white-margin spurge

commutata Engelm. - tinted spurge, wood spurge

corollata L. - flowering spurge, poison milkweed, tramp's spurge, white-flowered milkweed, wild hippo

cyathophora Murray - fire-on-the-mountain, painted poinsettia

cyparissias L. - cypress spurge, graveyard-weed, quack salvers-grass, salvers spurge

dentata Michx. - toothed spurge

esula L. - faitours-grass, leafy spurge, wolf's-milk

glyptosperma Engelm. - ridge-seed spurge

helioscopia L. - cat-milk, sun spurge, wart spurge, wart-weed

heterophylla L. - annual poinsettia, Japanese poinsettia, Mexican fire-plant, mole-plant, paint-leaf, painted spurge, wild poinsettia

hirta L. - garden spurge

humistrata Engelm. - prostrate spurge

hyssopifolia L. - hyssop spurge

ipecacuanhae L. - Carolina ipecac, ipecac spurge, wild ipecac

lathyris L. - caper spurge, gopher spurge, gopher-plant, mole-plant, myrtle spurge

lucida Waldst. & Kit. - broad-leaf spurge, shining spurge

maculata L. - eye-bane, eye-bright, milk-purslane, nodding spurge, slobber-weed, spotted spurge, stubble spurge

marginata Pursh - ghost-weed, snow-on-the-mountain

micromera Boiss. - little-leaf spurge

milii Des Moul. - Christ-plant, Christ-thorn, crown-of-thorns

nutans Lag. - nodding spurge
peplus L. - petty spurge
polygonifolia L. - seaside spurge
prostrata Aiton - ground spurge
prunifolia Jacq. - painted euphorbia
pulcherrima Willd. ex Klotzsch - Christmas-flower,
  Christmas-star, lobster-plant, Mexican flame-leaf, painted-
  leaf, poinsettia
serpens Kunth - creeping spurge, round-leaf spurge
serpyllifolia Pers. - thyme-leaved spurge
serrata L. - saw-tooth spurge, toothed spurge
spathulata Lam. - net-seed spurge, prairie spurge
vermiculata Raf. - hairy spurge

Euphrasia L. (Scrophulariaceae)
canadensis Towns. - Canadian eye-bright

Euryops (Cass.) Cass. (Asteraceae)
speciosissimus DC. - Clanwilliam daisy

Eustoma Salisb. (Gentianaceae)
russellianum (Hook.) G. Don - prairie-gentian

Euterpe Mart. (Arecaceae)
edulis Mart. - assai palm
oleracea Mart. - assai palm

Exacum L. (Gentianaceae)
affine Balf.f. - German violet, Persian violet

Exochorda Lindl. (Rosaceae)
racemosa (Lindl.) Rehder - pearl-bush

Exothea Macfad. (Sapindaceae)
paniculata Walp. ex Radlk. - butter-bough, inkwood

Eysenhardtia Kunth (Fabaceae)
polystachya (Ortega) Sarg. - kidney-wood

Fabiana Ruiz & Pav. (Solanaceae)
imbricata Ruiz & Pav. - pichi, pichi-pichi

Fagopyrum Mill. (Polygonaceae)
  buckwheat
esculentum Moench - buckwheat, notch-seeded buckwheat
tataricum (L.) Gaertn. - Indian wheat, Tatarian buckwheat

Fagus L. (Fagaceae)
  beech
grandifolia Ehrh. - American beech
sylvatica L. - European beech

Falcaria Bernh. (Apiaceae)
vulgaris Bernh. - sickle-weed

Fatshedera Guillaumin (Araliaceae)
  Note = *Fatsia* × *Hedera*.
×*lizei* Guillaumin - aralia-ivy, botanical-wonder, tree-ivy

Fatsia Decne. & Planch. (Araliaceae)
japonica (Thunb.) Decne. & Planch. - Formosan rice-tree,
  glossy-leaved paper-plant, Japanese fatsia, paper-plant

Felicia Cass. (Asteraceae)
bergeriana (Spreng.) O. Hoffm. - kingfisher daisy

Festuca L. (Poaceae)
  fescue
arizonica Vasey - Arizona fescue
arundinacea Schreb. - tall fescue

brevipila Tracey - hard fescue
idahoensis Elmer - blue bunch grass, blue-bunch fescue,
  Idaho fescue
obtusa Biehler - nodding fescue
occidentalis Hook. - western fescue
ovina L. - sheep fescue
pratensis Huds. - English bluegrass, meadow fescue
rubra L. - chewing festuca, red fescue
rubra L. var. *heterophylla* (Lam.) Mutel - shade fescue
scabrella Torr. ex Hook. - rough fescue
tenuifolia Sibth. - hair fescue, slender fescue
thurberi Vasey - Thurber's fescue
viridula Vasey - green fescue, green-leaf fescue, mountain
  bunch grass

Ficus L. (Moraceae)
  fig
altissima Blume - council-tree, false banyan, lofty fig
aurea Nutt. - Florida strangler fig, golden fig
benghalensis L. - banyan fig, East Indian fig-tree, Indian
  banyan
benjamina L. - Benjamin's-tree, Java fig, small-leaf rubber-
  plant, tropic-laurel, weeping fig
carica L. - edible fig, fig
citrifolia Mill. - short-leaf fig
deltoidea Jack - mistletoe ficus, mistletoe rubber-plant
elastica Roxb. ex Hornem. - Assam rubber, Indian rubber fig,
  Indian rubber-tree, rubber-plant
lyrata Warb. - banjo fig, fiddle-leaf, fiddle-leaf ficus
microcarpa L.f. - Indian laurel fig
pumila L. - climbing ficus, creeping ficus, creeping rubber-
  plant

Filago L. (Asteraceae)
  cotton-rose
germanica (L.) Huds. - herba impia
verna (Raf.) Shinners - many-stemmed evax

Filipendula Mill. (Rosaceae)
  meadowsweet
rubra (Hill) B. L. Rob. - queen-of-the-prairie
ulmaria (L.) Maxim. - queen-of-the-meadow
vulgaris Moench - dropwort

Fimbristylis Vahl (Cyperaceae)
annua (All.) Roem. & Schult. - annual fringe-rush
autumnalis (L.) Roem. & Schult. - slender fringe-rush
dichotoma (L.) Vahl - forked fringe-rush
littoralis Gaud. - globe fringe-rush
miliacea (L.) Vahl - globe fringe-rush

Firmiana Marsili (Sterculiaceae)
simplex (L.) W. Wight - Chinese parasol-tree, Japanese
  varnish-tree, Phoenix-tree

Fittonia Coem. (Acanthaceae)
verschaffeltii (Lem.) Coem. - mosaic-plant, nerve-plant, silver
  fittonia, silver nerve-plant, silver-net-plant, silver-threads,
  white-leaf fittonia

Flacourtia Comm. ex L'Hér. (Flacourtiaceae)
indica (Burm.) Merr. - Batoko-plum, governor's-plum,
  Madagascar plum, ramontchi

Flaveria Juss. (Asteraceae)
trinervia (Spreng.) C. Mohr - cluster-flower

Foeniculum Mill. (Apiaceae)
  vulgare Mill. - fennel, finocchio, Florence fennel

Forestiera Poir. (Oleaceae)
  acuminata (Michx.) Poir. - swamp-privet
  phillyreoides (Benth.) Torr. - desert-olive forestiera
  pubescens Nutt. - desert-olive

Forsythia Vahl (Oleaceae)
  viridissima Lind. - golden-bells

Fortunella Swingle (Rutaceae)
    kumquat
  hindsii (Champ. ex Benth.) Swingle - Hong Kong kumquat
  margarita (Lour.) Swingle - oval kumquat

Fouquieria Kunth (Fouquieriaceae)
  splendens Engelm. - candlewood, coach-whip, Jacob's-staff, ocotillo, vine cactus

Fragaria L. (Rosaceae)
    strawberry
  ×*ananassa* Duchesne - cultivated strawberry, garden strawberry
    Note = *F. chiloensis* × *F. virginiana.*
  chiloensis (L.) Duchesne - Chilean strawberry
  vesca L. - alpine strawberry, European strawberry, sow-teat strawberry, wood strawberry, woodland strawberry
  vesca L. subsp. *americana* (Porter) Staudt - woodland strawberry
  virginiana Duchesne - mountain strawberry, Virginia strawberry, wild strawberry

Franklinia Marshall (Theaceae)
  alatamaha Bartram - franklinia

Frasera Walter (Gentianaceae)
  speciosa Douglas ex Griseb. - columbo, green-gentian

Fraxinus L. (Oleaceae)
    ash
  americana L. - white ash
  anomala Torr. - single-leaf ash
  caroliniana Mill. - Carolina ash, pop ash, water ash
  dipetala Hook. & Arn. - California ash, two-petal ash
  excelsior L. - European ash
  latifolia Benth. - Oregon ash
  nigra Marsh. - black ash
  pennsylvanica Marsh. - green ash, red ash
  quadrangulata Michx. - blue ash
  velutina Torr. - velvet ash

Fremontodendron Coville (Sterculiaceae)
  californica (Torr.) Cov. - California fremontia, flannel-bush

Fritillaria L. (Liliaceae)
    fritillary
  affines (Schult.) Sealy - checker-lily, narrow-leaf fritillary, rice-grain fritillary
  camschatcensis (L.) Ker Gawl. - Kamchatka lily

Froelichia Moench (Amaranthaceae)
  floridana (Nutt.) Moq. - cotton-weed

Fuchsia L. (Onagraceae)
  triphylla L. - lady's-eardrops

Fuirena Rottb. (Cyperaceae)
  breviseta (Coville) Coville - short-bristled umbrella-grass

  pumila (Torr.) Spreng. - dwarf umbrella-grass
  scirpoidea Michx. - rush fuirena
  simplex Vahl - unbranched umbrella-grass
  squarrosa Michx. - squarrose umbrella-grass

Fumaria L. (Papaveraceae)
  officinalis L. - earth-smoke, fumitory

Furcraea Vent. (Agavaceae)
  selloa K. Koch - maguey

Gaillardia Foug. (Asteraceae)
    blanket-flower
  pulchella Foug. - fire-wheel, rose-ring blanket-flower, rose-ring gaillardia

Galactia P. Browne (Fabaceae)
    milk pea
  volubilis (L.) Britton - hairy milk pea

Galanthus L. (Liliaceae)
  nivalis L. - snowdrop

Galax Sims (Diapensiaceae)
  aphylla L. - beetle-wood, galax

Galega L. (Fabaceae)
  officinalis L. - goat's-rue

Galenia L. (Aizoaceae)
  pubescens (Eckl. & Zeyh.) Druce - green galenia

Galeopsis L. (Lamiaceae)
  ladanum L. - red hemp-nettle
  tetrahit L. - bee-nettle, dog-nettle, flowering-nettle, hemp-nettle, ironweed, wild hemp

Galinsoga Ruiz & Pav. (Asteraceae)
  parviflora Cav. - quick-weed, small-flowered galinsoga
  quadriradiata Ruiz & Pav. - French-weed, hairy galinsoga

Galium L. (Rubiaceae)
    bedstraw, cleavers
  aparine L. - catch-weed bedstraw, cleavers, goose-grass
  asprellum Michx. - kidney-vine, rough bedstraw
  boreale L. - chicken-weed, northern bedstraw
  circaezans Michx. - forest bedstraw, wild licorice, wild white licorice
  lanceolatum Torr. - wild licorice, yellow wild licorice
  mollugo L. - hedge bedstraw, smooth bedstraw, whip-tongue, white hedge bedstraw, wild madder
  odoratum (L.) Scop. - sweet woodruff, waldmeister
  saxatile L. - heath bedstraw
  sylvaticum L. - baby's-breath, Scotch-mist
  tricorne L. - corn bedstraw
  tricornutum Dandy - three-horn bedstraw
  triflorum Michx. - sweet-scented bedstraw
  verum L. - bed-flower, cheese-rennet, curd-wort, lady's bedstraw, yellow bedstraw

Garcinia L. (Clusiaceae)
  mangostana L. - mangosteen

Gardenia J. Ellis (Rubiaceae)
  jasminoides J. Ellis - Cape jasmine, gardenia

Garrya Douglas ex Lindl. (Garryaceae)
    silk-tassel-bush
  elliptica Douglas ex Lindl. - wavy-leaf silk-tassel

Gaultheria L. (Ericaceae)
hispidula (L.) Muhl. - capillaire, creeping snowberry,
maidenhair-berry, moxie, moxie-plum
humifusa (R. J. D. Graham) Rydb. - alpine wintergreen
procumbens L. - checkerberry, mountain-tea, teaberry,
wintergreen
shallon Pursh - salal, shallon, union-leaf

Gaura L. (Onagraceae)
biennis L. - biennial gaura
coccinea Pursh - linda-tarde, scarlet gaura, wild honeysuckle
parviflora Douglas ex Lehm. - lizard's-tail, velvet-weed
sinuata Nutt. ex Ser. - wavy-leaf gaura
villosa Torr. - hairy gaura

Gaylussacia Kunth (Ericaceae)
huckleberry
baccata (Wangenh.) K. Koch - black huckleberry, gueules-
noires
brachycera (Michx.) A. Gray - box huckleberry
dumosa (Andr.) Torr. & A. Gray - dwarf huckleberry
frondosa (L.) Torr. & A. Gray ex Torr. - blue-tangle,
dangleberry, dwarf huckleberry

Gazania Gaertn. (Asteraceae)
rigens (L.) Gaertn. - treasure-flower

Gelsemium Juss. (Loganiaceae)
sempervirens (L.) W. T. Aiton - Carolina jasmine, evening
trumpet-flower, yellow jessamine

Genipa L. (Rubiaceae)
genip
americana L. - genipap, marmalade, marmalade-box

Genista L. (Fabaceae)
monspessulanus (L.) L. A. S. Johnson - French broom
tinctoria L. - dyer's broom, dyer's greenwccd, dyer's
greenwood, dyeweed, woad-waxen

Gentiana L. (Gentianaceae)
gentian
andrewsii Griseb. - bottle gentian, closed gentian, prairie
bottle gentian
autumnalis L. - one-flowered gentian, pine-barren gentian
catesbaei Walter - Catesby's gentian, coastal plain gentian
clausa Raf. - blind gentian, bottle gentian, closed gentian,
meadow closed gentian
linearis Froel. - closed gentian, narrow-leaf gentian
newberryi A. Gray - alpine gentian
rubricaulis Schwein. - closed gentian, Great Lakes gentian
saponaria L. - downy gentian, soapwort gentian
setigera A. Gray - Mendocino gentian
villosa L. - Sampson's snakeroot, striped gentian

Gentianella Moench (Gentianaceae)
gentian
quinquefolia (L.) Small - ague-weed, stiff gentian

Gentianopsis Ma (Gentianaceae)
crinita (Froel.) Ma - fringed gentian
procera (T. Holm) Ma - fringed gentian, smaller fringed
gentian

Geocaulon Fernald (Santalaceae)
lividum (Richardson) Fernald - northern comandra

Geranium L. (Geraniaceae)
crane's-bill
bicknellii Britton - Bicknell's crane's-bill
carolinianum L. - Carolina geranium
columbinum L. - long-stalked crane's-bill
dissectum L. - cut-leaf geranium
ibericum Cav. - Iberian crane's-bill
maculatum L. - alumroot, spotted crane's-bill, spotted
geranium, wild geranium
molle L. - dove's-foot geranium
pratense L. - meadow geranium
pusillum Burm. - small-flowered crane's-bill, small-flowered
geranium
robertianum L. - herb-Robert, red-robin, Robert's geranium
sanguineum L. - blood-red geranium, bloody geranium
sibiricum L. - Siberian crane's-bill

Gerbera L. (Asteraceae)
jamesonii Bolus ex Hook. f. - African daisy, Barberton's
daisy, Transvaal daisy, veldt daisy

Geum L. (Rosaceae)
avens
aleppicum Jacq. - avens
canadense Jacq. - white avens
macrophyllum Willd. - big-leaf avens, large-leaf avens
rivale L. - chocolate-root, Indian chocolate, purple avens,
water avens
triflorum Pursh - long-plumed purple avens, old-man's-
whiskers
virginianum L. - cream-colored avens, rough avens

Gilia Ruiz & Pav. (Polemoniaceae)
capitata Sims - field gilia, gillyflower
tricolor Benth. - bird's-eyes

Gillenia Moench (Rosaceae)
stipulata (Muhl.) Baill. - American ipecac
trifoliata (L.) Moench - bowman's-root, Indian physic

Ginkgo L. (Ginkgoaceae)
biloba L. - ginkgo, maidenhair tree

Gladiolus L. (Iridaceae)
×*hortulanus* L. H. Bailey - garden gladiola
Note = Perhaps descended more or less directly from *G.
natalensis*.

Glaux L. (Primulaceae)
maritima L. - sea-milkwort

Glechoma L. (Lamiaceae)
hederacea L. - alehoof, field-balm, gill-over-the-ground,
ground-ivy, runaway-robin

Gleditsia L. (Fabaceae)
aquatica Marsh. - water locust
japonica Miq. - Japanese honey locust
triacanthos L. - honey locust, honey-shuck, sweet locust
triacanthos L. var. *inermis* (L.) C.K. Schneid. - thornless
honey locust

Gloriosa L. (Liliaceae)
superba L. - gloriosa-lily

Glyceria R. Br. (Poaceae)
borealis (Nash) Batsch - float grass, northern manna grass,
small floating manna grass

canadensis (Michx.) Trin. - rattlesnake manna grass
elata (Nash) M. E. Jones - fowl manna grass, tall manna grass
fluitans (L.) R. Br. - float grass, floating manna grass, floating sweet grass, water manna grass
grandis S. Watson ex A. Gray - American manna grass
septentrionalis Hitchc. - eastern manna grass, floating manna grass, tufted manna grass
striata (Lam.) Hitchc. - fowl manna grass

Glycine Willd. (Fabaceae)
max (L.) Merr. - sojabean, soybean

Glycyrrhiza L. (Fabaceae)
glabra L. - licorice, licorice-root
lepidota Pursh - American licorice, sweet-root, wild licorice

Gnaphalium L. (Asteraceae)
cudweed, everlasting
obtusifolium L. - cat's-foot, chafe-weed, cudweed, everlasting, fragrant cudweed, fragrant everlasting, old-field-balsam, rabbit-tobacco, sweet everlasting
pensylvanicum Willd. - wandering cudweed
purpureum L. - purple cudweed
purpureum L. var. *falcatum* (Lam.) Torr. & A. Gray - narrow-leaf cudweed
stramineum Kunth - cotton-batting cudweed
supinum L. - alpine cudweed
uliginosum L. - low cudweed
viscosum Kunth - clammy cudweed, clammy everlasting

Gomphrena L. (Amaranthaceae)
globosa L. - globe-amaranth, perpetua

Goodyera R. Br. (Orchidaceae)
oblongifolia Raf. - western rattlesnake plaintain

Gordonia J. Ellis (Theaceae)
lasianthus (L.) J. Ellis - black-laurel, loblolly-bay

Gossypium L. (Malvaceae)
cotton
arboreum L. - Ceylon cotton, tree cotton
barbadense L. - American-Egyptian cotton, extra-long staple cotton, Sea Island cotton, tree cotton
herbaceum L. - cultivated cotton, Levant cotton
hirsutum L. - upland cotton
sturtianum J. H. Willis - Sturt's desert-rose
thurberi Tod. - Arizona wild cotton

Gratiola L. (Scrophulariaceae)
aurea Pursh. - golden hedge-hyssop, golden-pert, yellow hedge-hyssop

Grevillea R. Br. ex Knight (Proteaceae)
spider-flower
robusta A. Cunn. ex R. Br. - silk-oak

Grindelia Willd. (Asteraceae)
squarrosa (Pursh) Dunal - curly-cup gum-weed, gum-plant, gum-weed, rosinweed, tarweed

Guajacum L. (Zygophyllaceae)
sanctum L. - holy-wood lignum-vitae, lignum-vitae

Guettarda L. (Rubiaceae)
scabra (L.) Vent. - rough-leaf velvet-seed

Guizotia Cass. (Asteraceae)

abyssinica (L.f.) Cass. - guizotia, Niger-seed, noog, nug, ramtilla

Gutierrezia Lag. (Asteraceae)
broom-weed , match-brush, match-weed, resin-weed, snakeweed, turpentine-weed
microcephala (DC.) A. Gray - sticky snakeweed, thread-leaf snakeweed
sarothrae (Pursh) Britton & Rusby - broom snakeroot, broom snakeweed

Gymnanthes Sw. (Euphorbiaceae)
lucida Sw. - crab-wood, oyster-wood, poison-wood

Gymnocarpium Newman (Pteridophyta)
dryopteris (L.) Newman - oak fern

Gymnocladus Lam. (Fabaceae)
dioica (L.) K. Koch - chicot, Kentucky coffee-tree, nickers-tree

Gynura Cass. (Asteraceae)
aurantiaca (Blume) DC. - purple-passion-vine, royal-vine-plant, velvet-plant

Gypsophila L. (Caryophyllaceae)
elegans M. Bieb. - baby's-breath
paniculata L. - baby's-breath, old-maid's-pink

Habenaria Willd. (Orchidaceae)
fringe orchid, rein orchid
dilatata (Pursh) Hook. - bog-candle, leaf white orchid, scented-bottle, tall white bog orchid
hyperborea (L.) R. Br. - northern green orchid
viridis (L.) R. Br. var. *bracteata* (Muhl. ex Willd.) A. Gray - bracted green-onion

Hackelia Opiz (Boraginaceae)
beggar's-lice, stickseed
diffusa (Lehm.) Johnst. - stickweed
floribunda (Lehm.) I. M. Johnst. - stickseed, western stickseed
virginiana (L.) I. M. Johnst. - wild comfrey

Hackelochloa Kuntze (Poaceae)
granularis (L.) Kuntze - pit-scale grass

Haemanthus L. (Liliaceae)
coccineus L. - African blood-lily, blood-lily

Halenia Borkh. (Gentianaceae)
elliptica D. Don - spurred-gentian

Halesia J. Ellis ex L. (Styracaceae)
silver-bell, snowdrop-tree
diptera J. Ellis - two-winged silver-bell
tetraptera J. Ellis - little silver-bell, opossum-wood, shittim-wood, wild olive

Halogeton C. A. Mey. (Chenopodiaceae)
glomeratus (Stephen ex M. Bieb.) C. A. Mey. - barilla

Hamamelis L. (Hamamelidaceae)
virginiana L. - witch-hazel

Hamelia Jacq. (Rubiaceae)
patens Jacq. - fire-bush, scarlet-bush

Haplopappus Cass. (Asteraceae)
golden-weed

drummondii (Torr. & A. Gray) Blake - Drummond's golden-weed

heterophyllus (A. Gray) S. F. Blake - jimmy-weed, rayless goldenrod

tenuisectus (Greene) S. F. Blake ex L. D. Benson - burro-weed

Hatiora Britton & Rose (Cactaceae)
gaertneri (Regel) Barthlott - Easter cactus

Hebe Comm. ex Juss. (Scrophulariaceae)
speciosa (A. Cunn.) Cockayne & Allan - showy speedwell

Hedeoma Pers. (Lamiaceae)
mock pennyroyal
hispidum Pursh - rough false pennyroyal
pulegioides (L.) Pers. - American false pennyroyal, American pennyroyal, mock pennyroyal, mosquito-plant, pudding-grass, squaw-mint, stinking-palm

Hedera L. (Araliaceae)
ivy
helix L. - English ivy

Hedychium J. König (Zingiberaceae)
garland-lily, ginger-lily
coronarium J. König ex Retz. - butterfly-ginger, butterfly-lily, cinnamon-jasmine, garland-flower, ginger-lily, white-ginger
gardnerianum Roscoe - Kahili ginger

Hedyotis L. (Rubiaceae)
caerulea (L.) Hook. - bluets, innocence, Quaker-ladies
corymbosa (L.) Lam. - Old World diamond-flower
procumbens (J. F. Gmel.) Fosberg - fairy-footprints
purpurea (L.) Torr. & A. Gray - large houstonia

Helenium L. (Asteraceae)
sneezeweed
amarum (Raf.) H. Rock - bitter sneezeweed, bitter-weed, fennel, yellow dog-fennel
autumnale L. - sneezeweed, stagger-wort, swamp sunflower, yellow-star
flexuosum Raf. - purple-head sneezeweed, southern sneezeweed
hoopesii A. Gray - orange sneezeweed
microcephalum DC. - sneezeweed

Helianthemum Mill. (Cistaceae)
rock-rose, sun-rose
bicknellii Fernald - frostweed
canadense (L.) Michx. - frost-wort, frostweed
nummularium (L.) Mill. - sun-rose

Helianthus L. (Asteraceae)
sunflower
angustifolius L. - narrow-leaf sunflower, swamp sunflower
annuus L. - Hopi sunflower, mirasol, sunflower, wild sunflower
argophyllus Torr. & A. Gray - silver-leaf sunflower
atrorubens L. - Appalachian sunflower, dark-eye sunflower, dark-red sunflower, hairy-wood sunflower
ciliaris DC. - blue-weed, Texas blue-weed
debilis Nutt. - beach sunflower, weak sunflower
decapetalus L. - forest sunflower, giant sunflower, ten-petals sunflower, thin-leaf sunflower, woodland sunflower
divaricatus L. - divaricate sunflower, divergent sunflower, woodland sunflower

giganteus L. - giant sunflower, swamp sunflower, tall sunflower
grosseserratus G. Martens - saw-tooth sunflower
heterophyllus Nutt. - different-leaf sunflower, hairy sunflower
hirsutus Raf. - hairy sunflower, rough sunflower, stiff-haired sunflower
kellermanii Britton - Kellerman's sunflower
×*laetiflorus* Pers. - cheerful sunflower, showy sunflower
Note = *H. rigidus* × *H. tuberosus* var. *rigidus.*
laetiflorus Pers. - bright sunflower
maximilianii Schrad. - Judge Daly's sunflower, Maximilian's sunflower
microcephalus Torr. & A. Gray - small-headed sunflower, small-wood sunflower
mollis Lam. - ashy sunflower, hairy sunflower, soft sunflower
nuttallii Torr. & A. Gray - Nuttall's sunflower
occidentalis Riddell - naked-stemmed sunflower, western sunflower
pauciflorus Nutt. - stiff sunflower
petiolaris Nutt. - Kansas sunflower, petioled sunflower, prairie sunflower
strumosus L. - pale-leaf sunflower, rough-leaf sunflower, swollen sunflower, woodland sunflower
tuberosus L. - earth-apple, girasole, Jerusalem artichoke, sun-choke, woodland sunflower

Helichrysum Mill. (Asteraceae)
everlasting, immortelle
bracteatum (Vent.) Andr. - straw-flower

Heliconia L. (Heliconiaceae)
psittacorum L.f. - parakeet-flower, parrot-flower, parrot-plantain

Heliopsis Pers. (Asteraceae)
ox-eye
helianthoides (L.) Sweet - ox-eye, sunflower-everlasting

Heliotropium L. (Boraginaceae)
heliotrope, turnsole
amplexicaule Vahl - clasping heliotrope
arborescens L. - cherry-pie, heliotrope
curassavicum L. - salt heliotrope, seaside heliotrope, white-weed, wild heliotrope
curassavicum L. var. *obovatum* DC. - seaside heliotrope, spatulate-leaved heliotrope
curassavicum L. var. *oculatum* (A. Heller) I. M. Johnst. - alkali heliotrope
europaeum L. - European heliotrope
indicum L. - Indian heliotrope, turnsole

Helipterum DC. (Asteraceae)
everlasting, straw-flower
splendidum Hemsl. - showy sunray, silky-white everlasting, splendid everlasting

Helleborus L. (Ranunculaceae)
niger L. - Christmas-rose

Hemarthria R. Br. (Poaceae)
altissima (Poir.) Stapf & F. T. Hubb. - limpo grass

Hemerocallis L. (Liliaceae)
daylily
fulva (L.) L. - fulvous daylily, orange daylily, tawny daylily
lilioasphodelus L. - lemon-lily, yellow daylily

Hemigraphis Nees (Acanthaceae)
  alternata (Burm. f.) T. Anderson - red-flame-ivy, red-ivy

Hemizonia DC. (Asteraceae)
  congesta DC. - hayfield tarweed
  pungens (Hook. & Arn.) Torr. & A. Gray - spike-weed

Hepatica Mill. (Ranunculaceae)
  hepatica, liver-leaf, noble-liverwort
  americana (DC.) Ker Gawl. - round-lobed hepatica

Heracleum L. (Apiaceae)
  cow-parsnip
  sphondylium L. subsp. *montanum* (Schleich. ex Gaudin) Briq.
    - American cow-parsnip, cow-parsnip, heltrot, hogweed,
    masterwort

Hesperis L. (Brassicaceae)
  rocket
  matronalis L. - dame's-rocket, dame's-violet, mother-of-the-
    evening, sweet-rocket

Heteranthera Ruiz & Pav. (Pontederiaceae)
  dubia (Jacq.) MacMill. - water star grass
  limosa (Sw.) Willd. - duck-salad
  reniformis Ruiz & Pav. - round-leaf mud-plantain

Heteromeles M. Roem. (Rosaceae)
  arbutifolia (Lindl.) Roem. - Christmas-berry

Heteropogon Pers. (Poaceae)
  contortus (L.) P. Beauv. ex Roem. & Schult. - tangle-head
  melanocarpus (Elliott) Benth. - sweet tangle-head

Heterotheca Cass. (Asteraceae)
  grandiflora Nutt. - telegraph-plant, telegraph-plant-weed

Heuchera L. (Saxifragaceae)
  alumroot
  americana L. - black-geranium, rock-geranium
  sanguinea Engelm. - coral-bells

Hevea Aubl. (Euphorbiaceae)
  brasiliensis (Willd. ex A. Juss.) Müll. Arg. - caouthchouc-
    tree, Pará rubber-tree, rubber-tree

Hibiscus L. (Malvaceae)
  giant mallow, rose mallow
  cannabinus L. - bastard-jute, bimli-jute, Deccan's-hemp,
    Indian hemp, kenaf
  grandiflorus Michx. - great rose mallow
  laevis All. - halberd-leaf mallow, smooth rose mallow, soldier
    rose mallow
  lasiocarpus Cav. - rose mallow, woolly rose mallow
  moscheutos L. - hibiscus, mallow-rose, rose mallow, swamp
    rose mallow, wild cotton
  moscheutos L. subsp. *palustris* (L.) R. T. Clausen - crimson-
    eyed rose mallow, mallow-rose, marsh mallow, sea
    hollyhock, swamp rose mallow, swamp-rose, wild cotton
  rosa-sinensis L. - blackening-plant, China rose, Chinese
    hibiscus, Hawaiian hibiscus, rose-of-China
  sabdariffa L. - Indian sorrel, Jamaican sorrel, roselle
  syriacus L. - rose-of-Sharon, shrub althea
  tiliaceus L. - mahoe, majagua, sea hibiscus
  trionum L. - bladder ketmia, flower-of-an-hour, modesty,
    shoofly, Venice mallow

Hieracium L. (Asteraceae)
  hawkweed
  aurantiacum L. - devil's-paintbrush, king-devil, orange
    hawkweed, orange-paintbrush
  caespitosum Dumort. - field hawkweed, yellow hawkweed,
    yellow king-devil, yellow-devil, yellow-paintbrush
  canadense Michx. - Canadian hawkweed
  floribundum Wimm. & Grab. - glaucous hawkweed, king-
    devil, smooth hawkweed, yellow-devil hawkweed
  gronovii L. - beaked hawkweed, Gronovius's hawkweed,
    hairy hawkweed
  lachenalii C. C. Gmel. - hawkweed
  murorum L. - golden lungwort, wall hawkweed
  pilosella L. - felon-herb, mouse-bloodwort, mouse-ear
    hawkweed
  piloselloides Vill. - glaucous king-devil, king-devil hawkweed
  praealtum Vill. ex Gochnat var. *decipiens* W.D.J. Koch - tall
    hawkweed
  scabrum Michx. - rough hawkweed, sticky hawkweed
  umbellatum L. - narrow-leaf hawkweed, northern hawkweed
  venosum L. - poor-Robin's-plantain, rattlesnake-weed, veiny
    hawkweed

Hierochloe R. Br. (Poaceae)
  occidentalis Buckley - California sweet grass
  odorata (L.) P. Beauv. - sweet grass, vanilla grass

Hilaria Kunth (Poaceae)
  belangeri (Steud.) Nash - curly-mesquite grass
  jamesii (Torr.) Benth. - galleta grass
  mutica (Buckley) Benth. - tobosa grass
  rigida (Thurb.) Benth. ex Scribn. - big galleta

Hippeastrum Herb. (Liliaceae)
  puniceum (Lam.) Urb. - amaryllis, Barbados lily

Hippuris L. (Hippuridaceae)
  vulgaris L. - mare's-tail

Hirschfeldia Moench (Brassicaceae)
  incana (L.) Lag.-Foss. - short-pod mustard

Hoffmannseggia Cav. (Fabaceae)
  glauca (Ortega) Eifert - hog-potato, pig-nut

Holcus L. (Poaceae)
  lanatus L. - velvet grass, Yorkshire fog
  mollis L. - creeping soft grass, creeping velvet grass, German
    velvet grass

Holocarpha Greene (Asteraceae)
  virgata (A. Gray) Keck - virgate tarweed

Holodiscus (K. Koch) Maxim. (Rosaceae)
  discolor (Pursh) Maxim. - cream-bush, ocean-spray, rock-
    spiraea
  dumosus (Nutt. ex Hook.) A. Heller - mountain-glory, rock-
    spiraea

Holosteum L. (Caryophyllaceae)
  umbellatum L. - jagged chickweed, umbrella-spurry

Homalocladium (F. Muell.) L. H. Bailey (Polygonaceae)
  platycladum (F. Muell.) L. H. Bailey - centipede-plant,
    ribbon-bush, tapeworm-plant

Honckenya Ehrh. (Caryophyllaceae)
peploides (L.) Ehrh. - sea beach sandwort, sea-chickweed, sea-purslane

Hordeum L. (Poaceae)
barley
arizonicum Covas - Arizona barley
brachyantherum Nevski - meadow barley
brevisubulatum (Trin.) Link - short-awn barley
bulbosum L. - bulbous barley
geniculatum All. - Mediterranean barley
glaucum Steud. - flicker-tail grass, foxtail grass, tickle grass, wall barley
jubatum L. - foxtail barley, skunk-tail grass, squirrel-tail barley, wild barley
leporinum Link - hare barley, wild barley
murinum L. - barley grass, mouse barley, wall barley
pusillum Nutt. - little barley
vulgare L. - barley, Nepal barley

Hosta Tratt. (Liliaceae)
lancifolia (Thunb.) Engl. - narrow-leaf plantain-lily
plantaginea (Lam.) Asch. - fragrant plantain-lily
ventricosa (Salisb.) Stearn - blue plantain-lily, muraski-giboshi

Hottonia L. (Primulaceae)
inflata Elliott - feather-foil, water-violet

Hovenia Thunb. (Rhamnaceae)
dulcis Thunb. - Japanese raisin-tree

Howeia Becc. (Arecaceae)
sentry palm
belmoreana (C. Moore & F. Muell.) Becc. - Belmore's palm, Belmore's sentry palm, curly palm
forsteriana (C. Moore & F. Muell.) Becc. - Forster's palm, Forster's sentry palm, kentia palm, sentry palm, thatch palm, thatch-leaf palm

Hoya R. Br. (Asclepiadaceae)
carnosa (L.f.) R. Br. - honey-plant, wax-plant

Hudsonia L. (Cistaceae)
beach-heather
ericoides L. - golden-heather
tomentosa Nutt. - beach-heath, false heather, hudsonia, poverty-grass

Humulus L. (Cannabaceae)
hop
japonicus Siebold & Zucc. - Japanese hop
lupulus L. - bine, European hop, hop

Hura L. (Euphorbiaceae)
crepitans L. - javillo, monkey's-dinner-bell, monkey-pistol, sandbox-tree

Hyacinthoides Heist. ex Fabr. (Liliaceae)
hispanica (Mill.) Rothm. - blue-flower squill

Hyacinthus L. (Liliaceae)
orientalis L. - Dutch hyacinth, garden hyacinth, hyacinth

Hybanthus Jacq. (Violaceae)
concolor (T. F. Forst.) Spreng. - green-violet

Hydrangea L. (Hydrangeaceae)
anomala D. Don subsp. *petiolaris* (Siebold & Zucc.) D. C. McClint. - climbing hydrangea
arborescens L. - American hydrangea, hills-of-snow, seven-bark, smooth hydrangea, wild hydrangea
macrophylla (Thunb.) Ser. - French hydrangea, hortensia, house hydrangea
paniculata Siebold - panicled hydrangea

Hydrastis L. (Ranunculaceae)
canadensis L. - goldenseal, orange-root, tumeric

Hydrilla Rich. (Hydrocharitaceae)
verticillata (L.f.) Royle - hydrilla

Hydrocotyle L. (Apiaceae)
navelwort, pennywort, water pennywort
americana L. - marsh pennywort, water pennywort
bonariensis Comm. ex Lam. - coastal plain pennywort
sibthorpioides Lam. - lawn pennywort
umbellata L. - water pennywort
verticillata Thunb. - whorled pennywort

Hydrolea L. (Hydrophyllaceae)
uniflora Raf. - one-flowered hydrolea

Hydrophyllum L. (Hydrophyllaceae)
water-leaf
canadense L. - broad-leaf water-leaf, hairy water-leaf, maple-leaf water-leaf
capitatum Douglas ex Benth. - cat-breeches, woolen-breeches
macrophyllum Nutt. - hairy water-leaf, large-leaf water-leaf
virginianum L. - eastern water-leaf, Indian-salad, John's-cabbage, Shawnee-salad, Virginia water-leaf

Hygrophila R. Br. (Acanthaceae)
difformis (L.f.) Blume - water-wisteria
lacustris (Schltdl. & Cham.) Nees - lake hygrophila
polysperma (Roxb.) T. Anderson - Indian hygrophila

Hylocereus A. Berger (Cactaceae)
undatus (Haw.) Britton & Rose - Honolulu-queen, night-blooming cereus, queen-of-the-night

Hymenocallis Salisb. (Liliaceae)
basket-flower, crown-beauty, sea-daffodil, spider-lily
narcissiflora (Jacq.) J. F. Macbr. - basket-flower, Peruvian daffodil

Hymenopappus L'Hér. (Asteraceae)
scabiosaeus L'Hér. - white-bracted hymenopappus
scabiosaeus L'Hér. var. *corymbosus* (Torr. & A. Gray) B. L. Turner - old-plainsman

Hymenoxys Cass. (Asteraceae)
odorata DC. - bitter rubber-weed
richardsonii Cockll. var. *floribunda* (A.Gray) Parker - pingue

Hyoscyamus L. (Solanaceae)
niger L. - black henbane, henbane, stinking nightshade

Hyparrhenia Andersson ex E. Fourn. (Poaceae)
rufa (Nees) Stapf - jaragua grass

Hypericum L. (Clusiaceae)
St. John's-wort
adpressum Barton - creeping St. John's-wort
calycinum L. - Aaron's-beard, creeping St. John's-wort, golden-flower, rose-of-Sharon

canadense L. - Canadian St. John's-wort
crux-andreae (L.) Crantz - St. Peter's-wort
drummondii (Grev. & Hook.) Torr. & A. Gray - nits-and-lice
frondosum Michx. - golden St. John's-wort
gentianoides (L.) Britton, et al. - orange-grass, pine-weed
hypericoides (L.) Crantz - St. Andrew's-cross
kalmianum L. - Kalm's St. John's-wort
mitchellianum Rydb. - Blue Ridge St. John's-wort, mountain St. John's-wort
mutilum L. - dwarf St. John's-wort
perforatum L. - eola-weed, goat-weed, Klamath-weed, rosin-rose, St. John's-wort, Tipton's-weed
prolificum L. - broom-brush, shrubby St. John's-wort
punctatum Lam. - spotted St. John's-wort
pyramidatum Aiton - great St. John's-wort
tubulosum Walter - marsh St. John's-wort
virginicum L. - marsh St. John's-wort

Hyphaene Gaertn. (Arecaceae)
thebaica (L.) Mart. - doum palm, Egyptian doum, ginberbread-tree, gingerbread palm

Hypochaeris L. (Asteraceae)
glabra L. - smooth cat's-ear, smooth false dandelion
radicata L. - cat's-ear, coast-dandelion, false dandelion, flat-weed, gosmore, rough cat's-ear, spotted cat's-ear

Hypoxis L. (Liliaceae)
hirsuta (L.) Cov. - star-grass

Hyptis Jacq. (Lamiaceae)
emoryi Torr. - desert lavender
mutabilis (A. Rich.) Briq. - bitter mint

Hyssopus L. (Lamiaceae)
officinalis L. - hyssop

Iberis L. (Brassicaceae)
candytuft
amara L. - rocket candytuft
umbellata L. - globe candytuft

Idesia Maxim. (Flacourtiaceae)
polycarpa Maxim. - iigiri-tree

Ilex L. (Aquifoliaceae)
holly, inkberry, winterberry
ambigua (Michx.) Torr. - Carolina holly
ambigua (Michx.) Torr. var. *montana* (A. Gray) Ahles - big-leaf holly, halver, large-leaf holly, mountain holly, mountain winterberry
amelanchier M. A. Curtis - Sarvis's holly
anomala Hook. & Arn. f. *sandwicensis* (Endl.) Loes. - kawa'ii
aquifolium L. - English holly, European holly, Oregon holly
cassine L. - cassina, dahoon, dahoon holly, yaupon
coriacea (Pursh) Chapm. - bay-gall-bush, gallberry, large gallberry, sweet gallberry
cornuta Lindl. & Paxton - Chinese holly, horned holly
crenata Thunb. - box-leaf holly, Japanese holly
decidua Walter - possum-haw
glabra (L.) A. Gray - Appalachian tea, bitter gallberry, gallberry, inkberry, winterberry
laevigata (Pursh) A. Gray - smooth winterberry
myrtifolia Walter - myrtle dahoon
opaca Aiton - American holly
verticillata (L.) A. Gray - black-alder, winterberry

vomitoria Aiton - cassina, yaupon

Illicium L. (Illiciaceae)
anise-tree
floridanum J. Ellis - Florida anise-tree, purple anise
parviflorum Vent. - yellow anise-tree

Impatiens L. (Balsaminaceae)
balsam, jewelweed, snap-weed, touch-me-not
balsamina L. - balsam, garden balsam, rose-balsam
capensis Meerb. - jewelweed, lady's-earrings, orange touch-me-not, spotted snap-weed, spotted touch-me-not
noli-tangere L. - touch-me-not
pallida Nutt. - jewelweed, pale touch-me-not, pale-snapdragon, yellow touch-me-not
wallerana Hook. f. - busy-Lizzy, patience-plant, patient-Lucy, sultana, Zanzibar balsam

Imperata Cirillo (Poaceae)
brasiliensis Trin. - Brazilian satin-tail
cylindrica (L.) Raeusch. - cogon grass

Indigofera L. (Fabaceae)
indigo
hirsuta L. - hairy indigo
pilosa Poir. - blanket indigo
spicata Forssk. - creeping indigo, spicate indigo
suffruticosa Mill. - anil indigo

Inula L. (Asteraceae)
helenium L. - elecampane, elf-dock, elf-wort, horse-elder, horse-heal, scab-wort, yellow starwort

Ipomoea L. (Convolvulaceae)
morning-glory
alba L. - moonflower
aquatica Forssk. - kangkong, swamp morning-glory, water-spinach
batatas (L.) Lam. - sweet-potato, yam
cairica (L.) Sweet - Cairo morning-glory
coccinea L. - red morning-glory, small red morning-glory, star ipomoea
cordatotriloba Dennst. - sharp-pod morning-glory
cordatotriloba Dennst. var. *torreyana* (A. Gray) D. F. Austin - cotton morning-glory
hederacea Jacq. - ivy-leaf morning-glory
hederifolia L. - ivy-leaf red morning-glory
indica (Burm.) Merr. - blue dawn-flower, blue morning-glory
lacunosa L. - pitted morning-glory, small white morning-glory, white morning-glory
leptophylla Torr. - bush moonflower, bush morning-glory, man-of-the-earth, man-root
nil (L.) Roth - Japanese morning-glory, white-edge morning-glory
pandurata (L.) G. Mey. - big-root morning-glory, man-of-the-earth, man-root, wild potato, wild sweet-potato-vine
pes-caprae (L.) R. Br. - beach morning-glory, railroad-vine
purpurea (L.) Roth - morning-glory, tall morning-glory
quamoclit L. - cardinal-climber, cypress-vine, cypress-vine morning-glory, star-glory
tricolor Cav. - multicolored morning-glory
triloba L. - three-lobed morning-glory
turbinata Lag. - purple moonflower
wrightii A. Gray - palm-leaf morning-glory

Iresine P. Browne (Amaranthaceae)
diffusa Humb. & Bonpl. ex Willd. - bloodleaf, Jubas-bush
herbstii Hook. - beef-plant, beefsteak-plant, chicken's-gizzard

Iris L. (Iridaceae)
  flag, fleur-de-lis
cristata Aiton - dwarf-crested iris
ensata Thunb. - Japanese iris, Japanese water iris
fulva Ker Gawl. - copper iris, red iris
×*germanica* L. - flag, fleur-de-lis
latifolia Mill. - English iris
missouriensis Nutt. - Rocky Mountain iris, western blue flag
pseudacorus L. - yellow flag iris
verna L. - dwarf iris, vernal iris, violet iris
versicolor L. - blue flag iris, larger blue flag, northern blue flag, poison flag, wild iris
virginica L. var. *schrevei* (Small) E. S. Anderson - southern blue flag
xiphium L. - Spanish iris

Isatis L. (Brassicaceae)
tinctoria L. - asp-of-Jerusalem, dyer's woad

Ischaemum L. (Poaceae)
rugosum Salisb. - Saramolla grass

Isocoma Nutt. (Asteraceae)
coronopifolia (A. Gray) Greene - golden-weed

Isotria Raf. (Orchidaceae)
medeoloides (Pursh) Raf. - little five-leaves, little whorled pogonia, small whorled pogonia

Itea L. (Grossulariaceae)
virginica L. - sweet-spire, tassel-white, Virginia willow

Iva L. (Asteraceae)
  marsh-elder
acerosa (Nutt.) R. C. Jacks. - copper-weed
annua L. - annual marsh-elder, rough marsh-elder, rough sump-weed
axillaris Pursh - death-weed, poverty sump-weed, poverty-weed
xanthifolia Nutt. - big marsh-elder, marsh-elder

Ixiolirion Herb. (Liliaceae)
tataricum (Pall.) Herb. - Siberian lily, tartar-lily

Ixora L. (Rubiaceae)
coccinea L. - flame-of-the-woods, scarlet ixora

Jacquemontia Choisy (Convolvulaceae)
tamnifolia (L.) Griseb. - small-flowered morning-glory

Jamesia Torr. & A. Gray (Hydrangeaceae)
americana Torr. & A. Gray - cliff-bush

Jasione L. (Campanulaceae)
montana L. - sheep's-bit

Jasminum L. (Oleaceae)
  jasmine
multiflorum (Burm.) Andr. - star jasmine
nitidum Skan - angel-wing jasmine, star jasmine, windmill jasmine
nudiflorum Lindl. - winter jasmine
sambac (L.) Aiton - Arabian jasmine

Jatropha L. (Euphorbiaceae)
curcas L. - Barbados-nut, physic-nut

Jeffersonia Barton (Berberidaceae)
diphylla (L.) Pers. - twinleaf

Juglans L. (Juglandaceae)
  butternut, walnut
ailantifolia Carrière - Japanese walnut, onigurumi
californica S. Watson - California black walnut, California walnut, southern California walnut
cinerea L. - butternut, white walnut
cordiformis Maxim. - heart-nut, himegurumi
hindsii (Jeps.) Rehder - Hinds' walnut, northern California black walnut
major (Torr.) A. Heller - Arizona walnut, nogal
microcarpa Berland. - little walnut, nogal, river walnut, Texas black walnut, Texas walnut
nigra L. - black walnut
regia L. - English walnut, Madeira-nut, Persian walnut

Juncus L. (Juncaceae)
  bog rush, rush
acuminatus Michx. - tufted rush
balticus Willd. - Baltic rush, wire rush
bufonius L. - toad rush
effusus L. - bog rush, Japanese mat rush, soft rush
gerardii Loisel. - black-grass, hog rush
lesueurii Bol. - salt rush
roemerianus Scheele - needle rush
tenuis Willd. - field rush, path rush, poverty rush, slender rush, slender yard rush, wiregrass

Juniperus L. (Cupressaceae)
  juniper
ashei Buchholz - Ashe's juniper, Ozark white-cedar
californica Carrière - California juniper
chinensis L. - Chinese juniper
chinensis L. var. *sargentii* A. Henry - Sargent's juniper
communis L. - juniper, mountain juniper
communis L. subsp. *depressa* (Pursh) Franco - low juniper
conferta Parl. - shore juniper
deppeana Steud. - alligator juniper, sweet-fruited juniper
excelsa M. Bieb. - Greek juniper
flaccida Schltdl. - Mexican drooping juniper
horizontalis Moench - creeping juniper, creeping savin juniper, creeping-cedar, shrubby red-cedar
monosperma (Engelm.) Sarg. - cherry-stone juniper, one-seeded juniper
occidentalis Hook. - California juniper, sierra juniper, western juniper
osteosperma (Torr.) Little - Utah juniper
oxycedrus L. - prickly juniper
pinchotii Sudw. - Pinchot's juniper, red-berry juniper
sabina L. - savin
scopulorum Sarg. - Colorado red-cedar, Rocky Mountain juniper
silicicola (Small) L. H. Bailey - southern red-cedar
virginiana L. - eastern red-cedar, pencil-cedar, red-cedar

Justicia L. (Acanthaceae)
americana (L.) Vahl - American water-willow
brandegeana Wassh. & L. B. Sm. - false hop-plant, junta-de-cobra-pintada, shrimp-bush, shrimp-plant

carnea Lindl. - balsamo-cor-de-carne, Brazilian-plume, flamingo-plant, king's-crown, paradise-plant, plume-flower, plume-plant

Kalanchoe Adans. (Crassulaceae)
Palm Beach bells
fedtschenkoi Raym.-Hamet & E. Perrier - lavender-scallops, South American air-plant
laciniata (L.) DC. - Christmas-tree kalanchoe, fig-tree kalanchoe
marmorata Baker - pen-wiper
pinnata (Lam.) Pers. - air-plant, curtain-plant, floppers, good-luck-leaf, life-plant, Mexican love-plant, miracle-leaf, sprouting-leaf
tomentosa Baker - panda-bear-plant, plush-plant, pussy's-ears

Kalmia L. (Ericaceae)
American laurel, laurel
angustifolia L. - calf-kill, dwarf-laurel, lambkill, narrow-leaf-laurel, pig-laurel, sheep-laurel, sheep-poison, wicky
latifolia L. - calico-bush, ivy, ivy-bush, mountain-laurel, poison-laurel, spoonwood
polifolia Wangenh. - alpine-laurel, bog kalmia, bog-laurel, pale-laurel, swamp-laurel

Kerria DC. (Rosaceae)
japonica (L.) DC. - globeflower, Japanese rose, kerria

Kickxia Dumort. (Scrophulariaceae)
elatine (L.) Dumort. - sharp-point fluvellin
spuria (L.) Dumort. - female fluvellin

Kigelia DC. (Bignoniaceae)
africana (Lam.) Benth. - sausage-tree

Knautia L. (Dipsacaceae)
arvensis (L.) Coult. - blue-buttons, field scabious

Kniphofia Moench (Aloaceae)
poker-plant, red-hot-poker, torch-lily, tritoma
uvaria (L.) Hook. - flame-flower, poker-plant, torch-lily

Kochia Roth (Chenopodiaceae)
americana S. Watson - green-molly

Koeleria Pers. (Poaceae)
pyrimidata (Lam.) P. Beauv. - June grass

Koelreuteria Laxm. (Sapindaceae)
golden-rain-tree
elegans (Seem.) A. C. Sm. - flame-gold
paniculata Laxm. - China-tree, golden-rain-tree, pride-of-India, varnish-tree

Kokia Lewton (Malvaceae)
drynarioides (Seem.) Lewton - kokio

Kolkwitzia Graebn. (Caprifoliaceae)
amabilis Graebn. - beauty-bush

Krascheninnikovia Gueldenst. (Chenopodiaceae)
lanata (Pursh) A. Meeuse & Smit - winter-fat

Krigia Schreb. (Asteraceae)
dwarf-dandelion
biflora (Walter) S. F. Blake - orange dwarf-dandelion, two-flowered cynthia
virginica (L.) Willd. - dwarf-dandelion, Virginia dwarf-dandelion

Krugiodendron Urb. (Rhamnaceae)
ferreum (Vahl) Urb. - lead-wood

Kummerowia Schindl. (Fabaceae)
stipulacea (Maxim.) Makino - Korean bush-clover, Korean clover, Korean lespedeza

Lablab Adans. (Fabaceae)
purpureus (L.) Sweet - bonavist bean, hyacinth bean, lablab bean

Laburnum Medik. (Fabaceae)
bean-tree
anagyroides Medik. - golden-chain

Lachnanthes Elliott (Haemodoraceae)
caroliniana (Lam.) Dandy - dye-root, paint-root, red-root

Lactuca L. (Asteraceae)
lettuce
biennis (Moench) Fernald - biennial lettuce, wild lettuce
canadensis L. - Canadian wild lettuce, tall lettuce
floridana (L.) Gaertn. - Florida wild lettuce, woodland lettuce
graminifolia Michx. - grass-leaf lettuce
muralis (L.) Fresen. - wall lettuce
saligna L. - willow-leaf lettuce
sativa L. - celtuce, garden lettuce, lettuce
sativa L. var. *capitata* L. - head lettuce
sativa L. var. *longifolia* Lam. - Cos lettuce, Romaine lettuce
serriola L. - compass-plant, horse-thistle, milk-thistle, prickly lettuce, wild opium
tatarica (L.) C. A. Mey. subsp. *pulchella* (Pursh) Stebbins - blue lettuce

Lagenaria Ser. (Cucurbitaceae)
siceraria (Molina) Standl. - bottle gourd, calabash gourd, white-flowered gourd

Lagerstroemia L. (Lythraceae)
indica L. - crape-myrtle

Laguncularia C. F. Gaertn. (Combretaceae)
racemosa (L.) C. F. Gaertn. - white buttonwood, white mangrove

Lagurus L. (Poaceae)
ovatus L. - hare's-tail, rabbit-tail grass

Lamarckia Moench (Poaceae)
aurea (L.) Moench - golden-top

Lamium L. (Lamiaceae)
dead-nettle
album L. - snowflake, white dead-nettle
amplexicaule L. - henbit
maculatum L. - spotted dead-nettle
purpureum L. - purple dead-nettle, red dead-nettle

Lampranthus N. E. Br. (Aizoaceae)
glomeratus (L.) N. E. Br. - akulikuli

Lansium M. P. Corréa (Meliaceae)
domesticum M. P. Corréa - langsat

Lantana L. (Verbenaceae)
shrub-verbena
camara L. - lantana, large-leaf lantana, yellow-sage
montevidensis (Spreng.) Briq. - pole-cat-geranium, trailing lantana, weeping lantana

Lappula Moench (Boraginaceae)
occidentalis (S. Watson) Greene - western sticktight
redowskii (Hornem.) Greene - stickseed

Lapsana L. (Asteraceae)
communis L. - ballogan, nipplewort, succory-dock

Larix Mill. (Pinaceae)
larch
decidua Mill. - European larch
kaempferi (Lamb.) Carrière - Japanese larch
laricina (Du Roi) K. Koch - American larch, black larch, hackmatack, tamarack
lyallii Parl. - subalpine larch
occidentalis Nutt. - western larch

Larrea Cav. (Zygophyllaceae)
tridentata (Sessé & Moç ex DC.) J. M. Coult. - creosote-bush

Lasiacis (Griseb.) Hitchc. (Poaceae)
divaricata (L.) Hitchc. - tibisee

Lasthenia Cass. (Asteraceae)
chrysostoma (Fisch. & C. A. Mey.) Greene - gold-fields

Lathyrus L. (Fabaceae)
vetchling, wild pea
cicera L. - flat-pod pea-vine
hirsutus L. - Austrian winter pea, Caley's pea, rough pea-vine, singletary pea
japonicus Willd. - beach pea, heath pea, sea pea, seaside pea
latifolius L. - everlasting pea, everlasting pea-vine, perennial pea, perennial sweet pea, wild sweet pea
littoralis (Nutt.) Endl. - beach pea
ochroleucus Hook. - white pea, yellow vetchling
odoratus L. - sweet pea
palustris L. - marsh pea-vine, marsh vetchling, wild pea, wing-stcmmcd wild pea-vine
pratensis L. - craw pea, meadow pea, meadow pea-vine, meadow vetch, tar-fitch, yellow tare, yellow vetchling
sativus L. - chickling pea, chickling vetchling, dog-tooth pea, grass pea-vine, Indian pea, Khesari, riga pea
sylvestris L. - everlasting pea, flat pea, flat pea-vine, perennial pea
tingitanus L. - Tangier pea
tuberosus L. - earth-chestnut, earthnut pea, groundnut pea-vine, tuberous vetch, tuberous vetchling

Laurus L. (Lauraceae)
nobilis L. - bay, Grecian laurel, laurel, sweet-bay

Lavandula L. (Lamiaceae)
angustifolia Mill. - English lavender, lavender

Lavatera L. (Malvaceae)
arborea L. - tree mallow

Ledum L. (Ericaceae)
Labrador tea
glandulosum Nutt. - glandular Labrador tea, trapper's-tea, western Labrador tea
groenlandicum Oeder - Labrador tea
palustre L. - crystal-tea, wild rosemary

Leea D. Royen ex L. (Leeaceae)
coccinea Planch. - West Indian holly

Leersia Sw. (Poaceae)
cut grass, white grass
hexandra Sw. - southern cut grass
lenticularis Michx. - catchfly grass
oryzoides (L.) Sw. - rice cut grass

Leiophyllum R. Hedw. (Ericaceae)
buxifolium (Berg) Elliot - box sand-myrtle

Leitneria Chap. (Leitneriaceae)
floridana Chap. - corkwood

Lemna L. (Lemnaceae)
gibba L. - inflated duckweed
minor L. - duckweed, lesser duckweed
minuta Kunth - minute duckweed
obscura (Austin) Daubs - obscure duckweed
perpusilla Torr. - duck-meal
trinervis (Austin) Small - three-nerved duckweed
trisulca L. - star duckweed
valdiviana Phil. - small duckweed

Lens Mill. (Fabaceae)
culinaris Medik. - lentil

Leonotis (Pers.) R. Br. (Lamiaceae)
lion's-ear, lion's-tail
nepetifolia (L.) R. Br. - lion's-ear

Leontodon L. (Asteraceae)
autumnalis L. - fall hawk's-bit, fall-dandelion, hawk's-bit, lion's-tooth
hirtus L. - hairy hawk's-bit, rough hawk's-bit
taraxacoides (Vill.) Mérat - rough hawk's-bit

Leonurus L. (Lamiaceae)
cardiaca L. - lion's-ear, lion's-tail, motherwort
marrubiastrum L. - horehound motherwort
sibiricus L. - Siberian motherwort

Lepechinia Willd. (Lamiaceae)
calycina (Benth.) Epling - pitcher-sage

Lepidium L. (Brassicaceae)
peppergrass, pepperwort, tongue-grass
campestre (L.) R. Br. - cow cress, downy pcppergrass, field pepper-weed, field peppergrass, fieldcress
densiflorum Schrad. - green-flowered pepper-weed, green-flowered peppergrass, wild tongue-grass
latifolium L. - perennial pepper-weed, perennial pepper-weed, perennial peppergrass
nitidum Nutt. - tongue pepper-weed
perfoliatum L. - clasping pepper-weed, clasping-leaf peppergrass
ruderale L. - narrow-leaf pepper-weed, roadside peppergrass, stinking pepper-weed
sativum L. - garden cress
virginicum L. - bird-pepper, peppergrass, poor-man's-pepper, tongue-grass, Virginia pepper-weed

Leptochloa P. Beauv. (Poaceae)
sprangle-top
chinensis (L.) Nees - Chinese sprangle-top
dubia (Kunth) Nees - green sprangle-top
fascicularis (Lam.) A. Gray - bearded sprangle-top
filiformis (Lam.) P. Beauv. - red sprangle-top

panicoides (C. Presl & J. Presl) Hitchc. - Amazon sprangle-top

uninervia (J. Presl) Hitchc. & Chase - Mexican sprangle-top

Leptodactylon Hook. & Arn. (Polemoniaceae)
californicum Hook. & Arn. - prickly phlox
pungens (Torr.) Rydb. - granite-gilia

Leptospermum J. R. Forst. & G. Forst. (Myrtaceae)
scoparium J. R. Forst. & G. Forst. - manuka, New Zealand tea-tree

Lepyrodiclis Fenzl (Caryophyllaceae)
holosteoides (C. A. Mey.) Fisch. & C. A. Mey. - lepyrodiclis

Lespedeza Michx. (Fabaceae)
bush-clover
bicolor Turcz. - bicolored lespedeza, shrub bush-clover, shrub lespedeza
capitata Michx. - bush-clover, round-head bush-clover, round-head lespedeza
cuneata (Dum. Cours.) G. Don - Chinese bush-clover, Chinese lespedeza, perennial lespedeza, Sericea lespedeza
intermedia (S. Watson) Britton - wand lespedeza, wand-like bush-clover
procumbens Michx. - downy trailing lespedeza, trailing bush-clover
repens (L.) Barton - creeping bush-clover, smooth trailing lespedeza
striata (Thunb. ex Js. Murray) Hook. & Arn. - annual lespedeza, Japanese bush-clover, Japanese clover, Japanese lespedeza, lespedeza, striate lespedeza
stuevei Nutt. - Stueve's bush-clover, velvety lespedeza
thunbergii (DC.) Nakai - Thunberg's lespedeza
violacea (L.) Pers. - slender lespedeza, violet lespedeza
virginica (L.) Britton - slender bush-clover, slender lespedeza, Virginia lespedeza

Lesquerella S. Watson (Brassicaceae)
bladder-pod
gordonii (A. Gray) S. Watson - Gordon's bladder-pod

Leucaena Benth. (Fabaceae)
leucocephala (Lam.) de Wit - ipil, ipil-ipil, jumbie-bean, lead-tree, leucaena, white papinac
pulverulenta (Schltdl.) Benth. - great lead-tree, tepeguaje
retusa Benth. - little-leaf lead-tree

Leucanthemum Mill. (Asteraceae)
maximum (Ramond) DC. - daisy chrysanthemum, Max's chrysanthemum, Shasta daisy
vulgare Lam. - field daisy, marguerite, ox-eye daisy, poorland-flower, white daisy, white-weed

Leucocoryne Lindl. (Liliaceae)
ixioides (Hook.) Lindl. - glory-of-the-sun

Leucocrinum Nutt. ex A. Gray (Liliaceae)
mountain-lily, sand-lily, star-lily
montanum Nutt. ex A. Gray - mountain-lily, sand-lily, star-lily

Leucojum L. (Liliaceae)
aestivum L. - snowflake, summer-snowflake
vernum L. - spring-snowflake

Leucophyllum Bonpl. (Scrophulariaceae)
frutescens (Berland.) I. M. Johnst. - barometer-bush, cenizo

Leucothoe D. Don (Ericaceae)
fetterbush
axillaris (Lam.) D. Don - swamp dog-laurel
fontanesiana (Steud.) Sleumer - dog's-hobble, drooping-laurel, mountain dog-laurel, switch-ivy
racemosa (L.) A. Gray - sweet-bells

Levisticum Hill (Apiaceae)
officinale W. D. J. Koch - lovage

Lewisia Pursh (Portulacaceae)
rediviva Pursh - bitter-root

Leymus Hochst. (Poaceae)
angustus (Trin.) Pilg. - Altai wild rye
arenarius (L.) Hochst. - European dune grass, Lyme grass
chinensis (Trin.) Tzvelev - Chinese wild rice, false wheat grass
cinereus (Scribn. & Merr.) A. Löve - basin wild rye, giant wild rye
mollis (Trin.) Pilg. - American dune grass
racemosus (Lam.) Tzvelev - mammoth wild rye
salinus (M. E. Jones) A. Löve - Salina wild rye
triticoides (Buckley) Pilg. - beardless wild rye

Liatris Gaertn. ex Schreb. (Asteraceae)
blazing-star, button snakeroot, gay-feather
cylindracea Michx. - cylindric blazing-star, few-headed blazing-star
punctata Hook. - dotted blazing-star, dotted gay-feather
pycnostachya Michx. - prairie blazing-star, thick-spike blazing-star
spicata (L.) Willd. - blazing-star, dense blazing-star, sessile blazing-star
squarrosa (L.) Michx. - plains blazing-star, scaly blazing-star

Licania Aubl. (Chrysobalanaceae)
rigida Benth. - oiticica

Licaria Aubl. (Lauraceae)
triandra (Sw.) Kosterm. - gulf licaria

Ligusticum L. (Apiaceae)
scothicum L. - Scotch lovage, sea-lovage

Ligustrum L. (Oleaceae)
hedge-plant, privet
amurense Carrière - Amur privet
japonicum Thunb. - Japanese privet, wax-leaf privet
lucidum Aiton - Chinese privet, glossy privet, Nepal privet, wax-leaf privet, white wax-tree
obtusifolium Siebold & Zucc. - California privet
ovalifolium Hassk. - California privet
sinense Lour. - Chinese privet
vulgare L. - prim, privet

Lilium L. (Liliaceae)
auratum Lindl. - gold-banded lily, golden-rayed lily, mountain lily
bulbiferum L. - orange lily
canadense L. - Canadian lily, wild yellow lily, yellow bell lily
candidum L. - Madonna-lily
catesbaei Walter - leopard lily, pine lily, southern red lily
columbianum Baker - Columbia lily, Oregon lily
dauricum Ker Gawl. - candlestick lily
hansonii Moore - Japanese Turk's-cap lily

×*hollandicum* Bergmans - candlestick lily
  Note = A group of lilies derived from crossing *L.*
    bulbiferum or *L. bulbiferum* var. *croceum* with forms of
    L. × maculatum.
lancifolium Thunb. - tiger lily
longiflorum Thunb. - trumpet lily, white-trumpet lily
longiflorum Thunb. var. *eximium* (Courtois) Baker - Bermuda
  Easter lily, Easter lily
martagon L. - Martagon lily, turban lily, Turk's-cap lily
michiganense Farw. - Michigan lily
monadelphum M. Bieb. - Caucasian lily
pardalinum Kellogg - leopard lily, panther lily
philadelphicum L. - orange-cup lily, wild orange-red lily,
  wood lily
philadelphicum L. var. *andinum* (Nutt.) Ker Gawl. - western
  orange-cup lily
regale Wilson - regal lily, royal lily
speciosum Thunb. - showy Japanese lily, showy lily
superbum L. - American Turk's-cap lily, lily-royal, swamp
  lily, Turk's-cap lily
×*testaceum* Lindl. - Nankeen lily
  Note = *L. candidum × L. chalcedonicum.*
washingtonianum Kellogg - Washington's lily

Limnobium Rich. (Hydrocharitaceae)
spongia (Bosc) Rich. ex Steud. - American frog's-bit

Limnophila R. Br. (Scrophulariaceae)
sessiliflora (Vahl) Blume - limnophila

Limonia L. (Rutaceae)
acidissima L. - elephant-apple, Indian wood-apple, wood-
  apple

Limonium Mill. (Plumbaginaceae)
  marsh-rosemary, sea-lavender, statice
californicum (Boiss.) A. Heller - western marsh-rosemary

Linaria Mill. (Scrophulariaceae)
  spurred snapdragon, toadflax
canadensis (L.) Dum. Cours. - blue toadflax, old-field
  toadflax
floridana Chapm. - Florida toadflax
genistifolia (L.) Mill. - broad-leaf toadflax
genistifolia (L.) Mill. subsp. *dalmatica* (L.) Maire & Petitm. -
  Dalmatian toadflax
repens (L.) Mill. - striped toadflax
texana Scheele - Texas toadflax
vulgaris Mill. - butter-and-eggs, eggs-and-bacon, flax-weed,
  ramsted, wild snapdragon, yellow toadflax

Lindera Thunb. (Lauraceae)
benzoin (L.) Blume - Benjamin's-bush, fever-bush, spicebush,
  wild allspice
melissaefolium (Walter) Blume - Jove's-fruit

Lindernia All. (Scrophulariaceae)
anagallidea (Michx.) Pennell - false pimpernel
dubia (L.) Pennell - low false pimpernel
grandiflora Nutt. - round-leaf false pimpernel
procumbens (Krock.) Philcox - false pimpernel

Linnaea L. (Caprifoliaceae)
borealis L. - twinflower

Linum L. (Linaceae)
  flax

catharticum L. - fairy flax, white flax
grandiflorum Desf. - flowering flax, wild blue flax
perenne L. - perennial flax, wild blue flax
perenne L. subsp. *lewisii* (Pursh) Hultén - prairie flax
usitatissimum L. - flax, linseed
virginianum L. - Virginia yellow flax, woodland flax, yellow
  flax

Liparis Rich. (Orchidaceae)
liliifolia (L.) Lindl. - large twayblade, mauve sleekwort

Lippia L. (Verbenaceae)
micromera Schauer - Spanish thyme

Liquidambar L. (Hamamelidaceae)
formosana Hance - feng-hsiang-shu, Formosan gum
styraciflua L. - American sweet-gum, bilsted, red gum, sweet-
  gum

Liriodendron L. (Magnoliaceae)
tulipifera L. - tulip-poplar, tulip-tree, whitewood, yellow-
  poplar

Liriope Lour. (Liliaceae)
  lily-turf
muscari (Decne.) L. H. Bailey - big blue lily-turf

Litchi Sonn. (Sapindaceae)
chinensis Sonn. - leechee, litchi, lychee

Lithocarpus Blume (Fagaceae)
densiflora (Hook. & Arn.) Rehder - tan oak, tan-bark oak

Lithophragma (Nutt.) Torr. & A. Gray (Saxifragaceae)
  woodland-star
affine A. Gray - woodland star

Lithops N. E. Br. (Aizoaceae)
hookeri (Berg) Schwantes - living-stone

Lithospermum L. (Boraginaceae)
  gromwell, puccoon
arvense L. - bastard-alkanet, corn gromwell, pigeon-weed,
  puccoon, red-root, stone-seed, wheat-thief
canescens (Michx.) Lehm. - hoary puccoon, Indian-paint
caroliniense (J. F. Gmel.) MacMill. - plains puccoon, puccoon
incisum Lehm. - narrow-leaf puccoon
officinale L. - gray-mile, gromwell, little wale, pearl
  gromwell, pearl-plant
ruderale Douglas ex Lehm. - western gromwell

Litsea Lam. (Lauraceae)
aestivalis (L.) Fernald - pond-spice

Livistona R. Br. (Arecaceae)
  fan palm
chinensis (Jacq.) R. Br. ex Mart. - fan palm

Lloydia Rchb. (Liliaceae)
serotina (L.) Salisb. ex Rchb. - snowdon-lily

Lobelia L. (Campanulaceae)
cardinalis L. - cardinal-flower, Indian pink
dortmanna L. - water lobelia
erinus L. - edging lobelia
inflata L. - asthma-weed, bladder-pod, emetic-weed, eye-
  bright, gag-root, Indian tobacco, lobelia, puke-weed
kalmii L. - brook lobelia
puberula Michx. - downy lobelia

siphilitica L. - blue lobelia, cardinal-flower, great lobelia
spicata Lam. - pale-spike lobelia, spiked lobelia

Lobularia Desv. (Brassicaceae)
maritima (L.) Desv. - sweet-alyssum

Loiseleuria Desv. (Ericaceae)
procumbens (L.) Desv. - alpine-azalea

Lolium L. (Poaceae)
darnel, rye grass
multiflorum Lam. - annual rye grass, Australian rye grass, Italian rye grass
perenne L. - darnel, English rye grass, Lyme rye grass, perennial ray grass, perennial rye grass, strand-wheat, Terrell's grass
persicum Boiss. & Hohen. ex Boiss. - Persian darnel
rigidum Gaudin - rigid rye grass, Wimmera rye grass
temulentum L. - bearded darnel, darnel, poison darnel, poison rye grass

Lomatium Raf. (Apiaceae)
biscuit-root
bicolor (S. Watson) J. M. Coult. & Rose - biscuit-root
bicolor (S. Watson) J. M. Coult. & Rose var. *leptocarpum* (Torr. & A. Gray) Schlessman - slender-fruit lomatium
nudicaule (Pursh) J. M. Coult. & Rose - pestle parsnip
triternatum (Pursh) J. M. Coult. & Rose - buck-parsnip
utriculatum (Torr. & A. Gray) J. M. Coult. & Rose - bladder-parsnip, spring-gold

Lonicera L. (Caprifoliaceae)
honeysuckle
caerulea L. - sweet-berry honeysuckle, water-berry
canadensis Marsh. - American honeysuckle, fly honeysuckle
caprifolium L. - Italian woodbine
dioica L. - limber honeysuckle, wild honeysuckle
flava Sims - yellow honeysuckle
hirsuta Eaton - hairy honeysuckle
hispidula Douglas - pink honeysuckle
involucrata (Richardson) Banks ex Spreng. - twinberry
japonica Thunb. - gold-and-silver-flower, Japanese honeysuckle
morrowii A. Gray - Morrow's honeysuckle
oblongifolia (Goldie) Hook. - swamp fly honeysuckle
periclymenum L. - European woodbine, woodbine
prolifera (Kirchn.) Rehder - grape honeysuckle
sempervirens L. - coral honeysuckle, trumpet-honeysuckle
tatarica L. - Tatarian honeysuckle
villosa (Michx.) Roem. & Schult. - mountain fly honeysuckle, northern fly honeysuckle
xylosteum L. - European fly honeysuckle, fly honeysuckle

Lophostemon Schott (Myrtaceae)
conferta (R. Br.) Peter G. Wilson & J. T. Waterh. - Brisbane-box

Lotus L. (Fabaceae)
corniculatus L. - bird's-foot trefoil
micranthus Benth. - small-flowered bird's-foot trefoil
pedunculatus Cav. - greater bird's-foot trefoil
tenuis Waldst. & Kit. ex Willd. - narrow-leaf trefoil, slender trefoil
uliginosus Schkuhr - big trefoil
unifoliatus (Hook.) Benth. - deer-vetch, prairie trefoil

Ludwigia L. (Onagraceae)
false loosestrife
alternifolia L. - bushy water-primrose, rattle-box, seed-box, square-pod water-primrose
decurrens Walter - bushy water-primrose, wing-stemmed water-primrose, winged water-primrose
octovalvis (Jacq.) Raven - long-fruited primrose-willow, primrose-willow
palustris (L.) Elliott - water-purslane
peploides (Kunth) Raven - creeping water-primrose
polycarpa Short & Peter - many-fruited false loosestrife, top-podded water-primrose
repens J. R. Forst. - floating water-primrose
uruguayensis (Cambess.) Hara - showy water-primrose, Uruguay water-primrose

Luffa Mill. (Cucurbitaceae)
acutangula (L.) Roxb. - angled luffa, sing-kwa, towel gourd
aegyptiaca Mill. - dishcloth gourd, loofah, luffa, smooth loofah, sponge gourd, vegetable-sponge

Lunaria L. (Brassicaceae)
annua L. - bolbonac, honesty-plant, money-plant, moonwort, penny-flower, silver-dollar

Lupinus L. (Fabaceae)
lupine
albicaulis Hook. - sickle-keeled lupine
albus L. - Egyptian lupine, field lupine, white lupine, wolf bean
angustifolius L. - blue lupine, European blue lupine
arboreus Sims - tree lupine, yellow bush lupine
arcticus S. Watson - arctic lupine
argenteus Pursh - silver lupine
argenteus Pursh var. *heteranthus* (S. Watson) Barneby - tail-cup lupine
benthamii A. Heller - Bentham's annual lupine, spider lupine
bicolor Lindl. - bicolored lupine, miniature lupine
kingii S. Watson - king's lupine
leucophyllus Lindl. - velvet lupine, woolly-leaf lupine
littoralis Douglas - Chinook licorice, seashore lupine
luteus L. - European yellow lupine, yellow lupine
mutabilis Sweet - pearl lupine, tarhui, tarwi
nootkatensis Donn ex Sims - Nootka lupine
perennis L. - blue pea, Indian beans, perennial lupine, Quaker's-bonnet, sundial, sundial lupine, wild lupine
pilosus L. - blue lupine
polyphyllus Lindl. - large-leaf lupine, Washington's lupine
pusillus Pursh - low lupine
rivularis Lindl. - river bank lupine
sericeus Pursh - silky lupine
subcarnosus Hook. - Texas bluebonnet, Texas lupine
texensis Hook. - Texas bluebonnet

Luziola Juss. (Poaceae)
fluitans (Michx.) Terrell & H. Rob. - southern water grass

Luzula DC. (Juncaceae)
wood rush
campestris (L.) DC. - field wood rush, wood rush
parviflora (Ehrh.) Desv. - millet wood rush

Lychnis L. (Caryophyllaceae)
campion, catchfly
chalcedonica L. - Jerusalem-cross, London-pride, Maltese-cross, scarlet lychnis, scarlet-lightening

coronaria (L.) Desr. - dusty-miller, mullein-pink, rose campion

flos-cuculi L. - crow-flower, cuckoo-flower, Indian pink, meadow campion, meadow-pink, ragged-jade, ragged-robin

viscaria L. - German catchfly

Lycium L. (Solanaceae)
boxthorn, matrimony-vine
barbarum L. - barbary matrimony-vine, Chinese matrimony-vine, gow, gow-kee
berlandieri Dunal - Berlandier's wolfberry
carolinianum Walter - Christmas-berry
pallidum Miers - desert-thorn

Lycopersicon Mill. (Solanaceae)
tomato
esculentum Mill. - gold-apple, love-apple, tomato
esculentum Mill. var. *cerasiforme* (Dunal) A. Gray - cherry tomato
esculentum Mill. var. *pyriforme* (Dunal) L. H. Bailey - pear tomato
pimpinellifolium (L.) Mill. - currant tomato

Lycopodium L. (Lycopodiaceae)
club-moss
annotinum L. - bristly club-moss, interrupted club-moss, stiff club-moss
clavatum L. - buck-horn, club-moss, coral-evergreen, elk-moss, ground-pine, running club-moss, running-pine, staghorn-evergreen, wolf's-claws
complanatum L. - Christmas-green, ground-cedar, running-evergreen, trailing-evergreen
inundatum L. - bog club-moss, marsh club-moss
lucidulum Michx. - shining club-moss
obscurum L. - bunch-evergreen, flat-branch ground-pine, princess-pine, tree club-moss
selago L. - fir club-moss
tristachyum Pursh - ground-cedar, ground-pine

Lycopus L. (Lamiaceae)
bugleweed, gypsy-wort, water-horehound
americanus Muhl. ex Barton - American bugleweed, cut-leaf bugleweed, cut-leaf water-horehound, water-horehound
asper Greene - rough bugleweed
europaeus L. - European bugleweed, European water horehound
uniflorus Michx. - northern bugleweed, slender bugleweed

Lycoris Herb. (Liliaceae)
radiata (L'Hér.) Herb. - red spider-lily
squamigera Maxim. - magic-lily, resurrection-lily

Lygodesmia D. Don (Asteraceae)
juncea (Pursh) D. Don ex Hook. - devil's-shoestring, rush-pink, skeleton-weed

Lygodium Sw. (Pteridophyta)
climbing fern
japonicum (Thunb.) Sw. - Japanese climbing fern

Lyonia Nutt. (Ericaceae)
ferruginea (Walter) Nutt. - stagger-bush, tree lyonia
ligustrina (L.) DC. - he-huckleberry, male-berry, male-blueberry
lucida (Lam.) K. Koch - evergreen swamp fetterbush, fetterbush, letterfush

mariana (L.) D. Don - stagger-bush

Lyonothamnus A. Gray (Rosaceae)
Catalina ironwood
floribundus A. Gray - Lyon-tree

Lysichiton Schott (Araceae)
americanum Hultén & H. St. John - western skunk-cabbage, yellow skunk-cabbage

Lysiloma Benth. (Fabaceae)
latisiliqua (L.) Benth. - sabicu, wild tamarind

Lysimachia L. (Primulaceae)
loosestrife
ciliata L. - fringed loosestrife
lanceolata Walter - lance-leaf loosestrife
nummularia L. - creeping loosestrife, creeping-Charlie, creeping-Jennie, moneywort, yellow-myrtle
punctata L. - dotted loosestrife, garden loosestrife, spotted loosestrife
quadrifolia L. - prairie loosestrife, smooth loosestrife, whorled loosestrife
terrestris (L.) Britton, et al. - swamp loosestrife, swamp-candles, yellow loosestrife
thyrsiflora L. - swamp loosestrife, tufted loosestrife
vulgaris L. - garden loosestrife

Lythrum L. (Lythraceae)
loosestrife
alatum Pursh - wing-angled loosestrife, winged loosestrife
alatum Pursh var. *lanceolatum* (Elliott) Torr. & A. Gray ex Rothr. - lance-leaf loosestrife
hyssopifolium L. - hyssop loosestrife, hyssop-leaf loosestrife
lineare L. - narrow-leaf loosestrife
salicaria L. - bouquet-violet, purple loosestrife, purple lythrum, spiked loosestrife

Maackia Rupr. (Fabaceae)
amurensis Rupr. & Maxim. - Amur maackia

Macadamia F. Muell. (Proteaceae)
integrifolia Maiden & Betcke - Australian-nut, macadamia-nut, Queensland-nut
tetraphylla L. A. S. Johnson - small-fruited Queensland-nut

Macfadyena DC. (Bignoniaceae)
unguis-cati (L.) A. H. Gentry - cat-claw trumpet, cat-claw-vine, funnel-creeper

Machaeranthera Nees (Asteraceae)
tanacetifolia (Kunth) Nees - Tahoka daisy

Maclura Nutt. (Moraceae)
pomifera (Raf.) C. K. Schneid. - bow-wood, osage-orange

Macroptilium (Benth.) Urb. (Fabaceae)
atropurpureum (Moç & Sessé ex DC.) Urb. - conchito, purple bean, siratro
lathyroides (L.) Urb. - one-leaf-clover, phasey bean

Macrotyloma (Wight & Arn.) Verdc. (Fabaceae)
uniflorum (Lam.) Verdc. - horse-gram

Madia Molina (Asteraceae)
tarweed
elegans Lindl. - madia, showy tarweed, tarweed
glomerata Hook. - cluster tarweed, mountain tarweed, stinking tarweed

sativa Molina - Chilean tarweed, coast tarweed, madia-oil-plant

Magnolia L. (Magnoliaceae)
acuminata (L.) L. - cucumber-tree
acuminata (L.) L. var. *subcordata* (Spach) Dandy - yellow cucumber-tree
fraseri Walter - ear-leaved umbrella-tree, Fraser's magnolia, mountain magnolia
grandiflora L. - bull-bay, southern magnolia
macrophylla Michx. - big-leaf magnolia, great-leaf magnolia, large-leaf cucumber-tree, umbrella-tree
×*soulangiana* Soul.-Bod. - Chinese magnolia, saucer magnolia
Note = *M. heptapeta* × *M. quinquepeta.*
stellata (Siebold & Zucc.) Maxim. - star magnolia
tripetala (L.) L. - umbrella magnolia, umbrella-tree
virginiana L. - beaver-tree-laurel, laurel magnolia, small magnolia, swamp-bay, sweet-bay, sweet-bay magnolia

Maianthemum F. H. Wigg. (Liliaceae)
false lily-of-the-valley

canadense Desf. - Canadian mayflower, false lily-of-the-valley, muguet, two-leaf Solomon's-seal, wild lily-of-the-valley

Malachra L. (Malvaceae)
alceifolia Jacq. - malachra

Malcolmia R. Br. (Brassicaceae)
africana (L.) R. Br. - malcolm stock

Malosma Nutt. (Anacardiaceae)
laurina (Nutt.) Abrams - laurel sumac

Malpighia L. (Malpighiaceae)
coccigera L. - dwarf-holly, miniature holly, Singapore holly
glabra L. - acerola, Barbados cherry, huesito

Malus Mill. (Rosaceae)
apple
angustifolia (Aiton) Michx. - American crabapple, southern crabapple, southern wild crabapple
baccata (L.) Borkh. - Siberian crabapple
coronaria (L.) Mill. - American crabapple, garland crabapple, sweet crabapple, sweet-scented crabapple, wild sweet crabapple
domestica Borkh. - apple
floribunda Siebold ex Van Houtte - Japanese flowering crabapple, purple chokeberry, showy crabapple
fusca (Raf.) C. K. Schneid. - Oregon crabapple
ioensis (A. Wood) Britton - prairie crabapple, wild crabapple
prunifolia (Willd.) Borkh. - Chinese apple, plum-leaf apple
pumila (L.) Mill. - paradise apple
sieboldii (Regel) Rehder - Toringo crabapple
spectabilis (Aiton) Borkh. - Chinese flowering crabapple
sylvestris Mill. - apple, crabapple, paradise apple, wild apple

Malva L. (Malvaceae)
mallow, musk mallow
alcea L. - European mallow, hollyhock mallow, verrain mallow
moschata L. - musk, musk mallow, musk-plant
neglecta Wallr. - cheese-plant, cheeses, mallow
nicaeensis All. - bull mallow
parviflora L. - cheese-weed, little mallow
sylvestris L. - high mallow

verticillata L. var. *crispa* L. - curled mallow

Malvastrum A. Gray (Malvaceae)
false mallow
hispidum Hochr. - yellow false mallow

Malvaviscus Fabr. (Malvaceae)
sleepy mallow
arboreus Cav. - Turk's-cap, wax mallow
arboreus Cav. var. *drummondii* (Torr. & A. Gray) Schery - Drummond's wax mallow, Texas mallow, Turk's-cap

Malvella Jaub. & Spach (Malvaceae)
leprosa (Ortega) Krapov. - alkali mallow, alkali-sida, white mallow, white-weed

Mammillaria Haw. (Cactaceae)
fish-hook cactus, pincushion, strawberry cactus
hahniana Werderm. - lady-of-Mexico cactus

Mandragora L. (Solanaceae)
officinarum L. - mandrake

Manfreda Salisb. (Agavaceae)
maculosa (Hook.) Rose - spice-lily, wild tuberose
virginica (L.) Salisb. ex Rose - false aloe, rattlesnake-master

Mangifera L. (Anacardiaceae)
indica L. - mango

Manihot Mill. (Euphorbiaceae)
esculenta Crantz - bitter cassava, cassava, manioc, sweet-potato-tree, tapioca-plant, yuca

Manilkara Adans. (Sapotaceae)
jaimiqui (W. Wright ex Griseb.) Dubard subsp. *emarginata* (L.) Cronquist - wild dilly, wild sapodilla
zapota (L.) D. Royen - chicle, chicozapote, naseberry, nispero, sapodilla, sapotilla

Marah Kellogg (Cucurbitaceae)
big-root, man-root
oreganus (Torr. & A. Gray) Howell - coast man-root, old-man-in-the-ground, western wild cucumber

Maranta L. (Marantaceae)
arundinacea L. - arrowroot, obedience-plant
leuconeura E. Morris - prayer-plant, ten-commandments
leuconeura E. Morris var. *kerchoveana* E. Morris - rabbit's-foot, rabbit-tracks

Marrubium L. (Lamiaceae)
horehound
vulgare L. - horehound, hound-bane, marrube, marvel, white horehound

Marsilea L. (Pteridophyta)
pepperwort, water-clover
quadrifolia L. - European pepperwort, European water-clover
vestita Hook. & Grev. - hairy pepperwort

Matricaria L. (Asteraceae)
matricary
chamomilla L. - wild chamomile
maritima L. - false chamomile, scentless-chamomile
matricarioides (Less.) Porter - pineapple-weed, rayless dog-fennel
perforata Mérat - scentless-chamomile

Matteuccia Tod. (Pteridophyta)
struthiopteris (L.) Tod. - ostrich fern

Matthiola R. Br. (Brassicaceae)
  stock
incana (L.) R. Br. - Brampton's stock, gillyflower, imperial stock, stock
incana (L.) R. Br. var. *annua* (Sweet) Voss - ten-weeks stock
longipetala (Vent.) DC. - evening stock, evening-scented stock, night-scented stock

Mayaca Aubl. (Mayacaceae)
fluviatilis Aubl. - bog-moss

Mazus Lour. (Scrophulariaceae)
pumilus (Burm. f.) Steenis - Asian mazus

Meconopsis Vig. (Papaveraceae)
betonicifolia Franch. - blue poppy

Medeola L. (Liliaceae)
virginica L. - Indian cucumber-root

Medicago L. (Fabaceae)
  medic
arabica (L.) Huds. - bur-clover, spotted bur-clover, spotted medic
lupulina L. - black medic, hop-clover, nonesuch clover, yellow-trefoil
minima (L.) Bartal. - bur-clover, little bur-clover
polymorpha L. - California bur-clover, toothed bur-clover, toothed medic
sativa L. - alfalfa, lucerne
sativa L. subsp. *falcata* (L.) Arcang. - sickle alfalfa, sickle medic, yellow alfalfa, yellow lucerne, yellow-flowered alfalfa

Megalodonta Greene (Asteraceae)
beckii (Torr.) Greene - water-marigold

Melaleuca L. (Myrtaceae)
  bottlebrush, honey-myrtle
leucadendra (L.) L. - cajeput, river tea-tree, weeping tea-tree
quinquenervia (Cav.) S. T. Blake - broad-leaf tea-tree, cajeput, cajeput-tree, paper-bark-tree, punk-tree, swamp tea-tree

Melampyrum L. (Scrophulariaceae)
lineare Desr. - cow-wheat

Melastoma L. (Melastomataceae)
malabathricum L. - Bank's melastoma, Indian rhododendron

Melia L. (Meliaceae)
  bead-tree
azedarach L. - China-berry, China-tree, Indian lilac, Japanese bead-tree, paradise-tree, Persian lilac, pride-of-China, pride-of-India, Syrian bead-tree

Melica L. (Poaceae)
  melic
bulbosa Geyer ex Porter & J. M. Coult. - onion grass
geyeri Munro - Geyer's onion grass
harfordii Bol. - Harford's melic
imperfecta Trin. - California melic
mutica Walter - two-flowered melic
nitens (Scribn.) Nutt. ex Piper - three-flowered melic
porteri Scribn. - Porter's melic

smithii (Porter ex A. Gray) Vasey - awned melic, Smith's melic
spectabilis Scribn. - purple onion grass
subulata (Griseb.) Scribn. - Alaska onion grass

Melicoccus P. Browne (Sapindaceae)
bijugatus Jacq. - genip, genipe, honey-berry, mamoncillo, Spanish lime

Melilotus Mill. (Fabaceae)
  melilot, sweet-clover
alba Medik. - Bukhara clover, honey-clover, hubam, hubam clover, melilot, tree-clover, white melilot, white sweet-clover
altissima Thuill. - tall yellow sweet-clover
indica (L.) All. - Indian sweet-clover, senji, sour-clover
officinalis (L.) Lam. - melist, yellow melilot, yellow sweet-clover

Melinis P. Beauv. (Poaceae)
minutiflora P. Beauv. - molasses grass

Melissa L. (Lamiaceae)
  balm
officinalis L. - bee balm, lemon balm, sweet balm

Melochia L. (Sterculiaceae)
corchorifolia L. - red-weed

Melothria L. (Cucurbitaceae)
pendula L. - creeping cucumber

Menispermum L. (Menispermaceae)
canadense L. - Canadian moonseed, yellow parilla

Mentha L. (Lamiaceae)
  mint
aquatica L. - water mint
arvensis L. - corn mint, field mint, Japanese mint, wild mint
crispa L. - curled mint
×*gentilis* L. - red mint, Scotch mint, Scotch spearmint
  Note = *M. arvensis* × *M. spicata*.
longifolia (L.) Huds. - European horsemint, horsemint
×*piperita* L. - brandy mint, lamb mint, peppermint
  Note = *M. aquatica* × *M. spicata*.
×*piperita* L. var. *citrata* (Ehrh.) Briq. - bergamot mint, lemon mint
rotundifolia (L.) Huds. - apple mint, round-leaf mint
spicata L. - spearmint
uaveolens Ehrh. - apple mint

Mentzelia L. (Loasaceae)
  blazing-star
albicaulis (Hook.) Torr. & A. Gray - white-stem stick-leaf
decapetala (Pursh ex Sims) Urb. & Gilg - ten-petal stick-leaf
laevicaulis (Douglas ex Hook.) Torr. & A. Gray - blazing-star
multiflora (Nutt.) A. Gray - desert stick-leaf
oligosperma Nutt. - stick-leaf

Menyanthes L. (Menyanthaceae)
trifoliata L. - bogbean, buckbean, marsh-trefoil

Menziesia Sm. (Ericaceae)
ferruginea Sm. - mock azalea, rusty-leaf
pilosa (Michx. ex Lam.) Juss. ex Pers. - minnie-bush

Mertensia Roth (Boraginaceae)
  bluebells, lungwort

maritima (L.) S. F. Gray - oyster-leaf, sea lungwort, sea mertensia, seaside bluebells
paniculata (Aiton) G. Don - northern bluebells
virginica (L.) Pers. - eastern bluebells, Roanoke bells, Virginia bluebells, Virginia cowslip

Mesembryanthemum L. (Aizoaceae)
crystallinum L. - ice-plant

Mespilus L. (Rosaceae)
germanica L. - medlar

Metasequoia Miki ex Hu & W. C. Cheng (Taxodiaceae)
glyptostroboides Hu & W. C. Cheng - dawn redwood

Metrosideros Banks & Gaertn. (Myrtaceae)
bottlebrush, iron-tree
polymorpha Gaudich. - 'ohi'a lehua

Michelia L. (Magnoliaceae)
figo (Lour.) Spreng. - banana-shrub

Micranthemum Michx. (Scrophulariaceae)
glomeratum (Chapm.) Shinners - hemianthus
umbrosum (Walter) S. F. Blake - globifera

Microsorium Link (Pteridophyta)
punctatum (L.) H. F. Copel. - climbing bird's nest fern, crested fern

Microstegium Nees (Poaceae)
vimineum (Trin.) A. Camus var. *imberbe* (Nees) Honda - Mary's grass

Mikania Willd. (Asteraceae)
cordata (Burm. f.) B. L. Rob. - African mile-a-minute
cordifolia (L.f.) Willd. - hairy hempweed
micrantha Kunth - mile-a-minute
scandens (L.) Willd. - climbing hempweed

Milium L. (Poaceae)
vernale M. Bieb. - spring millet

Mimosa L. (Fabaceae)
invisa C. Mart. - giant sensitive-plant
pigra L. - cat-claw mimosa
pudica L. - sensitive-plant
quadrivalvis L. var. *angustata* (Torr. & A. Gray) Barnaby - cat-claw sensitive-brier, little-leaf sensitive-brier
quadrivalvis L. var. *nuttallii* (DC.) Barnaby - cat's-claw, cat-claw schrankia, sensitive-brier

Mimulus L. (Scrophulariaceae)
monkey-flower
alatus Aiton - sharp-winged monkey-flower
cardinalis Benth. - scarlet monkey-flower
guttatus DC. - monkey-flower
moschatus Lindl. - musk-flower, musk-plant
ringens L. - Allegheny monkey-flower, square-stemmed monkey-flower

Mimusops L. (Sapotaceae)
elengi L. - medlar, Spanish cherry

Mirabilis L. (Nyctaginaceae)
umbrella-wort
jalapa L. - beauty-of-the-night, four-o'clock, marvel-of-Peru
multiflora (Torr.) A. Gray - Colorado four-o'clock

nyctaginea (Michx.) MacMill. - umbrella-wort, wild four-o'clock

Miscanthus Andersson (Poaceae)
sinensis Andersson - eulalia, eulalia grass, Japanese plume grass

Mitchella L. (Rubiaceae)
repens L. - partridgeberry, squaw-berry

Mitella L. (Saxifragaceae)
bishop's-cap, miterwort
diphylla L. - cool-wort, miterwort
nuda L. - naked miterwort

Modiola Moench (Malvaceae)
caroliniana (L.) G. Don - bristly mallow

Moehringia L. (Caryophyllaceae)
lateriflora (L.) Fenzl - grove sandwort

Molinia Schrank (Poaceae)
caerulea (L.) Moench - moor grass

Mollugo L. (Molluginaceae)
verticillata L. - carpetweed, devil's-grip, Indian chickweed, kurumaba-zakuro-so, whorled carpetweed

Moluccella L. (Lamiaceae)
laevis L. - bells-of-Ireland, Molucca balm, shell-flower

Momordica L. (Cucurbitaceae)
balsamina L. - balsam-apple, wonder-apple
charantia L. - balsam-apple, balsam-pear, bitter cucumber, bitter gourd, la-kwa

Monarda L. (Lamiaceae)
horsemint, wild bergamot
citriodora Cerv. ex Lag. - lemon bee balm, lemon mint
didyma L. - bee balm, Oswego tea
fistulosa L. - wild bergamot
punctata L. - dotted mint, horsemint, spotted bee balm

Monardella Benth. (Lamiaceae)
odoratissima Benth. - mountain mint
villosa Benth. - coyote mint

Moneses Salisb. ex A. Gray (Ericaceae)
uniflora (L.) A. Gray - one-flowered shin-leaf, one-flowered wintergreen, one-flowered-pyrola, wood-nymph

Monochoria C. Presl (Pontederiaceae)
hastata (L.) Solms - arrow-leaf monochoria
vaginalis (Burm. f.) Kunth - monochoria

Monodora Dunal (Annonaceae)
myristica (Gaertn.) Dunal - African nutmeg, calabash nutmeg, Jamaican nutmeg

Monolepis Schrad. (Chenopodiaceae)
nuttalliana (B. Schütt) Greene - Nuttall's poverty-weed

Monotropa L. (Ericaceae)
hypopithys L. - false beech-drops, pinesap
uniflora L. - convulsion-root, corpse-plant, fit's-root, Indian-pipe, pine-sap

Monotropsis Schwein. (Ericaceae)
odorata Elliott - pygmy-pipes, sweet pinesap

Monstera Adans. (Araceae)
  window-leaf
  deliciosa Liebm. - breadfruit-vine, ceriman, cut-leaf
    philodendron, fruit-salad-plant, hurricane-plant, monstera,
    Swiss-cheese-plant

Montia L. (Portulacaceae)
  miner's-lettuce
  linearis (Douglas ex Hook.) Greene - Indian lettuce, narrow-
    leaf miner's-lettuce

Morinda L. (Rubiaceae)
  citrifolia L. - Indian mulberry

Morrenia Lindl. (Asclepiadaceae)
  odorata (Hook. & Arn.) Lindl. - strangler-vine, tasi

Morus L. (Moraceae)
  mulberry
  alba L. - white mulberry
  alba L. var. *multicaulis* (Perr.) Loudon - silkworm mulberry
  microphylla Buckley - Texas mulberry
  nigra L. - black mulberry
  rubra L. - American mulberry, red mulberry

Mucuna Adans. (Fabaceae)
  novaguineensis Scheff. - New Guinea creeper
  pruriens (L.) DC. var. *utilis* (Wall. ex Wight) Baker ex Burck
    - Bengal velvet bean, cowage velvet bean, Florida velvet
    bean, Lyon bean, Mauritius velvet bean, velvet bean,
    Yokohama velvet bean

Muhlenbergia Schreb. (Poaceae)
  muhly, wire-plant
  andina (Nutt.) Hitchc. - foxtail muhly
  asperifolia (Nees & C. A. Mey.) Parodi - alkali muhly, scratch
    grass
  capillaris (Lam.) Trin. - hair grass, hairy-awn muhly
  cuspidata (Torr. ex Hook.) Rydb. - plains muhly
  emersleyi Vasey - bullgrass
  filiculmis Vasey - slim-stem muhly
  filiformis (Thurb. ex S. Watson) Rydb. - pull-up muhly
  frondosa (Poir.) Fernald - wire-stem muhly
  mexicana (L.) Trin. - Mexican muhly
  montana (Nutt.) Hitchc. - mountain muhly
  pauciflora Buckley - few-flowered muhly, New Mexico
    muhly
  porteri Scribn. ex Beal - bush muhly
  repens (J. Presl) Hitchc. - creeping muhly
  rigens (Benth.) Hitchc. - deer grass
  schreberi J. F. Gmel. - drop-seed wiregrass, nimble-will
  torreyi (Kunth) Hitchc. - ring grass
  utilis (Torr.) Hitchc. - aparejo grass
  wrightii Vasey - spike muhly

Munroa Torr. (Poaceae)
  squarrosa (Nutt.) Torr. - false buffalo grass

Muntingia L. (Tiliaceae)
  calabura L. - calabur, Jamaican cherry, Panama-berry

Murdannia Royle (Commelinaceae)
  keisak (Hassk.) Hand.-Mazz. - marsh-dayflower
  nudiflora (L.) Brenan - dove-weed

Murraya J. König ex L. (Rutaceae)
  paniculata (L.) Jack - China-box, cosmetic-bark-tree, orange
    jessamine, orange-jasmine, satinwood

Musa L. (Musaceae)
  acuminata Colla - dwarf banana, edible banana, plantain
  ×*paradisiaca* L. - banana, plantain
  Note = *M. acuminata* × *M. balbisiana*.

Muscari Mill. (Liliaceae)
  botryoides (L.) Mill. - grape-hyacinth
  comosum Mill. - tassel-hyacinth

Myoporum Banks & Sol. ex G. Forst. (Myoporaceae)
  sandwicense (A. DC.) A. Gray - bastard-sandle-wood, naio

Myosotis L. (Boraginaceae)
  forget-me-not, scorpion-grass
  arvensis (L.) Hill - field forget-me-not, field scorpion-grass
  laxa Lehm. - smaller forget-me-not
  scorpioides L. - forget-me-not, true forget-me-not
  sylvatica Ehrh. ex Hoffm. - garden forget-me-not
  verna Nutt. - spring forget-me-not

Myosurus L. (Ranunculaceae)
  minimus L. - mouse's-tail

Myrciaria O. Berg (Myrtaceae)
  cauliflora (DC.) Bergius - Brazilian grape tree, jaboticaba

Myrica L. (Myricaceae)
  californica Cham. & Schltdl. - California bayberry, California
    wax myrtle, Pacific bayberry, Pacific wax myrtle, wax
    myrtle
  cerifera L. - candleberry, southern bayberry, southern wax
    myrtle, wax myrtle, waxberry
  gale L. - bog myrtle, meadow-fern, sweet gale
  heterophylla Raf. - evergreen bayberry, southern bayberry,
    wax myrtle
  inodora Bartram - odorless bayberry
  pensylvanica Loisel. - bayberry, candleberry, northern
    bayberry
  pusilla Raf. - dwarf wax myrtle

Myriophyllum L. (Haloragaceae)
  aquaticum (Vell.) Verdc. - parrot's-feather, water-feather
  exalbescens Fernald - northern water-milfoil
  heterophyllum Michx. - variable water-milfoil
  laxum Shuttlew. ex Chapm. - lax water-milfoil
  pinnatum (Walter) Britton, et al. - eastern water-milfoil
  spicatum L. - Eurasian milfoil, Eurasian water-milfoil
  verticillatum L. - whorled water-milfoil

Myrsine L. (Myrsinaceae)
  africana L. - African boxwood, Cape myrtle
  guianensis (Aubl.) Kuntze - Guianan rapanea

Myrtus L. (Myrtaceae)
  communis L. - Greek myrtle, Indian buchu, myrtle, Swedish
    myrtle

Najas L. (Najadaceae)
  flexilis (Willd.) Rostk. & Schmidt - northern water-nymph,
    slender naiad
  graminea Delile - grassy naiad
  guadalupensis (Spreng.) Magnus - southern naiad, southern
    water-nymph
  marina L. - alkaline water-nymph, holly-leaf naiad

minor All. - brittle-leaf naiad, eutrophic water-nymph

Nandina Thunb. (Berberidaceae)
domestica Thunb. - heavenly-bamboo, sacred-bamboo

Napaea L. (Malvaceae)
dioica L. - glade mallow

Narcissus L. (Liliaceae)
daffodil, narcissus
jonquilla L. - jonquil
poeticus L. - narcissus, pheasant's-eye, poet's narcissus
pseudonarcissus L. - daffodil, trumpet narcissus
tazetta L. - polyanthus narcissus

Nassella Desv. (Poaceae)
trichotoma (Nees) Hack. - serrated tussock

Navarretia Ruiz & Pav. (Polemoniaceae)
intertexta (Benth.) Hook. - needle-weed, woolly-gilia
squarrosa (Eschsch.) Hook. & Arn. - skunk-weed, skunk-weed-gilia

Nelumbo Adans. (Nelumbonaceae)
sacred-bean, water lotus
lutea (Willd.) Pers. - American lotus, pond-nuts, water-chinquapin, wonkapin, yanquapin, yellow nelumbo
nucifera Gaertn. - East Indian lotus, Hindu lotus, Oriental lotus, sacred lotus

Nemopanthus Raf. (Aquifoliaceae)
mountain holly
mucronatus (L.) Trel. - cat-berry, mountain-holly

Nemophila Nutt. (Hydrophyllaceae)
menziesii Hook. & Arn. - baby-blue-eyes
menziesii Hook. & Arn. var. *atomaria* (Fisch. & C. A. Mey.) H. P. Chandler - white baby-blue-eyes

Nepeta L. (Lamiaceae)
cataria L. - cat-mint, catnip

Nephelium L. (Sapindaceae)
lappaceum L. - rambutan

Nephrolepis Schott (Pteridophyta)
biserrata (Sw.) Schott - sword fern

Nerine Herb. (Liliaceae)
sarniensis (L.) Herb. - Guernsey lily

Nerium L. (Apocynaceae)
oleander L. - oleander, rose-bay

Neslia Desv. (Brassicaceae)
paniculata (L.) Desv. - ball mustard

Neyraudia Hook. f. (Poaceae)
reynaudiana (Kunth) Keng ex Hitchc. - Burma reed

Nicandra Adans. (Solanaceae)
physalodes (L.) Gaertn. - apple-of-Peru, shoofly-plant

Nicotiana L. (Solanaceae)
alata Link & Otto - flowering tobacco, jasmine tobacco, winged tobacco
glauca Graham - mustard-tree, tree tobacco
repanda Willd. ex Lehm. - fiddle-leaf tobacco, tobasco cimarron, wild tobacco
rustica L. - Aztec tobacco

tabacum L. - tobacco
trigonophylla Dunal - desert tobacco

Nolina Michx. (Agavaceae)
bear-grass
microcarpa S. Watson - bear-grass, sacahuista
recurvata (Lem.) Hemsl. - bottle-palm, elephant-foot-tree, ponytail

Notholaena R. Br. (Pteridophyta)
sinuata (Lag. ex Sw.) Kaulf. - waxy cloak fern

Nothoscordum Kunth (Liliaceae)
bivalve (L.) Britton - false garlic

Nuphar Sm. (Nymphaeaceae)
cow-lily, marsh-collard, spatter-dock, water-collard, yellow pond-lily
lutea (L.) Sm. - yellow water-lily
lutea (L.) Sm. subsp. *advena* (Aiton) Kartesz & Gandhi - spatter-dock
lutea (L.) Sm. subsp. *polysepala* (Engelm.) E.O.Beal - Rocky Mountain spatter-dock

Nymphaea L. (Nymphaeaceae)
water-lily
alba L. - European white water-lily
ampla (Salisb.) DC. - large water-lily
capensis Thunb. - Cape blue water-lily
elegans Hook. - blue water-lily
glandulifera Rodschied - Fenzel's water-lily
mexicana Zucc. - banana water-lily, Mexican water-lily, yellow water-lily
odorata Aiton - American water-lily, fragrant water-lily, pond-lily, white water-lily
tuberosa Paine - magnolia water-lily, tuberous water-lily

Nymphoides Hill (Menyanthaceae)
floating-heart
aquatica (Walter) Kuntze - banana-plant, big floating-heart, fairy water-lily
peltata (S. Gmel.) Kuntze - water fringe, yellow floating-heart

Nyssa L. (Cornaceae)
aquatica L. - cotton-gum, large tupelo, water tupelo, wild olive
ogeche J. Bartram ex Marsh. - Ogeche tupelo, Ogeechee lime tupelo
sylvatica Marsh. - black gum, black tupelo, pepperidge, sour-gum, upland tupelo
sylvatica Marsh. var. *biflora* (Walter) Sarg. - swamp blackgum, swamp tupelo

Obolaria L. (Gentianaceae)
virginica L. - pennywort

Ocimum L. (Lamiaceae)
basil
basilicum L. - basil, sweet basil
sanctum L. - holy basil

Ocotea Aubl. (Lauraceae)
coriacea (Sw.) Britton - Jamaican nectandra

Odontites Ludw. (Scrophulariaceae)
verna (Bellardi) Dumort. subsp. *serotina* (Dumort.) Corb. - red bartsia

Oemleria Rchb. (Rosaceae)
cerasiformis (Hook. & Arn.) J. W. Landon - osoberry

Oenanthe L. (Apiaceae)
water-dropwort
sarmentosa J. Presl - water-celery

Oenothera L. (Onagraceae)
evening-primrose, sundrops
albicaulis Pursh - prairie evening-primrose
biennis L. - evening-primrose, fever-plant, field-primrose, German rampion, tree-primrose
deltoides Torr. & Frém. - desert evening-primrose
fruticosa L. - southern sundrops, sundrops
humifusa Nutt. - sea beach evening-primrose, spreading evening-primrose
laciniata Hill - cut-leaf evening-primrose
nuttallii Sweet - small-flowered evening-primrose, white-stem evening-primrose
parviflora L. - small-flowered evening-primrose
perennis L. - little sundrops, perennial sundrops
speciosa Nutt. - showy evening-primrose, white evening-primrose

Olea L. (Oleaceae)
europaea L. - olive

Olneya A. Gray (Fabaceae)
tesota A. Gray - desert ironweed, ironwood, tesota

Onobrychis Mill. (Fabaceae)
viciifolia Scop. - esparcet, holy-clover, sainfoin, sanfoin

Onoclea L. (Pteridophyta)
sensibilis L. - sensitive fern

Onopordum L. (Asteraceae)
acanthium L. - cotton thistle, Scotch thistle

Onosmodium Michx. (Boraginaceae)
molle Michx. - marble-seed, western false gromwell

Ophioglossum L. (Pteridophyta)
adder's-tongue fern
pusillum Raf. - northern adder's-tongue fern

Oplismenus P. Beauv. (Poaceae)
hirtellus (L.) P. Beauv. - basket grass

Oplopanax (Torr. & A. Gray) Miq. (Araliaceae)
horridum (Sm.) Miq. - devil's-club

Opuntia Mill. (Cactaceae)
cholla cactus, prickly-pear, tuna-plant
austrina Small - eastern prickly-pear
cochenillifera (L.) Mill. - cochineal cactus, cochineal-plant
compressa (Salisb.) J. F. Macbr. - eastern prickly-pear, prickly-pear, spreading prickly-pear
ficus-indica (L.) Mill. - Indian fig, spineless cactus
fragilis (Nutt.) Haw. - brittle prickly-pear
fulgida Engelm. - jumping cactus, jumping cholla
imbricata (Haw.) DC. - walking-stick cholla
leptocaulis DC. - tasajillo
lindheimeri Engelm. - Lindheimer's prickly-pear, Texas prickly-pear
polyacantha Haw. - plains prickly-pear
spinosior (Engelm.) Toumey - spiny cholla

stricta Haw. var. *dillenii* (Ker Gawl.) L. D. Benson - prickly-pear cactus
versicolor Engelm. ex J. M. Coult. - staghorn cholla

Orchis L. (Orchidaceae)
rotundifolia Banks - one-leaf orchis, small round-leaf orchis
spectabilis L. - showy orchis

Origanum L. (Lamiaceae)
majorana L. - annual marjoram, sweet marjoram
vulgare L. - marjoram, oregano, organy, origano, pot marjoram, wild marjoram

Ornithogalum L. (Liliaceae)
nutans L. - nodding star-of-Bethlehem
umbellatum L. - dove's-dung, nap-at-noon, star-of-Bethlehem, summer-snowflake

Ornithopus L. (Fabaceae)
sativus Brot. - serradella

Orobanche L. (Orobanchaceae)
cernua Loeffler - nodding broomrape
crenata Forssk. - crenate broomrape
ludoviciana Nutt. - Louisiana broomrape, prairie broomrape
minor Sm. - lesser broomrape
ramosa L. - branched broomrape
uniflora L. - naked broomrape, one-flowered cancer-root

Orontium L. (Araceae)
aquaticum L. - golden-club

Orthilia Raf. (Ericaceae)
secunda (L.) House - one-sided pyrola, one-sided-wintergreen

Orthocarpus Nutt. (Scrophulariaceae)
luteus Nutt. - owl-clover

Oryza L. (Poaceae)
sativa L. - red rice, rice, upland rice

Oryzopsis Michx. (Poaceae)
mountain grass, rice grass
exigua Thurb. ex Torr. - little rice grass
hymenoides (Roem. & Schult.) Ricker - Indian mullet, Indian rice, silk grass
micrantha (Trin. & Rupr.) Thurb. - little-seed rice grass
miliacea (L.) Benth. & Hook. f. ex Asch. & Schweinf. - smilo grass

Osmanthus Lour. (Oleaceae)
devil-weed
americanus (L.) Benth. & Hook. f. - American olive, devil-wood, wild olive
fragrans (Thunb.) Lour. - fragrant-olive, sweet osmanthus, sweet-olive, tea-olive
heterophyllus (G. Don) P. S. Green - Chinese olive, false holly, holly osmanthus, holly-olive

Osmorhiza Raf. (Apiaceae)
chilensis Hook. & Arn. - spreading sweet-root, tapering sweet-root
claytonii (Michx.) C. B. Clarke - bland sweet cicely, hairy sweet cicely, sweet jarvil, woolly sweet cicely
longistylis (Torr.) DC. - anise-root, long-styled anise-root

Osmunda L. (Pteridophyta)
flowering fern
cinnamomea L. - buck-horn, cinnamon fern, fiddleheads

claytoniana L. - interrupted fern
regalis L. - flowering fern, royal fern

Osteospermum L. (Asteraceae)
ecklonis (DC.) Norl. - African daisy, Vanstadens River daisy

Ostrya Scop. (Betulaceae)
hop-hornbeam
knowltonii Cov. - Knowlton's hop-hornbeam
virginiana (Mill.) K. Koch - American hop-hornbeam, eastern hop-hornbeam, ironwood, leverwood

Oxalis L. (Oxalidaceae)
lady's-sorrel, wood-sorrel
acetosella L. - European wood-sorrel, Irish shamrock, northern wood-sorrel
corniculata L. - creeping lady's-sorrel, creeping wood-sorrel, creeping yellow wood-sorrel
corymbosa DC. - violet wood-sorrel
dillenii Jacq. - Florida yellow wood-sorrel, southern yellow wood-sorrel
europaea Jord. - European wood-sorrel, upright yellow sorrel, yellow wood-sorrel
montana Raf. - wood shamrock
oregana Nutt. - red wood-sorrel
pes-caprae L. - Bermuda buttercup, buttercup oxalis
stricta L. - yellow wood-sorrel
violacea L. - violet wood-sorrel

Oxydendrum DC. (Ericaceae)
arboreum (L.) DC. - sorrel-tree, sourwood, titi

Oxypolis Raf. (Apiaceae)
rigidior (L.) Raf. - cowbane, water-dropwort

Oxyria Hill (Polygonaceae)
digyna (L.) Hill - mountain-sorrel

Oxytropis DC. (Fabaceae)
crazyweed, locoweed
lambertii Pursh - Lambert's crazyweed, Lambert's loco, locoweed, purple loco, white loco
sericea Nutt. - silky crazyweed
splendens Douglas - showy crazyweed, showy locoweed

Pachira Aubl. (Bombacaceae)
aquatica Aubl. - Guianan chestnut, provision-tree, water-chestnut, wild cocoa

Pachycereus (A. Berger) Britton & Rose (Cactaceae)
schottii (Engelm.) D. R. Hunt - senita, whisker cactus

Pachyrhizus Rich. ex DC. (Fabaceae)
erosus (L.) Urb. - jicama, yam bean

Pachysandra Michx. (Buxaceae)
procumbens Michx. - Allegheny pachysandra, Allegheny-spurge
terminalis Siebold & Zucc. - Japanese pachysandra, Japanese-spurge

Paederia L. (Rubiaceae)
foetida L. - skunk-vine

Paeonia L. (Paeoniaceae)
peony
lactiflora Pall. - Chinese peony, garden peony
officinalis L. - peony
suffruticosa Andr. - tree peony

Panax L. (Araliaceae)
ginseng
quinquefolius L. - American ginseng, ginseng, sang
trifolius L. - dwarf ginseng, groundnut

Pandanus Parkinson (Pandanaceae)
screw-pine
odoratissimus L.f. - breadfruit, pandang
tectorius Parkinson ex Du Roi - pandanus-palm, thatch screw-pine
utilis Bory - screw-pine
veitchii J. H. Veitch ex Mast. & T. Moore - Veitch's screw-pine

Panicum L. (Poaceae)
panic grass
adspersum Trin. - broad-leaf panicum
anceps Michx. - panic grass
bulbosum Kunth - bulb panic grass
capillare L. - fool-hay, old-witch grass, tickle grass, tumbleweed grass, witch grass, witch's-hair
ciliatum Elliott - fringed panicum
clandestinum L. - deer's-tongue
coloratum L. - Klein's grass
dichotomiflorum Michx. - fall panic grass, fall panicum, spreading witch grass, sprouting-crabgrass
gattingeri Nash - Gattinger's witch grass
hemitomon Schult. - maiden-cane
maximum Jacq. - guinea grass, panic grass
miliaceum L. - broom corn millet, broom millet, broom-corn, hog millet, millet, proso millet, wild proso millet
obtusum Kunth - vine mesquite, vine mesquite grass
purpurascens Raddi - Pará grass
repens L. - torpedo grass
rigidulum Bosc - red-top panicum
texanum Buckley - Texas millet, Texas panicum
virgatum L. - switch grass

Papaver L. (Papaveraceae)
poppy
argemone L. - pinnate poppy
dubium L. - field poppy
nudicaule L. - arctic poppy, Iceland poppy
orientale L. - Oriental poppy
rhoeas L. - corn poppy, field poppy, Flander's poppy, red poppy, Shirley's poppy
somniferum L. - opium poppy

Parapholis C. E. Hubb. (Poaceae)
incurva (L.) C. E. Hubb. - sickle grass

Parentucellia Viv. (Scrophulariaceae)
viscosa (L.) Caruel - eyebright, yellow parentucellia

Parietaria L. (Urticaceae)
pellitory
floridana Nutt. - Florida pellitory
pensylvanica Willd. - Pennsylvania pellitory

Parkinsonia L. (Fabaceae)
aculeata L. - horse bean, Jerusalem thorn, Mexican palo-verde, retaima
florida (Benth. ex A. Gray) S. Watson - blue palo-verde, palo-verde

Parnassia L. (Saxifragaceae)
bog-stars, grass-of-Parnassus
californica (A. Gray) Greene - bog-stars, grass-of-Parnassus
glauca Raf. - American grass-of-Parnassus

Paronychia Mill. (Caryophyllaceae)
chickweed, nail-wort, whitlow-wort
canadensis (L.) A. Wood - forked chickweed

Parthenium L. (Asteraceae)
argentatum A. Gray - guayule
hysterophorus L. - quinine-weed, ragweed parthenium, Santa Maria
integrifolium L. - American feverfew, eastern parthenium, wild quinine

Parthenocissus Planch. (Vitaceae)
quinquefolia (L.) Planch. - American ivy, five-leaved ivy, Virginia creeper, woodbine
tricuspidata (Siebold & Zucc.) Planch. - Boston ivy, Japanese ivy

Paspalum L. (Poaceae)
almum Chase - Comb's paspalum
boscianum Flügge - bull paspalum
ciliatifolium Michx. - fringe-leaf paspalum
conjugatum Bergius - buffalo grass, sour grass, sour paspalum
dilatatum Poir. - Dallis grass
distichum L. - knotgrass
floridanum Michx. - Florida paspalum
fluitans (Elliott) Kunth - water paspalum
laeve Michx. - field paspalum
notatum Flügge - Bahia grass
notatum Flügge var. *saurae* Parodi - Pensacola Bahia grass
plicatulum Michx. - brown-seed paspalum
setaceum Michx. - thin paspalum
setaceum Michx. var. *ciliatifolium* (Michx.) Vasey - fringe-leaf paspalum
setaceum Michx. var. *longipedunculatum* (Leconte) Wood - bare-stem paspalum
setaceum Michx. var. *supinum* (Bosc ex Poir.) Trin. - supine paspalum
urvillei Steud. - Vasey's grass
vaginatum Sw. - seashore paspalum

Passiflora L. (Passifloraceae)
passionflower
caerulea L. - blue passionflower
edulis Sims - passion-fruit, purple granadilla
foetida L. - love-in-a-mist, red-fruit passionflower, running-pop, wild water-lemon
incarnata L. - apricot-vine, may-pop passionflower, may-pops, wild passionflower
laurifolia L. - belle-apple, Jamaican honeysuckle, pomme-de-liane, vinegar-pear, water-lemon, yellow granadilla
quadrangularis L. - giant granadilla
suberosa L. - corky-stem passionflower

Pastinaca L. (Apiaceae)
sativa L. - bird's-nest, Hart's-eye, madnip, parsnip, wild parsnip

Paulownia Siebold & Zucc. (Scrophulariaceae)
tomentosa (Thunb.) Steud. - empress-tree, karri-tree, kiri-tree, princess-tree, royal paulownia

Paxistima Raf. (Celastraceae)
canbyi A. Gray - cliff-green, mountain-lover
myrsinites (Pursh) Raf. - myrtle box-leaf, Oregon boxwood

Pedicularis L. (Scrophulariaceae)
lousewort, wood-betony
canadensis L. - forest lousewort, lousewort, wood-betony
densiflora Hook. - Indian-warrior
groenlandica Retz. - elephant's-head
lanceolata Michx. - swamp lousewort
palustris L. - swamp lousewort
sylvatica L. - small lousewort

Pedilanthus Neck. ex Poit. (Euphorbiaceae)
tithymaloides (L.) Poit. - devil's-backbone, Japanese poinsettia, red-bird-cactus, red-bird-flower, ribbon-cactus, slipper-flower, slipper-plant

Peganum L. (Zygophyllaceae)
harmala L. - African rue, harmel peganum

Pelargonium L'Hér. (Geraniaceae)
geranium, stork's-bill
×*domesticum* L. H. Bailey - fancy geranium, Lady Washington's geranium, Martha Washington's geranium, pansy-flowered geranium, regal geranium, show geranium, summer-azalea
Note = A cultigen of complex hybrid origin.
graveolens L'Hér. ex Aiton - rose geranium, sweet-scented geranium
×*hortorum* L. H. Bailey - bedding geranium, fish geranium, horseshoe geranium, house geranium, zonal geranium
Note = A cultigen of complex hybrid origin.
peltatum (L.) L'Hér. ex Aiton - hanging geranium, ivy geranium

Pellaea Link (Pteridophyta)
cliff brake
andromedifolia (Kaulf.) Fée - coffee fern
glabella Mett. - smooth cliff brake
rotundifolia (G. Forst.) Hook. - button fern, New Zealand cliff brake

Pellionia Gaudich. (Urticaceae)
pulchra N. E. Br. - rainbow-vine, satin pellionia
repens (Lour.) Merr. - trailing watermelon-begonia

Peltandra Raf. (Araceae)
arrow-arum
sagittifolia (Michx.) Morong - spoonflower
virginica (L.) Schott & Endl. - arrow-arum, green arrow-arum, tuckahoe, Virginia wake-robin

Pennisetum Rich. ex Pers. (Poaceae)
alopecuroides (L.) Spreng. - Chinese pennisetum
ciliare (L.) Link - buffalo grass
clandestinum Hochst. ex Chiov. - Kikuyu grass
glaucum (L.) R. Br. - bulrush millet, cat-tail millet, pearl millet
pedicellatum Trin. - kyasuma grass
polystachyon (L.) Schult. - mission grass
purpureum Schumach. - elephant grass, napier grass
setaceum (Forssk.) Chiov. - crimson fountain grass

Penstemon Schmidel (Scrophulariaceae)
    beard-tongue
    canescens (Britton) Britton - Appalachian beard-tongue, gray
      beard-tongue
    centranthifolius (Benth.) Benth. - scarlet bugler
    eatonii A. Gray - Eaton's firecracker, Eaton's penstemon
    grandiflorus Nutt. - large beard-tongue, large-flowered beard-
      tongue
    hirsutus (L.) Willd. - hairy beard-tongue, northeastern beard-
      tongue
    newberryi A. Gray - mountain-pride

Pentas Benth. (Rubiaceae)
    lanceolata (Forssk.) Deflers - Egyptian star-cluster, star-
      cluster

Penthorum L. (Saxifragaceae)
    sedoides L. - ditch-stonecrop

Peperomia Ruiz & Pav. (Piperaceae)
    radiator-plant
    caperata Yunck. - emerald-ripple peperomia, green-ripple
      peperomia, little-fantasy peperomia
    obtusifolia (L.) A. Dietr. - American radiator-plant, baby-
      rubber-plant, oval-leaf peperomia, pepper-face

Peraphyllum Nutt. (Rosaceae)
    ramosissimum Nutt. - squaw-apple, wild crabapple

Perideridia Rchb. (Apiaceae)
    gairdneri (Hook. & Arn.) Mathias - false caraway, Indian
      caraway, squawroot

Perilla L. (Lamiaceae)
    frutescens (L.) Britton - mint perilla, perilla, perilla-mint

Periploca L. (Asclepiadaceae)
    graeca L. - silk-vine

Persea Mill. (Lauraceae)
    americana Mill. - alligator-pear, aquacate, avocado, palta
    borbonia (L.) Spreng. - Florida mahogany, laurel-tree, red-
      bay, swamp red-bay, sweet-bay, tiss-wood
    palustris (Raf.) Sarg. - swamp-bay

Petasites Mill. (Asteraceae)
    butter-bur, sweet-colt's-foot
    hybridus (L.) Gaertn., C. A. Mey & Scherb. - butter-bur,
      butterfly-dock

Petroselinum Hill (Apiaceae)
    crispum (Mill.) Nyman ex A. W. Hill - parsley

Petunia Juss. (Solanaceae)
    axillaris (Lam.) Britton, et al. - large white petunia, white-
      moon petunia
    ×*hybrida* Vilm. - garden petunia, petunia
      Note = A cultigen of complex hybrid origin.
    parviflora Juss. - seaside petunia, wild petunia

Peumus Molina (Monimiaceae)
    boldus Molina - boldo

Phacelia Juss. (Hydrophyllaceae)
    scorpion-weed
    purshii Buckley - Miami-mist, scorpion-weed

Phalaris L. (Poaceae)
    canary grass

angusta Nees ex Trin. - timothy canary grass
aquatica L. - bulbous canary grass, Harding's grass
arundinacea L. - reed canary grass
canariensis L. - bird-seed grass, canary grass
caroliniana Walter - Carolina canary grass, May grass
minor Retz. - little-seed canary grass
paradoxa L. - hood canary grass
stenoptera Hack. - Harding's grass

Phaseolus L. (Fabaceae)
    bean
    acutifolius A. Gray - Tepary bean
    coccineus L. - Dutch-case-knife bean, scarlet runner bean
    lunatus L. - butter bean, Carolina bean, civet bean, garden
      bean, lima bean, sewee bean, sieva bean
    polystachios (L.) Britton, et al. - bean-vine, thicket bean, wild
      bean
    vulgaris L. - bean, French bean, frijol, garden bean, green
      bean, haricot bean, kidney bean, navy bean, runner bean,
      salad bean, snap bean, string bean, wax bean

Philadelphus L. (Hydrangeaceae)
    coronarius L. - sweet mock-orange

Philodendron Schott (Araceae)
    cordatum (Vell.) Kunth - heart-leaf philodendron
    scandens K. Koch & Sello - heart-leaf philodendron
    scandens K. Koch & Sello f. *micans* (K. Koch) Bunting -
      velvet-leaf philodendron

Phleum L. (Poaceae)
    timothy
    alpinum L. - alpine cat timothy, alpine timothy
    pratense L. - cultivated timothy, head's grass, mountain
      timothy, timothy

Phlomis L. (Lamiaceae)
    tuberosa L. - Jerusalem sage

Phlox L. (Polemoniaceae)
    amplifolia Britton - broad-leaf phlox, wide-leaf phlox
    buckleyi Wherry - Buckley's phlox, shale-barren phlox,
      sword-leaf phlox
    carolina L. - thick-leaf phlox
    divaricata L. - blue phlox, forest phlox, wild sweet-William
    drummondii Hook. - annual phlox, Drummond's phlox
    glaberrima L. - smooth phlox
    maculata L. - meadow phlox, wild sweet-William
    nivalis Lodd. - spotted phlox, trailing phlox
    ovata L. - Allegheny phlox, mountain phlox
    paniculata L. - fall phlox, perennial phlox, summer perennial
      phlox
    pilosa L. - downy phlox, prairie phlox
    stolonifera Sims - crawling phlox, creeping phlox
    subulata L. - ground-pink, moss-pink

Phoenix L. (Arecaceae)
    date palm
    canariensis hort. ex Chabaud - Canary Island date palm
    dactylifera L. - date palm
    reclinata Jacq. - Senegal date palm, wild date palm
    roebelenii O'Brien - dwarf date palm, miniature date palm,
      pygmy date palm, Roebelin's palm

Phoradendron Nutt. (Viscaceae)
false mistletoe, mistletoe
serotinum (Raf.) M. C. Johnst. - American Christmas mistletoe, American mistletoe

Phormium J. R. Forst. & G. Forst. (Agavaceae)
tenax J. R. Forst. & G. Forst. - harakeke, korari, New Zealand flax, New Zealand hemp

Photinia Lindl. (Rosaceae)
glabra (Thunb.) Maxim. - Japanese photinia

Phragmites Adans. (Poaceae)
australis (Cav.) Trin. ex Steud. - carrizo, reed

Phryma L. (Verbenaceae)
leptostachya L. - lop-seed

Phyla Lour. (Verbenaceae)
cuneifolia (Torr.) Greene - wedge-leaf fog-fruit
lanceolata (Michx.) Greene - northern fog-fruit
nodiflora (L.) Greene - Cape-weed, fog-fruit, mat lippia, mat-grass, turkey-tangle
nodiflora (L.) Greene var. *canescens* (Kunth) Moldenke - creeping fog-fruit
nodiflora (L.) Greene var. *incisa* (Small) Moldenke - saw-tooth fog-fruit
nodiflora (L.) Greene var. *rosea* (D.Don) Moldenke - garden fog-fruit

Phyllanthus L. (Euphorbiaceae)
niruri L. - niruri

Phyllodoce Salisb. (Ericaceae)
mountain-heather
breweri (A. Gray) A. Heller - mountain-heather, red-heather

Phyllostachys Siebold & Zucc. (Poaceae)
bamboo
aurea Rivière & C. Rivière - fish-pole bamboo, golden bamboo
bambusoides Siebold & Zucc. - hardy timber bamboo, Japanese timber bamboo, madake
nigra (Lodd.) Munro - black bamboo

Physalis L. (Solanaceae)
ground-cherry, husk-tomato
alkekengi L. - alkekengi, Chinese lantern-plant, Japanese-lantern, strawberry ground-cherry, winter-cherry
angulata L. - cut-leaf ground-cherry, lance-leaf ground-cherry
heterophylla Nees - clammy ground-cherry
lobata Torr. - lobed ground-cherry, purple ground-cherry
longifolia Nutt. - long-leaf ground-cherry
longifolia Nutt. var. *subglabrata* (Mack. & Bush) Cronquist - perennial ground-cherry, smooth ground-cherry
peruviana L. - Barbados gooseberry, Cape gooseberry, cherry tomato, gooseberry-tomato, Peruvian cherry, Peruvian ground-cherry, poha, strawberry tomato, winter-cherry
philadelphica Lam. - tomatillo
pubescens L. - downy ground-cherry, husk-tomato, strawberry tomato
pubescens L. var. *grisea* Waterf. - dwarf Cape gooseberry, hairy ground-cherry, strawberry tomato
virginiana Mill. - Virginia ground-cherry

Physocarpus Maxim. (Rosaceae)
ninebark
opulifolius (L.) Maxim. - ninebark

Phytolacca L. (Phytolaccaceae)
americana L. - garget, pigeon-berry, pocan, pokeberry, pokeweed, scoke, Virginia poke

Picea A. Dietr. (Pinaceae)
spruce
abies (L.) H. Karst. - Norway spruce
breweriana S. Watson - Brewer's spruce, weeping spruce
engelmannii Parry ex Engelm. - Engelmann's spruce
glauca (Moench) Voss - cat spruce, white spruce
mariana (Mill.) Britton, et al. - black spruce, bog spruce, double spruce
pungens Engelm. - Colorado blue spruce, Colorado spruce
rubens Sarg. - he-balsam, red spruce
sitchensis (Bong.) Carrière - Sitka spruce

Picramnia Sw. (Simaroubaceae)
pentandra Sw. - bitter-bush

Picris L. (Asteraceae)
echioides L. - bristly ox's-tongue
hieracioides L. - bugloss, hawkweed ox-tongue, ox-tongue

Pieris D. Don (Ericaceae)
floribunda (Pursh) Benth. & Hook. - evergreen mountain fetterbush, fetterbush, mountain andromeda
japonica (Thunb.) D. Don ex G. Don - Japanese andromeda, lily-of-the-valley-bush

Pilea Lindl. (Urticaceae)
cadierei Gagnep. & Guillaumin - aluminum-plant, watermelon pilea
involucrata (Sims) Urb. - friendship-plant, panamica
microphylla (L.) Liebm. - artillery-plant, artillery-weed
nummariifolia (Sw.) Wedd. - creeping-Charlie
pumila (L.) A. Gray - clearweed, cool-wort, richweed

Pimenta Lindl. (Myrtaceae)
dioica (L.) Merr. - allspice, pimento

Pimpinella L. (Apiaceae)
anisum L. - anise
saxifraga L. - burnet-saxifrage

Pinckneya Michx. (Rubiaceae)
pubens Michx. - fever-tree

Pinguicula L. (Lentibulariaceae)
vulgaris L. - butterwort

Pinus L. (Pinaceae)
pine
albicaulis Engelm. - white-bark pine
attenuata Lemmon - knob-cone pine
balfouriana Grev. & Balf. - foxtail pine
banksiana Lamb. - gray pine, jack pine, Labrador pine, scrub pine
canariensis C. Sm. - Canary Island pine
caribaea Morelet - Caribbean pine, Cuban pine
cembra L. - Russian cedar, Swiss stone pine
cembroides Zucc. - Mexican pinyon, Mexican stone pine
clausa (Chapm. ex Engelm.) Vasey ex Sarg. - sand pine
contorta Loudon - beach pine, lodgepole pine, shore pine
contorta Loudon subsp. *murrayana* (Grev. & Balf.) Critchf. - lodgepole pine
coulteri D. Don - big-cone pine, Coulter's pine

densiflora Siebold & Zucc. - Japanese red pine
echinata Mill. - long-tag pine, short-leaf pine, spruce pine, yellow pine
edulis Engelm. - Colorado pine, Colorado pinyon, pinyon pine, two-leaf nut pine
elliottii Engelm. - slash pine
engelmannii Carrière - Apache pine, Arizona long-leaf pine
flexilis E. James - limber pine
glabra Walter - cedar pine, spruce pine
halepensis Mill. - Aleppo pine, Jerusalem pine
jeffreyi Grev. & Balf. - Jeffrey's pine
lambertiana Douglas - giant pine, sugar pine
leiophylla Schltdl. & Cham. - Chihuahuan pine
longaeva D. K. Bailey - Great Basin bristle-cone pine, hickory pine, western bristle-cone pine
monophylla Torr. & Frém. - nut pine, single-leaf pinyon pine, stone pine
montezumae Lamb. - rough-bark Mexican pine
monticola Douglas - western white pine
mugo Turra - Swiss mountain pine
muricata D. Don - bishop's pine
nigra Arnold - Austrian pine, black pine, Corsican pine
palustris Mill. - Georgia pine, long-leaf pine, southern pine, southern yellow pine
patula Schiede ex Schltdl. & Cham. - Mexican yellow pine
pinaster Aiton - cluster pine, maritime pine
pinea L. - Italian stone pine, umbrella pine
ponderosa Douglas ex Lawson & C. Lawson - ponderosa pine, western yellow pine
ponderosa Douglas ex Lawson & C. Lawson var. *scopulorum* Engelm. - Rocky Mountain yellow pine
pungens Lamb. - hickory pine, prickly pine, table mountain pine
quadrifolia Parl. - Parry's pinyon pine
radiata D. Don - Monterey pine
resinosa Aiton - Canadian pine, Norway pine, red pine
rigida Mill. - pitch pine, torch pine
sabiniana Douglas - digger pine, foothill pine, gray pine
serotina Michx. - Pocosin pine, pond pine
strobus L. - eastern white pine, white pine
sylvestris L. - Scot's pine, Scotch pine
taeda L. - frankincense pine, loblolly pine, old-field pine
thunbergiana Franco - Japanese black pine
torreyana Carrière - soledad pine, Torrey's pine
virginiana Mill. - Jersey pine, poverty pine, scrub pine, spruce pine, Virginia pine
wallichiana A. B. Jacks. - Bhutan pine, blue pine, Himalayan white pine

Piper L. (Piperaceae)
pepper
nigrum L. - black pepper, pepper-plant, white pepper
ornatum N. E. Br. - Celebes pepper

Piptochaetium J. Presl (Poaceae)
fimbriatum (Kunth) Hitchc. - pinyon rice grass

Piqueria Cav. (Asteraceae)
trinervia Cav. - stevia

Piscidia L. (Fabaceae)
piscipula (L.) Sarg. - fish-poison-tree, Jamaican dogwood, West Indian dogwood

Pisonia L. (Nyctaginaceae)
umbellifera (J. Presl & G. Forst.) Seem. - bird-catcher-tree, para-para

Pistacia L. (Anacardiaceae)
pistachio
chinensis Bunge - Chinese pistachio
vera L. - green-almond, pistachio-nut

Pistia L. (Araceae)
stratiotes L. - water-lettuce

Pisum L. (Fabaceae)
pea
sativum L. - English pea, field pea, garden pea, pea, snap pea, sugar pea
sativum L. var. *arvense* (L.) Poir. - field pea
sativum L. var. *macrocarpon* Ser. - edible-podded pea, snow pea, sugar pea

Pithecellobium Mart. (Fabaceae)
dulce (Roxb.) Benth. - guaymochil, huamuchil, Madras thorn, manila, Manila tamarind, opiuma
flexicaule (Benth.) J. M. Coult. - ebony black-bead, Texas ebony
keyense Britton ex Britton & Rose - black-bead
unguis-cati (L.) Benth. - black-bead, bread-and-cheeses, cat's-claw

Pittosporum Banks ex Sol. (Pittosporaceae)
crassifolium A. Cunn. - karo
tobira (Thunb.) W. T. Aiton - Australian laurel, house blooming mock-orange, Japanese pittosporum

Pityopsis Nutt. (Asteraceae)
graminifolia (Michx.) Nutt. - silk-grass

Pityrogramma Link (Pteridophyta)
calomelanos (L.) Link - silver fern
sulphurea (Sw.) Maxon - Jamaican gold fern
triangularis (Kaulf.) Maxon - California gold-bark fern

Plagiobothrys Fisch. & C. A. Mey. (Boraginaceae)
popcorn-flower
figuratus (Piper) Johnst. - popcorn-flower, scorpion-grass

Planera J. F. Gmel. (Ulmaceae)
aquatica (Walter) J. F. Gmel. - planer-tree, water-elm

Plantago L. (Plantaginaceae)
plantain, ribwort
aristata Michx. - bracted plantain
elongata Pursh - slender plantain
lanceolata L. - buck-horn plantain, English plantain, narrow-leaf plantain, ribgrass, ripple-grass
major L. - broad-leaf plantain, cart-track-plant, great plantain, plantain, white-man's-foot
maritima L. var. *juncoides* (Lam.) A. Gray - seaside plantain
media L. - hoary plantain, lamb's-tongue
patagonica Jacq. - woolly plantain
psyllium L. - leafy-stemmed plantain, psyllium
rugelii Decne. - American plantain, black-seed plantain, Rugel's plantain
virginica L. - pale-seed plantain

Platanus L. (Platanaceae)
×*acerifolia* (Aiton) Willd. - London plane-tree
Note = *P. occidentalis* × *P. orientalis.*

occidentalis L. - American sycamore, buttonwood, eastern
   sycamore, plane-tree
orientalis L. - Oriental plane-tree
racemosa Nutt. - California sycamore, western sycamore
wrightii S. Watson - Arizona sycamore

Platycerium Desv. (Pteridophyta)
bifurcatum (Cav.) C. Chr. - elkhorn fern, staghorn fern

Platycladus Spach (Cupressaceae)
orientalis (L.) Franco - Oriental arborvitae

Platycodon A. DC. (Campanulaceae)
grandiflorus (Jacq.) A. DC. - balloon-flower, Chinese bellflower

Plectranthus L'Hér. (Lamiaceae)
   prostrate-coleus, spur-flower, Swedish begonia, Swedish ivy
amboinicus (Lour.) Spreng. - country borage, French thyme,
   Indian mint, Mexican mint, soup mint, Spanish thyme

Pleomele Salisb. (Agavaceae)
aurea (H. Mann) N. E. Br. - golden dracaena, halapepe

Pleuropogon R. Br. (Poaceae)
refractus (A. Gray) Benth. ex Vasey - nodding semaphore
   grass

Pluchea Cass. (Asteraceae)
camphorata (L.) DC. - marsh fleabane, stinkweed
foetida (L.) DC. - stinking fleabane

Plumbago L. (Plumbaginaceae)
auriculata Lam. - Cape leadwort

Plumeria L. (Apocynaceae)
   frangipani, temple-tree
rubra L. - nosegay
rubra L. f. *acutifolia* (Poir.) Woodson - pagoda-tree

Poa L. (Poaceae)
   bluegrass
alpina L. - alpine bluegrass
annua L. - annual bluegrass, dwarf meadow grass, low spear
   grass, six-weeks grass
arachnifera Torr. - Texas bluegrass
arctica R. Br. - arctic bluegrass
arida Vasey - plains bluegrass
bigelovii Vasey & Scribn. - Bigelow's bluegrass
bulbosa L. - bulbous bluegrass
compressa L. - Canadian bluegrass, wiregrass
cusickii Vasey - Cusick's bluegrass
epilis Scribn. - skyline bluegrass
fendleriana (Steud.) Vasey - long-tongue mutton grass,
   mutton bluegrass
interior Rydb. - inland bluegrass
leibergii Scribn. - Leiberg's bluegrass
leptocoma Trin. - bog bluegrass
nemoralis L. - wood bluegrass
nervosa (Hook.) Vasey - Wheeler's bluegrass
palustris L. - fowl bluegrass, fowl meadow grass, rough-stalk
   bluegrass
pratensis L. - June grass, Kentucky bluegrass, spear grass
rupicola Nash - timberline bluegrass
scabrella (Thurb.) Benth. ex Vasey - pine bluegrass
secunda J. Presl - Canby's bluegrass, Pacific bluegrass, pine
   bluegrass, Sandberg's bluegrass, slender bluegrass

secunda J. Presl subsp. *nevadensis* (Scribn.) Soreng - alkali
   bluegrass, big bluegrass, Nevada bluegrass
trivialis L. - rough bluegrass, rough-stalk bluegrass

Podocarpus L'Hér. ex Pers. (Podocarpaceae)
macrophyllus (Thunb.) D. Don - yew podocarpus

Podophyllum L. (Berberidaceae)
peltatum L. - mandrake, May-apple, raccoon-berry, wild jalap,
   wild lemon

Pogonia Juss. (Orchidaceae)
ophioglossoides (L.) Juss. - rose pogonia, snakemouth

Polanisia Raf. (Capparaceae)
dodecandra (L.) DC. - rough-seed clammy-weed
dodecandra (L.) DC. subsp. *trachysperma* (Torr. & A. Gray)
   Iltis - western clammy-weed

Polemonium L. (Polemoniaceae)
   Greek valerian, Jacob's-ladder
caeruleum L. - blue Jacob's-ladder
caeruleum L. subsp. *villosum* (Rudolph ex Georgi) Brand -
   Greek valerian, Jacob's-ladder
micranthum Benth. - annual polemonium
reptans L. - Greek valerian, spreading Jacob's-ladder

Polianthes L. (Agavaceae)
tuberosa L. - tuberose

Polygala L. (Polygalaceae)
   milkwort
cruciata L. - cross-leaf milkwort, drum-heads
lutea L. - candy-weed, orange milkwort, yellow bachelor's-
   button, yellow milkwort
paucifolia Willd. - bird-on-the-wing, flowering-wintergreen,
   fringed polygala
sanguinea L. - blood milkwort, field milkwort
senega L. - Seneca snakeroot
verticillata L. - whorled milkwort

Polygonatum Mill. (Liliaceae)
   Solomon's-seal
biflorum (Walter) Elliott - great Solomon's-seal, Solomon's-
   seal
commutatum (Schult. & Schult. f.) A. Dietr. - great
   Solomon's-seal

Polygonella Michx. (Polygonaceae)
polygama (Vent.) Engelm. & A. Gray - joint-weed

Polygonum L. (Polygonaceae)
   fleece-flower, knotweed, smartweed
achoreum S. F. Blake - striate knotweed
amphibium L. - water smartweed
argyrocoleon Kunze - silver-sheath knotweed
arifolium L. - halberd-leaf tear-thumb
aviculare L. - prostrate knotweed
bistortoides Pursh - bistort, snakeweed, western bistort
bungeanum Turcz. - prickly smartweed
caespitosum Blume var *longisetum* (de Bruyn) A. Stewart -
   tufted knotweed
cilinode Michx. - black-fringe knotweed
coccineum Muhl. ex Willd. - swamp smartweed
convolvulus L. - black-bindweed, corn-bindweed, wild
   buckwheat

cuspidatum Siebold & Zucc. - fleece-flower, Japanese knotweed, Mexican bamboo
douglasii Greene - Douglas's knotweed
erectum L. - erect knotweed
glaucum Nutt. - sea beach knotweed
hydropiper L. - marsh-pepper smartweed, smartweed, water-pepper
hydropiperoides Michx. - mild smartweed, mild water-pepper
lapathifolium L. - knotweed, pale persicaria, pale smartweed, willow-weed
orientale L. - kiss-me-over-the-garden-gate, prince's-feather, princess-feather
pensylvanicum L. - glandular persicary, Pennsylvania smartweed, pink-weed, purple-head, swamp persicary
persicaria L. - heart's-ease, lady's-thumb
punctatum Elliott - dotted smartweed, water smartweed
ramosissimum Michx. - bushy knotweed
sachalinense F. Schmidt ex Maxim. - giant knotweed, sachaline
sagittatum L. - arrow-leaf tear-thumb, arrow-vine
scandens L. - climbing false buckwheat, hedge smartweed
viviparum L. - alpine bistort, alpine smartweed, serpent-grass

Polymnia L. (Asteraceae)
canadensis L. - pale-flowered leaf-cup, small-flowered leaf-cup
uvedalia L. - large-flowered leaf-cup, yellow-flowered leaf-cup

Polypodium L. (Pteridophyta)
californicum Kaulf. - California polypody
glycyrrhiza D. C. Eaton - licorice fern
hesperium Maxon - western polypody
phyllitidis L. - ribbon fern, strap fern
polypodioides (L.) Watt - resurrection fern
virginianum L. - American wall fern, polypody, rock polypody

Polypogon Desf. (Poaceae)
beard grass
interruptus Kunth - ditch beard grass, ditch polypogon
monspeliensis (L.) Desf. - annual beard grass, rabbit's-foot grass, rabbit's-foot polypogon
semiverticillatus (Forssk.) Hyl. - water-bent

Polypremum L. (Loganiaceae)
procumbens L. - rust-weed

Polyscias J. R. Forst. & G. Forst. (Araliaceae)
fruticosa (L.) Harms - Ming aralia
guilfoylei (W. Bull) L. H. Bailey - coffee-tree, geranium-leaf aralia, wild coffee
scutellaria (Burm. f.) Fosberg - Balfour's aralia

Polystichum Roth (Pteridophyta)
shield fern
acrostichoides (Michx.) Schott - canker-brake, Christmas fern, dagger fern
lonchitis (L.) Roth - mountain holly fern, northern holly fern
munitum (Kaulf.) C. Presl - giant holly fern, Pacific Christmas fern, western sword fern

Polytaenia DC. (Apiaceae)
nuttallii DC. - prairie-parsley

Poncirus Raf. (Rutaceae)
trifoliata (L.) Raf. - hardy-orange, trifoliate-orange

Pongamia Vent. (Fabaceae)
pinnata (L.) Pierre - karum-tree, pongam, poonga oil-tree

Pontederia L. (Pontederiaceae)
cordata L. - pickerelweed
cordata L. var. *lancifolia* (Muhl.) Torr. - lance-leaf pickerelweed

Populus L. (Salicaceae)
aspen, cottonwood, poplar
alba L. - abele, silver-leaf poplar, white poplar
angustifolia E. James - narrow-leaf cottonwood
balsamifera L. - balsam poplar, hackmatack, tacamahac
canescens (Aiton) Sm. - gray poplar
deltoides Marsh. - eastern cottonwood, necklace poplar, southern cottonwood
deltoides Marsh. var. *occidentalis* Rydb. - Great Plains cottonwood, plains cottonwood
fremontii S. Watson - Alamo cottonwood, Fremont's cottonwood
×*gileadensis* Rouleau - balm-of-Gilead
grandidentata Michx. - big-tooth aspen, large-tooth aspen
heterophylla L. - black cottonwood, downy poplar, swamp cottonwood
nigra L. - black poplar, Lombardy poplar
simonii Carrière - Simon's poplar
tomentosa Carrière - Chinese white poplar
tremuloides Michx. - quaking aspen, quiver-leaf, trembling aspen
trichocarpa Torr. & A. Gray - black cottonwood, western balsam poplar

Porlieria Ruiz & Pav. (Zygophyllaceae)
angustifolia (Engelm.) A. Gray - guayacan, Texas porlieria

Portulaca L. (Portulacaceae)
purslane
grandiflora Hook. - eleven-o'clock, moss-rose, rose-moss, sun-plant
oleracea L. - purslane, pusley, wild portulaca
pilosa L. - pink purslane

Portulacaria Jacq. (Portulacaceae)
afra (L.) Jacq. - elephant-bush

Potamogeton L. (Potamogetonaceae)
pondweed
amplifolius Tuck. - big-leaf pondweed, broad-leaf pondweed, large-leaf pondweed
crispus L. - crispate-leaf pondweed, curly muck-weed, curly pondweed, curly-leaf pondweed
diversifolius Raf. - diverse-leaf pondweed, snail-seed pondweed, water-leaf pondweed
epihydrus Raf. - Nuttall's pondweed, ribbon-leaf pondweed
filiformis Pers. - fine-leaf pondweed, thread-leaf pondweed
foliosus Raf. - leafy pondweed
friesii Rupr. - Fries's pondweed
gramineus L. - grass-leaf pondweed, variable pondweed
illinoensis Morong - Illinois pondweed, shining pondweed
natans L. - floating pondweed, floating-leaf pondweed
nodosus Poir. - American pondweed, long-leaf pondweed
pectinatus L. - fennel-leaf pondweed, sago pondweed

perfoliatus L. - clasping-leaf pondweed, perfoliate pondweed, red-head-grass
praelongus Wulfen - white-stem pondweed
pusillus L. - slender pondweed, small pondweed
zosteriformis Fernald - eel-grass pondweed, flat-stem pondweed

Potentilla L. (Rosaceae)
cinquefoil, five-fingers
anserina L. - goose-grass, goose-tansy, silver-weed, silver-weed cinquefoil
argentea L. - hoary cinquefoil, silvery cinquefoil
arguta Pursh - tall cinquefoil, tall potentilla, white cinquefoil
biennis Greene - biennial cinquefoil
canadensis L. - cinquefoil, dwarf cinquefoil, running five-fingers
fruticosa L. - bush cinquefoil, golden hard-hack, shrubby cinquefoil, widdy
intermedia L. - downy cinquefoil
norvegica L. - rough cinquefoil, tall five-finger
pacifica Howell - coastal silver-weed, Pacific silver-weed
palustris (L.) Scop. - marsh cinquefoil, marsh fire-finger
paradoxa Nutt. - bushy cinquefoil, diffuse potentilla
pensylvanica L. - prairie cinquefoil
recta L. - rough-fruited cinquefoil, sulphur cinquefoil, upright cinquefoil
reptans L. - creeping cinquefoil
simplex Michx. - old-field cinquefoil
sterilis (L.) Garcke - strawberry potentilla, strawberry-leaf cinquefoil
tridentata Sol. - mountain white potentilla, three-tooth cinquefoil

Pouteria Aubl. (Sapotaceae)
campechiana (Kunth) Baehni - canistel, egg-fruit-tree, sapote-amarillo, sapote-borracho, ti-es
sapota (Jacq.) H. E. Moore & Stearn - mamey-colorado, mammee sapote, marmalade-fruit, marmalade-plum, sapota, sapote

Prenanthes L. (Asteraceae)
rattlesnake-root
alba L. - rattlesnake-root, white-lettuce
altissima L. - tall white-lettuce
autumnalis Walter - slender rattlesnake-root
racemosa Michx. - glaucous white-lettuce, smooth white-lettuce
serpentaria Pursh - gall-of-the-earth, lion's-foot
trifoliolata (Cass.) Fernald - gall-of-the-earth

Primula L. (Primulaceae)
cowslip, primrose
egaliksensis Wormsk. - Greenland primrose
elatior (L.) Hill - oxlip
laurentiana Fernald - bird's-eye primrose
malacoides Franch. - baby primrose, fairy primrose
obconica Hance - German primrose, poison primrose
×*polyantha* hort. - polyanthus
Note = A hybrid group probably having the parentage of *P. veris, P. elatior* and *P. vulgaris.*
sinensis Sabine ex Lindl. - Chinese primrose
veris L. - cowslip
vulgaris Huds. - English primrose

Proboscidea Schmidel (Pedaliaceae)
louisianica (Mill.) Thell. - devil's-claw, proboscis-flower, ram's-horn, unicorn-flower, unicorn-plant

Proserpinaca L. (Haloragaceae)
mermaid-weed
palustris L. - marsh mermaid-weed, mermaid-weed
pectinata Lam. - cut-leaf mermaid-weed

Prosopis L. (Fabaceae)
chilensis (Molina) Stuntz - algarroba, algarroba de Chile, mesquite
farcta (Banks & Sol.) J. F. Macbr. - Syrian mesquite
glandulosa Torr. - glandular mesquite, honey mesquite, mesquite
glandulosa Torr. var. *torreyana* (L. D. Benson) M. C. Johnst. - mesquite, western honey mesquite
juliflora (Sw.) DC. - algarroba, mesquite
pallida (Humb. & Bonpl. ex Willd.) Kunth - algarroba, algarrobo
velutina Wooton - velvet mesquite

Protea L. (Proteaceae)
cynaroides L. - giant protea, king protea

Prunella L. (Lamiaceae)
vulgaris L. - carpenter-weed, heal-all, self-heal

Prunus L. (Rosaceae)
alleghaniensis Porter - Alleghany plum, sloe
americana Marsh. - American plum, August plum, goose plum, hog plum, sloe, wild plum
andersonii A. Gray - desert peach
angustifolia Marsh. - Chickasaw plum, sand plum
armeniaca L. - apricot
avium (L.) L. - bird cherry, gean, mazzard cherry, sweet cherry
caroliniana (Mill.) Aiton - Carolina cherry-laurel, mock-orange, wild orange
cerasifera Ehrh. - cherry plum, Myrobalan plum
cerasus L. - pie cherry, sour cherry
cerasus L. var. *austera* L. - Morello's cherry
davidiana (Carrière) Franch. - Chinese wild peach, David's peach
domestica L. - European plum, garden plum, plum, prune plum
domestica L. subsp. *insititia* (L.) C. K. Schneid. - bullace plum, Damson's plum
dulcis (Mill.) D. A. Webb - almond
emarginata (Hook.) Walp. - bitter cherry, Oregon cherry
fremontii S. Watson - desert apricot
fruticosa Pall. - European dwarf cherry, European ground cherry
hortulana L. H. Bailey - hortulan plum, wild goose plum
ilicifolia (Nutt.) Walp. - evergreen cherry, holly-leaf cherry, islay, mountain holly, wild cherry
ilicifolia (Nutt.) Walp. subsp. *lyonii* (Eastw.) Raven - Catalina Island cherry
laurocerasus L. - cherry-laurel, English cherry-laurel
lusitanica L. - Portuguese cherry-laurel
mahaleb L. - Mahaleb cherry, perfumed cherry, St. Lucie cherry
maritima Marsh. - beach plum, shore plum
mexicana S. Watson - big tree plum, Mexican plum
munsoniana W. Wight & Hedrick - wild goose plum

nigra Aiton - Canadian plum
padus L. - European bird cherry, hag-berry
pensylvanica L.f. - bird cherry, fire cherry, pin cherry, wild red cherry
persica (L.) Batsch - peach
persica (L.) Batsch var. *nucipersica* (Suckow) C. K. Schneid. - nectarine
pumila L. - dwarf cherry, sand cherry
pumila L. var. *besseyi* (L. H. Bailey) Gleason - Bessey's cherry, western sand cherry
salicina Lindl. - Japanese plum
serotina Ehrh. - rum cherry, wild black cherry
serotina Ehrh. subsp. *virens* (Wooton & Standl.) McVaugh - southwestern chokecherry
serrulata Lindl. - Japanese flowering cherry, Oriental cherry
simonii Carrière - apricot plum, Simon's plum
spinosa L. - black-thorn, sloe
subcordata Benth. - Klamath plum, Pacific plum, sierra plum
susquehanae Willd. - sand cherry
tenella Batsch - dwarf Russian almond
triloba Lindl. - flowering almond
umbellata Elliott - flat-woods plum, hog plum, sloe plum
virginiana L. - chokecherry
virginiana L. var. *demissa* (Nutt.) Torr. - western chokecherry
virginiana L. var. *melanocarpa* (A. Nelson) Sarg. - black chokecherry

Pseudobombax Dugand (Bombacaceae)
ellipticum (Kunth) Dugand - shaving-brush-tree

Pseudocydonia (C. K. Schneid.) C. K. Schneid. (Rosaceae)
sinensis (Dum. Cours.) C. K. Schneid. - Chinese quince

Pseudolarix Gordon (Pinaceae)
amabilis (J. Nels.) Rehd. - golden larch

Pseudopanax K. Koch (Araliaceae)
crassifolia (Sol. ex A. Cunn.) K. Koch - horoeka, lancewood

Pseudoroegneria (Nevski) A. Löve (Poaceae)
spicata (Pursh) A. Löve - beardless wheat grass, blue-bunch wheat grass

Pseudotsuga Carrière (Pinaceae)
macrocarpa (Vasey) Mayr - big-cone Douglas fir, big-cone-spruce
menziesii (Mirb.) Franco - Douglas fir
menziesii (Mirb.) Franco var. *glauca* (Beissn.) Franco - Rocky Mountain Douglas fir

Psidium L. (Myrtaceae)
guajava L. - apple guava, guava, yellow guava
littorale Raddi var. *longipes* (O. Berg) Fosberg - Cattley's guava, purple strawberry guava

Psilotum Sw. (Psilotaceae)
nudum (L.) P. Beauv. - whisk-fern

Psophocarpus Neck. ex DC. (Fabaceae)
tetragonolobus (L.) DC. - asparagus pea, Goa bean, winged bean

Psoralea L. (Fabaceae)
  scurfy pea
canescens Michx. - buckroot
esculenta Pursh - breadroot, Indian turnip, prairie-potato, prairie-turnip, shaggy prairie-turnip

macrostachya DC. - leather-root

Psoralidium Rydb. (Fabaceae)
lanceolatum (Pursh) Rydb. - lemon scurf pea
tenuiflorum (Pursh) Rydb. - gray scurf-pea, scurfy psoralea

Psorothamnus Rydb. (Fabaceae)
spinosa (A. Gray) Barneby - desert smoke-tree

Psychotria L. (Rubiaceae)
kaduana (Cham. & Schltdl.) Fosberg - kopiko-tea
nervosa Sw. - Seminole balsam

Ptelea L. (Rutaceae)
  hop-tree, shrubby trefoil
crenulata Greene - California hop-tree
trifoliata L. - hop-tree, stinking-ash, water-ash

Pteridium Gled. ex Scop. (Pteridophyta)
  brake, braken
aquilinum (L.) Kühn - bracken fern, pasture brake
aquilinum (L.) Kühn var. *pubescens* Underw. - bracken fern, western bracken, western bracken fern

Pteris L. (Pteridophyta)
  brake
cretica L. - Cretan brake
ensiformis Burm. - sword brake

Pterocarpus Jacq. (Fabaceae)
indicus Willd. - amboyna-wood, Burmese rosewood, narra, padauk, padouk

Pterocarya Kunth (Juglandaceae)
  wing-nut
stenoptera C. DC. - Chinese wing-nut

Pterospora Nutt. (Ericaceae)
andromedea Nutt. - giant bird's-nest, pinedrops

Ptilimnium Raf. (Apiaceae)
capillaceum (Michx.) Raf. - Atlantic mock bishop's-weed, bishop's-weed, mock bishop's-weed

Ptychosperma Labill. (Arecaceae)
macarthurii H. Wendl. - cluster palm, hurricane palm, MacArthur's feather palm

Puccinellia Parl. (Poaceae)
  alkali grass
distans (L.) Parl. - alkali grass, European alkali grass, weeping alkali grass
nuttalliana (Schult.) Hitchc. - Nuttall's alkali grass

Pueraria DC. (Fabaceae)
lobata (Willd.) Ohwi - kudzu, kudzu-vine
phaseoloides (Roxb.) Benth. - puero, tropical kudzu

Pulicaria Gaertn. (Asteraceae)
dysenterica (L.) Bernh. - fleabane

Punica L. (Punicaceae)
granatum L. - pomegranate

Pycnanthemum Michx. (Lamiaceae)
  mountain mint
pilosum Nutt. - hairy mountain mint
virginianum (L.) T. Durand & B. D. Jacks. ex B. L. Rob. & Fernald - Virginia mountain mint

Pyracantha M. Roem. (Rosaceae)
fire-thorn
angustifolia (Franch.) C. K. Schneid. - narrow-leaf fire-thorn
coccinea M. Roem. - scarlet fire-thorn

Pyrola L. (Ericaceae)
pyrola, shin-leaf, wintergreen
asarifolia Michx. - pink pyrola, pink wintergreen
chlorantha Sw. - green-flowered wintergreen, greenish-flowered pyrola
elliptica Nutt. - shin-leaf, wild lily-of-the-valley
glandiflora Radius - arctic pyrola, arctic wintergreen
minor L. - lesser pyrola, lesser wintergreen
picta Sm. - white-veined shin-leaf, white-veined wintergreen
rotundifolia L. - round-leaf pyrola, rounded shin-leaf, shin-leaf, wild lily-of-the-valley

Pyrostegia C. Presl (Bignoniaceae)
venusta (Ker Gawl.) Miers - flame-flower, flame-vine, flaming-trumpet, golden-shower

Pyrrhopappus DC. (Asteraceae)
carolinianus (Walter) DC. - Carolina false dandelion

Pyrus L. (Rosaceae)
pear
calleryana Decne. - Bradford's pear, Callery's pear
communis L. - pear
pyrifolia (Burm. f.) Nakai - Asian pear, Chinese pear, Japanese pear, Oriental pear, sand pear

Quercus L. (Fagaceae)
oak
acutissima Carruth. - saw-tooth oak
agrifolia Née - California field oak, California live oak, coast live oak, encina
ajoensis C. H. Müll. - Ajo oak
alba L. - white oak
arizonica Sarg. - Arizona white oak
arkansana Sarg. - Arkansas oak
austrina Small - bastard oak, bluff oak
bicolor Willd. - swamp white oak
cerris L. - European turkey oak
chapmanii Sarg. - Chapman's oak
chrysolepis Liebm. - canyon live oak, canyon oak, maul oak
coccinea Münchh. - scarlet oak
douglasii Hook. & Arn. - blue oak
dumosa Nutt. - California scrub oak, Nuttall's scrub oak
durandii Buckley - Durand's oak
durata Jeps. - leather oak
ellipsoidalis E. J. Hill - jack oak, northern pin oak
emoryi Torr. - Emory's oak
engelmannii Greene - Engelmann's oak, mesa oak
falcata Michx. - southern red oak, Spanish red oak
falcata Michx. var. *pagodifolia* Elliott - cherry-bark oak, swamp red oak
gambelii Nutt. - Gambel's oak
garryana Douglas - Oregon oak, Oregon white oak, western oak
gravesii Sudw. - Graves's oak
grisea Liebm. - gray oak
harvardii Rydb. - Harvard's shin oak, shinnery oak
hemisphaerica Bartram - Darlington's oak
hypoleucoides A. Camus - silver-leaf oak
ilex L. - holly oak, Holm's oak

ilicifolia Wangenh. - bear oak, scrub oak
imbricaria Michx. - jack oak, laurel oak, shingle oak
incana Roxb. - bluejack oak, high-ground willow oak, sand-jack, turkey oak
kelloggii Newb. - California black oak, Kellogg's oak
laevis Walter - Catesby's oak, turkey oak
laurifolia Michx. - laurel oak, laurel-leaved oak
lobata Née - California white oak, roble, valley oak
lyrata Walter - over-cup oak, swamp post oak
macdonaldii Greene - MacDonald's oak
macrocarpa Michx. - bur oak, mossy-cup oak
marilandica Münchh. - blackjack oak, jack oak
michauxii Nutt. - basket oak, cow oak, swamp chestnut oak, swamp white oak
minima (Sarg.) Small - dwarf live oak
mohriana Buckley ex Rydb. - Mohr's oak
mongolica Fisch. ex Turcz. - Mongolian oak
muehlenbergii Engelm. - chinkapin oak, yellow chestnut oak, yellow oak
myrtifolia Willd. - myrtle oak
nigra L. - possum oak, water oak
nuttallii E. J. Palmer - Nuttall's oak
oblongifolia Torr. - Mexican blue oak
oglethorpensis W. H. Duncan - Oglethorpe's oak
palustris Münchh. - pin oak, Spanish oak
phellos L. - willow oak
prinoides Willd. - chinquapin oak, dwarf chestnut oak, dwarf chinquapin oak
prinus L. - basket oak, chestnut oak, rock chestnut oak, swamp chestnut oak
pumila Walter - running oak
pungens Liebm. - sandpaper oak
robur L. - English oak, truffle oak
rubra L. - northern red oak
sadleriana R. Br. - deer oak
shumardii Buckley - Shumard's red oak
stellata Wangenh. - post oak
suber L. - cork oak
texana Buckley - Texas red oak
tomentella Engelm. - island live oak, island oak
toumeyi Sarg. - Toumey's oak
turbinella Greene - shrub live oak, turbinella oak
undulata Torr. - Rocky Mountain scrub oak, wavy-leaf oak
vaccinifolia Kellogg - huckleberry oak
velutina Lam. - black oak, quercitron, yellow-bark oak
virginiana Mill. - live oak, southern live oak
virginiana Mill. var. *fusiformis* (Small) Sarg. - plateau oak
wislizenii A. DC. - interior live oak

Quisqualis L. (Combretaceae)
indica L. - Rangoon creeper

Randia L. (Rubiaceae)
aculeata L. - inkberry

Ranunculus L. (Ranunculaceae)
buttercup, crowfoot
abortivus L. - kidney-leaf, small-flowered buttercup, small-flowered crowfoot
acris L. - meadow buttercup, tall buttercup
aquatilis L. var. *capillaceus* (Thuill.) DC. - white water buttercup
arvensis L. - corn buttercup, field buttercup
asiaticus L. - Persian buttercup, turban buttercup

bulbosus L. - bulbous buttercup
californicus Benth. - California buttercup
cymbalaria Pursh - seaside crowfoot, shore buttercup
fascicularis Muhl. ex Bigelow - early buttercup
ficaria L. - lesser celandine, pilewort, small celandine
flabellaris Raf. - yellow water buttercup, yellow water crowfoot
flammula L. - spearwort
gmelinii DC. - small yellow water-crowfoot
hispidus Michx. - hispid buttercup, swamp buttercup
muricatus L. - rough-seed buttercup, spiny-fruit buttercup
occidentalis Nutt. - western field buttercup
orthorhynchus Hook. - bird's-foot buttercup
pensylvanicus L.f. - bristly crowfoot
pygmaeus Wahlenb. - dwarf buttercup
recurvatus Poir. - hooked buttercup
repens L. - butter-daisy, creeping buttercup, yellow-gowan
reptans L. - creeping spearwort
rhomboideus Goldie - prairie buttercup, prairie crowfoot
sardous Crantz - hairy buttercup
sceleratus L. - bog buttercup, celery-leaf buttercup, crow-foot buttercup, cursed buttercup, cursed crowfoot
testiculatus Crantz - bur buttercup, horned-head buttercup, testiculate buttercup
trichophyllus Chaix - white water-crowfoot

Raphanus L. (Brassicaceae)
raphanistrum L. - jointed charlock, wild radish
sativus L. - radish
sativus L. cv.'longipinnatus' - Chinese radish, daikon

Rapistrum Crantz (Brassicaceae)
perenne (L.) All. - steppe-cabbage
rugosum (L.) All. - turnip-weed, wild turnip

Ratibida Raf. (Asteraceae)
prairie coneflower
columnifera (Nutt.) Wooton & Standl. - columnar prairie coneflower, upright prairie coneflower

Ravenala Adans. (Strelitziaceae)
madagascariensis Sonn. - traveler's-palm

Reseda L. (Resedaceae)
alba L. - white mignonette
lutea L. - cut-leaf mignonette, dyer's-rocket, yellow mignonette
odorata L. - garden mignonette, mignonette

Rhamnus L. (Rhamnaceae)
buckthorn
alnifolia L'Hér. - alder buckthorn, alder-leaf buckthorn
betulifolia Greene - birch-leaf buckthorn
californica Eschsch. - California buckthorn, coffee-berry
caroliniana Walter - Carolina buckthorn, Indian cherry, yellow buckthorn
cathartica L. - buckthorn, European buckthorn
crocea Nutt. - red-berry, spiny red-berry
frangula L. - alder buckthorn, European alder buckthorn, glossy buckthorn
ilicifolia Kellogg - holly-leaf buckthorn, red-berried buckthorn
lanceolata Pursh - lance-leaf buckthorn
purshiana DC. - bearberry, cascara, cascara buckthorn, cascara sagrada

Rhaphiolepis Lindl. (Rosaceae)
indica (L.) Lindl. - Indian hawthorn
umbellata (Thunb.) Makino - Yedda's-hawthorn

Rhapis L.f. ex Aiton (Arecaceae)
lady palm
excelsa (Thunb.) A. Henry - bamboo palm, dwarf ground rattan, fern rhapis, lady palm, miniature fan palm, rattan palm, slender lady palm

Rheum L. (Polygonaceae)
officinale Baill. - Chinese rhubarb
rhabarbarum L. - pie-plant, rhubarb, wine-plant

Rhexia L. (Melastomataceae)
deer grass, meadow-berry
mariana L. - dull meadow-pitchers, Maryland meadow-beauty

Rhinanthus L. (Scrophulariaceae)
yellow-rattle
crista-galli L. - rattle-box
minor L. - yellow-rattle

Rhizophora L. (Rhizophoraceae)
mangle L. - American mangrove, mangrove, red mangrove

Rhododendron L. (Ericaceae)
azalea, rhododendron
arborescens (Pursh) Torr. - smooth azalea, sweet azalea
arboreum Sm. - farkleberry, tree huckleberry, tree rhododendron
atlanticum (Ashe) Rehder - dwarf azalea
austrinum Rehder - early azalea, election-pink
calendulaceum (Michx.) Torr. - flame azalea, yellow azalea
campylocarpum Hook. f. - honey-bell rhododendron
canadense (L.) Torr. - Canadian rhododendron, rhodora
canescens (Michx.) Sweet - Florida pinxter, hoary azalea, Piedmont azalea
carolinianum Rehder - Carolina rhododendron
catawbiense Michx. - catawba rhododendron, mountain rose-bay, purple laurel
×*gandavense* (K. Koch) Rehder - Ghent hybrid azalea
grande Wight - silvery rhododendron
indicum (L.) Sweet - macranthum azalea
japonicum (A. Gray) Suringar - Japanese azalea
kaempferi Planch. - torch azalea
lapponicum (L.) Wahlenb. - Lapland rhododendron, Lapland rose-bay
luteum (L.) Sweet - pontic azalea
macrophyllum G. Don - California rose-bay, Pacific rhododendron, West Coast rhododendron
maximum L. - great rhododendron, great-laurel, rose-bay rhododendron, white-laurel
minus Michx. - Piedmont rhododendron
molle (Blume) G. Don - Chinese azalea
mucronatum G. Don - snow azalea
obtusum (Lindl.) Planch. - Hiryu azalea
occidentale (Torr. & A. Gray) A. Gray - western azalea
periclymenoides (Michx.) Shinners - pinxter-bloom
prunifolium (Small) Millais - plum-leaf azalea
smirnowii Trautv. ex Regel - Smirnow's rhododendron
vaseyi A. Gray - pink-shell azalea
viscosum (L.) Torr. - clammy azalea, swamp azalea, swamp-honeysuckle, white swamp azalea

Rhodomyrtus (DC.) Rchb. (Myrtaceae)
  tomentosus (Aiton) Hassk. - downy rose-myrtle

Rhodotypos Siebold & Zucc. (Rosaceae)
  jet-bead, white kerria
  scandens (Thunb.) Makino - black jet-bead

Rhus L. (Anacardiaceae)
  aromatica Aiton - fragrant sumac, lemon sumac, polecat-bush, squaw-bush, sweet-scented sumac
  chinensis Mill. - Chinese sumac, nutgall-tree
  choriophylla Wooton & Standl. - Mearns' sumac, New Mexico evergreen sumac
  copallina L. - dwarf sumac, flame-tree sumac, mountain sumac, shining sumac, wing-rib sumac, winged sumac
  diversiloba Torr. & A. Gray - Pacific poison-oak, western poison-oak
  glabra L. - scarlet sumac, smooth sumac, vinegar-tree
  hirta (L.) Sudw. - staghorn sumac, velvet sumac, Virginia sumac
  integrifolia (Nutt.) Brewer & S. Watson - lemonade sumac, lemonade-berry, sour-berry
  lancea L. f. - bastard-willow, karee, karoo-tree, willow rhus
  lanceolata (A. Gray) Britton - prairie sumac
  ovata S. Watson - sugar sumac, sugar-bush sumac
  sempervirens Scheele - evergreen sumac
  trilobata Torr. & A. Gray - skunk-brush, skunk-bush
  virens Lindh. ex A. Gray - evergreen sumac, lentisco, tobacco sumac

Rhynchelytrum Nees (Poaceae)
  repens (Willd.) F. T. Hubb. - Natal grass

Rhynchospora Vahl (Cyperaceae)
  beak rush
  alba (L.) Vahl - white beak rush
  colorata (L.) H. Pfeiff. - white-top sedge
  corniculata (Lam.) A. Gray - horned rush
  fusca (L.) W. T. Aiton - brown beak rush
  globularis (Chapm.) Small - pine-hill beak rush
  inexpansa (Michx.) Vahl - nodding beak rush
  latifolia (Baldwin ex Elliott) W. W. Thomas - giant white-top sedge

Ribes L. (Grossulariaceae)
  currant, gooseberry
  alpinum L. - alpine currant, mountain currant
  americanum Mill. - American black currant, eastern black currant, wild black currant
  aureum Pursh - buffalo currant, golden currant, Missouri currant
  bracteosum Douglas - California black currant, stink currant
  cereum Douglas - squaw currant, wax currant, white-flowered currant
  cynosbati L. - dog-bramble, dogberry, pasture gooseberry, prickly gooseberry
  divaricatum Douglas - white-stem gooseberry
  glandulosum Grauer - fetid currant, skunk currant
  grossularis L. - European gooseberry
  hirtellum Michx. - hairy-stem gooseberry
  hudsonianum Richardson - Hudson Bay currant, northern black currant
  lacustre (Pers.) Poir. - bristly black currant, swamp currant, swamp gooseberry
  lobbii A. Gray - gummy gooseberry

  malvaceum Sm. - chaparral currant
  menziesii Pursh - canyon gooseberry
  missouriense Nutt. - Missouri gooseberry
  nevadense Kellogg - mountain pink currant, sierra currant
  nigrum L. - black currant, European black currant, garden black currant
  odoratum H. Wendl. - buffalo currant, Missouri currant
  oxyacanthoides L. - hawthorn-leaved gooseberry, northern gooseberry
  roezlii Regel - sierra gooseberry
  rubrum L. - garden currant, northern red currant, red currant, white currant
  sanguineum Pursh - red-flower currant
  setosum Lindl. - Missouri gooseberry
  speciosum Pursh - fuchsia-flowered gooseberry
  triste Pall. - swamp red currant
  uva-crispa L. - English gooseberry, European gooseberry, garden gooseberry
  viscosissimum Pursh - sticky currant

Richardia L. (Rubiaceae)
  brasiliensis Gomes - Brazilian pusley
  scabra L. - Florida pusley, Mexican clover

Ricinus L. (Euphorbiaceae)
  communis L. - castor bean, castor-oil-plant, palma-christi, wonder-tree

Rivina L. (Phytolaccaceae)
  humilis L. - baby-pepper, blood-berry, rouge-plant

Robinia L. (Fabaceae)
  locust
  hispida L. - bristly locust, moss locust, rose-acacia
  neomexicana A. Gray - desert locust, New Mexico locust
  pseudoacacia L. - black locust, false acacia, yellow locust
  viscosa Vent. - clammy locust, rose-acacia

Roemeria Medik. (Papaveraceae)
  refracta (Stev.) DC. - Roemer's poppy

Rollinia A. St.-Hil. (Annonaceae)
  pulchrinervis A. DC. - biriba

Rorippa Scop. (Brassicaceae)
  yellow-cress
  amphibia (L.) Besser - great yellow-cress
  austriaca (Crantz) Besser - Austrian field-cress
  nasturtium-aquaticum (L.) Hayek - watercress
  palustris (L.) Besser - marsh-cress
  sinuata (Nutt.) Hitchc. - spreading yellow-cress
  sylvestris (L.) Besser - creeping yellow-cress, yellow field-cress
  teres (Michx.) Stuckey - terete yellow-cress

Rosa L. (Rosaceae)
  brier, rose
  acicularis Lindl. - bristly rose, prickly rose
  arkansana Porter - Arkansas rose, dwarf prairie rose, prairie rose, prairie wild rose
  blanda Aiton - smooth rose
  bracteata H. Wendl. - Macartney's rose
  californica Cham. & Schltdl. - California rose, California wild rose
  canina L. - dog rose, dog-brier
  carolina L. - pasture rose

centifolia L. - cabbage rose
chinensis Jacq. - Bengal rose, China rose
eglanteria L. - eglantine, sweetbrier
foetida Herrm. - Austrian brier, Austrian yellow rose
gallica L. - French rose
gymnocarpa Nutt. - wood rose
laevigata Michx. - Cherokee rose
majalis Herrm. - cinnamon rose
multiflora Thunb. - baby rose, multiflora rose
×*noisettiana* Thory cv.'Manettii' - Manetti's rose
   Note = *R. chinensis* × *R. moschata*.
nutkana C. Presl - Nootka rose, Nutka rose
odorata (Andr.) Sweet - tea rose
palustris Marsh. - swamp rose
pimpinellifolia L. - Scotch rose
pisocarpa A. Gray - cluster rose
rugosa Thunb. - Japanese rose, rugosa rose, Turkestan rose
setigera Michx. - climbing prairie rose, climbing rose, prairie rose
spinosissima L. - Burnet's rose, Scotch rose
virginiana Mill. - Virginia rose
wichuraiana Crép. - memorial rose
woodsii Lindl. - western rose, Wood's rose

Rosmarinus L. (Lamiaceae)
officinalis L. - rosemary

Rostraria Trin. (Poaceae)
cristata (L.) Tzvelev - annual-cat-tail

Rotala L. (Lythraceae)
indica (Willd.) Koehne - Indian tooth-cup
ramosior (L.) Koehne - tooth-cup

Rottboellia L.f. (Poaceae)
cochinchinensis (Lour.) W. Clayton - itch grass

Roystonea Cook (Arecaceae)
   royal palm
elata (W. Bartram) F. Harper - Florida royal palm

Rubia L. (Rubiaceae)
tinctorum L. - madder

Rubus L. (Rosaceae)
   blackberry
allegheniensis Porter - Allegheny blackberry, sow-teat blackberry
arcticus L. - arctic blackberry, crimson blackberry, crimson-berry
argutus Link - high-bush blackberry, southern blackberry
canadensis L. - smooth blackberry, smooth bramble
chamaemorus L. - baked-apple-berry, cloudberry, malka, salmonberry, yellow-berry
cuneifolius Pursh - sand blackberry
deliciosus Torr. - Rocky Mountain raspberry
discolor Weihe & Nees - Himalayan blackberry
flagellaris Willd. - American dewberry, northern dewberry, running blackberry, trailing bramble
fruticosus L. - European blackberry
hispidus L. - bristly dewberry, running blackberry, swamp blackberry, swamp dewberry
idaeus L. - European red raspberry, framboise, red raspberry
laciniatus Willd. - cut-leaf blackberry, evergreen blackberry, parsley-leaf blackberry
loganobaccus L. H. Bailey - boysenberry, loganberry

moluccanus L. - Molucca raspberry
neomexicanus A. Gray - New Mexico raspberry
occidentalis L. - black raspberry, black-cap, thimbleberry
odoratus L. - flowering raspberry, purple-flowering raspberry, thimbleberry
parviflorus Nutt. - salmonberry, thimbleberry, western thimbleberry
phoenicolasius Maxim. - wine-berry
pubescens Raf. - dwarf raspberry, dwarf red raspberry
rosifolius Sm. - Mauritius raspberry
spectabilis Pursh - salmonberry
stellatus Sm. - knesheneka, nagoon-berry
strigosus Michx. - American red raspberry
trivialis Michx. - coastal plain dewberry, southern dewberry
ursinus Cham. & Schltdl. - California blackberry, Pacific blackberry, Pacific dewberry

Rudbeckia L. (Asteraceae)
   coneflower
hirta L. - black-eyed-Susan, hairy coneflower
hirta L. var. *pulcherrima* Farw. - black-eyed-Susan
laciniata L. - coneflower, cut-leaf coneflower, golden-glow
triloba L. - brown-eyed-Susan, thin-leaf coneflower, three-lobed coneflower

Ruellia L. (Acanthaceae)
caroliniensis (J. F. Gmel.) Steud. - wild petunia
makoyana Closon - monkey-plant, trailing velvet-plant
strepens L. - smooth ruellia
tuberosa L. - meadow-weed, menow-weed

Rumex L. (Polygonaceae)
   dock sorrel, sorrel
acetosa L. - garden sorrel, green sorrel, sour dock
acetosella L. - red sorrel, sheep sorrel, sorrel, sour dock
alpinus L. - monk's rhubarb, mountain rhubarb
altissimus A. Wood - pale dock, smooth dock, water dock
conglomeratus Murray - cluster dock
crispus L. - curly dock, yellow dock
hymenosepalus Torr. - canaigre, tanner's dock, wild rhubarb
longifolius DC. - long-leaf dock, yard dock
obtusifolius L. - bitter dock, blunt-leaf dock, broad-leaf dock, red-veined dock
occidentalis S. Watson - western dock
orbiculatus A. Gray - water dock
pallidus Bigelow - sea beach dock, seaside dock, white dock
patientia L. - herb-patience, monk's rhubarb, patience dock, spinach dock
pulcher L. - fiddle dock, fiddle-leaf dock
stenophyllus Ledeb. - narrow-leaf dock
venosus Pursh - sour-greens, veiny dock, wild begonia, wild hydrangea, winged dock
verticillatus L. - swamp dock, water dock

Rumohra Raddi (Pteridophyta)
adiantiformis (G. Forst.) Ching - iron fern, leather fern

Ruppia L. (Potamogetonaceae)
maritima L. - widgeon-grass

Russelia Jacq. (Scrophulariaceae)
equisetiformis Schltdl. & Cham. - coral-plant, fountain-bush, fountain-plant

Ruta L. (Rutaceae)
graveolens L. - herb-of-grace, rue

Sabal Adans. (Arecaceae)
  palmetto
etonia Swingle ex Nash - Etonia palmetto, scrub palmetto
mexicana Mart. - Mexican palmetto, Rio Grande palmetto,
  Texas palmetto, Victoria palmetto
minor (Jacq.) Pers. - blue-stem palmetto, bush palmetto, dwarf
  palmetto, scrub palmetto
palmetto (Walter) Schult. & Schult. f. - cabbage palmetto,
  Florida cabbage palm

Sabatia Adans. (Gentianaceae)
  rose-gentian
angularis (L.) Pursh - bitter-bloom, marsh-pink, rose-pink

Saccharum L. (Poaceae)
alopecuroideum (L.) Nutt. - silver plume grass
brevibarbe (Michx.) Pers. - brown plume grass
contortum (Baldwin ex Elliott) Nutt. - bent-awn plume grass
giganteum (Walter) Pers. - sugar-cane plume grass
officinarum L. - sugar-cane
ravennae (L.) L. - Ravenna's grass
spontaneum L. - wild sugar-cane

Sagina L. (Caryophyllaceae)
  pearl-wort
procumbens L. - arctic pearl-wort, bird's-eye pearl-wort

Sagittaria L. (Alismataceae)
  arrowhead, swamp-potato
cuneata E. Sheld. - arum-leaf arrowhead, wapato, wedge-leaf
  arrowhead
graminea Michx. - coastal arrowhead, grass-leaved sagittaria,
  slender arrowhead
graminea Michx. var. *platyphylla* Engelm. - delta arrowhead
lancifolia L. - arrowhead, lance-leaf arrowhead, lance-leaf
  sagittaria
latifolia Willd. - arrowhead, duck-potato, wapato
longiloba Engelm. - Gregg's arrowhead
montevidensis Cham. & Schltdl. - California arrowhead, giant
  arrowhead
sagittifolia L. - arrowhead, Chinese arrowhead, Old World
  arrowhead, swan-potato
subulata (L.) Buchenau - Hudson sagittaria, spring-tape

Saintpaulia H. Wendl. (Gesneriaceae)
ionantha H. Wendl. - African violet, usambava violet

Salicornia L. (Chenopodiaceae)
  glasswort, samphire
bigelovii Torr. - dwarf glasswort, dwarf saltwort
europaea L. - chicken's-claws, pigeon's-foot, samphire,
  slender glasswort
virginica L. - lead-grass, perennial glasswort, woody
  glasswort

Salix L. (Salicaceae)
  osier, willow
alaxensis (Andersson) Coville - felt-leaf willow
alba L. - white willow
alba L. var. *vitellina* (L.) Stokes - golden willow
amygdaloides Andersson - peach-leaved willow
arbusculoides Andersson - little-tree willow
arctica Pall. - arctic willow
babylonica L. - weeping willow
bebbiana Sarg. - Bebb's willow, gray willow, long-beaked
  willow

brachycarpa Nutt. subsp. *niphoclada* (Rydb.) Argus - barren
  ground willow
candida Flüggé - sage-leaf willow, silver willow
caprea L. - goat willow, hoary willow
caroliniana Michx. - Carolina willow, coastal plain willow,
  Ward's willow
chamissonis Andersson - Chamisso willow
cinerea L. - gray willow
commutata Bebb - under-green willow
cordata Michx. - dune willow, heart-leaf willow
discolor Muhl. - large pussy willow, pussy willow
eriocephala Michx. - diamond willow, heart-leaf willow,
  Missouri willow
exigua Nutt. - coyote willow, ditch bank willow, narrow-leaf
  willow, sandbar willow
floridana Chapm. - Florida willow
fluviatilis Nutt. - river willow
fragilis L. - brittle willow, crack willow
fuscescens Andersson - Alaska bog willow
glauca L. - gray-leaf willow
gooddingii C. R. Ball - Goodding's willow
hookeriana Hook. - coastal willow, Hooker's willow
humilis Marsh. - gray willow, prairie willow, small pussy
  willow, upland willow
laevigata Bebb - polished willow, red willow
lanata L. subsp. *richardsonii* (Hook.) Skvortsov -
  Richardson's willow
lasiolepis Benth. - arroyo willow, Tracy's willow
lucida Muhl. - shining willow
lucida Muhl. subsp. *lasiandra* (Benth.) E. Murray - Pacific
  willow
monticola Bebb - park willow, serviceberry willow
myrtillifolia Andersson - low blueberry willow
nigra Marsh. - black willow
novaeangliae Andersson - tall blueberry willow
ovalifolia Trautv. - oval-leaf willow
pentandra L. - bay willow, bay-leaved willow, laurel willow
petiolaris Sm. - meadow willow, skeleton-leaf willow
planifolia Cham. - diamond-leaf willow, tea-leaf willow
purpurea L. - basket willow, purple osier, purple willow
pyrifolia Andersson - balsam willow
reticulata L. - net-leaf willow
rotundifolia Trautv. - least willow
scouleriana Barratt ex Hook. - mountain willow, Scouler's
  willow
sericea Marsh. - silky willow
serissima (L. H. Bailey) Fernald - autumn willow
sessilifolia Nutt. - northwest willow, sandbar willow
sitchensis Bong. - Sitka willow
stolonifera Crép. - oval-leaf willow
syrticola Fernald - sand-dune willow
taxifolia Kunth - yew-leaf willow
viminalis L. - basket willow, osier, silky osier

Salpiglossis Ruiz & Pav. (Solanaceae)
sinuata Ruiz & Pav. - painted-tongue

Salsola L. (Chenopodiaceae)
australis R. Br. - Russian thistle, salt-bush
kali L. - tumbleweed
paulsenii Litv. - barb-wire Russian thistle
vermiculata L. - Mediterranean saltwort

Salvia L. (Lamiaceae)
ramona, sage
aethiopis L. - Mediterranean sage
apiana Jeps. - greasewood, white sage
azurea Lam. - blue sage
azurea Lam. var. *grandiflora* Benth. - pitcher's sage
coccinea L. - scarlet sage, Texas sage
dorrii (Kellogg) Abrams - gray-ball sage
farinacea Benth. - mealy-cup sage
greggii A. Gray - autumn sage
mellifera Greene - black sage
officinalis L. - garden sage, sage
pratensis L. - meadow sage
reflexa Hornem. - lance-leaf sage
sonomensis Greene - creeping sage
spathacea Greene - pitcher sage
splendens Sellow ex Roem. & Schult. - scarlet sage
verticillata L. - whorled sage

Salvinia Ség. (Salviniaceae)
auriculata Aubl. - giant salvinia, salvinia
minima Baker - water fern
molesta D. S. Mitch. - giant salvinia, kariba-weed

Sambucus L. (Caprifoliaceae)
elder, elderberry
callicarpa Greene - Pacific Coast red elder, Pacific red elder, red elderberry
canadensis L. - American elder, American elderberry, elder, sweet elder
cerulea Raf. - blue elder, blue elderberry, blueberry elder, western elderberry
ebulus L. - dane-wort, dwarf elder
melanocarpa A. Gray - black-bead elder
mexicana C. Presl ex DC. - blue elderberry, Mexican elder
nigra L. - European elder
racemosa L. - European red elderberry, red-berried elder
racemosa L. var. *microbotrys* (Rydb.) Kearney & Peebles - American red elderberry, red-berried elder

Samolus L. (Primulaceae)
parviflorus (Nees) Raf. - brookweed, water pimpernel

Sanguinaria L. (Papaveraceae)
canadensis L. - bloodroot, red-puccoon

Sanguisorba L. (Rosaceae)
burnet
canadensis L. - American burnet, Canadian burnet
minor Scop. - salad burnet
officinalis L. - burnet-bloodwort, great burnet
stipulata Raf. - Sitka burnet

Sanicula L. (Apiaceae)
black snakeroot, sanicle
bipinnatifida Hook. - purple sanicle, shoe-buttons
marilandica L. - black snakeroot

Sansevieria Thunb. (Agavaceae)
trifasciata Prain - mother-in-law's-tongue, snake-plant
zeylanica (L.) Willd. - Ceylon bowstring-hemp

Santolina L. (Asteraceae)
chamaecyparissus L. - lavender-cotton
rosmarinifolia L. - green santolina

Sapindus L. (Sapindaceae)
soapberry
oahuensis Hildebr. ex Radlk. - Hawaiian soap-tree
saponaria L. - false dogwood, jaboncillo, southern soapberry, wing-leaf soapberry
saponaria L. var. *drummondii* (Hook. & Arn.) L. D. Benson - jaboncillo, western soapberry

Sapium P. Browne (Euphorbiaceae)
sebiferum (L.) Roxb. - Chinese tallow-tree, tallow-tree

Saponaria L. (Caryophyllaceae)
officinalis L. - bouncing-bet, soapwort

Sarcobatus Nees (Chenopodiaceae)
vermiculatus (Hook.) Torr. - greasewood

Sarcostemma R. Br. (Asclepiadaceae)
cynanchoides Decne. - climbing milkweed

Sarracenia L. (Sarraceniaceae)
huntsman's-cup, pitcher-plant
alata (A. Wood) A. Wood - yellow trumpets
flava L. - huntsman's-horn, trumpet-leaf, trumpets, umbrella-trumpets, watches, yellow pitcher-plant
purpurea L. - huntsman's-cup, Indian cup-plant, pitcher-plant, side-saddle-flower, southern pitcher-plant, sweet pitcher-plant

Sassafras Nees & Eberm. (Lauraceae)
albidum (Nutt.) Nees - sassafras, white sassafras

Satureja L. (Lamiaceae)
calamint, savory
douglasii (Benth.) Briq. - yerba-buena
hortensis L. - summer savory

Saururus L. (Saururaceae)
cernuus L. - lizard's-tail, swamp-lily, water-dragon

Saxifraga L. (Saxifragaceae)
rock-foil
aizoides L. - yellow alpine saxifrage, yellow mountain saxifrage
micranthidifolia (Haw.) Steud. - lettuce saxifrage, mountain-lettuce
oppositifolia L. - purple alpine saxifrage, purple mountain saxifrage
pensylvanica L. - swamp saxifrage, wild beet
rivularis L. - alpine-brook saxifrage
stolonifera Meerb. - beefsteak-geranium, creeping-sailor, mother-of-thousands, strawberry-begonia, strawberry-geranium
virginiensis Michx. - early saxifrage

Scabiosa L. (Dipsacaceae)
pincushion-flower, scabious
atropurpurea L. - mourning-bride, pincushion, sweet scabious

Scandix L. (Apiaceae)
pecten-veneris L. - shepherd's-needle, Venus's-comb, Venus's-needle

Schedonnardus Steud. (Poaceae)
paniculatus (Nutt.) Trel. - tumble grass

Schefflera J. R. Forst. & G. Forst. (Araliaceae)
actinophylla (Endl.) Jaeger - Australian umbrella-tree, octopus-tree, Queensland umbrella-tree, rubber-tree, star-leaf

Schinus L. (Anacardiaceae)
molle L. - Australian pepper, California pepper-tree, molle, Peruvian mastic-tree, Peruvian pepper-tree, pirul
terebinthifolius Raddi - Brazilian pepper-tree, Christmas-berry-tree

Schismus P. Beauv. (Poaceae)
barbatus (L.) Thell. - Mediterranean grass

Schizachne Hack. (Poaceae)
purpurascens (Torr.) Swallen - false melic

Schizachyrium Nees (Poaceae)
scoparium (Michx.) Nash - blue-stem, broom beard grass, broom wiregrass, bunch grass, little blue-stem, prairie beard grass

Schlumbergera Lem. (Cactaceae)
×*buckleyi* (T. Moore) Tjaden - Christmas cactus
truncata (Haw.) Moran - claw cactus, crab cactus, link-leaf, Thanksgiving cactus, yoke cactus

Schwalbea L. (Scrophulariaceae)
americana L. - chaff-seed

Sciadopitys Siebold & Zucc. (Taxodiaceae)
verticillata (Thunb.) Siebold & Zucc. - Japanese umbrella-pine

Scilla L. (Liliaceae)
squill, wild hyacinth
sibirica Andr. - Siberian squill

Scirpus L. (Cyperaceae)
bulrush
acutus Bigelow var. *occidentalis* (S. Watson) Beetle - hard-stem bulrush, tule
americanus Pers. - American bulrush, chairmaker's rush, three-square
atrovirens Willd. - black bulrush, dark-green bulrush
caespitosus L. - tufted club rush
californicus (C. A. Mey.) Steud. - California bulrush, southern bulrush
congdonii Britton - wool grass
cyperinus (L.) Kunth - wool grass, wool grass bulrush
fluviatilis (Torr.) A. Gray - river bulrush
heterochaetus Chase - slender bulrush
koilolepis (Steud.) Gleason - hollow-scaled bulrush
lacustris L. - great bulrush, mat rush
maritimus L. - bayonet-grass, prairie bayonet-grass
mucronatus L. - rice-field bulrush
pendulus Muhl. - lined bulrush
subterminalis Torr. - swaying rush, water bulrush
validus Vahl - American great bulrush, soft-stem bulrush

Scleranthus L. (Caryophyllaceae)
annuus L. - German knotgrass, knawel

Scleria Bergius (Cyperaceae)
nut rush, razor-sedge, stone-rush
triglomerata Michx. - whip-grass

Sclerochloa P. Beauv. (Poaceae)
dura P. Beauv. - hard grass, tufted hard grass

Scolochloa Link (Poaceae)
festucacea (Willd.) Link - marsh grass, sprangle-top

Scorzonera L. (Asteraceae)
hispanica L. - black oyster-plant, black salsify, Spanish oyster-plant, Spanish salsify, viper's-grass

Scrophularia L. (Scrophulariaceae)
figwort
lanceolata Pursh - figwort
marilandica L. - carpenter's-square, pilewort

Scutellaria L. (Lamiaceae)
skullcap
elliptica Muhl. - hairy skullcap
galericulata L. - marsh skullcap, marsh skullwort
incana Biehler - downy skullcap, hoary skullcap
integrifolia L. - hyssop skullcap
lateriflora L. - blue skullcap, mad-dog skullcap
parvula Michx. - little skullcap, smaller skullcap

Secale L. (Poaceae)
cereale L. - rye
montanum Guss. - mountain rye

Sechium P. Browne (Cucurbitaceae)
edule (Jacq.) Sw. - chayote, choyote, christophine

Sedum L. (Crassulaceae)
orpine, stonecrop
acre L. - gold-moss, golden-carpet, mossy stonecrop
alboroseum Baker - garden orpine
glaucophyllum R. T. Clausen - cliff stonewort
pulchellum Michx. - rock-moss, widow's-cross
rosea (L.) Scop. - rose-root
spurium M. Bieb. - two-row stonecrop
telephium L. - live-forever, live-forever stonecrop, orpine

Selaginella P. Beauv. (Selaginellaceae)
little club-moss, spike-moss
oregana D. C. Eaton - Oregon spike-moss
rupestris (L.) Spring - dwarf lycopod, rock spike-moss
wallacei Hieron. - Wallace's spike-moss
watsonii Underw. - Watson's spike-moss

Sempervivum L. (Crassulaceae)
tectorum L. - hens-and-chickens, house-leek, live-forever

Senecio L. (Asteraceae)
groundsel, ragwort, squaw-weed
anonymus Wood - Appalachian groundsel, Small's groundsel
aureus L. - golden ragwort, heart-leaf groundsel, squaw-weed
chenopodioides Kunth - Mexican flame-vine, orange-glow vine
congestus (R. Br.) DC. - marsh fleabane, northern swamp groundsel
cruentus (Masson ex L'Hér.) DC. - florist's cineraria
douglasii DC. - bush senecio
glabellus Poir. - butterweed, cress-leaf groundsel, yellowtop
jacobaea L. - ragwort, stinking-Willie, tansy ragwort
longilobus Benth. - three-leaf groundsel
mikanioides Otto ex Walp. - Cape ivy, German ivy, parlor-ivy, water-ivy

pauperculus Michx. - balsam groundsel, northern meadow groundsel
plattensis Nutt. - Platte groundsel, prairie groundsel
riddellii Torr. & A. Gray - Riddell's groundsel
sylvaticus L. - woodland groundsel
viscosus L. - sticky groundsel
vulgaris L. - groundsel

Senna Mill. (Fabaceae)
alata (L.) Roxb. - candlestick senna, Christmas-candle, empress-candle-plant, ringworm senna, ringworm-bush, ringworm-shrub
alexandrina Mill. - Alexandrian senna, Indian senna, Tinnevelly senna
covesii (A. Gray) H. S. Irwin & Barneby - Cove's cassia, desert senna
hebecarpa (Fernald) H. S. Irwin & Barneby - wild senna
marilandica (L.) Link - southern wild senna, wild senna
obtusifolia (L.) H. S. Irwin & Barneby - sickle-pod
occidentalis (L.) Link - coffee senna, Negro coffee, styptic-weed
tora (L.) Roxb. - sickle senna, sickle-pod

Sequoia Endl. (Taxodiaceae)
sempervirens (D. Don) Endl. - coastal redwood, redwood

Sequoiadendron Buchholz (Taxodiaceae)
giganteum (Lindl.) Buchholz - big-tree, giant redwood, giant-sequoia

Serenoa Hook. f. (Arecaceae)
repens (W. Bartram) Small - saw palmetto, scrub palmetto

Sesamum L. (Pedaliaceae)
indicum L. - sesame

Sesbania Scop. (Fabaceae)
drummondii (Rydb.) Cory - Drummond's rattle-bush
exaltata (Raf.) Rydb. ex A. W. Hill - Colorado River hemp, pea-tree, sesban
punicea (Cav.) Benth. - rattle-bush, scarlet wisteria-tree
vesicaria (Jacq.) Elliott - bag-pod sesbania

Sesuvium L. (Aizoaceae)
portulacastrum (L.) L. - cenicilla, sea-purslane

Setaria P. Beauv. (Poaceae)
bristle grass
faberi Herrm. - giant foxtail, mutton bluegrass, nodding foxtail
geniculata (Lam.) P. Beauv. - knot-root bristle grass, knot-root fox-tail
grisebachii E. Fourn. - Grisebach's bristle grass
italica (L.) P. Beauv. - Bengal grass, foxtail millet, German millet, Hungarian millet, Italian millet, Japanese millet
macrostachya Kunth - plains bristle grass
magna Griseb. - giant bristle grass, giant foxtail, salt marsh foxtail
pallide-fusca (Schumacher) Stapf & C. E. Hubb. - cat-tail grass
palmifolia (J. König) Stapf - palm grass
pumila (Poir.) Roem. & Schult. - glaucous bristle grass, yellow bristle grass, yellow foxtail
verticillata (L.) P. Beauv. - bristly foxtail, bur bristle grass, pigeon grass
viridis (L.) P. Beauv. - green bristle grass, green foxtail

viridis (L.) P. Beauv. var. *major* (Gaudin) Pospischal - giant green foxtail
viridis (L.) P. Beauv. var. *robusta-alba* Schreib. - robust white foxtail
viridis (L.) P. Beauv. var. *robusta-purpurea* Schreib. - robust purple foxtail

Severinia Ten. & Endl. (Rutaceae)
buxifolia (Poir.) Ten. - Chinese box-orange

Seymeria Pursh (Scrophulariaceae)
cassioides (J. F. Gmel.) Blake - smooth seymeria
pectinata Pursh - sticky seymeria

Shepherdia Nutt. (Elaeagnaceae)
argentea (Pursh) Nutt. - silver buffalo-berry, silverberry
canadensis (L.) Nutt. - rabbit-berry, russet buffalo-berry, soapberry

Sherardia L. (Rubiaceae)
arvensis L. - blue field madder, field madder

Shortia Torr. & A. Gray (Diapensiaceae)
galacifolia Torr. & A. Gray - Oconee bells

Sibara Greene (Brassicaceae)
virginica (L.) Rollins - rock-cress, sibara

Sicana Naudin (Cucurbitaceae)
odorifera (Vell.) Naudin - casabanana, coroa, curua, curubá

Sicyos L. (Cucurbitaceae)
angulatus L. - bur cucumber, star-cucumber

Sida L. (Malvaceae)
acuta Burm. - broom-weed , southern sida
cordifolia L. - heart-leaf sida
hermaphrodita (L.) Rusby - Virginia mallow
rhombifolia L. - arrow-leaf sida, Cuban jute
spinosa L. - prickly mallow, prickly sida

Sidalcea A. Gray (Malvaceae)
checker mallow
malvaeflora (DC.) Benth. - checker mallow, checker-bloom

Sideroxylon L. (Sapotaceae)
foetidissimum Jacq. - false mastic
lanuginosa Michx. - chittamwood, false buckthorn, gum bumelia, shittim-wood
lycioides L. - ironwood, mock-orange, southern buckthorn
tenax L. - tough buckthorn

Silene L. (Caryophyllaceae)
campion, catchfly
acaulis L. - cushion-pink, moss campion
antirrhina L. - sleepy catchfly, tarry cockle
armeria L. - garden catchfly, none-so-pretty, sweet-William catchfly
californica Durand - California Indian pink
caroliniana Walter - wild pink
conica L. - sand catchfly
conoidea L. - cone catchfly
csereii Baumg. - biennial campion, bubble-poppy, campion, cow-bell, white-bottle
dichotoma Ehrh. - forking catchfly, hairy catchfly
dioica (L.) Clairv. - morning campion, red campion, red cockle
gallica L. - English catchfly, French catchfly, windmill pink

latiflora Poir. - evening campion, evening lychnis, white campion, white cockle

nivea (Nutt.) Otth - snowy campion

noctiflora L. - clammy cockle, night-flowering catchfly, snowy cockle, sticky cockle

pendula L. - nodding catchfly

regia Sims - royal catchfly, wild pink

rotundifolia Nutt. - round-leaf catchfly

stellata (L.) W. T. Aiton - starry campion, widow's-frill

virginica L. - fire-pink

vulgaris (Moench) Garcke - bladder campion

Silphium L. (Asteraceae)
rosinweed

integrifolium Michx. - prairie rosinweed, rosinweed, whole-leaf rosinweed

laciniatum L. - compass-plant

perfoliatum L. - cup rosinweed, cup-plant, Indian cup

radula Nutt. - showy rosinweed

terebinthinaceum Jacq. - basal-leaved rosinweed, prairie-dock

trifoliatum L. - whorled rosinweed

Silybum Adans. (Asteraceae)

marianum (L.) Gaertn. - blessed milk-thistle, blessed thistle, holy thistle, milk-thistle, St. Mary's thistle

Simarouba Aubl. (Simaroubaceae)

glauca DC. - aceituna, bitter damson, bitter-wood, paradise-tree

Simmondsia Nutt. (Simmondsiaceae)

chinensis (Link) C. K. Schneid. - deer-nut, goat-nut, jojoba, pig-nut

Sinapis L. (Brassicaceae)

alba L. - white mustard

arvensis L. - California rape, charlock

Sinningia Nees (Gesneriaceae)

speciosa (Lodd.) Hiern - Brazilian gloxinia, florist's gloxinia, violet-slipper gloxinia

Sisymbrium L. (Brassicaceae)

altissimum L. - hedge-mustard, Jim Hill's mustard, tall hedge mustard, tumble mustard

irio L. - London rocket

loeselii L. - tall hedge-mustard

officinale (L.) Scop. - hedge-mustard

orientale L. - Oriental mustard

Sisyrinchium L. (Iridaceae)
blue-eyed-grass

angustifolium Mill. - narrow-leaf blue-eyed-grass

douglasii A. Dietr. - grass-widow, purple-eyed-grass

montanum Greene - Montana blue-eyed-grass

Sium L. (Apiaceae)

suave Walter - water-parsnip

Smilacina Desf. (Liliaceae)

racemosa (L.) Link - false Solomon's-seal, false spikenard, fat-Solomon, Solomon's-zigzag, treacle-berry

stellata (L.) Desf. - star-flowered lily-of-the-valley, starflower, starry false Solomon's-seal

trifolia (L.) Desf. - starry false Solomon's-seal, three-leaf false Solomon's seal

Smilax L. (Smilacaceae)
cat-brier, greenbrier

auriculata Walter - wild bamboo-vine

bona-nox L. - bull-brier, cat-brier, China brier, tramp's-trouble

glauca Walter - cat greenbrier, cat-brier, saw-brier

herbacea L. - carrion-flower

hispida Muhl. - bristly greenbrier, hag-brier, hell-fetter

laurifolia L. - bamboo-vine, blaspheme-vine, laurel-leaved greenbrier

pumila Walter - sarsaparilla-vine

rotundifolia L. - bull-brier, cat-brier, greenbrier, horse-brier, round-leaf greenbrier

smallii Morong - Jackson's brier, lance-leaf greenbrier

tamnoides L. - China-root, hell-fetter

walteri Pursh - coral greenbrier, red-berried greenbrier

Solanum L. (Solanaceae)

aculeatissimum Jacq. - soda-apple nightshade

americanum Mill. - American black nightshade

capsicastrum Link ex Schauer - false Jerusalem cherry

carolinense L. - ball nightshade, Carolina horse-nettle, Carolina nettle, horse-nettle

citrullifolium A. Braun - melon-leaf nightshade

dimidiatum Raf. - robust horse-nettle, Torrey's nightshade

dulcamara L. - bittersweet nightshade, deadly nightshade, poisonous nightshade

elaeagnifolium Cav. - bull-nettle, silver horse-nettle, silver-leaf nightshade, silver-leaf-nettle, trompillo, white-horse-nettle

erianthum D. Don - mullein nightshade

integrifolium Poir. - Chinese scarlet eggplant, fruited eggplant, scarlet eggplant, tomato-fruited eggplant

jamesii Torr. - wild potato

melongena L. - aubergine, eggplant, mad-apple, melongene

nigrum L. - black nightshade, nightshade, poison-berry

pseudocapsicum L. - Jerusalem cherry

rostratum Dunal - buffalo-berry, buffalo-bur

sarrachoides Sendtn. - hairy nightshade

sisymbriifolium Lam. - sticky nightshade

torvum Sw. - terongan, turkey-berry

triflorum Nutt. - cut-leaf nettle, cut-leaf nightshade, three-flowered-nettle, wild tomato

tuberosum L. - Irish potato, potato, white potato

xanti A. Gray - purple nightshade

Soleirolia Gaudich. (Urticaceae)

soleirolii (Req.) Dandy - angel's-tears, baby's-tears, Corsican carpet-plant, Corsican-curse, Irish moss, Japanese moss, mind-your-own-business-plant, peace-in-the-home, Pollyana-vine

Solidago L. (Asteraceae)
goldenrod

bicolor L. - silver-rod, white goldenrod

boottii Hook. - Boott's goldenrod

caesia L. - axillary goldenrod, blue-stem goldenrod, wreath goldenrod

californica Nutt. - California goldenrod

canadensis L. - Canadian goldenrod, tall goldenrod

canadensis L. var. *scabra* (Muhl.) Torr. & A. Gray - tall goldenrod

elliottii Torr. & A. Gray - coastal swamp goldenrod, Elliott's goldenrod

erecta Pursh - erect goldenrod, slender goldenrod
fistulosa Mill. - hollow goldenrod, pine-barren goldenrod
flexicaulis L. - broad-leaf goldenrod, hairy piney-woods
  goldenrod, zigzag goldenrod
gigantea Aiton - late goldenrod
hispida Muhl. ex Willd. - hairy goldenrod
juncea Aiton - early goldenrod
leavenworthii Torr. & A. Gray - Leavenworth's goldenrod
minor (L.) Salisb. - small-headed goldenrod
missouriensis Nutt. - Missouri goldenrod
nemoralis Aiton - gray goldenrod
odora Aiton - licorice goldenrod, sweet goldenrod
ohioensis Riddell - Ohio goldenrod
patula Muhl. ex Willd. - rough-leaf goldenrod
ptarmicoides (Nees) Boivin - upland white aster, white upland
  aster
rigida L. - hard-leaf goldenrod, rigid goldenrod, stiff
  goldenrod
sempervirens L. - seaside goldenrod
speciosa Nutt. - showy goldenrod
uliginosa Nutt. - bog goldenrod, northern bog goldenrod
ulmifolia Muhl. ex Willd. - elm-leaf goldenrod

Soliva Ruiz & Pav. (Asteraceae)
sessilis Ruiz & Pav. - lawn burweed

Sonchus L. (Asteraceae)
  sow-thistle
arvensis L. - creeping sow-thistle, field sow-thistle, gut-weed,
  perennial sow-thistle
arvensis L. spp. *uliginosus* (M. Bieb.) Nyman - marsh sow-
  thistle
asper (L.) Hill - prickly sow-thistle, spiny sow-thistle, spiny-
  leaved sow-thistle
oleraceus L. - annual sow-thistle, sow-thistle

Sophora L. (Fabaceae)
affinis Torr. & A. Gray - Texas sophora
chrysophylla (Salisb.) Seem. - mamane
davidii (Franch.) Skeels - skeels, vetch-leaf sophora
japonica L. - Japanese pagoda-tree, pagoda-tree
nuttalliana B. L. Turner - silky sophora
secundiflora (Ortega) Lag. ex DC. - frijolito, mescal bean,
  Texas mountain-laurel
sericea Nutt. - silky sophora
tomentosa L. - silver-bush

Sorbus L. (Rosaceae)
  mountain-ash
americana Marsh. - American mountain-ash, dogberry,
  missey-moosey, round-wood
aucuparia L. - European mountain-ash, quick-beam, rowan
decora (Sarg.) C. K. Schneid. - showy mountain-ash
domestica L. - service-tree
sitchensis M. Roem. - Pacific mountain-ash

Sorghastrum Nash (Poaceae)
nutans (L.) Nash - Indian grass

Sorghum Moench (Poaceae)
almum Parodi - Columbus grass, sorghum-almum
bicolor (L.) Moench - broom-corn, durra, grain sorghum,
  kaffir, kaffir-corn, milo, shatter-cane, sorghum, sweet
  sorghum
×*drummondii* (Nees ex Steud.) Millsp. & Chase - Sudan grass
  Note = *S. bicolor* × *S. arundinaceum*.

halepense (L.) Pers. - Aleppo grass, Egyptian millet, Johnson
  grass

Sparganium L. (Sparganiaceae)
  bur-reed
americanum Nutt. - three-square bur-reed
androcladum (Engelm.) Morong - branching bur-reed
chlorocarpum Rydb. - green-fruited bur-reed
emersum Rehmann - narrow-leaf bur-reed
erectum L. - branched bur-reed
eurycarpum Engelm. ex A. Gray - broad-fruited bur-reed,
  giant bur-reed
fluctuans (Morong) B. L. Rob. - water bur-reed

Spartina Schreb. (Poaceae)
  cord grass, marsh grass
alterniflora Loisel. - salt marsh cord grass, salt-water cord
  grass, smooth cord grass
cynosuroides (L.) Roth - big cord grass
gracilis Trin. - alkali cord grass
patens (Aiton) Muhl. - salt meadow cord grass
pectinata Link - freshwater cord grass, prairie cord grass,
  slough grass

Spartium L. (Fabaceae)
junceum L. - Spanish broom, weaver's-broom

Spathodea P. Beauv. (Bignoniaceae)
campanulata P. Beauv. - African tulip-tree, flame-of-the-
  forest

Spergula L. (Caryophyllaceae)
  spurry
arvensis L. - corn spurry, starwort, stickwort

Spergularia (Pers.) J. Presl & C. Presl (Caryophyllaceae)
rubra (L.) J. Presl & C. Presl - red-sands spurry

Spermacoce L. (Rubiaceae)
alata Aubl. - broad-leaf button-weed
assurgens Ruiz & Pav. - button-plant
glabra Michx. - button-weed, smooth button-weed

Spermolepis Raf. (Apiaceae)
divaricata (Walter) Britton - spreading scale-seed, western
  scale-seed

Sphaeralcea A. St.-Hil. (Malvaceae)
  false mallow, globe mallow
ambigua A. Gray - apricot mallow, desert hollyhock, desert
  mallow
angustifolia (Cav.) G. Don - narrow-leaf globe mallow
coccinea (Pursh) Rydb. - prairie mallow, red false mallow,
  scarlet globe mallow, scarlet mallow

Sphaerophysa DC. (Fabaceae)
salsula (Pall.) DC. - Austrian field pea, Austrian pea-weed

Sphenoclea Gaertn. (Sphenocleaceae)
zeylandica Gaertn. - goose-weed

Sphenomeris Maxon (Pteridophyta)
chinensis (L.) Maxon - lace fern

Sphenopholis Scribn. (Poaceae)
obtusata (Michx.) Scribn. - prairie wedge grass, wedge grass

Spigelia L. (Loganiaceae)
marilandica L. - Indian pink, pink-root, wormgrass

Spinacia L. (Chenopodiaceae)
oleracea L. - spinach

Spiraea L. (Rosaceae)
bridal-wreath
alba Du Roi - meadowsweet
alba Du Roi var. *latifolia* (Aiton) Dippel - hard-hack,
meadowsweet
douglasii Hook. - hack-brush, hard-hack
japonica L.f. - Japanese spiraea
prunifolia Siebold & Zucc. - bridal-wreath
tomentosa L. - hard-hack, steeplebush
×*vanhouttei* (Briot) Zabel - bridal-wreath
Note = *S. cantoniensis* × *S. trilobata*.

Spiranthes Rich. (Orchidaceae)
lacera Raf. - northern slender lady's-tresses, slender lady's-tresses

Spirodela Schleid. (Lemnaceae)
polyrhiza (L.) Schleid. - giant duckweed, great duckweed,
greater duckweed, water-flaxseed

Spondias L. (Anacardiaceae)
purpurea L. - hog-plum, jocote, purple mombin, red mombin,
Spanish plum

Sporobolus R. Br. (Poaceae)
drop-seed, rush grass
africanus (Poir.) Robyns & Tourn. - African drop-seed
airoides (Torr.) Torr. - alkali sacaton
asper (Michx.) Kunth var. *hookeri* (Trin.) Vasey - meadow
drop-seed
contractus Hitchc. - spike drop-seed
cryptandrus (Torr.) A. Gray - sand drop-seed
domingensis (Trin.) Kunth - coral drop-seed
giganteus Nash - giant drop-seed
heterolepis (A. Gray) A. Gray - northern drop-seed, prairie
drop-seed
junceus (Michx.) Kunth - piney-woods drop-seed
neglectus Nash - annual drop-seed
vaginiflorus (Torr.) A. Wood ex A. Gray - poverty drop-seed,
poverty grass
virginicus (L.) Kunth - seashore drop-seed

Stachys L. (Lamiaceae)
betony, hedge-nettle, woundwort
annua L. - hedge-nettle betony
arvensis L. - field-nettle betony, low hedge-nettle, low nettle
olympica Poir. - woolly hedge-nettle
palustris L. - woundwort
tenuifolia Willd. - smooth hedge-nettle

Stachytarpheta Vahl (Verbenaceae)
jamaicensis (L.) Vahl - Jamaican vervain

Stanleya Nutt. (Brassicaceae)
prince's-plume
pinnata (Pursh) Britton - desert-plume

Staphylea L. (Staphyleaceae)
bladder-nut
bolanderi A. Gray - sierra bladder-nut
trifolia L. - American bladder-nut

Stellaria L. (Caryophyllaceae)
chickweed, starwort

graminea L. - lesser starwort, lesser stitch-wort, little starwort
holostea L. - Easter-bell, Easter-bell starwort, greater stitch-wort
laeta Richardson - Alaska chickweed
longifolia Muhl. ex Willd. - long-leaf chickweed, long-leaf
starwort
longipes Goldie - starwort
media (L.) Vill. - chickweed, stitch-wort
pubera Michx. - great chickweed, star chickweed

Stenanthium (A. Gray) Kunth (Liliaceae)
gramineum (Ker Gawl.) Morong - feather-bells

Stenocarpus R. Br. (Proteaceae)
sinuatus Endl. - fire-wheel-tree, wheel-of-fire

Stenocereus (A. Berger) Riccob. (Cactaceae)
thurberi (Engelm.) Buxb. - organ-pipe cactus

Stenotaphrum Trin. (Poaceae)
secundatum (Walter) Kuntze - St. Augustine's grass

Stephanomeria Nutt. (Asteraceae)
wire-lettuce
pauciflora (Nutt.) A. Nelson - desert wire-lettuce, wire-lettuce
tenuifolia (Torr.) H. Hall - slender wire-lettuce

Sternbergia Waldst. & Kit. (Liliaceae)
lutea (L.) Spreng. - fall daffodil

Stewartia L. (Theaceae)
melachodendron L. - silky stewartia, silky-camellia, Virginia
stewartia
ovata (Cav.) Weath. - mountain camellia, mountain stewartia

Stigmaphyllon A. Juss. (Malpighiaceae)
ciliatum (Lam.) A. Juss. - Amazon-vine, Brazilian golden-vine, butterfly-vine, golden-creeper, orchid-vine

Stillingia Garden ex L. (Euphorbiaceae)
sylvatica L. - queen's-delight, queen's-root

Stipa L. (Poaceae)
feather grass, needle grass, spear grass
avenacea L. - black oat grass, black-seed needle grass
cernua Stebbins & D. Löve - California needle grass
columbiana Macoun - Columbia needle grass
comata Trin. & Rupr. - needle-and-thread, needle-and-thread
grass, western stipa
lemmonii (Vasey) Scribn. - Lemmon's needle grass
lepida Hitchc. - foothill needle grass
lettermanii Vasey - Letterman's needle grass
leucotricha Trin. & Rupr. - Texas needle grass
neomexicana (Thurb.) Scribn. - New Mexico feather grass
occidentalis Thurb. ex S. Watson - western needle grass
pringlei Scribn. - Pringle's needle grass
pulchra Hitchc. - California needle grass, purple needle grass
richardsonii Link - Richardson's needle grass
robusta (Vasey) Scribn. - sleepy grass
scribneri Vasey - Scribner's needle grass
sibirica (L.) Lam. - needle grass
spartea Trin. - auger-seed, porcupine grass, weather grass
speciosa Trin. & Rupr. - desert needle grass
thurberiana Piper - Thurber's needle grass
viridula Trin. - feather bunch grass, green needle grass

Stokesia L'Hér. (Asteraceae)
laevis (Hill) Greene - Stokes's aster

Stratiotes L. (Hydrocharitaceae)
  aloides L. - crab's-claw

Strelitzia Banks ex Dryand. (Strelitziaceae)
  reginae Banks ex Dryand. - bird-of-paradise, crane-flower,
    crane-lily, queen's bird-of-paradise

Streptocarpus Lindl. (Gesneriaceae)
  rexii Hook. - Cape primrose

Streptopus Michx. (Liliaceae)
  twisted-stalk
  amplexifolius (L.) DC. - twisted-stalk, white mandarin
  roseus Michx. - rose mandarin, rose twisted-stalk

Striga Lour. (Scrophulariaceae)
  asiatica (L.) Kuntze - witch-weed
  gesnerioides (Willd.) Vatke - cow pea witch-weed, tobacco
    witch-weed

Strongylodon Vogel (Fabaceae)
  macrobotrys A. Gray - jade-vine

Strophostyles Elliott (Fabaceae)
  helvola (L.) Elliott - amberique bean, annual wild bean,
    trailing wild bean, wild bean
  leiosperma (Torr. & A. Gray) Piper - small-flowered wild
    bean, smooth-seeded wild bean
  umbellata (Muhl. ex Willd.) Britton - perennial wild bean,
    pink wild bean

Stylophorum Nutt. (Papaveraceae)
  diphyllum (Michx.) Nutt. - celandine poppy, wood poppy

Stylosanthes Sw. (Fabaceae)
  erecta P. Beauv. - Nigerian stylo
  guianensis (Aubl.) Sw. - Brazilian lucerne, Brazilian stylo
  hamata (L.) Taub. - Caribbean stylo
  humilis Kunth - Townsville stylo

Styrax L. (Styracaceae)
  snowbell, storax
  americanus Lam. - American snowbell, mock-orange

grandifolus Aiton - big-leaf snowbell

Suaeda Forssk. ex Scop. (Chenopodiaceae)
  calceoliformis (Hook.) Moq. - western seep-weed

Succisa Neck. (Dipsacaceae)
  australis (Wulfen) Rchb. - devil's-bit, southern-scabious

Swietenia Jacq. (Meliaceae)
  mahagoni (L.) Jacq. - Madeira redwood, Spanish mahogany,
    West Indian mahogany

Syagrus Mart. (Arecaceae)
  inajai (Spruce) Becc. - pupunha palm
  romanzoffianum (Cham.) Glassman - queen palm

Symphoricarpos Duhamel (Caprifoliaceae)
  snowberry
  albus (L.) S. F. Blake - snowberry, waxberry
  albus (L.) S. F. Blake var. *laevigatus* (Fernald) S. F. Blake -
    snowberry, waxberry
  ×*chenaultii* Rehder - coral-berry, Indian currant
    Note = Probably a hybrid between *S. microphyllus* and *S.
      orbiculatus*.
  occidentalis Hook. - western snowberry, wolfberry
  orbiculatus Moench - buck-brush, coral-berry, Indian currant

Symphytum L. (Boraginaceae)
  comfrey
  asperum Lepech. - prickly comfrey
  officinale L. - back-wort, boneset, bruise-wort, comfrey,
    healing-herb
  ×*uplandicum* Nyman - blue comfrey, Quaker comfrey,
    Russian comfrey
    Note = *S. asperum* × *S. officinale*.

Symplocarpus Salisb. ex Nutt. (Araceae)
  foetidus (L.) Nutt. - polecat-weed, skunk-cabbage

Symplocos Jacq. (Symplocaceae)
  tinctoria (L.) L'Hér. - horse-sugar, sapphire-berry sweet-leaf,
    sweet-leaf, yellow-wood

Synadenium Boiss. (Euphorbiaceae)
  grantii Hook. f. - American milk-bush

Syngonium Schott (Araceae)
  podophyllum Schott - African evergreen, arrowhead-vine,
    nephthytis

Synsepalum (A. DC.) Baill. (Sapotaceae)
  dulcificum (Schumach. & Thonn.) Daniell - miraculous-berry

Syringa L. (Oleaceae)
  lilac
  amurensis Rupr. - Amur lilac
  ×*chinensis* Willd. - Chinese lilac
    Note = *S.* × *persica* × *S. vulgaris*.
  josikaea J. Jacq. - Hungarian lilac
  ×*persica* L. - Persian lilac
    Note = *S. afghanica* × *S. laciniata*.
  vulgaris L. - lilac

Syzygium Gaertn. (Myrtaceae)
  cumini (L.) Skeels - black-plum, jambolan, jambolan-plum,
    jambool, jambu, Java plum
  jambos (L.) Alston - jambos, Malabar plum, rose-apple
  malaccense (L.) Merr. & L. M. Perry - large-fruit rose-apple,
    Malay apple, pomerac jambos, rose-apple

Tabebuia B. A. Gomes ex DC. (Bignoniaceae)
  trumpet-tree
  argentea Britton - Paraguayan trumpet-tree, silver trumpet-
    tree, tree-of-gold
  rosea (Bertol.) DC. - whitewood

Tabernaemontana L. (Apocynaceae)
  divaricata (L.) R. Br. - Adam's-apple, broad-leaf rose-bay,
    crape-gardenia, crape-jasmine, East Indian rose-bay,
    flowers-of-love, moonbeam, pinwheel-flower, wax-flower

Taeniatherum Nevski (Poaceae)
  caput-medusae (L.) Nevski - Medusa's-head

Taenidia (Torr. & A. Gray) Drude (Apiaceae)
  integerrima (L.) Drude - yellow pimpernel

Tagetes L. (Asteraceae)
  marigold
  erecta L. - African marigold, Aztec marigold, big marigold,
    French marigold
  minuta L. - wild marigold

Talinum Adans. (Portulacaceae)
  paniculatum (Jacq.) Gaertn. - flame-flower, jewels-of-Opar

Tamarix L. (Tamaricaceae)
aphylla (L.) H. Karst. - Athel tamarisk
chinensis Lour. - Chinese tamarisk, five-stamen tamarisk
gallica L. - French tamarisk
hispida Willd. - Kashgar tamarisk
parviflora DC. - small-flowered tamarisk
ramosissima Ledeb. - salt-cedar

Tanacetum L. (Asteraceae)
tansy
parthenium (L.) Sch. Bip. - feverfew
vulgare L. - garden tansy, golden-buttons, tansy

Taraxacum F. H. Wigg. (Asteraccac)
blowballs, dandelion
erythrospermum Andrz. ex Besser - red-seed dandelion
kok-saghyz Rodin - Russian dandelion
officinale F. H. Wigg. - dandelion

Taxodium Rich. (Taxodiaceae)
cypress
distichum (L.) Rich. - bald cypress
distichum (L.) Rich. var. *imbricarium* (Nutt.) H. B. Croom - pond cypress

Taxus L. (Taxaceae)
yew
baccata L. - English yew
brevifolia Nutt. - Pacific yew, western yew
canadensis Marsh. - American yew, Canadian yew, ground-hemlock
cuspidata Siebold & Zucc. - Japanese yew
floridana Nutt. - Florida yew

Tecoma Juss. (Bignoniaceae)
capensis (Thunb.) Lindl. - Cape honeysuckle
stans (L.) Kunth - trumpet-flower, yellow elder, yellow-bells

Tectaria Cav. (Pteridophyta)
heracleifolia (Willd.) Underw. - halberd fern

Tectona L.f. (Verbenaceae)
grandis L.f. - teak

Teesdalia R. Br. (Brassicaceae)
nudicaulis (L.) R. Br. - shepherd's-cress

Tellima R. Br. (Saxifragaceae)
grandiflora (Pursh) Lindl. - fringe-cups

Tephrosia Pers. (Fabaceae)
hoary pea
virginiana (L.) Pers. - cat's-gut, rabbit pea

Teramnus P. Browne (Fabaceae)
labialis (L.f.) Spreng. - blue wiss

Terminalia L. (Combretaceae)
cattapa L. - Indian almond, tropical almond

Tetradymia DC. (Asteraceae)
canescens DC. - spineless horse-brush
glabrata Torr. & A. Gray - little-leaf horse-brush

Tetragonia L. (Aizoaceae)
tetragonioides (Pall.) Kuntze - New Zealand spinach, tsuru-na, Warrigal's cabbage

Tetrapanax (K. Koch) K. Koch (Araliaceae)
papyrifer (Hook.) K. Koch - Chinese rice-paper-plant, rice-paper-plant

Tetrastigma (Miq.) Planch. (Vitaceae)
voinieranum (Pierre ex Nichols & Mottet) Gagnep. - chestnut-vine, lizard-plant

Tetrazygia Rich. ex DC. (Melastomataceae)
bicolor (Mill.) Cogn. - Florida tetrazygia
elaeagnoides (Sw.) DC. - cenizo

Teucrium L. (Lamiaceae)
botrys L. - cut-leaf germander
canadense L. - American germander, wood-sage
scorodonia L. - germander-sage, wood germander, wood-sage

Thalictrum L. (Ranunculaceae)
meadow-rue
alpinum L. - alpine meadow-rue
clavatum DC. - lady-rue, mountain meadow-rue
dasycarpum Fisch. & Avé-Lall. - purple meadow-rue
dioicum L. - early meadow-rue, quick-silver-weed
pubescens Pursh. - king-of-the-meadow, muskrat-weed, tall meadow-rue
revolutum DC. - purple meadow-rue, skunk meadow-rue, wax-leaf meadow-rue
thalictroides (L.) Eames & Boivin - rue anemone

Thaspium Nutt. (Apiaceae)
trifoliatum (L.) A. Gray - meadow parsnip, smooth meadow parsnip

Thelypteris Schmidel (Pteridophyta)
noveboracensis (L.) Nieuwl. - New York fern
palustris Schott - marsh fern, meadow fern, snuffbox fern

Theobroma L. (Sterculiaceae)
cacao L. - cacao, cacao-bean, chocolate tree, cocoa-bean

Thermopsis R. Br. ex Aiton & W. T. Aiton (Fabaceae)
false lupine
mollis (Michx.) M. A. Curtis - bush pea, Piedmont buckbean

Thespesia Sol. ex M. P. Corréa (Malvaceae)
populnea (L.) Sol. - portia-tree

Thevetia L. (Apocynaceae)
peruviana (Pers.) K. Schum. - be-still-tree, lucky-nut, yellow oleander

Thlaspi L. (Brassicaceae)
penny-grass, pennycress
arvense L. - fan-weed, field pennycress, French-weed, mithridate mustard, stinkweed
perfoliatum L. - thoroughwort pennycress

Thuja L. (Cupressaceae)
arborvitae
occidentalis L. - American arborvitae, northern white-cedar
plicata D. Don - giant arborvitae, giant-cedar, western red-cedar

Thujopsis Siebold & Zucc. ex Endl. (Cupressaceae)
dolobrata (L.f.) Siebold & Zucc. - false arborvitae, hiba, hiba-arborvitae

Thymus L. (Lamiaceae)

serpyllum L. - creeping thyme, lemon thyme, mother-of-thyme, wild thyme
vulgaris L. - garden thyme, thyme

Thysanocarpus Hook. (Brassicaceae)
curvipes Hook. - fringe-pod, lace-pod

Tiarella L. (Saxifragaceae)
false miterwort
cordifolia L. - foamflower
unifoliata Hook. - sugar-scoop

Tibouchina Aubl. (Melastomataceae)
urvilleana (DC.) Cogn. - glory-bush, lasiandra, pleroma, princess-flower, purple glory-tree

Tidestromia Standl. (Amaranthaceae)
lanuginosa (Nutt.) Standl. - woolly tidestromia

Tigridia Juss. (Iridaceae)
pavonia (L.f.) DC. - tiger-flower

Tilia L. (Tiliaceae)
americana L. - American basswood, American linden, basswood, whitewood
caroliniana Mill. - basswood, Carolina basswood
cordata Mill. - European linden, little-leaf linden
heterophylla Vent. - white basswood
petiolaris DC. - pendant silver linden, pendant white-lime, weeping-lime
platyphyllos Scop. - big-leaf linden, broad-leaf linden, large-leaf linden

Tillandsia L. (Bromeliaceae)
fasciculata Sw. - wild pineapple
recurvata (L.) L. - ball-moss, bunch-moss
usneoides (L.) L. - gray-beard, Spanish-moss

Tipuana (Benth.) Benth. (Fabaceae)
tipu (Benth.) Kuntze - pride-of-Bolivia, rosewood, tipa, tipu-tree

Tithonia Desf. ex Juss. (Asteraceae)
rotundifolia (Mill.) S. F. Blake - Mexican sunflower

Tolmiea Torr. & A. Gray (Saxifragaceae)
menziesii (Pursh) Torr. & A. Gray - pick-a-back-plant, piggyback-plant, thousand-mothers, youth-on-age

Toona (Endl.) M. Roem. (Meliaceae)
ciliata M. Roem. var. *australis* F. Muell. - Australian cedar
sinensis (A. Juss.) M. Roem. - cedrela

Torilis Adans. (Apiaceae)
arvensis (Huds.) Link - hedge-parsley
japonica (Houtt.) DC. - Japanese hedge-parsley
nodosa (L.) Gaertn. - knotted hedge-parsley

Torreya Arn. (Taxaceae)
californica Torr. - California nutmeg, California torreya
taxifolia Arn. - Florida torreya, stinking-cedar

Toxicodendron Mill. (Anacardiaceae)
pubescens Mill. - eastern poison-oak, hiedra, poison-ivy, poison-oak
radicans (L.) Kuntze - cow-itch, markry, mercury, poison-ivy
rydbergii (Small) Greene - western poison-ivy
vernix (L.) Kuntze - poison sumac, poison-dogwood, poison-elder, swamp sumac, western poison-ivy

Trachelospermum Lem. (Apocynaceae)
difforme (Walter) A. Gray - climbing dogbane
jasminoides Lem. - Confederate-jasmine, star-jasmine

Trachycarpus H. Wendl. (Arecaceae)
fortunei (Hook.) H. Wendl. - Chinese windmill palm, hemp palm, windmill palm

Trachymene Rudge (Apiaceae)
coerulea Graham - blue lace-flower

Tradescantia L. (Commelinaceae)
albiflora Kunth - green wandering-Jew
fluminensis Vell. - wandering-Jew, white-flowered spiderwort
ohiensis Raf. - smooth spiderwort, spiderwort
spathacea Sw. - boat-lily, man-in-a-boat, Moses-in-a-boat, Moses-in-the-bulrushes, Moses-on-a-raft, oyster-plant, purple-leaved spiderwort, three-men-in-a-boat
virginiana L. - spiderwort, Virginia spiderwort, widow's-tears
zebrina hort. ex Bosse - inch-plant, wandering-Jew

Tragopogon L. (Asteraceae)
goat's-beard
dubius Scop. - western salsify, yellow salsify
porrifolius L. - oyster-plant, purple-flowered salsify, salsify, vegetable-oyster
pratensis L. - Jack-go-to-bed-at-noon, John-go-to-bed-at-noon, meadow salsify

Trapa L. (Trapaceae)
natans L. - caltrop, Jesuit-nut, ling, saligot, trapa-nut, water-chestnut, water-nut

Trautvetteria Fisch. & C. A. Mey. (Ranunculaceae)
false bugbane
carolinensis (Walter) Vail - tassel-rue

Trema Lour. (Ulmaceae)
lamarckiana (Roem. & Schult.) Blume - West Indian trema
micrantha (L.) Blume - Florida trema

Trianthema L. (Aizoaceae)
portulacastrum L. - horse-purslane

Tribulus L. (Zygophyllaceae)
caltrop
cistoides L. - Jamaican fever-plant
terrestris L. - bur-nut, puncture-vine

Trichosanthes L. (Cucurbitaceae)
anguina L. - club gourd, serpent-cucumber, snake gourd, viper's-gourd

Trichostema L. (Lamiaceae)
blue-curls
brachiatum L. - false-pennyroyal
dichotomum L. - bastard-pennyroyal
lanceolatum Benth. - vinegar-weed

Tridax L. (Asteraceae)
procumbens L. - coat-buttons

Tridens Roem. & Schult. (Poaceae)
flavus (L.) Hitchc. - purple-top

Trientalis L. (Primulaceae)
chickweed, starflower, wintergreen
borealis Raf. - starflower

Trifolium L. (Fabaceae)
clover, trefoil
alexandrinum L. - berseem clover, Egyptian clover
ambiguum M. Bieb. - Caucasian clover, clover, kura, kura clover
arvense L. - old-field clover, rabbit-foot clover, stone clover
aureum Pollich - hop clover, palmate hop clover
campestre Schreb. - hop clover, large hop clover
carolinianum Michx. - Carolina clover
cherleri L. - cupped clover
dubium Sibth. - cow-hop clover, Irish shamrock, small hop clover, yellow clover
fragiferum L. - strawberry clover
hirtum All. - rose clover
hybridum L. - Alsike clover
incarnatum L. - crimson clover, Italian clover
medium L. - zigzag clover
michelianum Savi - big-flower clover
pratense L. - red clover
pratense L. var. *sativum* Schreb. - cultivated red clover
reflexum L. - annual buffalo clover, buffalo clover
repens L. - ladino clover, white clover
resupinatum L. - Persian clover, reversed clover
semipilosum Fresen. - Kenyan clover, Kenyan wild white clover
stoloniferum Muhl. - running buffalo clover
striatum L. - knotted clover, striate clover, trillium
subterraneum L. - subterranean clover
variegatum Nutt. - white-tip clover
vesiculosum Savi - arrow-leaf clover
wormskioldii Lehm. - seaside clover, sierra clover

Triglochin L. (Juncaginaceae)
arrow grass
maritima L. - seaside arrow grass
palustris L. - marsh arrow grass

Trigonella L. (Fabaceae)
foenum-graecum L. - fenugreek

Trillium L. (Liliaceae)
birthroot, trillium, wake-robin
cernuum L. - nodding trillium
erectum L. - purple trillium, red trillium, squawroot, stinking-Benjamin, wake-robin
flexipes Raf. - bent trillium, drooping trillium
grandiflorum (Michx.) Salisb. - big white trillium, large-flowered trillium, white trillium
nivale Riddell - dwarf white trillium, snow trillium
recurvatum L. C. Beck - prairie trillium
undulatum Willd. - painted trillium

Triodanis Raf. ex Greene (Campanulaceae)
biflora (Ruiz & Pav.) Greene - small Venus's-looking-glass
perfoliata (L.) Nieuwl. - clasping-bellwort, Venus's-looking-glass

Triosteum L. (Caprifoliaceae)
feverwort, horse-gentian
aurantiacum E. P. Bicknell - wild coffee
perfoliatum L. - feverwort, perfoliate horse-gentian, tinker's-weed, wild coffee

Triphasia Lour. (Rutaceae)
trifolia (Burm.) P. Wilson - lime-berry

Triplasis P. Beauv. (Poaceae)
purpurea (Walter) Chapm. - purple sand grass, sand grass

Tripogandra Raf. (Commelinaceae)
multiflora (Sw.) Raf. - fern-leaf inch-plant

Tripsacum L. (Poaceae)
dactyloides (L.) L. - eastern gama grass
floridanum Porter ex Vasey - Florida gama grass

Trisetum Pers. (Poaceae)
canescens Buckley - tall trisetum
cernuum Trin. - nodding trisetum
flavescens (L.) P. Beauv. - golden oat grass, yellow oat grass
pensylvanicum (L.) P. Beauv. - swamp-oats
spicatum (L.) Richt. - spike trisetum
wolfii Vasey - Wolf's trisetum

Triteleia Douglas ex Lindl. (Liliaceae)
ixioides (S. Watson) Greene - golden-brodiaea, pretty-face
laxa Benth. - grass-nut, triplet-lily

Triticum L. (Poaceae)
wheat
aestivum L. - bread wheat, wheat
dicoccoides (Kornh. ex Asch. & Graebn.) Aarons. - wild emmer
dicoccon Schrank - emmer
durum Desf. - durum wheat
monococcum L. - einkorn
polonicum L. - Polish wheat
spelta L. - spelt
turgidum L. - Alaska wheat, cone wheat, English wheat, Mediterranean wheat, poulard wheat, rivet wheat

Triumfetta L. (Tiliaceae)
semitriloba (L.) Jacq. - Sacramento bur

Trollius L. (Ranunculaceae)
globeflower
laxus Salisb. - spreading globeflower

Tropaeolum L. (Tropaeolaceae)
majus L. - garden nasturtium, Indian cress, tall nasturtium

Tsuga Carrière (Pinaceae)
hemlock, hemlock-spruce
canadensis (L.) Carrière - Canadian hemlock, eastern hemlock
caroliniana Engelm. - Carolina hemlock
diversifolia (Maxim.) Mast. - northern Japanese hemlock
heterophylla (Raf.) Sarg. - western hemlock
mertensiana (Bong.) Carrière - mountain hemlock

Tulbaghia L. (Liliaceae)
violacea Harv. - society-garlic

Turnera L. (Turneraceae)
ulmifolia L. - yellow-alder

Tussilago L. (Asteraceae)
farfara L. - clay-weed, colt's-foot, cough-wort, dove-dock, ginger-root, horse's-hoof

Typha L. (Typhaceae)
angustifolia L. - narrow-leaf cat-tail, small-bulrush, soft-flag
domingensis Pers. - southern cat-tail
glauca Godr. - blue cat-tail
latifolia L. - broad-leaf cat-tail, bull-rush, cat-tail, Cossack asparagus, nail-rod

Ulex L. (Fabaceae)
europaeus L. - furze, gorse

Ulmus L. (Ulmaceae)
elm
alata Michx. - wahoo elm, winged elm
americana L. - American elm, water elm, white elm
americana L. cv. 'moline' - Moline's elm
crassifolia Nutt. - cedar elm
glabra Huds. - Scotch elm, Wych's elm
japonica (Rehder) Sarg. - Japanese elm
minor Mill. - English elm, smooth-leaved elm
parvifolia Jacq. - Chinese elm, leather-leaf elm
pumila L. - Chinese elm, dwarf elm, Siberian elm
rubra Muhl. - red elm, slippery elm
serotina Sarg. - red elm, September elm
thomasii Sarg. - cork elm, rock elm

Umbellularia (Nees) Nutt. (Lauraceae)
californica (Hook. & Arn.) Nutt. - California bay, California laurel, California olive, Oregon myrtle, pepper-wood

Ungnadia Endl. (Sapindaceae)
speciosa Endl. - Mexican buckeye

Uniola L. (Poaceae)
paniculata L. - sea-oats

Urena L. (Malvaceae)
lobata L. - aramina, bur mallow, cadillo, caesar-weed, Congo jute

Urochloa P. Beauv. (Poaceae)
panicoides P. Beauv. - liver-seed grass
reptans (L.) Stapf - sprawling-panicum

Urtica L. (Urticaceae)
nettle
chamaedryoides Pursh - southern nettle
dioica L. - European nettle, slender nettle, stinging nettle, tall nettle
dioica L. subsp. *gracilis* (Aiton) Selander - Lyall's nettle
urens L. - burning nettle, dog nettle, dwarf nettle

Utricularia L. (Lentibulariaceae)
biflora Lam. - tangled bladderwort
floridana Nash - Florida bladderwort
foliosa L. - leafy bladderwort
gibba L. - slender-stem bladderwort
inflata Walter - floating bladderwort, inflated bladderwort
olivacea Wright ex Griseb. - minute bladderwort
purpurea Walter - purple bladderwort, spotted bladderwort
radiata Small - floating bladderwort, swollen bladderwort
vulgaris L. - bladderwort

Uvularia L. (Liliaceae)
bellwort, merry-bells
grandiflora Sm. - large-flowered bellwort
perfoliata L. - perfoliate bellwort
sessilifolia L. - little bellwort, sessile bellwort, wild oats

Vaccaria Wolf (Caryophyllaceae)
hispanica (Mill.) Rauschert - cow cockle, cow soapwort, cow-herb, pink cockle, spring cockle

Vaccinium L. (Ericaceae)
bilberry, blueberry, cranberry, huckleberry
angustifolium Aiton - late sweet blueberry, low blueberry, low sweet blueberry, low-bush blueberry, sweet-hurts
arboreum Marsh. - farkleberry, sparkleberry
ashei Reade - rabbit's-eye blueberry
atrococcum (A. Gray) A. Heller - black high-bush blueberry, black huckleberry, downy swamp huckleberry
caesariense Mack. - high-bush blueberry, New Jersey blueberry
caesium Greene - deer-berry, squaw huckleberry
caespitosum Michx. - dwarf bilberry
corymbosum L. - high-bush blueberry, swamp blueberry, whortleberry
crassifolium Andr. - creeping blueberry
elliottii Chapm. - Elliott's blueberry, southern high-bush blueberry
erythrocarpum Michx. - bearberry, dingle-berry, mountain cranberry
macrocarpon Aiton - American cranberry, cultivated cranberry, large cranberry
membranaceum Douglas - blue huckleberry, mountain bilberry, mountain blueberry, thin-leaf huckleberry
myrtilloides Michx. - sour-top blueberry, velvet-leaf blueberry
myrtillus L. - bilberry, whin-berry, whortleberry
occidentalis A. Gray - western blueberry
ovalifolium Sm. - mathers, tall bilberry
ovatum Pursh - California huckleberry, evergreen huckleberry, shot huckleberry
oxycoccos L. - small cranberry
parvifolium Sm. - red huckleberry
scoparium Leiberg ex Crép. - grouseberry, little-leaf huckleberry
stamineum L. - deer-berry, squaw huckleberry
uliginosum L. - alpine blueberry, bog bilberry, moor-berry
vacillans Torr. - early sweet bilberry, hillside blueberry, low bilberry, low blueberry, sweet blueberry
vitis-idaea L. - cowberry, fox-berry, lingonberry, mountain cranberry, partridgeberry
vitis-idaea L. var. *minus* Lodd. - ling-berry, lingon, lingonberry, mountain cranberry, rock cranberry

Valeriana L. (Valerianaceae)
edulis Nutt. ex Torrey & A. Gray - edible valerian, tap-rooted valerian, tobacco-root
officinalis L. - garden-heliotrope, valerian

Valerianella Mill. (Valerianaceae)
corn-salad, lamb's-lettuce
eriocarpa Desv. - Italian corn-salad
locusta (L.) Betcke - corn-salad, fetticus, lamb's-lettuce, milk-grass
radiata (L.) Dufr. - beaked corn-salad

Vallisneria L. (Hydrocharitaceae)
americana Michx. - American eelgrass, water-celery
spiralis L. - tape-weed

Vancouveria C. Morren & Decne. (Berberidaceae)
planipetala Calloni - inside-out-flower, redwood-ivy

Vanilla Mill. (Orchidaceae)
planifolia Andr. - vanilla

Vauquelinia M. P. Corréa ex Humb. & Bonpl. (Rosaceae)
californica (Torr.) Sarg. - Torrey's vauquelinia

Veitchia H. Wendl. (Arecaceae)
merrillii (Becc.) H. E. Moore - Christmas palm, Manila palm, Merrill's palm

Ventenata Koeler (Poaceae)
dubia (Leers) Coss. - ventenata

Veratrum L. (Liliaceae)
false hellebore
californicum Durand - California false hellebore, western false hellebore
viride Aiton - Indian poke, itch-weed, white hellebore

Verbascum L. (Scrophulariaceae)
mullein
blattaria L. - moth mullein
blattaria L. var. *albiflora* (D. Don) House - white moth mullein
lychnitis L. - white mullein
phlomoides L. - clasping mullein, mullein
thapsus L. - flannel-plant, mullein, velvet-plant, woolly mullein
virgatum Stokes - purple-stamen mullein, wand mullein

Verbena L. (Verbenaceae)
bipinnatifida Nutt. - Dakota vervain
bonariensis L. - tall vervain
bracteata Lag. & Rodr. - bracted vervain, prostrate vervain, wild verbena
brasiliensis Vell. - Brazilian vervain
canadensis (L.) Britton - clump verbena, creeping vervain, rose verbena, rose vervain
hastata L. - blue verbena, blue vervain, simpler's-joy
×*hybrida* Voss - garden verbena
Note = Probably a hybrid of *V. peruviana* and other species.
litoralis Kunth - seashore vervain
menthifolia Benth. - mint vervain
officinalis L. - European vervain
simplex Lehm. - narrow-leaf vervain
stricta Vent. - hoary vervain, woolly verbena
tenuisecta Briq. - moss vervain
urticifolia L. - white verbena, white vervain

Verbesina L. (Asteraceae)
crown-beard
alternifolia (L.) Britton ex Kearney - wing-stem, yellow-ironweed
encelioides (Cav.) Benth. & Hook. f. ex A. Gray - butter-daisy, golden-crown daisy
occidentalis (L.) Walter - crown-beard, southern flat-seed-sunflower
virginica L. - frostweed, tick-weed

Vernicia Lour. (Euphorbiaceae)
fordii (Hemsl.) Airy Shaw - China-wood oil-tree, tung oil-tree
montana Lour. - mu, mu-oil-tree, mu-tree, tung

Vernonia Schreb. (Asteraceae)
ironweed
baldwinii Torr. - western ironweed
cinerea (L.) Less. - little ironweed
gigantea (Walter) Trel. ex F. V. Branner & Coville - tall ironweed
noveboracensis (L.) Michx. - ironweed, New York ironweed

Veronica L. (Scrophulariaceae)
brooklime, speedwell
agrestis L. - field speedwell
americana (Raf.) Schwein. - American brooklime
anagallis-aquatica L. - water speedwell
arvensis L. - corn speedwell
chamaedrys L. - germander speedwell
filiformis Sm. - creeping speedwell, slender speedwell
hederifolia L. - ivy-leaf speedwell
longifolia L. - beach speedwell, long-leaf speedwell
officinalis L. - gypsy-weed, speedwell
peregrina L. - neck-weed, purslane speedwell
peregrina L. subsp. *xalapensis* (Kunth) Pennell - purslane speedwell, western purslane speedwell
persica Poir. - bird's-eye, Persian purslane, Persian speedwell, winter speedwell
polita Fr. - wayside speedwell
scutellata L. - marsh speedwell
serpyllifolia L. subsp. *humifusa* (Dicks.) Syme - creeping speedwell, thyme-leaved speedwell
spicata L. - spike speedwell

Veronicastrum Heist. ex Fabr. (Scrophulariaceae)
virginicum (L.) Farw. - black-root, bowman's-root, Culver's-root

Vetiveria Bory (Poaceae)
zizanioides (L.) Nash - khus-khus, vetiver

Viburnum L. (Caprifoliaceae)
arrow-wood
acerifolium L. - arrow-wood, dockmackie, flowering-maple, maple-leaf viburnum, possum-haw
alnifolium Marsh. - American wayfaring-tree, devil's-shoestring, dog's-hobble, dogberry, hobblebush, moosewood, tangel-foot, White Mountains dogwood, witch's-hobble
cassinoides L. - Appalachian tea, swamp haw, teaberry, wild raisin, withe-rod
dentatum L. - arrow-wood, arrow-wood viburnum
dilatatum Thunb. - Linden's viburnum
edule (Michx.) Raf. - moose-berry, squash-berry
lantana L. - twist-wood, wayfaring-tree
lentago L. - black-haw, cowberry, nannyberry, sheepberry, sweet viburnum, tea-plant
nudum L. - black-alder, naked viburnum, possum-haw, smooth withe-rod, swamp-haw
obovatum Walter - Walter's viburnum
odoratissimum Ker Gawl. - sweet viburnum
opulus L. - European cranberry-bush, guelder-rose, snowball-bush, whitten-tree
opulus L. var. *roseum* L. - snowball-bush
plicatum Thunb. var. *tomentosum* Miq. - double-file viburnum

prunifolium L. - black-haw, black-haw viburnum, nannyberry, sheepberry, stag-bush, sweet-haw
rafinesquianum Schult. - downy arrow-wood, downy-leaf arrow-wood
recognitum Fernald - smooth arrow-wood
rufidulum Raf. - rusty black-haw, southern black-haw
tinus L. - laurustinus
trilobum Marsh. - cramp-bark, cranberry-bush, grouseberry, high-bush cranberry, pimbina, summer-berry

Vicia L. (Fabaceae)
tare, vetch
acutifolia Elliott - sand vetch
americana Willd. - American vetch, purple vetch
benghalensis L. - purple vetch
caroliniana Walter - Carolina vetch, pale vetch, wood vetch
cracca L. - bird vetch, boreal vetch, Canadian pea, cow vetch, tufted vetch
faba L. - broad bean, fava bean, horse bean, Windsor bean
gigantea Hook. - large vetch, Sitka vetch
hirsuta (L.) S. F. Gray - tiny vetch
pannonica Crantz - Hungarian vetch
sativa L. - spring-vetch, tare, vetch
sativa L. subsp. *nigra* (L.) Ehrh. - black-pod vetch, vetch
sepium L. - hedge vetch
tenuifolia Roth - bramble vetch
tetrasperma (L.) Schreb. - four-seeded vetch, lentil tare, slender vetch, smooth tare
villosa Roth - hairy vetch, large Russian vetch, winter vetch
villosa Roth subsp. *varia* (Host) Corb. - winter vetch

Victoria Lindl. (Nymphaeaceae)
giant water-lily, water-platter
amazonica (Poepp.) Sowerby - Amazon water-lily, Amazon water-platter, royal water-lily, water-maize

Vigna Savi (Fabaceae)
angularis (Willd.) Ohwi & H. Ohashi - azuki bean
mungo (L.) Hepper - black gram, urd
radiata (L.) R. Wilczek - golden gram, green gram, mung bean
unguiculata (L.) Walp. - black-eyed pea, cow pea, crowder pea, southern pea, yard-long bean
unguiculata (L.) Walp. subsp. *cylindrica* (L.) Verdc. - cat-jang, Jerusalem pea, marble pea
unguiculata (L.) Walp. subsp. *sesquipedalis* (L.) Verdc. - asparagus bean, yard-long bean

Viguiera Kunth (Asteraceae)
annua (Jones) S. F. Blake - goldeneye
stenoloba S. F. Blake - resin-bush

Vinca L. (Apocynaceae)
ground-myrtle, periwinkle
major L. - band-plant, big periwinkle, big-leaf periwinkle, blue-buttons, greater periwinkle
minor L. - lesser periwinkle, periwinkle, running-myrtle

Vincetoxicum Wolf (Asclepiadaceae)
hirundinaria Medik. - swallow-wort, white swallow-wort

Viola L. (Violaceae)
violet
adunca Sm. - hook-spur violet, western dog violet
affinis Leconte - Leconte's violet
arvensis Js. Murray - European field pansy, field violet

blanda Willd. - sweet white violet
canadensis L. - Canadian violet, tall white violet
conspersa Rchb. - American dog violet
cornuta L. - bedding pansy, horned violet, tufted pansy, viola
cucullata Aiton - blue marsh violet, marsh blue violet
glabella Nutt. - stream violet
hastata Michx. - halberd-leaf violet, spear-leaf violet
incognita Brainerd - large-leaf white violet
lanceolata L. - lance-leaf violet, lance-leaf violet, strap-leaf violet, water violet
langsdorfii Fisch. ex Ging. - Alaska violet
lobata Benth. - pine violet, yellow wood violet
macloskeyi F. E. Lloyd - western sweet white violet
macloskeyi F. E. Lloyd var. *pallens* (Banks) C. L. Hitchc. - northern white violet
missouriensis Greene - Missouri violet
nuttallii Pursh - yellow prairie violet
ocellata Torr. & A. Gray - two-eyed violet, western heart's-ease
odorata L. - English violet, florist's violet, sweet violet
orbiculata Geyer ex Holz. - western round-leaf violet
palmata L. - early blue violet, wood violet
palmata L. var. *pedatifida* (G. Don) Cronquist - larkspur violet, purple prairie violet
palustris L. - alpine marsh violet, marsh violet
papilionacea Pursh - blue violet
pedata L. - bird's-foot violet
primulifolia L. - primrose-leaf violet
pubescens Aiton - downy yellow violet, smooth yellow violet
rafinesquii Greene - field pansy
renifolia A. Gray - kidney-leaf violet, northern white violet
rostrata Pursh - long-spurred violet
rotundifolia Michx. - early yellow violet, round-leaf violet, round-leaf yellow violet
sagittata Aiton - arrow-leaf violet, arrowhead violet
selkirkii Pursh - great-spurred violet
sororia Willd. - woolly blue violet
sororia Willd. subsp. *affinis* (Leconte) R. J. Little - northern bog violet
striata Aiton - cream violet, creamy violet, pale violet, striped violet
tricolor L. - European wild pansy, field pansy, heart's-ease, Johnny-jump-up, miniature pansy, pansy, wild violet

Vitex L. (Verbenaceae)
agnus-castus L. - chaste-tree, hemp-tree, Indian spice, lilac chaste-tree, monk's-pepper-tree, sage-tree, wild pepper

Vitis L. (Vitaceae)
grape
acerifolia Raf. - bush grape
aestivalis Michx. - bunch grape, pigeon grape, summer grape
aestivalis Michx. var. *argentifolia* (Munson) Fernald - silver-leaf grape, summer grape
arizonica Engelm. - canyon grape
baileyana Munson - possum grape
berlandieri Planch. - Spanish grape, winter grape
californica Benth. - California wild grape, western wild grape
cinerea Engelm. - gray-black grape, pigeon grape
labrusca L. - American grape, fox grape, skunk grape, stunt grape
lincecumii Buckley - post-oak grape
munsoniana Planch. - bird grape
mustangensis Buckley - mustang grape

novae-angliae Fernald - New England grape
palmata Vahl - cat grape, catbird grape, red grape
riparia Michx. - frost grape, river bank grape
rotundifolia Michx. - bullace grape, muscadine grape,
scuppernong grape, southern fox grape
rupestris Scheele - bush grape, mountain grape, rock grape,
sand grape, sugar grape
vinifera L. - cultivated grape, European grape, wine grape
vulpina L. - chicken grape, frost grape, winter grape

Vulpia C. C. Gmel. (Poaceae)
bromoides (L.) A. Gray - rat's-tail, squirrel-tail fescue
microstachys (Nutt.) Benth. var. *ciliata* (Beal) Lonard &
Gould - gray fescue
microstachys (Nutt.) Benth. var. *pauciflora* (Beal) Lonard &
Gould - Pacific fescue
myuros (L.) C. C. Gmel. - foxtail fescue, rat-tail fescue
myuros (L.) C. C. Gmel. var. *hirsuta* Hack. - foxtail fescue

Wahlenbergia Schrad. ex Roth (Campanulaceae)
marginata (Thunb.) DC. - Asiatic bellflower

Waldsteinia Willd. (Rosaceae)
fragarioides (Michx.) Tratt. - barren-strawberry

Waltheria L. (Sterculiaceae)
indica L. - Indian waltheria

Washingtonia H. Wendl. (Arecaceae)
Washington palm
filifera H. Wendl. - California fan palm, California palm,
California Washington palm, desert fan palm, petticoat palm
robusta H. Wendl. - desert palm, Mexican Washington palm,
thread palm

Wedelia Jacq. (Asteraceae)
trilobata (L.) Hitchc. - creeping-Charlie, porch-vine, trailing
wedelia, wedelia

Weigela Thunb. (Caprifoliaceae)
florida (Bunge) A. DC. - pink weigela

Whipplea Torr. (Hydrangeaceae)
modesta Torr. - modesty, yerba-de-selva

Wisteria Nutt. (Fabaceae)
floribunda (Willd.) DC. - Japanese wisteria
sinensis (Sims) Sweet - Chinese wisteria

Wolffia Horkel ex Schleid. (Lemnaceae)
columbiana H. Karst. - water-meal
papulifera C. H. Thomps. - pimpled water-meal
punctata Griseb. - spotted water-meal

Wolffiella (Hegelm.) Hegelm. (Lemnaceae)
floridana (J. D. Sm. ex Hegelm.) C. H. Thomps. - flat bog-
mat
gladiata Hegelm. - bog-mat

Woodsia R. Br. (Pteridophyta)
glabella R. Br. - smooth cliff fern, smooth woodsia
oregana D. C. Eaton - Oregon woodsia, western cliff fern
scopulina D. C. Eaton - mountain cliff fern, Rocky Mountain
woodsia

Woodwardia Sm. (Pteridophyta)
chain fern
areolata (L.) T. Moore - nettle chain fern
fimbriata Sm. - giant chain fern

radicans (L.) Sm. - European chain fern
virginica (L.) Sm. - Virginia chain fern

Xanthium L. (Asteraceae)
spinosum L. - spiny cocklebur
strumarium L. - broad-leaf cocklebur, cocklebur, heart-leaf
cocklebur
strumarium L. var. *canadense* (Mill.) Torr. & A. Gray -
cocklebur

Xanthorhiza Marsh. (Ranunculaceae)
simplicissima Marsh. - shrub yellow-root

Xanthosoma Schott (Araceae)
sagittifolium (L.) Schott - malanga, ocumo, tannia, tanyah,
yautia

Xerophyllum Michx. (Liliaceae)
asphodeloides (L.) Nutt. - turkey-beard
tenax (Pursh) Nutt. - bear-grass, squaw-grass

Ximenia L. (Olacaceae)
americana L. - hog-plum, tallow-wood

Xyris L. (Xyridaceae)
yellow-eyed-grass
caroliniana Walter - yellow-eyed-grass

Youngia Cass. (Asteraceae)
japonica (L.) DC. - Asiatic hawk's-beard

Yucca L. (Agavaceae)
aloifolia L. - aloe yucca, dagger-plant, Spanish bayonet
baccata Torr. - banana yucca, blue yucca, datil, Spanish
bayonet
brevifolia Engelm. - Joshua-tree
elata Engelm. - palmella, soap-tree, soap-tree yucca, soap-
weed
elephantipes Regel - izote, spineless yucca
filamentosa L. - Adam's-needle, bear-grass, needle-palm, silk-
grass, spoon-leaf yucca
glauca Nutt. ex J. Fraser - soap-weed, soap-well
gloriosa L. - Lord's-candlestick, palm-lily, Roman-candle,
Spanish-dagger
rupicola Scheele - twisted-leaf yucca
schottii Engelm. - Schott's yucca
whipplei Torr. - Our-Lord's-candle

Zamia L. (Zamiaceae)
pumila L. - comptie, coontie, Florida arrowroot, sago cycas,
Seminole bread

Zannichellia L. (Zannichelliaceae)
palustris L. - horned-pondweed

Zantedeschia Spreng. (Araceae)
calla-lily
aethiopica (L.) Spreng. - arum-lily, calla-lily, florist's calla,
garden calla, pig-lily, trumpet-lily, white calla-lily
elliottiana (W. Watson) Engl. - golden calla-lily, yellow calla-
lily
rehmannii Engl. - pink calla-lily, red calla-lily

Zanthoxylum L. (Rutaceae)
americanum Mill. - northern prickly-ash, prickly-ash,
toothache-tree
clava-herculis L. - Hercules's-club, pepper-wood, sea-ash,
southern prickly-ash

fagara (L.) Sarg. - lime prickly-ash, wild lime
hirsutum Buckley - Texas Hercules's-club

Zea L. (Poaceae)
mays L. - Dent's corn, field corn, flint corn, maize, pod corn, popcorn, sweet corn, volunteer corn
mays L. subsp. *mexicana* (Schrad.) Iltis - teosinte
perennis (Hitchc.) Reeves & Mangelsd. - perennial teosinte

Zelkova Spach (Ulmaceae)
carpinifolia (Pall.) K. Koch - Caucasian elm, elm-leaf zelkova
serrata (Thunb.) Makino - Japanese zelkova, saw-leaf zelkova

Zephyranthes Herb. (Liliaceae)
atamasco (L.) Herb. - atamasco-lily
candida (Lindl.) Herb. - zephyr-lily

Zigadenus Michx. (Liliaceae)
densus (Desr.) Fernald - black snakeroot, crow-poison
elegans Pursh - alkali-grass, mountain death camass, white camass
nuttallii (A. Gray) S. Watson - death camass, Nuttall's death camass, poison camass
paniculatus (Nutt.) S. Watson - foothill death camass
venenosus S. Watson - alkali-grass, meadow death camass, soap-plant

Zingiber Boehm. (Zingiberaceae)
ginger
officinale Roscoe - Canton ginger, ginger, true ginger
zerumbet (L.) Sm. - zerumbet ginger

Zinnia L. (Asteraceae)
violacea Cav. - zinnia

Zizania L. (Poaceae)
water-oats, wild rice
aquatica L. - annual wild rice, eastern wild rice, Indian rice, wild rice
latifolia (Griseb.) Turcz. ex Stapf - Manchurian wild rice
palustris L. - cultivated northern wild rice

Zizaniopsis Döll & Asch. (Poaceae)
miliacea (Michx.) Döll & Asch. - giant cut grass, southern wild rice

Zizia W. Koch (Apiaceae)
aurea (L.) W. Koch - golden alexanders, meadow-parsnip

Ziziphus Mill. (Rhamnaceae)
jujuba Mill. - Chinese date, Chinese jujube, jujube
mauritiana Lam. - cottony jujube, Indian jujube

Zostera L. (Zosteraceae)
marina L. - eelgrass

Zoysia Willd. (Poaceae)
japonica Steud. - Japanese lawn grass, Korean grass, Korean lawn grass, zoysia
matrella (L.) Merr. - flawn, Japanese carpet grass, Manila grass, zoysia grass
tenuifolia Willd. ex Trin. - Mascarene grass

Zygophyllum L. (Zygophyllaceae)
fabago L. - bean-caper, Syrian bean-caper

Common Names

Aaron's-beard - *Hypericum calycinum*
abele - *Populus alba*
abelia, glossy abelia - *Abelia grandiflora*
absinth wormwood - *Artemisia absinthium*
absinthe - *Artemisia absinthium*
absinthium - *Artemisia absinthium*
Abyssinian tea - *Catha edulis*
acacia
   Berlandier's acacia - *Acacia berlandieri*
   black-brush acacia - *Acacia rigidula*
   black-wood acacia - *Acacia melanoxylon*
   cat-claw acacia - *Acacia greggii*
   cultivated acacia - *Acacia cyclopis*
   false acacia - *Robinia pseudoacacia*
   mescat acacia - *Acacia constricta*
   prairie acacia - *Acacia angustissima*
   rose-acacia - *Robinia hispida, R. viscosa*
   sweet acacia - *Acacia farnesiana*
   twisted acacia - *Acacia tortuosa*
   white-thorn acacia - *Acacia constricta*
aceituna - *Simarouba glauca*
acerola - *Malpighia glabra*
achiote - *Bixa orellana*
achira - *Canna indica*
aconite - *Aconitum napellus, Aconitum*
acre, staves-acre - *Delphinium menziesii*
Adam-and-Eve - *Aplectrum hyemale*
Adam's
   Adam's-apple - *Tabernaemontana divaricata*
   Adam's-needle - *Yucca filamentosa*
adder's
   adder's-tongue - *Erythronium*
   adder's-tongue fern - *Ophioglossum*
   northern adder's-tongue fern - *Ophioglossum pusillum*
   yellow adder's-tongue - *Erythronium americanum*
adlay
   adlay - *Coix lacryma-jobi*
   adlay millet - *Coix lacryma-jobi*
adonis
   fall adonis - *Adonis annua*
   pheasant-eye adonis - *Adonis annua*
   spring adonis - *Adonis vernalis*
aerial yam - *Dioscorea bulbifera*
African
   African Bermuda grass - *Cynodon transvaalensis*
   African blood-lily - *Haemanthus coccineus*
   African boxwood - *Myrsine africana*
   African daisy - *Arctotis, Gerbera jamesonii, Osteospermum ecklonis*
   African drop-seed - *Sporobolus africanus*
   African evergreen - *Syngonium podophyllum*
   African iris - *Dietes iridioides*
   African lily - *Agapanthus africanus*
   African marigold - *Tagetes erecta*
   African mile-a-minute - *Mikania cordata*
   African millet - *Eleusine coracana*
   African mustard - *Brassica tournefortii*
   African nutmeg - *Monodora myristica*
   African rue - *Peganum harmala*
   African tulip-tree - *Spathodea campanulata*
   African violet - *Saintpaulia ionantha*
   blue-eyed African daisy - *Arctotis stoechadifolia*
agapanthus, blue agapanthus - *Agapanthus africanus*
age, youth-on-age - *Tolmiea menziesii*
ageratum
   ageratum - *Ageratum houstonianum*
   hardy ageratum - *Conoclinium coelestinum*
   tropic ageratum - *Ageratum conyzoides*
agrimony
   agrimony - *Agrimonia eupatoria, A. gryposepala, Agrimonia*
   medicinal agrimony - *Agrimonia eupatoria*
   roadside agrimony - *Agrimonia striata*
ague
   ague-root - *Aletris farinosa*
   ague-weed - *Eupatorium perfoliatum, Gentianella quinquefolia*
ailanthus - *Ailanthus altissima*
Ajo oak - *Quercus ajoensis*
akebia, five-leaf akebia - *Akebia quinata*
akee - *Blighia sapida*
akulikuli - *Lampranthus glomeratus*
Alabama supple-jack - *Berchemia scandens*
Alamo cottonwood - *Populus fremontii*
Alaska
   Alaska bog willow - *Salix fuscescens*
   Alaska cedar - *Chamaecyparis nootkatensis*
   Alaska chickweed - *Stellaria laeta*
   Alaska onion grass - *Melica subulata*
   Alaska violet - *Viola langsdorfii*
   Alaska wheat - *Triticum turgidum*
   Alaska yellow-cedar - *Chamaecyparis nootkatensis*
albizia, silk-tree albizia - *Albizia julibrissin*
alder
   alder - *Alnus*
   alder buckthorn - *Rhamnus alnifolia, R. frangula*
   alder-leaf buckthorn - *Rhamnus alnifolia*
   alder-leaf pepper-bush - *Clethra alnifolia*
   American green alder - *Alnus viridis* subsp. *crispa*
   Arizona alder - *Alnus oblongifolia*
   black alder - *Alnus glutinosa*
   brook alder - *Alnus maritima*
   coast white-alder - *Clethra alnifolia*
   European alder - *Alnus glutinosa*
   European alder buckthorn - *Rhamnus frangula*
   European green alder - *Alnus viridis*
   gray alder - *Alnus incana*
   green alder - *Alnus sinuata, A. viridis* subsp. *crispa*
   hazel alder - *Alnus rugosa, A. serrulata*
   Japanese alder - *Alnus japonica*
   mountain alder - *Alnus tenuifolia, A. viridis* subsp. *crispa*

mountain white-alder - *Clethra acuminata*
New Mexico alder - *Alnus oblongifolia*
Oregon alder - *Alnus rubra*
red alder - *Alnus rubra*
seaside alder - *Alnus maritima*
sierra alder - *Alnus rhombifolia*
Sitka alder - *Alnus sinuata*
smooth alder - *Alnus rugosa, A. serrulata*
speckled alder - *Alnus incana, A. rugosa*
thin-leaf alder - *Alnus tenuifolia*
white alder - *Alnus incana, A. rhombifolia*
yellow-alder - *Turnera ulmifolia*
alehoof - *Glechoma hederacea*
Aleppo
  Aleppo grass - *Sorghum halepense*
  Aleppo pine - *Pinus halepensis*
Alexander
  Alexander grass - *Brachiaria plantaginea*
  small-flowered Alexander grass - *Brachiaria subquadripara*
alexanders
  alexanders - *Angelica atropurpurea*
  golden alexanders - *Zizia aurea*
Alexandra's palm - *Archontophoenix alexandrae*
Alexandrian
  Alexandrian-laurel - *Calophyllum inophyllum*
  Alexandrian senna - *Senna alexandrina*
alfalfa
  alfalfa - *Medicago sativa*
  alfalfa dodder - *Cuscuta suaveolens*
  sickle alfalfa - *Medicago sativa* subsp. *falcata*
  small-seeded alfalfa dodder - *Cuscuta approximata*
  yellow alfalfa - *Medicago sativa* subsp. *falcata*
  yellow-flowered alfalfa - *Medicago sativa* subsp. *falcata*
alfilaria - *Erodium cicutarium*
alfombrilla - *Drymaria arenarioides*
algarroba
  algarroba - *Prosopis chilensis, P. juliflora, P. pallida*
  algarroba bean - *Ceratonia siliqua*
  algarroba de Chile - *Prosopis chilensis*
algarrobo - *Prosopis pallida*
alkali
  alkali bluegrass - *Poa secunda* subsp. *nevadensis*
  alkali cord grass - *Spartina gracilis*
  alkali grass - *Puccinellia distans, Puccinellia*
  alkali heliotrope - *Heliotropium curassavicum* var. *oculatum*
  alkali mallow - *Malvella leprosa*
  alkali muhly - *Muhlenbergia asperifolia*
  alkali sacaton - *Sporobolus airoides*
  alkali-sida - *Malvella leprosa*
  alkali-weed - *Cressa truxillensis*
  European alkali grass - *Puccinellia distans*
  Nuttall's alkali grass - *Puccinellia nuttalliana*
  weeping alkali grass - *Puccinellia distans*
alkaline water-nymph - *Najas marina*
alkanet
  alkanet - *Anchusa officinalis*
  bastard-alkanet - *Lithospermum arvense*
alkekengi - *Physalis alkekengi*
all
  all-good - *Chenopodium bonus-henricus*
  all-scale - *Atriplex polycarpa*

heal-all - *Prunella vulgaris*
allamanda
  allamanda - *Allamanda cathartica*
  bush allamanda - *Allamanda schottii*
Alleghany plum - *Prunus alleghaniensis*
Allegheny
  Allegheny barberry - *Berberis canadensis*
  Allegheny blackberry - *Rubus allegheniensis*
  Allegheny chinquapin - *Castanea pumila*
  Allegheny monkey-flower - *Mimulus ringens*
  Allegheny pachysandra - *Pachysandra procumbens*
  Allegheny phlox - *Phlox ovata*
  Allegheny serviceberry - *Amelanchier laevis*
  Allegheny-spurge - *Pachysandra procumbens*
  Allegheny-vine - *Adlumia fungosa*
alligator
  alligator juniper - *Juniperus deppeana*
  alligator-pear - *Persea americana*
  alligator-weed - *Alternanthera philoxeroides*
allspice
  allspice - *Pimenta dioica*
  Carolina allspice - *Calycanthus floridus*
  wild allspice - *Lindera benzoin*
almond
  almond - *Prunus dulcis*
  dwarf Russian almond - *Prunus tenella*
  earth-almond - *Cyperus esculentus*
  flowering almond - *Prunus triloba*
  green-almond - *Pistacia vera*
  Indian almond - *Terminalia cattapa*
  tropical almond - *Terminalia cattapa*
almum, sorghum-almum - *Sorghum almum*
aloe
  aloe yucca - *Yucca aloifolia*
  Barbados aloe - *Aloe vera*
  bitter aloe - *Aloe vera*
  Curaçao aloe - *Aloe vera*
  false aloe - *Manfreda virginica*
  kanniedood aloe - *Aloe variegata*
  medicinal aloe - *Aloe vera*
  tiger aloe - *Aloe variegata*
  true aloe - *Aloe vera*
alpine
  alpine-azalea - *Loiseleuria procumbens*
  alpine bartsia - *Bartsia alpina*
  alpine bearberry - *Arctostaphylos alpina*
  alpine bistort - *Polygonum viviparum*
  alpine blueberry - *Vaccinium uliginosum*
  alpine bluegrass - *Poa alpina*
  alpine-brook saxifrage - *Saxifraga rivularis*
  alpine calamint - *Acinos arvensis*
  alpine cat timothy - *Phleum alpinum*
  alpine cudweed - *Gnaphalium supinum*
  alpine currant - *Ribes alpinum*
  alpine fir - *Abies lasiocarpa*
  alpine foxtail - *Alopecurus alpinus*
  alpine gentian - *Gentiana newberryi*
  alpine lady's-mantle - *Alchemilla alpina*
  alpine-laurel - *Kalmia polifolia*
  alpine marsh violet - *Viola palustris*
  alpine meadow-rue - *Thalictrum alpinum*
  alpine milk-vetch - *Astragalus alpinus*

alpine shooting-star - *Dodecatheon alpinum*
alpine smartweed - *Polygonum viviparum*
alpine strawberry - *Fragaria vesca*
alpine timothy - *Phleum alpinum*
alpine wintergreen - *Gaultheria humifusa*
loose-flowered alpine sedge - *Carex rariflora*
purple alpine saxifrage - *Saxifraga oppositifolia*
yellow alpine saxifrage - *Saxifraga aizoides*
Alsike clover - *Trifolium hybridum*
Altai wild rye - *Leymus angustus*
alternate-leaf dogwood - *Cornus alternifolia*
althea, shrub althea - *Hibiscus syriacus*
aluminum-plant - *Pilea cadierei*
alumroot - *Geranium maculatum, Heuchera*
Alyce clover - *Alysicarpus vaginalis*
alyssum
  dwarf alyssum - *Alyssum desertorum*
  field alyssum - *Alyssum minus*
  golden-tuft alyssum - *Alyssum saxatile*
  hoary-alyssum - *Berteroa incana*
  sweet-alyssum - *Lobularia maritima*
  yellow alyssum - *Alyssum alyssoides*
amaranth
  amaranth - *Amaranthus*
  amaranth pigweed - *Amaranthus hybridus, A. retroflexus*
  Chinese amaranth - *Amaranthus tricolor*
  Ganges amaranth - *Amaranthus tricolor*
  giant amaranth - *Amaranthus australis*
  globe-amaranth - *Gomphrena globosa*
  green amaranth - *Amaranthus hybridus, A. retroflexus, A. viridis*
  livid amaranth - *Amaranthus lividus*
  mat amaranth - *Amaranthus graecizans*
  Palmer's amaranth - *Amaranthus palmeri*
  Powell's amaranth - *Amaranthus powellii*
  prostrate amaranth - *Amaranthus albus, A. blitoides, A. graecizans*
  purple amaranth - *Amaranthus cruentus*
  sand-hills amaranth - *Amaranthus arenicola*
  slender amaranth - *Amaranthus viridis*
  spiny amaranth - *Amaranthus spinosus*
  thorny amaranth - *Amaranthus spinosus*
  Torrey's amaranth - *Amaranthus bigelovii*
amaranthus
  red amaranthus - *Amaranthus hybridus*
  spleen amaranthus - *Amaranthus hybridus*
amarillo, sapote-amarillo - *Pouteria campechiana*
amaryllis - *Hippeastrum puniceum*
amatungulu - *Carissa macrocarpa*
Amazon
  Amazon lily - *Eucharis grandiflora*
  Amazon sprangle-top - *Leptochloa panicoides*
  Amazon-vine - *Stigmaphyllon ciliatum*
  Amazon water-lily - *Victoria amazonica*
  Amazon water-platter - *Victoria amazonica*
  lily-of-the-Amazon - *Eucharis grandiflora*
amber-bell - *Erythronium americanum*
amberique bean - *Strophostyles helvola*
amboyna-wood - *Pterocarpus indicus*
ambrosia, tall ambrosia - *Ambrosia trifida*

American
  American angelica - *Angelica atropurpurea*
  American arborvitae - *Thuja occidentalis*
  American barberry - *Berberis canadensis*
  American basswood - *Tilia americana*
  American beach grass - *Ammophila breviligulata*
  American beauty-berry - *Callicarpa americana*
  American beech - *Fagus grandifolia*
  American bellflower - *Campanula americana*
  American bittersweet - *Celastrus scandens*
  American black currant - *Ribes americanum*
  American black nightshade - *Solanum americanum*
  American bladder-nut - *Staphylea trifolia*
  American brooklime - *Veronica americana*
  American bugbane - *Cimicifuga americana*
  American bugleweed - *Lycopus americanus*
  American bulrush - *Scirpus americanus*
  American burn-weed - *Erechtites hieracifolia*
  American burnet - *Sanguisorba canadensis*
  American century-plant - *Agave americana*
  American chestnut - *Castanea dentata*
  American Christmas mistletoe - *Phoradendron serotinum*
  American cow-parsnip - *Heracleum sphondylium* subsp. *montanum*
  American cowslip - *Dodecatheon conjugens, D. meadia, Dodecatheon*
  American crabapple - *Malus angustifolia, M. coronaria*
  American cranberry - *Vaccinium macrocarpon*
  American dewberry - *Rubus flagellaris*
  American dittany - *Cunila origanoides*
  American dog violet - *Viola conspersa*
  American dogwood - *Cornus sericea*
  American dragonhead - *Dracocephalum parviflorum*
  American dune grass - *Leymus mollis*
  American eelgrass - *Vallisneria americana*
  American-Egyptian cotton - *Gossypium barbadense*
  American elder - *Sambucus canadensis*
  American elderberry - *Sambucus canadensis*
  American elm - *Ulmus americana*
  American false pennyroyal - *Hedeoma pulegioides*
  American feverfew - *Parthenium integrifolium*
  American filbert - *Corylus americana*
  American frog's-bit - *Limnobium spongia*
  American germander - *Teucrium canadense*
  American ginseng - *Panax quinquefolius*
  American grape - *Vitis labrusca*
  American grass-of-Parnassus - *Parnassia glauca*
  American great bulrush - *Scirpus validus*
  American green alder - *Alnus viridis* subsp. *crispa*
  American hazelnut - *Corylus americana*
  American hemp - *Apocynum cannabinum*
  American holly - *Ilex opaca*
  American honeysuckle - *Lonicera canadensis*
  American hop-hornbeam - *Ostrya virginiana*
  American hornbeam - *Carpinus caroliniana*
  American hydrangea - *Hydrangea arborescens*
  American ipecac - *Gillenia stipulata*
  American ivy - *Parthenocissus quinquefolia*
  American joint-vetch - *Aeschynomene americana*
  American knapweed - *Centaurea americana*
  American larch - *Larix laricina*
  American laurel - *Kalmia*
  American licorice - *Glycyrrhiza lepidota*

American linden - *Tilia americana*
American otus - *Nelumbo lutea*
American maidenhair fern - *Adiantum pedatum*
American mangrove - *Rhizophora mangle*
American manna grass - *Glyceria grandis*
American milk-bush - *Synadenium grantii*
American mistletoe - *Phoradendron serotinum*
American mountain-ash - *Sorbus americana*
American mulberry - *Morus rubra*
American olive - *Osmanthus americanus*
American pasqueflower - *Anemone patens*
American pawpaw - *Asimina triloba*
American pennyroyal - *Hedeoma pulegioides*
American persimmon - *Diospyros virginiana*
American plantain - *Plantago rugelii*
American plum - *Prunus americana*
American pondweed - *Potamogeton nodosus*
American potato bean - *Apios americana*
American radiator-plant - *Peperomia obtusifolia*
American red elderberry - *Sambucus racemosa* var.
  microbotrys
American red raspberry - *Rubus strigosus*
American sea-rocket - *Cakile edentula*
American slough grass - *Beckmannia syzigachne*
American smoke-tree - *Cotinus obovatus*
American snowbell - *Styrax americanus*
American spikenard - *Aralia racemosa*
American sweet-gum - *Liquidambar styraciflua*
American sycamore - *Platanus occidentalis*
American Turk's-cap lily - *Lilium superbum*
American valerian - *Cypripedium calceolus* var. *pubescens*
American vetch - *Vicia americana*
American wall fern - *Polypodium virginianum*
American water-lily - *Nymphaea odorata*
American water-willow - *Justicia americana*
American wayfaring-tree - *Viburnum alnifolium*
American worm-seed - *Chenopodium ambrosioides*
American yew - *Taxus canadensis*
South American air-plant - *Kalanchoe fedtschenkoi*
South American jelly palm - *Butia capitata*
South American waterweed - *Egeria densa*
ammannia
pink ammannia - *Ammannia latifolia*
purple ammannia - *Ammannia coccinea*
ammi
greater ammi - *Ammi majus*
toothpick ammi - *Ammi visnaga*
amole - *Chlorogalum pomeridianum*
ampelopsis, heart-leaf ampelopsis - *Ampelopsis cordata*
amsonia, willow amsonia - *Amsonia tabernaemontana*
Amur
Amur lilac - *Syringa amurensis*
Amur maackia - *Maackia amurensis*
Amur maple - *Acer ginnala*
Amur privet - *Ligustrum amurense*
anacahuita - *Cordia boissieri*
anchored water-hyacinth - *Eichhornia azurea*
Andrew's, St. Andrew's-cross - *Hypericum hypericoides*
andromeda
Japanese andromeda - *Pieris japonica*
mountain andromeda - *Pieris floribunda*

androsace
slender-stem androsace - *Androsace filiformis*
western androsace - *Androsace occidentalis*
anemone
anemone - *Anemone*
Canadian anemone - *Anemone canadensis*
candle anemone - *Anemone cylindrica*
European wood anemone - *Anemone nemorosa*
Japanese anemone - *Anemone hupehensis* var. *japonica*
long-headed anemone - *Anemone cylindrica*
meadow anemone - *Anemone canadensis*
poppy anemone - *Anemone coronaria*
rue anemone - *Thalictrum thalictroides*
tall anemone - *Anemone virginiana*
wood anemone - *Anemone quinquefolia*
angel-wing jasmine - *Jasminum nitidum*
angel's
angel's-tears - *Soleirolia soleirolii*
angel's-trumpet - *Datura innoxia*
angelica
American angelica - *Angelica atropurpurea*
angelica-tree - *Aralia spinosa*
filmy angelica - *Angelica triquinata*
garden angelica - *Angelica archangelica*
great angelica - *Angelica atropurpurea*
mountain angelica - *Angelica triquinata*
purple-stem angelica - *Angelica atropurpurea*
angle-pod - *Cynanchum laeve*
angled
angled luffa - *Luffa acutangula*
wing-angled loosestrife - *Lythrum alatum*
Angola pea - *Cajanus cajan*
anil indigo - *Indigofera suffruticosa*
animated oat - *Avena sterilis*
anise
anise - *Pimpinella anisum*
anise-hyssop - *Agastache foeniculum*
anise-root - *Osmorhiza longistylis*
anise-tree - *Illicium*
Florida anise-tree - *Illicium floridanum*
long-styled anise-root - *Osmorhiza longistylis*
purple anise - *Illicium floridanum*
yellow anise-tree - *Illicium parviflorum*
annatto - *Bixa orellana*
Anne's, Queen Anne's-lace - *Daucus carota*
Annie, sweet Annie - *Artemisia annua*
annual
annual aster - *Callistephus chinensis*
annual beard grass - *Polypogon monspeliensis*
annual bluegrass - *Poa annua*
annual broom-weed - *Amphiachyris dracunculoides*
annual buffalo clover - *Trifolium reflexum*
annual bur-sage - *Ambrosia acanthicarpa*
annual burweed - *Ambrosia acanthicarpa*
annual-cat-tail - *Rostraria cristata*
annual drop-seed - *Sporobolus neglectus*
annual fleabane - *Erigeron annuus*
annual fringe-rush - *Fimbristylis annua*
annual hair grass - *Deschampsia danthonioides*
annual lespedeza - *Lespedeza striata*
annual marjoram - *Origanum majorana*

annual marsh-elder - *Iva annua*
annual phlox - *Phlox drummondii*
annual poinsettia - *Euphorbia heterophylla*
annual polemonium - *Polemonium micranthum*
annual prickly poppy - *Argemone polyanthemos*
annual rye grass - *Lolium multiflorum*
annual salt marsh aster - *Aster subulatus*
annual sedge - *Cyperus compressus*
annual sow-thistle - *Sonchus oleraceus*
annual wild bean - *Strophostyles helvola*
annual wild rice - *Zizania aquatica*
annual wormwood - *Artemisia annua*
Bentham's annual lupine - *Lupinus benthamii*
anoda, spurred anoda - *Anoda cristata*
anthericum, walking anthericum - *Chlorophytum comosum*
Antwerp hollyhock - *Alcea ficifolia*
Apache pine - *Pinus engelmannii*
aparajita - *Clitoria ternatea*
aparejo grass - *Muhlenbergia utilis*
Appalachian
  Appalachian beard-tongue - *Penstemon canescens*
  Appalachian bellflower - *Campanula divaricata*
  Appalachian groundsel - *Senecio anonymus*
  Appalachian sunflower - *Helianthus atrorubens*
  Appalachian tea - *Ilex glabra, Viburnum cassinoides*
apple
  Adam's-apple - *Tabernaemontana divaricata*
  apple - *Malus domestica, M. sylvestris, Malus*
  apple guava - *Psidium guajava*
  apple mint - *Mentha rotundifolia, M. suaveolens*
  apple-of-Peru - *Nicandra physalodes*
  baked-apple-berry - *Rubus chamaemorus*
  balsam-apple - *Clusia rosea, Echinocystis lobata, Momordica balsamina, M. charantia*
  belle-apple - *Passiflora laurifolia*
  bitter-apple - *Citrullus colocynthis*
  Chinese apple - *Malus prunifolia*
  custard-apple - *Annona cherimola, A. reticulata, A. squamosa*
  downy thorn-apple - *Datura innoxia, D. metel*
  earth-apple - *Helianthus tuberosus*
  elephant-apple - *Limonia acidissima*
  gold-apple - *Lycopersicon esculentum*
  Indian apple - *Datura innoxia*
  Indian wood-apple - *Limonia acidissima*
  kai-apple - *Dovyalis caffra*
  kau-apple - *Dovyalis caffra*
  kei-apple - *Dovyalis caffra*
  large-fruit rose-apple - *Syzygium malaccense*
  love-apple - *Lycopersicon esculentum*
  mad-apple - *Datura stramonium, Solanum melongena*
  Malay apple - *Syzygium malaccense*
  May-apple - *Podophyllum peltatum*
  Mexican apple - *Casimiroa edulis*
  mock-apple - *Echinocystis lobata*
  paradise apple - *Malus pumila, M. sylvestris*
  Peruvian apple cactus - *Cereus uruguayanus*
  plum-leaf apple - *Malus prunifolia*
  pond-apple - *Annona glabra*
  possum-apple - *Diospyros virginiana*
  rose-apple - *Syzygium jambos, S. malaccense*
  soda-apple nightshade - *Solanum aculeatissimum*

  squaw-apple - *Peraphyllum ramosissimum*
  stramonium thorn-apple - *Datura stramonium*
  sugar-apple - *Annona squamosa*
  thorn-apple - *Crataegus, Datura stramonium, Datura*
  wild apple - *Malus sylvestris*
  wild balsam-apple - *Echinocystis lobata*
  wonder-apple - *Momordica balsamina*
  wood-apple - *Limonia acidissima*
apples, cane-apples - *Arbutus unedo*
apricot
  apricot - *Prunus armeniaca*
  apricot mallow - *Sphaeralcea ambigua*
  apricot plum - *Prunus simonii*
  apricot-vine - *Passiflora incarnata*
  desert apricot - *Prunus fremontii*
aquacate - *Persea americana*
Arabian
  Arabian jasmine - *Jasminum sambac*
  Arabian tea - *Catha edulis*
arabica coffee - *Coffea arabica*
aralia
  aralia-ivy - *Fatshedera lizei*
  Balfour's aralia - *Polyscias scutellaria*
  geranium-leaf aralia - *Polyscias guilfoylei*
  Ming aralia - *Polyscias fruticosa*
aramina - *Urena lobata*
arborvitae
  American arborvitae - *Thuja occidentalis*
  arborvitae - *Thuja*
  false arborvitae - *Thujopsis dolobrata*
  giant arborvitae - *Thuja plicata*
  hiba-arborvitae - *Thujopsis dolobrata*
  Oriental arborvitae - *Platycladus orientalis*
arbutus, trailing arbutus - *Epigaea repens*
arctic
  arctic blackberry - *Rubus arcticus*
  arctic bluegrass - *Poa arctica*
  arctic lupine - *Lupinus arcticus*
  arctic pearl-wort - *Sagina procumbens*
  arctic poppy - *Papaver nudicaule*
  arctic pyrola - *Pyrola glandiflora*
  arctic willow - *Salix arctica*
  arctic willow-weed - *Epilobium davuricum*
  arctic wintergreen - *Pyrola glandiflora*
  long-awn arctic sedge - *Carex podocarpa*
  weak arctic sedge - *Carex supina*
ardisia - *Ardisia crispa*
areca palm - *Chrysalidocarpus lutescens*
areng palm - *Arenga pinnata*
argemony - *Argemone*
Arizona
  Arizona alder - *Alnus oblongifolia*
  Arizona barley - *Hordeum arizonicum*
  Arizona brome - *Bromus arizonicus*
  Arizona cotton-top - *Digitaria californica*
  Arizona cypress - *Cupressus arizonica*
  Arizona fescue - *Festuca arizonica*
  Arizona-giant - *Carnegiea gigantea*
  Arizona long-leaf pine - *Pinus engelmannii*
  Arizona sycamore - *Platanus wrightii*
  Arizona three-awn grass - *Aristida arizonica*

Arizona walnut - *Juglans major*
Arizona white oak - *Quercus arizonica*
Arizona wild cotton - *Gossypium thurberi*
rough-bark Arizona cypress - *Cupressus arizonica*
ark, Noah's-ark - *Cypripedium calceolus* var. *pubescens*
Arkansas
    Arkansas oak - *Quercus arkansana*
    Arkansas rose - *Rosa arkansana*
aromo - *Azara microphylla*
arrow
    arrow-arum - *Peltandra virginica, Peltandra*
    arrow-feather - *Aristida purpurascens*
    arrow-feather three-awn - *Aristida purpurascens*
    arrow grass - *Triglochin*
    arrow-leaf aster - *Aster sagittifolius*
    arrow-leaf clover - *Trifolium vesiculosum*
    arrow-leaf monochoria - *Monochoria hastata*
    arrow-leaf sida - *Sida rhombifolia*
    arrow-leaf tear-thumb - *Polygonum sagittatum*
    arrow-leaf violet - *Viola sagittata*
    arrow-vine - *Polygonum sagittatum*
    arrow-wood - *Viburnum acerifolium, V. dentatum, Viburnum*
    arrow-wood viburnum - *Viburnum dentatum*
    downy arrow-wood - *Viburnum rafinesquianum*
    downy-leaf arrow-wood - *Viburnum rafinesquianum*
    green arrow-arum - *Peltandra virginica*
    marsh arrow grass - *Triglochin palustris*
    seaside arrow grass - *Triglochin maritima*
    smooth arrow-wood - *Viburnum recognitum*
arrowhead
    arrowhead - *Sagittaria lancifolia, S. latifolia, S. sagittifolia, Sagittaria*
    arrowhead-vine - *Syngonium podophyllum*
    arrowhead violet - *Viola sagittata*
    arum-leaf arrowhead - *Sagittaria cuneata*
    California arrowhead - *Sagittaria montevidensis*
    Chinese arrowhead - *Sagittaria sagittifolia*
    coastal arrowhead - *Sagittaria graminea*
    delta arrowhead - *Sagittaria graminea* var. *platyphylla*
    giant arrowhead - *Sagittaria montevidensis*
    Gregg's arrowhead - *Sagittaria longiloba*
    lance-leaf arrowhead - *Sagittaria lancifolia*
    Old World arrowhead - *Sagittaria sagittifolia*
    slender arrowhead - *Sagittaria graminea*
    wedge-leaf arrowhead - *Sagittaria cuneata*
arrowroot
    arrowroot - *Maranta arundinacea*
    Florida arrowroot - *Zamia pumila*
    Queensland arrowroot - *Canna indica*
    Spanish arrowroot - *Canna indica*
arroyo willow - *Salix lasiolepis*
arthraxon, joint-head arthraxon - *Arthraxon hispidus*
artichoke
    globe artichoke - *Cynara scolymus*
    Jerusalem artichoke - *Helianthus tuberosus*
artillery
    artillery-plant - *Pilea microphylla*
    artillery-weed - *Pilea microphylla*
arum
    arrow-arum - *Peltandra virginica, Peltandra*
    arum-leaf arrowhead - *Sagittaria cuneata*
    arum-lily - *Zantedeschia aethiopica*

    dragon-arum - *Arisaema dracontium*
    green arrow-arum - *Peltandra virginica*
    ivy-arum - *Epipremnum aureum*
    umbrella arum - *Amorphophallus rivieri*
    water-arum - *Calla palustris*
asarabacca - *Asarum*
ash
    American mountain-ash - *Sorbus americana*
    ash - *Fraxinus*
    ash gourd - *Benincasa hispida*
    ash-leaf maple - *Acer negundo*
    black ash - *Fraxinus nigra*
    black mountain-ash - *Eucalyptus sieber*
    blue ash - *Fraxinus quadrangulata*
    California ash - *Fraxinus dipetala*
    Carolina ash - *Fraxinus caroliniana*
    European ash - *Fraxinus excelsior*
    European mountain-ash - *Sorbus aucuparia*
    green ash - *Fraxinus pennsylvanica*
    lime prickly-ash - *Zanthoxylum fagara*
    mountain-ash - *Sorbus*
    northern prickly-ash - *Zanthoxylum americanum*
    Oregon ash - *Fraxinus latifolia*
    Pacific mountain-ash - *Sorbus sitchensis*
    pop ash - *Fraxinus caroliniana*
    prickly-ash - *Aralia spinosa, Zanthoxylum americanum*
    red ash - *Fraxinus pennsylvanica*
    sea-ash - *Zanthoxylum clava-herculis*
    showy mountain-ash - *Sorbus decora*
    silvertop-ash - *Eucalyptus sieber*
    single-leaf ash - *Fraxinus anomala*
    southern prickly-ash - *Zanthoxylum clava-herculis*
    stinking-ash - *Ptelea trifoliata*
    two-petal ash - *Fraxinus dipetala*
    velvet ash - *Fraxinus velutina*
    water ash - *Fraxinus caroliniana*
    white ash - *Fraxinus americana*
Ashe's
    Ashe's birch - *Betula uber*
    Ashe's juniper - *Juniperus ashei*
ashy sunflower - *Helianthus mollis*
Asian
    Asian mazus - *Mazus pumilus*
    Asian pear - *Pyrus pyrifolia*
    Asian pigeon-wings - *Clitoria ternatea*
Asiatic
    Asiatic bellflower - *Wahlenbergia marginata*
    Asiatic dayflower - *Commelina communis*
    Asiatic hawk's-beard - *Youngia japonica*
    Asiatic pennywort - *Centella asiatica*
asp-of-Jerusalem - *Isatis tinctoria*
asparagus
    asparagus - *Asparagus officinalis*
    asparagus bean - *Vigna unguiculata* subsp. *sesquipedalis*
    asparagus broccoli - *Brassica oleracea* var. *italica*
    asparagus-fern - *Asparagus densiflorus, A. setaceus*
    asparagus pea - *Psophocarpus tetragonolobus*
    Cossack asparagus - *Typha latifolia*
    fern asparagus - *Asparagus setaceus*
    garden asparagus - *Asparagus officinalis*
    smilax asparagus - *Asparagus asparagoides*

Sprenger's asparagus - *Asparagus densiflorus*
aspen
  aspen - *Populus*
  big-tooth aspen - *Populus grandidentata*
  large-tooth aspen - *Populus grandidentata*
  quaking aspen - *Populus tremuloides*
  trembling aspen - *Populus tremuloides*
assai palm - *Euterpe edulis, E. oleracea*
Assam rubber - *Ficus elastica*
aster
  annual aster - *Callistephus chinensis*
  annual salt marsh aster - *Aster subulatus*
  arrow-leaf aster - *Aster sagittifolius*
  aster - *Aster*
  awl aster - *Aster pilosus*
  azure aster - *Aster oolentangiensis*
  big-leaf aster - *Aster macrophyllus*
  blue heart-leaf aster - *Aster cordifolius*
  blue-wood aster - *Aster cordifolius*
  bog aster - *Aster nemoralis*
  bristly aster - *Aster puniceus*
  bushy aster - *Aster dumosus*
  calico aster - *Aster lateriflorus*
  China aster - *Callistephus chinensis*
  clasping aster - *Aster patens*
  clasping heart-leaf aster - *Aster undulatus*
  falcate golden aster - *Chrysopsis falcata*
  flat-top white aster - *Aster umbellatus*
  goblet aster - *Aster lateriflorus*
  golden aster - *Chrysopsis*
  grass-leaf golden aster - *Chrysopsis graminifolia*
  heart-leaf aster - *Aster cordifolius*
  heath aster - *Aster ericoides*
  large-leaf aster - *Aster macrophyllus*
  large purple aster - *Aster patens*
  leafy bog aster - *Aster nemoralis*
  long-stalked aster - *Aster dumosus*
  Lowrie's aster - *Aster lowrieanus*
  Maryland golden aster - *Chrysopsis mariana*
  midwestern blue heart-leaf aster - *Aster shortii*
  New England aster - *Aster novae-angliae*
  New York aster - *Aster novi-belgii*
  panicled aster - *Aster lanceolatus*
  perennial salt marsh aster - *Aster tenuifolius*
  prairie golden aster - *Chrysopsis camporum*
  purple-stem aster - *Aster puniceus*
  rock aster - *Aster alpinus*
  savory-leaf aster - *Aster linariifolius*
  Schreber's aster - *Aster schreberi*
  shaggy golden aster - *Chrysopsis mariana*
  shore aster - *Aster tradescantii*
  Short's aster - *Aster shortii*
  sickle-leaf golden aster - *Chrysopsis falcata*
  small white aster - *Aster laterifolius*
  smooth aster - *Aster laevis*
  southern aster - *Aster hemisphericus*
  spiny aster - *Chloranthus spinosa*
  squarrose white aster - *Aster ericoides*
  starved aster - *Aster lateriflorus*
  stiff aster - *Aster linariifolius*
  Stokes's aster - *Stokesia laevis*
  tall flat-top white aster - *Aster umbellatus*

  Tradescant's aster - *Aster tradescantii*
  upland white aster - *Solidago ptarmicoides*
  wavy-leaf aster - *Aster undulatus*
  white field aster - *Aster lanceolatus*
  white heart-leaf aster - *Aster divaricatus*
  white heath aster - *Aster pilosus*
  white upland aster - *Solidago ptarmicoides*
  white-wood aster - *Aster divaricatus*
  whorled aster - *Aster acuminatus*
  whorled wood aster - *Aster acuminatus*
asthma-weed - *Lobelia inflata*
atamasco-lily - *Zephyranthes atamasco*
Athel tamarisk - *Tamarix aphylla*
ati - *Calophyllum inophyllum*
Atlantic
  Atlantic mock bishop's-weed - *Ptilimnium capillaceum*
  Atlantic white-cedar - *Chamaecyparis thyoides*
Atlas cedar - *Cedrus atlantica*
atriplex, tumbling atriplex - *Atriplex rosea*
attorney, Scotch-attorney - *Clusia rosea*
aubergine - *Solanum melongena*
aucuba, Japanese aucuba - *Aucuba japonica*
auger-seed - *Stipa spartea*
August plum - *Prunus americana*
Augustine's, St. Augustine's grass - *Stenotaphrum secundatum*
auricula-tree - *Calotropis procera*
Australian
  Australian black-wood - *Acacia melanoxylon*
  Australian burn-weed - *Erechtites minima*
  Australian cedar - *Toona ciliata* var. *australis*
  Australian chess - *Bromus arenarius*
  Australian chestnut - *Castanospermum australe*
  Australian dodder - *Cuscuta obtusiflora*
  Australian flame-tree - *Brachychiton acerifolius*
  Australian gum - *Eucalyptus*
  Australian laurel - *Pittosporum tobira*
  Australian-nut - *Macadamia integrifolia*
  Australian pepper - *Schinus molle*
  Australian pine - *Casuarina equisetifolia, Casuarina*
  Australian rye grass - *Lolium multiflorum*
  Australian saltbush - *Atriplex semibaccata*
  Australian tree-fern - *Cyathea australis, C. cooperi*
  Australian umbrella-tree - *Schefflera actinophylla*
Austrian
  Austrian brier - *Rosa foetida*
  Austrian field-cress - *Rorippa austriaca*
  Austrian field pea - *Sphaerophysa salsula*
  Austrian pea-weed - *Sphaerophysa salsula*
  Austrian pine - *Pinus nigra*
  Austrian winter pea - *Lathyrus hirsutus*
  Austrian yellow rose - *Rosa foetida*
autumn
  autumn bent grass - *Agrostis perennans*
  autumn coralroot - *Corallorrhiza odontorhiza*
  autumn crocus - *Colchicum autumnale*
  autumn elaeagnus - *Elaeagnus umbellata*
  autumn-olive - *Elaeagnus umbellata*
  autumn sage - *Salvia greggii*
  autumn squash - *Cucurbita maxima*
  autumn willow - *Salix serissima*

avalanche-lily - *Erythronium grandiflorum*
avens
  avens - *Geum aleppicum, Geum*
  big-leaf avens - *Geum macrophyllum*
  cream-colored avens - *Geum virginianum*
  large-leaf avens - *Geum macrophyllum*
  long-plumed purple avens - *Geum triflorum*
  mountain-avens - *Dryas octopetala, D. octopetala* subsp.
   alaskensis
  purple avens - *Geum rivale*
  rough avens - *Geum virginianum*
  water avens - *Geum rivale*
  white avens - *Geum canadense*
avocado - *Persea americana*
awl aster - *Aster pilosus*
awn
  Arizona three-awn grass - *Aristida arizonica*
  arrow-feather three-awn - *Aristida purpurascens*
  awn-fruited sedge - *Carex stipata*
  bent-awn plume grass - *Saccharum contortum*
  church-mouse three-awn - *Aristida dichotoma*
  hairy-awn muhly - *Muhlenbergia capillaris*
  long-awn arctic sedge - *Carex podocarpa*
  pine-land three-awn - *Aristida stricta*
  poverty three-awn - *Aristida divaricata*
  prairie three-awn - *Aristida adscensionis*
  purple three-awn - *Aristida purpurea*
  red three-awn - *Aristida purpurea*
  Reverchon's three-awn - *Aristida purpurea*
  short-awn barley - *Hordeum brevisubulatum*
  short-awn foxtail - *Alopecurus aequalis*
  six-weeks three-awn - *Aristida adscensionis*
  slim-spike three-awn - *Aristida longespica*
  three-awn - *Aristida*
  Wooton's three-awn - *Aristida pansa*
awned
  awned cyperus - *Cyperus squarrosus*
  awned melic - *Melica smithii*
awnless brome grass - *Bromus inermis*
ax-seed - *Coronilla varia*
axillary goldenrod - *Solidago caesia*
azalea
  alpine-azalea - *Loiseleuria procumbens*
  azalea - *Rhododendron*
  Chinese azalea - *Rhododendron molle*
  clammy azalea - *Rhododendron viscosum*
  dwarf azalea - *Rhododendron atlanticum*
  early azalea - *Rhododendron austrinum*
  flame azalea - *Rhododendron calendulaceum*
  Ghent hybrid azalea - *Rhododendron gandavense*
  Hiryu azalea - *Rhododendron obtusum*
  hoary azalea - *Rhododendron canescens*
  Japanese azalea - *Rhododendron japonicum*
  macranthum azalea - *Rhododendron indicum*
  mock azalea - *Menziesia ferruginea*
  Piedmont azalea - *Rhododendron canescens*
  pink-shell azalea - *Rhododendron vaseyi*
  plum-leaf azalea - *Rhododendron prunifolium*
  pontic azalea - *Rhododendron luteum*
  smooth azalea - *Rhododendron arborescens*
  snow azalea - *Rhododendron mucronatum*
  summer-azalea - *Pelargonium domesticum*

  swamp azalea - *Rhododendron viscosum*
  sweet azalea - *Rhododendron arborescens*
  torch azalea - *Rhododendron kaempferi*
  western azalea - *Rhododendron occidentale*
  white swamp azalea - *Rhododendron viscosum*
  yellow azalea - *Rhododendron calendulaceum*
azolla - *Azolla caroliniana*
Aztec
  Aztec marigold - *Tagetes erecta*
  Aztec tobacco - *Nicotiana rustica*
azuki bean - *Vigna angularis*
azure
  azure aster - *Aster oolentangiensis*
  azure monk's-hood - *Aconitum carmichaelii*
baby
  baby-blue-eyes - *Nemophila menziesii*
  baby-pepper - *Rivina humilis*
  baby primrose - *Primula malacoides*
  baby rose - *Rosa multiflora*
  baby-rubber-plant - *Peperomia obtusifolia*
  cry-baby-tree - *Erythrina crista-galli*
  Texas mud-baby - *Echinodorus cordifolius*
  white baby-blue-eyes - *Nemophila menziesii* var. *atomaria*
baby's
  baby's-breath - *Galium sylvaticum, Gypsophila elegans,*
   G. paniculata
  baby's-tears - *Soleirolia soleirolii*
baccharis
  dwarf baccharis - *Baccharis pilularis*
  eastern baccharis - *Baccharis halimifolia*
  kidney-wort baccharis - *Baccharis pilularis* var. *consanguinea*
  sticky baccharis - *Baccharis salicifolia*
  willow baccharis - *Baccharis salicina*
baccharus, broom baccharus - *Baccharis sarothroides*
bachelor's
  bachelor's-button - *Centaurea cyanus*
  yellow bachelor's-button - *Polygala lutea*
back
  back-wort - *Symphytum officinale*
  pick-a-back-plant - *Tolmiea menziesii*
backbone, devil's-backbone - *Pedilanthus tithymaloides*
bacon
  bacon-weed - *Chenopodium album*
  eggs-and-bacon - *Linaria vulgaris*
bag
  bag-flower - *Clerodendrum thompsoniae*
  bag-pod sesbania - *Sesbania vesicaria*
Bahia
  Bahia grass - *Paspalum notatum*
  Bahia love grass - *Eragrostis bahiensis*
  Pensacola Bahia grass - *Paspalum notatum* var. *saurae*
baked-apple-berry - *Rubus chamaemorus*
Baker's cypress - *Cupressus bakeri*
bald cypress - *Taxodium distichum*
Balfour's aralia - *Polyscias scutellaria*
ball
  ball fern - *Davallia trichomanoides*
  ball-moss - *Tillandsia recurvata*
  ball mustard - *Neslia paniculata*
  ball nightshade - *Solanum carolinense*
  cotton-ball - *Espostoa lanata*

gray-ball sage - *Salvia dorrii*
ballogan - *Lapsana communis*
balloon
    balloon-flower - *Platycodon grandiflorus*
    balloon-vine - *Cardiospermum halicacabum*
balls, wax-balls - *Acalypha virginica*
balm
    balm - *Melissa*
    balm-of-Gilead - *Populus gileadensis*
    bee balm - *Melissa officinalis, Monarda didyma*
    field-balm - *Glechoma hederacea*
    horse-balm - *Collinsonia canadensis, Collinsonia*
    lemon balm - *Melissa officinalis*
    lemon bee balm - *Monarda citriodora*
    Molucca balm - *Moluccella laevis*
    northern horse-balm - *Collinsonia canadensis*
    spotted bee balm - *Monarda punctata*
    sweet balm - *Melissa officinalis*
balmony - *Chelone glabra*
balsam
    balsam - *Impatiens balsamina, Impatiens*
    balsam-apple - *Clusia rosea, Echinocystis lobata, Momordica balsamina, M. charantia*
    balsam fir - *Abies balsamea, A. grandis*
    balsam groundsel - *Senecio pauperculus*
    balsam-pear - *Momordica charantia*
    balsam poplar - *Populus balsamifera*
    balsam-root - *Balsamorhiza*
    balsam willow - *Salix pyrifolia*
    fir-balsam - *Abies balsamea*
    Fraser's balsam fir - *Abies fraseri*
    garden balsam - *Impatiens balsamina*
    he-balsam - *Picea rubens*
    old-field-balsam - *Gnaphalium obtusifolium*
    rose-balsam - *Impatiens balsamina*
    Seminole balsam - *Psychotria nervosa*
    she-balsam - *Abies fraseri*
    western balsam poplar - *Populus trichocarpa*
    wild balsam-apple - *Echinocystis lobata*
    Zanzibar balsam - *Impatiens wallerana*
balsamo-cor-de-carne - *Justicia carnea*
Baltic rush - *Juncus balticus*
bamboo
    bamboo - *Arundinaria, Bambusa vulgaris, Bambusa, Phyllostachys*
    bamboo palm - *Chamaedorea erumpens, Rhapis excelsa*
    bamboo-vine - *Smilax laurifolia*
    black bamboo - *Phyllostachys nigra*
    cane-break bamboo - *Arundinaria gigantea*
    fish-pole bamboo - *Phyllostachys aurea*
    golden bamboo - *Phyllostachys aurea*
    hardy timber bamboo - *Phyllostachys bambusoides*
    heavenly-bamboo - *Nandina domestica*
    hedge bamboo - *Bambusa multiplex*
    Japanese timber bamboo - *Phyllostachys bambusoides*
    Mexican bamboo - *Polygonum cuspidatum*
    sacred-bamboo - *Nandina domestica*
    wild bamboo-vine - *Smilax auriculata*
banana
    banana - *Musa paradisiaca*
    banana-plant - *Nymphoides aquatica*
    banana-shrub - *Michelia figo*

banana water-lily - *Nymphaea mexicana*
    banana yucca - *Yucca baccata*
    dwarf banana - *Musa acuminata*
    edible banana - *Musa acuminata*
band-plant - *Vinca major*
banded
    gold-banded lily - *Lilium auratum*
    green-banded mariposa - *Calochortus macrocarpus*
bane
    children's-bane - *Cicuta maculata*
    eye-bane - *Euphorbia maculata*
    garden wolf-bane - *Aconitum napellus*
    hound-bane - *Marrubium vulgare*
    leopard-bane - *Doronicum plantagineum*
    sow-bane - *Chenopodium hybridum, C. murale*
    swine-bane - *Chenopodium murale*
    trailing wolf-bane - *Aconitum reclinatum*
    wolf-bane - *Aconitum*
baneberry
    baneberry - *Actaea rubra, Actaea*
    red baneberry - *Actaea rubra*
    white baneberry - *Actaea alba*
Bangalow, northern Bangalow palm - *Archontophoenix alexandrae*
banjo fig - *Ficus lyrata*
bank
    ditch bank willow - *Salix exigua*
    river bank grape - *Vitis riparia*
    river bank lupine - *Lupinus rivularis*
    river bank sedge - *Carex riparia*
Bank's melastoma - *Melastoma malabathricum*
banyan
    banyan fig - *Ficus benghalensis*
    false banyan - *Ficus altissima*
    Indian banyan - *Ficus benghalensis*
barb-wire Russian thistle - *Salsola paulsenii*
Barbados
    Barbados aloe - *Aloe vera*
    Barbados cherry - *Malpighia glabra*
    Barbados gooseberry - *Physalis peruviana*
    Barbados lily - *Hippeastrum puniceum*
    Barbados-nut - *Jatropha curcas*
    pride-of-Barbados - *Caesalpinia pulcherrima*
Barbara's, St. Barbara's cress - *Barbarea vulgaris*
barbary matrimony-vine - *Lycium barbarum*
barbe-de-capuchin - *Cichorium intybus*
barbed goat grass - *Aegilops triuncialis*
barberry
    Allegheny barberry - *Berberis canadensis*
    American barberry - *Berberis canadensis*
    barberry - *Berberis vulgaris, Berberis*
    blue barberry - *Berberis aquifolium*
    Colorado barberry - *Berberis fendleri*
    creeping barberry - *Berberis repens*
    European barberry - *Berberis vulgaris*
    holly barberry - *Berberis aquifolium*
    Japanese barberry - *Berberis thunbergii*
    red barberry - *Berberis haematocarpa*
Barberton's daisy - *Gerbera jamesonii*

bare-stem paspalum - *Paspalum setaceum* var.
  longipedunculatum
barilla - *Halogeton glomeratus*
bark
  California gold-bark fern - *Pityrogramma triangularis*
  cherry-bark oak - *Quercus falcata* var. *pagodifolia*
  cinnamon-bark - *Canella winteriana*
  cork-bark fir - *Abies lasiocarpa* var. *arizonica*
  cosmetic-bark-tree - *Murraya paniculata*
  cramp-bark - *Viburnum trilobum*
  green-bark ceanothus - *Ceanothus spinosus*
  mealy stringy-bark - *Eucalyptus cinerea*
  messmate stringy-bark - *Eucalyptus obliqua*
  paper-bark-tree - *Melaleuca quinquenervia*
  rope-bark - *Dirca palustris*
  rough-bark Arizona cypress - *Cupressus arizonica*
  rough-bark Mexican pine - *Pinus montezumae*
  scaly-bark beefwood - *Casuarina glauca*
  seven-bark - *Hydrangea arborescens*
  stringy-bark - *Eucalyptus*
  tan-bark oak - *Lithocarpus densiflora*
  white-bark maple - *Acer leucoderme*
  white-bark pine - *Pinus albicaulis*
  yellow-bark oak - *Quercus velutina*
barley
  Arizona barley - *Hordeum arizonicum*
  barley - *Hordeum vulgare, Hordeum*
  barley grass - *Hordeum murinum*
  bulbous barley - *Hordeum bulbosum*
  foxtail barley - *Hordeum jubatum*
  hare barley - *Hordeum leporinum*
  little barley - *Hordeum pusillum*
  meadow barley - *Hordeum brachyantherum*
  Mediterranean barley - *Hordeum geniculatum*
  mouse barley - *Hordeum murinum*
  Nepal barley - *Hordeum vulgare*
  short-awn barley - *Hordeum brevisubulatum*
  squirrel-tail barley - *Hordeum jubatum*
  wall barley - *Hordeum glaucum, H. murinum*
  wild barley - *Hordeum jubatum, H. leporinum*
barn grass - *Echinochloa crus-galli*
barnyard grass - *Echinochloa crus-galli*
barometer-bush - *Leucophyllum frutescens*
barren
  barren brome - *Bromus sterilis*
  barren ground willow - *Salix brachycarpa* subsp. *niphoclada*
  barren-strawberry - *Waldsteinia fragarioides*
  pine-barren cyperus - *Cyperus cylindricus*
  pine-barren gentian - *Gentiana autumnalis*
  pine-barren goldenrod - *Solidago fistulosa*
  pine-barren sandwort - *Arenaria caroliniana*
  pine-barren tick-trefoil - *Desmodium strictum*
  shale-barren phlox - *Phlox buckleyi*
barroom-plant - *Aspidistra elatior*
bartsia
  alpine bartsia - *Bartsia alpina*
  red bartsia - *Odontites verna* subsp. *serotina*
basal-leaved rosinweed - *Silphium terebinthinaceum*
basil
  basil - *Ocimum basilicum, Ocimum*
  holy basil - *Ocimum sanctum*

sweet basil - *Ocimum basilicum*
wild basil - *Clinopodium vulgare*
basin
  basin sagebrush - *Artemisia tridentata*
  basin wild rye - *Leymus cinereus*
Basin, Great Basin bristle-cone pine - *Pinus longaeva*
basket
  basket-flower - *Centaurea americana, Hymenocallis narcissiflora, Hymenocallis*
  basket grass - *Oplismenus hirtellus*
  basket oak - *Quercus michauxii, Q. prinus*
  basket-of-gold - *Alyssum saxatile*
  basket willow - *Salix purpurea, S. viminalis*
  scarlet basket-vine - *Aeschynanthus pulcher*
bassia
  five-hook bassia - *Bartonia virginica, Bassia hyssopifolia*
  hyssop bassia - *Bassia hyssopifolia*
basswood
  American basswood - *Tilia americana*
  basswood - *Tilia americana, T. caroliniana*
  Carolina basswood - *Tilia caroliniana*
  white basswood - *Tilia heterophylla*
bastard
  bastard-alkanet - *Lithospermum arvense*
  bastard-indigo - *Amorpha fruticosa*
  bastard-jute - *Hibiscus cannabinus*
  bastard oak - *Quercus austrina*
  bastard-pennyroyal - *Trichostema dichotomum*
  bastard-saffron - *Carthamus tinctorius*
  bastard-sandle-wood - *Myoporum sandwicense*
  bastard-toadflax - *Comandra umbellata, Comandra*
  bastard-willow - *Rhus lancea*
batino - *Alstonia macrophylla*
Batoko-plum - *Flacourtia indica*
batting, cotton-batting cudweed - *Gnaphalium stramineum*
bauhinia
  nasturtium bauhinia - *Bauhinia galpinii*
  red bauhinia - *Bauhinia galpinii*
bay
  bay - *Laurus nobilis*
  bay-gall-bush - *Ilex coriacea*
  bay-leaved willow - *Salix pentandra*
  bay willow - *Salix pentandra*
  broad-leaf rose-bay - *Tabernaemontana divaricata*
  bull-bay - *Magnolia grandiflora*
  California bay - *Umbellularia californica*
  California rose-bay - *Rhododendron macrophyllum*
  East Indian rose-bay - *Tabernaemontana divaricata*
  Hudson Bay currant - *Ribes hudsonianum*
  Lapland rose-bay - *Rhododendron lapponicum*
  loblolly-bay - *Gordonia lasianthus*
  Moreton Bay chestnut - *Castanospermum australe*
  mountain rose-bay - *Rhododendron catawbiense*
  red-bay - *Persea borbonia*
  rose-bay - *Nerium oleander*
  rose-bay rhododendron - *Rhododendron maximum*
  swamp-bay - *Magnolia virginiana, Persea palustris*
  swamp red-bay - *Persea borbonia*
  sweet-bay - *Laurus nobilis, Magnolia virginiana, Persea borbonia*
  sweet-bay magnolia - *Magnolia virginiana*

bayberry
  bayberry - *Myrica pensylvanica*
  California bayberry - *Myrica californica*
  evergreen bayberry - *Myrica heterophylla*
  northern bayberry - *Myrica pensylvanica*
  odorless bayberry - *Myrica inodora*
  Pacific bayberry - *Myrica californica*
  southern bayberry - *Myrica cerifera, M. heterophylla*
bayonet
  bayonet-grass - *Scirpus maritimus*
  prairie bayonet-grass - *Scirpus maritimus*
  Spanish bayonet - *Yucca aloifolia, Y. baccata*
beach
  American beach grass - *Ammophila breviligulata*
  beach-heath - *Hudsonia tomentosa*
  beach-heather - *Hudsonia*
  beach morning-glory - *Ipomoea pes-caprae*
  beach pea - *Lathyrus japonicus, L. littoralis*
  beach pine - *Pinus contorta*
  beach plum - *Prunus maritima*
  beach speedwell - *Veronica longifolia*
  beach sunflower - *Helianthus debilis*
  beach wormwood - *Artemisia stelleriana*
  European beach grass - *Ammophila arenaria*
  sea beach dock - *Rumex pallidus*
  sea beach evening-primrose - *Oenothera humifusa*
  sea beach knotweed - *Polygonum glaucum*
  sea beach orache - *Atriplex arenaria*
  sea beach sandwort - *Honckenya peploides*
Beach, Palm Beach bells - *Kalanchoe*
bead
  bead-podded mustard - *Chorispora tenella*
  bead-tree - *Melia*
  black-bead - *Pithecellobium keyense, P. unguis-cati*
  black-bead elder - *Sambucus melanocarpa*
  black jet-bead - *Rhodotypos scandens*
  blue-bead-lily - *Clintonia borealis*
  coral-bead-plant - *Abrus precatorius*
  ebony black-bead - *Pithecellobium flexicaule*
  Japanese bead-tree - *Melia azedarach*
  jet-bead - *Rhodotypos*
  red-bead-plant - *Abrus precatorius*
  Syrian bead-tree - *Melia azedarach*
beads
  coral-beads - *Cocculus carolinus*
  prayer-beads - *Abrus precatorius*
beak
  beak rush - *Rhynchospora*
  brown beak rush - *Rhynchospora fusca*
  bur-beak chervil - *Anthriscus caucalis*
  nodding beak rush - *Rhynchospora inexpansa*
  pine-hill beak rush - *Rhynchospora globularis*
  white beak rush - *Rhynchospora alba*
beaked
  beaked corn-salad - *Valerianella radiata*
  beaked filbert - *Corylus cornuta*
  beaked hawkweed - *Hieracium gronovii*
  beaked hazelnut - *Corylus cornuta*
  beaked sedge - *Carex rostrata*
  beaked spike rush - *Eleocharis rostellata*
  long-beaked willow - *Salix bebbiana*
beam, quick-beam - *Sorbus aucuparia*

bean
  algarroba bean - *Ceratonia siliqua*
  amberique bean - *Strophostyles helvola*
  American potato bean - *Apios americana*
  annual wild bean - *Strophostyles helvola*
  asparagus bean - *Vigna unguiculata* subsp. *sesquipedalis*
  azuki bean - *Vigna angularis*
  bean - *Phaseolus vulgaris, Phaseolus*
  bean-caper - *Zygophyllum fabago*
  bean-tree - *Laburnum*
  bean-vine - *Phaseolus polystachios*
  Bengal velvet bean - *Mucuna pruriens* var. *utilis*
  black bean - *Castanospermum australe*
  bonavist bean - *Lablab purpureus*
  broad bean - *Vicia faba*
  buck bean - *Menyanthes trifoliata*
  butter bean - *Phaseolus lunatus*
  cacao-bean - *Theobroma cacao*
  Carolina bean - *Phaseolus lunatus*
  castor bean - *Ricinus communis*
  Cherokee bean - *Erythrina herbacea*
  civet bean - *Phaseolus lunatus*
  cluster bean - *Cyamopsis tetragonoloba*
  cocoa-bean - *Theobroma cacao*
  coral bean - *Erythrina herbacea*
  cowage velvet bean - *Mucuna pruriens* var. *utilis*
  Dutch-case-knife bean - *Phaseolus coccineus*
  eastern coral bean - *Erythrina herbacea*
  fava bean - *Vicia faba*
  Florida velvet bean - *Mucuna pruriens* var. *utilis*
  French bean - *Phaseolus vulgaris*
  garden bean - *Phaseolus lunatus, P. vulgaris*
  giant stock bean - *Canavalia ensiformis*
  Goa bean - *Psophocarpus tetragonolobus*
  green bean - *Phaseolus vulgaris*
  haricot bean - *Phaseolus vulgaris*
  horse bean - *Canavalia ensiformis, Parkinsonia aculeata, Vicia faba*
  hyacinth bean - *Lablab purpureus*
  Indian bean - *Catalpa speciosa*
  Indian bean catalpa - *Catalpa bignonioides*
  jack bean - *Canavalia ensiformis*
  jequirity bean - *Abrus precatorius*
  jumbie-bean - *Leucaena leucocephala*
  kidney bean - *Phaseolus vulgaris*
  lablab bean - *Lablab purpureus*
  lima bean - *Phaseolus lunatus*
  locust bean - *Ceratonia siliqua*
  Lyon bean - *Mucuna pruriens* var. *utilis*
  Mauritius velvet bean - *Mucuna pruriens* var. *utilis*
  mescal bean - *Sophora secundiflora*
  mung bean - *Vigna radiata*
  navy bean - *Phaseolus vulgaris*
  perennial wild bean - *Strophostyles umbellata*
  phasey bean - *Macroptilium lathyroides*
  pink wild bean - *Strophostyles umbellata*
  precatory bean - *Abrus precatorius*
  purple bean - *Macroptilium atropurpureum*
  runner bean - *Phaseolus vulgaris*
  sacred-bean - *Nelumbo*
  salad bean - *Phaseolus vulgaris*
  scarlet runner bean - *Phaseolus coccineus*

sewee bean - *Phaseolus lunatus*
sieva bean - *Phaseolus lunatus*
small-flowered wild bean - *Strophostyles leiosperma*
smooth-seeded wild bean - *Strophostyles leiosperma*
snap bean - *Phaseolus vulgaris*
southwestern coral bean - *Erythrina flabelliformis*
string bean - *Phaseolus vulgaris*
sword bean - *Canavalia ensiformis, C. gladiata*
Syrian bean-caper - *Zygophyllum fabago*
Tepary bean - *Phaseolus acutifolius*
thicket bean - *Phaseolus polystachios*
trailing wild bean - *Strophostyles helvola*
velvet bean - *Mucuna pruriens* var. *utilis*
wax bean - *Phaseolus vulgaris*
western coral bean - *Erythrina flabelliformis*
wild bean - *Apios americana, Phaseolus polystachios,*
    Strophostyles helvola
Windsor bean - *Vicia faba*
winged bean - *Psophocarpus tetragonolobus*
wolf bean - *Lupinus albus*
wonder bean - *Canavalia ensiformis*
yam bean - *Pachyrhizus erosus*
yard-long bean - *Vigna unguiculata, V. unguiculata* subsp.
    sesquipedalis
Yokohama velvet bean - *Mucuna pruriens* var. *utilis*
beans, Indian beans - *Lupinus perennis*
bear's
    bear's-foot - *Aconitum napellus*
    bear's-paw fern - *Aglaomorpha meyeniana*
bearberry
    alpine bearberry - *Arctostaphylos alpina*
    bearberry - *Arctostaphylos uva-ursi, Arctostaphylos, Rhamnus*
        purshiana, Vaccinium erythrocarpum
    black bearberry - *Arctostaphylos alpina*
beard
    Aaron's-beard - *Hypericum calycinum*
    annual beard grass - *Polypogon monspeliensis*
    Appalachian beard-tongue - *Penstemon canescens*
    Asiatic hawk's-beard - *Youngia japonica*
    beard grass - *Andropogon gerardii, A. virginicus,*
        Andropogon, Polypogon
    beard-tongue - *Penstemon*
    bristly hawk's-beard - *Crepis setosa*
    broom beard grass - *Schizachyrium scoparium*
    bushy beard grass - *Andropogon glomeratus*
    bushy beard-stem - *Andropogon glomeratus*
    cane beard grass - *Bothriochloa barbinodis*
    crown-beard - *Verbesina occidentalis, Verbesina*
    ditch beard grass - *Polypogon interruptus*
    Elliott's beard grass - *Andropogon elliottii*
    European hawk's-beard - *Crepis vesicaria* subsp. *haenseleri*
    false goat's-beard - *Astilbe biternata*
    goat's-beard - *Aruncus dioicus, A. sylvester, Tragopogon*
    gray-beard - *Tillandsia usneoides*
    gray beard-tongue - *Penstemon canescens*
    hairy beard-tongue - *Penstemon hirsutus*
    hawk's-beard - *Crepis*
    Jupiter's-beard - *Centranthus ruber*
    large beard-tongue - *Penstemon grandiflorus*
    large-flowered beard-tongue - *Penstemon grandiflorus*
    lion's-beard - *Anemone patens*
    narrow-leaf hawk's-beard - *Crepis tectorum*
    northeastern beard-tongue - *Penstemon hirsutus*
    old-man's-beard - *Chionanthus virginicus*
    pitted beard grass - *Bothriochloa pertusa*
    prairie beard grass - *Schizachyrium scoparium*
    ringed beard grass - *Dichanthium annulatum*
    rough hawk's-beard - *Crepis biennis*
    silver beard grass - *Bothriochloa saccharoides*
    smooth hawk's-beard - *Crepis capillaris*
    turkey-beard - *Xerophyllum asphodeloides*
    Virginia beard grass - *Andropogon virginicus*
    western hawk's-beard - *Crepis occidentalis*
bearded
    bearded beggar-ticks - *Bidens aristosa*
    bearded couch - *Elymus trachycaulus*
    bearded-creeper - *Crupina vulgaris*
    bearded darnel - *Lolium temulentum*
    bearded sprangle-top - *Leptochloa fascicularis*
beardless
    beardless wheat grass - *Pseudoroegneria spicata*
    beardless wild rye - *Leymus triticoides*
beauty
    American beauty-berry - *Callicarpa americana*
    beauty-bush - *Kolkwitzia amabilis*
    beauty-of-the-night - *Mirabilis jalapa*
    broad-leaf spring-beauty - *Claytonia caroliniana*
    Carolina spring-beauty - *Claytonia caroliniana*
    Chinese beauty-berry - *Callicarpa dichotoma*
    climbing-beauty - *Aeschynanthus pulcher*
    crown-beauty - *Hymenocallis*
    Maryland meadow-beauty - *Rhexia mariana*
    narrow-leaf spring-beauty - *Claytonia virginica*
    purple beauty-berry - *Callicarpa dichotoma*
    river-beauty - *Epilobium latifolium*
    spring-beauty - *Claytonia megarrhiza, C. virginica, Claytonia*
beaver
    beaver-poison - *Cicuta maculata*
    beaver-tree-laurel - *Magnolia virginiana*
Bebb's willow - *Salix bebbiana*
bed
    bed-flower - *Galium verum*
    Jack-go-to-bed-at-noon - *Tragopogon pratensis*
    John-go-to-bed-at-noon - *Tragopogon pratensis*
bedding
    bedding geranium - *Pelargonium hortorum*
    bedding pansy - *Viola cornuta*
bedstraw
    bedstraw - *Galium*
    bedstraw bellflower - *Campanula aparinoides*
    catch-weed bedstraw - *Galium aparine*
    corn bedstraw - *Galium tricorne*
    forest bedstraw - *Galium circaezans*
    heath bedstraw - *Galium saxatile*
    hedge bedstraw - *Galium mollugo*
    lady's bedstraw - *Galium verum*
    northern bedstraw - *Galium boreale*
    rough bedstraw - *Galium asprellum*
    smooth bedstraw - *Galium mollugo*
    sweet-scented bedstraw - *Galium triflorum*
    three-horn bedstraw - *Galium tricornutum*
    white hedge bedstraw - *Galium mollugo*
    yellow bedstraw - *Galium verum*

bee
  bee balm - *Melissa officinalis, Monarda didyma*
  bee-nettle - *Galeopsis tetrahit*
  lemon bee balm - *Monarda citriodora*
  Rocky Mountain bee-plant - *Cleome serrulata*
  spotted bee balm - *Monarda punctata*
  yellow bee-plant - *Cleome lutea*
beech
  American beech - *Fagus grandifolia*
  beech - *Fagus*
  beech-drops - *Epifagus virginiana*
  blue-beech - *Carpinus caroliniana*
  European beech - *Fagus sylvatica*
  false beech-drops - *Monotropa hypopithys*
  water-beech - *Carpinus caroliniana*
beef-plant - *Iresine herbstii*
beefsteak
  beefsteak-geranium - *Begonia rex-cultorum, Saxifraga stolonifera*
  beefsteak-plant - *Acalypha wilkesiana, Iresine herbstii*
beefwood
  beefwood - *Casuarina cunninghamiana*
  beefwood-tree - *Casuarina equisetifolia, Casuarina*
  Cunningham's beefwood - *Casuarina cunninghamiana*
  scaly-bark beefwood - *Casuarina glauca*
beet
  beet - *Beta vulgaris, Beta*
  garden beet - *Beta vulgaris*
  sea beet - *Beta vulgaris*
  spinach beet - *Beta vulgaris* subsp. *cicla*
  sugar beet - *Beta vulgaris*
  wild beet - *Amaranthus hybridus, A. retroflexus, Saxifraga pensylvanica*
beetle-wood - *Galax aphylla*
beetroot - *Beta vulgaris*
beggar
  bearded beggar-ticks - *Bidens aristosa*
  beggar-tick grass - *Aristida orcuttiana*
  beggar-ticks - *Agrimonia gryposepala, Bidens bipinnata, B. pilosa, Bidens, Desmodium*
  beggar-weed - *Desmodium canadense, D. tortuosum*
  Bigelow's beggar-ticks - *Bidens bigelovii*
  connate beggar-ticks - *Bidens connata*
  coreopsis beggar-ticks - *Bidens polylepis*
  creeping beggar-weed - *Desmodium incanum*
  devil's beggar-ticks - *Bidens frondosa*
  estuary beggar-ticks - *Bidens hyperborea*
  Florida beggar-weed - *Desmodium tortuosum*
  hairy beggar-ticks - *Bidens pilosa*
  marsh beggar-ticks - *Bidens mitis*
  nodding beggar-ticks - *Bidens cernua*
  northern estuarine beggar-ticks - *Bidens hyperborea*
  purple-stem beggar-ticks - *Bidens connata*
  tall beggar-ticks - *Bidens vulgata*
  three-flowered beggar-weed - *Desmodium triflorum*
  trifid beggar-ticks - *Bidens tripartita*
  western beggar-ticks - *Bidens vulgata*
beggar's
  beggar's-button - *Arctium lappa, Arctium*
  beggar's-lice - *Cynoglossum, Hackelia*
begonia
  hybrid tuberous begonia - *Begonia tuberhybrida*

  rex begonia - *Begonia rex-cultorum*
  strawberry-begonia - *Saxifraga stolonifera*
  Swedish begonia - *Plectranthus*
  trailing watermelon-begonia - *Pellionia repens*
  wild begonia - *Rumex venosus*
Belgian evergreen - *Dracaena sanderana*
bell
  amber-bell - *Erythronium americanum*
  bell pepper - *Capsicum annuum, C. frutescens* cv. 'grossum'
  Chinese bell-flower - *Platycodon grandiflorus*
  cow-bell - *Silene csereii*
  Easter-bell - *Stellaria holostea*
  Easter-bell starwort - *Stellaria holostea*
  honey-bell rhododendron - *Rhododendron campylocarpum*
  little silver-bell - *Halesia tetraptera*
  monkey's-dinner-bell - *Hura crepitans*
  purple bell - *Campanula rapunculoides*
  silver-bell - *Halesia*
  two-winged silver-bell - *Halesia diptera*
  yellow bell lily - *Lilium canadense*
belladonna
  belladonna - *Atropa belladonna*
  belladonna-lily - *Amaryllis belladonna*
  Cape belladonna - *Amaryllis belladonna*
belle-apple - *Passiflora laurifolia*
bellflower
  American bellflower - *Campanula americana*
  Appalachian bellflower - *Campanula divaricata*
  Asiatic bellflower - *Wahlenbergia marginata*
  bedstraw bellflower - *Campanula aparinoides*
  bellflower - *Campanula rapunculoides, Campanula*
  clustered bellflower - *Campanula glomerata*
  creeping bellflower - *Campanula rapunculoides*
  marsh bellflower - *Campanula aparinoides*
  nettle-leaf bellflower - *Campanula trachelium*
  rover bellflower - *Campanula rapunculoides*
  tall bellflower - *Campanula americana*
  tussock bellflower - *Campanula carpatica*
  willow bellflower - *Campanula persicifolia*
bells
  bells-of-Ireland - *Moluccella laevis*
  Canterbury bells - *Campanula medium*
  coral-bells - *Heuchera sanguinea*
  fairy-bells - *Disporum hookeri*
  feather-bells - *Stenanthium gramineum*
  golden-bells - *Forsythia viridissima*
  merry-bells - *Uvularia*
  Oconee bells - *Shortia galacifolia*
  Palm Beach bells - *Kalanchoe*
  peach-bells - *Campanula persicifolia*
  rabbit-bells - *Crotalaria rotundifolia*
  Roanoke bells - *Mertensia virginica*
  sweet-bells - *Leucothoe racemosa*
  velvet-bells - *Bartsia alpina*
  yellow-bells - *Tecoma stans*
bellwort
  bellwort - *Uvularia*
  clasping-bellwort - *Triodanis perfoliata*
  large-flowered bellwort - *Uvularia grandiflora*
  little bellwort - *Uvularia sessilifolia*
  perfoliate bellwort - *Uvularia perfoliata*
  sessile bellwort - Uvularia sessilifolia

Belmore's
    Belmore's palm - *Howeia belmoreana*
    Belmore's sentry palm - *Howeia belmoreana*
belvedere - *Bassia scoparia*
Bend, Big Bend loco - *Astragalus earlei*
Bengal
    Bengal gram - *Cicer arietinum*
    Bengal grass - *Setaria italica*
    Bengal rose - *Rosa chinensis*
    Bengal velvet bean - *Mucuna pruriens* var. *utilis*
Benjamin, stinking-Benjamin - *Trillium erectum*
Benjamin's
    Benjamin's-bush - *Lindera benzoin*
    Benjamin's-tree - *Ficus benjamina*
bent
    autumn bent grass - *Agrostis perennans*
    bent-awn plume grass - *Saccharum contortum*
    bent foxtail - *Alopecurus geniculatus*
    bent grass - *Agrostis capillaris, Agrostis*
    bent trillium - *Trillium flexipes*
    black bent grass - *Agrostis gigantea*
    brown bent grass - *Agrostis canina, A. perennans*
    colonial bent grass - *Agrostis capillaris*
    creeping bent grass - *Agrostis stolonifera*
    dune bent grass - *Agrostis pallens*
    Oregon bent grass - *Agrostis oregonensis*
    Rhode Island bent grass - *Agrostis capillaris*
    rough bent grass - *Agrostis scabra*
    upland bent grass - *Agrostis perennans*
    velvet bent grass - *Agrostis canina*
    water-bent - *Polypogon semiverticillatus*
    winter bent grass - *Agrostis hyemalis*
Bentham's annual lupine - *Lupinus benthamii*
bergamot
    bergamot mint - *Mentha piperita* var. *citrata*
    wild bergamot - *Monarda fistulosa, Monarda*
Berlandier's
    Berlandier's acacia - *Acacia berlandieri*
    Berlandier's wolfberry - *Lycium berlandieri*
Bermuda
    African Bermuda grass - *Cynodon transvaalensis*
    Bermuda buttercup - *Oxalis pes-caprae*
    Bermuda Easter lily - *Lilium longiflorum* var. *eximium*
    Bermuda grass - *Cynodon dactylon*
berried
    red-berried buckthorn - *Rhamnus ilicifolia*
    red-berried elder - *Sambucus racemosa, S. racemosa*
        var. *microbotrys*
    red-berried greenbrier - *Smilax walteri*
berry
    American beauty-berry - *Callicarpa americana*
    baked-apple-berry - *Rubus chamaemorus*
    berry bladder fern - *Cystopteris bulbifera*
    berry saltbush - *Atriplex semibaccata*
    blood-berry - *Rivina humilis*
    buffalo-berry - *Solanum rostratum*
    cat-berry - *Nemopanthus mucronatus*
    China-berry - *Melia azedarach*
    Chinese beauty-berry - *Callicarpa dichotoma*
    Christmas-berry - *Crossopetalum ilicifolia, Heteromeles*
        arbutifolia, Lycium carolinianum

Christmas-berry-tree - *Schinus terebinthifolius*
coffee-berry - *Rhamnus californica*
coral-berry - *Ardisia crenata, Symphoricarpos chenaultii,*
    S. orbiculatus
cracker-berry - *Cornus canadensis*
crimson-berry - *Rubus arcticus*
curlew-berry - *Empetrum nigrum*
deer-berry - *Vaccinium caesium, V. stamineum*
dingle-berry - *Vaccinium erythrocarpum*
fox-berry - *Vaccinium vitis-idaea*
goar-berry gourd - *Cucumis anguria*
hag-berry - *Prunus padus*
honey-berry - *Melicoccus bijugatus*
jaundice-berry - *Berberis vulgaris*
lemonade-berry - *Rhus integrifolia*
lime-berry - *Triphasia trifolia*
ling-berry - *Vaccinium vitis-idaea* var. *minus*
maidenhair-berry - *Gaultheria hispidula*
male-berry - *Lyonia ligustrina*
meadow-berry - *Rhexia*
meal-berry - *Arctostaphylos uva-ursi*
miraculous-berry - *Synsepalum dulcificum*
moor-berry - *Vaccinium uliginosum*
moose-berry - *Viburnum edule*
nagoon-berry - *Rubus stellatus*
Panama-berry - *Muntingia calabura*
pigeon-berry - *Duranta erecta, Phytolacca americana*
poison-berry - *Solanum nigrum*
pudding-berry - *Cornus canadensis*
purple beauty-berry - *Callicarpa dichotoma*
rabbit-berry - *Shepherdia canadensis*
raccoon-berry - *Podophyllum peltatum*
red-berry - *Rhamnus crocea*
red-berry juniper - *Juniperus pinchotii*
red-berry moonseed - *Cocculus carolinus*
russet buffalo-berry - *Shepherdia canadensis*
sand-berry - *Arctostaphylos uva-ursi*
sapphire-berry sweet-leaf - *Symplocos tinctoria*
silver buffalo-berry - *Shepherdia argentea*
sour-berry - *Rhus integrifolia*
spiny red-berry - *Rhamnus crocea*
squash-berry - *Viburnum edule*
squaw-berry - *Mitchella repens*
summer-berry - *Viburnum trilobum*
sweet-berry honeysuckle - *Lonicera caerulea*
treacle-berry - *Smilacina racemosa*
turkey-berry - *Solanum torvum*
water-berry - *Lonicera caerulea*
whin-berry - *Vaccinium myrtillus*
wine-berry - *Rubus phoenicolasius*
yellow-berry - *Rubus chamaemorus*
berseem clover - *Trifolium alexandrinum*
Bessey's cherry - *Prunus pumila* var. *besseyi*
bet, bouncing-bet - *Saponaria officinalis*
Bethlehem
    nodding star-of-Bethlehem - *Ornithogalum nutans*
    star-of-Bethlehem - *Ornithogalum umbellatum*
betony
    betony - *Stachys*
    field-nettle betony - *Stachys arvensis*
    hedge-nettle betony - *Stachys annua*
    wood-betony - *Pedicularis canadensis, Pedicularis*

Bhutan pine - *Pinus wallichiana*
Bicknell's crane's-bill - *Geranium bicknellii*
bicolored
  bicolored lespedeza - *Lespedeza bicolor*
  bicolored lupine - *Lupinus bicolor*
biennial
  biennial campion - *Silene csereii*
  biennial cinquefoil - *Potentilla biennis*
  biennial gaura - *Gaura biennis*
  biennial lettuce - *Lactuca biennis*
  biennial wormwood - *Artemisia biennis*
Big Bend loco - *Astragalus earlei*
big
  big blue lily-turf - *Liriope muscari*
  big blue-stem - *Andropogon gerardii*
  big bluegrass - *Poa secunda* subsp. *nevadensis*
  big carpet grass - *Axonopus furcatus*
  big-cone Douglas fir - *Pseudotsuga macrocarpa*
  big-cone pine - *Pinus coulteri*
  big-cone-spruce - *Pseudotsuga macrocarpa*
  big cord grass - *Spartina cynosuroides*
  big floating-heart - *Nymphoides aquatica*
  big-flower clover - *Trifolium michelianum*
  big galleta - *Hilaria rigida*
  big-head knapweed - *Centaurea macrocephala*
  big-leaf aster - *Aster macrophyllus*
  big-leaf avens - *Geum macrophyllum*
  big-leaf holly - *Ilex ambigua* var. *montana*
  big-leaf linden - *Tilia platyphyllos*
  big-leaf magnolia - *Magnolia macrophylla*
  big-leaf maple - *Acer macrophyllum*
  big-leaf periwinkle - *Vinca major*
  big-leaf pondweed - *Potamogeton amplifolius*
  big-leaf snowbell - *Styrax grandifolus*
  big marigold - *Tagetes erecta*
  big marsh-elder - *Iva xanthifolia*
  big num-num - *Carissa macrocarpa*
  big periwinkle - *Vinca major*
  big-root - *Marah*
  big-root morning-glory - *Ipomoea pandurata*
  big sagebrush - *Artemisia tridentata*
  big tick-trefoil - *Desmodium cuspidatum*
  big-tooth aspen - *Populus grandidentata*
  big-tooth maple - *Acer grandidentatum*
  big-top love grass - *Eragrostis hirsuta*
  big-tree - *Sequoiadendron giganteum*
  big tree plum - *Prunus mexicana*
  big trefoil - *Lotus uliginosus*
  big white trillium - *Trillium grandiflorum*
bigarade - *Citrus aurantium*
Bigelow's
  Bigelow's beggar-ticks - *Bidens bigelovii*
  Bigelow's bluegrass - *Poa bigelovii*
bignay - *Antidesma bunius*
bilberry
  bilberry - *Vaccinium myrtillus, Vaccinium*
  bog bilberry - *Vaccinium uliginosum*
  dwarf bilberry - *Vaccinium caespitosum*
  early sweet bilberry - *Vaccinium vacillans*
  low bilberry - *Vaccinium vacillans*
  mountain bilberry - *Vaccinium membranaceum*
  tall bilberry - *Vaccinium ovalifolium*

bill
  Bicknell's crane's-bill - *Geranium bicknellii*
  crane's-bill - *Geranium*
  heron's-bill - *Erodium*
  Iberian crane's-bill - *Geranium ibericum*
  long-stalked crane's-bill - *Geranium columbinum*
  Siberian crane's-bill - *Geranium sibiricum*
  small-flowered crane's-bill - *Geranium pusillum*
  spotted crane's-bill - *Geranium maculatum*
  stork's-bill - *Erodium, Pelargonium*
billion-dollar grass - *Echinochloa frumentacea*
bills, mosquito-bills - *Dodecatheon hendersonii*
bilsted - *Liquidambar styraciflua*
Biltmore hawthorn - *Crataegus intricata*
bimble-box - *Eucalyptus populnea*
bimli-jute - *Hibiscus cannabinus*
bind
  bear-bind - *Convolvulus arvensis*
  corn-bind - *Convolvulus arvensis*
bindweed
  bindweed - *Calystegia, Convolvulus arvensis*
  black-bindweed - *Polygonum convolvulus*
  bracted bindweed - *Calystegia sepium*
  corn-bindweed - *Polygonum convolvulus*
  field bindweed - *Convolvulus arvensis*
  great bindweed - *Calystegia sepium*
  hedge bindweed - *Calystegia sepium*
  Japanese bindweed - *Calystegia hederacea*
  low bindweed - *Calystegia spithamaea*
  small bindweed - *Convolvulus arvensis*
bine
  bine - *Humulus lupulus*
  tree-bine - *Cissus*
  Venezuelan tree-bine - *Cissus rhombifolia*
birch
  Ashe's birch - *Betula uber*
  birch - *Betula*
  birch-lead mountain-mahogany - *Cercocarpus betuloides*
  birch-leaf buckthorn - *Rhamnus betulifolia*
  black birch - *Betula lenta, B. nigra*
  blue birch - *Betula caerulea*
  bog birch - *Betula glandulosa*
  canoe birch - *Betula papyrifera*
  cherry birch - *Betula lenta*
  dwarf birch - *Betula glandulosa*
  dwarf white birch - *Betula minor*
  European white birch - *Betula pendula*
  fire birch - *Betula populifolia*
  gray birch - *Betula alleghaniensis, B. populifolia, B. pumila*
  Kenai birch - *Betula papyrifera* var. *kenaica*
  low birch - *Betula pumila*
  mahogany birch - *Betula lenta*
  mountain birch - *Betula occidentalis*
  old-field birch - *Betula populifolia*
  paper birch - *Betula papyrifera*
  red birch - *Betula nigra*
  resin birch - *Betula glandulosa*
  river birch - *Betula nigra*
  Sandberg's birch - *Betula sandbergii*
  swamp birch - *Betula pumila*
  sweet birch - *Betula lenta*
  water birch - *Betula occidentalis*

West Indian birch - *Bursera simaruba*
white birch - *Betula papyrifera, B. populifolia*
yellow birch - *Betula alleghaniensis, B. pumila*
bird
   bird-catcher-tree - *Pisonia umbellifera*
   bird cherry - *Prunus avium, P. pensylvanica*
   bird grape - *Vitis munsoniana*
   bird-of-paradise - *Caesalpinia gilliesii, Strelitzia reginae*
   bird-on-the-wing - *Polygala paucifolia*
   bird pepper - *Capsicum frutescens*
   bird-seed grass - *Phalaris canariensis*
   bird vetch - *Vicia cracca*
   European bird cherry - *Prunus padus*
   queen's bird-of-paradise - *Strelitzia reginae*
   red-bird-cactus - *Pedilanthus tithymaloides*
   red-bird-flower - *Pedilanthus tithymaloides*
bird's
   bird's-eye - *Adonis annua, Veronica persica*
   bird's-eye pearl-wort - *Sagina procumbens*
   bird's-eye primrose - *Primula laurentiana*
   bird's-eyes - *Gilia tricolor*
   bird's-foot buttercup - *Ranunculus orthorhynchus*
   bird's-foot trefoil - *Lotus corniculatus*
   bird's-foot violet - *Viola pedata*
   bird's-nest - *Daucus carota, Pastinaca sativa*
   bird's-nest fern - *Asplenium nidus*
   bird's-rape - *Brassica rapa*
   bird's-rape mustard - *Brassica rapa*
   climbing bird's nest fern - *Microsorium punctatum*
   giant bird's-nest - *Pterospora andromedea*
   greater bird's-foot trefoil - *Lotus pedunculatus*
   small-flowered bird's-foot trefoil - *Lotus micranthus*
biriba - *Rollinia pulchrinervis*
birthroot - *Trillium*
birthwort - *Aristolochia*
biscuit-root - *Lomatium bicolor, Lomatium*
bishop's
   Atlantic mock bishop's-weed - *Ptilimnium capillaceum*
   bishop's-cap - *Astrophytum myriostigma, A. ornatum, Mitella*
   bishop's gout-weed - *Aegopodium podagraria*
   bishop's-hood - *Astrophytum myriostigma*
   bishop's pine - *Pinus muricata*
   bishop's-weed - *Ptilimnium capillaceum*
   mock bishop's-weed - *Ptilimnium capillaceum*
bisnaga - *Ammi visnaga*
bistort
   alpine bistort - *Polygonum viviparum*
   bistort - *Polygonum bistortoides*
   western bistort - *Polygonum bistortoides*
bit
   American frog's-bit - *Limnobium spongia*
   devil's-bit - *Succisa australis*
   fall hawk's-bit - *Leontodon autumnalis*
   hairy hawk's-bit - *Leontodon hirtus*
   hawk's-bit - *Leontodon autumnalis*
   rough hawk's-bit - *Leontodon hirtus, L. taraxacoides*
   sheep's-bit - *Jasione montana*
bitter
   bitter aloe - *Aloe vera*
   bitter-apple - *Citrullus colocynthis*
   bitter-bloom - *Sabatia angularis*
   bitter-bush - *Chromolaena odorata, Picramnia pentandra*

bitter cassava - *Manihot esculenta*
bitter cherry - *Prunus emarginata*
bitter-cress - *Barbarea vulgaris, Cardamine*
bitter cucumber - *Momordica charantia*
bitter damson - *Simarouba glauca*
bitter dock - *Rumex obtusifolius*
bitter dogbane - *Apocynum androsaemifolium*
bitter gallberry - *Ilex glabra*
bitter gourd - *Momordica charantia*
bitter hickory - *Carya aquatica*
bitter mint - *Hyptis mutabilis*
bitter-nut hickory - *Carya cordiformis*
bitter orange - *Citrus aurantium*
bitter pecan - *Carya aquatica*
bitter-root - *Lewisia rediviva*
bitter rubber-weed - *Hymenoxys odorata*
bitter sneezeweed - *Helenium amarum*
bitter-weed - *Ambrosia artemisiifolia, A. trifida, Artemisia biennis, Conyza canadensis, Helenium amarum*
bitter-wood - *Simarouba glauca*
cuckoo bitter-cress - *Cardamine pratensis*
dry-land bitter-cress - *Cardamine parviflora*
hairy bitter-cress - *Cardamine hirsuta*
little bitter-cress - *Cardamine oligosperma*
Pennsylvania bitter-cress - *Cardamine pensylvanica*
small-flowered bitter-cress - *Cardamine parviflora*
trailing bitter-cress - *Cardamine rotundifolia*
bittersweet
   American bittersweet - *Celastrus scandens*
   bittersweet - *Celastrus*
   bittersweet nightshade - *Solanum dulcamara*
   climbing bittersweet - *Celastrus scandens*
   false bittersweet - *Celastrus scandens*
   Oriental bittersweet - *Celastrus orbiculatus*
   shrubby bittersweet - *Celastrus scandens, Celastrus*
black
   American black currant - *Ribes americanum*
   American black nightshade - *Solanum americanum*
   Australian black-wood - *Acacia melanoxylon*
   black alder - *Alnus glutinosa*
   black ash - *Fraxinus nigra*
   black bamboo - *Phyllostachys nigra*
   black-bead - *Pithecellobium keyense, P. unguis-cati*
   black-bead elder - *Sambucus melanocarpa*
   black bean - *Castanospermum australe*
   black bearberry - *Arctostaphylos alpina*
   black bent grass - *Agrostis gigantea*
   black-bindweed - *Polygonum convolvulus*
   black birch - *Betula lenta, B. nigra*
   black-brush acacia - *Acacia rigidula*
   black bulrush - *Scirpus atrovirens*
   black-butt - *Eucalyptus pilularis*
   black-cap - *Rubus occidentalis*
   black chokeberry - *Aronia melanocarpa*
   black chokecherry - *Prunus virginiana* var. *melanocarpa*
   black cohosh - *Cimicifuga racemosa*
   black cottonwood - *Populus heterophylla, P. trichocarpa*
   black crowberry - *Empetrum nigrum*
   black currant - *Ribes nigrum*
   black dalea - *Dalea frutescens*
   black-edged sedge - *Carex nigromarginata* var. *elliptica*
   black-eyed pea - *Vigna unguiculata*
   black-eyed-Susan - *Rudbeckia hirta, R. hirta* var. *pulcherrima*

black fiber palm - *Arenga pinnata*
black-fringe knotweed - *Polygonum cilinode*
black-geranium - *Heuchera americana*
black gram - *Vigna mungo*
black grama - *Bouteloua eriopoda*
black grass - *Alopecurus myosuroides*
black gum - *Nyssa sylvatica*
black-haw - *Viburnum lentago, V. prunifolium*
black-haw viburnum - *Viburnum prunifolium*
black hawthorn - *Crataegus douglasii*
black henbane - *Hyoscyamus niger*
black hickory - *Carya texana*
black high-bush blueberry - *Vaccinium atrococcum*
black horehound - *Ballota nigra*
black huckleberry - *Gaylussacia baccata, Vaccinium atrococcum*
black jet-bead - *Rhodotypos scandens*
black knapweed - *Centaurea nigra*
black larch - *Larix laricina*
black-laurel - *Gordonia lasianthus*
black locust - *Robinia pseudoacacia*
black mangrove - *Avicennia germinans*
black maple - *Acer nigrum*
black medic - *Medicago lupulina*
black mountain-ash - *Eucalyptus sieber*
black mulberry - *Morus nigra*
black mustard - *Brassica nigra*
black nightshade - *Solanum nigrum*
black oak - *Quercus velutina*
black oat grass - *Stipa avenacea*
black-olive - *Bucida buceras*
black oyster-plant - *Scorzonera hispanica*
black pepper - *Piper nigrum*
black persimmon - *Diospyros texana*
black pine - *Pinus nigra*
black-plum - *Syzygium cumini*
black-pod vetch - *Vicia sativa* subsp. *nigra*
black poplar - *Populus nigra*
black raspberry - *Rubus occidentalis*
black-root - *Veronicastrum virginicum*
black sage - *Salvia mellifera*
black salsify - *Scorzonera hispanica*
black sedge - *Carex atratiformis*
black-seed needle grass - *Stipa avenacea*
black-seed plantain - *Plantago rugelii*
black snakeroot - *Cimicifuga racemosa, Sanicula marilandica, Sanicula, Zigadenus densus*
black spruce - *Picea mariana*
black sugar maple - *Acer nigrum*
black swallow-wort - *Cynanchum louiseae*
black-thorn - *Crataegus calpodendron, Prunus spinosa*
black titi - *Cliftonia monophylla, Cyrilla racemiflora*
black tupelo - *Nyssa sylvatica*
black walnut - *Juglans nigra*
black-weed - *Ambrosia acanthicarpa*
black willow - *Salix nigra*
black-wood - *Acacia melanoxylon, Avicennia germinans*
black-wood acacia - *Acacia melanoxylon*
bristly black currant - *Ribes lacustre*
California black currant - *Ribes bracteosum*
California black oak - *Quercus kelloggii*
California black walnut - *Juglans californica*
eastern black currant - *Ribes americanum*

ebony black-bead - *Pithecellobium flexicaule*
European black currant - *Ribes nigrum*
garden black currant - *Ribes nigrum*
gray-black grape - *Vitis cinerea*
Japanese black pine - *Pinus thunbergiana*
northern black currant - *Ribes hudsonianum*
northern California black walnut - *Juglans hindsii*
rusty black-haw - *Viburnum rufidulum*
southern black-haw - *Viburnum rufidulum*
Texas black walnut - *Juglans microcarpa*
West Indian black-thorn - *Acacia farnesiana*
wild black cherry - *Prunus serotina*
wild black currant - *Ribes americanum*
blackberry
  Allegheny blackberry - *Rubus allegheniensis*
  arctic blackberry - *Rubus arcticus*
  blackberry - *Rubus*
  blackberry-lily - *Belamcanda chinensis*
  California blackberry - *Rubus ursinus*
  crimson blackberry - *Rubus arcticus*
  cut-leaf blackberry - *Rubus laciniatus*
  European blackberry - *Rubus fruticosus*
  evergreen blackberry - *Rubus laciniatus*
  high-bush blackberry - *Rubus argutus*
  Himalayan blackberry - *Rubus discolor*
  Pacific blackberry - *Rubus ursinus*
  parsley-leaf blackberry - *Rubus laciniatus*
  running blackberry - *Rubus flagellaris, R. hispidus*
  sand blackberry - *Rubus cuneifolius*
  smooth blackberry - *Rubus canadensis*
  southern blackberry - *Rubus argutus*
  sow-teat blackberry - *Rubus allegheniensis*
  swamp blackberry - *Rubus hispidus*
blackening-plant - *Hibiscus rosa-sinensis*
blackgum, swamp blackgum - *Nyssa sylvatica* var. *biflora*
blackjack oak - *Quercus marilandica*
bladder
  American bladder-nut - *Staphylea trifolia*
  berry bladder fern - *Cystopteris bulbifera*
  bladder campion - *Silene vulgaris*
  bladder ketmia - *Hibiscus trionum*
  bladder-nut - *Staphylea*
  bladder-parsnip - *Lomatium utriculatum*
  bladder-pod - *Lesquerella, Lobelia inflata*
  bladder sedge - *Carex intumescens*
  bladder-senna - *Colutea arborescens*
  brittle bladder fern - *Cystopteris fragilis*
  bulblet bladder fern - *Cystopteris bulbifera*
  Gordon's bladder-pod - *Lesquerella gordonii*
  mountain bladder fern - *Cystopteris montana*
  sierra bladder-nut - *Staphylea bolanderi*
bladderwort
  bladderwort - *Utricularia vulgaris*
  floating bladderwort - *Utricularia inflata, U. radiata*
  Florida bladderwort - *Utricularia floridana*
  inflated bladderwort - *Utricularia inflata*
  leafy bladderwort - *Utricularia foliosa*
  minute bladderwort - *Utricularia olivacea*
  purple bladderwort - *Utricularia purpurea*
  slender-stem bladderwort - *Utricularia gibba*
  spotted bladderwort - *Utricularia purpurea*
  swollen bladderwort - *Utricularia radiata*

tangled bladderwort - *Utricularia biflora*
blanco, zapote-blanco - *Casimiroa edulis*
bland sweet cicely - *Osmorhiza claytonii*
blanket
    blanket-flower - *Gaillardia*
    blanket indigo - *Indigofera pilosa*
    rose-ring blanket-flower - *Gaillardia pulchella*
blaspheme-vine - *Smilax laurifolia*
blazing
    blazing-star - *Liatris spicata, Liatris, Mentzelia laevicaulis,*
      Mentzelia
    cylindric blazing-star - *Liatris cylindracea*
    dense blazing-star - *Liatris spicata*
    dotted blazing-star - *Liatris punctata*
    few-headed blazing-star - *Liatris cylindracea*
    plains blazing-star - *Liatris squarrosa*
    prairie blazing-star - *Liatris pycnostachya*
    scaly blazing-star - *Liatris squarrosa*
    sessile blazing-star - *Liatris spicata*
    thick-spike blazing-star - *Liatris pycnostachya*
bleed, nose-bleed - *Achillea millefolium*
bleeding
    bleeding-glory-bower - *Clerodendrum thompsoniae*
    bleeding-heart - *Dicentra spectabilis*
    love-lies-bleeding - *Amaranthus caudatus*
    tropical bleeding-heart - *Clerodendrum thompsoniae*
    western bleeding-heart - *Dicentra formosa*
blessed
    blessed milk-thistle - *Silybum marianum*
    blessed thistle - *Cnicus benedictus, Silybum marianum*
blimbing - *Averrhoa carambola*
blind gentian - *Gentiana clausa*
blister-cress - *Erysimum*
blite
    blite goosefoot - *Chenopodium capitatum*
    blite-mulberry - *Chenopodium capitatum*
    coast-blite - *Chenopodium rubrum*
    frost-blite - *Chenopodium album*
    strawberry-blite - *Chenopodium capitatum*
blob, May-blob - *Caltha palustris*
blond tree-fern - *Cibotium splendens*
blonde-Lilian - *Erythronium albidum*
blood
    African blood-lily - *Haemanthus coccineus*
    blood-berry - *Rivina humilis*
    blood-flower milkweed - *Asclepias curassavica*
    blood-lily - *Haemanthus coccineus*
    blood milkwort - *Polygala sanguinea*
    blood ragweed - *Ambrosia trifida* var. *texana*
    blood-red geranium - *Geranium sanguineum*
    blood-red netbush - *Calothamnus sanguineus*
    blood-stanch - *Conyza canadensis*
    blood-twig dogwood - *Cornus sanguinea*
bloodleaf - *Iresine diffusa*
bloodroot - *Sanguinaria canadensis*
bloodwort
    bloodwort - *Achillea millefolium*
    burnet-bloodwort - *Sanguisorba officinalis*
    mouse-bloodwort - *Hieracium pilosella*
bloody geranium - *Geranium sanguineum*

bloom
    bitter-bloom - *Sabatia angularis*
    checker-bloom - *Sidalcea malvaeflora*
    honey-bloom - *Apocynum androsaemifolium*
    pinxter-bloom - *Rhododendron periclymenoides*
blooming
    day-blooming cestrum - *Cestrum diurnum*
    house blooming mock-orange - *Pittosporum tobira*
    night-blooming cereus - *Hylocereus undatus*
blossum, blue-blossum - *Ceanothus thyrsiflorus*
blowballs - *Taraxacum*
Blue Ridge St. John's-wort - *Hypericum mitchellianum*
blue
    baby-blue-eyes - *Nemophila menziesii*
    big blue lily-turf - *Liriope muscari*
    big blue-stem - *Andropogon gerardii*
    blue agapanthus - *Agapanthus africanus*
    blue ash - *Fraxinus quadrangulata*
    blue barberry - *Berberis aquifolium*
    blue-bead-lily - *Clintonia borealis*
    blue-beech - *Carpinus caroliniana*
    blue birch - *Betula caerulea*
    blue-blossom - *Ceanothus thyrsiflorus*
    blue boneset - *Conoclinium coelestinum*
    blue-brush - *Ceanothus thyrsiflorus*
    blue-bunch fescue - *Festuca idahoensis*
    blue bunch grass - *Festuca idahoensis*
    blue-bunch wheat grass - *Pseudoroegneria spicata*
    blue-buttons - *Knautia arvensis, Vinca major*
    blue cat-tail - *Typha glauca*
    blue cohosh - *Caulophyllum thalictroides*
    blue comfrey - *Symphytum uplandicum*
    blue-couch - *Digitaria scalarum*
    blue-curls - *Trichostema*
    blue daisy - *Cichorium intybus*
    blue dawn-flower - *Ipomoea indica*
    blue-devil - *Echium vulgare*
    blue-dicks - *Dichelostemma pulchellum*
    blue dogbane - *Amsonia tabernaemontana*
    blue elder - *Sambucus cerulea*
    blue elderberry - *Sambucus cerulea, S. mexicana*
    blue-eyed African daisy - *Arctotis stoechadifolia*
    blue-eyed Cape-marigold - *Dimorphotheca sinuata*
    blue-eyed-grass - *Sisyrinchium*
    blue-eyed-Mary - *Collinsia verna*
    blue false indigo - *Baptisia australis*
    blue field madder - *Sherardia arvensis*
    blue flag iris - *Iris versicolor*
    blue-flower squill - *Hyacinthoides hispanica*
    blue giant hyssop - *Agastache foeniculum*
    blue grama - *Bouteloua gracilis*
    blue gum - *Eucalyptus globulus*
    blue heart-leaf aster - *Aster cordifolius*
    blue-hearts - *Buchnera*
    blue huckleberry - *Vaccinium membranaceum*
    blue Jacob's-ladder - *Polemonium caeruleum*
    blue-jasmine - *Clematis crispa*
    blue-joint - *Calamagrostis canadensis*
    blue lace-flower - *Trachymene coerulea*
    blue lettuce - *Lactuca tatarica* subsp. *pulchella*
    blue-lips - *Collinsia grandiflora*
    blue lobelia - *Lobelia siphilitica*

blue lupine - *Lupinus angustifolius, L. pilosus*
blue marsh violet - *Viola cucullata*
blue morning-glory - *Ipomoea indica*
blue mustard - *Chorispora tenella*
blue oak - *Quercus douglasii*
blue palo-verde - *Parkinsonia florida*
blue passionflower - *Passiflora caerulea*
blue pea - *Clitoria ternatea, Lupinus perennis*
blue phlox - *Phlox divaricata*
blue pine - *Pinus wallichiana*
blue plantain-lily - *Hosta ventricosa*
blue poppy - *Meconopsis betonicifolia*
blue sage - *Salvia azurea*
blue-sailors - *Cichorium intybus*
blue skullcap - *Scutellaria lateriflora*
blue spring daisy - *Erigeron pulchellus*
blue-star - *Amsonia tabernaemontana, Amsonia*
blue-stem - *Andropogon glomeratus, Andropogon,*
    Schizachyrium scoparium
blue-stem goldenrod - *Solidago caesia*
blue-stem palmetto - *Sabal minor*
blue-stem prickle poppy - *Argemone albiflora*
blue-tangle - *Gaylussacia frondosa*
blue-thistle - *Echium vulgare*
blue toadflax - *Linaria canadensis*
blue verbena - *Verbena hastata*
blue vervain - *Verbena hastata*
blue-vine - *Cynanchum laeve*
blue violet - *Viola papilionacea*
blue water-lily - *Nymphaea elegans*
blue wax-weed - *Cuphea viscosissima*
blue-weed - *Helianthus ciliaris*
blue wild indigo - *Baptisia australis*
blue wild rye - *Elymus glaucus*
blue wiss - *Teramnus labialis*
blue-wood aster - *Aster cordifolius*
blue yucca - *Yucca baccata*
broom sedge blue-stem - *Andropogon virginicus*
bushy blue-stem - *Andropogon glomeratus*
cane blue-stem - *Bothriochloa barbinodis*
Cape blue water-lily - *Nymphaea capensis*
Colorado blue spruce - *Picea pungens*
Diaz's blue-stem - *Dichanthium annulatum*
early blue violet - *Viola palmata*
eastern blue-eyed-Mary - *Collinsia verna*
Elliott's blue-stem - *Andropogon elliottii*
European blue lupine - *Lupinus angustifolius*
Hall's blue-stem - *Andropogon hallii*
King Ranch blue-stem - *Bothriochloa ischaemum,*
    B. ischaemum var. *songarica*
Kleberg's blue-stem - *Dichanthium annulatum*
larger blue flag - *Iris versicolor*
little blue-stem - *Schizachyrium scoparium*
marsh blue violet - *Viola cucullata*
Mexican blue oak - *Quercus oblongifolia*
Mexican blue-wood - *Condalia mexicana*
midwestern blue heart-leaf aster - *Aster shortii*
Montana blue-eyed-grass - *Sisyrinchium montanum*
narrow-leaf blue-eyed-grass - *Sisyrinchium angustifolium*
northern blue flag - *Iris versicolor*
pitted blue-stem - *Bothriochloa pertusa*
plains blue-stem - *Bothriochloa ischaemum*
red-white-and-blue-flower - *Cuphea ignea*

sand blue-stem - *Andropogon hallii*
silver blue-stem - *Bothriochloa saccharoides*
southern blue flag - *Iris virginica* var. *schrevei*
Sydney blue gum - *Eucalyptus saligna*
Tasmanian blue gum - *Eucalyptus globulus*
Texas blue-weed - *Helianthus ciliaris*
Turkestan blue-stem - *Bothriochloa ischaemum*
western blue flag - *Iris missouriensis*
white baby-blue-eyes - *Nemophila menziesii* var. *atomaria*
wild blue flax - *Linum grandiflorum, L. perenne*
woolly blue violet - *Viola sororia*
yellow blue-stem - *Bothriochloa ischaemum*
bluebell - *Campanula rotundifolia*
bluebells
    bluebells - *Mertensia*
    eastern bluebells - *Mertensia virginica*
    northern bluebells - *Mertensia paniculata*
    seaside bluebells - *Mertensia maritima*
    Virginia bluebells - *Mertensia virginica*
blueberry
    alpine blueberry - *Vaccinium uliginosum*
    black high-bush blueberry - *Vaccinium atrococcum*
    blueberry - *Vaccinium*
    blueberry elder - *Sambucus cerulea*
    creeping blueberry - *Vaccinium crassifolium*
    Elliott's blueberry - *Vaccinium elliottii*
    high-bush blueberry - *Vaccinium caesariense, V. corymbosum*
    hillside blueberry - *Vaccinium vacillans*
    late sweet blueberry - *Vaccinium angustifolium*
    low blueberry - *Vaccinium angustifolium, V. vacillans*
    low blueberry willow - *Salix myrtillifolia*
    low-bush blueberry - *Vaccinium angustifolium*
    low sweet blueberry - *Vaccinium angustifolium*
    male-blueberry - *Lyonia ligustrina*
    mountain blueberry - *Vaccinium membranaceum*
    New Jersey blueberry - *Vaccinium caesariense*
    rabbit's-eye blueberry - *Vaccinium ashei*
    sour-top blueberry - *Vaccinium myrtilloides*
    southern high-bush blueberry - *Vaccinium elliottii*
    swamp blueberry - *Vaccinium corymbosum*
    sweet blueberry - *Vaccinium vacillans*
    tall blueberry willow - *Salix novaeangliae*
    velvet-leaf blueberry - *Vaccinium myrtilloides*
    western blueberry - *Vaccinium occidentalis*
    bluebonnet, Texas bluebonnet - *Lupinus subcarnosus,*
        L. texensis
bluebottle - *Centaurea cyanus*
bluegrass
    alkali bluegrass - *Poa secunda* subsp. *nevadensis*
    alpine bluegrass - *Poa alpina*
    annual bluegrass - *Poa annua*
    arctic bluegrass - *Poa arctica*
    big bluegrass - *Poa secunda* subsp. *nevadensis*
    Bigelow's bluegrass - *Poa bigelovii*
    bluegrass - *Poa*
    bog bluegrass - *Poa leptocoma*
    bulbous bluegrass - *Poa bulbosa*
    Canadian bluegrass - *Poa compressa*
    Canby's bluegrass - *Poa secunda*
    Cusick's bluegrass - *Poa cusickii*
    English bluegrass - *Festuca pratensis*
    fowl bluegrass - *Poa palustris*

inland bluegrass - *Poa interior*
Kentucky bluegrass - *Poa pratensis*
Leiberg's bluegrass - *Poa leibergii*
mutton bluegrass - *Poa fendleriana, Setaria faberi*
Nevada bluegrass - *Poa secunda* subsp. *nevadensis*
Pacific bluegrass - *Poa secunda*
pine bluegrass - *Poa scabrella, P. secunda*
plains bluegrass - *Poa arida*
rough bluegrass - *Poa trivialis*
rough-stalk bluegrass - *Poa palustris, P. trivialis*
Sandberg's bluegrass - *Poa secunda*
skyline bluegrass - *Poa epilis*
slender bluegrass - *Poa secunda*
Texas bluegrass - *Poa arachnifera*
timberline bluegrass - *Poa rupicola*
Wheeler's bluegrass - *Poa nervosa*
wood bluegrass - *Poa nemoralis*
bluejack oak - *Quercus incana*
bluet, mountain bluet - *Centaurea montana*
bluets - *Hedyotis caerulea*
blueweed - *Echium vulgare*
bluff oak - *Quercus austrina*
blunt
blunt-broom sedge - *Carex tribuloides*
blunt-leaf dock - *Rumex obtusifolius*
blunt-leaf milkweed - *Asclepias amplexicaulis*
blunt spike rush - *Eleocharis ovata*
boat
boat-lily - *Tradescantia spathacea*
man-in-a-boat - *Tradescantia spathacea*
Moses-in-a-boat - *Tradescantia spathacea*
three-men-in-a-boat - *Tradescantia spathacea*
Boer's love grass - *Eragrostis curvula*
bog
Alaska bog willow - *Salix fuscescens*
bog aster - *Aster nemoralis*
bog bilberry - *Vaccinium uliginosum*
bog birch - *Betula glandulosa*
bog bluegrass - *Poa leptocoma*
bog buttercup - *Ranunculus sceleratus*
bog-candle - *Habenaria dilatata*
bog club-moss - *Lycopodium inundatum*
bog goldenrod - *Solidago uliginosa*
bog-hemp - *Boehmeria cylindrica*
bog kalmia - *Kalmia polifolia*
bog-laurel - *Kalmia polifolia*
bog-mat - *Wolffiella gladiata*
bog-moss - *Mayaca fluviatilis*
bog myrtle - *Myrica gale*
bog-rosemary - *Andromeda glaucophylla, Andromeda*
bog rush - *Juncus effusus, Juncus*
bog spruce - *Picea mariana*
bog-stars - *Parnassia californica, Parnassia*
flat bog-mat - *Wolffiella floridana*
leafy bog aster - *Aster nemoralis*
northern bog goldenrod - *Solidago uliginosa*
northern bog violet - *Viola sororia* subsp. *affinis*
tall white bog orchid - *Habenaria dilatata*
bogbean - *Menyanthes trifoliata*
bolbonac - *Lunaria annua*
boldo - *Peumus boldus*

Bolivia, pride-of-Bolivia - *Tipuana tipu*
bonavist bean - *Lablab purpureus*
boneset
blue boneset - *Conoclinium coelestinum*
boneset - *Eupatorium perfoliatum, Eupatorium, Symphytum officinale*
false boneset - *Brickellia eupatorioides*
hyssop-leaf boneset - *Eupatorium hyssopifolium*
late boneset - *Eupatorium serotinum*
purple boneset - *Eupatorium maculatum, E. perfoliatum*
tall boneset - *Eupatorium maculatum*
upland boneset - *Eupatorium sessilifolium*
bonnet
bonnet grass - *Danthonia spicata*
bride's-bonnet - *Clintonia uniflora*
Quaker's-bonnet - *Lupinus perennis*
Boott's goldenrod - *Solidago boottii*
boquet, Kashmir boquet - *Clerodendrum*
borage
borage - *Borago officinalis, Borago*
country borage - *Plectranthus amboinicus*
bore cole - *Brassica oleracea* var. *acephala*
boreal vetch - *Vicia cracca*
borracho, sapote-borracho - *Pouteria campechiana*
Boston ivy - *Parthenocissus tricuspidata*
botanical-wonder - *Fatshedera lizei*
bottle
bottle gentian - *Gentiana andrewsii, G. clausa*
bottle gourd - *Lagenaria siceraria*
bottle-palm - *Nolina recurvata*
flame bottle-tree - *Brachychiton acerifolius*
prairie bottle gentian - *Gentiana andrewsii*
scented-bottle - *Habenaria dilatata*
white-bottle - *Silene csereii*
bottlebrush
bottlebrush - *Elymus hystrix, Equisetum arvense, Melaleuca, Metrosideros*
bottlebrush buckeye - *Aesculus parviflora*
bottlebrush sedge - *Carex hystricina*
crimson bottlebrush - *Callistemon citrinus*
one-sided bottlebrush - *Calothamnus sanguineus*
weeping bottlebrush - *Callistemon viminalis*
bough, butter-bough - *Exothea paniculata*
boulder fern - *Dennstaedtia punctilobula*
bouncing-bet - *Saponaria officinalis*
bouquet
bouquet larkspur - *Delphinium grandiflorum*
bouquet-violet - *Lythrum salicaria*
bow-wood - *Maclura pomifera*
bower
bleeding-glory-bower - *Clerodendrum thompsoniae*
glory-bower - *Clerodendrum*
virgin's-bower - *Clematis virginiana, Clematis*
bowman's-root - *Apocynum cannabinum, Gillenia trifoliata, Veronicastrum virginicum*
bowstring
bowstring-hemp - *Calotropis gigantea*
Ceylon bowstring-hemp - *Sansevieria zeylanica*
box
bimble-box - *Eucalyptus populnea*

box elder - *Acer negundo*
box huckleberry - *Gaylussacia brachycera*
box-leaf eugenia - *Eugenia foetida*
box-leaf holly - *Ilex crenata*
box sand-myrtle - *Leiophyllum buxifolium*
Brisbane-box - *Lophostemon conferta*
California box elder - *Acer negundo* subsp. *californicum*
China-box - *Murraya paniculata*
Chinese box-orange - *Severinia buxifolia*
low rattle-box - *Crotalaria rotundifolia*
marmalade-box - *Genipa americana*
mountain-box - *Arctostaphylos uva-ursi*
myrtle box-leaf - *Paxistima myrsinites*
rattle-box - *Crotalaria sagittalis, Crotalaria, Ludwigia alternifolia, Rhinanthus crista-galli*
red box - *Eucalyptus polyanthemos*
seed-box - *Ludwigia alternifolia*
showy rattle-box - *Crotalaria spectabilis*
water-leaf rattle-box - *Crotalaria retusa*
wedge-leaf rattle-box - *Crotalaria retusa*
weedy rattle-box - *Crotalaria sagittalis*
boxthorn - *Lycium*
boxwood
African boxwood - *Myrsine africana*
boxwood - *Buxus sempervirens, Buxus*
Japanese boxwood - *Buxus microphylla* var. *japonica*
Korean boxwood - *Buxus microphylla* var. *koreana*
little-leaf boxwood - *Buxus microphylla*
Oregon boxwood - *Paxistima myrsinites*
Turkish boxwood - *Buxus sempervirens*
boysenberry - *Rubus loganobaccus*
bracken
bracken fern - *Pteridium aquilinum, P. aquilinum* var. *pubescens*
western bracken - *Pteridium aquilinum* var. *pubescens*
western bracken fern - *Pteridium aquilinum* var. *pubescens*
Bradford's pear - *Pyrus calleryana*
Bragg's manzanita - *Arctostaphylos nummularia*
Brahman grass - *Dichanthium annulatum*
Brainerd's hawthorn - *Crataegus brainerdii*
brake
brake - *Pteridium, Pteris*
canker-brake - *Polystichum acrostichoides*
cliff brake - *Pellaea*
Cretan brake - *Pteris cretica*
fragile cliff brake - *Cryptogramma stelleri*
New Zealand cliff brake - *Pellaea rotundifolia*
pasture brake - *Pteridium aquilinum*
rock brake - *Cryptogramma crispa, Cryptogramma*
slender cliff brake - *Cryptogramma stelleri*
smooth cliff brake - *Pellaea glabella*
Steller's rock brake - *Cryptogramma stelleri*
sword brake - *Pteris ensiformis*
braken - *Pteridium*
bramble
bramble vetch - *Vicia tenuifolia*
dog-bramble - *Ribes cynosbati*
smooth bramble - *Rubus canadensis*
trailing bramble - *Rubus flagellaris*
Brampton's stock - *Matthiola incana*
branch, flat-branch ground-pine - *Lycopodium obscurum*

branched
branched broomrape - *Orobanche ramosa*
branched bur-reed - *Sparganium erectum*
branching bur-reed - *Sparganium androcladum*
branco, veludo-branco - *Celosia argentea*
brandy mint - *Mentha piperita*
braschette - *Brassica oleracea* var. *acephala*
brasil - *Condalia hookeri*
brass
brass-buttons - *Cotula coronopifolia*
southern brass-buttons - *Cotula australis*
brava, cana-brava - *Aristida longespica, Arundo donax*
Brazilian
Brazilian cherry - *Eugenia uniflora*
Brazilian elodea - *Egeria densa*
Brazilian gloxinia - *Sinningia speciosa*
Brazilian golden-vine - *Stigmaphyllon ciliatum*
Brazilian grape tree - *Myrciaria cauliflora*
Brazilian lucerne - *Stylosanthes guianensis*
Brazilian oak - *Casuarina glauca*
Brazilian pepper-tree - *Schinus terebinthifolius*
Brazilian pine - *Araucaria angustifolia*
Brazilian-plume - *Justicia carnea*
Brazilian pusley - *Richardia brasiliensis*
Brazilian satin-tail - *Imperata brasiliensis*
Brazilian sky-flower - *Duranta erecta*
Brazilian stylo - *Stylosanthes guianensis*
Brazilian vervain - *Verbena brasiliensis*
Brazilian waterweed - *Egeria densa*
bread
bread-and-cheeses - *Pithecellobium unguis-cati*
bread-palm - *Cycas*
bread wheat - *Triticum aestivum*
Seminole bread - *Zamia pumila*
St. John's-bread - *Ceratonia siliqua*
breadfruit
breadfruit - *Artocarpus altilis, Pandanus odoratissimus*
breadfruit-vine - *Monstera deliciosa*
breadroot - *Psoralea esculenta*
break, cane-break bamboo - *Arundinaria gigantea*
breast, partridge-breast - *Aloe variegata*
breath, baby's-breath - *Galium sylvaticum, Gypsophila elegans, G. paniculata*
breeches
cat-breeches - *Hydrophyllum capitatum*
Dutchman's-breeches - *Dicentra cucullaria*
woolen-breeches - *Hydrophyllum capitatum*
Brewer's spruce - *Picea breweriana*
brickel-bush - *Brickellia*
bridal-wreath - *Spiraea prunifolia, S. vanhouttei, Spiraea*
bride, mourning-bride - *Scabiosa atropurpurea*
bride's-bonnet - *Clintonia uniflora*
brier
Austrian brier - *Rosa foetida*
brier - *Rosa*
bull-brier - *Smilax bona-nox, S. rotundifolia*
cat-brier - *Smilax bona-nox, S. glauca, S. rotundifolia, Smilax*
cat-claw sensitive-brier - *Mimosa quadrivalvis* var. *angustata*
China brier - *Smilax bona-nox*
dog-brier - *Rosa canina*
hag-brier - *Smilax hispida*

horse-brier - *Smilax rotundifolia*
Jackson's brier - *Smilax smallii*
little-leaf sensitive-brier - *Mimosa quadrivalvis* var. *angustata*
saw-brier - *Smilax glauca*
sensitive-brier - *Mimosa quadrivalvis* var. *nuttallii*

bright
bright-eyes - *Catharanthus roseus*
bright sunflower - *Helianthus laetiflorus*
Canadian eye-bright - *Euphrasia canadensis*
eye-bright - *Anagallis arvensis, Euphorbia maculata, Lobelia inflata*
meadow-bright - *Caltha palustris*

Brisbane-box - *Lophostemon conferta*

bristle
bristle-cone fir - *Abies bracteata*
bristle grass - *Setaria*
bristle oat - *Avena strigosa*
bristle-stalk sedge - *Carex leptalea*
bur bristle grass - *Setaria verticillata*
giant bristle grass - *Setaria magna*
glaucous bristle grass - *Setaria pumila*
Great Basin bristle-cone pine - *Pinus longaeva*
green bristle grass - *Setaria viridis*
Grisebach's bristle grass - *Setaria grisebachii*
knot-root bristle grass - *Setaria geniculata*
plains bristle grass - *Setaria macrostachya*
western bristle-cone pine - *Pinus longaeva*
yellow bristle grass - *Setaria pumila*

bristled, short-bristled umbrella-grass - *Fuirena breviseta*

bristly
bristly aster - *Aster puniceus*
bristly black currant - *Ribes lacustre*
bristly club-moss - *Lycopodium annotinum*
bristly crowfoot - *Ranunculus pensylvanicus*
bristly dewberry - *Rubus hispidus*
bristly foxtail - *Setaria verticillata*
bristly greenbrier - *Smilax hispida*
bristly hawk's-beard - *Crepis setosa*
bristly locust - *Robinia hispida*
bristly mallow - *Modiola caroliniana*
bristly ox's-tongue - *Picris echioides*
bristly rose - *Rosa acicularis*
bristly sarsaparilla - *Aralia hispida*
bristly sedge - *Carex comosa*
bristly star-bur - *Acanthospermum hispidum*

brittle
brittle bladder fern - *Cystopteris fragilis*
brittle fern - *Cystopteris fragilis*
brittle-leaf naiad - *Najas minor*
brittle maidenhair fern - *Adiantum tenerum*
brittle prickly-pear - *Opuntia fragilis*
brittle willow - *Salix fragilis*

broad
broad bean - *Vicia faba*
broad-fruited bur-reed - *Sparganium eurycarpum*
broad-leaf button-weed - *Spermacoce alata*
broad-leaf cat-tail - *Typha latifolia*
broad-leaf cocklebur - *Xanthium strumarium*
broad-leaf dock - *Rumex obtusifolius*
broad-leaf filaree - *Erodium botrys*
broad-leaf goldenrod - *Solidago flexicaulis*
broad-leaf linden - *Tilia platyphyllos*

broad-leaf milkweed - *Asclepias latifolia*
broad-leaf panicum - *Panicum adspersum*
broad-leaf phlox - *Phlox amplifolia*
broad-leaf plantain - *Plantago major*
broad-leaf pondweed - *Potamogeton amplifolius*
broad-leaf rose-bay - *Tabernaemontana divaricata*
broad-leaf signal grass - *Brachiaria platyphylla*
broad-leaf spring-beauty - *Claytonia caroliniana*
broad-leaf spurge - *Euphorbia lucida*
broad-leaf tea-tree - *Melaleuca quinquenervia*
broad-leaf toadflax - *Linaria genistifolia*
broad-leaf water-leaf - *Hydrophyllum canadense*

broccoli
asparagus broccoli - *Brassica oleracea* var. *italica*
broccoli - *Brassica oleracea* var. *botrytis*
Italian broccoli - *Brassica oleracea* var. *italica*
sprouting broccoli - *Brassica oleracea* var. *italica*

brodiaea
golden-brodiaea - *Triteleia ixioides*
harvest brodiaea - *Brodiaea coronaria*

brome
Arizona brome - *Bromus arizonicus*
awnless brome grass - *Bromus inermis*
barren brome - *Bromus sterilis*
brome grass - *Bromus*
California brome - *Bromus carinatus*
Canadian brome - *Bromus purgans*
compact brome - *Bromus madritensis*
downy brome - *Bromus tectorum*
downy brome grass - *Bromus tectorum*
drooping brome - *Bromus tectorum*
false brome - *Brachypodium distachyon*
field brome - *Bromus arvensis*
foxtail brome - *Bromus rubens*
foxtail brome grass - *Bromus rubens*
fringed brome - *Bromus ciliatus*
Hungarian brome - *Bromus inermis*
Japanese brome - *Bromus japonicus*
Japanese false brome grass - *Brachypodium pinnatum*
Madrid brome - *Bromus madritensis*
meadow brome - *Bromus erectus*
mountain brome - *Bromus carinatus*
nodding brome - *Bromus anomalus*
poverty brome - *Bromus sterilis*
rattlesnake brome - *Bromus briziformis*
red brome - *Bromus rubens*
rip-gut brome - *Bromus diandrus, B. rigidus*
rip-gut brome grass - *Bromus rigidus*
rye brome - *Bromus secalinus*
sand brome - *Bromus arenarius*
Schrader's brome - *Bromus catharticus*
slender false brome grass - *Brachypodium sylvaticum*
smooth brome - *Bromus inermis*
smooth brome grass - *Bromus commutatus*
soft brome - *Bromus hordeaceus*
Spanish brome - *Bromus madritensis*
Spanish brome grass - *Bromus madritensis*
sterile brome - *Bromus sterilis*
upright brome - *Bromus erectus*

brook
alpine-brook saxifrage - *Saxifraga rivularis*
brook alder - *Alnus maritima*

brook grass - *Catabrosa*
brook lobelia - *Lobelia kalmii*
brook saxifrage - *Boykinia aconitifolia*
brooklime
  American brooklime - *Veronica americana*
  brooklime - *Veronica*
brookweed - *Samolus parviflorus*
broom
  annual broom-weed - *Amphiachyris dracunculoides*
  blunt-broom sedge - *Carex tribuloides*
  broom - *Cytisus scoparius, Cytisus*
  broom baccharus - *Baccharis sarothroides*
  broom beard grass - *Schizachyrium scoparium*
  broom-brush - *Hypericum prolificum*
  broom-corn - *Panicum miliaceum, Sorghum bicolor*
  broom corn millet - *Panicum miliaceum*
  broom hickory - *Carya glabra*
  broom millet - *Panicum miliaceum*
  broom-sedge - *Andropogon virginicus*
  broom sedge blue-stem - *Andropogon virginicus*
  broom snakeroot - *Gutierrezia sarothrae*
  broom snakeweed - *Gutierrezia sarothrae*
  broom-weed - *Amphiachyris dracunculoides, Gutierrezia, Sida acuta*
  broom wiregrass - *Schizachyrium scoparium*
  chaparral-broom - *Baccharis pilularis*
  desert-broom - *Baccharis sarothroides*
  dyer's broom - *Genista tinctoria*
  European broom - *Cytisus scoparius*
  false broom-weed - *Ericameria austrotexana*
  French broom - *Genista monspessulanus*
  pointed broom-sedge - *Carex scoparia*
  Scotch broom - *Cytisus scoparius*
  Spanish broom - *Spartium junceum*
  weaver's-broom - *Spartium junceum*
broomrape
  branched broomrape - *Orobanche ramosa*
  crenate broomrape - *Orobanche crenata*
  lesser broomrape - *Orobanche minor*
  Louisiana broomrape - *Orobanche ludoviciana*
  naked broomrape - *Orobanche uniflora*
  nodding broomrape - *Orobanche cernua*
  prairie broomrape - *Orobanche ludoviciana*
brown
  brown beak rush - *Rhynchospora fusca*
  brown bent grass - *Agrostis canina, A. perennans*
  brown dogwood - *Cornus glabrata*
  brown-eyed-Susan - *Rudbeckia triloba*
  brown knapweed - *Centaurea jacea*
  brown mustard - *Brassica juncea, B. nigra*
  brown plume grass - *Saccharum brevibarbe*
  brown-seed paspalum - *Paspalum plicatulum*
  brown-top - *Agrostis capillaris*
  brown-top millet - *Brachiaria fasciculata, B. ramosa*
  brown-top panicum - *Brachiaria fasciculata*
browned sedge - *Carex adusta*
brownish sedge - *Carex brunnescens*
bruise-wort - *Symphytum officinale*
brush
  black-brush acacia - *Acacia rigidula*
  blue-brush - *Ceanothus thyrsiflorus*
  broom-brush - *Hypericum prolificum*

buck-brush - *Ceanothus cuneatus, Symphoricarpos orbiculatus*
coyote-brush - *Baccharis pilularis*
Cupid's-shaving-brush - *Emilia fosbergii*
deer-brush - *Ceanothus integerrimus*
desert rabbit-brush - *Chrysothamnus paniculatus*
Douglas's rabbit-brush - *Chrysothamnus viscidiflorus*
fox-brush - *Centranthus ruber*
gray rabbit-brush - *Chrysothamnus nauseosus*
green-plume rabbit-brush - *Chrysothamnus nauseosus* subsp. *graveolens*
hack-brush - *Spiraea douglasii*
little-leaf horse-brush - *Tetradymia glabrata*
match-brush - *Gutierrezia*
Parry's rabbit-brush - *Chrysothamnus parryi*
rabbit-brush - *Chrysothamnus*
rosin-brush - *Baccharis sarothroides*
rubber rabbit-brush - *Chrysothamnus nauseosus*
shaving-brush-tree - *Pseudobombax ellipticum*
skunk-brush - *Rhus trilobata*
southwestern rabbit-brush - *Chrysothamnus pulchellus*
spineless horse-brush - *Tetradymia canescens*
Texas white-brush - *Aloysia gratissima*
white-brush - *Aloysia gratissima*
yellow rabbit-brush - *Chrysothamnus viscidiflorus*
Brussels sprouts - *Brassica oleracea* var. *gemmifera*
bryony, white bryony - *Bryonia alba*
bubble-poppy - *Silene csereii*
buchu, Indian buchu - *Myrtus communis*
bucida, ox-horn bucida - *Bucida buceras*
buck
  buck-brush - *Ceanothus cuneatus, Symphoricarpos orbiculatus*
  buck-horn - *Lycopodium clavatum, Osmunda cinnamomea*
  buck-horn plantain - *Plantago lanceolata*
  buck-parsnip - *Lomatium triternatum*
buckbean
  buckbean - *Menyanthes trifoliata*
  Piedmont buckbean - *Thermopsis mollis*
buckeye
  bottlebrush buckeye - *Aesculus parviflora*
  buckeye - *Aesculus*
  California buckeye - *Aesculus californica*
  dwarf buckeye - *Aesculus sylvatica*
  fetid buckeye - *Aesculus glabra*
  Georgia buckeye - *Aesculus sylvatica*
  Mexican buckeye - *Ungnadia speciosa*
  Ohio buckeye - *Aesculus glabra*
  painted buckeye - *Aesculus sylvatica*
  red buckeye - *Aesculus pavia*
  sweet buckeye - *Aesculus flava*
  Texas buckeye - *Aesculus glabra* var. *arguta*
  yellow buckeye - *Aesculus flava*
Buckley's
  Buckley's hickory - *Carya texana*
  Buckley's phlox - *Phlox buckleyi*
buckroot - *Psoralea canescens*
buckthorn
  alder buckthorn - *Rhamnus alnifolia, R. frangula*
  alder-leaf buckthorn - *Rhamnus alnifolia*
  birch-leaf buckthorn - *Rhamnus betulifolia*
  buckthorn - *Rhamnus cathartica, Rhamnus*
  buckthorn-weed - *Amsinckia intermedia*

California buckthorn - *Rhamnus californica*
Carolina buckthorn - *Rhamnus caroliniana*
cascara buckthorn - *Rhamnus purshiana*
European alder buckthorn - *Rhamnus frangula*
European buckthorn - *Rhamnus cathartica*
false buckthorn - *Sideroxylon lanuginosa*
glossy buckthorn - *Rhamnus frangula*
holly-leaf buckthorn - *Rhamnus ilicifolia*
lance-leaf buckthorn - *Rhamnus lanceolata*
red-berried buckthorn - *Rhamnus ilicifolia*
southern buckthorn - *Sideroxylon lycioides*
tough buckthorn - *Sideroxylon tenax*
yellow buckthorn - *Rhamnus caroliniana*

buckwheat
buckwheat - *Fagopyrum esculentum, Fagopyrum*
buckwheat-bush - *Cliftonia monophylla*
buckwheat-tree - *Cliftonia monophylla*
buckwheat-vine - *Brunnichia ovata*
California buckwheat - *Eriogonum fasciculatum*
climbing false buckwheat - *Polygonum scandens*
long-leaf wild buckwheat - *Eriogonum longifolium*
notch-seeded buckwheat - *Fagopyrum esculentum*
Tatarian buckwheat - *Fagopyrum tataricum*
wild buckwheat - *Eriogonum, Polygonum convolvulus*

buddleja, orange-eyed buddleja - *Buddleja davidii*
buena, yerba-buena - *Satureja douglasii*

buffalo
annual buffalo clover - *Trifolium reflexum*
buffalo-berry - *Solanum rostratum*
buffalo-bur - *Solanum rostratum*
buffalo clover - *Trifolium reflexum*
buffalo currant - *Ribes aureum, R. odoratum*
buffalo gourd - *Cucurbita foetidissima*
buffalo grass - *Buchloe dactyloides, Paspalum conjugatum,*
    Pennisetum ciliare
buffalo-weed - *Ambrosia trifida*
false buffalo grass - *Munroa squarrosa*
running buffalo clover - *Trifolium stoloniferum*
russet buffalo-berry - *Shepherdia canadensis*
silver buffalo-berry - *Shepherdia argentea*

bugbane
American bugbane - *Cimicifuga americana*
bugbane - *Cimicifuga racemosa, Cimicifuga*
false bugbane - *Trautvetteria*
mountain bugbane - *Cimicifuga americana*

bugle
bugle - *Ajuga reptans*
carpet bugle - *Ajuga reptans*
erect bugle - *Ajuga genevensis*
Geneva bugle - *Ajuga genevensis*
red bugle-vine - *Aeschynanthus pulcher*
standing bugle - *Ajuga genevensis*

bugler
royal-red-bugler - *Aeschynanthus pulcher*
scarlet bugler - *Penstemon centranthifolius*

bugleweed
American bugleweed - *Lycopus americanus*
bugleweed - *Lycopus*
carpet bugleweed - *Ajuga reptans*
creeping bugleweed - *Ajuga reptans*
cut-leaf bugleweed - *Lycopus americanus*
European bugleweed - *Lycopus europaeus*

northern bugleweed - *Lycopus uniflorus*
rough bugleweed - *Lycopus asper*
slender bugleweed - *Lycopus uniflorus*
yellow bugleweed - *Ajuga chamaepitys*

bugloss
bugloss - *Anchusa officinalis, Picris hieracioides*
Italian bugloss - *Anchusa azurea*
small bugloss - *Anchusa arvensis*
viper-bugloss - *Echium vulgare*

Bukhara clover - *Melilotus alba*

bulb
bulb-bearing water-hemlock - *Cicuta bulbifera*
bulb panic grass - *Panicum bulbosum*
wonder-bulb - *Colchicum autumnale*

bulbil-bearing yam - *Dioscorea bulbifera*
bulblet bladder fern - *Cystopteris bulbifera*

bulbous
bulbous barley - *Hordeum bulbosum*
bulbous bluegrass - *Poa bulbosa*
bulbous buttercup - *Ranunculus bulbosus*
bulbous canary grass - *Phalaris aquatica*

bull
bull-bay - *Magnolia grandiflora*
bull-brier - *Smilax bona-nox, S. rotundifolia*
bull mallow - *Malva nicaeensis*
bull-nettle - *Cnidoscolus stimulosus, Solanum elaeagnifolium*
bull paspalum - *Paspalum boscianum*
bull-rush - *Typha latifolia*
bull thistle - *Cirsium pumilim, C. vulgare*
Texas bull-nettle - *Cnidoscolus texanus*

bullace
bullace grape - *Vitis rotundifolia*
bullace plum - *Prunus domestica* subsp. *insititia*

bullgrass - *Muhlenbergia emersleyi*
bullock's-heart - *Annona reticulata*

bulrush
American bulrush - *Scirpus americanus*
American great bulrush - *Scirpus validus*
black bulrush - *Scirpus atrovirens*
bulrush - *Scirpus*
bulrush millet - *Pennisetum glaucum*
California bulrush - *Scirpus californicus*
dark-green bulrush - *Scirpus atrovirens*
great bulrush - *Scirpus lacustris*
hard-stem bulrush - *Scirpus acutus* var. *occidentalis*
hollow-scaled bulrush - *Scirpus koilolepis*
lined bulrush - *Scirpus pendulus*
rice-field bulrush - *Scirpus mucronatus*
river bulrush - *Scirpus fluviatilis*
slender bulrush - *Scirpus heterochaetus*
small-bulrush - *Typha angustifolia*
soft-stem bulrush - *Scirpus validus*
southern bulrush - *Scirpus californicus*
water bulrush - *Scirpus subterminalis*
wool grass bulrush - *Scirpus cyperinus*
bulrushes, Moses-in-the-bulrushes - *Tradescantia spathacea*

bumelia, gum bumelia - *Sideroxylon lanuginosa*

bunch
blue-bunch fescue - *Festuca idahoensis*
blue bunch grass - *Festuca idahoensis*
blue-bunch wheat grass - *Pseudoroegneria spicata*

bunch-evergreen - *Lycopodium obscurum*
bunch grape - *Vitis aestivalis*
bunch grass - *Schizachyrium scoparium*
bunch-moss - *Tillandsia recurvata*
feather bunch grass - *Stipa viridula*
mountain bunch grass - *Festuca viridula*
bunchberry - *Cornus canadensis*
bunching
   bunching onion - *Allium fistulosum*
   Japanese bunching onion - *Allium fistulosum*
bundle
   bundle-flower - *Desmanthus illinoensis, Desmanthus*
   Illinois bundle-flower - *Desmanthus illinoensis*
   velvet bundle-flower - *Desmanthus velutinus*
bunias, crested bunias - *Bunias erucago*
bunk - *Cichorium intybus*
bur
   annual bur-sage - *Ambrosia acanthicarpa*
   branched bur-reed - *Sparganium erectum*
   branching bur-reed - *Sparganium androcladum*
   bristly star-bur - *Acanthospermum hispidum*
   broad-fruited bur-reed - *Sparganium eurycarpum*
   buffalo-bur - *Solanum rostratum*
   bur-beak chervil - *Anthriscus caucalis*
   bur bristle grass - *Setaria verticillata*
   bur buttercup - *Ranunculus testiculatus*
   bur chervil - *Anthriscus caucalis*
   bur-clover - *Medicago arabica, M. minima*
   bur cucumber - *Sicyos angulatus*
   bur gherkin - *Cucumis anguria*
   bur grass - *Cenchrus incertus*
   bur-head - *Echinodorus cordifolius*
   bur mallow - *Urena lobata*
   bur-marigold - *Bidens pilosa, Bidens*
   bur-nut - *Tribulus terrestris*
   bur oak - *Quercus macrocarpa*
   bur ragweed - *Ambrosia tomentosa*
   bur-reed - *Sparganium*
   butter-bur - *Petasites hybridus, Petasites*
   California bur-clover - *Medicago polymorpha*
   creeping bur-head - *Echinodorus cordifolius*
   giant bur-reed - *Sparganium eurycarpum*
   green-fruited bur-reed - *Sparganium chlorocarpum*
   little bur-clover - *Medicago minima*
   narrow-leaf bur-reed - *Sparganium emersum*
   nodding bur-marigold - *Bidens cernua*
   Paraguay bur - *Acanthospermum australe*
   Paraguay star-bur - *Acanthospermum australe*
   Sacramento bur - *Triumfetta semitriloba*
   skeleton-leaf bur - *Ambrosia tomentosa*
   skeleton-leaf bur-sage - *Ambrosia tomentosa*
   slim-leaf bur-sage - *Ambrosia confertiflora*
   smaller bur-marigold - *Bidens cernua*
   spotted bur-clover - *Medicago arabica*
   three-square bur-reed - *Sparganium americanum*
   toothed bur-clover - *Medicago polymorpha*
   water bur-reed - *Sparganium fluctuans*
   woolly-leaf bur-sage - *Ambrosia grayi*
burdock
   burdock - *Arctium minus, Arctium*
   cotton burdock - *Arctium tomentosum*
   edible burdock - *Arctium lappa*

   great burdock - *Arctium lappa*
   lesser burdock - *Arctium minus*
   smaller burdock - *Arctium minus*
   woolly burdock - *Arctium tomentosum*
Burma reed - *Neyraudia reynaudiana*
Burmese
   Burmese fishtail palm - *Caryota mitis*
   Burmese rosewood - *Pterocarpus indicus*
burn
   American burn-weed - *Erechtites hieracifolia*
   Australian burn-weed - *Erechtites minima*
   burn-weed - *Erechtites hieracifolia*
   yellow burn-weed - *Amsinckia lycopsoides*
burnet
   American burnet - *Sanguisorba canadensis*
   burnet - *Sanguisorba*
   burnet-bloodwort - *Sanguisorba officinalis*
   burnet-saxifrage - *Pimpinella saxifraga*
   Canadian burnet - *Sanguisorba canadensis*
   great burnet - *Sanguisorba officinalis*
   salad burnet - *Sanguisorba minor*
   Sitka burnet - *Sanguisorba stipulata*
Burnet's rose - *Rosa spinosissima*
burning
   burning-bush - *Bassia scoparia, Dictamnus albus, Euonymus atropurpureus, Euonymus*
   burning nettle - *Urtica urens*
   western burning-bush - *Euonymus occidentalis*
   winged burning-bush - *Euonymus alatus*
burnt-weed - *Epilobium angustifolium*
burro-weed - *Haplopappus tenuisectus*
bursting-heart - *Euonymus americanus*
burweed
   annual burweed - *Ambrosia acanthicarpa*
   lawn burweed - *Soliva sessilis*
bush
   alder-leaf pepper-bush - *Clethra alnifolia*
   American milk-bush - *Synadenium grantii*
   barometer-bush - *Leucophyllum frutescens*
   bay-gall-bush - *Ilex coriacea*
   beauty-bush - *Kolkwitzia amabilis*
   Benjamin's-bush - *Lindera benzoin*
   bitter-bush - *Chromolaena odorata, Picramnia pentandra*
   black high-bush blueberry - *Vaccinium atrococcum*
   brickel-bush - *Brickellia*
   buckwheat-bush - *Cliftonia monophylla*
   burning-bush - *Bassia scoparia, Dictamnus albus, Euonymus atropurpureus, Euonymus*
   bush allamanda - *Allamanda schottii*
   bush chinquapin - *Castanopsis sempervirens*
   bush cinquefoil - *Potentilla fruticosa*
   bush-clover - *Lespedeza capitata, Lespedeza*
   bush grape - *Vitis acerifolia, V. rupestris*
   bush-honeysuckle - *Diervilla lonicera*
   bush moonflower - *Ipomoea leptophylla*
   bush morning-glory - *Ipomoea leptophylla*
   bush muhly - *Muhlenbergia porteri*
   bush palmetto - *Sabal minor*
   bush pea - *Thermopsis mollis*
   bush poppy - *Dendromecon rigida*
   bush pumpkin - *Cucurbita pepo* var. *melopepo*
   bush senecio - *Senecio douglasii*

bush squash - *Cucurbita pepo* var. *melopepo*
butterfly-bush - *Buddleja davidii, Buddleja*
calico-bush - *Kalmia latifolia*
caper-bush - *Capparis*
Chinese bush-clover - *Lespedeza cuneata*
cliff-bush - *Jamesia americana*
cranberry-bush - *Viburnum trilobum*
cream-bush - *Holodiscus discolor*
creeping bush-clover - *Lespedeza repens*
creosote-bush - *Larrea tridentata*
deer-bush - *Ceanothus integerrimus*
Drummond's rattle-bush - *Sesbania drummondii*
elephant-bush - *Portulacaria afra*
European cranberry-bush - *Viburnum opulus*
false indigo-bush - *Amorpha fruticosa*
fern-bush - *Chamaebatiaria millefolium*
fever-bush - *Lindera benzoin*
fire-bush - *Hamelia patens*
flannel-bush - *Fremontodendron californica*
Florida hop-bush - *Dodonaea viscosa*
fountain-bush - *Russelia equisetiformis*
glory-bush - *Tibouchina urvilleana*
groundsel-bush - *Baccharis halimifolia*
high-bush blackberry - *Rubus argutus*
high-bush blueberry - *Vaccinium caesariense, V. corymbosum*
high-bush cranberry - *Viburnum trilobum*
indigo-bush - *Amorpha fruticosa, Dalea*
ivy-bush - *Kalmia latifolia*
Japanese bush-clover - *Lespedeza striata*
Jubas-bush - *Iresine diffusa*
Korean bush-clover - *Kummerowia stipulacea*
lily-of-the-valley-bush - *Pieris japonica*
low-bush blueberry - *Vaccinium angustifolium*
minnie-bush - *Menziesia pilosa*
northern bush-honeysuckle - *Diervilla lonicera*
pearl-bush - *Exochorda racemosa*
poison-bush - *Acokanthera*
polecat-bush - *Rhus aromatica*
rattle-bush - *Sesbania punicea*
resin-bush - *Viguiera stenoloba*
ribbon-bush - *Homalocladium platycladum*
ringworm-bush - *Senna alata*
round-head bush-clover - *Lespedeza capitata*
running strawberry-bush - *Euonymus obovatus*
salt-bush - *Salsola australis*
scarlet-bush - *Hamelia patens*
shrimp-bush - *Justicia brandegeana*
shrub bush-clover - *Lespedeza bicolor*
silk-tassel-bush - *Garrya*
silver-bush - *Convolvulus cneorum, Sophora tomentosa*
skunk-bush - *Rhus trilobata*
slender bush-clover - *Lespedeza virginica*
smoke-bush - *Cotinus coggygria*
snowball-bush - *Viburnum opulus, V. opulus* var. *roseum*
soap-bush - *Clidemia hirta*
southern high-bush blueberry - *Vaccinium elliottii*
spleenwort-bush - *Comptonia peregrina*
squaw-bush - *Condalia spathulata, Rhus aromatica*
stag-bush - *Viburnum prunifolium*
stagger-bush - *Lyonia ferruginea, L. mariana*
strawberry-bush - *Euonymus americanus*
Stueve's bush-clover - *Lespedeza stuevei*
sugar-bush sumac - *Rhus ovata*
sweet pepper-bush - *Clethra acuminata, C. alnifolia*

trailing bush-clover - *Lespedeza procumbens*
wand-like bush-clover - *Lespedeza intermedia*
western burning-bush - *Euonymus occidentalis*
winged burning-bush - *Euonymus alatus*
yellow bush lupine - *Lupinus arboreus*
bushman's-poison - *Acokanthera oppositifolia,*
    Acokanthera
bushy
bushy aster - *Aster dumosus*
bushy beard grass - *Andropogon glomeratus*
bushy beard-stem - *Andropogon glomeratus*
bushy blue-stem - *Andropogon glomeratus*
bushy cinquefoil - *Potentilla paradoxa*
bushy knotweed - *Polygonum ramosissimum*
bushy wallflower - *Erysimum repandum*
bushy water-primrose - *Ludwigia alternifolia, L. decurrens*
bushy yate - *Eucalyptus lehmannii*
business, mind-your-own-business-plant - *Soleirolia*
    soleirolii
busy-Lizzy - *Impatiens wallerana*
butt, black-butt - *Eucalyptus pilularis*
butter
butter-and-eggs - *Linaria vulgaris*
butter bean - *Phaseolus lunatus*
butter-bough - *Exothea paniculata*
butter-bur - *Petasites hybridus, Petasites*
butter-daisy - *Ranunculus repens, Verbesina encelioides*
butter-print - *Abutilon theophrasti*
butter-weed - *Abutilon theophrasti*
buttercup
Bermuda buttercup - *Oxalis pes-caprae*
bird's-foot buttercup - *Ranunculus orthorhynchus*
bog buttercup - *Ranunculus sceleratus*
bulbous buttercup - *Ranunculus bulbosus*
bur buttercup - *Ranunculus testiculatus*
buttercup - *Ranunculus*
buttercup oxalis - *Oxalis pes-caprae*
California buttercup - *Ranunculus californicus*
celery-leaf buttercup - *Ranunculus sceleratus*
corn buttercup - *Ranunculus arvensis*
creeping buttercup - *Ranunculus repens*
crow-foot buttercup - *Ranunculus sceleratus*
cursed buttercup - *Ranunculus sceleratus*
dwarf buttercup - *Ranunculus pygmaeus*
early buttercup - *Ranunculus fascicularis*
field buttercup - *Ranunculus arvensis*
hairy buttercup - *Ranunculus sardous*
hispid buttercup - *Ranunculus hispidus*
hooked buttercup - *Ranunculus recurvatus*
horned-head buttercup - *Ranunculus testiculatus*
meadow buttercup - *Ranunculus acris*
Persian buttercup - *Ranunculus asiaticus*
prairie buttercup - *Ranunculus rhomboideus*
rough-seed buttercup - *Ranunculus muricatus*
shore buttercup - *Ranunculus cymbalaria*
small-flowered buttercup - *Ranunculus abortivus*
spiny-fruit buttercup - *Ranunculus muricatus*
swamp buttercup - *Ranunculus hispidus*
tall buttercup - *Ranunculus acris*
testiculate buttercup - *Ranunculus testiculatus*
turban buttercup - *Ranunculus asiaticus*
western field buttercup - *Ranunculus occidentalis*
white water buttercup - *Ranunculus aquatilis* var. *capillaceus*

yellow water buttercup - *Ranunculus flabellaris*
butterfly
    butterfly-bush - *Buddleja davidii, Buddleja*
    butterfly-dock - *Petasites hybridus*
    butterfly-ginger - *Hedychium coronarium*
    butterfly-lily - *Hedychium coronarium*
    butterfly milkweed - *Asclepias tuberosa*
    butterfly palm - *Chrysalidocarpus lutescens*
    butterfly pea - *Centrosema pubescens, Centrosema, Clitoria mariana, C. ternatea*
    butterfly-tulip - *Calochortus*
    butterfly-vine - *Stigmaphyllon ciliatum*
    butterfly-weed - *Asclepias tuberosa*
    spurred butterfly pea - *Centrosema virginianum*
    yellow butterfly palm - *Chrysalidocarpus lutescens*
butternut - *Juglans cinerea, Juglans*
butterweed - *Conyza canadensis, Senecio glabellus*
butterwort - *Pinguicula vulgaris*
button
    bachelor's-button - *Centaurea cyanus*
    beggar's-button - *Arctium lappa, Arctium*
    broad-leaf button-weed - *Spermacoce alata*
    button fern - *Pellaea rotundifolia*
    button mangrove - *Conocarpus erectus*
    button-plant - *Spermacoce assurgens*
    button snakeroot - *Eryngium aquaticum, E. yuccifolium, Liatris*
    button-weed - *Diodia teres, Spermacoce glabra*
    cockle-button - *Arctium lappa, A. minus*
    cuckoo-button - *Arctium minus*
    rough button-weed - *Diodia teres*
    silver button - *Anaphalis margaritacea*
    silver button mangrove - *Conocarpus erectus* var. *sericeus*
    smooth button-weed - *Spermacoce glabra*
    Virginia button-weed - *Diodia virginiana*
    yellow bachelor's-button - *Polygala lutea*
buttonbush - *Cephalanthus occidentalis*
buttons
    blue-buttons - *Knautia arvensis, Vinca major*
    brass-buttons - *Cotula coronopifolia*
    coat-buttons - *Tridax procumbens*
    golden-buttons - *Tanacetum vulgare*
    shoe-buttons - *Sanicula bipinnatifida*
    southern brass-buttons - *Cotula australis*
buttonwood
    buttonwood - *Conocarpus erectus, Platanus occidentalis*
    silver buttonwood - *Conocarpus erectus* var. *sericeus*
    white buttonwood - *Laguncularia racemosa*
cabbage
    cabbage - *Brassica oleracea, B. oleracea* var. *capitata*
    cabbage gum - *Eucalyptus amplifolia, E. pauciflora*
    cabbage palmetto - *Sabal palmetto*
    cabbage rose - *Rosa centifolia*
    cabbage-tree - *Cordyline, Cussonia spicata*
    celery cabbage - *Brassica pekinensis*
    Chinese cabbage - *Brassica pekinensis*
    Florida cabbage palm - *Sabal palmetto*
    flowering cabbage - *Brassica oleracea* var. *acephala*
    John's-cabbage - *Hydrophyllum virginianum*
    mustard cabbage - *Brassica juncea*
    savoy cabbage - *Brassica oleracea* var. *capitata*
    saw cabbage palm - *Acoelorraphe wrightii*

    Shantung cabbage - *Brassica pekinensis*
    skunk-cabbage - *Symplocarpus foetidus*
    steppe-cabbage - *Rapistrum perenne*
    Warrigal's cabbage - *Tetragonia tetragonioides*
    western skunk-cabbage - *Lysichiton americanum*
    wild cabbage - *Brassica oleracea*
    yellow skunk-cabbage - *Lysichiton americanum*
cacao
    cacao - *Theobroma cacao*
    cacao-bean - *Theobroma cacao*
cactus
    cholla cactus - *Opuntia*
    Christmas cactus - *Schlumbergera buckleyi*
    claw cactus - *Schlumbergera truncata*
    cochineal cactus - *Opuntia cochenillifera*
    crab cactus - *Schlumbergera truncata*
    Easter cactus - *Hatiora gaertneri*
    fish-hook cactus - *Mammillaria*
    giant cactus - *Carnegiea gigantea*
    hedgehog cactus - *Echinocereus viridiflorus*
    jumping cactus - *Opuntia fulgida*
    lady-of-Mexico cactus - *Mammillaria hahniana*
    new old-man cactus - *Espostoa lanata*
    organ-pipe cactus - *Stenocereus thurberi*
    Peruvian apple cactus - *Cereus uruguayanus*
    Peruvian old-man cactus - *Espostoa lanata*
    Peruvian snowball cactus - *Espostoa lanata*
    prickly-pear cactus - *Opuntia stricta* var. *dillenii*
    red-bird-cactus - *Pedilanthus tithymaloides*
    ribbon-cactus - *Pedilanthus tithymaloides*
    spineless cactus - *Opuntia ficus-indica*
    star cactus - *Astrophytum ornatum, Astrophytum*
    strawberry cactus - *Mammillaria*
    Thanksgiving cactus - *Schlumbergera truncata*
    torch cactus - *Echinopsis spachiana*
    Unguentine-cactus - *Aloe vera*
    vine cactus - *Fouquieria splendens*
    whisker cactus - *Pachycereus schottii*
    white torch cactus - *Echinopsis spachiana*
    yoke cactus - *Schlumbergera truncata*
cadillo - *Urena lobata*
cadlock - *Brassica nigra*
caesar-weed - *Urena lobata*
cafta - *Catha edulis*
Cairo morning-glory - *Ipomoea cairica*
cajeput
    cajeput - *Melaleuca leucadendra, M. quinquenervia*
    cajeput-tree - *Melaleuca quinquenervia*
caju - *Anacardium occidentale*
calabash
    calabash - *Crescentia cujete*
    calabash gourd - *Lagenaria siceraria*
    calabash nutmeg - *Monodora myristica*
    calabash-tree - *Crescentia cujete*
calabazilla - *Cucurbita foetidissima*
calabur - *Muntingia calabura*
caladium, fancy-leaf caladium - *Caladium bicolor, C. hortulanum*
calamint
    alpine calamint - *Acinos arvensis*
    calamint - *Calamintha nepeta, Satureja*

calamondin - *Citrofortunella mitis*
calamus - *Acorus calamus*
Caledonia, New Caledonia pine - *Araucaria columnaris*
Caley's pea - *Lathyrus hirsutus*
calf-kill - *Kalmia angustifolia*
calico
    calico aster - *Aster lateriflorus*
    calico-bush - *Kalmia latifolia*
    calico-plant - *Alternanthera bettzichiana*
California
    California arrowhead - *Sagittaria montevidensis*
    California ash - *Fraxinus dipetala*
    California bay - *Umbellularia californica*
    California bayberry - *Myrica californica*
    California black currant - *Ribes bracteosum*
    California black oak - *Quercus kelloggii*
    California black walnut - *Juglans californica*
    California blackberry - *Rubus ursinus*
    California box elder - *Acer negundo* subsp. *californicum*
    California brome - *Bromus carinatus*
    California buckeye - *Aesculus californica*
    California buckthorn - *Rhamnus californica*
    California buckwheat - *Eriogonum fasciculatum*
    California bulrush - *Scirpus californicus*
    California bur-clover - *Medicago polymorpha*
    California buttercup - *Ranunculus californicus*
    California chinquapin - *Castanopsis sempervirens*
    California false hellebore - *Veratrum californicum*
    California fan palm - *Washingtonia filifera*
    California fern - *Conium maculatum*
    California field oak - *Quercus agrifolia*
    California fremontia - *Fremontodendron californica*
    California fuchsia - *Epilobium canum*
    California gold-bark fern - *Pityrogramma triangularis*
    California goldenrod - *Solidago californica*
    California hazelnut - *Corylus cornuta* var. *californica*
    California hop-tree - *Ptelea crenulata*
    California horse-chestnut - *Aesculus californica*
    California huckleberry - *Vaccinium ovatum*
    California incense-cedar - *Calocedrus decurrens*
    California Indian pink - *Silene californica*
    California juniper - *Juniperus californica, J. occidentalis*
    California laurel - *Umbellularia californica*
    California live oak - *Quercus agrifolia*
    California melic - *Melica imperfecta*
    California mugwort - *Artemisia douglasiana*
    California needle grass - *Stipa cernua, S. pulchra*
    California nutmeg - *Torreya californica*
    California oat grass - *Danthonia californica*
    California olive - *Umbellularia californica*
    California palm - *Washingtonia filifera*
    California pepper-tree - *Schinus molle*
    California pitcher-plant - *Darlingtonia californica*
    California polypody - *Polypodium californicum*
    California poppy - *Eschscholzia californica*
    California privet - *Ligustrum obtusifolium, L. ovalifolium*
    California rape - *Sinapis arvensis*
    California red fir - *Abies magnifica*
    California redbud - *Cercis occidentalis*
    California rose - *Rosa californica*
    California rose-bay - *Rhododendron macrophyllum*
    California sagebrush - *Artemisia californica*

    California scrub oak - *Quercus dumosa*
    California sun-cup - *Camissonia bistorta*
    California sweet grass - *Hierochloe occidentalis*
    California sweet-shrub - *Calycanthus occidentalis*
    California sycamore - *Platanus racemosa*
    California torreya - *Torreya californica*
    California walnut - *Juglans californica*
    California Washington palm - *Washingtonia filifera*
    California wax myrtle - *Myrica californica*
    California white oak - *Quercus lobata*
    California wild grape - *Vitis californica*
    California wild rose - *Rosa californica*
    northern California black walnut - *Juglans hindsii*
    southern California walnut - *Juglans californica*
calla
    calla-lily - *Zantedeschia aethiopica, Zantedeschia*
    florist's calla - *Zantedeschia aethiopica*
    garden calla - *Zantedeschia aethiopica*
    golden calla-lily - *Zantedeschia elliottiana*
    pink calla-lily - *Zantedeschia rehmannii*
    red calla-lily - *Zantedeschia rehmannii*
    white calla-lily - *Zantedeschia aethiopica*
    wild calla - *Calla palustris*
    yellow calla-lily - *Zantedeschia elliottiana*
Callery's pear - *Pyrus calleryana*
calliopsis - *Coreopsis tinctoria*
calopo - *Calopogonium mucunoides*
caltrop - *Trapa natans, Tribulus*
camagueyana - *Bothriochloa pertusa*
camasey - *Clidemia hirta*
camass
    camass - *Camassia quamash*
    death camass - *Zigadenus nuttallii*
    eastern camass - *Camassia scilloides*
    foothill death camass - *Zigadenus paniculatus*
    meadow death camass - *Zigadenus venenosus*
    mountain death camass - *Zigadenus elegans*
    Nuttall's death camass - *Zigadenus nuttallii*
    poison camass - *Zigadenus nuttallii*
    white camass - *Zigadenus elegans*
camel-thorn - *Alhagi maurorum, A. pseudalhagi*
camellia
    camellia - *Camellia japonica*
    mountain camellia - *Stewartia ovata*
    Sasanqua camellia - *Camellia sasanqua*
    silky-camellia - *Stewartia melachodendron*
camosh - *Camassia quamash*
camphor-tree - *Cinnamomum camphora*
campion
    biennial campion - *Silene csereii*
    bladder campion - *Silene vulgaris*
    campion - *Lychnis, Silene csereii, Silene*
    corn-campion - *Agrostemma githago*
    evening campion - *Silene latiflora*
    meadow campion - *Lychnis flos-cuculi*
    morning campion - *Silene dioica*
    moss campion - *Silene acaulis*
    red campion - *Silene dioica*
    rose campion - *Lychnis coronaria*
    snowy campion - *Silene nivea*
    starry campion - *Silene stellata*

white campion - *Silene latiflora*
cana-brava - *Aristida longespica, Arundo donax*
Canadian
  Canadian anemone - *Anemone canadensis*
  Canadian bluegrass - *Poa compressa*
  Canadian brome - *Bromus purgans*
  Canadian burnet - *Sanguisorba canadensis*
  Canadian columbine - *Aquilegia canadensis*
  Canadian eye-bright - *Euphrasia canadensis*
  Canadian goldenrod - *Solidago canadensis*
  Canadian hawkweed - *Hieracium canadense*
  Canadian hemlock - *Tsuga canadensis*
  Canadian lily - *Lilium canadense*
  Canadian mayflower - *Maianthemum canadense*
  Canadian moonseed - *Menispermum canadense*
  Canadian pea - *Vicia cracca*
  Canadian pine - *Pinus resinosa*
  Canadian plum - *Prunus nigra*
  Canadian pumpkin - *Cucurbita moschata*
  Canadian pussy-toes - *Antennaria neglecta* var. *canadensis*
  Canadian rhododendron - *Rhododendron canadense*
  Canadian St. John's-wort - *Hypericum canadense*
  Canadian thistle - *Cirsium arvense*
  Canadian tick-trefoil - *Desmodium canadense*
  Canadian violet - *Viola canadensis*
  Canadian wild lettuce - *Lactuca canadensis*
  Canadian wild rye - *Elymus canadensis*
  Canadian yew - *Taxus canadensis*
canaigre - *Rumex hymenosepalus*
Canary
  Canary Island date palm - *Phoenix canariensis*
  Canary Island pine - *Pinus canariensis*
canary
  bulbous canary grass - *Phalaris aquatica*
  canary grass - *Phalaris canariensis, Phalaris*
  Carolina canary grass - *Phalaris caroliniana*
  hood canary grass - *Phalaris paradoxa*
  little-seed canary grass - *Phalaris minor*
  reed canary grass - *Phalaris arundinacea*
  timothy canary grass - *Phalaris angusta*
Canby's bluegrass - *Poa secunda*
cancer
  cancer-root - *Epifagus virginiana*
  one-flowered cancer-root - *Orobanche uniflora*
candle
  bog-candle - *Habenaria dilatata*
  candle anemone - *Anemone cylindrica*
  candle larkspur - *Delphinium elatum*
  Christmas-candle - *Senna alata*
  desert-candle - *Eremurus robustus*
  empress-candle-plant - *Senna alata*
  Our-Lord's-candle - *Yucca whipplei*
  Roman-candle - *Yucca gloriosa*
candleberry
  candleberry - *Aleurites moluccana, Myrica cerifera,*
    M. pensylvanica
  candleberry-tree - *Aleurites moluccana*
candlenut
  candlenut - *Aleurites moluccana*
  candlenut-tree - *Aleurites moluccana*
candles, swamp-candles - *Lysimachia terrestris*

candlestick
  candlestick lily - *Lilium dauricum, L. hollandicum*
  candlestick senna - *Senna alata*
  Lord's-candlestick - *Yucca gloriosa*
candlewood - *Fouquieria splendens*
candy-weed - *Polygala lutea*
candytuft
  candytuft - *Iberis*
  globe candytuft - *Iberis umbellata*
  rocket candytuft - *Iberis amara*
cane
  cane - *Arundinaria*
  cane-apples - *Arbutus unedo*
  cane beard grass - *Bothriochloa barbinodis*
  cane blue-stem - *Bothriochloa barbinodis*
  cane-break bamboo - *Arundinaria gigantea*
  cane palm - *Chrysalidocarpus lutescens*
  dumb-cane - *Dieffenbachia seguine, Dieffenbachia*
  giant cane - *Arundinaria gigantea, A. gigantea* subsp. *tecta*
  golden cane palm - *Chrysalidocarpus lutescens*
  large cane - *Arundinaria gigantea*
  maiden-cane - *Panicum hemitomon*
  shatter-cane - *Sorghum bicolor*
  southern cane - *Arundinaria gigantea*
  spotted dumb-cane - *Dieffenbachia maculata*
  sugar-cane - *Saccharum officinarum*
  sugar-cane plume grass - *Saccharum giganteum*
  switch-cane - *Arundinaria gigantea* subsp. *tecta*
  wild sugar-cane - *Saccharum spontaneum*
canella - *Canella winteriana*
canistel - *Pouteria campechiana*
canker
  canker-brake - *Polystichum acrostichoides*
  canker-root - *Coptis groenlandica, C. trifolia*
canna
  edible canna - *Canna indica*
  garden canna - *Canna generalis*
canoe birch - *Betula papyrifera*
canotia - *Canotia holocantha*
cantaloupe - *Cucumis melo* var. *cantalupensis*
Canterbury bells - *Campanula medium*
Canton ginger - *Zingiber officinale*
canyon
  canyon gooseberry - *Ribes menziesii*
  canyon grape - *Vitis arizonica*
  canyon live oak - *Quercus chrysolepis*
  canyon maple - *Acer macrophyllum*
  canyon oak - *Quercus chrysolepis*
caouthchouc-tree - *Hevea brasiliensis*
cap
  American Turk's-cap lily - *Lilium superbum*
  bishop's-cap - *Astrophytum myriostigma, A. ornatum, Mitella*
  black-cap - *Rubus occidentalis*
  dunce-cap larkspur - *Delphinium occidentale*
  friar's-cap - *Aconitum napellus*
  Japanese Turk's-cap lily - *Lilium hansonii*
  soldier's-cap - *Aconitum napellus*
  Turk's-cap - *Aconitum napellus, Malvaviscus arboreus,*
    M. arboreus var. *drummondii*
  Turk's-cap lily - *Lilium martagon, L. superbum*

Cape
 blue-eyed Cape-marigold - *Dimorphotheca sinuata*
 Cape belladonna - *Amaryllis belladonna*
 Cape blue water-lily - *Nymphaea capensis*
 Cape gooseberry - *Physalis peruviana*
 Cape honeysuckle - *Tecoma capensis*
 Cape ivy - *Senecio mikanioides*
 Cape jasmine - *Gardenia jasminoides*
 Cape leadwort - *Plumbago auriculata*
 Cape-marigold - *Dimorphotheca*
 Cape myrtle - *Myrsine africana*
 Cape primrose - *Streptocarpus rexii*
 Cape-weed - *Phyla nodiflora*
 dwarf Cape gooseberry - *Physalis pubescens* var. *grisea*
caper
 bean-caper - *Zygophyllum fabago*
 caper-bush - *Capparis*
 caper spurge - *Euphorbia lathyris*
 Jamaican caper - *Capparis cynophallophora*
 Jamaican caper-tree - *Capparis cynophallophora*
 Syrian bean-caper - *Zygophyllum fabago*
capillaire - *Gaultheria hispidula*
caps, sailor-caps - *Dodecatheon hendersonii*
capuchin, barbe-de-capuchin - *Cichorium intybus*
caramba - *Averrhoa carambola*
carambola - *Averrhoa carambola*
caraway
 caraway - *Carum carvi*
 false caraway - *Perideridia gairdneri*
 Indian caraway - *Perideridia gairdneri*
cardamom
 Ceylon cardamom - *Elettaria cardamomum*
 Malabar cardamom - *Elettaria cardamomum*
cardamon - *Elettaria cardamomum*
cardinal
 cardinal-climber - *Ipomoea quamoclit*
 cardinal-flower - *Lobelia cardinalis, L. siphilitica*
 cardinal larkspur - *Delphinium cardinale*
 cardinal-spear - *Erythrina herbacea*
 red cardinal - *Erythrina herbacea*
cardo-del-valle - *Centaurea americana*
cardoon - *Cynara cardunculus*
careless-weed - *Amaranthus retroflexus*
carib grass - *Eriochloa polystachya*
Caribbean
 Caribbean pine - *Pinus caribaea*
 Caribbean stylo - *Stylosanthes hamata*
Carmel
 Carmel ceanothus - *Ceanothus griseus*
 Carmel creeper - *Ceanothus griseus* var. *horizontalis*
carnation - *Dianthus caryophyllus*
carne, balsamo-cor-de-carne - *Justicia carnea*
carob - *Ceratonia siliqua*
Carolina
 Carolina allspice - *Calycanthus floridus*
 Carolina ash - *Fraxinus caroliniana*
 Carolina basswood - *Tilia caroliniana*
 Carolina bean - *Phaseolus lunatus*
 Carolina buckthorn - *Rhamnus caroliniana*
 Carolina canary grass - *Phalaris caroliniana*
 Carolina cherry-laurel - *Prunus caroliniana*

Carolina clover - *Trifolium carolinianum*
Carolina dichondra - *Dichondra carolinensis*
Carolina false dandelion - *Pyrrhopappus carolinianus*
Carolina foxtail - *Alopecurus carolinianus*
Carolina geranium - *Geranium carolinianum*
Carolina hemlock - *Tsuga caroliniana*
Carolina holly - *Ilex ambigua*
Carolina horse-nettle - *Solanum carolinense*
Carolina ipecac - *Euphorbia ipecacuanhae*
Carolina jasmine - *Gelsemium sempervirens*
Carolina moonseed - *Cocculus carolinus*
Carolina mosquito-fern - *Azolla caroliniana*
Carolina nettle - *Solanum carolinense*
Carolina rhododendron - *Rhododendron carolinianum*
Carolina spring-beauty - *Claytonia caroliniana*
Carolina vetch - *Vicia caroliniana*
Carolina water-hyssop - *Bacopa caroliniana*
Carolina whitlow-grass - *Draba reptans*
Carolina willow - *Salix caroliniana*
carpenter-weed - *Prunella vulgaris*
carpenter's-square - *Scrophularia marilandica*
carpet
 big carpet grass - *Axonopus furcatus*
 carpet bugle - *Ajuga reptans*
 carpet bugleweed - *Ajuga reptans*
 carpet grass - *Axonopus affinis*
 Corsican carpet-plant - *Soleirolia soleirolii*
 golden-carpet - *Sedum acre*
 Japanese carpet grass - *Zoysia matrella*
 tropical carpet grass - *Axonopus compressus*
carpetweed
 carpetweed - *Mollugo verticillata*
 whorled carpetweed - *Mollugo verticillata*
carrion-flower - *Smilax herbacea*
carrizo - *Arundo donax, Phragmites australis*
carrot
 carrot - *Daucus carota* subsp. *sativus*
 southwestern carrot - *Daucus pusillus*
 wild carrot - *Daucus carota*
cart-track-plant - *Plantago major*
carthamus - *Carthamus oxyacantha*
casaba melon - *Cucumis melo* var. *inodorus*
casabanana - *Sicana odorifera*
Cascade fir - *Abies amabilis*
Cascades mahonia - *Berberis nervosa*
cascara
 cascara - *Rhamnus purshiana*
 cascara buckthorn - *Rhamnus purshiana*
 cascara sagrada - *Rhamnus purshiana*
case, Dutch-case-knife bean - *Phaseolus coccineus*
cashew - *Anacardium occidentale*
cassandra - *Chamaedaphne calyculata*
cassava
 bitter cassava - *Manihot esculenta*
 cassava - *Manihot esculenta*
cassia
 Cove's cassia - *Senna covesii*
 feathery cassia - *Cassia artemisioides*
 purging cassia - *Cassia fistula*
 wormwood cassia - *Cassia artemisioides*
cassie - *Acacia farnesiana*

cassina - *Ilex cassine, I. vomitoria*
cast-iron-plant - *Aspidistra elatior*
castor
  castor bean - *Ricinus communis*
  castor-oil-plant - *Ricinus communis*
casuarina, horsetail casuarina - *Casuarina equisetifolia*
cat
  alpine cat timothy - *Phleum alpinum*
  annual-cat-tail - *Rostraria cristata*
  blue cat-tail - *Typha glauca*
  broad-leaf cat-tail - *Typha latifolia*
  cat-berry - *Nemopanthus mucronatus*
  cat-breeches - *Hydrophyllum capitatum*
  cat-brier - *Smilax bona-nox, S. glauca, S. rotundifolia, Smilax*
  cat-claw acacia - *Acacia greggii*
  cat-claw mimosa - *Mimosa pigra*
  cat-claw schrankia - *Mimosa quadrivalvis* var. *nuttallii*
  cat-claw sensitive-brier - *Mimosa quadrivalvis* var. *angustata*
  cat-claw trumpet - *Macfadyena unguis-cati*
  cat-claw-vine - *Macfadyena unguis-cati*
  cat grape - *Vitis palmata*
  cat greenbrier - *Smilax glauca*
  cat-jang - *Vigna unguiculata* subsp. *cylindrica*
  cat-jang pea - *Cajanus cajan*
  cat-milk - *Euphorbia helioscopia*
  cat-mint - *Nepeta cataria*
  cat spruce - *Picea glauca*
  cat-tail - *Typha latifolia*
  cat-tail grass - *Setaria pallide-fusca*
  cat-tail millet - *Pennisetum glaucum*
  narrow-leaf cat-tail - *Typha angustifolia*
  pole-cat-geranium - *Lantana montevidensis*
  southern cat-tail - *Typha domingensis*
cat's
  cat's-claw - *Acacia tortuosa, Mimosa quadrivalvis* var.
    nuttallii, Pithecellobium unguis-cati
  cat's-ear - *Hypochaeris radicata*
  cat's-foot - *Gnaphalium obtusifolium*
  cat's-gut - *Tephrosia virginiana*
  Gregg's cat's-claw - *Acacia greggii*
  long-flowered cat's-claw - *Acacia greggii*
  rough cat's-ear - *Hypochaeris radicata*
  smooth cat's-ear - *Hypochaeris glabra*
  spotted cat's-ear - *Hypochaeris radicata*
Catalina
  Catalina ceanothus - *Ceanothus arboreus*
  Catalina ironwood - *Lyonothamnus*
  Catalina Island cherry - *Prunus ilicifolia* subsp. *lyonii*
  Catalina mountain-lilac - *Ceanothus arboreus*
catalpa
  catalpa - *Catalpa bignonioides*
  Chinese catalpa - *Catalpa ovata*
  Indian bean catalpa - *Catalpa bignonioides*
  northern catalpa - *Catalpa speciosa*
  southern catalpa - *Catalpa bignonioides*
  western catalpa - *Catalpa speciosa*
catawba
  catawba - *Catalpa bignonioides*
  catawba rhododendron - *Rhododendron catawbiense*
  catawba-tree - *Catalpa speciosa*
catbird grape - *Vitis palmata*

catch
  catch-weed - *Asperugo procumbens*
  catch-weed bedstraw - *Galium aparine*
catcher, bird-catcher-tree - *Pisonia umbellifera*
catchfly
  catchfly - *Lychnis, Silene*
  catchfly grass - *Leersia lenticularis*
  cone catchfly - *Silene conoidea*
  English catchfly - *Silene gallica*
  forking catchfly - *Silene dichotoma*
  French catchfly - *Silene gallica*
  garden catchfly - *Silene armeria*
  German catchfly - *Lychnis viscaria*
  hairy catchfly - *Silene dichotoma*
  night-flowering catchfly - *Silene noctiflora*
  nodding catchfly - *Silene pendula*
  round-leaf catchfly - *Silene rotundifolia*
  royal catchfly - *Silene regia*
  sand catchfly - *Silene conica*
  sleepy catchfly - *Silene antirrhina*
  sweet-William catchfly - *Silene armeria*
Catesby's
  Catesby's gentian - *Gentiana catesbaei*
  Catesby's oak - *Quercus laevis*
Catharine's, St. Catharine's-lace - *Eriogonum giganteum*
catnip
  catnip - *Agastache nepetoides, Nepeta cataria*
  catnip giant hyssop - *Agastache nepetoides*
Cattley's guava - *Psidium littorale* var. *longipes*
Caucasian
  Caucasian clover - *Trifolium ambiguum*
  Caucasian elm - *Zelkova carpinifolia*
  Caucasian lily - *Lilium monadelphum*
cauliflower
  cauliflower - *Brassica oleracea* var. *botrytis*
  cauliflower-ears - *Crassula ovata*
Cayenne pepper - *Capsicum annuum, C. frutescens*
ceanothus
  Carmel ceanothus - *Ceanothus griseus*
  Catalina ceanothus - *Ceanothus arboreus*
  felt-leaf ceanothus - *Ceanothus arboreus*
  green-bark ceanothus - *Ceanothus spinosus*
  Jersey tea ceanothus - *Ceanothus americanus*
  Monterey ceanothus - *Ceanothus cuneatus* var. *rigidus*
  Point Reyes ceanothus - *Ceanothus gloriosus*
  red-stem ceanothus - *Ceanothus sanguineus*
  spiny ceanothus - *Ceanothus spinosus*
  Yankee Point ceanothus - *Ceanothus griseus* var. *horizontalis*
cedar
  Alaska cedar - *Chamaecyparis nootkatensis*
  Alaska yellow-cedar - *Chamaecyparis nootkatensis*
  Atlantic white-cedar - *Chamaecyparis thyoides*
  Atlas cedar - *Cedrus atlantica*
  Australian cedar - *Toona ciliata* var. *australis*
  California incense-cedar - *Calocedrus decurrens*
  cedar - *Cedrus*
  cedar elm - *Ulmus crassifolia*
  cedar-of-Goa - *Cupressus lusitanica*
  cedar pine - *Pinus glabra*
  Colorado red-cedar - *Juniperus scopulorum*
  creeping-cedar - *Juniperus horizontalis*
  deodara cedar - *Cedrus deodara*

eastern red-cedar - *Juniperus virginiana*
giant-cedar - *Thuja plicata*
ground-cedar - *Lycopodium complanatum, L. tristachyum*
incense-cedar - *Calocedrus decurrens*
Japanese cedar - *Cryptomeria japonica*
Nootka yellow-cedar - *Chamaecyparis nootkatensis*
northern white-cedar - *Thuja occidentalis*
Ozark white-cedar - *Juniperus ashei*
pencil-cedar - *Juniperus virginiana*
Port Orford cedar - *Chamaecyparis lawsoniana*
red-cedar - *Juniperus virginiana*
Russian cedar - *Pinus cembra*
salt-cedar - *Tamarix ramosissima*
shrubby red-cedar - *Juniperus horizontalis*
southern red-cedar - *Juniperus silicicola*
southern white-cedar - *Chamaecyparis thyoides*
stinking-cedar - *Torreya taxifolia*
swamp white-cedar - *Chamaecyparis thyoides*
western red-cedar - *Thuja plicata*
cedrela - *Toona sinensis*
ceiba - *Ceiba pentandra*
celandine
celandine - *Chelidonium majus*
celandine poppy - *Stylophorum diphyllum*
greater celandine - *Chelidonium majus*
lesser celandine - *Ranunculus ficaria*
small celandine - *Ranunculus ficaria*
Celebes pepper - *Piper ornatum*
celeriac - *Apium graveolens* var. *rapaceum*
celery
celery - *Apium graveolens, A. graveolens* var. *dulce*
celery cabbage - *Brassica pekinensis*
celery-leaf buttercup - *Ranunculus sceleratus*
knob celery - *Apium graveolens* var. *rapaceum*
turnip-rooted celery - *Apium graveolens* var. *rapaceum*
water-celery - *Oenanthe sarmentosa, Vallisneria americana*
wild celery - *Ciclospermum leptophyllum*
celosia - *Celosia argentea*
celtuce - *Lactuca sativa*
cenicilla - *Sesuvium portulacastrum*
cenizo - *Atriplex canescens, Leucophyllum frutescens, Tetrazygia elaeagnoides*
centaury - *Centaurium erythraea*
centipede
centipede grass - *Eremochloa ophiuroides*
centipede-plant - *Homalocladium platycladum*
century
American century-plant - *Agave americana*
century-plant - *Agave americana*
cereus, night-blooming cereus - *Hylocereus undatus*
ceriman - *Monstera deliciosa*
cerro hawthorn - *Crataegus erythropoda*
cestrum, day-blooming cestrum - *Cestrum diurnum*
Ceylon
Ceylon bowstring-hemp - *Sansevieria zeylanica*
Ceylon cardamom - *Elettaria cardamomum*
Ceylon cinnamon - *Cinnamomum verum*
Ceylon cotton - *Gossypium arboreum*
golden Ceylon creeper - *Epipremnum aureum*
chafe-weed - *Gnaphalium obtusifolium*

chaff
chaff-seed - *Schwalbea americana*
mat chaff-flower - *Alternanthera caracasana*
chain
chain fern - *Woodwardia*
European chain fern - *Woodwardia radicans*
giant chain fern - *Woodwardia fimbriata*
golden-chain - *Laburnum anagyroides*
nettle chain fern - *Woodwardia areolata*
Virginia chain fern - *Woodwardia virginica*
chairmaker's rush - *Scirpus americanus*
chalice, fire-chalice - *Epilobium canum*
chalk maple - *Acer leucoderme*
chamise - *Adenostoma fasciculatum*
Chamisso willow - *Salix chamissonis*
chamiza - *Atriplex canescens*
chamomile
chamomile - *Anthemis, Chamaemelum*
corn-chamomile - *Anthemis arvensis*
dog-chamomile - *Anthemis cotula*
false chamomile - *Matricaria maritima*
fetid-chamomile - *Anthemis cotula*
field-chamomile - *Anthemis arvensis*
garden chamomile - *Chamaemelum nobile*
mayweed-chamomile - *Anthemis cotula*
Roman chamomile - *Chamaemelum nobile*
scentless-chamomile - *Matricaria maritima, M. perforata*
stinking-chamomile - *Anthemis cotula*
wild chamomile - *Matricaria chamomilla*
yellow-chamomile - *Anthemis tinctoria*
chaparral
chaparral-broom - *Baccharis pilularis*
chaparral currant - *Ribes malvaceum*
Chapman's oak - *Quercus chapmanii*
chapote - *Diospyros texana*
chard
chard - *Beta vulgaris* subsp. *cicla*
Swiss-chard - *Beta vulgaris* subsp. *cicla*
Charlie, creeping-Charlie - *Lysimachia nummularia, Pilea nummulariifolia, Wedelia trilobata*
charlock
charlock - *Sinapis arvensis*
jointed charlock - *Raphanus raphanistrum*
chaste
chaste-tree - *Vitex agnus-castus*
lilac chaste-tree - *Vitex agnus-castus*
chat - *Catha edulis*
chayote - *Sechium edule*
cheat
cheat - *Bromus secalinus*
cheat grass - *Bromus secalinus, B. tectorum*
checker
checker-bloom - *Sidalcea malvaeflora*
checker-lily - *Fritillaria affines*
checker mallow - *Sidalcea malvaeflora, Sidalcea*
checkerberry - *Gaultheria procumbens*
cheerful sunflower - *Helianthus laetiflorus*
cheese
cheese-plant - *Malva neglecta*
cheese-rennet - *Galium verum*

cheese-weed - *Malva parviflora*
Swiss-cheese-plant - *Monstera deliciosa*
cheeses
bread-and-cheeses - *Pithecellobium unguis-cati*
cheeses - *Malva neglecta*
cherimalla - *Annona cherimola*
cherimola - *Annona cherimola*
cherimoya - *Annona cherimola*
Cherokee
Cherokee bean - *Erythrina herbacea*
Cherokee rose - *Rosa laevigata*
cherry
Barbados cherry - *Malpighia glabra*
Bessey's cherry - *Prunus pumila* var. *besseyi*
bird cherry - *Prunus avium, P. pensylvanica*
bitter cherry - *Prunus emarginata*
Brazilian cherry - *Eugenia uniflora*
Carolina cherry-laurel - *Prunus caroliniana*
Catalina Island cherry - *Prunus ilicifolia* subsp. *lyonii*
cherry-bark oak - *Quercus falcata* var. *pagodifolia*
cherry birch - *Betula lenta*
cherry-laurel - *Prunus laurocerasus*
cherry-pie - *Heliotropium arborescens*
cherry plum - *Prunus cerasifera*
cherry-stone juniper - *Juniperus monosperma*
cherry tomato - *Lycopersicon esculentum* var. *cerasiforme, Physalis peruviana*
clammy ground-cherry - *Physalis heterophylla*
cut-leaf ground-cherry - *Physalis angulata*
downy ground-cherry - *Physalis pubescens*
dwarf cherry - *Prunus pumila*
English cherry-laurel - *Prunus laurocerasus*
European bird cherry - *Prunus padus*
European dwarf cherry - *Prunus fruticosa*
European ground cherry - *Prunus fruticosa*
evergreen cherry - *Prunus ilicifolia*
false Jerusalem cherry - *Solanum capsicastrum*
fire cherry - *Prunus pensylvanica*
ground-cherry - *Physalis*
hairy ground-cherry - *Physalis pubescens* var. *grisea*
holly-leaf cherry - *Prunus ilicifolia*
Indian cherry - *Rhamnus caroliniana*
Jamaican cherry - *Muntingia calabura*
Japanese flowering cherry - *Prunus serrulata*
Jerusalem cherry - *Solanum pseudocapsicum*
lance-leaf ground-cherry - *Physalis angulata*
lobed ground-cherry - *Physalis lobata*
long-leaf ground-cherry - *Physalis longifolia*
Mahaleb cherry - *Prunus mahaleb*
mazzard cherry - *Prunus avium*
Morello's cherry - *Prunus cerasus* var. *austera*
Oregon cherry - *Prunus emarginata*
Oriental cherry - *Prunus serrulata*
perennial ground-cherry - *Physalis longifolia* var. *subglabrata*
perfumed cherry - *Prunus mahaleb*
Peruvian cherry - *Physalis peruviana*
Peruvian ground-cherry - *Physalis peruviana*
pie cherry - *Prunus cerasus*
pin cherry - *Prunus pensylvanica*
Portuguese cherry-laurel - *Prunus lusitanica*
purple ground-cherry - *Physalis lobata*
rum cherry - *Prunus serotina*

sand cherry - *Prunus pumila, P. susquehanae*
smooth ground-cherry - *Physalis longifolia* var. *subglabrata*
sour cherry - *Prunus cerasus*
Spanish cherry - *Mimusops elengi*
St. Lucie cherry - *Prunus mahaleb*
strawberry ground-cherry - *Physalis alkekengi*
Surinam cherry - *Eugenia uniflora*
sweet cherry - *Prunus avium*
Virginia ground-cherry - *Physalis virginiana*
western sand cherry - *Prunus pumila* var. *besseyi*
wild black cherry - *Prunus serotina*
wild cherry - *Prunus ilicifolia*
wild red cherry - *Prunus pensylvanica*
winter-cherry - *Physalis alkekengi, P. peruviana*
chervil
bur-beak chervil - *Anthriscus caucalis*
bur chervil - *Anthriscus caucalis*
chervil - *Anthriscus cerefolium*
dwarf ground-chervil - *Chamaesaracha coronopus*
hairy-fruit chervil - *Chaerophyllum tainturieri*
southern chervil - *Chaerophyllum tainturieri*
white chervil - *Cryptotaenia canadensis*
wild chervil - *Anthriscus sylvestris, Cryptotaenia canadensis*
chess
Australian chess - *Bromus arenarius*
chess - *Bromus secalinus, Bromus*
Chilean chess - *Bromus trinii*
downy chess - *Bromus tectorum*
early chess - *Bromus tectorum*
foxtail chess - *Bromus madritensis, B. rubens*
hairy chess - *Bromus commutatus*
Japanese chess - *Bromus japonicus*
rattlesnake chess - *Bromus briziformis*
slender chess - *Bromus tectorum*
soft chess - *Bromus hordeaceus*
chestnut
American chestnut - *Castanea dentata*
Australian chestnut - *Castanospermum australe*
California horse-chestnut - *Aesculus californica*
chestnut - *Castanea dentata, Castanea*
chestnut oak - *Quercus prinus*
chestnut-vine - *Tetrastigma voinieranum*
Chinese chestnut - *Castanea mollissima*
dwarf chestnut oak - *Quercus prinoides*
dwarf horse-chestnut - *Aesculus parviflora*
earth-chestnut - *Lathyrus tuberosus*
Eurasian chestnut - *Castanea sativa*
European chestnut - *Castanea sativa*
European horse-chestnut - *Aesculus hippocastanum*
Guianan chestnut - *Pachira aquatica*
horse-chestnut - *Aesculus hippocastanum, Aesculus*
Japanese chestnut - *Castanea crenata*
Japanese horse-chestnut - *Aesculus turbinata*
Moreton Bay chestnut - *Castanospermum australe*
red horse-chestnut - *Aesculus carnea*
rock chestnut oak - *Quercus prinus*
Spanish chestnut - *Castanea sativa*
swamp chestnut oak - *Quercus michauxii, Q. prinus*
water-chestnut - *Pachira aquatica, Trapa natans*
yellow chestnut oak - *Quercus muehlenbergii*
chewing festuca - *Festuca rubra*
Chickasaw plum - *Prunus angustifolia*

chicken
  chicken grape - *Vitis vulpina*
  chicken-weed - *Galium boreale*
chicken's
  chicken's-claws - *Salicornia europaea*
  chicken's-gizzard - *Iresine herbstii*
chickens, hens-and-chickens - *Sempervivum tectorum*
chickling
  chickling pea - *Lathyrus sativus*
  chickling vetchling - *Lathyrus sativus*
chickpea - *Cicer arietinum*
chickweed
  Alaska chickweed - *Stellaria laeta*
  chickweed - *Paronychia, Stellaria media, Stellaria, Trientalis*
  field chickweed - *Cerastium arvense*
  forked chickweed - *Paronychia canadensis*
  great chickweed - *Stellaria pubera*
  Indian chickweed - *Mollugo verticillata*
  jagged chickweed - *Holosteum umbellatum*
  long-leaf chickweed - *Stellaria longifolia*
  meadow chickweed - *Cerastium arvense*
  mouse-ear chickweed - *Cerastium glomeratum, C. vulgatum, Cerastium*
  nodding chickweed - *Cerastium nutans*
  poison-chickweed - *Anagallis arvensis*
  red-chickweed - *Anagallis arvensis*
  sea-chickweed - *Honckenya peploides*
  star chickweed - *Stellaria pubera*
  sticky chickweed - *Cerastium glomeratum*
  water-chickweed - *Callitriche*
chicle - *Manilkara zapota*
chicory - *Cichorium intybus, Cichorium*
chicot - *Gymnocladus dioica*
chicozapote - *Manilkara zapota*
chigger-flower - *Asclepias tuberosa*
Chihuahuan pine - *Pinus leiophylla*
childing pink - *Dianthus prolifer*
children's-bane - *Cicuta maculata*
Chile, algarroba de Chile - *Prosopis chilensis*
Chilean
  Chilean chess - *Bromus trinii*
  Chilean pine - *Araucaria araucana*
  Chilean strawberry - *Fragaria chiloensis*
  Chilean tarweed - *Madia sativa*
chili pepper - *Capsicum annuum, C. frutescens*
chilicote - *Erythrina flabelliformis*
chin-chin - *Azara microphylla*
China
  China aster - *Callistephus chinensis*
  China-berry - *Melia azedarach*
  China-box - *Murraya paniculata*
  China brier - *Smilax bona-nox*
  China fir - *Cunninghamia lanceolata*
  China grass - *Boehmeria nivea*
  China jute - *Abutilon theophrasti*
  China-root - *Smilax tamnoides*
  China rose - *Hibiscus rosa-sinensis, Rosa chinensis*
  China-tree - *Koelreuteria paniculata, Melia azedarach*
  China-wood oil-tree - *Vernicia fordii*
  pride-of-China - *Melia azedarach*
  rose-of-China - *Hibiscus rosa-sinensis*

Chinese
  Chinese amaranth - *Amaranthus tricolor*
  Chinese apple - *Malus prunifolia*
  Chinese arrowhead - *Sagittaria sagittifolia*
  Chinese azalea - *Rhododendron molle*
  Chinese beauty-berry - *Callicarpa dichotoma*
  Chinese bell-flower - *Platycodon grandiflorus*
  Chinese box-orange - *Severinia buxifolia*
  Chinese bush-clover - *Lespedeza cuneata*
  Chinese cabbage - *Brassica pekinensis*
  Chinese catalpa - *Catalpa ovata*
  Chinese chestnut - *Castanea mollissima*
  Chinese chives - *Allium tuberosum*
  Chinese cup grass - *Eriochloa villosa*
  Chinese date - *Ziziphus jujuba*
  Chinese elm - *Ulmus parvifolia, U. pumila*
  Chinese ephedra - *Ephedra sinica*
  Chinese evergreen - *Aglaonema modestum, A. simplex*
  Chinese flowering crabapple - *Malus spectabilis*
  Chinese forget-me-not - *Cynoglossum amabile*
  Chinese fuzzy gourd - *Benincasa hispida*
  Chinese gooseberry - *Actinidia deliciosa*
  Chinese hibiscus - *Hibiscus rosa-sinensis*
  Chinese holly - *Ilex cornuta*
  Chinese jade-plant - *Crassula arborescens*
  Chinese jujube - *Ziziphus jujuba*
  Chinese juniper - *Juniperus chinensis*
  Chinese lantern-plant - *Physalis alkekengi*
  Chinese laurel - *Antidesma bunius*
  Chinese lespedeza - *Lespedeza cuneata*
  Chinese lilac - *Syringa chinensis*
  Chinese magnolia - *Magnolia soulangiana*
  Chinese matrimony-vine - *Lycium barbarum*
  Chinese mustard - *Brassica juncea*
  Chinese olive - *Osmanthus heterophyllus*
  Chinese parasol-tree - *Firmiana simplex*
  Chinese parsley - *Coriandrum sativum*
  Chinese pear - *Pyrus pyrifolia*
  Chinese pennisetum - *Pennisetum alopecuroides*
  Chinese peony - *Paeonia lactiflora*
  Chinese pistachio - *Pistacia chinensis*
  Chinese preserving melon - *Benincasa hispida*
  Chinese primrose - *Primula sinensis*
  Chinese privet - *Ligustrum lucidum, L. sinense*
  Chinese quince - *Pseudocydonia sinensis*
  Chinese radish - *Raphanus sativus* cv. 'longipinnatus'
  Chinese redbud - *Cercis chinensis*
  Chinese rhubarb - *Rheum officinale*
  Chinese rice-paper-plant - *Tetrapanax papyrifer*
  Chinese rubber-plant - *Crassula ovata*
  Chinese scarlet eggplant - *Solanum integrifolium*
  Chinese silk-plant - *Boehmeria nivea*
  Chinese sprangle-top - *Leptochloa chinensis*
  Chinese sumac - *Rhus chinensis*
  Chinese tallow-tree - *Sapium sebiferum*
  Chinese tamarisk - *Tamarix chinensis*
  Chinese taro - *Alocasia cucullata*
  Chinese watermelon - *Benincasa hispida*
  Chinese white poplar - *Populus tomentosa*
  Chinese wild peach - *Prunus davidiana*
  Chinese wild rice - *Leymus chinensis*
  Chinese windmill palm - *Trachycarpus fortunei*
  Chinese wing-nut - *Pterocarya stenoptera*

Chinese winter melon - *Benincasa hispida*
Chinese wisteria - *Wisteria sinensis*
Chinese yam - *Dioscorea batatas*
chinkapin
  chinkapin oak - *Quercus muehlenbergii*
  Florida chinkapin - *Castanea alnifolia*
Chinook licorice - *Lupinus littoralis*
chinquapin
  Allegheny chinquapin - *Castanea pumila*
  bush chinquapin - *Castanopsis sempervirens*
  California chinquapin - *Castanopsis sempervirens*
  chinquapin - *Castanopsis*
  chinquapin oak - *Quercus prinoides*
  dwarf chinquapin oak - *Quercus prinoides*
  giant chinquapin - *Castanopsis chrysophylla*
  golden chinquapin - *Castanopsis chrysophylla*
  sierra chinquapin - *Castanopsis sempervirens*
  water-chinquapin - *Nelumbo lutea*
chirimoya - *Annona cherimola*
chittamwood - *Cotinus obovatus, Sideroxylon lanuginosa*
chives
  Chinese chives - *Allium tuberosum*
  chives - *Allium schoenoprasum*
chloris, fringed chloris - *Chloris ciliata*
chocolate
  chocolate-root - *Geum rivale*
  chocolate tree - *Theobroma cacao*
  chocolate-vine - *Akebia quinata*
  Indian chocolate - *Geum rivale*
Choctaw-root - *Apocynum cannabinum*
choi
  gai-choi - *Brassica juncea*
  pak-choi - *Brassica chinensis*
choke, sun-choke - *Helianthus tuberosus*
chokeberry
  black chokeberry - *Aronia melanocarpa*
  chokeberry - *Aronia*
  Florida chokeberry - *Aronia prunifolia*
  purple chokeberry - *Aronia prunifolia, Malus floribunda*
  red chokeberry - *Aronia arbutifolia*
chokecherry
  black chokecherry - *Prunus virginiana* var. *melanocarpa*
  chokecherry - *Prunus virginiana*
  southwestern chokecherry - *Prunus serotina* subsp. *virens*
  western chokecherry - *Prunus virginiana* var. *demissa*
cholla
  cholla cactus - *Opuntia*
  jumping cholla - *Opuntia fulgida*
  spiny cholla - *Opuntia spinosior*
  staghorn cholla - *Opuntia versicolor*
  walking-stick cholla - *Opuntia imbricata*
chonta - *Bactris gasipaes*
choyote - *Sechium edule*
Christ
  Christ-plant - *Euphorbia milii*
  Christ-thorn - *Euphorbia milii*
christi, palma-christi - *Ricinus communis*
Christmas
  American Christmas mistletoe - *Phoradendron serotinum*
  Christmas-berry - *Crossopetalum ilicifolia, Heteromeles arbutifolia, Lycium carolinianum*

Christmas-berry-tree - *Schinus terebinthifolius*
Christmas cactus - *Schlumbergera buckleyi*
Christmas-candle - *Senna alata*
Christmas fern - *Polystichum acrostichoides*
Christmas-flower - *Euphorbia pulcherrima*
Christmas-green - *Lycopodium complanatum*
Christmas palm - *Veitchia merrillii*
Christmas-rose - *Helleborus niger*
Christmas-star - *Euphorbia pulcherrima*
Christmas-tree kalanchoe - *Kalanchoe laciniata*
Pacific Christmas fern - *Polystichum munitum*
christophine - *Sechium edule*
chrysanthemum
  corn chrysanthemum - *Chrysanthemum segetum*
  costmary chrysanthemum - *Chrysanthemum balsamita*
  daisy chrysanthemum - *Leucanthemum maximum*
  florist's chrysanthemum - *Chrysanthemum morifolium*
  garland chrysanthemum - *Chrysanthemum coronarium*
  Max's chrysanthemum - *Leucanthemum maximum*
chufa - *Cyperus esculentus*
chuparosa - *Anisacanthus thurberi*
church-mouse three-awn - *Aristida dichotoma*
ciboule - *Allium fistulosum*
cicely
  bland sweet cicely - *Osmorhiza claytonii*
  hairy sweet cicely - *Osmorhiza claytonii*
  woolly sweet cicely - *Osmorhiza claytonii*
cicer milk-vetch - *Astragalus cicer*
cigar
  cigar-flower - *Cuphea ignea*
  cigar-tree - *Catalpa speciosa*
cilantro - *Coriandrum sativum*
cimarron, tobasco cimarron - *Nicotiana repanda*
cineraria, florist's cineraria - *Senecio cruentus*
cinnamon
  Ceylon cinnamon - *Cinnamomum verum*
  cinnamon - *Cinnamomum verum*
  cinnamon-bark - *Canella winteriana*
  cinnamon fern - *Osmunda cinnamomea*
  cinnamon-jasmine - *Hedychium coronarium*
  cinnamon rose - *Rosa majalis*
  cinnamon-vine - *Dioscorea batatas*
  wild cinnamon - *Canella winteriana*
cinquefoil
  biennial cinquefoil - *Potentilla biennis*
  bush cinquefoil - *Potentilla fruticosa*
  bushy cinquefoil - *Potentilla paradoxa*
  cinquefoil - *Potentilla canadensis, Potentilla*
  creeping cinquefoil - *Potentilla reptans*
  downy cinquefoil - *Potentilla intermedia*
  dwarf cinquefoil - *Potentilla canadensis*
  hoary cinquefoil - *Potentilla argentea*
  marsh cinquefoil - *Potentilla palustris*
  old-field cinquefoil - *Potentilla simplex*
  prairie cinquefoil - *Potentilla pensylvanica*
  rough cinquefoil - *Potentilla norvegica*
  rough-fruited cinquefoil - *Potentilla recta*
  shrubby cinquefoil - *Potentilla fruticosa*
  silver-weed cinquefoil - *Potentilla anserina*
  silvery cinquefoil - *Potentilla argentea*
  strawberry-leaf cinquefoil - *Potentilla sterilis*

sulphur cinquefoil - *Potentilla recta*
tall cinquefoil - *Potentilla arguta*
three-tooth cinquefoil - *Potentilla tridentata*
upright cinquefoil - *Potentilla recta*
white cinquefoil - *Potentilla arguta*
cistus, St. Vincent's cistus - *Cistus palhinhai*
citron
 citron - *Citrullus lanatus* var. *citroides, Citrus medica*
 citron-melon - *Citrullus lanatus* var. *citroides*
citronella - *Collinsonia canadensis*
city goosefoot - *Chenopodium urbicum*
civet bean - *Phaseolus lunatus*
clammy
 clammy azalea - *Rhododendron viscosum*
 clammy cockle - *Silene noctiflora*
 clammy cudweed - *Gnaphalium viscosum*
 clammy cuphea - *Cuphea viscosissima*
 clammy everlasting - *Gnaphalium viscosum*
 clammy ground-cherry - *Physalis heterophylla*
 clammy locust - *Robinia viscosa*
 rough-seed clammy-weed - *Polanisia dodecandra*
 western clammy-weed - *Polanisia dodecandra* subsp.
   trachysperma
Clanwilliam daisy - *Euryops speciosissimus*
clasping
 clasping aster - *Aster patens*
 clasping-bellwort - *Triodanis perfoliata*
 clasping dogbane - *Apocynum sibiricum*
 clasping heart-leaf aster - *Aster undulatus*
 clasping heliotrope - *Heliotropium amplexicaule*
 clasping-leaf peppergrass - *Lepidium perfoliatum*
 clasping-leaf pondweed - *Potamogeton perfoliatus*
 clasping milkweed - *Asclepias amplexicaulis*
 clasping mullein - *Verbascum phlomoides*
 clasping pepper-weed - *Lepidium perfoliatum*
claw
 cat-claw acacia - *Acacia greggii*
 cat-claw mimosa - *Mimosa pigra*
 cat-claw schrankia - *Mimosa quadrivalvis* var. *nuttallii*
 cat-claw sensitive-brier - *Mimosa quadrivalvis* var. *angustata*
 cat-claw trumpet - *Macfadyena unguis-cati*
 cat-claw-vine - *Macfadyena unguis-cati*
 cat's-claw - *Acacia tortuosa, Mimosa quadrivalvis* var.
   nuttallii, Pithecellobium unguis-cati
 claw cactus - *Schlumbergera truncata*
 crab's-claw - *Stratiotes aloides*
 devil's-claw - *Proboscidea louisianica*
 Gregg's cat's-claw - *Acacia greggii*
 long-flowered cat's-claw - *Acacia greggii*
claws
 chicken's-claws - *Salicornia europaea*
 wolf's-claws - *Lycopodium clavatum*
clay-weed - *Tussilago farfara*
clearweed - *Pilea pumila*
cleavers - *Galium aparine, Galium*
clematis - *Clematis*
cleome
 golden cleome - *Cleome lutea*
 yellow cleome - *Cleome lutea*
cliff
 cliff brake - *Pellaea*

cliff-bush - *Jamesia americana*
cliff-green - *Paxistima canbyi*
cliff-rose - *Cowania mexicana*
cliff stonewort - *Sedum glaucophyllum*
fragile cliff brake - *Cryptogramma stelleri*
mountain cliff fern - *Woodsia scopulina*
New Zealand cliff brake - *Pellaea rotundifolia*
slender cliff brake - *Cryptogramma stelleri*
smooth cliff brake - *Pellaea glabella*
smooth cliff fern - *Woodsia glabella*
western cliff fern - *Woodsia oregana*
climber
 cardinal-climber - *Ipomoea quamoclit*
 elephant-climber - *Argyreia nervosa*
 love grass petticoat-climber - *Eragrostis tenella*
climbing
 climbing-beauty - *Aeschynanthus pulcher*
 climbing bird's nest fern - *Microsorium punctatum*
 climbing bittersweet - *Celastrus scandens*
 climbing dogbane - *Trachelospermum difforme*
 climbing euonymus - *Euonymus fortunei*
 climbing false buckwheat - *Polygonum scandens*
 climbing fern - *Lygodium*
 climbing ficus - *Ficus pumila*
 climbing fumitory - *Adlumia fungosa*
 climbing hempweed - *Mikania scandens*
 climbing hydrangea - *Decumaria barbara, Hydrangea*
   anomala subsp. *petiolaris*
 climbing milkweed - *Cynanchum louiseae, Sarcostemma*
   cynanchoides
 climbing prairie rose - *Rosa setigera*
 climbing rose - *Rosa setigera*
 Japanese climbing fern - *Lygodium japonicum*
clintonia, yellow clintonia - *Clintonia borealis*
cloak, waxy cloak fern - *Notholaena sinuata*
clock
 shepherd's-clock - *Anagallis arvensis*
 townhall's-clock - *Adoxa moschatellina*
closed
 closed gentian - *Gentiana andrewsii, G. clausa, G. linearis,*
   G. rubricaulis
 meadow closed gentian - *Gentiana clausa*
clotbur - *Arctium minus, Arctium*
cloth
 oil-cloth-flower - *Anthurium andraeanum*
 tapa-cloth-tree - *Broussonetia papyrifera*
cloudberry - *Rubus chamaemorus*
clove pink - *Dianthus caryophyllus*
clover
 Alsike clover - *Trifolium hybridum*
 Alyce clover - *Alysicarpus vaginalis*
 annual buffalo clover - *Trifolium reflexum*
 arrow-leaf clover - *Trifolium vesiculosum*
 berseem clover - *Trifolium alexandrinum*
 big-flower clover - *Trifolium michelianum*
 buffalo clover - *Trifolium reflexum*
 Bukhara clover - *Melilotus alba*
 bur-clover - *Medicago arabica, M. minima*
 bush-clover - *Lespedeza capitata, Lespedeza*
 California bur-clover - *Medicago polymorpha*
 Carolina clover - *Trifolium carolinianum*
 Caucasian clover - *Trifolium ambiguum*

Chinese bush-clover - *Lespedeza cuneata*
clover - *Trifolium ambiguum, Trifolium*
clover dodder - *Cuscuta epithymum*
cow-hop clover - *Trifolium dubium*
creeping bush-clover - *Lespedeza repens*
crimson clover - *Trifolium incarnatum*
cultivated red clover - *Trifolium pratense* var. *sativum*
cupped clover - *Trifolium cherleri*
downy prairie-clover - *Dalea villosa*
Egyptian clover - *Trifolium alexandrinum*
elk-clover - *Aralia californica*
European water-clover - *Marsilea quadrifolia*
hoary tick-clover - *Desmodium canescens*
holy-clover - *Onobrychis viciifolia*
honey-clover - *Melilotus alba*
hop clover - *Trifolium aureum, T. campestre*
hubam clover - *Melilotus alba*
Indian sweet-clover - *Melilotus indica*
Italian clover - *Trifolium incarnatum*
Japanese bush-clover - *Lespedeza striata*
Japanese clover - *Lespedeza striata*
Kenyan clover - *Trifolium semipilosum*
Kenyan wild white clover - *Trifolium semipilosum*
knotted clover - *Trifolium striatum*
Korean bush-clover - *Kummerowia stipulacea*
Korean clover - *Kummerowia stipulacea*
kura clover - *Trifolium ambiguum*
ladino clover - *Trifolium repens*
large hop clover - *Trifolium campestre*
little bur-clover - *Medicago minima*
Mexican clover - *Richardia scabra*
musk-clover - *Erodium moschatum*
nonesuch clover - *Medicago lupulina*
old-field clover - *Trifolium arvense*
one-leaf-clover - *Alysicarpus vaginalis, Macroptilium lathyroides*
owl-clover - *Orthocarpus luteus*
palmate hop clover - *Trifolium aureum*
Persian clover - *Trifolium resupinatum*
pin-clover - *Erodium cicutarium*
purple prairie-clover - *Dalea purpurea*
rabbit-foot clover - *Trifolium arvense*
red clover - *Trifolium pratense*
reversed clover - *Trifolium resupinatum*
rose clover - *Trifolium hirtum*
round-head bush-clover - *Lespedeza capitata*
running buffalo clover - *Trifolium stoloniferum*
seaside clover - *Trifolium wormskioldii*
sessile tick-clover - *Desmodium sessilifolium*
shrub bush-clover - *Lespedeza bicolor*
sierra clover - *Trifolium wormskioldii*
silky prairie-clover - *Dalea villosa*
slender bush-clover - *Lespedeza virginica*
small hop clover - *Trifolium dubium*
sour-clover - *Melilotus indica*
Spanish tick-clover - *Desmodium uncinatum*
spotted bur-clover - *Medicago arabica*
stinking-clover - *Cleome serrulata*
stone clover - *Trifolium arvense*
strawberry clover - *Trifolium fragiferum*
striate clover - *Trifolium striatum*
Stueve's bush-clover - *Lespedeza stuevei*
subterranean clover - *Trifolium subterraneum*

sweet-clover - *Melilotus*
tall yellow sweet-clover - *Melilotus altissima*
toothed bur-clover - *Medicago polymorpha*
trailing bush-clover - *Lespedeza procumbens*
tree-clover - *Melilotus alba*
wand-like bush-clover - *Lespedeza intermedia*
water-clover - *Marsilea*
white clover - *Trifolium repens*
white prairie-clover - *Dalea candida*
white sweet-clover - *Melilotus alba*
white-tip clover - *Trifolium variegatum*
yellow clover - *Trifolium dubium*
yellow sweet-clover - *Melilotus officinalis*
zigzag clover - *Trifolium medium*
club
    bog club-moss - *Lycopodium inundatum*
    bristly club-moss - *Lycopodium annotinum*
    club gourd - *Trichosanthes anguina*
    club-moss - *Lycopodium clavatum, Lycopodium*
    devil's-club - *Oplopanax horridum*
    fir club-moss - *Lycopodium selago*
    golden-club - *Orontium aquaticum*
    Hercules's-club - *Aralia spinosa, Zanthoxylum clava-herculis*
    interrupted club-moss - *Lycopodium annotinum*
    little club-moss - *Selaginella*
    marsh club-moss - *Lycopodium inundatum*
    running club-moss - *Lycopodium clavatum*
    shining club-moss - *Lycopodium lucidulum*
    spiny-club palm - *Bactris*
    stiff club-moss - *Lycopodium annotinum*
    Texas Hercules's-club - *Zanthoxylum hirsutum*
    tree club-moss - *Lycopodium obscurum*
    tufted club rush - *Scirpus caespitosus*
clump verbena - *Verbena canadensis*
cluster
    cluster bean - *Cyamopsis tetragonoloba*
    cluster dock - *Rumex conglomeratus*
    cluster-flower - *Flaveria trinervia*
    cluster-leaf tick-trefoil - *Desmodium glutinosum*
    cluster mahonia - *Berberis pinnata*
    cluster palm - *Ptychosperma macarthurii*
    cluster pine - *Pinus pinaster*
    cluster rose - *Rosa pisocarpa*
    cluster serviceberry - *Amelanchier alnifolia* var. *pumila*
    cluster tarweed - *Madia glomerata*
    Egyptian star-cluster - *Pentas lanceolata*
    star-cluster - *Pentas lanceolata*
clustered
    clustered bellflower - *Campanula glomerata*
    clustered fishtail palm - *Caryota mitis*
    northern clustered sedge - *Carex arcta*
coach-whip - *Fouquieria splendens*
coast
    coast-blite - *Chenopodium rubrum*
    coast-dandelion - *Hypochaeris radicata*
    coast fiddle-neck - *Amsinckia lycopsoides*
    coast live oak - *Quercus agrifolia*
    coast man-root - *Marah oreganus*
    coast sandbur - *Cenchrus incertus*
    coast tarweed - *Madia sativa*
    coast wallflower - *Erysimum capitatum*
    coast white-alder - *Clethra alnifolia*

Gulf Coast spike rush - *Eleocharis cellulosa*
Pacific Coast red elder - *Sambucus callicarpa*
West Coast rhododendron - *Rhododendron macrophyllum*
coastal
coastal arrowhead - *Sagittaria graminea*
coastal plain dewberry - *Rubus trivialis*
coastal plain gentian - *Gentiana catesbaei*
coastal plain pennywort - *Hydrocotyle bonariensis*
coastal plain tickseed - *Bidens mitis*
coastal plain willow - *Salix caroliniana*
coastal redwood - *Sequoia sempervirens*
coastal shield fern - *Dryopteris arguta*
coastal silver-weed - *Potentilla pacifica*
coastal swamp goldenrod - *Solidago elliottii*
coastal wattle - *Acacia cyclopis*
coastal willow - *Salix hookeriana*
coastal wood fern - *Dryopteris arguta*
coat
coat-buttons - *Tridax procumbens*
Jacob's-coat - *Acalypha wilkesiana*
Joseph's-coat - *Amaranthus tricolor*
cobra
cobra-lily - *Darlingtonia californica*
junta-de-cobra-pintada - *Justicia brandegeana*
cochineal
cochineal cactus - *Opuntia cochenillifera*
cochineal-plant - *Opuntia cochenillifera*
cochita - *Centrosema*
cock
cock-foot panicum - *Echinochloa crus-galli*
cock grass - *Bromus secalinus*
cock's
cock's-comb - *Celosia argentea* var. *cristata*
cock's-foot - *Dactylis glomerata*
cockle
clammy cockle - *Silene noctiflora*
cockle-button - *Arctium lappa, A. minus*
corn cockle - *Agrostemma githago*
cow cockle - *Vaccaria hispanica*
pink cockle - *Vaccaria hispanica*
red cockle - *Silene dioica*
snowy cockle - *Silene noctiflora*
spring cockle - *Vaccaria hispanica*
sticky cockle - *Silene noctiflora*
tarry cockle - *Silene antirrhina*
white cockle - *Silene latiflora*
cocklebur
broad-leaf cocklebur - *Xanthium strumarium*
cocklebur - *Agrimonia, Xanthium strumarium, X. strumarium*
var. *canadense*
heart-leaf cocklebur - *Xanthium strumarium*
spiny cocklebur - *Xanthium spinosum*
cockspur
cockspur - *Crataegus crus-galli*
cockspur coral-tree - *Erythrina crista-galli*
cockspur grass - *Echinochloa crus-galli*
cockspur hawthorn - *Crataegus crus-galli*
cockspur-thorn - *Crataegus crus-galli*
gulf cockspur - *Echinochloa crus-pavonis*
coco
coco - *Cyperus esculentus*
coco-grass - *Cyperus rotundus*

coco-plum - *Chrysobalanus icaco*
coco sedge - *Cyperus esculentus, C. rotundus*
coco-yam - *Colocasia esculenta*
cocoa
cocoa-bean - *Theobroma cacao*
wild cocoa - *Pachira aquatica*
coconut
coconut - *Cocos nucifera*
coconut palm - *Cocos nucifera*
coffee
arabica coffee - *Coffea arabica*
coffee - *Coffea arabica*
coffee-berry - *Rhamnus californica*
coffee colubrina - *Colubrina arborescens*
coffee fern - *Pellaea andromedifolia*
coffee senna - *Senna occidentalis*
coffee-tree - *Polyscias guilfoylei*
coffee-weed - *Cichorium intybus*
Kentucky coffee-tree - *Gymnocladus dioica*
Negro coffee - *Senna occidentalis*
robusta coffee - *Coffea canephora*
wild coffee - *Colubrina arborescens, Polyscias guilfoylei,*
Triosteum aurantiacum, T. perfoliatum
wild robusta coffee - *Coffea canephora*
cogon grass - *Imperata cylindrica*
cohosh
black cohosh - *Cimicifuga racemosa*
blue cohosh - *Caulophyllum thalictroides*
cohosh - *Actaea*
summer cohosh - *Cimicifuga americana*
white cohosh - *Actaea alba*
cole
bore cole - *Brassica oleracea* var. *acephala*
cole - *Brassica oleracea* var. *acephala*
red-cole - *Armoracia rusticana*
southern cole - *Brassica juncea*
coleus, prostrate-coleus - *Plectranthus*
colewort - *Brassica oleracea* var. *acephala, Crambe*
abyssinica
colic
colic-root - *Aletris, Dioscorea villosa*
yellow colic-root - *Aletris aurea*
coliseum-ivy - *Cymbalaria muralis*
collard
marsh-collard - *Nuphar*
water-collard - *Nuphar*
collards - *Brassica oleracea* var. *acephala*
collared dodder - *Cuscuta indecora*
colonial bent grass - *Agrostis capillaris*
Colorado
Colorado barberry - *Berberis fendleri*
Colorado blue spruce - *Picea pungens*
Colorado four-o'clock - *Mirabilis multiflora*
Colorado pine - *Pinus edulis*
Colorado pinyon - *Pinus edulis*
Colorado red-cedar - *Juniperus scopulorum*
Colorado River hemp - *Sesbania exaltata*
Colorado spruce - *Picea pungens*
colorado, mamey-colorado - *Pouteria sapota*

colt's
  colt's-foot - *Tussilago farfara*
  sweet-colt's-foot - *Petasites*
colubrina, coffee colubrina - *Colubrina arborescens*
Columbia
  Columbia hawthorn - *Crataegus columbiana*
  Columbia lily - *Lilium columbianum*
  Columbia milk-vetch - *Astragalus miser* var. *serotinus*
  Columbia needle grass - *Stipa columbiana*
columbine
  Canadian columbine - *Aquilegia canadensis*
  columbine - *Aquilegia canadensis, Aquilegia*
  European columbine - *Aquilegia vulgaris*
  garden columbine - *Aquilegia vulgaris*
  wild columbine - *Aquilegia canadensis*
columbo - *Frasera speciosa*
Columbus grass - *Sorghum almum*
column, golden-column - *Echinopsis spachiana*
columnar prairie coneflower - *Ratibida columnifera*
colza - *Brassica napus*
comagueyana - *Bothriochloa pertusa*
comandra, northern comandra - *Geocaulon lividum*
comb
  cock's-comb - *Celosia argentea* var. *cristata*
  Venus's-comb - *Scandix pecten-veneris*
Comb's paspalum - *Paspalum almum*
comfrey
  blue comfrey - *Symphytum uplandicum*
  comfrey - *Symphytum officinale, Symphytum*
  northern wild comfrey - *Cynoglossum boreale*
  prickly comfrey - *Symphytum asperum*
  Quaker comfrey - *Symphytum uplandicum*
  Russian comfrey - *Symphytum uplandicum*
  wild comfrey - *Cynoglossum virginianum, Hackelia virginiana*
commandments, ten-commandments - *Maranta leuconeura*
compact brome - *Bromus madritensis*
compass-plant - *Lactuca serriola, Silphium laciniatum*
comptie - *Zamia pumila*
conchitas - *Clitoria ternatea*
conchito - *Macroptilium atropurpureum*
condalia, Mexican condalia - *Condalia mexicana*
cone
  big-cone Douglas fir - *Pseudotsuga macrocarpa*
  big-cone pine - *Pinus coulteri*
  big-cone-spruce - *Pseudotsuga macrocarpa*
  bristle-cone fir - *Abies bracteata*
  cone catchfly - *Silene conoidea*
  cone wheat - *Triticum turgidum*
  Great Basin bristle-cone - *Pinus longaeva*
  knob-cone pine - *Pinus attenuata*
  western bristle-cone pine - *Pinus longaeva*
coneflower
  columnar prairie coneflower - *Ratibida columnifera*
  coneflower - *Dracopis amplexicaulis, Rudbeckia laciniata, Rudbeckia*
  cut-leaf coneflower - *Rudbeckia laciniata*
  hairy coneflower - *Rudbeckia hirta*
  pale purple coneflower - *Echinacea pallida*

  prairie coneflower - *Ratibida*
  purple coneflower - *Echinacea purpurea, Echinacea*
  thin-leaf coneflower - *Rudbeckia triloba*
  three-lobed coneflower - *Rudbeckia triloba*
  upright prairie coneflower - *Ratibida columnifera*
conehead
  conehead - *Cycas*
  sago conehead - *Cycas*
Confederate-jasmine - *Trachelospermum jasminoides*
Congo
  Congo jute - *Urena lobata*
  Congo pea - *Cajanus cajan*
connate beggar-ticks - *Bidens connata*
consumption-weed - *Baccharis halimifolia*
convulsion-root - *Monotropa uniflora*
cool
  cool-tankard - *Borago officinalis*
  cool-wort - *Mitella diphylla, Pilea pumila*
coon's-tail - *Ceratophyllum demersum*
coontie - *Zamia pumila*
copa-tree - *Ailanthus altissima*
copey - *Clusia rosea*
copper
  copper iris - *Iris fulva*
  copper-weed - *Iva acerosa*
  hop-hornbeam copper-leaf - *Acalypha ostryifolia*
  painted copper-leaf - *Acalypha wilkesiana*
  rhombic copper-leaf - *Acalypha rhomboidea*
  rough-pod copper-leaf - *Acalypha ostryifolia*
  short-stalk copper-leaf - *Acalypha gracilens*
  slender copper-leaf - *Acalypha gracilens*
  Virginia copper-leaf - *Acalypha virginica*
cor, balsamo-cor-de-carne - *Justicia carnea*
coral
  cockspur coral-tree - *Erythrina crista-galli*
  coral-bead-plant - *Abrus precatorius*
  coral-beads - *Cocculus carolinus*
  coral bean - *Erythrina herbacea*
  coral-bells - *Heuchera sanguinea*
  coral-berry - *Ardisia crenata, Symphoricarpos chenaultii, S. orbiculatus*
  coral drop-seed - *Sporobolus domingensis*
  coral-evergreen - *Lycopodium clavatum*
  coral greenbrier - *Smilax walteri*
  coral honeysuckle - *Lonicera sempervirens*
  coral-plant - *Russelia equisetiformis*
  coral-tree - *Erythrina*
  eastern coral bean - *Erythrina herbacea*
  southwestern coral bean - *Erythrina flabelliformis*
  western coral bean - *Erythrina flabelliformis*
coralroot
  autumn coralroot - *Corallorrhiza odontorhiza*
  coralroot - *Corallorrhiza*
  early coralroot - *Corallorrhiza trifida*
  large coralroot - *Corallorrhiza maculata*
  late coralroot - *Corallorrhiza odontorhiza*
  northern coralroot - *Corallorrhiza trifida*
  pale coralroot - *Corallorrhiza trifida*
  spotted coralroot - *Corallorrhiza maculata*
corazon - *Annona reticulata*

cord
   alkali cord grass - *Spartina gracilis*
   big cord grass - *Spartina cynosuroides*
   cord grass - *Spartina*
   freshwater cord grass - *Spartina pectinata*
   prairie cord grass - *Spartina pectinata*
   salt marsh cord grass - *Spartina alterniflora*
   salt meadow cord grass - *Spartina patens*
   salt-water cord grass - *Spartina alterniflora*
   smooth cord grass - *Spartina alterniflora*
coreopsis
   coreopsis beggar-ticks - *Bidens polylepis*
   garden coreopsis - *Coreopsis lanceolata*
   lance-leaf coreopsis - *Coreopsis lanceolata*
   plains coreopsis - *Coreopsis tinctoria*
   stiff coreopsis - *Coreopsis palmata*
   tall coreopsis - *Coreopsis tripteris*
   whorled coreopsis - *Coreopsis verticillata*
coriander - *Coriandrum sativum*
cork
   cork-bark fir - *Abies lasiocarpa* var. *arizonica*
   cork elm - *Ulmus thomasii*
   cork fir - *Abies lasiocarpa* var. *arizonica*
   cork oak - *Quercus suber*
corkwood - *Leitneria floridana*
corky-stem passionflower - *Passiflora suberosa*
corn
   beaked corn-salad - *Valerianella radiata*
   broom-corn - *Panicum miliaceum, Sorghum bicolor*
   broom corn millet - *Panicum miliaceum*
   corn bedstraw - *Galium tricorne*
   corn-bind - *Convolvulus arvensis*
   corn-bindweed - *Polygonum convolvulus*
   corn buttercup - *Ranunculus arvensis*
   corn-campion - *Agrostemma githago*
   corn-chamomile - *Anthemis arvensis*
   corn chrysanthemum - *Chrysanthemum segetum*
   corn cockle - *Agrostemma githago*
   corn gromwell - *Lithospermum arvense*
   corn-lily - *Clintonia borealis*
   corn-marigold - *Chrysanthemum segetum*
   corn mint - *Mentha arvensis*
   corn-mullein - *Agrostemma githago*
   corn-plant - *Dracaena fragrans* cv. 'massangeana'
   corn poppy - *Papaver rhoeas*
   corn-rose - *Agrostemma githago*
   corn-salad - *Valerianella locusta, Valerianella*
   corn speedwell - *Veronica arvensis*
   corn spurry - *Spergula arvensis*
   crow-corn - *Aletris farinosa*
   Dent's corn - *Zea mays*
   field corn - *Zea mays*
   flint corn - *Zea mays*
   Italian corn-salad - *Valerianella eriocarpa*
   kaffir-corn - *Sorghum bicolor*
   pod corn - *Zea mays*
   squirrel-corn - *Dicentra canadensis*
   sweet corn - *Zea mays*
   volunteer corn - *Zea mays*
cornel
   cornel - *Cornus*
   dwarf cornel - *Cornus canadensis*

   small rough-leaf cornel - *Cornus asperifolia*
   stiff-cornel dogwood - *Cornus stricta*
   western cornel - *Cornus glabrata*
cornered, three-cornered-jack - *Emex australis*
cornflower - *Centaurea cyanus*
Cornish heath - *Erica vagans*
coroa - *Sicana odorifera*
corpse-plant - *Monotropa uniflora*
Corsican
   Corsican carpet-plant - *Soleirolia soleirolii*
   Corsican-curse - *Soleirolia soleirolii*
   Corsican pine - *Pinus nigra*
corydalis
   golden corydalis - *Corydalis aurea*
   pale corydalis - *Corydalis sempervirens*
   slender corydalis - *Corydalis micrantha*
   yellow corydalis - *Corydalis flavula*
Cos lettuce - *Lactuca sativa* var. *longifolia*
cosmetic-bark-tree - *Murraya paniculata*
cosmos
   cosmos - *Cosmos bipinnatus*
   garden cosmos - *Cosmos bipinnatus*
   orange cosmos - *Cosmos sulphureus*
   yellow cosmos - *Cosmos sulphureus*
Cossack asparagus - *Typha latifolia*
costmary
   costmary - *Chrysanthemum balsamita*
   costmary chrysanthemum - *Chrysanthemum balsamita*
cottage pink - *Dianthus plumarius*
cotton
   American-Egyptian cotton - *Gossypium barbadense*
   Arizona cotton-top - *Digitaria californica*
   Arizona wild cotton - *Gossypium thurberi*
   Ceylon cotton - *Gossypium arboreum*
   cotton - *Gossypium*
   cotton-ball - *Espostoa lanata*
   cotton-batting cudweed - *Gnaphalium stramineum*
   cotton burdock - *Arctium tomentosum*
   cotton grass - *Eriophorum*
   cotton-gum - *Nyssa aquatica*
   cotton morning-glory - *Ipomoea cordatotriloba* var. *torreyana*
   cotton-rose - *Filago*
   cotton thistle - *Onopordum acanthium*
   cotton weed - *Anaphalis margaritacea, Asclepias syriaca*
   cultivated cotton - *Gossypium herbaceum*
   dumb-cotton - *Calotropis procera*
   extra-long staple cotton - *Gossypium barbadense*
   lavender-cotton - *Santolina chamaecyparissus*
   Levant cotton - *Gossypium herbaceum*
   red silk cotton-tree - *Bombax ceiba*
   russet cotton grass - *Eriophorum russeolum*
   Sea Island cotton - *Gossypium barbadense*
   sheathed cotton grass - *Eriophorum vaginatum*
   silk cotton-tree - *Ceiba pentandra*
   tree cotton - *Gossypium arboreum, G. barbadense*
   tussock cotton grass - *Eriophorum vaginatum*
   upland cotton - *Gossypium hirsutum*
   Virginia cotton grass - *Eriophorum virginicum*
   white silk cotton-tree - *Ceiba pentandra*
   wild cotton - *Hibiscus moscheutos, H. moscheutos* subsp. *palustris*

cottonwood
  Alamo cottonwood - *Populus fremontii*
  black cottonwood - *Populus heterophylla, P. trichocarpa*
  cottonwood - *Populus*
  eastern cottonwood - *Populus deltoides*
  Fremont's cottonwood - *Populus fremontii*
  Great Plains cottonwood - *Populus deltoides* var. *occidentalis*
  narrow-leaf cottonwood - *Populus angustifolia*
  plains cottonwood - *Populus deltoides* var. *occidentalis*
  southern cottonwood - *Populus deltoides*
  swamp cottonwood - *Populus heterophylla*
cottony jujube - *Ziziphus mauritiana*
couch
  bearded couch - *Elymus trachycaulus*
  blue-couch - *Digitaria scalarum*
couchgrass - *Elytrigia repens*
cough-wort - *Tussilago farfara*
Coulter's
  Coulter's pine - *Pinus coulteri*
  Coulter's spiderling - *Boerhavia coulteri*
council-tree - *Ficus altissima*
country
  country borage - *Plectranthus amboinicus*
  country gooseberry - *Averrhoa carambola*
  country-spinach - *Basella alba*
  country-walnut - *Aleurites moluccana*
Cove's cassia - *Senna covesii*
cow
  American cow-parsnip - *Heracleum sphondylium* subsp. *montanum*
  cow-bell - *Silene csereii*
  cow cockle - *Vaccaria hispanica*
  cow cress - *Lepidium campestre*
  cow-herb - *Vaccaria hispanica*
  cow-hop clover - *Trifolium dubium*
  cow-itch - *Campsis radicans, Toxicodendron radicans*
  cow-lily - *Nuphar*
  cow oak - *Quercus michauxii*
  cow-parsnip - *Heracleum sphondylium* subsp. *montanum, Heracleum*
  cow pea - *Vigna unguiculata*
  cow pea witch-weed - *Striga gesnerioides*
  cow-poison - *Delphinium menziesii, D. trolliifolium*
  cow soapwort - *Vaccaria hispanica*
  cow vetch - *Vicia cracca*
  cow-wheat - *Melampyrum lineare*
  cow's-tail-pine - *Cephalotaxus harringtonia* var. *drupacea*
cowage velvet bean - *Mucuna pruriens* var. *utilis*
cowbane
  cowbane - *Oxypolis rigidior*
  spotted cowbane - *Cicuta maculata*
cowberry - *Vaccinium vitis-idaea, Viburnum lentago*
cowslip
  American cowslip - *Dodecatheon conjugens, D. meadia, Dodecatheon*
  cowslip - *Caltha palustris, Primula veris, Primula*
  Virginia cowslip - *Mertensia virginica*
coyote
  coyote-brush - *Baccharis pilularis*
  coyote mint - *Monardella villosa*
  coyote willow - *Salix exigua*

crab
  crab cactus - *Schlumbergera truncata*
  crab-wood - *Gymnanthes lucida*
crab's
  crab's-claw - *Stratiotes aloides*
  crab's-eye - *Abrus precatorius*
crabapple
  American crabapple - *Malus angustifolia, M. coronaria*
  Chinese flowering crabapple - *Malus spectabilis*
  crabapple - *Malus sylvestris*
  garland crabapple - *Malus coronaria*
  Japanese flowering crabapple - *Malus floribunda*
  Oregon crabapple - *Malus fusca*
  prairie crabapple - *Malus ioensis*
  showy crabapple - *Malus floribunda*
  Siberian crabapple - *Malus baccata*
  southern crabapple - *Malus angustifolia*
  southern wild crabapple - *Malus angustifolia*
  sweet crabapple - *Malus coronaria*
  sweet-scented crabapple - *Malus coronaria*
  Toringo crabapple - *Malus sieboldii*
  wild crabapple - *Malus ioensis, Peraphyllum ramosissimum*
  wild sweet crabapple - *Malus coronaria*
crabgrass
  crabgrass - *Digitalis*
  hairy crabgrass - *Digitaria sanguinalis*
  Indian crabgrass - *Digitaria longiflora*
  large crabgrass - *Digitaria sanguinalis*
  purple crabgrass - *Digitaria sanguinalis*
  slender crabgrass - *Digitaria filiformis*
  small crabgrass - *Digitaria ischaemum*
  smooth crabgrass - *Digitaria ischaemum*
  southern crabgrass - *Digitaria ciliaris*
  sprouting-crabgrass - *Panicum dichotomiflorum*
  tropical crabgrass - *Digitaria bicornis*
  violet crabgrass - *Digitaria violascens*
crack willow - *Salix fragilis*
cracker-berry - *Cornus canadensis*
crakeberry - *Empetrum nigrum*
crambe - *Crambe abyssinica*
cramp-bark - *Viburnum trilobum*
cranberry
  American cranberry - *Vaccinium macrocarpon*
  cranberry - *Vaccinium*
  cranberry-bush - *Viburnum trilobum*
  cultivated cranberry - *Vaccinium macrocarpon*
  European cranberry-bush - *Viburnum opulus*
  high-bush cranberry - *Viburnum trilobum*
  hog cranberry - *Arctostaphylos uva-ursi*
  large cranberry - *Vaccinium macrocarpon*
  mountain cranberry - *Vaccinium erythrocarpum, V. vitis-idaea, V. vitis-idaea* var. *minus*
  rock cranberry - *Vaccinium vitis-idaea* var. *minus*
  small cranberry - *Vaccinium oxycoccos*
crane
  crane-flower - *Strelitzia reginae*
  crane-lily - *Strelitzia reginae*
crane's
  Bicknell's crane's-bill - *Geranium bicknellii*
  crane's-bill - *Geranium*
  Iberian crane's-bill - *Geranium ibericum*
  long-stalked crane's-bill - *Geranium columbinum*

Siberian crane's-bill - *Geranium sibiricum*
small-flowered crane's-bill - *Geranium pusillum*
spotted crane's-bill - *Geranium maculatum*

crape
crape-gardenia - *Tabernaemontana divaricata*
crape-jasmine - *Tabernaemontana divaricata*
crape-myrtle - *Lagerstroemia indica*

craw pea - *Lathyrus pratensis*

crawling phlox - *Phlox stolonifera*

crazyweed
crazyweed - *Oxytropis*
Lambert's crazyweed - *Oxytropis lambertii*
showy crazyweed - *Oxytropis splendens*
silky crazyweed - *Oxytropis sericea*

cream
cream-bush - *Holodiscus discolor*
cream-colored avens - *Geum virginianum*
cream violet - *Viola striata*

creamy violet - *Viola striata*

creashak - *Arctostaphylos uva-ursi*

creeper
bearded-creeper - *Crupina vulgaris*
Carmel creeper - *Ceanothus griseus* var. *horizontalis*
funnel-creeper - *Macfadyena unguis-cati*
golden Ceylon creeper - *Epipremnum aureum*
golden-creeper - *Stigmaphyllon ciliatum*
New Guinea creeper - *Mucuna novaguineensis*
Point Reyes creeper - *Ceanothus gloriosus*
Rangoon creeper - *Quisqualis indica*
trumpet-creeper - *Campsis radicans*
Virginia creeper - *Parthenocissus quinquefolia*

creeping
creeping barberry - *Berberis repens*
creeping beggar-weed - *Desmodium incanum*
creeping bellflower - *Campanula rapunculoides*
creeping bent grass - *Agrostis stolonifera*
creeping blueberry - *Vaccinium crassifolium*
creeping bugleweed - *Ajuga reptans*
creeping bur-head - *Echinodorus cordifolius*
creeping bush-clover - *Lespedeza repens*
creeping buttercup - *Ranunculus repens*
creeping-cedar - *Juniperus horizontalis*
creeping-Charlie - *Lysimachia nummularia, Pilea nummulariifolia, Wedelia trilobata*
creeping cinquefoil - *Potentilla reptans*
creeping cucumber - *Melothria pendula*
creeping dayflower - *Commelina diffusa*
creeping ficus - *Ficus pumila*
creeping fog-fruit - *Phyla nodiflora* var. *canescens*
creeping foxtail - *Alopecurus arundinaceus*
creeping indigo - *Indigofera spicata*
creeping-Jennie - *Convolvulus arvensis, Lysimachia nummularia*
creeping juniper - *Juniperus horizontalis*
creeping knapweed - *Centaurea repens*
creeping lady's-sorrel - *Oxalis corniculata*
creeping loosestrife - *Lysimachia nummularia*
creeping muhly - *Muhlenbergia repens*
creeping phlox - *Phlox stolonifera*
creeping rubber-plant - *Ficus pumila*
creeping sage - *Salvia sonomensis*
creeping-sailor - *Saxifraga stolonifera*

creeping savin juniper - *Juniperus horizontalis*
creeping sedge - *Carex chordorrhiza*
creeping snowberry - *Gaultheria hispidula*
creeping soft grass - *Holcus mollis*
creeping sow-thistle - *Sonchus arvensis*
creeping spearwort - *Ranunculus reptans*
creeping speedwell - *Veronica filiformis, V. serpyllifolia* subsp. *humifusa*
creeping spike rush - *Eleocharis palustris*
creeping spurge - *Euphorbia serpens*
creeping St. John's-wort - *Hypericum adpressum, H. calycinum*
creeping thistle - *Cirsium arvense*
creeping thyme - *Thymus serpyllum*
creeping velvet grass - *Holcus mollis*
creeping vervain - *Verbena canadensis*
creeping water-primrose - *Ludwigia peploides*
creeping wood-sorrel - *Oxalis corniculata*
creeping yellow-cress - *Rorippa sylvestris*
creeping yellow wood-sorrel - *Oxalis corniculata*

crenate broomrape - *Orobanche crenata*

creosote-bush - *Larrea tridentata*

cress
Austrian field-cress - *Rorippa austriaca*
bitter-cress - *Barbarea vulgaris, Cardamine*
blister-cress - *Erysimum*
cow cress - *Lepidium campestre*
creeping yellow-cress - *Rorippa sylvestris*
cress-leaf groundsel - *Senecio glabellus*
cuckoo bitter-cress - *Cardamine pratensis*
dry-land bitter-cress - *Cardamine parviflora*
early winter-cress - *Barbarea verna*
garden cress - *Lepidium sativum*
globe-podded hoary-cress - *Cardaria pubescens*
great yellow-cress - *Rorippa amphibia*
green rock-cress - *Arabis missouriensis*
hairy bitter-cress - *Cardamine hirsuta*
hairy rock-cress - *Arabis hirsuta*
heart-padded hoary-cress - *Cardaria draba*
hoary-cress - *Cardaria draba*
Indian cress - *Tropaeolum majus*
little bitter-cress - *Cardamine oligosperma*
marsh-cress - *Rorippa palustris*
mountain winter-cress - *Cardamine rotundifolia*
mouse-ear-cress - *Arabidopsis thaliana*
Pennsylvania bitter-cress - *Cardamine pensylvanica*
purple-cress - *Cardamine douglasii*
rock-cress - *Arabis, Sibara virginica*
rocket-cress - *Barbarea vulgaris*
shepherd's-cress - *Teesdalia nudicaulis*
small-flowered bitter-cress - *Cardamine parviflora*
smooth rock-cress - *Arabis laevigata*
spreading yellow-cress - *Rorippa sinuata*
spring-cress - *Cardamine bulbosa*
St. Barbara's cress - *Barbarea vulgaris*
swine-cress - *Coronopus didymus*
terete yellow-cress - *Rorippa teres*
thale-cress - *Arabidopsis thaliana*
trailing bitter-cress - *Cardamine rotundifolia*
upland-cress - *Barbarea vulgaris*
wart-cress - *Coronopus*
winter-cress - *Barbarea orthoceras, B. vulgaris*
yellow-cress - *Rorippa*

yellow field-cress - *Rorippa sylvestris*
crested
  crested bunias - *Bunias erucago*
  crested dog-tail grass - *Cynosurus cristatus*
  crested dog's-tail - *Cynosurus cristatus*
  crested fern - *Microsorium punctatum*
  crested prickle poppy - *Argemone platyceras*
  crested sedge - *Carex cristatella*
  crested wheat grass - *Agropyron cristatum*
  desert crested wheat grass - *Agropyron desertorum*
  dwarf-crested iris - *Iris cristata*
  fairway crested wheat grass - *Agropyron cristatum*
  standard crested wheat grass - *Agropyron desertorum*
Cretan brake - *Pteris cretica*
crimson
  crimson-berry - *Rubus arcticus*
  crimson blackberry - *Rubus arcticus*
  crimson bottlebrush - *Callistemon citrinus*
  crimson clover - *Trifolium incarnatum*
  crimson-eyed rose mallow - *Hibiscus moscheutos* subsp.
   palustris
  crimson fountain grass - *Pennisetum setaceum*
crinkle-root - *Cardamine diphylla*
crinkled hair grass - *Deschampsia flexuosa*
crinum
  crinum-lily - *Crinum*
  southern swamp crinum - *Crinum americanum*
crispate-leaf pondweed - *Potamogeton crispus*
crista-de-galo - *Celosia argentea* var. *cristata*
crocus
  autumn crocus - *Colchicum autumnale*
  wild crocus - *Anemone patens*
crofton-weed - *Ageratina adenophora*
crook-neck squash - *Cucurbita moschata*
cross
  cross-leaf heath - *Erica tetralix*
  cross-leaf milkwort - *Polygala cruciata*
  cross-vine - *Bignonia capreolata*
  Jerusalem-cross - *Lychnis chalcedonica*
  Maltese-cross - *Lychnis chalcedonica*
  St. Andrew's-cross - *Hypericum hypericoides*
  widow's-cross - *Sedum pulchellum*
crotalaria
  lance-leaf crotalaria - *Crotalaria lanceolata*
  showy crotalaria - *Crotalaria spectabilis*
  slender-leaf crotalaria - *Crotalaria brevidens*
  smooth crotalaria - *Crotalaria pallida*
  striped crotalaria - *Crotalaria pallida*
croton
  croton - *Codiaeum, Croton*
  garden-croton - *Codiaeum variegatum*
  one-seeded croton - *Croton monanthogynus*
  prairie-tea croton - *Croton monanthogynus*
  Texas croton - *Croton texensis*
  three-seeded croton - *Croton lindheimerianus*
  tooth-leaved croton - *Croton glandulosa*
  tropical croton - *Croton glandulosus* var. *septentrionalis*
  woolly croton - *Croton capitatus*
crow
  crow-corn - *Aletris farinosa*
  crow-flower - *Lychnis flos-cuculi*

crow-foot buttercup - *Ranunculus sceleratus*
crow garlic - *Allium vineale*
crow-poison - *Zigadenus densus*
crowberry
  black crowberry - *Empetrum nigrum*
  crowberry - *Empetrum nigrum*
  purple crowberry - *Empetrum atropurpureum*
crowder pea - *Vigna unguiculata*
crowfoot
  bristly crowfoot - *Ranunculus pensylvanicus*
  crowfoot - *Ranunculus*
  crowfoot grass - *Dactyloctenium aegyptium, Digitaria*
   sanguinalis, Eleusine indica
  cursed crowfoot - *Ranunculus sceleratus*
  European crowfoot - *Aquilegia vulgaris*
  garden crowfoot - *Aquilegia vulgaris*
  prairie crowfoot - *Ranunculus rhomboideus*
  seaside crowfoot - *Ranunculus cymbalaria*
  small-flowered crowfoot - *Ranunculus abortivus*
  small yellow water-crowfoot - *Ranunculus gmelinii*
  white water-crowfoot - *Ranunculus trichophyllus*
  yellow water crowfoot - *Ranunculus flabellaris*
crown
  crown-beard - *Verbesina occidentalis, Verbesina*
  crown-beauty - *Hymenocallis*
  crown daisy - *Chrysanthemum coronarium*
  crown-flower - *Calotropis gigantea*
  crown-of-the-field - *Agrostemma githago*
  crown-of-thorns - *Euphorbia milii*
  crown-plant - *Calotropis gigantea*
  crown-vetch - *Coronilla varia*
  crown-weed - *Ambrosia trifida*
  golden-crown daisy - *Verbesina encelioides*
  king's-crown - *Justicia carnea*
  small crown-flower - *Calotropis procera*
  trailing crown-vetch - *Coronilla varia*
crupina - *Crupina vulgaris*
cry-baby-tree - *Erythrina crista-galli*
cryptanthus, pink cryptanthus - *Cryptanthus bromelioides*
crystal-tea - *Ledum palustre*
Cuban
  Cuban jute - *Sida rhombifolia*
  Cuban pine - *Pinus caribaea*
cuckold - *Arctium lappa, Bidens bipinnata*
cuckoo
  cuckoo bitter-cress - *Cardamine pratensis*
  cuckoo-button - *Arctium minus*
  cuckoo-flower - *Cardamine pratensis, Lychnis flos-cuculi*
cucumber
  bitter cucumber - *Momordica charantia*
  bur cucumber - *Sicyos angulatus*
  creeping cucumber - *Melothria pendula*
  cucumber - *Cucumis sativus*
  cucumber-tree - *Magnolia acuminata*
  Indian cucumber-root - *Medeola virginica*
  large-leaf cucumber-tree - *Magnolia macrophylla*
  mock cucumber - *Echinocystis lobata*
  prickly cucumber - *Echinocystis lobata*
  serpent-cucumber - *Trichosanthes anguina*
  star-cucumber - *Sicyos angulatus*
  western wild cucumber - *Marah oreganus*
  wild cucumber - *Echinocystis lobata*

yellow cucumber-tree - *Magnolia acuminata* var. *subcordata*
cudweed
  alpine cudweed - *Gnaphalium supinum*
  clammy cudweed - *Gnaphalium viscosum*
  cotton-batting cudweed - *Gnaphalium stramineum*
  cudweed - *Artemisia ludoviciana, Gnaphalium obtusifolium,*
    Gnaphalium
  fragrant cudweed - *Gnaphalium obtusifolium*
  low cudweed - *Gnaphalium uliginosum*
  narrow-leaf cudweed - *Gnaphalium purpureum* var. *falcatum*
  purple cudweed - *Gnaphalium purpureum*
  wandering cudweed - *Gnaphalium pensylvanicum*
cultivated
  cultivated acacia - *Acacia cyclopis*
  cultivated cotton - *Gossypium herbaceum*
  cultivated cranberry - *Vaccinium macrocarpon*
  cultivated grape - *Vitis vinifera*
  cultivated northern wild rice - *Zizania palustris*
  cultivated oat - *Avena sativa*
  cultivated red clover - *Trifolium pratense* var. *sativum*
  cultivated strawberry - *Fragaria ananassa*
  cultivated timothy - *Phleum pratense*
Culver's-root - *Veronicastrum virginicum*
cumin - *Cuminum cyminum*
Cunningham's beefwood - *Casuarina cunninghamiana*
cup
  California sun-cup - *Camissonia bistorta*
  Chinese cup grass - *Eriochloa villosa*
  cup fern - *Dennstaedtia*
  cup grass - *Echinochloa*
  cup-plant - *Silphium perfoliatum*
  cup rosinweed - *Silphium perfoliatum*
  curly-cup gum-weed - *Grindelia squarrosa*
  huntsman's-cup - *Sarracenia purpurea, Sarracenia*
  Indian cup - *Silphium perfoliatum*
  Indian cup-plant - *Sarracenia purpurea*
  Indian tooth-cup - *Rotala indica*
  king's-cup - *Caltha palustris*
  large-flowered leaf-cup - *Polymnia uvedalia*
  mealy-cup sage - *Salvia farinacea*
  mossy-cup oak - *Quercus macrocarpa*
  orange-cup lily - *Lilium philadelphicum*
  over-cup oak - *Quercus lyrata*
  painted-cup - *Castilleja coccinea, Castilleja*
  pale-flowered leaf-cup - *Polymnia canadensis*
  prairie cup grass - *Eriochloa contracta*
  queen's-cup - *Clintonia uniflora*
  seaside painted-cup - *Castilleja latifolia*
  small-flowered leaf-cup - *Polymnia canadensis*
  sun-cup - *Camissonia bistorta, C. subacaulis*
  tail-cup lupine - *Lupinus argenteus* var. *heteranthus*
  tooth-cup - *Rotala ramosior*
  western orange-cup lily - *Lilium philadelphicum* var. *andinum*
  woolly cup grass - *Eriochloa villosa*
  woolly painted-cup - *Castilleja foliolosa*
  yellow-flowered leaf-cup - *Polymnia uvedalia*
cupey - *Clusia rosea*
cuphea
  clammy cuphea - *Cuphea viscosissima*
  tarweed cuphea - *Cuphea carthagenensis*
Cupid's-shaving-brush - *Emilia fosbergii*
cupped clover - *Trifolium cherleri*

cups, fringe-cups - *Tellima grandiflora*
Curaçao aloe - *Aloe vera*
curd-wort - *Galium verum*
curled
  curled mallow - *Malva verticillata* var. *crispa*
  curled mint - *Mentha crispa*
curlew-berry - *Empetrum nigrum*
curls, blue-curls - *Trichostema*
curly
  curly-cup gum-weed - *Grindelia squarrosa*
  curly dock - *Rumex crispus*
  curly-heads - *Clematis ochroleuca*
  curly-indigo - *Aeschynomene virginica*
  curly-leaf pondweed - *Potamogeton crispus*
  curly-mesquite grass - *Hilaria belangeri*
  curly muck-weed - *Potamogeton crispus*
  curly palm - *Howeia belmoreana*
  curly pondweed - *Potamogeton crispus*
currant
  alpine currant - *Ribes alpinum*
  American black currant - *Ribes americanum*
  black currant - *Ribes nigrum*
  bristly black currant - *Ribes lacustre*
  buffalo currant - *Ribes aureum, R. odoratum*
  California black currant - *Ribes bracteosum*
  chaparral currant - *Ribes malvaceum*
  currant - *Ribes*
  currant tomato - *Lycopersicon pimpinellifolium*
  eastern black currant - *Ribes americanum*
  European black currant - *Ribes nigrum*
  fetid currant - *Ribes glandulosum*
  garden black currant - *Ribes nigrum*
  garden currant - *Ribes rubrum*
  golden currant - *Ribes aureum*
  Hudson Bay currant - *Ribes hudsonianum*
  Indian currant - *Symphoricarpos chenaultii, S. orbiculatus*
  Missouri currant - *Ribes aureum, R. odoratum*
  mountain currant - *Ribes alpinum*
  mountain pink currant - *Ribes nevadense*
  northern black currant - *Ribes hudsonianum*
  northern red currant - *Ribes rubrum*
  red currant - *Ribes rubrum*
  red-flower currant - *Ribes sanguineum*
  sierra currant - *Ribes nevadense*
  skunk currant - *Ribes glandulosum*
  squaw currant - *Ribes cereum*
  sticky currant - *Ribes viscosissimum*
  stink currant - *Ribes bracteosum*
  swamp currant - *Ribes lacustre*
  swamp red currant - *Ribes triste*
  wax currant - *Ribes cereum*
  white currant - *Ribes rubrum*
  white-flowered currant - *Ribes cereum*
  wild black currant - *Ribes americanum*
curse, Corsican-curse - *Soleirolia soleirolii*
cursed
  cursed buttercup - *Ranunculus sceleratus*
  cursed crowfoot - *Ranunculus sceleratus*
curtain-plant - *Kalanchoe pinnata*
curua - *Sicana odorifera*
curubá - *Sicana odorifera*

cushion-pink - *Silene acaulis*
Cusick's
  Cusick's bluegrass - *Poa cusickii*
  Cusick's serviceberry - *Amelanchier alnifolia* var. *cusickii*
  custard-apple - *Annona cherimola, A. reticulata, A. squamosa*
cut
  cut grass - *Leersia*
  cut-leaf blackberry - *Rubus laciniatus*
  cut-leaf bugleweed - *Lycopus americanus*
  cut-leaf coneflower - *Rudbeckia laciniata*
  cut-leaf evening-primrose - *Oenothera laciniata*
  cut-leaf geranium - *Geranium dissectum*
  cut-leaf germander - *Teucrium botrys*
  cut-leaf ground-cherry - *Physalis angulata*
  cut-leaf mermaid-weed - *Proserpinaca pectinata*
  cut-leaf mignonette - *Reseda lutea*
  cut-leaf nettle - *Solanum triflorum*
  cut-leaf nightshade - *Solanum triflorum*
  cut-leaf philodendron - *Monstera deliciosa*
  cut-leaf teasel - *Dipsacus laciniatus*
  cut-leaf toothwort - *Cardamine concatenata*
  cut-leaf water-horehound - *Lycopus americanus*
  giant cut grass - *Zizaniopsis miliacea*
  rice cut grass - *Leersia oryzoides*
  southern cut grass - *Leersia hexandra*
cycas, sago cycas - *Zamia pumila*
cyclamen, florist's cyclamen - *Cyclamen persicum*
cylindric
  cylindric blazing-star - *Liatris cylindracea*
  cylindric sedge - *Cyperus retrorsus*
cynanchum, leafless cynanchum - *Cynanchum scoparium*
cynthia, two-flowered cynthia - *Krigia biflora*
cyperus
  awned cyperus - *Cyperus squarrosus*
  globose cyperus - *Cyperus ovularis*
  marsh cyperus - *Cyperus pseudovegetus*
  pine-barren cyperus - *Cyperus cylindricus*
  red-rooted cyperus - *Cyperus erythrorhizos*
  shining cyperus - *Cyperus rivularis*
  slender cyperus - *Cyperus filiculmis*
  straw-colored cyperus - *Cyperus strigosus*
cypress
  Arizona cypress - *Cupressus arizonica*
  Baker's cypress - *Cupressus bakeri*
  bald cypress - *Taxodium distichum*
  cypress - *Cupressus, Taxodium*
  cypress spurge - *Euphorbia cyparissias*
  cypress-vine - *Ipomoea quamoclit*
  cypress-vine morning-glory - *Ipomoea quamoclit*
  false cypress - *Chamaecyparis*
  Guadalupe cypress - *Cupressus guadalupensis*
  Hinoki cypress - *Chamaecyparis obtusa*
  Hinoki false cypress - *Chamaecyparis obtusa*
  Italian cypress - *Cupressus sempervirens*
  Japanese false cypress - *Chamaecyparis obtusa*
  Lawson's cypress - *Chamaecyparis lawsoniana*
  McNab's cypress - *Cupressus macnabiana*
  Mendocino cypress - *Cupressus pygmaea*
  Mexican cypress - *Cupressus lusitanica*
  modoc cypress - *Cupressus bakeri*
  Monterey cypress - *Cupressus macrocarpa*
  pond cypress - *Taxodium distichum* var. *imbricarium*

Portuguese cypress - *Cupressus lusitanica*
  rough-bark Arizona cypress - *Cupressus arizonica*
  Sargent's cypress - *Cupressus sargentii*
  Sawara cypress - *Chamaecyparis pisifera*
  summer-cypress - *Bassia scoparia*
  tecate cypress - *Cupressus guadalupensis*
daffodil
  daffodil - *Narcissus pseudonarcissus, Narcissus*
  fall daffodil - *Sternbergia lutea*
  Peruvian daffodil - *Hymenocallis narcissiflora*
  sea-daffodil - *Hymenocallis*
dagger
  dagger fern - *Polystichum acrostichoides*
  dagger-plant - *Yucca aloifolia*
  Spanish-dagger - *Yucca gloriosa*
dahl - *Cajanus cajan*
Dahlberg's daisy - *Dyssodia tenuiloba*
dahoon
  dahoon - *Ilex cassine*
  dahoon holly - *Ilex cassine*
  myrtle dahoon - *Ilex myrtifolia*
daikon - *Raphanus sativus* cv. 'longipinnatu'
daisy
  African daisy - *Arctotis, Gerbera jamesonii, Osteospermum ecklonis*
  Barberton's daisy - *Gerbera jamesonii*
  blue daisy - *Cichorium intybus*
  blue-eyed African daisy - *Arctotis stoechadifolia*
  blue spring daisy - *Erigeron pulchellus*
  butter-daisy - *Ranunculus repens, Verbesina encelioides*
  Clanwilliam daisy - *Euryops speciosissimus*
  crown daisy - *Chrysanthemum coronarium*
  Dahlberg's daisy - *Dyssodia tenuiloba*
  daisy - *Bellis*
  daisy chrysanthcmum - *Leucanthemum maximum*
  daisy fleabane - *Erigeron annuus, E. philadelphicus, E. strigosus*
  Engelmann's daisy - *Engelmannia pinnatifida*
  English daisy - *Bellis perennis*
  European daisy - *Bellis perennis*
  field daisy - *Leucanthemum vulgare*
  golden-crown daisy - *Verbesina encelioides*
  kingfisher daisy - *Felicia bergeriana*
  mountain daisy - *Arenaria groenlandica*
  ox-eye daisy - *Leucanthemum vulgare*
  painted daisy - *Chrysanthemum coccineum*
  Paris daisy - *Chrysanthemum frutescens*
  rough daisy fleabane - *Erigeron strigosus*
  Shasta daisy - *Leucanthemum maximum*
  stinking-daisy - *Anthemis cotula*
  Tahoka daisy - *Machaeranthera tanacetifolia*
  Transvaal daisy - *Gerbera jamesonii*
  Vanstadens River daisy - *Osteospermum ecklonis*
  veldt daisy - *Gerbera jamesonii*
  western daisy - *Astranthium integrifolium*
  white daisy - *Chrysanthemum frutescens, Leucanthemum vulgare*
Dakota vervain - *Verbena bipinnatifida*
dalea
  black dalea - *Dalea frutescens*
  foxtail dalea - *Dalea leporina*
  hare-foot dalea - *Dalea leporina*

Dallis grass - *Paspalum dilatatum*
Dalmatian
    Dalmatian insect-flower - *Chrysanthemum cinerariifolium*
    Dalmatian pyrethrum - *Chrysanthemum cinerariifolium*
    Dalmatian toadflax - *Linaria genistifolia* subsp. *dalmatica*
Daly's, Judge Daly's sunflower - *Helianthus maximilianii*
dame's
    dame's-rocket - *Hesperis matronalis*
    dame's-violet - *Hesperis matronalis*
damson, bitter damson - *Simarouba glauca*
Damson's plum - *Prunus domestica* subsp. *insititia*
dandelion
    Carolina false dandelion - *Pyrrhopappus carolinianus*
    coast-dandelion - *Hypochaeris radicata*
    dandelion - *Taraxacum officinale, Taraxacum*
    dwarf-dandelion - *Krigia virginica, Krigia*
    fall-dandelion - *Leontodon autumnalis*
    false dandelion - *Hypochaeris radicata*
    orange dwarf-dandelion - *Krigia biflora*
    red-seed dandelion - *Taraxacum erythrospermum*
    Russian dandelion - *Taraxacum kok-saghyz*
    smooth false dandelion - *Hypochaeris glabra*
    Virginia dwarf-dandelion - *Krigia virginica*
dane-wort - *Sambucus ebulus*
dangleberry - *Gaylussacia frondosa*
daphne
    February daphne - *Daphne mezereum*
    rose daphne - *Daphne cneorum*
    winter daphne - *Daphne odora*
dark
    dark-eye sunflower - *Helianthus atrorubens*
    dark-green bulrush - *Scirpus atrovirens*
    dark-red sunflower - *Helianthus atrorubens*
Darlington's oak - *Quercus hemisphaerica*
darnel
    bearded darnel - *Lolium temulentum*
    darnel - *Lolium perenne, L. temulentum, Lolium*
    Persian darnel - *Lolium persicum*
    poison darnel - *Lolium temulentum*
darning, devil's-darning-needle - *Clematis virginiana*
dasheen - *Colocasia esculenta*
date
    Canary Island date palm - *Phoenix canariensis*
    Chinese date - *Ziziphus jujuba*
    date palm - *Phoenix dactylifera, Phoenix*
    date-plum - *Diospyros kaki, D. virginiana*
    dwarf date palm - *Phoenix roebelenii*
    miniature date palm - *Phoenix roebelenii*
    pygmy date palm - *Phoenix roebelenii*
    Senegal date palm - *Phoenix reclinata*
    wild date palm - *Phoenix reclinata*
datil - *Yucca baccata*
datura
    Hindu datura - *Datura metel*
    sacred datura - *Datura innoxia*
    small datura - *Datura discolor*
David's peach - *Prunus davidiana*
dawn
    blue dawn-flower - *Ipomoea indica*
    dawn redwood - *Metasequoia glyptostroboides*

day
    day-blooming cestrum - *Cestrum diurnum*
    day jessamine - *Cestrum diurnum*
dayflower
    Asiatic dayflower - *Commelina communis*
    creeping dayflower - *Commelina diffusa*
    dayflower - *Commelina communis*
    marsh-dayflower - *Murdannia keisak*
    spreading dayflower - *Commelina diffusa*
    Virginia dayflower - *Commelina virginica*
daylily
    daylily - *Hemerocallis*
    fulvous daylily - *Hemerocallis fulva*
    orange daylily - *Hemerocallis fulva*
    tawny daylily - *Hemerocallis fulva*
    yellow daylily - *Hemerocallis lilioasphodelus*
dead
    dead-nettle - *Lamium*
    purple dead-nettle - *Lamium purpureum*
    red dead-nettle - *Lamium purpureum*
    spotted dead-nettle - *Lamium maculatum*
    white dead-nettle - *Lamium album*
deadly
    deadly-hemlock - *Conium maculatum*
    deadly nightshade - *Atropa belladonna, Solanum dulcamara*
death
    death camass - *Zigadenus nuttallii*
    death-weed - *Iva axillaris*
    foothill death camass - *Zigadenus paniculatus*
    meadow death camass - *Zigadenus venenosus*
    mountain death camass - *Zigadenus elegans*
    Nuttall's death camass - *Zigadenus nuttallii*
Deccan's-hemp - *Hibiscus cannabinus*
deer
    deer-berry - *Vaccinium caesium, V. stamineum*
    deer-brush - *Ceanothus integerrimus*
    deer-bush - *Ceanothus integerrimus*
    deer fern - *Blechnum spicant*
    deer grass - *Muhlenbergia rigens, Rhexia*
    deer-nut - *Simmondsia chinensis*
    deer oak - *Quercus sadleriana*
    deer-vetch - *Lotus unifoliatus*
    deer-wort - *Ageratina altissima*
deer's
    deer's-foot - *Achlys triphylla*
    deer's-tongue - *Panicum clandestinum*
del, cardo-del-valle - *Centaurea americana*
delight, queen's-delight - *Stillingia sylvatica*
delta
    delta arrowhead - *Sagittaria graminea* var. *platyphylla*
    delta maidenhair fern - *Adiantum raddianum* cv. 'decorum'
dense blazing-star - *Liatris spicata*
Dent's corn - *Zea mays*
deodar - *Cedrus deodara*
deodara cedar - *Cedrus deodara*
Deptford pink - *Dianthus armeria*
desert
    desert apricot - *Prunus fremontii*
    desert-broom - *Baccharis sarothroides*
    desert-candle - *Eremurus robustus*
    desert crested wheat grass - *Agropyron desertorum*

desert evening-primrose - *Oenothera deltoides*
desert fan palm - *Washingtonia filifera*
desert hackberry - *Celtis pallida*
desert hollyhock - *Sphaeralcea ambigua*
desert-honeysuckle - *Anisacanthus thurberi*
desert ironweed - *Olneya tesota*
desert larkspur - *Delphinium andersonii*
desert lavender - *Hyptis emoryi*
desert locust - *Robinia neomexicana*
desert mallow - *Sphaeralcea ambigua*
desert-marigold - *Baileya multiradiacata*
desert mariposa - *Calochortus kennedyi*
desert needle grass - *Stipa speciosa*
desert-olive - *Forestiera pubescens*
desert-olive forestiera - *Forestiera phillyreoides*
desert palm - *Washingtonia robusta*
desert peach - *Prunus andersonii*
desert-plume - *Stanleya pinnata*
desert rabbit-brush - *Chrysothamnus paniculatus*
desert-rose - *Adenium obesum*
desert salt grass - *Distichlis stricta*
desert saltbush - *Atriplex polycarpa*
desert senna - *Senna covesii*
desert smoke-tree - *Psorothamnus spinosa*
desert stick-leaf - *Mentzelia multiflora*
desert-thorn - *Lycium pallidum*
desert tobacco - *Nicotiana trigonophylla*
desert-willow - *Chilopsis linearis*
desert wire-lettuce - *Stephanomeria pauciflora*
Sturt's desert-rose - *Gossypium sturtianum*
desmodium, silver-leaf desmodium - *Desmodium uncinatum*
devil
  blue-devil - *Echium vulgare*
  devil grass - *Cynodon dactylon*
  devil-weed - *Osmanthus*
  devil-wood - *Osmanthus americanus*
  glaucous king-devil - *Hieracium piloselloides*
  king-devil - *Hieracium aurantiacum, H. floribundum*
  king-devil hawkweed - *Hieracium piloselloides*
  yellow-devil - *Hieracium caespitosum*
  yellow-devil hawkweed - *Hieracium floribundum*
  yellow king-devil - *Hieracium caespitosum*
devil's
  devil's-backbone - *Pedilanthus tithymaloides*
  devil's beggar-ticks - *Bidens frondosa*
  devil's-bit - *Succisa australis*
  devil's-claw - *Proboscidea louisianica*
  devil's-club - *Oplopanax horridum*
  devil's-darning-needle - *Clematis virginiana*
  devil's-grandmother - *Elephantopus tomentosus*
  devil's grass - *Elytrigia repens*
  devil's-grip - *Mollugo verticillata*
  devil's-hair - *Cuscuta epilinum*
  devil's-ivy - *Epipremnum aureum*
  devil's-lettuce - *Amsinckia tessellata*
  devil's-paintbrush - *Hieracium aurantiacum*
  devil's-plague - *Daucus carota*
  devil's-shoestring - *Lygodesmia juncea, Viburnum alnifolium*
  devil's-tongue - *Amorphophallus rivieri, Amorphophallus*
  devil's-vine - *Calystegia sepium*
  devil's-walking-stick - *Aralia spinosa*

dew-thread - *Drosera filiformis*
dewberry
  American dewberry - *Rubus flagellaris*
  bristly dewberry - *Rubus hispidus*
  coastal plain dewberry - *Rubus trivialis*
  northern dewberry - *Rubus flagellaris*
  Pacific dewberry - *Rubus ursinus*
  southern dewberry - *Rubus trivialis*
  swamp dewberry - *Rubus hispidus*
dewdrop
  dewdrop - *Dalibarda repens*
  garden dewdrop - *Duranta erecta*
  golden dewdrop - *Duranta erecta*
diamond
  diamond-leaf willow - *Salix planifolia*
  diamond willow - *Salix eriocephala*
  Old World diamond-flower - *Hedyotis corymbosa*
Diaz's blue-stem - *Dichanthium annulatum*
dichondra, Carolina dichondra - *Dichondra carolinensis*
dicks, blue-dicks - *Dichelostemma pulchellum*
dicksonia, hairy dicksonia - *Dennstaedtia punctilobula*
different-leaf sunflower - *Helianthus heterophyllus*
diffuse
  diffuse knapweed - *Centaurea diffusa*
  diffuse potentilla - *Potentilla paradoxa*
digger pine - *Pinus sabiniana*
digitalis - *Digitalis lanata*
dill
  dill - *Anethum graveolens*
  dill-weed - *Anthemis cotula*
  dilly, wild dilly - *Manilkara jaimiqui* subsp. *emarginata*
dingle-berry - *Vaccinium erythrocarpum*
dinner, monkey's-dinner-bell - *Hura crepitans*
disc water-hyssop - *Bacopa rotundifolia*
dishcloth gourd - *Luffa aegyptiaca*
distaff
  distaff thistle - *Carthamus lanatus*
  woolly distaff thistle - *Carthamus lanatus*
ditch
  ditch bank willow - *Salix exigua*
  ditch beard grass - *Polypogon interruptus*
  ditch polypogon - *Polypogon interruptus*
  ditch-stonecrop - *Penthorum sedoides*
dittany
  American dittany - *Cunila origanoides*
  dittany - *Cunila origanoides, Dictamnus albus*
divaricate sunflower - *Helianthus divaricatus*
divergent sunflower - *Helianthus divaricatus*
diverse-leaf pondweed - *Potamogeton diversifolius*
divine-flower - *Dianthus caryophyllus*
dock
  bitter dock - *Rumex obtusifolius*
  blunt-leaf dock - *Rumex obtusifolius*
  broad-leaf dock - *Rumex obtusifolius*
  butterfly-dock - *Petasites hybridus*
  cluster dock - *Rumex conglomeratus*
  curly dock - *Rumex crispus*
  dock sorrel - *Rumex*
  dove-dock - *Tussilago farfara*
  elf-dock - *Inula helenium*

fiddle dock - *Rumex pulcher*
fiddle-leaf dock - *Rumex pulcher*
long-leaf dock - *Rumex longifolius*
narrow-leaf dock - *Rumex stenophyllus*
pale dock - *Rumex altissimus*
patience dock - *Rumex patientia*
prairie-dock - *Silphium terebinthinaceum*
purple wan-dock - *Brasenia schreberi*
red-veined dock - *Rumex obtusifolius*
Rocky Mountain spatter-dock - *Nuphar lutea* subsp.
  polysepala
sea beach dock - *Rumex pallidus*
seaside dock - *Rumex pallidus*
smooth dock - *Rumex altissimus*
sour dock - *Rumex acetosa, R. acetosella*
spatter-dock - *Nuphar lutea* subsp. *advena, Nuphar*
spinach dock - *Rumex patientia*
succory-dock - *Lapsana communis*
swamp dock - *Rumex verticillatus*
tanner's dock - *Rumex hymenosepalus*
veiny dock - *Rumex venosus*
water dock - *Rumex altissimus, R. orbiculatus, R. verticillatus*
western dock - *Rumex occidentalis*
white dock - *Rumex pallidus*
winged dock - *Rumex venosus*
yard dock - *Rumex longifolius*
yellow dock - *Rumex crispus*
dockmackie - *Viburnum acerifolium*
dodder
alfalfa dodder - *Cuscuta suaveolens*
Australian dodder - *Cuscuta obtusiflora*
clover dodder - *Cuscuta epithymum*
collared dodder - *Cuscuta indecora*
dodder - *Cuscuta gronovii*
field dodder - *Cuscuta campestris, C. pentagona*
flax dodder - *Cuscuta epilinum*
hazel dodder - *Cuscuta coryli*
large-fruit dodder - *Cuscuta umbrosa*
large-seeded dodder - *Cuscuta indecora, C. pentagona*
lespedeza dodder - *Cuscuta pentagona*
onion dodder - *Cuscuta gronovii*
polygonum dodder - *Cuscuta polygonorum*
Sandwich's dodder - *Cuscuta sandwichiana*
small-seeded alfalfa dodder - *Cuscuta approximata*
small-seeded dodder - *Cuscuta planiflora*
smartweed dodder - *Cuscuta polygonorum*
southern dodder - *Cuscuta obtusiflora*
swamp dodder - *Cuscuta gronovii*
thyme dodder - *Cuscuta epithymum*
umbrella dodder - *Cuscuta umbellata*
western flax dodder - *Cuscuta campestris*
dog
American dog violet - *Viola conspersa*
crested dog-tail grass - *Cynosurus cristatus*
dog-bramble - *Ribes cynosbati*
dog-brier - *Rosa canina*
dog-chamomile - *Anthemis cotula*
dog-fennel - *Anthemis cotula, Anthemis, Eupatorium capillifolium*
dog-mustard - *Erucastrum gallicum*
dog nettle - *Urtica urens*
dog rose - *Rosa canina*
dog-tooth grass - *Cynodon dactylon*

dog-tooth pea - *Lathyrus sativus*
dog-tooth-violet - *Erythronium*
dog-weed - *Dyssodia*
hedgehog dog-tail grass - *Cynosurus echinatus*
mad-dog skullcap - *Scutellaria lateriflora*
mountain dog-laurel - *Leucothoe fontanesiana*
rayless dog-fennel - *Matricaria matricarioides*
swamp dog-laurel - *Leucothoe axillaris*
western dog violet - *Viola adunca*
white dog-tooth-violet - *Erythronium albidum*
yellow dog-fennel - *Helenium amarum*
dog's
crested dog's-tail - *Cynosurus cristatus*
dog's-hobble - *Leucothoe fontanesiana, Viburnum alnifolium*
rough dog's-tail - *Cynosurus echinatus*
dogbane
bitter dogbane - *Apocynum androsaemifolium*
blue dogbane - *Amsonia tabernaemontana*
clasping dogbane - *Apocynum sibiricum*
climbing dogbane - *Trachelospermum difforme*
dogbane - *Apocynum androsaemifolium, Apocynum*
hemp dogbane - *Apocynum cannabinum*
prairie dogbane - *Apocynum sibiricum*
spreading dogbane - *Apocynum androsaemifolium*
dogberry - *Cornus sanguinea, Ribes cynosbati, Sorbus americana, Viburnum alnifolium*
dogwood
alternate-leaf dogwood - *Cornus alternifolia*
American dogwood - *Cornus sericea*
blood-twig dogwood - *Cornus sanguinea*
brown dogwood - *Cornus glabrata*
dogwood - *Cornus*
eastern flowering dogwood - *Cornus florida*
false dogwood - *Sapindus saponaria*
flowering dogwood - *Cornus florida*
gray dogwood - *Cornus racemosa*
Jamaican dogwood - *Piscidia piscipula*
knob-styled dogwood - *Cornus amomum*
mountain dogwood - *Cornus nuttallii*
northern swamp dogwood - *Cornus racemosa*
Pacific dogwood - *Cornus nuttallii*
pagoda dogwood - *Cornus alternifolia*
panicled dogwood - *Cornus racemosa*
poison-dogwood - *Toxicodendron vernix*
red osier dogwood - *Cornus sericea*
rough-leaf dogwood - *Cornus drummondii*
round-leaf dogwood - *Cornus rugosa*
silky dogwood - *Cornus amomum, C. obliqua*
southern swamp dogwood - *Cornus stricta*
stiff-cornel dogwood - *Cornus stricta*
stiff dogwood - *Cornus stricta*
West Indian dogwood - *Piscidia piscipula*
white dogwood - *Cornus florida*
White Mountains dogwood - *Viburnum alnifolium*
doll's-eyes - *Actaea alba*
dollar
billion-dollar grass - *Echinochloa frumentacea*
dollar-plant - *Crassula ovata*
silver-dollar - *Crassula arborescens, Lunaria annua*
silver-dollar gum - *Eucalyptus polyanthemos*
silver-dollar-tree - *Eucalyptus cinerea*
dopatrium - *Dopatrium junceum*

dotted
  dotted blazing-star - *Liatris punctata*
  dotted gay-feather - *Liatris punctata*
  dotted hawthorn - *Crataegus punctata*
  dotted loosestrife - *Lysimachia punctata*
  dotted mint - *Monarda punctata*
  dotted smartweed - *Polygonum punctatum*
double
  double-file viburnum - *Viburnum plicatum* var. *tomentosum*
  double spruce - *Picea mariana*
Douglas
  big-cone Douglas fir - *Pseudotsuga macrocarpa*
  Douglas fir - *Pseudotsuga menziesii*
  Rocky Mountain Douglas fir - *Pseudotsuga menziesii* var. *glauca*
Douglas's
  Douglas's fiddlehead - *Amsinckia douglasiana*
  Douglas's knotweed - *Polygonum douglasii*
  Douglas's maple - *Acer glabrum* subsp. *douglasii*
  Douglas's rabbit-brush - *Chrysothamnus viscidiflorus*
doum
  doum palm - *Hyphaene thebaica*
  Egyptian doum - *Hyphaene thebaica*
dove
  dove-dock - *Tussilago farfara*
  dove-plum - *Coccoloba diversifolia*
  dove-weed - *Eremocarpus setigerus, Murdannia nudiflora*
dove's
  dove's-dung - *Ornithogalum umbellatum*
  dove's-foot geranium - *Geranium molle*
downy
  downy arrow-wood - *Viburnum rafinesquianum*
  downy brome - *Bromus tectorum*
  downy brome grass - *Bromus tectorum*
  downy chess - *Bromus tectorum*
  downy cinquefoil - *Potentilla intermedia*
  downy false foxglove - *Aureolaria virginica*
  downy gentian - *Gentiana saponaria*
  downy ground-cherry - *Physalis pubescens*
  downy hawthorn - *Crataegus mollis*
  downy-leaf arrow-wood - *Viburnum rafinesquianum*
  downy lobelia - *Lobelia puberula*
  downy oat grass - *Danthonia sericea*
  downy paintbrush - *Castilleja sessiliflora*
  downy peppergrass - *Lepidium campestre*
  downy phlox - *Phlox pilosa*
  downy poplar - *Populus heterophylla*
  downy prairie-clover - *Dalea villosa*
  downy rose-myrtle - *Rhodomyrtus tomentosus*
  downy serviceberry - *Amelanchier arborea, A. canadensis*
  downy skullcap - *Scutellaria incana*
  downy swamp huckleberry - *Vaccinium atrococcum*
  downy thorn-apple - *Datura innoxia, D. metel*
  downy trailing lespedeza - *Lespedeza procumbens*
  downy wild rye - *Elymus villosus*
  downy woodmint - *Blephilia ciliata*
  downy yellow violet - *Viola pubescens*
dracaena
  gold-dust dracaena - *Dracaena surculosa*
  golden dracaena - *Pleomele aurea*
  spotted dracaena - *Dracaena surculosa*

dragon
  dragon-arum - *Arisaema dracontium*
  dragon-root - *Arisaema triphyllum*
  dragon-tree - *Dracaena draco*
  fire-dragon - *Acalypha wilkesiana*
  green-dragon - *Arisaema dracontium*
  water-dragon - *Calla palustris, Saururus cernuus*
dragonhead
  American dragonhead - *Dracocephalum parviflorum*
  dragonhead - *Dracocephalum parviflorum*
dragonroot - *Arisaema dracontium*
drake - *Avena fatua*
drooping
  drooping brome - *Bromus tectorum*
  drooping-laurel - *Leucothoe fontanesiana*
  drooping she-oak - *Casuarina stricta*
  drooping trillium - *Trillium flexipes*
  drooping wood-reed - *Cinna latifolia*
  drooping wood sedge - *Carex arctata*
  Mexican drooping juniper - *Juniperus flaccida*
drop
  African drop-seed - *Sporobolus africanus*
  annual drop-seed - *Sporobolus neglectus*
  coral drop-seed - *Sporobolus domingensis*
  drop-seed - *Sporobolus*
  drop-seed wiregrass - *Muhlenbergia schreberi*
  giant drop-seed - *Sporobolus giganteus*
  hairy-drop-seed - *Blepharoneuron tricholepis*
  meadow drop-seed - *Sporobolus asper* var. *hookeri*
  northern drop-seed - *Sporobolus heterolepis*
  piney-woods drop-seed - *Sporobolus junceus*
  poverty drop-seed - *Sporobolus vaginiflorus*
  prairie drop-seed - *Sporobolus heterolepis*
  sand drop-seed - *Sporobolus cryptandrus*
  seashore drop-seed - *Sporobolus virginicus*
  spike drop-seed - *Sporobolus contractus*
drops
  beech-drops - *Epifagus virginiana*
  false beech-drops - *Monotropa hypopithys*
dropwort
  dropwort - *Filipendula vulgaris*
  water-dropwort - *Oenanthe, Oxypolis rigidior*
drought, woolly white drought-weed - *Eremocarpus setigerus*
drum-heads - *Polygala cruciata*
Drummond's
  Drummond's golden-weed - *Haplopappus drummondii*
  Drummond's phlox - *Phlox drummondii*
  Drummond's rattle-bush - *Sesbania drummondii*
  Drummond's wax mallow - *Malvaviscus arboreus* var. *drummondii*
dry
  dry-land bitter-cress - *Cardamine parviflora*
  dry-spiked sedge - *Carex siccata*
drymary, heart-leaf drymary - *Drymaria cordata*
duck
  duck-meal - *Lemna perpusilla*
  duck-potato - *Sagittaria latifolia*
  duck-salad - *Heteranthera limosa*
duckweed
  duckweed - *Lemna minor*
  giant duckweed - *Spirodela polyrhiza*

great duckweed - *Spirodela polyrhiza*
greater duckweed - *Spirodela polyrhiza*
inflated duckweed - *Lemna gibba*
lesser duckweed - *Lemna minor*
minute duckweed - *Lemna minuta*
obscure duckweed - *Lemna obscura*
small duckweed - *Lemna valdiviana*
star duckweed - *Lemna trisulca*
three-nerved duckweed - *Lemna trinervis*

dull meadow-pitchers - *Rhexia mariana*

dumb
dumb-cane - *Dieffenbachia seguine, Dieffenbachia*
dumb-cotton - *Calotropis procera*
spotted dumb-cane - *Dieffenbachia maculata*

dunce-cap larkspur - *Delphinium occidentale*

dune
American dune grass - *Leymus mollis*
dune bent grass - *Agrostis pallens*
dune manzanita - *Arctostaphylos pumila*
dune sandbur - *Cenchrus tribuloides*
dune willow - *Salix cordata*
European dune grass - *Leymus arenarius*
sand-dune willow - *Salix syrticola*

dung, dove's-dung - *Ornithogalum umbellatum*

Durand's oak - *Quercus durandii*

durra - *Sorghum bicolor*

durum wheat - *Triticum durum*

dust
gold-dust - *Alyssum saxatile*
gold-dust dracaena - *Dracaena surculosa*

duster, fairy-duster - *Calliandra eriophylla*

dusty-miller - *Centaurea cineraria, Lychnis coronaria*

Dutch
Dutch-case-knife bean - *Phaseolus coccineus*
Dutch flax - *Camelina microcarpa*
Dutch hyacinth - *Hyacinthus orientalis*

Dutchman's
Dutchman's-breeches - *Dicentra cucullaria*
Dutchman's-pipe - *Aristolochia macrophyllum*

dwarf
dwarf alyssum - *Alyssum desertorum*
dwarf azalea - *Rhododendron atlanticum*
dwarf baccharis - *Baccharis pilularis*
dwarf banana - *Musa acuminata*
dwarf bilberry - *Vaccinium caespitosum*
dwarf birch - *Betula glandulosa*
dwarf buckeye - *Aesculus sylvatica*
dwarf buttercup - *Ranunculus pygmaeus*
dwarf Cape gooseberry - *Physalis pubescens* var. *grisea*
dwarf cherry - *Prunus pumila*
dwarf chestnut oak - *Quercus prinoides*
dwarf chinquapin oak - *Quercus prinoides*
dwarf cinquefoil - *Potentilla canadensis*
dwarf cornel - *Cornus canadensis*
dwarf-crested iris - *Iris cristata*
dwarf-dandelion - *Krigia virginica, Krigia*
dwarf date palm - *Phoenix roebelenii*
dwarf elder - *Sambucus ebulus*
dwarf elm - *Ulmus pumila*
dwarf fleabane - *Conyza ramosissima*
dwarf ginseng - *Panax trifolius*

dwarf glasswort - *Salicornia bigelovii*
dwarf ground-chervil - *Chamaesaracha coronopus*
dwarf ground rattan - *Rhapis excelsa*
dwarf-holly - *Malpighia coccigera*
dwarf horse-chestnut - *Aesculus parviflora*
dwarf huckleberry - *Gaylussacia dumosa, G. frondosa*
dwarf iris - *Iris verna*
dwarf larkspur - *Delphinium nuttallianum, D. tricorne*
dwarf-laurel - *Kalmia angustifolia*
dwarf live oak - *Quercus minima*
dwarf lycopod - *Selaginella rupestris*
dwarf maple - *Acer glabrum*
dwarf meadow grass - *Poa annua*
dwarf mistletoe - *Arceuthobium pusillum*
dwarf nettle - *Urtica urens*
dwarf nipplewort - *Arnoseris minima*
dwarf palmetto - *Sabal minor*
dwarf pawpaw - *Asimina parviflora*
dwarf poinciana - *Caesalpinia pulcherrima*
dwarf prairie rose - *Rosa arkansana*
dwarf raspberry - *Rubus pubescens*
dwarf red raspberry - *Rubus pubescens*
dwarf rubber-plant - *Crassula ovata*
dwarf Russian almond - *Prunus tenella*
dwarf saltwort - *Salicornia bigelovii*
dwarf snapdragon - *Chaenorhinum minus, Chaenorhinum*
dwarf spike rush - *Eleocharis parvula*
dwarf St. John's-wort - *Hypericum mutilum*
dwarf sumac - *Rhus copallina*
dwarf umbrella-grass - *Fuirena pumila*
dwarf wax myrtle - *Myrica pusilla*
dwarf white birch - *Betula minor*
dwarf white trillium - *Trillium nivale*
dwarf wild indigo - *Amorpha nana*
eastern dwarf mistletoe - *Arceuthobium pusillum*
European dwarf cherry - *Prunus fruticosa*
orange dwarf-dandelion - *Krigia biflora*
Virginia dwarf-dandelion - *Krigia virginica*

dye-root - *Lachnanthes caroliniana*

dyer's
dyer's broom - *Genista tinctoria*
dyer's greenweed - *Genista tinctoria*
dyer's greenwood - *Genista tinctoria*
dyer's-rocket - *Reseda lutea*
dyer's woad - *Isatis tinctoria*

dyeweed - *Genista tinctoria*

eardrops, lady's-eardrops - *Brunnichia ovata, Fuchsia triphylla*

early
early azalea - *Rhododendron austrinum*
early blue violet - *Viola palmata*
early buttercup - *Ranunculus fascicularis*
early chess - *Bromus tectorum*
early coralroot - *Corallorrhiza trifida*
early everlasting - *Antennaria plantaginifolia*
early goldenrod - *Solidago juncea*
early meadow-rue - *Thalictrum dioicum*
early saxifrage - *Saxifraga virginiensis*
early sweet bilberry - *Vaccinium vacillans*
early water grass - *Echinochloa oryzoides*
early winter-cress - *Barbarea verna*
early yellow violet - *Viola rotundifolia*

earrings, lady's-earrings - *Impatiens capensis*
ears
    cauliflower-ears - *Crassula ovata*
    mouse's-ears - *Antennaria plantaginifolia*
    pussy's-ears - *Calochortus tolmiei, Kalanchoe tomentosa*
earth
    earth-almond - *Cyperus esculentus*
    earth-apple - *Helianthus tuberosus*
    earth-chestnut - *Lathyrus tuberosus*
    earth-nut - *Arachis hypogaea*
    earth-smoke - *Fumaria officinalis*
    earth-star - *Cryptanthus*
    gall-of-the-earth - *Prenanthes serpentaria, P. trifoliolata*
    man-of-the-earth - *Ipomoea leptophylla, I. pandurata*
earthnut pea - *Lathyrus tuberosus*
ease
    heart's-ease - *Polygonum persicaria, Viola tricolor*
    western heart's-ease - *Viola ocellata*
East
    East Indian fig-tree - *Ficus benghalensis*
    East Indian Jew's mallow - *Corchorus aestuans*
    East Indian lemongrass - *Cymbopogon flexuosus*
    East Indian lotus - *Nelumbo nucifera*
    East Indian rose-bay - *Tabernaemontana divaricata*
Easter
    Bermuda Easter lily - *Lilium longiflorum* var. *eximium*
    Easter-bell - *Stellaria holostea*
    Easter-bell starwort - *Stellaria holostea*
    Easter cactus - *Hatiora gaertneri*
    Easter lily - *Lilium longiflorum* var. *eximium*
eastern
    eastern baccharis - *Baccharis halimifolia*
    eastern black currant - *Ribes americanum*
    eastern blue-eyed-Mary - *Collinsia verna*
    eastern bluebells - *Mertensia virginica*
    eastern camass - *Camassia scilloides*
    eastern coral bean - *Erythrina herbacea*
    eastern cottonwood - *Populus deltoides*
    eastern dwarf mistletoe - *Arceuthobium pusillum*
    eastern flowering dogwood - *Cornus florida*
    eastern gama grass - *Tripsacum dactyloides*
    eastern hemlock - *Tsuga canadensis*
    eastern hop-hornbeam - *Ostrya virginiana*
    eastern manna grass - *Glyceria septentrionalis*
    eastern parthenium - *Parthenium integrifolium*
    eastern pasqueflower - *Anemone patens*
    eastern poison-oak - *Toxicodendron pubescens*
    eastern prickly-pear - *Opuntia austrina, O. compressa*
    eastern red-cedar - *Juniperus virginiana*
    eastern redbud - *Cercis canadensis*
    eastern serviceberry - *Amelanchier canadensis*
    eastern shooting-star - *Dodecatheon meadia*
    eastern sycamore - *Platanus occidentalis*
    eastern wahoo - *Euonymus atropurpureus*
    eastern water-leaf - *Hydrophyllum virginianum*
    eastern water-milfoil - *Myriophyllum pinnatum*
    eastern white pine - *Pinus strobus*
    eastern whorled milkweed - *Asclepias verticillata*
    eastern wild rice - *Zizania aquatica*
    eastern willow-herb - *Epilobium coloratum*
Eastwood manzanita - *Arctostaphylos glandulosa*

Eaton's
    Eaton's firecracker - *Penstemon eatonii*
    Eaton's penstemon - *Penstemon eatonii*
ebony
    ebony black-bead - *Pithecellobium flexicaule*
    mountain-ebony - *Bauhinia hookeri, B. variegata*
    Texas ebony - *Pithecellobium flexicaule*
echinacea, pale echinacea - *Echinacea pallida*
eddo - *Colocasia esculenta*
edge, white-edge morning-glory - *Ipomoea nil*
edged, black-edged sedge - *Carex nigromarginata* var.
    elliptica
edging lobelia - *Lobelia erinus*
edible
    edible banana - *Musa acuminata*
    edible burdock - *Arctium lappa*
    edible canna - *Canna indica*
    edible fig - *Ficus carica*
    edible galingale - *Cyperus esculentus*
    edible-podded pea - *Pisum sativum* var. *macrocarpon*
    edible valerian - *Valeriana edulis*
eel-grass pondweed - *Potamogeton zosteriformis*
eelgrass
    American eelgrass - *Vallisneria americana*
    eelgrass - *Zostera marina*
egeria - *Egeria densa*
egg-fruit-tree - *Pouteria campechiana*
eggplant
    Chinese scarlet eggplant - *Solanum integrifolium*
    eggplant - *Solanum melongena*
    fruited eggplant - *Solanum integrifolium*
    scarlet eggplant - *Solanum integrifolium*
    tomato-fruited eggplant - *Solanum integrifolium*
eggs
    butter-and-eggs - *Linaria vulgaris*
    eggs-and-bacon - *Linaria vulgaris*
eglantine - *Rosa eglanteria*
Egyptian
    American-Egyptian cotton - *Gossypium barbadense*
    Egyptian clover - *Trifolium alexandrinum*
    Egyptian doum - *Hyphaene thebaica*
    Egyptian grass - *Dactyloctenium aegyptium*
    Egyptian lupine - *Lupinus albus*
    Egyptian millet - *Sorghum halepense*
    Egyptian onion - *Allium cepa*
    Egyptian pea - *Cicer arietinum*
    Egyptian star-cluster - *Pentas lanceolata*
einkorn - *Triticum monococcum*
Eisen's water-hyssop - *Bacopa eisenii*
elaeagnus
    autumn elaeagnus - *Elaeagnus umbellata*
    thorny elaeagnus - *Elaeagnus pungens*
elder
    American elder - *Sambucus canadensis*
    annual marsh-elder - *Iva annua*
    big marsh-elder - *Iva xanthifolia*
    black-bead elder - *Sambucus melanocarpa*
    blue elder - *Sambucus cerulea*
    blueberry elder - *Sambucus cerulea*
    box elder - *Acer negundo*
    California box elder - *Acer negundo* subsp. *californicum*

dwarf elder - *Sambucus ebulus*
elder - *Sambucus canadensis, Sambucus*
European elder - *Sambucus nigra*
horse-elder - *Inula helenium*
marsh-elder - *Iva xanthifolia, Iva*
Mexican elder - *Sambucus mexicana*
Pacific Coast red elder - *Sambucus callicarpa*
Pacific red elder - *Sambucus callicarpa*
poison-elder - *Toxicodendron vernix*
red-berried elder - *Sambucus racemosa, S. racemosa*
  var. *microbotrys*
rough marsh-elder - *Iva annua*
sweet elder - *Sambucus canadensis*
yellow elder - *Tecoma stans*
elderberry
  American elderberry - *Sambucus canadensis*
  American red elderberry - *Sambucus racemosa* var.
   microbotrys
  blue elderberry - *Sambucus cerulea, S. mexicana*
  elderberry - *Sambucus*
  European red elderberry - *Sambucus racemosa*
  red elderberry - *Sambucus callicarpa*
  western elderberry - *Sambucus cerulea*
elecampane - *Inula helenium*
election-pink - *Rhododendron austrinum*
elemi, gum-elemi - *Bursera simaruba*
elephant
  elephant-apple - *Limonia acidissima*
  elephant-bush - *Portulacaria afra*
  elephant-climber - *Argyreia nervosa*
  elephant-ear-plant - *Alocasia, Colocasia esculenta*
  elephant-foot-tree - *Nolina recurvata*
  elephant garlic - *Allium ampeloprasum*
  elephant grass - *Pennisetum purpureum*
elephant's
  elephant's-ear - *Caladium bicolor*
  elephant's-head - *Pedicularis groenlandica*
  Florida elephant's-foot - *Elephantopus elatus*
  purple elephant's-foot - *Elephantopus nudatus*
eleven-o'clock - *Portulaca grandiflora*
elf
  elf-dock - *Inula helenium*
  elf-wort - *Inula helenium*
elfin-herb - *Cuphea hyssopifolia*
elk
  elk-clover - *Aralia californica*
  elk-moss - *Lycopodium clavatum*
  elk sedge - *Carex geyeri*
elkhorn fern - *Platycerium bifurcatum*
Elliott's
  Elliott's beard grass - *Andropogon elliottii*
  Elliott's blue-stem - *Andropogon elliottii*
  Elliott's blueberry - *Vaccinium elliottii*
  Elliott's goldenrod - *Solidago elliottii*
  Elliott's love grass - *Eragrostis elliottii*
ellisia - *Ellisia nyctelea*
elm
  American elm - *Ulmus americana*
  Caucasian elm - *Zelkova carpinifolia*
  cedar elm - *Ulmus crassifolia*
  Chinese elm - *Ulmus parvifolia, U. pumila*

cork elm - *Ulmus thomasii*
dwarf elm - *Ulmus pumila*
elm - *Ulmus*
elm-leaf goldenrod - *Solidago ulmifolia*
elm-leaf zelkova - *Zelkova carpinifolia*
English elm - *Ulmus minor*
Japanese elm - *Ulmus japonica*
leather-leaf elm - *Ulmus parvifolia*
Moline's elm - *Ulmus americana* cv. 'moline'
red elm - *Ulmus rubra, U. serotina*
rock elm - *Ulmus thomasii*
Scotch elm - *Ulmus glabra*
September elm - *Ulmus serotina*
Siberian elm - *Ulmus pumila*
slippery elm - *Ulmus rubra*
smooth-leaved elm - *Ulmus minor*
wahoo elm - *Ulmus alata*
water elm - *Ulmus americana*
white elm - *Ulmus americana*
winged elm - *Ulmus alata*
Wych's elm - *Ulmus glabra*
elodea
  Brazilian elodea - *Egeria densa*
  elodea - *Elodea canadensis*
  long-sheath elodea - *Elodea longivaginata*
  western elodea - *Elodea nuttallii*
emerald-ripple peperomia - *Peperomia caperata*
emetic-weed - *Lobelia inflata*
emex, spiny emex - *Emex spinosa*
emmer
  emmer - *Triticum dicoccon*
  wild emmer - *Triticum dicoccoides*
Emory's oak - *Quercus emoryi*
empress
  empress-candle-plant - *Senna alata*
  empress-tree - *Paulownia tomentosa*
enchanter's
  enchanter's-nightshade - *Circaea quadrisulcata, Circaea*
  small enchanter's-nightshade - *Circaea alpina*
encina - *Quercus agrifolia*
endive - *Cichorium endivia*
Engelmann's
  Engelmann's daisy - *Engelmannia pinnatifida*
  Engelmann's oak - *Quercus engelmannii*
  Engelmann's spruce - *Picea engelmannii*
England
  New England aster - *Aster novae-angliae*
  New England grape - *Vitis novae-angliae*
  New England serviceberry - *Amelanchier sanguinea*
English
  English bluegrass - *Festuca pratensis*
  English catchfly - *Silene gallica*
  English cherry-laurel - *Prunus laurocerasus*
  English daisy - *Bellis perennis*
  English elm - *Ulmus minor*
  English gooseberry - *Ribes uva-crispa*
  English hawthorn - *Crataegus laevigata, C. monogyna*
  English holly - *Ilex aquifolium*
  English iris - *Iris latifolia*
  English ivy - *Hedera helix*
  English lavender - *Lavandula angustifolia*

English oak - *Quercus robur*
English pea - *Pisum sativum*
English plantain - *Plantago lanceolata*
English primrose - *Primula vulgaris*
English rye grass - *Lolium perenne*
English violet - *Viola odorata*
English wallflower - *Erysimum cheiri*
English walnut - *Juglans regia*
English wheat - *Triticum turgidum*
English yew - *Taxus baccata*
entangled hawthorn - *Crataegus intricata*
eola-weed - *Hypericum perforatum*
ephedra
 Chinese ephedra - *Ephedra sinica*
 gray ephedra - *Ephedra nevadensis*
 green ephedra - *Ephedra viridis*
 long-leaf ephedra - *Ephedra trifurca*
 Nevada ephedra - *Ephedra nevadensis*
erect
 erect bugle - *Ajuga genevensis*
 erect goldenrod - *Solidago erecta*
 erect knotweed - *Polygonum erectum*
 erect spiderling - *Boerhavia erecta*
eryngo
 eryngo - *Eryngium aquaticum*
 marsh eryngo - *Eryngium aquaticum*
escarole - *Cichorium endivia*
eschalot - *Allium cepa*
esparcet - *Onobrychis viciifolia*
estafiata - *Artemisia frigida*
estragon - *Artemisia dracunculus*
estuarine, northern estuarine beggar-ticks - *Bidens hyperborea*
estuary beggar-ticks - *Bidens hyperborea*
Etonia palmetto - *Sabal etonia*
eucalypt
 eucalypt - *Eucalyptus*
 round-leaf eucalypt - *Eucalyptus polyanthemos*
eucalyptus, spiral eucalyptus - *Eucalyptus cinerea*
Eucharist-lily - *Eucharis grandiflora*
eugenia, box-leaf eugenia - *Eugenia foetida*
eulalia
 eulalia - *Miscanthus sinensis*
 eulalia grass - *Miscanthus sinensis*
euonymus
 climbing euonymus - *Euonymus fortunei*
 European euonymus - *Euonymus europaeus*
 evergreen euonymus - *Euonymus japonicus*
 winged euonymus - *Euonymus alatus*
eupatorium
 late eupatorium - *Eupatorium serotinum*
 round-leaf eupatorium - *Eupatorium rotundifolium*
 white-bracted eupatorium - *Eupatorium album*
euphorbia, painted euphorbia - *Euphorbia prunifolia*
Eurasian
 Eurasian chestnut - *Castanea sativa*
 Eurasian milfoil - *Myriophyllum spicatum*
 Eurasian water-milfoil - *Myriophyllum spicatum*
European
 European alder - *Alnus glutinosa*

European alder buckthorn - *Rhamnus frangula*
European alkali grass - *Puccinellia distans*
European ash - *Fraxinus excelsior*
European barberry - *Berberis vulgaris*
European beach grass - *Ammophila arenaria*
European beech - *Fagus sylvatica*
European bird cherry - *Prunus padus*
European black currant - *Ribes nigrum*
European blackberry - *Rubus fruticosus*
European blue lupine - *Lupinus angustifolius*
European broom - *Cytisus scoparius*
European buckthorn - *Rhamnus cathartica*
European bugleweed - *Lycopus europaeus*
European chain fern - *Woodwardia radicans*
European chestnut - *Castanea sativa*
European columbine - *Aquilegia vulgaris*
European cranberry-bush - *Viburnum opulus*
European crowfoot - *Aquilegia vulgaris*
European daisy - *Bellis perennis*
European dune grass - *Leymus arenarius*
European dwarf cherry - *Prunus fruticosa*
European elder - *Sambucus nigra*
European euonymus - *Euonymus europaeus*
European field pansy - *Viola arvensis*
European filbert - *Corylus avellana*
European fly honeysuckle - *Lonicera xylosteum*
European gooseberry - *Ribes grossularis, R. uva-crispa*
European grape - *Vitis vinifera*
European green alder - *Alnus viridis*
European ground cherry - *Prunus fruticosa*
European hawk's-beard - *Crepis vesicaria* subsp. *haenseleri*
European hazelnut - *Corylus avellana*
European heliotrope - *Heliotropium europaeum*
European holly - *Ilex aquifolium*
European hop - *Humulus lupulus*
European hornbeam - *Carpinus betulus*
European horse-chestnut - *Aesculus hippocastanum*
European horsemint - *Mentha longifolia*
European larch - *Larix decidua*
European linden - *Tilia cordata*
European mallow - *Malva alcea*
European mountain-ash - *Sorbus aucuparia*
European nettle - *Urtica dioica*
European parsley fern - *Cryptogramma crispa*
European pepperwort - *Marsilea quadrifolia*
European plum - *Prunus domestica*
European red elderberry - *Sambucus racemosa*
European red raspberry - *Rubus idaeus*
European spindle-tree - *Euonymus europaeus*
European strawberry - *Fragaria vesca*
European turkey oak - *Quercus cerris*
European vervain - *Verbena officinalis*
European water-clover - *Marsilea quadrifolia*
European water horehound - *Lycopus europaeus*
European water-starwort - *Callitriche stagnalis*
European white birch - *Betula pendula*
European white water-lily - *Nymphaea alba*
European wild pansy - *Viola tricolor*
European wood anemone - *Anemone nemorosa*
European wood-sorrel - *Oxalis acetosella, O. europaea*
European woodbine - *Lonicera periclymenum*
European yellow lupine - *Lupinus luteus*
eutrophic water-nymph - *Najas minor*
evax, many-stemmed evax - *Filago verna*

evening
  cut-leaf evening-primrose - *Oenothera laciniata*
  desert evening-primrose - *Oenothera deltoides*
  evening campion - *Silene latiflora*
  evening lychnis - *Silene latiflora*
  evening-primrose - *Oenothera biennis, Oenothera*
  evening-scented stock - *Matthiola longipetala*
  evening stock - *Matthiola longipetala*
  evening trumpet-flower - *Gelsemium sempervirens*
  mother-of-the-evening - *Hesperis matronalis*
  prairie evening-primrose - *Oenothera albicaulis*
  sea beach evening-primrose - *Oenothera humifusa*
  showy evening-primrose - *Oenothera speciosa*
  small-flowered evening-primrose - *Oenothera nuttallii,*
    O. parviflora
  spreading evening-primrose - *Oenothera humifusa*
  white evening-primrose - *Oenothera speciosa*
  white-stem evening-primrose - *Oenothera nuttallii*
Everglades palm - *Acoelorraphe wrightii*
evergreen
  African evergreen - *Syngonium podophyllum*
  Belgian evergreen - *Dracaena sanderana*
  bunch-evergreen - *Lycopodium obscurum*
  Chinese evergreen - *Aglaonema modestum, A. simplex*
  coral-evergreen - *Lycopodium clavatum*
  evergreen bayberry - *Myrica heterophylla*
  evergreen blackberry - *Rubus laciniatus*
  evergreen cherry - *Prunus ilicifolia*
  evergreen euonymus - *Euonymus japonicus*
  evergreen huckleberry - *Vaccinium ovatum*
  evergreen mountain fetterbush - *Pieris floribunda*
  evergreen sumac - *Rhus sempervirens, R. virens*
  evergreen swamp fetterbush - *Lyonia lucida*
  New Mexico evergreen sumac - *Rhus choriophylla*
  running-evergreen - *Lycopodium complanatum*
  staghorn-evergreen - *Lycopodium clavatum*
  trailing-evergreen - *Lycopodium complanatum*
everlasting
  clammy everlasting - *Gnaphalium viscosum*
  early everlasting - *Antennaria plantaginifolia*
  everlasting - *Anaphalis, Antennaria, Gnaphalium*
    obtusifolium, Gnaphalium, Helichrysum, Helipterum
  everlasting pea - *Lathyrus latifolius, L. sylvestris*
  everlasting pea-vine - *Lathyrus latifolius*
  fragrant everlasting - *Gnaphalium obtusifolium*
  life-everlasting - *Anaphalis*
  pearly everlasting - *Anaphalis margaritacea*
  plantain-leaf everlasting - *Antennaria plantaginifolia*
  silky-white everlasting - *Helipterum splendidum*
  splendid everlasting - *Helipterum splendidum*
  sunflower-everlasting - *Heliopsis helianthoides*
  sweet everlasting - *Gnaphalium obtusifolium*
extra-long staple cotton - *Gossypium barbadense*
eye
  bird's-eye - *Adonis annua, Veronica persica*
  bird's-eye pearl-wort - *Sagina procumbens*
  bird's-eye primrose - *Primula laurentiana*
  Canadian eye-bright - *Euphrasia canadensis*
  crab's-eye - *Abrus precatorius*
  dark-eye sunflower - *Helianthus atrorubens*
  eye-bane - *Euphorbia maculata*

eye-bright - *Anagallis arvensis, Euphorbia maculata, Lobelia*
    inflata
  Hart's-eye - *Pastinaca sativa*
  ox-eye - *Heliopsis helianthoides, Heliopsis*
  ox-eye daisy - *Leucanthemum vulgare*
  pheasant-eye adonis - *Adonis annua*
  pheasant's-eye - *Adonis annua, Narcissus poeticus*
  rabbit's-eye blueberry - *Vaccinium ashei*
eyebright - *Parentucellia viscosa*
eyed
  black-eyed pea - *Vigna unguiculata*
  black-eyed-Susan - *Rudbeckia hirta, R. hirta* var. *pulcherrima*
  blue-eyed African daisy - *Arctotis stoechadifolia*
  blue-eyed Cape-marigold - *Dimorphotheca sinuata*
  blue-eyed-grass - *Sisyrinchium*
  blue-eyed-Mary - *Collinsia verna*
  brown-eyed-Susan - *Rudbeckia triloba*
  crimson-eyed rose mallow - *Hibiscus moscheutos* subsp.
    palustris
  eastern blue-eyed-Mary - *Collinsia verna*
  Montana blue-eyed-grass - *Sisyrinchium montanum*
  narrow-leaf blue-eyed-grass - *Sisyrinchium angustifolium*
  orange-eyed buddleja - *Buddleja davidii*
  purple-eyed-grass - *Sisyrinchium douglasii*
  two-eyed violet - *Viola ocellata*
  yellow-eyed-grass - *Xyris caroliniana, Xyris*
eyes
  baby-blue-eyes - *Nemophila menziesii*
  bird's-eyes - *Gilia tricolor*
  bright-eyes - *Catharanthus roseus*
  doll's-eyes - *Actaea alba*
  white baby-blue-eyes - *Nemophila menziesii* var. *atomaria*
ezo-negi - *Allium schoenoprasum*
face
  pepper-face - *Peperomia obtusifolia*
  pretty-face - *Triteleia ixioides*
fairway crested wheat grass - *Agropyron cristatum*
fairy
  fairy-bells - *Disporum hookeri*
  fairy-duster - *Calliandra eriophylla*
  fairy flax - *Linum catharticum*
  fairy-footprints - *Hedyotis procumbens*
  fairy primrose - *Primula malacoides*
  fairy-slipper - *Calypso bulbosa*
  fairy water-lily - *Nymphoides aquatica*
faitours-grass - *Euphorbia esula*
falcate golden aster - *Chrysopsis falcata*
fall
  fall adonis - *Adonis annua*
  fall daffodil - *Sternbergia lutea*
  fall-dandelion - *Leontodon autumnalis*
  fall hawk's-bit - *Leontodon autumnalis*
  fall panic grass - *Panicum dichotomiflorum*
  fall panicum - *Panicum dichotomiflorum*
  fall phlox - *Phlox paniculata*
false
  American false pennyroyal - *Hedeoma pulegioides*
  blue false indigo - *Baptisia australis*
  California false hellebore - *Veratrum californicum*
  Carolina false dandelion - *Pyrrhopappus carolinianus*
  climbing false buckwheat - *Polygonum scandens*
  downy false foxglove - *Aureolaria virginica*

false acacia - *Robinia pseudoacacia*
false aloe - *Manfreda virginica*
false arborvitae - *Thujopsis dolobrata*
false banyan - *Ficus altissima*
false beech-drops - *Monotropa hypopithys*
false bittersweet - *Celastrus scandens*
false boneset - *Brickellia eupatorioides*
false brome - *Brachypodium distachyon*
false broom-weed - *Ericameria austrotexana*
false buckthorn - *Sideroxylon lanuginosa*
false buffalo grass - *Munroa squarrosa*
false bugbane - *Trautvetteria*
false caraway - *Perideridia gairdneri*
false chamomile - *Matricaria maritima*
false cypress - *Chamaecyparis*
false dandelion - *Hypochaeris radicata*
false dogwood - *Sapindus saponaria*
false flax - *Camelina microcarpa, Camelina*
false foxglove - *Aureolaria pedicularia*
false garlic - *Nothoscordum bivalve*
false goat's-beard - *Astilbe biternata*
false heather - *Cuphea hyssopifolia, Hudsonia tomentosa*
false hellebore - *Veratrum*
false holly - *Osmanthus heterophyllus*
false hop-plant - *Justicia brandegeana*
false indigo - *Amorpha fruticosa, Amorpha, Baptisia*
false indigo-bush - *Amorpha fruticosa*
false ipecac - *Asclepias curassavica*
false Jerusalem cherry - *Solanum capsicastrum*
false lily-of-the-valley - *Maianthemum canadense,
    Maianthemum*
false loosestrife - *Ludwigia*
false lupine - *Thermopsis*
false mallow - *Malvastrum, Sphaeralcea*
false mastic - *Sideroxylon foetidissimum*
false mayweed - *Dyssodia papposa*
false melic - *Schizachne purpurascens*
false mesquite - *Calliandra*
false mistletoe - *Phoradendron*
false miterwort - *Tiarella*
false nettle - *Boehmeria cylindrica*
false nut sedge - *Cyperus strigosus*
false-pennyroyal - *Trichostema brachiatum*
false pimpernel - *Lindernia anagallidea, L. procumbens*
false ragweed - *Ambrosia tenuifolia*
false saffron - *Carthamus tinctorius*
false shagbark - *Carya ovalis*
false Solomon's-seal - *Smilacina racemosa*
false spikenard - *Smilacina racemosa*
false strawberry - *Duchesnea indica*
false tansy - *Artemisia biennis*
false wheat grass - *Leymus chinensis*
flat-seeded false flax - *Camelina microcarpa*
fragrant false indigo - *Amorpha nana*
Hinoki false cypress - *Chamaecyparis obtusa*
Japanese false brome grass - *Brachypodium pinnatum*
Japanese false cypress - *Chamaecyparis obtusa*
large-seeded false flax - *Camelina sativa*
low false pimpernel - *Lindernia dubia*
many-fruited false loosestrife - *Ludwigia polycarpa*
red false mallow - *Sphaeralcea coccinea*
rough false pennyroyal - *Hedeoma hispidum*
round-leaf false pimpernel - *Lindernia grandiflora*

slender false brome grass - *Brachypodium sylvaticum*
small-seeded false flax - *Camelina microcarpa*
smooth false dandelion - *Hypochaeris glabra*
starry false Solomon's-seal - *Smilacina stellata, S. trifolia*
three-leaf false Solomon's seal - *Smilacina trifolia*
western false gromwell - *Onosmodium molle*
western false hellebore - *Veratrum californicum*
yellow false mallow - *Malvastrum hispidum*
fan
    California fan palm - *Washingtonia filifera*
    desert fan palm - *Washingtonia filifera*
    fan-leaf hawthorn - *Crataegus flabellata*
    fan maidenhair fern - *Adiantum tenerum*
    fan palm - *Coccothrinax, Livistona chinensis, Livistona*
    fan-weed - *Thlaspi arvense*
    miniature fan palm - *Rhapis excelsa*
fancy
    fancy fern - *Dryopteris carthusiana*
    fancy geranium - *Pelargonium domesticum*
    fancy-leaf caladium - *Caladium bicolor, C. hortulanum*
fantasy, little-fantasy peperomia - *Peperomia caperata*
fanwort
    green fanwort - *Cabomba caroliniana* cv. 'multipartita'
    purple fanwort - *Cabomba caroliniana*
farewell-to-spring - *Clarkia amoena, Clarkia*
farkleberry - *Rhododendron arboreum, Vaccinium
    arboreum*
fat
    fat-hen - *Chenopodium album, C. bonus-henricus*
    fat-Solomon - *Smilacina racemosa*
    mule's-fat - *Baccharis viminea*
    sheep's-fat - *Atriplex confertifolia*
    winter-fat - *Krascheninnikovia lanata*
fatsia, Japanese fatsia - *Fatsia japonica*
fava bean - *Vicia faba*
fawn-lily - *Erythronium*
feather
    arrow-feather - *Aristida purpurascens*
    arrow-feather three-awn - *Aristida purpurascens*
    dotted gay-feather - *Liatris punctata*
    feather-bells - *Stenanthium gramineum*
    feather bunch grass - *Stipa viridula*
    feather finger grass - *Chloris virgata*
    feather-foil - *Hottonia inflata*
    feather-geranium - *Chenopodium botrys*
    feather grass - *Stipa*
    feather-head knapweed - *Centaurea trichocephala*
    feather-plume - *Dalea formosa*
    feather tumble grass - *Eragrostis tenella*
    gay-feather - *Liatris*
    golden feather palm - *Chrysalidocarpus lutescens*
    MacArthur's feather palm - *Ptychosperma macarthurii*
    New Mexico feather grass - *Stipa neomexicana*
    parrot's-feather - *Myriophyllum aquaticum*
    prince's-feather - *Polygonum orientale*
    princess-feather - *Polygonum orientale*
    water-feather - *Myriophyllum aquaticum*
feathery cassia - *Cassia artemisioides*
February daphne - *Daphne mezereum*
Fee's lip fern - *Cheilanthes feei*
feijoa - *Acca sellowiana*
felon-herb - *Artemisia vulgaris, Hieracium pilosella*

felt
    felt-leaf ceanothus - *Ceanothus arboreus*
    felt-leaf willow - *Salix alaxensis*
female fluvellin - *Kickxia spuria*
fence
    flower-fence poinciana - *Caesalpinia pulcherrima*
    flowering-fence - *Caesalpinia pulcherrima*
feng-hsiang-shu - *Liquidambar formosana*
fennel
    dog-fennel - *Anthemis cotula, Anthemis, Eupatorium capillifolium*
    fennel - *Foeniculum vulgare, Helenium amarum*
    fennel giant hyssop - *Agastache foeniculum*
    fennel-leaf pondweed - *Potamogeton pectinatus*
    Florence fennel - *Foeniculum vulgare*
    rayless dog-fennel - *Matricaria matricarioides*
    yellow dog-fennel - *Helenium amarum*
fenugreek - *Trigonella foenum-graecum*
Fenzel's water-lily - *Nymphaea glandulifera*
fern
    adder's-tongue fern - *Ophioglossum*
    American maidenhair fern - *Adiantum pedatum*
    American wall fern - *Polypodium virginianum*
    asparagus-fern - *Asparagus densiflorus, A. setaceus*
    Australian tree-fern - *Cyathea australis, C. cooperi*
    ball fern - *Davallia trichomanoides*
    bear's-paw fern - *Aglaomorpha meyeniana*
    berry bladder fern - *Cystopteris bulbifera*
    bird's-nest fern - *Asplenium nidus*
    blond tree-fern - *Cibotium splendens*
    boulder fern - *Dennstaedtia punctilobula*
    bracken fern - *Pteridium aquilinum, P. aquilinum* var. *pubescens*
    brittle bladder fern - *Cystopteris fragilis*
    brittle fern - *Cystopteris fragilis*
    brittle maidenhair fern - *Adiantum tenerum*
    bulblet bladder fern - *Cystopteris bulbifera*
    button fern - *Pellaea rotundifolia*
    California fern - *Conium maculatum*
    California gold-bark fern - *Pityrogramma triangularis*
    Carolina mosquito-fern - *Azolla caroliniana*
    chain fern - *Woodwardia*
    Christmas fern - *Polystichum acrostichoides*
    cinnamon fern - *Osmunda cinnamomea*
    climbing bird's nest fern - *Microsorium punctatum*
    climbing fern - *Lygodium*
    coastal shield fern - *Dryopteris arguta*
    coastal wood fern - *Dryopteris arguta*
    coffee fern - *Pellaea andromedifolia*
    crested fern - *Microsorium punctatum*
    cup fern - *Dennstaedtia*
    dagger fern - *Polystichum acrostichoides*
    deer fern - *Blechnum spicant*
    delta maidenhair fern - *Adiantum raddianum* cv. 'decorum'
    elkhorn fern - *Platycerium bifurcatum*
    European chain fern - *Woodwardia radicans*
    European parsley fern - *Cryptogramma crispa*
    fan maidenhair fern - *Adiantum tenerum*
    fancy fern - *Dryopteris carthusiana*
    Fee's lip fern - *Cheilanthes feei*
    fern asparagus - *Asparagus setaceus*
    fern-bush - *Chamaebatiaria millefolium*

    fern-gale - *Comptonia peregrina*
    fern-leaf inch-plant - *Tripogandra multiflora*
    fern-palm - *Cycas circinalis*
    fern rhapis - *Rhapis excelsa*
    five-finger fern - *Adiantum pedatum*
    floating fern - *Ceratopteris pteridoides*
    floating water fern - *Ceratopteris thalictroides*
    florist's fern - *Dryopteris carthusiana*
    flowering fern - *Osmunda regalis, Osmunda*
    fragile fern - *Cystopteris fragilis*
    giant chain fern - *Woodwardia fimbriata*
    giant holly fern - *Polystichum munitum*
    grape fern - *Botrychium*
    halberd fern - *Tectaria heracleifolia*
    hand fern - *Dryopteris pedata*
    hard fern - *Blechnum spicant*
    Hawaiian tree-fern - *Cibotium chamissoi, C. splendens*
    hay-scented fern - *Dennstaedtia punctilobula*
    horsetail-fern - *Equisetum arvense*
    house holly fern - *Cyrtomium falcata*
    interrupted fern - *Osmunda claytoniana*
    iron fern - *Rumohra adiantiformis*
    Jamaican gold fern - *Pityrogramma sulphurea*
    Japanese climbing fern - *Lygodium japonicum*
    Japanese fern-palm - *Cycas revoluta*
    Japanese holly fern - *Cyrtomium falcata*
    lace fern - *Sphenomeris chinensis*
    lace lip fern - *Cheilanthes gracillima*
    lady fern - *Athyrium felix-femina*
    lanate lip fern - *Cheilanthes lanosa*
    lance-leaf grape fern - *Botrychium lanceolatum*
    leather fern - *Acrostichum aureum, Rumohra adiantiformis*
    leather wood fern - *Dryopteris marginalis*
    licorice fern - *Polypodium glycyrrhiza*
    lip fern - *Cheilanthes*
    little grape fern - *Botrychium simplex*
    maidenhair fern - *Adiantum*
    man tree-fern - *Cibotium splendens*
    marginal shield fern - *Dryopteris marginalis*
    marginal wood fern - *Dryopteris marginalis*
    marsh fern - *Thelypteris palustris*
    meadow fern - *Thelypteris palustris*
    mountain bladder fern - *Cystopteris montana*
    mountain cliff fern - *Woodsia scopulina*
    mountain holly fern - *Polystichum lonchitis*
    mountain parsley fern - *Cryptogramma crispa*
    mountain wood fern - *Dryopteris campyloptera*
    Nebraska fern - *Conium maculatum*
    nettle chain fern - *Woodwardia areolata*
    New York fern - *Thelypteris noveboracensis*
    northern adder's-tongue fern - *Ophioglossum pusillum*
    northern holly fern - *Polystichum lonchitis*
    northern lady fern - *Athyrium felix-femina* var. *michauxii*
    northern maidenhair fern - *Adiantum pedatum*
    oak fern - *Gymnocarpium dryopteris*
    ostrich fern - *Matteuccia struthiopteris*
    Pacific Christmas fern - *Polystichum munitum*
    parsley fern - *Cryptogramma crispa, Cryptogramma*
    pinnate mosquito-fern - *Azolla pinnata*
    rattlesnake fern - *Botrychium virginianum*
    resurrection fern - *Polypodium polypodioides*
    ribbon fern - *Polypodium phyllitidis*
    rough tree-fern - *Cyathea australis*

royal fern - *Osmunda regalis*
savannah fern - *Dicranopteris linearis*
sensitive fern - *Onoclea sensibilis*
shield fern - *Dryopteris, Polystichum*
shrubby-fern - *Comptonia peregrina*
silver fern - *Pityrogramma calomelanos*
slender lip fern - *Cheilanthes feei*
smooth cliff fern - *Woodsia glabella*
snuffbox fern - *Thelypteris palustris*
spinulose wood fern - *Dryopteris carthusiana*
spreading wood fern - *Dryopteris campyloptera*
squirrel-foot fern - *Davallia trichomanoides*
staghorn fern - *Platycerium bifurcatum*
strap fern - *Polypodium phyllitidis*
swamp fern - *Acrostichum*
sweet-fern - *Comptonia asplenifolia, C. peregrina*
sword fern - *Nephrolepis biserrata*
toothed wood fern - *Dryopteris carthusiana*
tree-fern - *Cyathea*
Venus's-hair fern - *Adiantum capillaris-veneris*
Virginia chain fern - *Woodwardia virginica*
walking fern - *Asplenium rhizophyllum*
water fern - *Ceratopteris thalictroides, Salvinia minima*
waxy cloak fern - *Notholaena sinuata*
West Indian tree-fern - *Cyathea arborea*
western bracken fern - *Pteridium aquilinum* var. *pubescens*
western cliff fern - *Woodsia oregana*
western sword fern - *Polystichum munitum*
whisk-fern - *Psilotum nudum*
winter-fern - *Conium maculatum*
wood fern - *Dryopteris*
woolly lip fern - *Cheilanthes lanosa*
fescue
  Arizona fescue - *Festuca arizonica*
  blue-bunch fescue - *Festuca idahoensis*
  fescue - *Festuca*
  foxtail fescue - *Vulpia myuros, V. myuros* var. *hirsuta*
  gray fescue - *Vulpia microstachys* var. *ciliata*
  green fescue - *Festuca viridula*
  green-leaf fescue - *Festuca viridula*
  hair fescue - *Festuca tenuifolia*
  hard fescue - *Festuca brevipila*
  Idaho fescue - *Festuca idahoensis*
  meadow fescue - *Festuca pratensis*
  nodding fescue - *Festuca obtusa*
  Pacific fescue - *Vulpia microstachys* var. *pauciflora*
  rat-tail fescue - *Vulpia myuros*
  red fescue - *Festuca rubra*
  rough fescue - *Festuca scabrella*
  shade fescue - *Festuca rubra* var. *heterophylla*
  sheep fescue - *Festuca ovina*
  slender fescue - *Festuca tenuifolia*
  squirrel-tail fescue - *Vulpia bromoides*
  tall fescue - *Festuca arundinacea*
  Thurber's fescue - *Festuca thurberi*
  western fescue - *Festuca occidentalis*
festuca, chewing festuca - *Festuca rubra*
fetid
  fetid buckeye - *Aesculus glabra*
  fetid-chamomile - *Anthemis cotula*
  fetid currant - *Ribes glandulosum*
  fetid horehound - *Ballota*
  fetid-marigold - *Dyssodia papposa, Dyssodia*

fetid wild pumpkin - *Cucurbita foetidissima*
fetter, hell-fetter - *Smilax hispida, S. tamnoides*
fetterbush
  evergreen mountain fetterbush - *Pieris floribunda*
  evergreen swamp fetterbush - *Lyonia lucida*
  fetterbush - *Leucothoe, Lyonia lucida, Pieris floribunda*
fetticus - *Valerianella locusta*
fever
  fever-bush - *Lindera benzoin*
  fever grass - *Cymbopogon citratus*
  fever-plant - *Oenothera biennis*
  fever-tree - *Pinckneya pubens*
  hay-fever-weed - *Ambrosia acanthicarpa*
  Jamaican fever-plant - *Tribulus cistoides*
feverfew
  American feverfew - *Parthenium integrifolium*
  feverfew - *Agrimonia gryposepala, Tanacetum parthenium*
feverweed - *Eupatorium perfoliatum*
feverwort - *Triosteum perfoliatum, Triosteum*
few
  few-flowered milkweed - *Asclepias lanceolata*
  few-flowered muhly - *Muhlenbergia pauciflora*
  few-flowered sedge - *Carex pauciflora*
  few-headed blazing-star - *Liatris cylindracea*
  few-seeded sedge - *Carex oligosperma*
fiber, black fiber palm - *Arenga pinnata*
ficus
  climbing ficus - *Ficus pumila*
  creeping ficus - *Ficus pumila*
  fiddle-leaf ficus - *Ficus lyrata*
  mistletoe ficus - *Ficus deltoidea*
fiddle
  coast fiddle-neck - *Amsinckia lycopsoides*
  fiddle dock - *Rumex pulcher*
  fiddle-leaf - *Ficus lyrata*
  fiddle-leaf dock - *Rumex pulcher*
  fiddle-leaf ficus - *Ficus lyrata*
  fiddle-leaf tobacco - *Nicotiana repanda*
  fiddle-necks - *Amsinckia lycopsoides*
  rigid fiddle-neck - *Amsinckia retrorsa*
  small-flowered fiddle-neck - *Amsinckia menziesii*
  tarweed fiddle-neck - *Amsinckia lycopsoides*
  western fiddle-neck - *Amsinckia tessellata*
fiddlehead, Douglas's fiddlehead - *Amsinckia douglasiana*
fiddleheads - *Osmunda cinnamomea*
field
  Austrian field-cress - *Rorippa austriaca*
  Austrian field pea - *Sphaerophysa salsula*
  blue field madder - *Sherardia arvensis*
  California field oak - *Quercus agrifolia*
  crown-of-the-field - *Agrostemma githago*
  European field pansy - *Viola arvensis*
  field alyssum - *Alyssum minus*
  field-balm - *Glechoma hederacea*
  field bindweed - *Convolvulus arvensis*
  field brome - *Bromus arvensis*
  field buttercup - *Ranunculus arvensis*
  field-chamomile - *Anthemis arvensis*
  field chickweed - *Cerastium arvense*
  field corn - *Zea mays*

field daisy - *Leucanthemum vulgare*
field dodder - *Cuscuta campestris, C. pentagona*
field forget-me-not - *Myosotis arvensis*
field garlic - *Allium vineale*
field gilia - *Gilia capitata*
field hawkweed - *Hieracium caespitosum*
field horsetail - *Equisetum arvense*
field larkspur - *Delphinium consolida, D. menziesii*
field lupine - *Lupinus albus*
field madder - *Sherardia arvensis*
field maple - *Acer campestre*
field milk-vetch - *Astragalus agrestis*
field milkwort - *Polygala sanguinea*
field mint - *Mentha arvensis*
field mustard - *Brassica rapa*
field-nettle betony - *Stachys arvensis*
field pansy - *Viola rafinesquii, V. tricolor*
field paspalum - *Paspalum laeve*
field pea - *Pisum sativum, P. sativum* var. *arvense*
field pennycress - *Thlaspi arvense*
field pepper-weed - *Lepidium campestre*
field peppergrass - *Lepidium campestre*
field poppy - *Papaver dubium, P. rhoeas*
field-primrose - *Oenothera biennis*
field pussy-toes - *Antennaria neglecta*
field rush - *Juncus tenuis*
field sage-wort - *Artemisia campestris* subsp. *caudata*
field sandbur - *Cenchrus incertus*
field scabious - *Knautia arvensis*
field scorpion-grass - *Myosotis arvensis*
field sedge - *Carex conoidea*
field sow-thistle - *Sonchus arvensis*
field speedwell - *Veronica agrestis*
field thistle - *Cirsium discolor*
field violet - *Viola arvensis*
field wood rush - *Luzula campestris*
lily-of-the-field - *Anemone*
old-field-balsam - *Gnaphalium obtusifolium*
old-field birch - *Betula populifolia*
old-field cinquefoil - *Potentilla simplex*
old-field clover - *Trifolium arvense*
old-field pine - *Pinus taeda*
old-field toadflax - *Linaria canadensis*
rice-field bulrush - *Scirpus mucronatus*
western field buttercup - *Ranunculus occidentalis*
white field aster - *Aster lanceolatus*
yellow field-cress - *Rorippa sylvestris*
fieldcress - *Lepidium campestre*
fields, gold-fields - *Lasthenia chrysostoma*
fig
  banjo fig - *Ficus lyrata*
  banyan fig - *Ficus benghalensis*
  East Indian fig-tree - *Ficus benghalensis*
  edible fig - *Ficus carica*
  fig - *Ficus carica, Ficus*
  fig-leaved goosefoot - *Chenopodium ficifolium*
  fig-tree kalanchoe - *Kalanchoe laciniata*
  Florida strangler fig - *Ficus aurea*
  golden fig - *Ficus aurea*
  hottentot-fig - *Carpobrotus edulis*
  Indian fig - *Opuntia ficus-indica*
  Indian laurel fig - *Ficus microcarpa*
  Indian rubber fig - *Ficus elastica*

  Java fig - *Ficus benjamina*
  keg-fig - *Diospyros kaki*
  lofty fig - *Ficus altissima*
  sea-fig - *Carpobrotus chilensis*
  short-leaf fig - *Ficus citrifolia*
  weeping fig - *Ficus benjamina*
figwort - *Scrophularia lanceolata, Scrophularia*
filaree
  broad-leaf filaree - *Erodium botrys*
  red-stem filaree - *Erodium cicutarium*
  Texas filaree - *Erodium texanum*
  white-stem filaree - *Erodium moschatum*
filbert
  American filbert - *Corylus americana*
  beaked filbert - *Corylus cornuta*
  European filbert - *Corylus avellana*
  filbert - *Corylus*
  giant filbert - *Corylus maxima*
filmy angelica - *Angelica triquinata*
fine-leaf pondweed - *Potamogeton filiformis*
finger
  feather finger grass - *Chloris virgata*
  finger grass - *Chloris, Digitalis, Digitaria ischaemum*
  finger-leaf gourd - *Cucurbita digitata*
  finger millet - *Eleusine coracana*
  finger-of-God - *Aechmea orlandiana*
  finger tickseed - *Coreopsis palmata*
  finger-weed - *Amsinckia intermedia*
  five-finger fern - *Adiantum pedatum*
  lady's-finger - *Abelmoschus esculentus*
  marsh fire-finger - *Potentilla palustris*
  radiate finger grass - *Chloris radiata*
  rock finger grass - *Chloris petraea*
  tall five-finger - *Potentilla norvegica*
  velvet finger grass - *Digitaria velutina*
  woolly finger grass - *Digitaria eriantha* subsp. *pentzii*
fingers
  five-fingers - *Potentilla*
  lady's-fingers - *Anthyllis vulneraria*
  running five-fingers - *Potentilla canadensis*
finocchio - *Foeniculum vulgare*
fir
  alpine fir - *Abies lasiocarpa*
  balsam fir - *Abies balsamea, A. grandis*
  big-cone Douglas fir - *Pseudotsuga macrocarpa*
  bristle-cone fir - *Abies bracteata*
  California red fir - *Abies magnifica*
  Cascade fir - *Abies amabilis*
  China fir - *Cunninghamia lanceolata*
  cork-bark fir - *Abies lasiocarpa* var. *arizonica*
  cork fir - *Abies lasiocarpa* var. *arizonica*
  Douglas fir - *Pseudotsuga menziesii*
  fir - *Abies*
  fir-balsam - *Abies balsamea*
  fir club-moss - *Lycopodium selago*
  Fraser's balsam fir - *Abies fraseri*
  Fraser's fir - *Abies fraseri*
  grand fir - *Abies grandis*
  Greek fir - *Abies cephalonica*
  joint-fir - *Ephedra*
  lowland fir - *Abies grandis*
  lowland white fir - *Abies grandis*

noble fir - *Abies procera*
Pacific silver fir - *Abies amabilis*
red fir - *Abies magnifica*
Rocky Mountain Douglas fir - *Pseudotsuga menziesii* var.
  glauca
Santa Lucia fir - *Abies bracteata*
silver fir - *Abies alba, A. amabilis, A. grandis*
subalpine fir - *Abies lasiocarpa*
white fir - *Abies amabilis, A. concolor*
fire
  fire birch - *Betula populifolia*
  fire-bush - *Hamelia patens*
  fire-chalice - *Epilobium canum*
  fire cherry - *Prunus pensylvanica*
  fire-dragon - *Acalypha wilkesiana*
  fire-lily - *Cyrtanthus angusitfolius*
  fire-on-the-mountain - *Euphorbia cyathophora*
  fire-pink - *Silene virginica*
  fire-thorn - *Pyracantha*
  fire-wheel - *Gaillardia pulchella*
  fire-wheel-tree - *Stenocarpus sinuatus*
  marsh fire-finger - *Potentilla palustris*
  Mexican fire-plant - *Euphorbia heterophylla*
  narrow-leaf fire-thorn - *Pyracantha angustifolia*
  rancher's fire-weed - *Amsinckia menziesii*
  scarlet fire-thorn - *Pyracantha coccinea*
  wheel-of-fire - *Stenocarpus sinuatus*
fireberry hawthorn - *Crataegus chrysocarpa*
firecracker
  Eaton's firecracker - *Penstemon eatonii*
  firecracker-flower - *Crossandra infundibuliformis*
  firecracker-plant - *Cuphea ignea*
fireweed
  fireweed - *Epilobium angustifolium, Epilobium, Erechtites*
    hieracifolia
  New Zealand fireweed - *Erechtites glomerata*
fish
  fish geranium - *Pelargonium hortorum*
  fish-hook cactus - *Mammillaria*
  fish-poison-tree - *Piscidia piscipula*
  fish-pole bamboo - *Phyllostachys aurea*
fishtail
  Burmese fishtail palm - *Caryota mitis*
  clustered fishtail palm - *Caryota mitis*
  fishtail palm - *Caryota urens, Caryota*
  tufted fishtail palm - *Caryota mitis*
fisolilla - *Calopogonium mucunoides*
fistula, purging fistula - *Cassia fistula*
  fit's-root - *Astragalus glycyphyllos, Monotropa uniflora*
fitch, tar-fitch - *Lathyrus pratensis*
fittonia
  silver fittonia - *Fittonia verschaffeltii*
  white-leaf fittonia - *Fittonia verschaffeltii*
five
  five-finger fern - *Adiantum pedatum*
  five-fingers - *Potentilla*
  five-hook bassia - *Bartonia virginica, Bassia hyssopifolia*
  five-leaf akebia - *Akebia quinata*
  five-leaved ivy - *Parthenocissus quinquefolia*
  five-stamen tamarisk - *Tamarix chinensis*
  little five-leaves - *Isotria medeoloides*
  running five-fingers - *Potentilla canadensis*

  tall five-finger - *Potentilla norvegica*
flaccid sedge - *Carex leptalea*
flag
  blue flag iris - *Iris versicolor*
  flag - *Iris germanica, Iris*
  flag pawpaw - *Asimina incana*
  flag-root - *Acorus calamus*
  larger blue flag - *Iris versicolor*
  northern blue flag - *Iris versicolor*
  poison flag - *Iris versicolor*
  soft-flag - *Typha angustifolia*
  southern blue flag - *Iris virginica* var. *schrevei*
  spiral-flag - *Costus malortieanus*
  sweet-flag - *Acorus calamus*
  western blue flag - *Iris missouriensis*
  yellow flag iris - *Iris pseudacorus*
flamboyant-tree - *Delonix regia*
flame
  Australian flame-tree - *Brachychiton acerifolius*
  flame azalea - *Rhododendron calendulaceum*
  flame bottle-tree - *Brachychiton acerifolius*
  flame-flower - *Kniphofia uvaria, Pyrostegia venusta, Talinum*
    paniculatum
  flame-gold - *Koelreuteria elegans*
  flame-of-the-forest - *Spathodea campanulata*
  flame-of-the-woods - *Ixora coccinea*
  flame-tree - *Delonix regia*
  flame-tree sumac - *Rhus copallina*
  flame-vine - *Pyrostegia venusta*
  Mexican flame-leaf - *Euphorbia pulcherrima*
  Mexican flame-vine - *Senecio chenopodioides*
  red-flame-ivy - *Hemigraphis alternata*
flaming-trumpet - *Pyrostegia venusta*
flamingo
  flamingo-flower - *Anthurium andraeanum*
  flamingo-lily - *Anthurium andraeanum*
  flamingo-plant - *Justicia carnea*
Flander's poppy - *Papaver rhoeas*
flannel
  flannel-bush - *Fremontodendron californica*
  flannel-plant - *Verbascum thapsus*
flat
  flat bog-mat - *Wolffiella floridana*
  flat-branch ground-pine - *Lycopodium obscurum*
  flat pea - *Lathyrus sylvestris*
  flat pea-vine - *Lathyrus sylvestris*
  flat-pod pea-vine - *Lathyrus cicera*
  flat sedge - *Cyperus odoratus, Cyperus*
  flat-seeded false flax - *Camelina microcarpa*
  flat-stem pondweed - *Potamogeton zosteriformis*
  flat-top white aster - *Aster umbellatus*
  flat-weed - *Hypochaeris radicata*
  flat-woods plum - *Prunus umbellata*
  jointed flat sedge - *Cyperus articulatus*
  red-root flat sedge - *Cyperus erythrorhizos*
  rice flat sedge - *Cyperus iria*
  southern flat-seed-sunflower - *Verbesina occidentalis*
  tall flat-top white aster - *Aster umbellatus*
flaver - *Avena fatua*
flawn - *Zoysia matrella*
flax
  Dutch flax - *Camelina microcarpa*

fairy flax - *Linum catharticum*
false flax - *Camelina microcarpa, Camelina*
flat-seeded false flax - *Camelina microcarpa*
flax - *Linum usitatissimum, Linum*
flax dodder - *Cuscuta epilinum*
flax-weed - *Linaria vulgaris*
flowering flax - *Linum grandiflorum*
large-seeded false flax - *Camelina sativa*
New Zealand flax - *Phormium tenax*
perennial flax - *Linum perenne*
prairie flax - *Linum perenne* subsp. *lewisii*
small-seeded false flax - *Camelina microcarpa*
Virginia yellow flax - *Linum virginianum*
western flax - *Camelina microcarpa*
western flax dodder - *Cuscuta campestris*
white flax - *Linum catharticum*
wild blue flax - *Linum grandiflorum, L. perenne*
woodland flax - *Linum virginianum*
yellow flax - *Linum virginianum*
flaxseed, water-flaxseed - *Spirodela polyrhiza*
fleabane
annual fleabane - *Erigeron annuus*
daisy fleabane - *Erigeron annuus, E. philadelphicus, E. strigosus*
dwarf fleabane - *Conyza ramosissima*
fleabane - *Conyza canadensis, Erigeron philadelphicus, Erigeron, Pulicaria dysenterica*
hairy fleabane - *Conyza bonariensis*
marsh fleabane - *Pluchea camphorata, Senecio congestus*
Oregon fleabane - *Erigeron speciosus*
Philadelphia fleabane - *Erigeron philadelphicus*
poor-Robin's fleabane - *Erigeron pulchellus*
rough daisy fleabane - *Erigeron strigosus*
rough fleabane - *Erigeron strigosus*
stinking fleabane - *Pluchea foetida*
tall fleabane - *Conyza floribunda*
fleckled milk-vetch - *Astragalus lentiginosus*
fleece
fleece-flower - *Polygonum cuspidatum, Polygonum*
golden-fleece - *Dyssodia tenuiloba*
fleshy hawthorn - *Crataegus succulenta*
fleur-de-lis - *Iris germanica, Iris*
flexible milk-vetch - *Astragalus flexuosus*
flicker-tail grass - *Hordeum glaucum*
flint corn - *Zea mays*
flix-weed - *Descurainia sophia*
float grass - *Glyceria borealis, G. fluitans*
floating
big floating-heart - *Nymphoides aquatica*
floating bladderwort - *Utricularia inflata, U. radiata*
floating fern - *Ceratopteris pteridoides*
floating foxtail - *Alopecurus geniculatus*
floating-heart - *Nymphoides*
floating-leaf pondweed - *Potamogeton natans*
floating manna grass - *Glyceria fluitans, G. septentrionalis*
floating pondweed - *Potamogeton natans*
floating sweet grass - *Glyceria fluitans*
floating water fern - *Ceratopteris thalictroides*
floating water-primrose - *Ludwigia repens*
small floating manna grass - *Glyceria borealis*
yellow floating-heart - *Nymphoides peltata*

Flodman's thistle - *Cirsium flodmanii*
floppers - *Kalanchoe pinnata*
Flora's paintbrush - *Emilia javanica*
Florence fennel - *Foeniculum vulgare*
Florida
Florida anise-tree - *Illicium floridanum*
Florida arrowroot - *Zamia pumila*
Florida beggar-weed - *Desmodium tortuosum*
Florida bladderwort - *Utricularia floridana*
Florida cabbage palm - *Sabal palmetto*
Florida chinkapin - *Castanea alnifolia*
Florida chokeberry - *Aronia prunifolia*
Florida elephant's-foot - *Elephantopus elatus*
Florida gama grass - *Tripsacum floridanum*
Florida hickory - *Carya floridana*
Florida hop-bush - *Dodonaea viscosa*
Florida mahogany - *Persea borbonia*
Florida maple - *Acer barbatum*
Florida paspalum - *Paspalum floridanum*
Florida pellitory - *Parietaria floridana*
Florida pinxter - *Rhododendron canescens*
Florida pusley - *Richardia scabra*
Florida royal palm - *Roystonea elata*
Florida silver palm - *Coccothrinax argentata*
Florida strangler fig - *Ficus aurea*
Florida tetrazygia - *Tetrazygia bicolor*
Florida toadflax - *Linaria floridana*
Florida torreya - *Torreya taxifolia*
Florida trema - *Trema micrantha*
Florida velvet bean - *Mucuna pruriens* var. *utilis*
Florida wild lettuce - *Lactuca floridana*
Florida willow - *Salix floridana*
Florida yellow wood-sorrel - *Oxalis dillenii*
Florida yew - *Taxus floridana*
florist's
florist's calla - *Zantedeschia aethiopica*
florist's chrysanthemum - *Chrysanthemum morifolium*
florist's cineraria - *Senecio cruentus*
florist's cyclamen - *Cyclamen persicum*
florist's fern - *Dryopteris carthusiana*
florist's gloxinia - *Sinningia speciosa*
florist's violet - *Viola odorata*
floss
floss-flower - *Ageratum*
floss-silk-tree - *Chorisia speciosa*
flowering
Chinese flowering crabapple - *Malus spectabilis*
eastern flowering dogwood - *Cornus florida*
flowering almond - *Prunus triloba*
flowering cabbage - *Brassica oleracea* var. *acephala*
flowcring dogwood - *Cornus florida*
flowering-fence - *Caesalpinia pulcherrima*
flowering fern - *Osmunda regalis, Osmunda*
flowering flax - *Linum grandiflorum*
flowering-maple - *Abutilon pictum, Abutilon, Viburnum acerifolium*
flowering-nettle - *Galeopsis tetrahit*
flowering quince - *Chaenomeles japonica*
flowering raspberry - *Rubus odoratus*
flowering-rush - *Butomus umbellatus*
flowering spurge - *Euphorbia corollata*
flowering tobacco - *Nicotiana alata*

flowering-willow - *Chilopsis linearis, Epilobium angustifolium*
flowering-wintergreen - *Polygala paucifolia*
Japanese flowering cherry - *Prunus serrulata*
Japanese flowering crabapple - *Malus floribunda*
late-flowering goosefoot - *Chenopodium strictum* var. *glaucophyllum*
late-flowering thoroughwort - *Eupatorium serotinum*
night-flowering catchfly - *Silene noctiflora*
purple-flowering raspberry - *Rubus odoratus*

flowers-of-love - *Tabernaemontana divaricata*

fluvellin
female fluvellin - *Kickxia spuria*
sharp-point fluvellin - *Kickxia elatine*

fly
European fly honeysuckle - *Lonicera xylosteum*
fly-away grass - *Agrostis scabra*
fly honeysuckle - *Lonicera canadensis, L. xylosteum*
mountain fly honeysuckle - *Lonicera villosa*
northern fly honeysuckle - *Lonicera villosa*
swamp fly honeysuckle - *Lonicera oblongifolia*

flytrap, Venus's-flytrap - *Dionaea muscipula*

foamflower - *Tiarella cordifolia*

fog
creeping fog-fruit - *Phyla nodiflora* var. *canescens*
fog-fruit - *Phyla nodiflora*
garden fog-fruit - *Phyla nodiflora* var. *rosea*
northern fog-fruit - *Phyla lanceolata*
old-fog - *Danthonia spicata*
saw-tooth fog-fruit - *Phyla nodiflora* var. *incisa*
wedge-leaf fog-fruit - *Phyla cuneifolia*
Yorkshire fog - *Holcus lanatus*

foil
feather-foil - *Hottonia inflata*
rock-foil - *Saxifraga*

fonio - *Digitaria exilis*

fool-hay - *Panicum capillare*

fool's-parsley - *Aethusa cynapium*

foot
bear's-foot - *Aconitum napellus*
bird's-foot buttercup - *Ranunculus orthorhynchus*
bird's-foot trefoil - *Lotus corniculatus*
bird's-foot violet - *Viola pedata*
cat's-foot - *Gnaphalium obtusifolium*
cock-foot panicum - *Echinochloa crus-galli*
cock's-foot - *Dactylis glomerata*
colt's-foot - *Tussilago farfara*
crow-foot buttercup - *Ranunculus sceleratus*
deer's-foot - *Achlys triphylla*
dove's-foot geranium - *Geranium molle*
elephant-foot-tree - *Nolina recurvata*
Florida elephant's-foot - *Elephantopus elatus*
greater bird's-foot trefoil - *Lotus pedunculatus*
hare-foot dalea - *Dalea leporina*
lion's-foot - *Prenanthes serpentaria*
pigeon's-foot - *Salicornia europaea*
purple elephant's-foot - *Elephantopus nudatus*
pussy's-foot - *Ageratum*
rabbit-foot clover - *Trifolium arvense*
rabbit's-foot - *Maranta leuconeura* var. *kerchoveana*
rabbit's-foot grass - *Polypogon monspeliensis*
rabbit's-foot polypogon - *Polypogon monspeliensis*

small-flowered bird's-foot trefoil - *Lotus micranthus*
squirrel-foot fern - *Davallia trichomanoides*
sweet-colt's-foot - *Petasites*
tangel-foot - *Viburnum alnifolium*
turkey-foot - *Andropogon gerardii, A. hallii*
white-man's-foot - *Plantago major*

foothill
foothill death camass - *Zigadenus paniculatus*
foothill needle grass - *Stipa lepida*
foothill pine - *Pinus sabiniana*

footprints, fairy-footprints - *Hedyotis procumbens*

forest
flame-of-the-forest - *Spathodea campanulata*
forest bedstraw - *Galium circaezans*
forest lousewort - *Pedicularis canadensis*
forest phlox - *Phlox divaricata*
forest red gum - *Eucalyptus tereticornis*
forest sunflower - *Helianthus decapetalus*

forestiera, desert-olive forestiera - *Forestiera phillyreoides*

forever
live-forever - *Sedum telephium, Sempervivum tectorum*
live-forever stonecrop - *Sedum telephium*

forget
Chinese forget-me-not - *Cynoglossum amabile*
field forget-me-not - *Myosotis arvensis*
forget-me-not - *Myosotis scorpioides, Myosotis*
garden forget-me-not - *Myosotis sylvatica*
smaller forget-me-not - *Myosotis laxa*
spring forget-me-not - *Myosotis verna*
true forget-me-not - *Myosotis scorpioides*
white forget-me-not - *Cryptantha intermedia*
yellow forget-me-not - *Amsinckia lycopsoides*

forked
forked chickweed - *Paronychia canadensis*
forked fringe-rush - *Fimbristylis dichotoma*

forking
forking catchfly - *Silene dichotoma*
forking larkspur - *Delphinium consolida*

Formosan
Formosan gum - *Liquidambar formosana*
Formosan rice-tree - *Fatsia japonica*

Forster's
Forster's palm - *Howeia forsteriana*
Forster's sentry palm - *Howeia forsteriana*

fountain
crimson fountain grass - *Pennisetum setaceum*
fountain-bush - *Russelia equisetiformis*
fountain-plant - *Russelia equisetiformis*

four
Colorado four-o'clock - *Mirabilis multiflora*
four-leaved milkweed - *Asclepias quadrifolia*
four-o'clock - *Mirabilis jalapa*
four-seeded vetch - *Vicia tetrasperma*
four-wing saltbush - *Atriplex canescens*
trailing four-o'clock - *Allionia incarnata*
wild four-o'clock - *Mirabilis nyctaginea*

fowl
fowl bluegrass - *Poa palustris*
fowl manna grass - *Glyceria elata, G. striata*
fowl meadow grass - *Poa palustris*

fox
  fox-berry - *Vaccinium vitis-idaea*
  fox-brush - *Centranthus ruber*
  fox-flower - *Armeria maritima* subsp. *sibirica*
  fox grape - *Vitis labrusca*
  fox sedge - *Carex vulpinoidea*
  knot-root fox-tail - *Setaria geniculata*
  southern fox grape - *Vitis rotundifolia*
foxglove
  downy false foxglove - *Aureolaria virginica*
  false foxglove - *Aureolaria pedicularia*
  foxglove - *Digitalis purpurea*
  Grecian foxglove - *Digitalis lanata*
  mullein-foxglove - *Dasistoma macrophylla*
  straw foxglove - *Digitalis lutea*
foxtail
  alpine foxtail - *Alopecurus alpinus*
  bent foxtail - *Alopecurus geniculatus*
  bristly foxtail - *Setaria verticillata*
  Carolina foxtail - *Alopecurus carolinianus*
  creeping foxtail - *Alopecurus arundinaceus*
  floating foxtail - *Alopecurus geniculatus*
  foxtail barley - *Hordeum jubatum*
  foxtail brome - *Bromus rubens*
  foxtail brome grass - *Bromus rubens*
  foxtail chess - *Bromus madritensis, B. rubens*
  foxtail dalea - *Dalea leporina*
  foxtail fescue - *Vulpia myuros, V. myuros* var. *hirsuta*
  foxtail grass - *Hordeum glaucum*
  foxtail millet - *Setaria italica*
  foxtail muhly - *Muhlenbergia andina*
  foxtail pine - *Pinus balfouriana*
  foxtail-rush - *Equisetum arvense*
  foxtail sedge - *Carex alopecoidea*
  giant foxtail - *Setaria faberi, S. magna*
  giant green foxtail - *Setaria viridis* var. *major*
  green foxtail - *Setaria viridis*
  marsh foxtail - *Alopecurus geniculatus*
  meadow foxtail - *Alopecurus pratensis*
  mouse foxtail - *Alopecurus myosuroides*
  nodding foxtail - *Setaria faberi*
  reed foxtail - *Alopecurus arundinaceus*
  Rendle's foxtail - *Alopecurus rendlei*
  robust purple foxtail - *Setaria viridis* var. *robusta-purpurea*
  robust white foxtail - *Setaria viridis* var. *robusta-alba*
  salt marsh foxtail - *Setaria magna*
  short-awn foxtail - *Alopecurus aequalis*
  slim-spike foxtail - *Alopecurus myosuroides*
  water foxtail - *Alopecurus geniculatus*
  yellow foxtail - *Setaria pumila*
fragile
  fragile cliff brake - *Cryptogramma stelleri*
  fragile fern - *Cystopteris fragilis*
fragrant
  fragrant cudweed - *Gnaphalium obtusifolium*
  fragrant everlasting - *Gnaphalium obtusifolium*
  fragrant false indigo - *Amorpha nana*
  fragrant giant hyssop - *Agastache foeniculum*
  fragrant-olive - *Osmanthus fragrans*
  fragrant plantain-lily - *Hosta plantaginea*
  fragrant sumac - *Rhus aromatica*
  fragrant thistle - *Cirsium pumilim*

fragrant water-lily - *Nymphaea odorata*
framboise - *Rubus idaeus*
frangipani - *Plumeria*
Frank's sedge - *Carex frankii*
frankincense pine - *Pinus taeda*
franklinia - *Franklinia alatamaha*
franseria, white-leaf franseria - *Ambrosia tomentosa*
Fraser's
  Fraser's balsam fir - *Abies fraseri*
  Fraser's fir - *Abies fraseri*
  Fraser's magnolia - *Magnolia fraseri*
  Fraser's sedge - *Cymophyllus fraseri*
fraxinella - *Dictamnus albus*
free-flowered waterweed - *Elodea nuttallii*
Fremont's
  Fremont's cottonwood - *Populus fremontii*
  Fremont's mahonia - *Berberis fremontii*
fremontia, California fremontia - *Fremontodendron californica*
French
  French bean - *Phaseolus vulgaris*
  French broom - *Genista monspessulanus*
  French catchfly - *Silene gallica*
  French hydrangea - *Hydrangea macrophylla*
  French marigold - *Tagetes erecta*
  French mulberry - *Callicarpa americana*
  French rose - *Rosa gallica*
  French tamarisk - *Tamarix gallica*
  French thyme - *Plectranthus amboinicus*
  French-weed - *Galinsoga quadriradiata, Thlaspi arvense*
freshwater cord grass - *Spartina pectinata*
friar's-cap - *Aconitum napellus*
friendship-plant - *Pilea involucrata*
Fries's pondweed - *Potamogeton friesii*
frijol - *Phaseolus vulgaris*
frijolito - *Sophora secundiflora*
frill, widow's-frill - *Silene stellata*
fringe
  annual fringe-rush - *Fimbristylis annua*
  black-fringe knotweed - *Polygonum cilinode*
  forked fringe-rush - *Fimbristylis dichotoma*
  fringe-cups - *Tellima grandiflora*
  fringe-leaf paspalum - *Paspalum ciliatifolium, P. setaceum* var. *ciliatifolium*
  fringe orchid - *Habenaria*
  fringe-pod - *Thysanocarpus curvipes*
  fringe sedge - *Carex crinita*
  fringe-tree - *Chionanthus virginicus*
  globe fringe-rush - *Fimbristylis littoralis, F. miliacea*
  slender fringe-rush - *Fimbristylis autumnalis*
  water fringe - *Nymphoides peltata*
fringed
  fringed brome - *Bromus ciliatus*
  fringed chloris - *Chloris ciliata*
  fringed gentian - *Gentianopsis crinita, G. procera*
  fringed loosestrife - *Lysimachia ciliata*
  fringed panicum - *Panicum ciliatum*
  fringed pigweed - *Amaranthus fimbriatus*
  fringed polygala - *Polygala paucifolia*
  fringed sagebrush - *Artemisia frigida*
  fringed signal grass - *Brachiaria ciliatissima*

smaller fringed gentian - *Gentianopsis procera*
fritillary
    fritillary - *Fritillaria*
    narrow-leaf fritillary - *Fritillaria affines*
    rice-grain fritillary - *Fritillaria affines*
frog's, American frog's-bit - *Limnobium spongia*
frost
    frost-blite - *Chenopodium album*
    frost-flower - *Aster lanceolatus, Aster*
    frost grape - *Vitis riparia, V. vulpina*
    frost-wort - *Helianthemum canadense*
frosted hawthorn - *Crataegus pruinosa*
frostweed - *Helianthemum bicknellii, H. canadense,*
    Verbesina virginica
fruit
    creeping fog-fruit - *Phyla nodiflora* var. *canescens*
    egg-fruit-tree - *Pouteria campechiana*
    fog-fruit - *Phyla nodiflora*
    fruit-salad-plant - *Monstera deliciosa*
    garden fog-fruit - *Phyla nodiflora* var. *rosea*
    hairy-fruit chervil - *Chaerophyllum tainturieri*
    Jove's-fruit - *Lindera melissaefolium*
    kiwi-fruit - *Actinidia deliciosa*
    large-fruit dodder - *Cuscuta umbrosa*
    large-fruit rose-apple - *Syzygium malaccense*
    marmalade-fruit - *Pouteria sapota*
    northern fog-fruit - *Phyla lanceolata*
    passion-fruit - *Passiflora edulis*
    red-fruit passionflower - *Passiflora foetida*
    saw-tooth fog-fruit - *Phyla nodiflora* var. *incisa*
    slender-fruit lomatium - *Lomatium bicolor* var. *leptocarpum*
    spiny-fruit buttercup - *Ranunculus muricatus*
    star-fruit - *Averrhoa carambola*
    wedge-leaf fog-fruit - *Phyla cuneifolia*
    wide-fruit sedge - *Carex angustata*
    wool-fruit sedge - *Carex lasiocarpa*
fruited
    awn-fruited sedge - *Carex stipata*
    broad-fruited bur-reed - *Sparganium eurycarpum*
    fruited eggplant - *Solanum integrifolium*
    golden-fruited sedge - *Carex aurea*
    green-fruited bur-reed - *Sparganium chlorocarpum*
    long-fruited primrose-willow - *Ludwigia octovalvis*
    many-fruited false loosestrife - *Ludwigia polycarpa*
    plum-fruited yew - *Cephalotaxus harringtonia* var. *drupacea*
    rough-fruited cinquefoil - *Potentilla recta*
    small-fruited hickory - *Carya glabra*
    small-fruited Queensland-nut - *Macadamia tetraphylla*
    sweet-fruited juniper - *Juniperus deppeana*
    three-fruited sedge - *Carex trisperma*
    tomato-fruited eggplant - *Solanum integrifolium*
    yellow-fruited-thorn - *Crataegus flava*
fry-wood - *Albizia lebbeck*
fuchsia
    California fuchsia - *Epilobium canum*
    fuchsia-flowered gooseberry - *Ribes speciosum*
fuirena, rush fuirena - *Fuirena scirpoidea*
Fuller's teasel - *Dipsacus sativus*
fulvous daylily - *Hemerocallis fulva*
fume
    slender fume-wort - *Corydalis micrantha*

yellow fume-wort - *Corydalis flavula*
fumitory
    climbing fumitory - *Adlumia fungosa*
    fumitory - *Fumaria officinalis*
fundi - *Digitaria exilis*
funeral-palm - *Cycas*
funnel-creeper - *Macfadyena unguis-cati*
furze - *Ulex europaeus*
fuzzy, Chinese fuzzy gourd - *Benincasa hispida*
gag-root - *Lobelia inflata*
gai-choi - *Brassica juncea*
gaillardia, rose-ring gaillardia - *Gaillardia pulchella*
galax - *Galax aphylla*
gale
    fern-gale - *Comptonia peregrina*
    sweet gale - *Myrica gale*
galenia, green galenia - *Galenia pubescens*
galingale
    edible galingale - *Cyperus esculentus*
    galingale - *Cyperus*
    sedge galingale - *Cyperus diandrus*
galinsoga
    hairy galinsoga - *Galinsoga quadriradiata*
    small-flowered galinsoga - *Galinsoga parviflora*
gall
    bay-gall-bush - *Ilex coriacea*
    gall-of-the-earth - *Prenanthes serpentaria, P. trifoliolata*
gallberry
    bitter gallberry - *Ilex glabra*
    gallberry - *Ilex coriacea, I. glabra*
    large gallberry - *Ilex coriacea*
    sweet gallberry - *Ilex coriacea*
galleta
    big galleta - *Hilaria rigida*
    galleta grass - *Hilaria jamesii*
gallow-grass - *Cannabis sativa*
galo, crista-de-galo - *Celosia argentea* var. *cristata*
gama
    eastern gama grass - *Tripsacum dactyloides*
    Florida gama grass - *Tripsacum floridanum*
Gambel's oak - *Quercus gambelii*
Ganges amaranth - *Amaranthus tricolor*
garbanzo - *Cicer arietinum*
garden
    garden angelica - *Angelica archangelica*
    garden asparagus - *Asparagus officinalis*
    garden balsam - *Impatiens balsamina*
    garden bean - *Phaseolus lunatus, P. vulgaris*
    garden beet - *Beta vulgaris*
    garden black currant - *Ribes nigrum*
    garden calla - *Zantedeschia aethiopica*
    garden canna - *Canna generalis*
    garden catchfly - *Silene armeria*
    garden chamomile - *Chamaemelum nobile*
    garden columbine - *Aquilegia vulgaris*
    garden coreopsis - *Coreopsis lanceolata*
    garden cosmos - *Cosmos bipinnatus*
    garden cress - *Lepidium sativum*
    garden-croton - *Codiaeum variegatum*
    garden crowfoot - *Aquilegia vulgaris*
    garden currant - *Ribes rubrum*

garden dewdrop - *Duranta erecta*
garden fog-fruit - *Phyla nodiflora* var. *rosea*
garden forget-me-not - *Myosotis sylvatica*
garden gladiola - *Gladiolus hortulanus*
garden gooseberry - *Ribes uva-crispa*
garden-heliotrope - *Valeriana officinalis*
garden hyacinth - *Hyacinthus orientalis*
garden lettuce - *Lactuca sativa*
garden loosestrife - *Lysimachia punctata, L. vulgaris*
garden mignonette - *Reseda odorata*
garden monk's-hood - *Aconitum napellus*
garden nasturtium - *Tropaeolum majus*
garden orache - *Atriplex hortensis*
garden orpine - *Sedum alboroseum*
garden pea - *Pisum sativum*
garden peony - *Paeonia lactiflora*
garden pepper - *Capsicum annuum*
garden petunia - *Petunia hybrida*
garden pink - *Dianthus plumarius*
garden plum - *Prunus domestica*
garden rocket - *Eruca vesicaria* subsp. *sativa*
garden sage - *Salvia officinalis*
garden snapdragon - *Antirrhinum majus*
garden sorrel - *Rumex acetosa*
garden spurge - *Euphorbia hirta*
garden strawberry - *Fragaria ananassa*
garden tansy - *Tanacetum vulgare*
garden thyme - *Thymus vulgaris*
garden verbena - *Verbena hybrida*
garden wolf-bane - *Aconitum napellus*
kiss-me-over-the-garden-gate - *Polygonum orientale*
gardenia
  crape-gardenia - *Tabernaemontana divaricata*
  gardenia - *Gardenia jasminoides*
garget - *Phytolacca americana*
garland
  garland chrysanthemum - *Chrysanthemum coronarium*
  garland crabapple - *Malus coronaria*
  garland-flower - *Daphne cneorum, Hedychium coronarium*
  garland-lily - *Hedychium*
garlic
  crow garlic - *Allium vineale*
  elephant garlic - *Allium ampeloprasum*
  false garlic - *Nothoscordum bivalve*
  field garlic - *Allium vineale*
  garlic - *Allium sativum*
  garlic mustard - *Alliaria petiolata*
  garlic-vine - *Cydista aequinoctialis*
  great-headed garlic - *Allium ampeloprasum*
  meadow garlic - *Allium canadense*
  Oriental garlic - *Allium tuberosum*
  society-garlic - *Tulbaghia violacea*
  stag's garlic - *Allium vineale*
  wild garlic - *Allium canadense, A. vineale*
gas-plant - *Dictamnus albus*
gate, kiss-me-over-the-garden-gate - *Polygonum orientale*
Gattinger's witch grass - *Panicum gattingeri*
gaura
  biennial gaura - *Gaura biennis*
  hairy gaura - *Gaura villosa*
  scarlet gaura - *Gaura coccinea*
  wavy-leaf gaura - *Gaura sinuata*

gay
  dotted gay-feather - *Liatris punctata*
  gay-feather - *Liatris*
gean - *Prunus avium*
Geneva bugle - *Ajuga genevensis*
genip - *Genipa, Melicoccus bijugatus*
genipap - *Genipa americana*
genipe - *Melicoccus bijugatus*
gentian
  alpine gentian - *Gentiana newberryi*
  blind gentian - *Gentiana clausa*
  bottle gentian - *Gentiana andrewsii, G. clausa*
  Catesby's gentian - *Gentiana catesbaei*
  closed gentian - *Gentiana andrewsii, G. clausa, G. linearis,*
    G. rubricaulis
  coastal plain gentian - *Gentiana catesbaei*
  downy gentian - *Gentiana saponaria*
  fringed gentian - *Gentianopsis crinita, G. procera*
  gentian - *Gentiana, Gentianella*
  Great Lakes gentian - *Gentiana rubricaulis*
  green-gentian - *Frasera speciosa*
  horse-gentian - *Triosteum*
  meadow closed gentian - *Gentiana clausa*
  Mendocino gentian - *Gentiana setigera*
  narrow-leaf gentian - *Gentiana linearis*
  one-flowered gentian - *Gentiana autumnalis*
  perfoliate horse-gentian - *Triosteum perfoliatum*
  pine-barren gentian - *Gentiana autumnalis*
  prairie bottle gentian - *Gentiana andrewsii*
  prairie-gentian - *Eustoma russellianum*
  rose-gentian - *Sabatia*
  smaller fringed gentian - *Gentianopsis procera*
  soapwort gentian - *Gentiana saponaria*
  spurred-gentian - *Halenia elliptica*
  stiff gentian - *Gentianella quinquefolia*
  striped gentian - *Gentiana villosa*
George's-lily - *Cyrtanthus elatus*
Georgia
  Georgia buckeye - *Aesculus sylvatica*
  Georgia pine - *Pinus palustris*
geranium
  bedding geranium - *Pelargonium hortorum*
  beefsteak-geranium - *Begonia rex-cultorum, Saxifraga*
    stolonifera
  black-geranium - *Heuchera americana*
  blood-red geranium - *Geranium sanguineum*
  bloody geranium - *Geranium sanguineum*
  Carolina geranium - *Geranium carolinianum*
  cut-leaf geranium - *Geranium dissectum*
  dove's-foot geranium - *Geranium molle*
  fancy geranium - *Pelargonium domesticum*
  feather-geranium - *Chenopodium botrys*
  fish geranium - *Pelargonium hortorum*
  geranium - *Pelargonium*
  geranium-leaf aralia - *Polyscias guilfoylei*
  hanging geranium - *Pelargonium peltatum*
  horseshoe geranium - *Pelargonium hortorum*
  house geranium - *Pelargonium hortorum*
  ivy geranium - *Pelargonium peltatum*
  Lady Washington's geranium - *Pelargonium domesticum*
  Martha Washington's geranium - *Pelargonium domesticum*
  meadow geranium - *Geranium pratense*

mint-geranium - *Chrysanthemum balsamita*
pansy-flowered geranium - *Pelargonium domesticum*
pole-cat-geranium - *Lantana montevidensis*
regal geranium - *Pelargonium domesticum*
Robert's geranium - *Geranium robertianum*
rock-geranium - *Heuchera americana*
rose geranium - *Pelargonium graveolens*
show geranium - *Pelargonium domesticum*
small-flowered geranium - *Geranium pusillum*
spotted geranium - *Geranium maculatum*
strawberry-geranium - *Saxifraga stolonifera*
sweet-scented geranium - *Pelargonium graveolens*
wild geranium - *Geranium maculatum*
zonal geranium - *Pelargonium hortorum*

gerardia - *Agalinis purpurea*

German
German catchfly - *Lychnis viscaria*
German ivy - *Senecio mikanioides*
German knotgrass - *Scleranthus annuus*
German millet - *Setaria italica*
German primrose - *Primula obconica*
German rampion - *Oenothera biennis*
German velvet grass - *Holcus mollis*
German violet - *Exacum affine*

germander
American germander - *Teucrium canadense*
cut-leaf germander - *Teucrium botrys*
germander-sage - *Teucrium scorodonia*
germander speedwell - *Veronica chamaedrys*
wood germander - *Teucrium scorodonia*

Geyer's
Geyer's larkspur - *Delphinium geyeri*
Geyer's onion grass - *Melica geyeri*

Ghent hybrid azalea - *Rhododendron gandavense*

gherkin
bur gherkin - *Cucumis anguria*
West Indian gherkin - *Cucumis anguria*

ghost
ghost gum - *Eucalyptus pauciflora*
ghost-weed - *Euphorbia marginata*

giant
Arizona-giant - *Carnegiea gigantea*
blue giant hyssop - *Agastache foeniculum*
catnip giant hyssop - *Agastache nepetoides*
fennel giant hyssop - *Agastache foeniculum*
fragrant giant hyssop - *Agastache foeniculum*
giant amaranth - *Amaranthus australis*
giant arborvitae - *Thuja plicata*
giant arrowhead - *Sagittaria montevidensis*
giant bird's-nest - *Pterospora andromedea*
giant bristle grass - *Setaria magna*
giant bur-reed - *Sparganium eurycarpum*
giant cactus - *Carnegiea gigantea*
giant cane - *Arundinaria gigantea, A. gigantea* subsp. *tecta*
giant-cedar - *Thuja plicata*
giant chain fern - *Woodwardia fimbriata*
giant chinquapin - *Castanopsis chrysophylla*
giant cut grass - *Zizaniopsis miliacea*
giant drop-seed - *Sporobolus giganteus*
giant duckweed - *Spirodela polyrhiza*
giant filbert - *Corylus maxima*
giant foxtail - *Setaria faberi, S. magna*

giant granadilla - *Passiflora quadrangularis*
giant green foxtail - *Setaria viridis* var. *major*
giant holly fern - *Polystichum munitum*
giant horsetail - *Equisetum telmateia*
giant hyssop - *Agastache nepetoides, Agastache*
giant knotweed - *Polygonum sachalinense*
giant mallow - *Hibiscus*
giant milkweed - *Calotropis gigantea*
giant pine - *Pinus lambertiana*
giant protea - *Protea cynaroides*
giant ragweed - *Ambrosia trifida*
giant redwood - *Sequoiadendron giganteum*
giant reed - *Arundo donax*
giant saguaro - *Carnegiea gigantea*
giant salvinia - *Salvinia auriculata, S. molesta*
giant sensitive-plant - *Mimosa invisa*
giant-sequoia - *Sequoiadendron giganteum*
giant stock bean - *Canavalia ensiformis*
giant sunflower - *Helianthus decapetalus, H. giganteus*
giant water-lily - *Victoria*
giant white-top sedge - *Rhynchospora latifolia*
giant wild rye - *Leymus cinereus*
lavender giant hyssop - *Agastache foeniculum*
nettle-leaf giant hyssop - *Agastache urticifolia*
purple giant hyssop - *Agastache scrophulariaefolia*
yellow giant hyssop - *Agastache nepetoides*

giboshi, muraski-giboshi - *Hosta ventricosa*

Gilead, balm-of-Gilead - *Populus gileadensis*

gilia
field gilia - *Gilia capitata*
granite-gilia - *Leptodactylon pungens*
skunk-weed-gilia - *Navarretia squarrosa*
woolly-gilia - *Navarretia intertexta*

gill-over-the-ground - *Glechoma hederacea*

gillyflower - *Gilia capitata, Matthiola incana*

ginberbread-tree - *Hyphaene thebaica*

ginger
butterfly-ginger - *Hedychium coronarium*
Canton ginger - *Zingiber officinale*
ginger - *Zingiber officinale, Zingiber*
ginger-lily - *Alpinia, Hedychium coronarium, Hedychium*
ginger-root - *Tussilago farfara*
Kahili ginger - *Hedychium gardnerianum*
red-ginger - *Alpinia purpurata*
shell-ginger - *Alpinia zerumbet*
spiral-ginger - *Costus malortieanus*
true ginger - *Zingiber officinale*
white-ginger - *Hedychium coronarium*
wild ginger - *Asarum canadense, Asarum*
zerumbet ginger - *Zingiber zerumbet*

gingerbread palm - *Hyphaene thebaica*

ginkgo - *Ginkgo biloba*

ginseng
American ginseng - *Panax quinquefolius*
dwarf ginseng - *Panax trifolius*
ginseng - *Panax quinquefolius, Panax*

girasole - *Helianthus tuberosus*

gizzard, chicken's-gizzard - *Iresine herbstii*

glacier-lily - *Erythronium grandiflorum*

glade mallow - *Napaea dioica*

gladiola, garden gladiola - *Gladiolus hortulanus*

gladiolus, water-gladiolus - *Butomus umbellatus*
glandular
  glandular Labrador tea - *Ledum glandulosum*
  glandular mesquite - *Prosopis glandulosa*
  glandular persicary - *Polygonum pensylvanicum*
glass
  looking-glass-plant - *Coprosma repens*
  small Venus's-looking-glass - *Triodanis biflora*
  Venus's-looking-glass - *Triodanis perfoliata*
glasswort
  dwarf glasswort - *Salicornia bigelovii*
  glasswort - *Salicornia*
  perennial glasswort - *Salicornia virginica*
  slender glasswort - *Salicornia europaea*
  starry glasswort - *Cerastium arvense*
  woody glasswort - *Salicornia virginica*
glaucous
  glaucous bristle grass - *Setaria pumila*
  glaucous hawkweed - *Hieracium floribundum*
  glaucous king-devil - *Hieracium piloselloides*
  glaucous white-lettuce - *Prenanthes racemosa*
globe
  globe-amaranth - *Gomphrena globosa*
  globe artichoke - *Cynara scolymus*
  globe candytuft - *Iberis umbellata*
  globe fringe-rush - *Fimbristylis littoralis, F. miliacea*
  globe mallow - *Sphaeralcea*
  globe-podded hoary-cress - *Cardaria pubescens*
  globe sedge - *Cyperus globulosus*
  globe thistle - *Echinops*
  globe-tulip - *Calochortus*
  great globe thistle - *Echinops sphaerocephalus*
  narrow-leaf globe mallow - *Sphaeralcea angustifolia*
  scarlet globe mallow - *Sphaeralcea coccinea*
globeflower
  globeflower - *Kerria japonica, Trollius*
  spreading globeflower - *Trollius laxus*
globifera - *Micranthemum umbrosum*
globose cyperus - *Cyperus ovularis*
gloriosa-lily - *Gloriosa superba*
glory
  beach morning-glory - *Ipomoea pes-caprae*
  big-root morning-glory - *Ipomoea pandurata*
  bleeding-glory-bower - *Clerodendrum thompsoniae*
  blue morning-glory - *Ipomoea indica*
  bush morning-glory - *Ipomoea leptophylla*
  Cairo morning-glory - *Ipomoea cairica*
  cotton morning-glory - *Ipomoea cordatotriloba* var. *torreyana*
  cypress-vine morning-glory - *Ipomoea quamoclit*
  glory-bower - *Clerodendrum*
  glory-bush - *Tibouchina urvilleana*
  glory-of-the-sun - *Leucocoryne ixioides*
  glory-tree - *Clerodendrum thompsoniae*
  ivy-leaf morning-glory - *Ipomoea hederacea*
  ivy-leaf red morning-glory - *Ipomoea hederifolia*
  Japanese morning-glory - *Ipomoea nil*
  morning-glory - *Ipomoea purpurea, Ipomoea*
  mountain-glory - *Holodiscus dumosus*
  multicolored morning-glory - *Ipomoea tricolor*
  orchard morning-glory - *Convolvulus arvensis*
  palm-leaf morning-glory - *Ipomoea wrightii*
  pitted morning-glory - *Ipomoea lacunosa*

  purple glory-tree - *Tibouchina urvilleana*
  red morning-glory - *Ipomoea coccinea*
  sharp-pod morning-glory - *Ipomoea cordatotriloba*
  small-flowered morning-glory - *Convolvulus arvensis, Jacquemontia tamnifolia*
  small red morning-glory - *Ipomoea coccinea*
  small white morning-glory - *Ipomoea lacunosa*
  star-glory - *Ipomoea quamoclit*
  swamp morning-glory - *Ipomoea aquatica*
  tall morning-glory - *Ipomoea purpurea*
  three-lobed morning-glory - *Ipomoea triloba*
  white-edge morning-glory - *Ipomoea nil*
  white morning-glory - *Ipomoea lacunosa*
  wild morning-glory - *Calystegia sepium, Convolvulus arvensis*
  woolly morning-glory - *Argyreia nervosa*
glossy
  glossy abelia - *Abelia grandiflora*
  glossy buckthorn - *Rhamnus frangula*
  glossy-leaved paper-plant - *Fatsia japonica*
  glossy privet - *Ligustrum lucidum*
glow
  golden-glow - *Rudbeckia laciniata*
  orange-glow vine - *Senecio chenopodioides*
gloxinia
  Brazilian gloxinia - *Sinningia speciosa*
  florist's gloxinia - *Sinningia speciosa*
  violet-slipper gloxinia - *Sinningia speciosa*
Goa
  cedar-of-Goa - *Cupressus lusitanica*
  Goa bean - *Psophocarpus tetragonolobus*
goar-berry gourd - *Cucumis anguria*
goat
  barbed goat grass - *Aegilops triuncialis*
  goat grass - *Aegilops*
  goat-nut - *Simmondsia chinensis*
  goat-weed - *Hypericum perforatum*
  goat willow - *Salix caprea*
  jointed goat grass - *Aegilops cylindrica*
goat's
  false goat's-beard - *Astilbe biternata*
  goat's-beard - *Aruncus dioicus, A. sylvester, Tragopogon*
  goat's-rue - *Galega officinalis*
goblet aster - *Aster lateriflorus*
gobo - *Arctium lappa*
God, finger-of-God - *Aechmea orlandiana*
godetia - *Clarkia*
gokorna - *Clitoria ternatea*
gold
  basket-of-gold - *Alyssum saxatile*
  California gold-bark fern - *Pityrogramma triangularis*
  flame-gold - *Koelreuteria elegans*
  gold-and-silver-flower - *Lonicera japonica*
  gold-apple - *Lycopersicon esculentum*
  gold-banded lily - *Lilium auratum*
  gold-dust - *Alyssum saxatile*
  gold-dust dracaena - *Dracaena surculosa*
  gold-fields - *Lasthenia chrysostoma*
  gold-moss - *Sedum acre*
  gold-thread - *Coptis groenlandica, Coptis*
  gold-thread-vine - *Cuscuta gronovii*

Jamaican gold fern - *Pityrogramma sulphurea*
spring-gold - *Lomatium utriculatum*
tree-of-gold - *Tabebuia argentea*
golden
  Brazilian golden-vine - *Stigmaphyllon ciliatum*
  Drummond's golden-weed - *Haplopappus drummondii*
  falcate golden aster - *Chrysopsis falcata*
  golden alexanders - *Zizia aurea*
  golden aster - *Chrysopsis*
  golden bamboo - *Phyllostachys aurea*
  golden-bells - *Forsythia viridissima*
  golden-brodiaea - *Triteleia ixioides*
  golden-buttons - *Tanacetum vulgare*
  golden calla-lily - *Zantedeschia elliottiana*
  golden cane palm - *Chrysalidocarpus lutescens*
  golden-carpet - *Sedum acre*
  golden Ceylon creeper - *Epipremnum aureum*
  golden-chain - *Laburnum anagyroides*
  golden chinquapin - *Castanopsis chrysophylla*
  golden cleome - *Cleome lutea*
  golden-club - *Orontium aquaticum*
  golden-column - *Echinopsis spachiana*
  golden corydalis - *Corydalis aurea*
  golden-creeper - *Stigmaphyllon ciliatum*
  golden-crown daisy - *Verbesina encelioides*
  golden currant - *Ribes aureum*
  golden dewdrop - *Duranta erecta*
  golden dracaena - *Pleomele aurea*
  golden feather palm - *Chrysalidocarpus lutescens*
  golden fig - *Ficus aurea*
  golden-fleece - *Dyssodia tenuiloba*
  golden-flower - *Hypericum calycinum*
  golden-fruited sedge - *Carex aurea*
  golden-glow - *Rudbeckia laciniata*
  golden gram - *Vigna radiata*
  golden hard-hack - *Potentilla fruticosa*
  golden-heather - *Hudsonia ericoides*
  golden hedge-hyssop - *Gratiola aurea*
  golden larch - *Pseudolarix amabilis*
  golden lungwort - *Hieracium murorum*
  golden marquerite - *Anthemis tinctoria*
  golden oat grass - *Trisetum flavescens*
  golden-pert - *Gratiola aurea*
  golden pothos - *Epipremnum aureum*
  golden ragwort - *Senecio aureus*
  golden-rain - *Cassia fistula*
  golden-rain-tree - *Koelreuteria paniculata, Koelreuteria*
  golden-rayed lily - *Lilium auratum*
  golden-shower - *Cassia fistula, Pyrostegia venusta*
  golden-slipper - *Cypripedium calceolus* var. *pubescens*
  golden St. John's-wort - *Hypericum frondosum*
  golden-top - *Lamarckia aurea*
  golden-trumpet - *Allamanda cathartica*
  golden-tuft alyssum - *Alyssum saxatile*
  golden-tuft madwort - *Alyssum saxatile*
  golden-weed - *Haplopappus, Isocoma coronopifolia*
  golden willow - *Salix alba* var. *vitellina*
  grass-leaf golden aster - *Chrysopsis graminifolia*
  Maryland golden aster - *Chrysopsis mariana*
  prairie golden aster - *Chrysopsis camporum*
  shaggy golden aster - *Chrysopsis mariana*
  sickle-leaf golden aster - *Chrysopsis falcata*
goldeneye - *Viguiera annua*

goldenrod
  axillary goldenrod - *Solidago caesia*
  blue-stem goldenrod - *Solidago caesia*
  bog goldenrod - *Solidago uliginosa*
  Boott's goldenrod - *Solidago boottii*
  broad-leaf goldenrod - *Solidago flexicaulis*
  California goldenrod - *Solidago californica*
  Canadian goldenrod - *Solidago canadensis*
  coastal swamp goldenrod - *Solidago elliottii*
  early goldenrod - *Solidago juncea*
  Elliott's goldenrod - *Solidago elliottii*
  elm-leaf goldenrod - *Solidago ulmifolia*
  erect goldenrod - *Solidago erecta*
  goldenrod - *Solidago*
  gray goldenrod - *Solidago nemoralis*
  hairy goldenrod - *Solidago hispida*
  hairy piney-woods goldenrod - *Solidago flexicaulis*
  hard-leaf goldenrod - *Solidago rigida*
  hollow goldenrod - *Solidago fistulosa*
  late goldenrod - *Solidago gigantea*
  Leavenworth's goldenrod - *Solidago leavenworthii*
  licorice goldenrod - *Solidago odora*
  Missouri goldenrod - *Solidago missouriensis*
  northern bog goldenrod - *Solidago uliginosa*
  Ohio goldenrod - *Solidago ohioensis*
  pine-barren goldenrod - *Solidago fistulosa*
  rayless goldenrod - *Haplopappus heterophyllus*
  rigid goldenrod - *Solidago rigida*
  rough-leaf goldenrod - *Solidago patula*
  seaside goldenrod - *Solidago sempervirens*
  showy goldenrod - *Solidago speciosa*
  slender goldenrod - *Solidago erecta*
  small-headed goldenrod - *Solidago minor*
  stiff goldenrod - *Solidago rigida*
  sweet goldenrod - *Solidago odora*
  tall goldenrod - *Solidago canadensis, S. canadensis* var.
    scabra
  white goldenrod - *Solidago bicolor*
  wreath goldenrod - *Solidago caesia*
  zigzag goldenrod - *Solidago flexicaulis*
goldenseal - *Hydrastis canadensis*
Gomuti palm - *Arenga pinnata*
goober - *Arachis hypogaea*
good
  all-good - *Chenopodium bonus-henricus*
  good-King-Henry - *Chenopodium bonus-henricus*
  good-luck-leaf - *Kalanchoe pinnata*
  good-luck palm - *Chamaedorea elegans*
  good-luck-plant - *Cordyline terminalis*
Goodding's willow - *Salix gooddingii*
goose
  goose grass - *Eleusine indica*
  goose plum - *Prunus americana*
  goose-tansy - *Potentilla anserina*
  goose-weed - *Sphenoclea zeylandica*
  wild goose plum - *Prunus hortulana, P. munsoniana*
gooseberry
  Barbados gooseberry - *Physalis peruviana*
  canyon gooseberry - *Ribes menziesii*
  Cape gooseberry - *Physalis peruviana*
  Chinese gooseberry - *Actinidia deliciosa*
  country gooseberry - *Averrhoa carambola*

dwarf Cape gooseberry - *Physalis pubescens* var. *grisea*
English gooseberry - *Ribes uva-crispa*
European gooseberry - *Ribes grossularis, R. uva-crispa*
fuchsia-flowered gooseberry - *Ribes speciosum*
garden gooseberry - *Ribes uva-crispa*
gooseberry - *Ribes*
gooseberry gourd - *Cucumis anguria*
gooseberry mallee - *Eucalyptus calycogona*
gooseberry-tomato - *Physalis peruviana*
gummy gooseberry - *Ribes lobbii*
hairy-stem gooseberry - *Ribes hirtellum*
hawthorn-leaved gooseberry - *Ribes oxyacanthoides*
Missouri gooseberry - *Ribes missouriense, R. setosum*
northern gooseberry - *Ribes oxyacanthoides*
pasture gooseberry - *Ribes cynosbati*
prickly gooseberry - *Ribes cynosbati*
sierra gooseberry - *Ribes roezlii*
swamp gooseberry - *Ribes lacustre*
white-stem gooseberry - *Ribes divaricatum*

goosefoot
blite goosefoot - *Chenopodium capitatum*
city goosefoot - *Chenopodium urbicum*
fig-leaved goosefoot - *Chenopodium ficifolium*
goosefoot - *Chenopodium album, Chenopodium*
Jerusalem oak goosefoot - *Chenopodium botrys*
late-flowering goosefoot - *Chenopodium strictum* var. *glaucophyllum*
many-seeded goosefoot - *Chenopodium polyspermum*
maple-leaf goosefoot - *Chenopodium hybridum, C. simplex*
mealy goosefoot - *Chenopodium incanum*
Missouri goosefoot - *Chenopodium missouriense*
nettle-leaf goosefoot - *Chenopodium murale*
oak-leaf goosefoot - *Chenopodium glaucum*
perennial goosefoot - *Chenopodium bonus-henricus*
pit-seed goosefoot - *Chenopodium berlandieri*
red goosefoot - *Chenopodium rubrum*
stinking goosefoot - *Chenopodium vulvaria*
turnpike goosefoot - *Chenopodium botrys*
white goosefoot - *Chenopodium album*

gopher
gopher-plant - *Euphorbia lathyris*
gopher spurge - *Euphorbia lathyris*
gopher-tail love grass - *Eragrostis ciliaris*

Gordon's bladder-pod - *Lesquerella gordonii*
gorse - *Ulex europaeus*
gosmore - *Hypochaeris radicata*

gourd
ash gourd - *Benincasa hispida*
bitter gourd - *Momordica charantia*
bottle gourd - *Lagenaria siceraria*
buffalo gourd - *Cucurbita foetidissima*
calabash gourd - *Lagenaria siceraria*
Chinese fuzzy gourd - *Benincasa hispida*
club gourd - *Trichosanthes anguina*
dishcloth gourd - *Luffa aegyptiaca*
finger-leaf gourd - *Cucurbita digitata*
goar-berry gourd - *Cucumis anguria*
gooseberry gourd - *Cucumis anguria*
gourd - *Cucurbita*
hedgehog gourd - *Cucumis dipsaceus*
Missouri gourd - *Cucurbita foetidissima*
snake gourd - *Trichosanthes anguina*
sponge gourd - *Luffa aegyptiaca*

teasel gourd - *Cucumis dipsaceus*
Texas gourd - *Cucurbita texana*
towel gourd - *Luffa acutangula*
viper's-gourd - *Trichosanthes anguina*
wax gourd - *Benincasa hispida*
white-flowered gourd - *Lagenaria siceraria*
white gourd - *Benincasa hispida*
wild gourd - *Cucumis dipsaceus, Cucurbita foetidissima*
yellow-flowered gourd - *Cucurbita pepo* var. *ovifera*

gout
bishop's gout-weed - *Aegopodium podagraria*
gout-weed - *Aegopodium podagraria*

governor's-plum - *Flacourtia indica*

gow
gow - *Lycium barbarum*
gow-kee - *Lycium barbarum*

gowan, yellow-gowan - *Ranunculus repens*
grace, herb-of-grace - *Ruta graveolens*
graceful sedge - *Carex gracillima*

grain
grain sorghum - *Sorghum bicolor*
rice-grain fritillary - *Fritillaria affines*

gram
Bengal gram - *Cicer arietinum*
black gram - *Vigna mungo*
golden gram - *Vigna radiata*
green gram - *Vigna radiata*
horse-gram - *Macrotyloma uniflorum*
yellow gram - *Cicer arietinum*

grama
black grama - *Bouteloua eriopoda*
blue grama - *Bouteloua gracilis*
grama - *Bouteloua*
hairy grama - *Bouteloua hirsuta*
mat grama - *Bouteloua simplex*
needle grama - *Bouteloua aristidoides*
purple grama - *Bouteloua radicosa*
Rothrock's grama - *Bouteloua rothrockii*
side-oats grama - *Bouteloua curtipendula*
six-weeks grama - *Bouteloua barbata*
six-weeks needle grama - *Bouteloua aristidoides*

granadilla
giant granadilla - *Passiflora quadrangularis*
purple granadilla - *Passiflora edulis*
yellow granadilla - *Passiflora laurifolia*

grand fir - *Abies grandis*
Grande, Rio Grande palmetto - *Sabal mexicana*
grandmother, devil's-grandmother - *Elephantopus tomentosus*
granite-gilia - *Leptodactylon pungens*
granjeno - *Celtis pallida*

grape
American grape - *Vitis labrusca*
bear-grape - *Arctostaphylos uva-ursi*
bird grape - *Vitis munsoniana*
Brazilian grape tree - *Myrciaria cauliflora*
bullace grape - *Vitis rotundifolia*
bunch grape - *Vitis aestivalis*
bush grape - *Vitis acerifolia, V. rupestris*
California wild grape - *Vitis californica*
canyon grape - *Vitis arizonica*

cat grape - *Vitis palmata*
catbird grape - *Vitis palmata*
chicken grape - *Vitis vulpina*
cultivated grape - *Vitis vinifera*
European grape - *Vitis vinifera*
fox grape - *Vitis labrusca*
frost grape - *Vitis riparia, V. vulpina*
grape - *Vitis*
grape fern - *Botrychium*
grape honeysuckle - *Lonicera prolifera*
grape-hyacinth - *Muscari botryoides*
grape ivy - *Cissus rhombifolia, Cissus*
gray-black grape - *Vitis cinerea*
lance-leaf grape fern - *Botrychium lanceolatum*
little grape fern - *Botrychium simplex*
mountain grape - *Vitis rupestris*
muscadine grape - *Vitis rotundifolia*
mustang grape - *Vitis mustangensis*
New England grape - *Vitis novae-angliae*
Oregon grape - *Berberis aquifolium*
pigeon grape - *Vitis aestivalis, V. cinerea*
possum grape - *Cissus verticillata, Vitis baileyana*
post-oak grape - *Vitis lincecumii*
racoon-grape - *Ampelopsis cordata*
red grape - *Vitis palmata*
river bank grape - *Vitis riparia*
rock grape - *Vitis rupestris*
sand grape - *Vitis rupestris*
scuppernong grape - *Vitis rotundifolia*
sea-grape - *Coccoloba uvifera*
silver-leaf grape - *Vitis aestivalis* var. *argentifolia*
skunk grape - *Vitis labrusca*
southern fox grape - *Vitis rotundifolia*
Spanish grape - *Vitis berlandieri*
stunt grape - *Vitis labrusca*
sugar grape - *Vitis rupestris*
summer grape - *Vitis aestivalis, V. aestivalis* var. *argentifolia*
western wild grape - *Vitis californica*
wine grape - *Vitis vinifera*
winter grape - *Vitis berlandieri, V. vulpina*

grapefruit - *Citrus paradisi*

grass
    African Bermuda grass - *Cynodon transvaalensis*
    Alaska onion grass - *Melica subulata*
    Aleppo grass - *Sorghum halepense*
    Alexander grass - *Brachiaria plantaginea*
    alkali cord grass - *Spartina gracilis*
    alkali grass - *Puccinellia distans, Puccinellia*
    American beach grass - *Ammophila breviligulata*
    American dune grass - *Leymus mollis*
    American grass-of-Parnassus - *Parnassia glauca*
    American manna grass - *Glyceria grandis*
    American slough grass - *Beckmannia syzigachne*
    annual beard grass - *Polypogon monspeliensis*
    annual hair grass - *Deschampsia danthonioides*
    annual rye grass - *Lolium multiflorum*
    aparejo grass - *Muhlenbergia utilis*
    Arizona three-awn grass - *Aristida arizonica*
    arrow grass - *Triglochin*
    Australian rye grass - *Lolium multiflorum*
    autumn bent grass - *Agrostis perennans*
    awnless brome grass - *Bromus inermis*
    Bahia grass - *Paspalum notatum*
    Bahia love grass - *Eragrostis bahiensis*
    barbed goat grass - *Aegilops triuncialis*
    barley grass - *Hordeum murinum*
    barn grass - *Echinochloa crus-galli*
    barnyard grass - *Echinochloa crus-galli*
    basket grass - *Oplismenus hirtellus*
    bayonet-grass - *Scirpus maritimus*
    bear-grass - *Dasylirion, Nolina microcarpa, Nolina, Xerophyllum tenax, Yucca filamentosa*
    beard grass - *Andropogon gerardii, A. virginicus, Andropogon, Polypogon*
    beardless wheat grass - *Pseudoroegneria spicata*
    beggar-tick grass - *Aristida orcuttiana*
    Bengal grass - *Setaria italica*
    bent-awn plume grass - *Saccharum contortum*
    bent grass - *Agrostis capillaris, Agrostis*
    Bermuda grass - *Cynodon dactylon*
    big carpet grass - *Axonopus furcatus*
    big cord grass - *Spartina cynosuroides*
    big-top love grass - *Eragrostis hirsuta*
    billion-dollar grass - *Echinochloa frumentacea*
    bird-seed grass - *Phalaris canariensis*
    black bent grass - *Agrostis gigantea*
    black grass - *Alopecurus myosuroides*
    black oat grass - *Stipa avenacea*
    black-seed needle grass - *Stipa avenacea*
    blue bunch grass - *Festuca idahoensis*
    blue-bunch wheat grass - *Pseudoroegneria spicata*
    blue-eyed-grass - *Sisyrinchium*
    Boer's love grass - *Eragrostis curvula*
    bonnet grass - *Danthonia spicata*
    Brahman grass - *Dichanthium annulatum*
    bristle grass - *Setaria*
    broad-leaf signal grass - *Brachiaria platyphylla*
    brome grass - *Bromus*
    brook grass - *Catabrosa*
    broom beard grass - *Schizachyrium scoparium*
    brown bent grass - *Agrostis canina, A. perennans*
    brown plume grass - *Saccharum brevibarbe*
    buffalo grass - *Buchloe dactyloides, Paspalum conjugatum, Pennisetum ciliare*
    bulb panic grass - *Panicum bulbosum*
    bulbous canary grass - *Phalaris aquatica*
    bunch grass - *Schizachyrium scoparium*
    bur bristle grass - *Setaria verticillata*
    bur grass - *Cenchrus incertus*
    bushy beard grass - *Andropogon glomeratus*
    California needle grass - *Stipa cernua, S. pulchra*
    California oat grass - *Danthonia californica*
    California sweet grass - *Hierochloe occidentalis*
    canary grass - *Phalaris canariensis, Phalaris*
    cane beard grass - *Bothriochloa barbinodis*
    carib grass - *Eriochloa polystachya*
    Carolina canary grass - *Phalaris caroliniana*
    Carolina whitlow-grass - *Draba reptans*
    carpet grass - *Axonopus affinis*
    cat-tail grass - *Setaria pallide-fusca*
    catchfly grass - *Leersia lenticularis*
    centipede grass - *Eremochloa ophiuroides*
    cheat grass - *Bromus secalinus, B. tectorum*
    China grass - *Boehmeria nivea*
    Chinese cup grass - *Eriochloa villosa*
    cock grass - *Bromus secalinus*

cockspur grass - *Echinochloa crus-galli*
coco-grass - *Cyperus rotundus*
cogon grass - *Imperata cylindrica*
colonial bent grass - *Agrostis capillaris*
Columbia needle grass - *Stipa columbiana*
Columbus grass - *Sorghum almum*
cord grass - *Spartina*
cotton grass - *Eriophorum*
creeping bent grass - *Agrostis stolonifera*
creeping soft grass - *Holcus mollis*
creeping velvet grass - *Holcus mollis*
crested dog-tail grass - *Cynosurus cristatus*
crested wheat grass - *Agropyron cristatum*
crimson fountain grass - *Pennisetum setaceum*
crinkled hair grass - *Deschampsia flexuosa*
crowfoot grass - *Dactyloctenium aegyptium, Digitaria
    sanguinalis, Eleusine indica*
cup grass - *Echinochloa*
curly-mesquite grass - *Hilaria belangeri*
cut grass - *Leersia*
Dallis grass - *Paspalum dilatatum*
deer grass - *Muhlenbergia rigens, Rhexia*
desert crested wheat grass - *Agropyron desertorum*
desert needle grass - *Stipa speciosa*
desert salt grass - *Distichlis stricta*
devil grass - *Cynodon dactylon*
devil's grass - *Elytrigia repens*
ditch beard grass - *Polypogon interruptus*
dog-tooth grass - *Cynodon dactylon*
downy brome grass - *Bromus tectorum*
downy oat grass - *Danthonia sericea*
dune bent grass - *Agrostis pallens*
dwarf meadow grass - *Poa annua*
dwarf umbrella-grass - *Fuirena pumila*
early water grass - *Echinochloa oryzoides*
eastern gama grass - *Tripsacum dactyloides*
eastern manna grass - *Glyceria septentrionalis*
eel-grass pondweed - *Potamogeton zosteriformis*
Egyptian grass - *Dactyloctenium aegyptium*
elephant grass - *Pennisetum purpureum*
Elliott's beard grass - *Andropogon elliottii*
Elliott's love grass - *Eragrostis elliottii*
English rye grass - *Lolium perenne*
eulalia grass - *Miscanthus sinensis*
European alkali grass - *Puccinellia distans*
European beach grass - *Ammophila arenaria*
European dune grass - *Leymus arenarius*
fairway crested wheat grass - *Agropyron cristatum*
faitours-grass - *Euphorbia esula*
fall panic grass - *Panicum dichotomiflorum*
false buffalo grass - *Munroa squarrosa*
false wheat grass - *Leymus chinensis*
feather bunch grass - *Stipa viridula*
feather finger grass - *Chloris virgata*
feather grass - *Stipa*
feather tumble grass - *Eragrostis tenella*
fever grass - *Cymbopogon citratus*
field scorpion-grass - *Myosotis arvensis*
finger grass - *Chloris, Digitalis, Digitaria ischaemum*
flicker-tail grass - *Hordeum glaucum*
float grass - *Glyceria borealis, G. fluitans*
floating manna grass - *Glyceria fluitans, G. septentrionalis*
floating sweet grass - *Glyceria fluitans*

Florida gama grass - *Tripsacum floridanum*
fly-away grass - *Agrostis scabra*
foothill needle grass - *Stipa lepida*
fowl manna grass - *Glyceria elata, G. striata*
fowl meadow grass - *Poa palustris*
foxtail brome grass - *Bromus rubens*
foxtail grass - *Hordeum glaucum*
freshwater cord grass - *Spartina pectinata*
fringed signal grass - *Brachiaria ciliatissima*
galleta grass - *Hilaria jamesii*
gallow-grass - *Cannabis sativa*
Gattinger's witch grass - *Panicum gattingeri*
German velvet grass - *Holcus mollis*
Geyer's onion grass - *Melica geyeri*
giant bristle grass - *Setaria magna*
giant cut grass - *Zizaniopsis miliacea*
glaucous bristle grass - *Setaria pumila*
goat grass - *Aegilops*
golden oat grass - *Trisetum flavescens*
goose grass - *Eleusine indica*
gopher-tail love grass - *Eragrostis ciliaris*
grass-leaf golden aster - *Chrysopsis graminifolia*
grass-leaf lettuce - *Lactuca graminifolia*
grass-leaf pondweed - *Potamogeton gramineus*
grass-leaved sagittaria - *Sagittaria graminea*
grass-nut - *Arachis hypogaea, Triteleia laxa*
grass-of-Parnassus - *Parnassia californica, Parnassia*
grass pea-vine - *Lathyrus sativus*
grass pink - *Dianthus armeria, D. plumarius*
grass-widow - *Sisyrinchium douglasii*
green bristle grass - *Setaria viridis*
green needle grass - *Stipa viridula*
Grisebach's bristle grass - *Setaria grisebachii*
guinea grass - *Panicum maximum*
hair grass - *Agrostis hyemalis, A. scabra, Deschampsia,
    Muhlenbergia capillaris*
hard grass - *Sclerochloa dura*
Harding's grass - *Phalaris aquatica, P. stenoptera*
head's grass - *Phleum pratense*
hedgehog dog-tail grass - *Cynosurus echinatus*
hedgehog grass - *Cenchrus echinatus*
hood canary grass - *Phalaris paradoxa*
hurricane grass - *Bothriochloa pertusa*
Indian grass - *Sorghastrum nutans*
Indian love grass - *Eragrostis pilosa*
intermediate oat grass - *Danthonia intermedia*
intermediate wheat grass - *Elytrigia intermedia*
Italian rye grass - *Lolium multiflorum*
itch grass - *Rottboellia cochinchinensis*
Japanese carpet grass - *Zoysia matrella*
Japanese false brome grass - *Brachypodium pinnatum*
Japanese lawn grass - *Zoysia japonica*
Japanese plume grass - *Miscanthus sinensis*
jaragua grass - *Hyparrhenia rufa*
Johnson grass - *Sorghum halepense*
jointed goat grass - *Aegilops cylindrica*
June grass - *Danthonia spicata, Koeleria pyrimidata, Poa
    pratensis*
jungle rice grass - *Echinochloa colona, E. crus-galli*
Kikuyu grass - *Pennisetum clandestinum*
Kleberg's grass - *Dichanthium annulatum*
Klein's grass - *Panicum coloratum*
knot-root bristle grass - *Setaria geniculata*

Korean grass - *Zoysia japonica*
Korean lawn grass - *Zoysia japonica*
kyasuma grass - *Pennisetum pedicellatum*
lace grass - *Eragrostis capillaris*
late water grass - *Echinochloa phyllopogon*
lazy-man's grass - *Eremochloa ophiuroides*
lead-grass - *Salicornia virginica*
Lehmann's love grass - *Eragrostis lehmanniana*
Lemmon's needle grass - *Stipa lemmonii*
Letterman's needle grass - *Stipa lettermanii*
limpo grass - *Hemarthria altissima*
little love grass - *Eragrostis poaeoides*
little quaking grass - *Briza minor*
little rice grass - *Oryzopsis exigua*
little-seed canary grass - *Phalaris minor*
little-seed rice grass - *Oryzopsis micrantha*
liver-seed grass - *Urochloa panicoides*
long-tongue mutton grass - *Poa fendleriana*
love grass - *Eragrostis*
love grass petticoat-climber - *Eragrostis tenella*
low spear grass - *Poa annua*
Lyme grass - *Elymus, Leymus arenarius*
Lyme rye grass - *Lolium perenne*
Malabar grass - *Cymbopogon flexuosus*
Manila grass - *Zoysia matrella*
marram grass - *Ammophila arenaria*
marsh arrow grass - *Triglochin palustris*
marsh grass - *Scolochloa festucacea, Spartina*
Mary's grass - *Microstegium vimineum* var. *imberbe*
Mascarene grass - *Zoysia tenuifolia*
mat-grass - *Phyla nodiflora*
May grass - *Phalaris caroliniana*
Mediterranean grass - *Schismus barbatus*
Mediterranean love grass - *Eragrostis barrelieri*
mesquite grass - *Bouteloua*
Mexican love grass - *Eragrostis mexicana*
milk-grass - *Valerianella locusta*
mission grass - *Pennisetum polystachyon*
molasses grass - *Melinis minutiflora*
Montana blue-eyed-grass - *Sisyrinchium montanum*
moor grass - *Molinia caerulea*
mountain bunch grass - *Festuca viridula*
mountain grass - *Oryzopsis*
mountain hair grass - *Deschampsia atropurpurea*
napier grass - *Pennisetum purpureum*
narrow-leaf blue-eyed-grass - *Sisyrinchium angustifolium*
narrow-leaf signal grass - *Brachiaria piligera*
narrow reed grass - *Calamagrostis neglecta*
Natal grass - *Rhynchelytrum repens*
needle-and-thread grass - *Stipa comata*
needle grass - *Stipa sibirica, Stipa*
New Mexico feather grass - *Stipa neomexicana*
nodding semaphore grass - *Pleuropogon refractus*
northern manna grass - *Glyceria borealis*
northern nut-grass - *Cyperus esculentus*
northern reed grass - *Calamagrostis neglecta*
nut-grass - *Cyperus rotundus*
Nuttall's alkali grass - *Puccinellia nuttalliana*
oat grass - *Danthonia*
old-witch grass - *Panicum capillare*
one-spike oat grass - *Danthonia unispicata*
onion grass - *Melica bulbosa*
orange-grass - *Hypericum gentianoides*

orchard grass - *Dactylis glomerata*
Oregon bent grass - *Agrostis oregonensis*
Pacific reed grass - *Calamagrostis nutkaensis*
palm grass - *Setaria palmifolia*
pampas grass - *Cortaderia selloana*
pangola grass - *Digitaria eriantha* subsp. *pentzii*
panic grass - *Echinochloa crus-galli, Panicum anceps,*
   P. maximum, Panicum
Pará grass - *Panicum purpurascens*
Para grass - *Brachiaria mutica*
Parry's oat grass - *Danthonia parryi*
penny-grass - *Thlaspi*
Pensacola Bahia grass - *Paspalum notatum* var. *saurae*
perennial quaking grass - *Briza media*
perennial ray grass - *Lolium perenne*
perennial rye grass - *Lolium perenne*
pigeon grass - *Digitaria sanguinalis, Setaria verticillata*
pin grass - *Erodium cicutarium*
pine grass - *Calamagrostis rubescens*
pinyon rice grass - *Piptochaetium fimbriatum*
pit-scale grass - *Hackelochloa granularis*
pitted beard grass - *Bothriochloa pertusa*
plains bristle grass - *Setaria macrostachya*
plains love grass - *Eragrostis intermedia*
plains reed grass - *Calamagrostis montanensis*
poison rye grass - *Lolium temulentum*
polar grass - *Arctagrostis latifolia*
porcupine grass - *Stipa spartea*
poverty grass - *Aristida dichotoma, Danthonia spicata,*
   Sporobolus vaginiflorus
poverty oat grass - *Danthonia spicata*
prairie bayonet-grass - *Scirpus maritimus*
prairie beard grass - *Schizachyrium scoparium*
prairie cord grass - *Spartina pectinata*
prairie cup grass - *Eriochloa contracta*
prairie grass - *Bromus catharticus*
prairie wedge grass - *Sphenopholis obtusata*
Pringle's needle grass - *Stipa pringlei*
pubescent wheat grass - *Elytrigia intermedia*
pudding-grass - *Hedeoma pulegioides*
purple-eyed-grass - *Sisyrinchium douglasii*
purple love grass - *Eragrostis spectabilis*
purple needle grass - *Stipa pulchra*
purple onion grass - *Melica spectabilis*
purple reed grass - *Calamagrostis purpurascens*
purple sand grass - *Triplasis purpurea*
quack salvers-grass - *Euphorbia cyparissias*
quaking grass - *Briza media*
quick grass - *Elytrigia repens*
quitch grass - *Elytrigia repens*
rabbit-tail grass - *Lagurus ovatus*
rabbit's-foot grass - *Polypogon monspeliensis*
radiate finger grass - *Chloris radiata*
rattlesnake manna grass - *Glyceria canadensis*
Ravenna's grass - *Saccharum ravennae*
red-head-grass - *Potamogeton perfoliatus*
red love grass - *Eragrostis secundiflora*
reed canary grass - *Phalaris arundinacea*
reed grass - *Calamagrostis*
rescue grass - *Bromus catharticus*
Rhode Island bent grass - *Agrostis capillaris*
Rhodes's grass - *Chloris gayana*
rice cut grass - *Leersia oryzoides*

rice grass - *Oryzopsis*
Richardson's needle grass - *Stipa richardsonii*
rigid rye grass - *Lolium rigidum*
ring grass - *Muhlenbergia torreyi*
ringed beard grass - *Dichanthium annulatum*
rip-gut brome grass - *Bromus rigidus*
rip-gut grass - *Bromus rigidus*
ripple-grass - *Plantago lanceolata*
rock finger grass - *Chloris petraea*
rough bent grass - *Agrostis scabra*
rush grass - *Sporobolus*
russet cotton grass - *Eriophorum russeolum*
rye grass - *Lolium*
salt grass - *Distichlis spicata, Distichlis*
salt marsh cord grass - *Spartina alterniflora*
salt meadow cord grass - *Spartina patens*
salt-water cord grass - *Spartina alterniflora*
sand grass - *Triplasis purpurea*
sand love grass - *Eragrostis trichodes*
Saramolla grass - *Ischaemum rugosum*
savanna grass - *Axonopus compressus*
scorpion-grass - *Myosotis, Plagiobothrys figuratus*
scratch grass - *Muhlenbergia asperifolia*
Scribner's needle grass - *Stipa scribneri*
Scribner's reed grass - *Calamagrostis scribneri*
scurvy-grass - *Barbarea verna, Cochlearia officinalis, Crambe maritima*
scutch grass - *Cynodon dactylon, Elytrigia repens*
seashore salt grass - *Distichlis spicata*
seaside arrow grass - *Triglochin maritima*
sedge grass - *Andropogon virginicus*
serpent-grass - *Polygonum viviparum*
sheathed cotton grass - *Eriophorum vaginatum*
Shelly's grass - *Elytrigia repens*
short-bristled umbrella-grass - *Fuirena breviseta*
Siberian wheat grass - *Agropyron fragile* subsp. *sibiricum*
sickle grass - *Parapholis incurva*
silk grass - *Oryzopsis hymenoides*
silver beard grass - *Bothriochloa saccharoides*
silver hair grass - *Aira caryophyllea*
silver plume grass - *Saccharum alopecuroideum*
six-weeks grass - *Poa annua*
skunk-tail grass - *Hordeum jubatum*
sleepy grass - *Stipa robusta*
slender false brome grass - *Brachypodium sylvaticum*
slender hair grass - *Deschampsia elongata*
slender wheat grass - *Elymus trachycaulus*
slough grass - *Spartina pectinata*
small floating manna grass - *Glyceria borealis*
small-flowered Alexander grass - *Brachiaria subquadripara*
smilo grass - *Oryzopsis miliacea*
smooth brome grass - *Bromus commutatus*
smooth cord grass - *Spartina alterniflora*
snake-grass - *Equisetum arvense*
sour grass - *Digitaria insularis, Paspalum conjugatum*
southern cut grass - *Leersia hexandra*
southern water grass - *Luziola fluitans*
Spanish brome grass - *Bromus madritensis*
spear grass - *Poa pratensis, Stipa*
spider grass - *Aristida ternipes*
spreading witch grass - *Panicum dichotomiflorum*
spring whitlow-grass - *Draba verna*
squarrose umbrella-grass - *Fuirena squarrosa*

squaw-grass - *Xerophyllum tenax*
St. Augustine's grass - *Stenotaphrum secundatum*
standard crested wheat grass - *Agropyron desertorum*
star grass - *Aletris*
stink grass - *Eragrostis cilianensis*
Sudan grass - *Sorghum drummondii*
sugar-cane plume grass - *Saccharum giganteum*
sugar-grass sedge - *Carex atherodes*
sweet grass - *Hierochloe odorata*
sweet vernal grass - *Anthoxanthum odoratum*
switch grass - *Panicum virgatum*
tall manna grass - *Glyceria elata*
tall oat grass - *Arrhenatherum elatius*
tall wheat grass - *Elytrigia elongata*
Terrell's grass - *Lolium perenne*
Texas needle grass - *Stipa leucotricha*
thick-spike wheat grass - *Elymus lanceolatus*
thin grass - *Agrostis pallens*
Thurber's needle grass - *Stipa thurberiana*
tickle grass - *Agrostis hyemalis, A. scabra, Hordeum glaucum, Panicum capillare*
timber oat grass - *Danthonia intermedia*
timothy canary grass - *Phalaris angusta*
tobosa grass - *Hilaria mutica*
tongue-grass - *Lepidium virginicum, Lepidium*
toothache grass - *Ctenium aromaticum*
torpedo grass - *Panicum repens*
tropical carpet grass - *Axonopus compressus*
tuber oat grass - *Arrhenatherum elatius*
tufted hair grass - *Deschampsia caespitosa*
tufted hard grass - *Sclerochloa dura*
tufted love grass - *Eragrostis pectinacea*
tufted manna grass - *Glyceria septentrionalis*
tufted spear grass - *Eragrostis pectinacea*
tumble grass - *Schedonnardus paniculatus*
tumble windmill grass - *Chloris verticillata*
tumbleweed grass - *Panicum capillare*
tussock cotton grass - *Eriophorum vaginatum*
unbranched umbrella-grass - *Fuirena simplex*
upland bent grass - *Agrostis perennans*
vanilla grass - *Hierochloe odorata*
Vasey's grass - *Paspalum urvillei*
velvet bent grass - *Agrostis canina*
velvet finger grass - *Digitaria velutina*
velvet grass - *Holcus lanatus*
vine mesquite grass - *Panicum obtusum*
viper's-grass - *Scorzonera hispanica*
Virginia beard grass - *Andropogon virginicus*
Virginia cotton grass - *Eriophorum virginicum*
water grass - *Echinochloa crus-galli*
water manna grass - *Glyceria fluitans*
water star grass - *Heteranthera dubia*
weather grass - *Stipa spartea*
wedge grass - *Sphenopholis obtusata*
weeping alkali grass - *Puccinellia distans*
weeping love grass - *Eragrostis curvula*
western needle grass - *Stipa occidentalis*
western wheat grass - *Elymus smithii*
wheat grass - *Agropyron, Elytrigia repens*
whip-grass - *Scleria triglomerata*
white grass - *Leersia*
white oat grass - *Danthonia spicata*
whitlow-grass - *Draba verna*

widgeon-grass - *Ruppia maritima*
wild tongue-grass - *Lepidium densiflorum*
Wimmera rye grass - *Lolium rigidum*
windmill grass - *Chloris verticillata, Chloris*
winter bent grass - *Agrostis hyemalis*
witch grass - *Elytrigia repens, Panicum capillare*
wood whitlow-grass - *Draba nemorosa*
wool grass - *Scirpus congdonii, S. cyperinus*
wool grass bulrush - *Scirpus cyperinus*
woolly cup grass - *Eriochloa villosa*
woolly finger grass - *Digitaria eriantha* subsp. *pentzii*
yard grass - *Eleusine indica*
yellow bristle grass - *Setaria pumila*
yellow-eyed-grass - *Xyris caroliniana, Xyris*
yellow nut-grass - *Cyperus esculentus*
yellow oat grass - *Trisetum flavescens*
zoysia grass - *Zoysia matrella*

grassy
  grassy naiad - *Najas graminea*
  grassy-rush - *Butomus umbellatus*

Graves's oak - *Quercus gravesii*

graveyard-weed - *Euphorbia cyparissias*

gray
  gray alder - *Alnus incana*
  gray-ball sage - *Salvia dorrii*
  gray-beard - *Tillandsia usneoides*
  gray beard-tongue - *Penstemon canescens*
  gray birch - *Betula alleghaniensis, B. populifolia, B. pumila*
  gray-black grape - *Vitis cinerea*
  gray dogwood - *Cornus racemosa*
  gray ephedra - *Ephedra nevadensis*
  gray fescue - *Vulpia microstachys* var. *ciliata*
  gray goldenrod - *Solidago nemoralis*
  gray gum - *Eucalyptus punctata*
  gray-leaf willow - *Salix glauca*
  gray-mile - *Lithospermum officinale*
  gray nickers - *Caesalpinia bonduc*
  gray oak - *Quercus grisea*
  gray pine - *Pinus banksiana, P. sabiniana*
  gray poplar - *Populus canescens*
  gray rabbit-brush - *Chrysothamnus nauseosus*
  gray scurf-pea - *Psoralidium tenuiflorum*
  gray willow - *Salix bebbiana, S. cinerea, S. humilis*
  greasewood - *Adenostoma fasciculatum, Salvia apiana, Sarcobatus vermiculatus*

great
  American great bulrush - *Scirpus validus*
  great angelica - *Angelica atropurpurea*
  Great Basin bristle-cone pine - *Pinus longaeva*
  great bindweed - *Calystegia sepium*
  great bulrush - *Scirpus lacustris*
  great burdock - *Arctium lappa*
  great burnet - *Sanguisorba officinalis*
  great chickweed - *Stellaria pubera*
  great duckweed - *Spirodela polyrhiza*
  great globe thistle - *Echinops sphaerocephalus*
  great-headed garlic - *Allium ampeloprasum*
  great Indian plantain - *Cacalia muhlenbergii*
  Great Lakes gentian - *Gentiana rubricaulis*
  great-laurel - *Rhododendron maximum*
  great lead-tree - *Leucaena pulverulenta*
  great-leaf magnolia - *Magnolia macrophylla*

  great lobelia - *Lobelia siphilitica*
  Great Plains cottonwood - *Populus deltoides* var. *occidentalis*
  great plantain - *Plantago major*
  great ragweed - *Ambrosia trifida*
  great rhododendron - *Rhododendron maximum*
  great rose mallow - *Hibiscus grandiflorus*
  great Solomon's-seal - *Polygonatum biflorum, P. commutatum*
  great-spurred violet - *Viola selkirkii*
  great St. John's-wort - *Hypericum pyramidatum*
  great willow-herb - *Epilobium angustifolium*
  great yellow-cress - *Rorippa amphibia*

greater
  greater ammi - *Ammi majus*
  greater bird's-foot trefoil - *Lotus pedunculatus*
  greater celandine - *Chelidonium majus*
  greater duckweed - *Spirodela polyrhiza*
  greater periwinkle - *Vinca major*
  greater stitch-wort - *Stellaria holostea*

Grecian
  Grecian foxglove - *Digitalis lanata*
  Grecian laurel - *Laurus nobilis*

Greek
  Greek fir - *Abies cephalonica*
  Greek juniper - *Juniperus excelsa*
  Greek milkweed - *Asclepias speciosa*
  Greek myrtle - *Myrtus communis*
  Greek valerian - *Polemonium caeruleum* subsp. *villosum, P. reptans, Polemonium*

green
  American green alder - *Alnus viridis* subsp. *crispa*
  bracted green-onion - *Habenaria viridis* var. *bracteata*
  Christmas-green - *Lycopodium complanatum*
  cliff-green - *Paxistima canbyi*
  dark-green bulrush - *Scirpus atrovirens*
  European green alder - *Alnus viridis*
  giant green foxtail - *Setaria viridis* var. *major*
  green alder - *Alnus sinuata, A. viridis* subsp. *crispa*
  green-almond - *Pistacia vera*
  green amaranth - *Amaranthus hybridus, A. retroflexus, A. viridis*
  green arrow-arum - *Peltandra virginica*
  green ash - *Fraxinus pennsylvanica*
  green-banded mariposa - *Calochortus macrocarpus*
  green-bark ceanothus - *Ceanothus spinosus*
  green bean - *Phaseolus vulgaris*
  green bristle grass - *Setaria viridis*
  green-dragon - *Arisaema dracontium*
  green ephedra - *Ephedra viridis*
  green fanwort - *Cabomba caroliniana* cv. 'multipartita'
  green fescue - *Festuca viridula*
  green-flowered pepper-weed - *Lepidium densiflorum*
  green-flowered peppergrass - *Lepidium densiflorum*
  green-flowered wintergreen - *Pyrola chlorantha*
  green foxtail - *Setaria viridis*
  green-fruited bur-reed - *Sparganium chlorocarpum*
  green galenia - *Galenia pubescens*
  green-gentian - *Frasera speciosa*
  green gram - *Vigna radiata*
  green hawthorn - *Crataegus viridis*
  green kyllinga - *Cyperus brevifolius*
  green-leaf fescue - *Festuca viridula*

green-leaf manzanita - *Arctostaphylos patula*
green milkweed - *Asclepias viridiflora*
green-molly - *Kochia americana*
green needle grass - *Stipa viridula*
green osier - *Cornus alternifolia*
green pepper - *Capsicum annuum*
green-plume rabbit-brush - *Chrysothamnus nauseosus* subsp.
  graveolens
green-ripple peperomia - *Peperomia caperata*
green rock-cress - *Arabis missouriensis*
green santolina - *Santolina rosmarinifolia*
green sedge - *Carex viridula, Cyperus virens*
green sorrel - *Rumex acetosa*
green sprangle-top - *Leptochloa dubia*
green-stemmed Joe-Pye-weed - *Eupatorium purpureum*
green tansy mustard - *Descurainia pinnata* var. *brachycarpa*
green thistle - *Cirsium arvense*
green-vine - *Convolvulus arvensis*
green-violet - *Hybanthus concolor*
green wandering-Jew - *Tradescantia albiflora*
northern green orchid - *Habenaria hyperborea*
under-green willow - *Salix commutata*
greenbrier
  bristly greenbrier - *Smilax hispida*
  cat greenbrier - *Smilax glauca*
  coral greenbrier - *Smilax walteri*
  greenbrier - *Smilax rotundifolia, Smilax*
  lance-leaf greenbrier - *Smilax smallii*
  laurel-leaved greenbrier - *Smilax laurifolia*
  red-berried greenbrier - *Smilax walteri*
  round-leaf greenbrier - *Smilax rotundifolia*
greenish-flowered pyrola - *Pyrola chlorantha*
Greenland primrose - *Primula egaliksensis*
greens
  mustard-greens - *Brassica juncea*
  sour-greens - *Rumex venosus*
greenweed, dyer's greenweed - *Genista tinctoria*
greenwood, dyer's greenwood - *Genista tinctoria*
Gregg's
  Gregg's arrowhead - *Sagittaria longiloba*
  Gregg's cat's-claw - *Acacia greggii*
grip, devil's-grip - *Mollugo verticillata*
Grisebach's bristle grass - *Setaria grisebachii*
gromwell
  corn gromwell - *Lithospermum arvense*
  gromwell - *Lithospermum officinale, Lithospermum*
  pearl gromwell - *Lithospermum officinale*
  western false gromwell - *Onosmodium molle*
  western gromwell - *Lithospermum ruderale*
Gronovius's hawkweed - *Hieracium gronovii*
grooved
  two-grooved milk-vetch - *Astragalus bisulcatus*
  two-grooved poison-vetch - *Astragalus bisulcatus*
ground
  barren ground willow - *Salix brachycarpa* subsp. *niphoclada*
  clammy ground-cherry - *Physalis heterophylla*
  cut-leaf ground-cherry - *Physalis angulata*
  downy ground-cherry - *Physalis pubescens*
  dwarf ground-chervil - *Chamaesaracha coronopus*
  dwarf ground rattan - *Rhapis excelsa*
  European ground cherry - *Prunus fruticosa*

flat-branch ground-pine - *Lycopodium obscurum*
gill-over-the-ground - *Glechoma hederacea*
ground-cedar - *Lycopodium complanatum, L. tristachyum*
ground-cherry - *Physalis*
ground-hemlock - *Taxus canadensis*
ground-ivy - *Glechoma hederacea*
ground-myrtle - *Vinca*
ground-pine - *Lycopodium clavatum, L. tristachyum*
ground-pink - *Phlox subulata*
ground spurge - *Euphorbia prostrata*
hairy ground-cherry - *Physalis pubescens* var. *grisea*
high-ground willow oak - *Quercus incana*
lance-leaf ground-cherry - *Physalis angulata*
lobed ground-cherry - *Physalis lobata*
long-leaf ground-cherry - *Physalis longifolia*
old-man-in-the-ground - *Marah oreganus*
perennial ground-cherry - *Physalis longifolia* var. *subglabrata*
Peruvian ground-cherry - *Physalis peruviana*
purple ground-cherry - *Physalis lobata*
smooth ground-cherry - *Physalis longifolia* var. *subglabrata*
strawberry ground-cherry - *Physalis alkekengi*
Virginia ground-cherry - *Physalis virginiana*
groundnut
  groundnut - *Apios americana, Arachis hypogaea, Panax*
    trifolius
  groundnut pea-vine - *Lathyrus tuberosus*
groundsel
  Appalachian groundsel - *Senecio anonymus*
  balsam groundsel - *Senecio pauperculus*
  cress-leaf groundsel - *Senecio glabellus*
  groundsel - *Senecio vulgaris, Senecio*
  groundsel-bush - *Baccharis halimifolia*
  groundsel-tree - *Baccharis halimifolia*
  heart-leaf groundsel - *Senecio aureus*
  northern meadow groundsel - *Senecio pauperculus*
  northern swamp groundsel - *Senecio congestus*
  Platte groundsel - *Senecio plattensis*
  prairie groundsel - *Senecio plattensis*
  Riddell's groundsel - *Senecio riddellii*
  Small's groundsel - *Senecio anonymus*
  sticky groundsel - *Senecio viscosus*
  three-leaf groundsel - *Senecio longilobus*
  woodland groundsel - *Senecio sylvaticus*
grouseberry - *Vaccinium scoparium, Viburnum trilobum*
grove sandwort - *Moehringia lateriflora*
gruya - *Canna indica*
Guadalupe cypress - *Cupressus guadalupensis*
guajillo - *Acacia berlandieri*
guar - *Cyamopsis tetragonoloba*
guava
  apple guava - *Psidium guajava*
  Cattley's guava - *Psidium littorale* var. *longipes*
  guava - *Psidium guajava*
  pineapple-guava - *Acca sellowiana*
  purple strawberry guava - *Psidium littorale* var. *longipes*
  yellow guava - *Psidium guajava*
guayacan - *Porlieria angustifolia*
guaymochil - *Pithecellobium dulce*
guayule - *Parthenium argentatum*
guelder-rose - *Viburnum opulus*
Guernsey lily - *Nerine sarniensis*

gueules-noires - *Gaylussacia baccata*
Guianan
    Guianan chestnut - *Pachira aquatica*
    Guianan rapanea - *Myrsine guianensis*
guinea grass - *Panicum maximum*
Guinea, New Guinea creeper - *Mucuna novaguineensis*
guizotia - *Guizotia abyssinica*
Gulf Coast spike rush - *Eleocharis cellulosa*
gulf
    gulf cockspur - *Echinochloa crus-pavonis*
    gulf licaria - *Licaria triandra*
gum
    American sweet-gum - *Liquidambar styraciflua*
    Australian gum - *Eucalyptus*
    black gum - *Nyssa sylvatica*
    blue gum - *Eucalyptus globulus*
    cabbage gum - *Eucalyptus amplifolia, E. pauciflora*
    cotton-gum - *Nyssa aquatica*
    curly-cup gum-weed - *Grindelia squarrosa*
    forest red gum - *Eucalyptus tereticornis*
    Formosan gum - *Liquidambar formosana*
    ghost gum - *Eucalyptus pauciflora*
    gray gum - *Eucalyptus punctata*
    gum bumelia - *Sideroxylon lanuginosa*
    gum-elemi - *Bursera simaruba*
    gum-plant - *Grindelia squarrosa*
    gum succory - *Chondrilla juncea*
    gum-weed - *Grindelia squarrosa*
    Lehmann's gum - *Eucalyptus lehmannii*
    lemon-scented gum - *Eucalyptus citriodora*
    manna gum - *Eucalyptus viminalis*
    Murray's red gum - *Eucalyptus camaldulensis*
    red gum - *Eucalyptus calophylla, E. camaldulensis,*
       Liquidambar styraciflua
    river red gum - *Eucalyptus camaldulensis*
    rose gum - *Eucalyptus grandis*
    silver-dollar gum - *Eucalyptus polyanthemos*
    silver-leaf gum - *Eucalyptus pulverulenta*
    silver-leaf mountain gum - *Eucalyptus pulverulenta*
    snow gum - *Eucalyptus niphophila, E. pauciflora*
    sour-gum - *Nyssa sylvatica*
    sweet-gum - *Liquidambar styraciflua*
    Sydney blue gum - *Eucalyptus saligna*
    Tasmanian blue gum - *Eucalyptus globulus*
gumbo
    gumbo - *Abelmoschus esculentus, Bursera simaruba*
    gumbo-limbo - *Bursera simaruba*
gummy gooseberry - *Ribes lobbii*
gut
    cat's-gut - *Tephrosia virginiana*
    gut-weed - *Sonchus arvensis*
    rip-gut brome - *Bromus diandrus, B. rigidus*
    rip-gut brome grass - *Bromus rigidus*
    rip-gut grass - *Bromus rigidus*
    rip-gut sedge - *Carex lacustris*
gypsy
    gypsy-weed - *Veronica officinalis*
    gypsy-wort - *Lycopus*
hack
    golden hard-hack - *Potentilla fruticosa*
    hack-brush - *Spiraea douglasii*

hard-hack - *Spiraea alba* var. *latifolia, S. douglasii,*
    S. tomentosa
hackberry
    desert hackberry - *Celtis pallida*
    hackberry - *Celtis occidentalis, Celtis*
    Mississippi hackberry - *Celtis laevigata*
    net-leaf hackberry - *Celtis reticulata*
    northern hackberry - *Celtis occidentalis*
    southern hackberry - *Celtis laevigata*
    spiny hackberry - *Celtis pallida*
hackmatack - *Larix laricina, Populus balsamifera*
hag
    hag-berry - *Prunus padus*
    hag-brier - *Smilax hispida*
hairbrush, water-hairbrush - *Catabrosa aquatica*
hairy
    hairy-awn muhly - *Muhlenbergia capillaris*
    hairy beard-tongue - *Penstemon hirsutus*
    hairy beggar-ticks - *Bidens pilosa*
    hairy bitter-cress - *Cardamine hirsuta*
    hairy buttercup - *Ranunculus sardous*
    hairy catchfly - *Silene dichotoma*
    hairy chess - *Bromus commutatus*
    hairy coneflower - *Rudbeckia hirta*
    hairy crabgrass - *Digitaria sanguinalis*
    hairy dicksonia - *Dennstaedtia punctilobula*
    hairy-drop-seed - *Blepharoneuron tricholepis*
    hairy fleabane - *Conyza bonariensis*
    hairy-fruit chervil - *Chaerophyllum tainturieri*
    hairy galinsoga - *Galinsoga quadriradiata*
    hairy gaura - *Gaura villosa*
    hairy goldenrod - *Solidago hispida*
    hairy grama - *Bouteloua hirsuta*
    hairy ground-cherry - *Physalis pubescens* var. *grisea*
    hairy hawk's-bit - *Leontodon hirtus*
    hairy hawkweed - *Hieracium gronovii*
    hairy hempweed - *Mikania cordifolia*
    hairy honeysuckle - *Lonicera hirsuta*
    hairy indigo - *Indigofera hirsuta*
    hairy manzanita - *Arctostaphylos columbiana*
    hairy milk pea - *Galactia volubilis*
    hairy mountain mint - *Pycnanthemum pilosum*
    hairy nightshade - *Solanum sarrachoides*
    hairy pepperwort - *Marsilea vestita*
    hairy piney-woods goldenrod - *Solidago flexicaulis*
    hairy rock-cress - *Arabis hirsuta*
    hairy skullcap - *Scutellaria elliptica*
    hairy spurge - *Euphorbia vermiculata*
    hairy-stem gooseberry - *Ribes hirtellum*
    hairy sunflower - *Helianthus heterophyllus, H. hirsutus,*
       H. mollis
    hairy sweet cicely - *Osmorhiza claytonii*
    hairy vetch - *Vicia villosa*
    hairy water-leaf - *Hydrophyllum canadense, H. macrophyllum*
    hairy white-top - *Cardaria pubescens*
    hairy wild rye - *Elymus villosus*
    hairy willow-herb - *Epilobium hirsutum*
    hairy willow-weed - *Epilobium hirsutum*
    hairy wood mint - *Blephilia hirsuta*
    hairy-wood sunflower - *Helianthus atrorubens*
halapepe - *Pleomele aurea*

halberd
  halberd fern - *Tectaria heracleifolia*
  halberd-leaf mallow - *Hibiscus laevis*
  halberd-leaf orache - *Atriplex patula* subsp. *hastata*
  halberd-leaf tear-thumb - *Polygonum arifolium*
  halberd-leaf violet - *Viola hastata*
half-moon loco - *Astragalus allochrous*
Hall's blue-stem - *Andropogon hallii*
halver - *Ilex ambigua* var. *montana*
hand fern - *Dryopteris pedata*
hanging geranium - *Pelargonium peltatum*
hapuu
  hapuu - *Cibotium splendens*
  hapuu-ii - *Cibotium chamissoi*
harakeke - *Phormium tenax*
harbinger-of-spring - *Erigenia bulbosa*
hard
  golden hard-hack - *Potentilla fruticosa*
  hard fern - *Blechnum spicant*
  hard fescue - *Festuca brevipila*
  hard grass - *Sclerochloa dura*
  hard-hack - *Spiraea alba* var. *latifolia, S. douglasii,*
    S. tomentosa
  hard-leaf goldenrod - *Solidago rigida*
  hard maple - *Acer saccharum*
  hard-stem bulrush - *Scirpus acutus* var. *occidentalis*
  tufted hard grass - *Sclerochloa dura*
Harding's grass - *Phalaris aquatica, P. stenoptera*
hardy
  hardy ageratum - *Conoclinium coelestinum*
  hardy-orange - *Poncirus trifoliata*
  hardy timber bamboo - *Phyllostachys bambusoides*
hare
  hare barley - *Hordeum leporinum*
  hare-ear mustard - *Conringia orientalis*
  hare-foot dalea - *Dalea leporina*
hare's-tail - *Lagurus ovatus*
harebell
  harebell - *Campanula rotundifolia*
  southern harebell - *Campanula divaricata*
Harford's melic - *Melica harfordii*
haricot bean - *Phaseolus vulgaris*
harlequin
  rock harlequin - *Corydalis sempervirens*
  yellow harlequin - *Corydalis flavula*
harlock - *Arctium lappa*
harmel peganum - *Peganum harmala*
Harrington's plum-yew - *Cephalotaxus harringtonia*
Hart's-eye - *Pastinaca sativa*
Hartshorn's-plant - *Anemone patens*
Harvard's shin oak - *Quercus harvardii*
harvest
  harvest brodiaea - *Brodiaea coronaria*
  harvest-lice - *Agrimonia*
hastate Indian plantain - *Cacalia suaveolens*
haw
  black-haw - *Viburnum lentago, V. prunifolium*
  black-haw viburnum - *Viburnum prunifolium*
  possum-haw - *Ilex decidua, Viburnum acerifolium, V. nudum*
  red haw - *Crataegus*

  rusty black-haw - *Viburnum rufidulum*
  southern black-haw - *Viburnum rufidulum*
  summer-haw - *Crataegus flava*
  swamp haw - *Viburnum cassinoides*
  sweet-haw - *Viburnum prunifolium*
Hawaiian
  Hawaiian hibiscus - *Hibiscus rosa-sinensis*
  Hawaiian soap-tree - *Sapindus oahuensis*
  Hawaiian tree-fern - *Cibotium chamissoi, C. splendens*
hawk's
  Asiatic hawk's-beard - *Youngia japonica*
  bristly hawk's-beard - *Crepis setosa*
  European hawk's-beard - *Crepis vesicaria* subsp. *haenseleri*
  fall hawk's-bit - *Leontodon autumnalis*
  hairy hawk's-bit - *Leontodon hirtus*
  hawk's-beard - *Crepis*
  hawk's-bit - *Leontodon autumnalis*
  narrow-leaf hawk's-beard - *Crepis tectorum*
  rough hawk's-beard - *Crepis biennis*
  rough hawk's-bit - *Leontodon hirtus, L. taraxacoides*
  smooth hawk's-beard - *Crepis capillaris*
  western hawk's-beard - *Crepis occidentalis*
hawkweed
  beaked hawkweed - *Hieracium gronovii*
  Canadian hawkweed - *Hieracium canadense*
  field hawkweed - *Hieracium caespitosum*
  glaucous hawkweed - *Hieracium floribundum*
  Gronovius's hawkweed - *Hieracium gronovii*
  hairy hawkweed - *Hieracium gronovii*
  hawkweed - *Hieracium lachenalii, Hieracium*
  hawkweed ox-tongue - *Picris hieracioides*
  king-devil hawkweed - *Hieracium piloselloides*
  mouse-ear hawkweed - *Hieracium pilosella*
  narrow-leaf hawkweed - *Hieracium umbellatum*
  northern hawkweed - *Hieracium umbellatum*
  orange hawkweed - *Hieracium aurantiacum*
  rough hawkweed - *Hieracium scabrum*
  smooth hawkweed - *Hieracium floribundum*
  sticky hawkweed - *Hieracium scabrum*
  tall hawkweed - *Hieracium praealtum* var. *decipiens*
  veiny hawkweed - *Hieracium venosum*
  wall hawkweed - *Hieracium murorum*
  yellow-devil hawkweed - *Hieracium floribundum*
  yellow hawkweed - *Hieracium caespitosum*
hawthorn
  Biltmore hawthorn - *Crataegus intricata*
  black hawthorn - *Crataegus douglasii*
  Brainerd's hawthorn - *Crataegus brainerdii*
  cerro hawthorn - *Crataegus erythropoda*
  cockspur hawthorn - *Crataegus crus-galli*
  Columbia hawthorn - *Crataegus columbiana*
  dotted hawthorn - *Crataegus punctata*
  downy hawthorn - *Crataegus mollis*
  English hawthorn - *Crataegus laevigata, C. monogyna*
  entangled hawthorn - *Crataegus intricata*
  fan-leaf hawthorn - *Crataegus flabellata*
  fireberry hawthorn - *Crataegus chrysocarpa*
  fleshy hawthorn - *Crataegus succulenta*
  frosted hawthorn - *Crataegus pruinosa*
  green hawthorn - *Crataegus viridis*
  hawthorn - *Crataegus aestivalis, Crataegus*
  hawthorn-leaved gooseberry - *Ribes oxyacanthoides*

Indian hawthorn - *Rhaphiolepis indica*
Kansas hawthorn - *Crataegus coccinioides*
little-hip hawthorn - *Crataegus spathulata*
May hawthorn - *Crataegus aestivalis*
one-seeded hawthorn - *Crataegus monogyna*
parsley hawthorn - *Crataegus marshallii*
pasture hawthorn - *Crataegus spathulata*
pear hawthorn - *Crataegus calpodendron*
river hawthorn - *Crataegus rivularis*
round-leaf hawthorn - *Crataegus chrysocarpa*
scarlet hawthorn - *Crataegus coccinea*
Washington's hawthorn - *Crataegus phaenopyrum*
Yedda's-hawthorn - *Rhaphiolepis umbellata*
yellow hawthorn - *Crataegus flava*

hay
fool-hay - *Panicum capillare*
hay-fever-weed - *Ambrosia acanthicarpa*
hay-scented fern - *Dennstaedtia punctilobula*
hay sedge - *Carex foenea*

hayfield tarweed - *Hemizonia congesta*

hazel
hazel - *Corylus*
hazel alder - *Alnus rugosa, A. serrulata*
hazel dodder - *Cuscuta coryli*
western hazel - *Corylus cornuta* var. *californica*
witch-hazel - *Hamamelis virginiana*

hazelnut
American hazelnut - *Corylus americana*
beaked hazelnut - *Corylus cornuta*
California hazelnut - *Corylus cornuta* var. *californica*
European hazelnut - *Corylus avellana*
hazelnut - *Corylus*

head's grass - *Phleum pratense*

heads
curly-heads - *Clematis ochroleuca*
drum-heads - *Polygala cruciata*

heal
heal-all - *Prunella vulgaris*
horse-heal - *Inula helenium*
self-heal - *Prunella vulgaris*

healing-herb - *Symphytum officinale*

heart
big floating-heart - *Nymphoides aquatica*
bleeding-heart - *Dicentra spectabilis*
blue heart-leaf aster - *Aster cordifolius*
bullock's-heart - *Annona reticulata*
bursting-heart - *Euonymus americanus*
clasping heart-leaf aster - *Aster undulatus*
floating-heart - *Nymphoides*
heart-leaf ampelopsis - *Ampelopsis cordata*
heart-leaf aster - *Aster cordifolius*
heart-leaf cocklebur - *Xanthium strumarium*
heart-leaf drymary - *Drymaria cordata*
heart-leaf groundsel - *Senecio aureus*
heart-leaf philodendron - *Philodendron cordatum, P. scandens*
heart-leaf sida - *Sida cordifolia*
heart-leaf willow - *Salix cordata, S. eriocephala*
heart-nut - *Juglans cordiformis*
heart-of-Jesus - *Caladium bicolor*
heart-padded hoary-cress - *Cardaria draba*
heart-seed - *Cardiospermum halicababum*

midwestern blue heart-leaf aster - *Aster shortii*
tropical bleeding-heart - *Clerodendrum thompsoniae*
western bleeding-heart - *Dicentra formosa*
white-heart hickory - *Carya tomentosa*
white heart-leaf aster - *Aster divaricatus*
yellow floating-heart - *Nymphoides peltata*

heart's
heart's-ease - *Polygonum persicaria, Viola tricolor*
western heart's-ease - *Viola ocellata*

hearts, blue-hearts - *Buchnera*

heath
beach-heath - *Hudsonia tomentosa*
Cornish heath - *Erica vagans*
cross-leaf heath - *Erica tetralix*
heath - *Erica*
heath aster - *Aster ericoides*
heath bedstraw - *Galium saxatile*
heath pea - *Lathyrus japonicus*
Scotch heath - *Erica cinerea*
spring heath - *Erica carnea*
twisted heath - *Erica cinerea*
white heath aster - *Aster pilosus*
winter heath - *Erica carnea*

heather
beach-heather - *Hudsonia*
false heather - *Cuphea hyssopifolia, Hudsonia tomentosa*
golden-heather - *Hudsonia ericoides*
heather - *Calluna vulgaris*
mountain-heather - *Phyllodoce breweri, Phyllodoce*
red-heather - *Phyllodoce breweri*
Scotch heather - *Calluna vulgaris*
snow-heather - *Erica carnea*

heaven, tree-of-heaven - *Ailanthus altissima*

heavenly-bamboo - *Nandina domestica*

heavy sedge - *Carex gravida*

hedge
golden hedge-hyssop - *Gratiola aurea*
hedge bamboo - *Bambusa multiplex*
hedge bedstraw - *Galium mollugo*
hedge bindweed - *Calystegia sepium*
hedge maple - *Acer campestre*
hedge-mustard - *Sisymbrium altissimum, S. officinale*
hedge-nettle - *Stachys*
hedge-nettle betony - *Stachys annua*
hedge-parsley - *Torilis arvensis*
hedge-plant - *Ligustrum*
hedge smartweed - *Polygonum scandens*
hedge vetch - *Vicia sepium*
Japanese hedge-parsley - *Torilis japonica*
knotted hedge-parsley - *Torilis nodosa*
low hedge-nettle - *Stachys arvensis*
smooth hedge-nettle - *Stachys tenuifolia*
tall hedge mustard - *Sisymbrium altissimum*
white hedge bedstraw - *Galium mollugo*
woolly hedge-nettle - *Stachys olympica*
yellow hedge-hyssop - *Gratiola aurea*

hedgehog
hedgehog cactus - *Echinocereus viridiflorus*
hedgehog dog-tail grass - *Cynosurus echinatus*
hedgehog gourd - *Cucumis dipsaceus*
hedgehog grass - *Cenchrus echinatus*

heliotrope
  alkali heliotrope - *Heliotropium curassavicum* var. *oculatum*
  clasping heliotrope - *Heliotropium amplexicaule*
  European heliotrope - *Heliotropium europaeum*
  garden-heliotrope - *Valeriana officinalis*
  heliotrope - *Heliotropium arborescens, Heliotropium*
  Indian heliotrope - *Heliotropium indicum*
  salt heliotrope - *Heliotropium curassavicum*
  seaside heliotrope - *Heliotropium curassavicum,*
    H. curassavicum var. *obovatum*
  spatulate-leaved heliotrope - *Heliotropium curassavicum* var.
    obovatum
  wild heliotrope - *Heliotropium curassavicum*
hell-fetter - *Smilax hispida, S. tamnoides*
hellebore
  California false hellebore - *Veratrum californicum*
  false hellebore - *Veratrum*
  western false hellebore - *Veratrum californicum*
  white hellebore - *Veratrum viride*
helmut-flower - *Aconitum napellus*
heltrot - *Heracleum sphondylium* subsp. *montanum*
hemianthus - *Micranthemum glomeratum*
hemlock
  bulb-bearing water-hemlock - *Cicuta bulbifera*
  Canadian hemlock - *Tsuga canadensis*
  Carolina hemlock - *Tsuga caroliniana*
  deadly-hemlock - *Conium maculatum*
  eastern hemlock - *Tsuga canadensis*
  ground-hemlock - *Taxus canadensis*
  hemlock - *Tsuga*
  hemlock-parsley - *Conioselinum pacificum*
  hemlock-spruce - *Tsuga*
  mountain hemlock - *Tsuga mertensiana*
  northern Japanese hemlock - *Tsuga diversifolia*
  poison-hemlock - *Conium maculatum*
  spotted-hemlock - *Conium maculatum*
  spotted water-hemlock - *Cicuta maculata*
  water-hemlock - *Cicuta maculata, Cicuta*
  western hemlock - *Tsuga heterophylla*
  western water-hemlock - *Cicuta douglasii*
hemp
  American hemp - *Apocynum cannabinum*
  bog-hemp - *Boehmeria cylindrica*
  bowstring-hemp - *Calotropis gigantea*
  Ceylon bowstring-hemp - *Sansevieria zeylanica*
  Colorado River hemp - *Sesbania exaltata*
  Deccan's-hemp - *Hibiscus cannabinus*
  hemp - *Cannabis*
  hemp dogbane - *Apocynum cannabinum*
  hemp-leaved mallow - *Althaea cannabina*
  hemp-nettle - *Galeopsis tetrahit*
  hemp palm - *Trachycarpus fortunei*
  hemp-plant - *Agave sisalana*
  hemp-tree - *Vitex agnus-castus*
  Indian hemp - *Abutilon theophrasti, Apocynum cannabinum,*
    Crotalaria juncea, Hibiscus cannabinus
  Madras hemp - *Crotalaria juncea*
  New Zealand hemp - *Phormium tenax*
  red hemp-nettle - *Galeopsis ladanum*
  sisal-hemp - *Agave sisalana*
  soft hemp - *Cannabis sativa*
  sunn hemp - *Crotalaria juncea*

  tall water-hemp - *Amaranthus tuberculatus*
  true hemp - *Cannabis sativa*
  water-hemp - *Amaranthus rudis*
  wild hemp - *Ambrosia trifida, Galeopsis tetrahit*
hempweed
  climbing hempweed - *Mikania scandens*
  hairy hempweed - *Mikania cordifolia*
hen, fat-hen - *Chenopodium album, C. bonus-henricus*
henbane
  black henbane - *Hyoscyamus niger*
  henbane - *Hyoscyamus niger*
henbit - *Lamium amplexicaule*
Henry, good-King-Henry - *Chenopodium bonus-henricus*
hens-and-chickens - *Sempervivum tectorum*
hepatica
  hepatica - *Hepatica*
  round-lobed hepatica - *Hepatica americana*
herb
  cow-herb - *Vaccaria hispanica*
  eastern willow-herb - *Epilobium coloratum*
  elfin-herb - *Cuphea hyssopifolia*
  felon-herb - *Artemisia vulgaris, Hieracium pilosella*
  great willow-herb - *Epilobium angustifolium*
  hairy willow-herb - *Epilobium hirsutum*
  healing-herb - *Symphytum officinale*
  herb-of-grace - *Ruta graveolens*
  herb-patience - *Rumex patientia*
  herb-Robert - *Geranium robertianum*
  herb-Sophia - *Descurainia sophia*
  pot-herb - *Barbarea vulgaris*
  purple-leaf willow-herb - *Epilobium coloratum*
  willow-herb - *Epilobium hirsutum, Epilobium*
herba impia - *Filago germanica*
Hercules's
  Hercules's-club - *Aralia spinosa, Zanthoxylum clava-herculis*
  Texas Hercules's-club - *Zanthoxylum hirsutum*
heron's-bill - *Erodium*
hiba
  hiba - *Thujopsis dolobrata*
  hiba-arborvitae - *Thujopsis dolobrata*
hibiscus
  Chinese hibiscus - *Hibiscus rosa-sinensis*
  Hawaiian hibiscus - *Hibiscus rosa-sinensis*
  hibiscus - *Hibiscus moscheutos*
  sea hibiscus - *Hibiscus tiliaceus*
hickory
  bitter hickory - *Carya aquatica*
  bitter-nut hickory - *Carya cordiformis*
  black hickory - *Carya texana*
  broom hickory - *Carya glabra*
  Buckley's hickory - *Carya texana*
  Florida hickory - *Carya floridana*
  hickory - *Carya*
  hickory pine - *Pinus longaeva, P. pungens*
  mocker-nut hickory - *Carya tomentosa*
  nutmeg hickory - *Carya myristiciformis*
  Ozark hickory - *Carya texana*
  pale hickory - *Carya pallida*
  pig-nut hickory - *Carya glabra*
  sand hickory - *Carya pallida*
  scrub hickory - *Carya floridana*

shagbark hickory - *Carya ovata*
shellbark hickory - *Carya laciniosa, C. ovata*
small-fruited hickory - *Carya glabra*
swamp hickory - *Carya cordiformis*
water hickory - *Carya aquatica*
white-heart hickory - *Carya tomentosa*
white hickory - *Carya tomentosa*
hiedra - *Toxicodendron pubescens*
high
black high-bush blueberry - *Vaccinium atrococcum*
high-bush blackberry - *Rubus argutus*
high-bush blueberry - *Vaccinium caesariense, V. corymbosum*
high-bush cranberry - *Viburnum trilobum*
high-ground willow oak - *Quercus incana*
high mallow - *Malva sylvestris*
southern high-bush blueberry - *Vaccinium elliottii*
hill
hill-mustard - *Bunias orientalis*
pine-hill beak rush - *Rhynchospora globularis*
Hill's, Jim Hill's mustard - *Sisymbrium altissimum*
hills
hills-of-snow - *Hydrangea arborescens*
sand-hills amaranth - *Amaranthus arenicola*
hillside
hillside blueberry - *Vaccinium vacillans*
hillside sedge - *Carex siccata*
Himalayan
Himalayan blackberry - *Rubus discolor*
Himalayan white pine - *Pinus wallichiana*
himegurumi - *Juglans cordiformis*
Hinds' walnut - *Juglans hindsii*
Hindu
Hindu datura - *Datura metel*
Hindu lotus - *Nelumbo nucifera*
Hinoki
Hinoki cypress - *Chamaecyparis obtusa*
Hinoki false cypress - *Chamaecyparis obtusa*
hip, little-hip hawthorn - *Crataegus spathulata*
hippo, wild hippo - *Euphorbia corollata*
Hiryu azalea - *Rhododendron obtusum*
hispid buttercup - *Ranunculus hispidus*
hoary
globe-podded hoary-cress - *Cardaria pubescens*
heart-padded hoary-cress - *Cardaria draba*
hoary-alyssum - *Berteroa incana*
hoary azalea - *Rhododendron canescens*
hoary cinquefoil - *Potentilla argentea*
hoary-cress - *Cardaria draba*
hoary pea - *Tephrosia*
hoary pepperwort - *Cardaria draba*
hoary plantain - *Plantago media*
hoary puccoon - *Lithospermum canescens*
hoary skullcap - *Scutellaria incana*
hoary tick-clover - *Desmodium canescens*
hoary tick-trefoil - *Desmodium canescens*
hoary vervain - *Verbena stricta*
hoary willow - *Salix caprea*
hobble
dog's-hobble - *Leucothoe fontanesiana, Viburnum alnifolium*
witch's-hobble - *Viburnum alnifolium*
hobblebush - *Viburnum alnifolium*

hog
hog cranberry - *Arctostaphylos uva-ursi*
hog millet - *Panicum miliaceum*
hog-peanut - *Amphicarpaea bracteata*
hog plum - *Prunus americana, P. umbellata*
hog-potato - *Hoffmannseggia glauca*
hog rush - *Juncus gerardii*
hog-wort - *Croton capitatus*
hogweed - *Ambrosia artemisiifolia, Conyza canadensis, Heracleum sphondylium* subsp. *montanum*
hollow
hollow goldenrod - *Solidago fistulosa*
hollow Joe-Pye-weed - *Eupatorium fistulosum*
hollow-scaled bulrush - *Scirpus koilolepis*
hollow-stemmed Joe-Pye-weed - *Eupatorium fistulosum*
holly
American holly - *Ilex opaca*
big-leaf holly - *Ilex ambigua* var. *montana*
box-leaf holly - *Ilex crenata*
Carolina holly - *Ilex ambigua*
Chinese holly - *Ilex cornuta*
dahoon holly - *Ilex cassine*
dwarf-holly - *Malpighia coccigera*
English holly - *Ilex aquifolium*
European holly - *Ilex aquifolium*
false holly - *Osmanthus heterophyllus*
giant holly fern - *Polystichum munitum*
holly - *Ilex*
holly barberry - *Berberis aquifolium*
holly-leaf buckthorn - *Rhamnus ilicifolia*
holly-leaf cherry - *Prunus ilicifolia*
holly-leaf naiad - *Najas marina*
holly mahonia - *Berberis aquifolium*
holly oak - *Quercus ilex*
holly-olive - *Osmanthus heterophyllus*
holly osmanthus - *Osmanthus heterophyllus*
horned holly - *Ilex cornuta*
house holly fern - *Cyrtomium falcata*
Japanese holly - *Ilex crenata*
Japanese holly fern - *Cyrtomium falcata*
large-leaf holly - *Ilex ambigua* var. *montana*
miniature holly - *Malpighia coccigera*
mountain holly - *Ilex ambigua* var. *montana, Nemopanthus, Prunus ilicifolia*
mountain holly fern - *Polystichum lonchitis*
northern holly fern - *Polystichum lonchitis*
Oregon holly - *Ilex aquifolium*
Sarvis's holly - *Ilex amelanchier*
Singapore holly - *Malpighia coccigera*
West Indian holly - *Leea coccinea*
hollyhock
Antwerp hollyhock - *Alcea ficifolia*
desert hollyhock - *Sphaeralcea ambigua*
hollyhock - *Alcea rosea*
hollyhock mallow - *Malva alcea*
sea hollyhock - *Hibiscus moscheutos* subsp. *palustris*
Holm's oak - *Quercus ilex*
holy
holy basil - *Ocimum sanctum*
holy-clover - *Onobrychis viciifolia*
holy thistle - *Silybum marianum*
holy-wood lignum-vitae - *Guajacum sanctum*

home, peace-in-the-home - *Soleirolia soleirolii*
honesty-plant - *Lunaria annua*
honewort - *Cryptotaenia canadensis*
honey
  honey-bell rhododendron - *Rhododendron campylocarpum*
  honey-berry - *Melicoccus bijugatus*
  honey-bloom - *Apocynum androsaemifolium*
  honey-clover - *Melilotus alba*
  honey locust - *Gleditsia triacanthos*
  honey mesquite - *Prosopis glandulosa*
  honey-myrtle - *Melaleuca*
  honey-plant - *Hoya carnosa*
  honey-shuck - *Gleditsia triacanthos*
  honey-vine - *Cynanchum laeve*
  Japanese honey locust - *Gleditsia japonica*
  thornless honey locust - *Gleditsia triacanthos* var. *inermis*
  western honey mesquite - *Prosopis glandulosa* var. *torreyana*
honeydew melon - *Cucumis melo* var. *inodorus*
honeysuckle
  American honeysuckle - *Lonicera canadensis*
  bush-honeysuckle - *Diervilla lonicera*
  Cape honeysuckle - *Tecoma capensis*
  coral honeysuckle - *Lonicera sempervirens*
  desert-honeysuckle - *Anisacanthus thurberi*
  European fly honeysuckle - *Lonicera xylosteum*
  fly honeysuckle - *Lonicera canadensis, L. xylosteum*
  grape honeysuckle - *Lonicera prolifera*
  hairy honeysuckle - *Lonicera hirsuta*
  honeysuckle - *Aquilegia canadensis, Lonicera*
  Jamaican honeysuckle - *Passiflora laurifolia*
  Japanese honeysuckle - *Lonicera japonica*
  limber honeysuckle - *Lonicera dioica*
  Morrow's honeysuckle - *Lonicera morrowii*
  mountain fly honeysuckle - *Lonicera villosa*
  northern bush-honeysuckle - *Diervilla lonicera*
  northern fly honeysuckle - *Lonicera villosa*
  pink honeysuckle - *Lonicera hispidula*
  swamp fly honeysuckle - *Lonicera oblongifolia*
  swamp-honeysuckle - *Rhododendron viscosum*
  sweet-berry honeysuckle - *Lonicera caerulea*
  Tatarian honeysuckle - *Lonicera tatarica*
  trumpet-honeysuckle - *Campsis radicans, Lonicera sempervirens*
  wild honeysuckle - *Gaura coccinea, Lonicera dioica*
  yellow honeysuckle - *Lonicera flava*
Hong Kong kumquat - *Fortunella hindsii*
Honolulu-queen - *Hylocereus undatus*
hood
  azure monk's-hood - *Aconitum carmichaelii*
  bishop's-hood - *Astrophytum myriostigma*
  garden monk's-hood - *Aconitum napellus*
  hood canary grass - *Phalaris paradoxa*
  monk's-hood - *Aconitum, Astrophytum myriostigma*
  monk's-hood-vine - *Ampelopsis aconitifolia*
  ornamental monk's-hood - *Astrophytum ornatum*
  southern monk's-hood - *Aconitum uncinatum*
  wild monk's-hood - *Aconitum uncinatum*
hoof, horse's-hoof - *Tussilago farfara*
hook
  fish-hook cactus - *Mammillaria*
  five-hook bassia - *Bartonia virginica, Bassia hyssopifolia*
  hook-spur violet - *Viola adunca*

hooked buttercup - *Ranunculus recurvatus*
Hooker's willow - *Salix hookeriana*
hop
  American hop-hornbeam - *Ostrya virginiana*
  California hop-tree - *Ptelea crenulata*
  cow-hop clover - *Trifolium dubium*
  eastern hop-hornbeam - *Ostrya virginiana*
  European hop - *Humulus lupulus*
  false hop-plant - *Justicia brandegeana*
  Florida hop-bush - *Dodonaea viscosa*
  hop - *Humulus lupulus, Humulus*
  hop clover - *Trifolium aureum, T. campestre*
  hop-hornbeam - *Ostrya*
  hop-hornbeam copper-leaf - *Acalypha ostryifolia*
  hop-like sedge - *Carex lupuliformis*
  hop sedge - *Carex lupulina*
  hop-tree - *Ptelea trifoliata, Ptelea*
  Japanese hop - *Humulus japonicus*
  Knowlton's hop-hornbeam - *Ostrya knowltonii*
  large hop clover - *Trifolium campestre*
  palmate hop clover - *Trifolium aureum*
  small hop clover - *Trifolium dubium*
Hopi sunflower - *Helianthus annuus*
horehound
  black horehound - *Ballota nigra*
  cut-leaf water-horehound - *Lycopus americanus*
  European water horehound - *Lycopus europaeus*
  fetid horehound - *Ballota*
  horehound - *Marrubium vulgare, Marrubium*
  horehound motherwort - *Leonurus marrubiastrum*
  water-horehound - *Lycopus americanus, Lycopus*
  white horehound - *Marrubium vulgare*
horn
  buck-horn - *Lycopodium clavatum, Osmunda cinnamomea*
  buck-horn plantain - *Plantago lanceolata*
  horn-of-plenty - *Datura metel*
  huntsman's-horn - *Sarracenia flava*
  ox-horn bucida - *Bucida buceras*
  ram's-horn - *Proboscidea louisianica*
  three-horn bedstraw - *Galium tricornutum*
hornbeam
  American hop-hornbeam - *Ostrya virginiana*
  American hornbeam - *Carpinus caroliniana*
  eastern hop-hornbeam - *Ostrya virginiana*
  European hornbeam - *Carpinus betulus*
  hop-hornbeam - *Ostrya*
  hop-hornbeam copper-leaf - *Acalypha ostryifolia*
  hornbeam - *Carpinus*
  Knowlton's hop-hornbeam - *Ostrya knowltonii*
horned
  horned-head buttercup - *Ranunculus testiculatus*
  horned holly - *Ilex cornuta*
  horned-pondweed - *Zannichellia palustris*
  horned rush - *Rhynchospora corniculata*
  horned violet - *Viola cornuta*
horoeka - *Pseudopanax crassifolia*
horse
  California horse-chestnut - *Aesculus californica*
  Carolina horse-nettle - *Solanum carolinense*
  dwarf horse-chestnut - *Aesculus parviflora*
  European horse-chestnut - *Aesculus hippocastanum*
  horse-balm - *Collinsonia canadensis, Collinsonia*

horse bean - *Canavalia ensiformis, Parkinsonia aculeata, Vicia faba*
horse-brier - *Smilax rotundifolia*
horse-chestnut - *Aesculus hippocastanum, Aesculus*
horse-elder - *Inula helenium*
horse-gentian - *Triosteum*
horse-gram - *Macrotyloma uniflorum*
horse-heal - *Inula helenium*
horse-nettle - *Solanum carolinense*
horse-pipes - *Equisetum arvense*
horse-purslane - *Trianthema portulacastrum*
horse-sugar - *Symplocos tinctoria*
horse-thistle - *Lactuca serriola*
horse-weed - *Ambrosia trifida*
Japanese horse-chestnut - *Aesculus turbinata*
little-leaf horse-brush - *Tetradymia glabrata*
northern horse-balm - *Collinsonia canadensis*
perfoliate horse-gentian - *Triosteum perfoliatum*
red horse-chestnut - *Aesculus carnea*
robust horse-nettle - *Solanum dimidiatum*
silver horse-nettle - *Solanum elaeagnifolium*
spineless horse-brush - *Tetradymia canescens*
white-horse - *Danthonia spicata*
white-horse-nettle - *Solanum elaeagnifolium*
horse's-hoof - *Tussilago farfara*
horsemint
European horsemint - *Mentha longifolia*
horsemint - *Mentha longifolia, Monarda punctata, Monarda*
sweet horsemint - *Cunila origanoides*
horseradish - *Armoracia rusticana*
horseshoe geranium - *Pelargonium hortorum*
horsetail
field horsetail - *Equisetum arvense*
giant horsetail - *Equisetum telmateia*
horsetail - *Equisetum arvense, Equisetum*
horsetail casuarina - *Casuarina equisetifolia*
horsetail-fern - *Equisetum arvense*
horsetail milkweed - *Asclepias verticillata*
horsetail-tree - *Casuarina equisetifolia*
marsh horsetail - *Equisetum palustre*
shade horsetail - *Equisetum palustre*
Sylvan's horsetail - *Equisetum sylvaticum*
variegated horsetail - *Equisetum variegatum*
water horsetail - *Equisetum fluviatile*
woodland horsetail - *Equisetum sylvaticum*
horseweed
horseweed - *Collinsonia, Conyza canadensis*
sprawling horseweed - *Calyptocarpus vialis*
hortensia - *Hydrangea macrophylla*
hortulan plum - *Prunus hortulana*
hot, red-hot-poker - *Kniphofia*
hottentot-fig - *Carpobrotus edulis*
hound-bane - *Marrubium vulgare*
hound's-tongue - *Cynoglossum officinale, Cynoglossum*
hour, flower-of-an-hour - *Hibiscus trionum*
house
house blooming mock-orange - *Pittosporum tobira*
house geranium - *Pelargonium hortorum*
house holly fern - *Cyrtomium falcata*
house hydrangea - *Hydrangea macrophylla*
house-leek - *Sempervivum tectorum*

house-pine - *Araucaria heterophylla*
houses, meeting-houses - *Aquilegia canadensis*
houstonia, large houstonia - *Hedyotis purpurea*
hsiang, feng-hsiang-shu - *Liquidambar formosana*
huamuchil - *Pithecellobium dulce*
hubam
hubam - *Melilotus alba*
hubam clover - *Melilotus alba*
huckleberry
black huckleberry - *Gaylussacia baccata, Vaccinium atrococcum*
blue huckleberry - *Vaccinium membranaceum*
box huckleberry - *Gaylussacia brachycera*
California huckleberry - *Vaccinium ovatum*
downy swamp huckleberry - *Vaccinium atrococcum*
dwarf huckleberry - *Gaylussacia dumosa, G. frondosa*
evergreen huckleberry - *Vaccinium ovatum*
he-huckleberry - *Cyrilla racemiflora, Lyonia ligustrina*
huckleberry - *Gaylussacia, Vaccinium*
huckleberry oak - *Quercus vaccinifolia*
little-leaf huckleberry - *Vaccinium scoparium*
red huckleberry - *Vaccinium parvifolium*
shot huckleberry - *Vaccinium ovatum*
squaw huckleberry - *Vaccinium caesium, V. stamineum*
thin-leaf huckleberry - *Vaccinium membranaceum*
tree huckleberry - *Rhododendron arboreum*
Hudson
Hudson Bay currant - *Ribes hudsonianum*
Hudson sagittaria - *Sagittaria subulata*
hudsonia - *Hudsonia tomentosa*
huesito - *Malpighia glabra*
huisache - *Acacia farnesiana*
Hungarian
Hungarian brome - *Bromus inermis*
Hungarian lilac - *Syringa josikaea*
Hungarian millet - *Setaria italica*
Hungarian vetch - *Vicia pannonica*
hungry-rice - *Digitaria exilis*
hunter's-robe - *Epipremnum aureum*
huntsman's
huntsman's-cup - *Sarracenia purpurea, Sarracenia*
huntsman's-horn - *Sarracenia flava*
hurricane
hurricane grass - *Bothriochloa pertusa*
hurricane palm - *Ptychosperma macarthurii*
hurricane-plant - *Monstera deliciosa*
hurts, sweet-hurts - *Vaccinium angustifolium*
husk-tomato - *Physalis pubescens, Physalis*
hyacinth
anchored water-hyacinth - *Eichhornia azurea*
Dutch hyacinth - *Hyacinthus orientalis*
garden hyacinth - *Hyacinthus orientalis*
grape-hyacinth - *Muscari botryoides*
hyacinth - *Hyacinthus orientalis*
hyacinth bean - *Lablab purpureus*
meadow-hyacinth - *Camassia scilloides*
tassel-hyacinth - *Muscari comosum*
water-hyacinth - *Eichhornia crassipes*
wild hyacinth - *Camassia scilloides, Dichelostemma pulchellum, Scilla*

hybrid
  Ghent hybrid azalea - *Rhododendron gandavense*
  hybrid tuberous begonia - *Begonia tuberhybrida*
hydrangea
  American hydrangea - *Hydrangea arborescens*
  climbing hydrangea - *Decumaria barbara, Hydrangea anomala* subsp. *petiolaris*
  French hydrangea - *Hydrangea macrophylla*
  house hydrangea - *Hydrangea macrophylla*
  panicled hydrangea - *Hydrangea paniculata*
  smooth hydrangea - *Hydrangea arborescens*
  wild hydrangea - *Hydrangea arborescens, Rumex venosus*
hydrilla - *Hydrilla verticillata*
hydrolea, one-flowered hydrolea - *Hydrolea uniflora*
hygrophila
  Indian hygrophila - *Hygrophila polysperma*
  lake hygrophila - *Hygrophila lacustris*
hymenopappus, white-bracted hymenopappus - *Hymenopappus scabiosaeus*
hyssop
  anise-hyssop - *Agastache foeniculum*
  blue giant hyssop - *Agastache foeniculum*
  Carolina water-hyssop - *Bacopa caroliniana*
  catnip giant hyssop - *Agastache nepetoides*
  disc water-hyssop - *Bacopa rotundifolia*
  Eisen's water-hyssop - *Bacopa eisenii*
  fennel giant hyssop - *Agastache foeniculum*
  fragrant giant hyssop - *Agastache foeniculum*
  giant hyssop - *Agastache nepetoides, Agastache*
  golden hedge-hyssop - *Gratiola aurea*
  hyssop - *Hyssopus officinalis*
  hyssop bassia - *Bassia hyssopifolia*
  hyssop-leaf boneset - *Eupatorium hyssopifolium*
  hyssop-leaf loosestrife - *Lythrum hyssopifolium*
  hyssop-leaf thoroughwort - *Eupatorium hyssopifolium*
  hyssop-leaf tickseed - *Corispermum hyssopifolium*
  hyssop loosestrife - *Lythrum hyssopifolium*
  hyssop skullcap - *Scutellaria integrifolia*
  hyssop spurge - *Euphorbia hyssopifolia*
  lavender giant hyssop - *Agastache foeniculum*
  nameless water-hyssop - *Bacopa innominata*
  nettle-leaf giant hyssop - *Agastache urticifolia*
  purple giant hyssop - *Agastache scrophulariaefolia*
  water-hyssop - *Bacopa monnieri*
  yellow giant hyssop - *Agastache nepetoides*
  yellow hedge-hyssop - *Gratiola aurea*
ibapah spring parsley - *Cymopterus ibapensis*
Iberian
  Iberian crane's-bill - *Geranium ibericum*
  Iberian star-thistle - *Centaurea iberica*
icaco - *Chrysobalanus icaco*
ice-plant - *Mesembryanthemum crystallinum*
Iceland poppy - *Papaver nudicaule*
Idaho
  Idaho fescue - *Festuca idahoensis*
  Idaho redtop - *Agrostis idahoensis*
iigiri-tree - *Idesia polycarpa*
Illinois
  Illinois bundle-flower - *Desmanthus illinoensis*
  Illinois pondweed - *Potamogeton illinoensis*
immortelle - *Helichrysum*

imperial stock - *Matthiola incana*
impia, herba impia - *Filago germanica*
Inca wheat - *Amaranthus caudatus*
incense
  California incense-cedar - *Calocedrus decurrens*
  incense-cedar - *Calocedrus decurrens*
inch
  fern-leaf inch-plant - *Tripogandra multiflora*
  inch-plant - *Tradescantia zebrina*
India, pride-of-India - *Koelreuteria paniculata, Melia azedarach*
Indian
  California Indian pink - *Silene californica*
  East Indian fig-tree - *Ficus benghalensis*
  East Indian Jew's mallow - *Corchorus aestuans*
  East Indian lemongrass - *Cymbopogon flexuosus*
  East Indian lotus - *Nelumbo nucifera*
  East Indian rose-bay - *Tabernaemontana divaricata*
  great Indian plantain - *Cacalia muhlenbergii*
  hastate Indian plantain - *Cacalia suaveolens*
  Indian almond - *Terminalia cattapa*
  Indian apple - *Datura innoxia*
  Indian banyan - *Ficus benghalensis*
  Indian bean - *Catalpa speciosa*
  Indian bean catalpa - *Catalpa bignonioides*
  Indian beans - *Lupinus perennis*
  Indian buchu - *Myrtus communis*
  Indian caraway - *Perideridia gairdneri*
  Indian cherry - *Rhamnus caroliniana*
  Indian chickweed - *Mollugo verticillata*
  Indian chocolate - *Geum rivale*
  Indian crabgrass - *Digitaria longiflora*
  Indian cress - *Tropaeolum majus*
  Indian cucumber-root - *Medeola virginica*
  Indian cup - *Silphium perfoliatum*
  Indian cup-plant - *Sarracenia purpurea*
  Indian currant - *Symphoricarpos chenaultii, S. orbiculatus*
  Indian fig - *Opuntia ficus-indica*
  Indian grass - *Sorghastrum nutans*
  Indian hawthorn - *Rhaphiolepis indica*
  Indian heliotrope - *Heliotropium indicum*
  Indian hemp - *Abutilon theophrasti, Apocynum cannabinum, Crotalaria juncea, Hibiscus cannabinus*
  Indian hygrophila - *Hygrophila polysperma*
  Indian joint-vetch - *Aeschynomene indica*
  Indian jujube - *Ziziphus mauritiana*
  Indian laburnum - *Cassia fistula*
  Indian laurel - *Calophyllum inophyllum*
  Indian laurel fig - *Ficus microcarpa*
  Indian lettuce - *Montia linearis*
  Indian licorice - *Abrus precatorius*
  Indian lilac - *Melia azedarach*
  Indian love grass - *Eragrostis pilosa*
  Indian mallow - *Abutilon incanum, A. theophrasti, Abutilon*
  Indian milk-vetch - *Astragalus aboriginorum*
  Indian mint - *Plectranthus amboinicus*
  Indian mock-strawberry - *Duchesnea indica*
  Indian mulberry - *Morinda citrifolia*
  Indian mullet - *Oryzopsis hymenoides*
  Indian mustard - *Brassica juncea*
  Indian-paint - *Chenopodium capitatum, Lithospermum canescens*

Indian paintbrush - *Asclepias tuberosa, Castilleja affinis, C. coccinea, Castilleja*
Indian pea - *Lathyrus sativus*
Indian physic - *Apocynum cannabinum, Gillenia trifoliata*
Indian pink - *Lobelia cardinalis, Lychnis flos-cuculi, Spigelia marilandica*
Indian-pipe - *Monotropa uniflora*
Indian poke - *Veratrum viride*
Indian rhododendron - *Melastoma malabathricum*
Indian rice - *Oryzopsis hymenoides, Zizania aquatica*
Indian rosewood - *Dalbergia sissoo*
Indian rubber fig - *Ficus elastica*
Indian rubber-tree - *Ficus elastica*
Indian-salad - *Hydrophyllum virginianum*
Indian sanicle - *Ageratina altissima*
Indian senna - *Senna alexandrina*
Indian-shot - *Canna indica*
Indian sorrel - *Hibiscus sabdariffa*
Indian spice - *Vitex agnus-castus*
Indian spinach - *Basella alba*
Indian strawberry - *Duchesnea indica*
Indian sweet-clover - *Melilotus indica*
Indian thistle - *Cirsium edule*
Indian tobacco - *Lobelia inflata*
Indian tooth-cup - *Rotala indica*
Indian turnip - *Arisaema triphyllum, Psoralea esculenta*
Indian walnut - *Aleurites moluccana*
Indian waltheria - *Waltheria indica*
Indian-warrior - *Pedicularis densiflora*
Indian wheat - *Fagopyrum tataricum*
Indian wickup - *Epilobium angustifolium*
Indian wood-apple - *Limonia acidissima*
Monterey Indian paintbrush - *Castilleja latifolia*
pale Indian plantain - *Cacalia atriplicifolia*
sweet-scented Indian plantain - *Cacalia suaveolens*
West Indian birch - *Bursera simaruba*
West Indian black-thorn - *Acacia farnesiana*
West Indian dogwood - *Piscidia piscipula*
West Indian gherkin - *Cucumis anguria*
West Indian holly - *Leea coccinea*
West Indian lemongrass - *Cymbopogon citratus*
West Indian lime - *Citrus aurantiifolia*
West Indian mahogany - *Swietenia mahagoni*
West Indian tree-fern - *Cyathea arborea*
West Indian trema - *Trema lamarckiana*
woolly Indian paintbrush - *Castilleja foliolosa*
yellow Indian-shoe - *Cypripedium calceolus* var. *pubescens*
indigo
anil indigo - *Indigofera suffruticosa*
bastard-indigo - *Amorpha fruticosa*
blanket indigo - *Indigofera pilosa*
blue false indigo - *Baptisia australis*
blue wild indigo - *Baptisia australis*
creeping indigo - *Indigofera spicata*
curly-indigo - *Aeschynomene virginica*
dwarf wild indigo - *Amorpha nana*
false indigo - *Amorpha fruticosa, Amorpha, Baptisia*
false indigo-bush - *Amorpha fruticosa*
fragrant false indigo - *Amorpha nana*
hairy indigo - *Indigofera hirsuta*
indigo - *Indigofera*
indigo-bush - *Amorpha fruticosa, Dalea*
indigo-squill - *Camassia scilloides*

spicate indigo - *Indigofera spicata*
inflated
inflated bladderwort - *Utricularia inflata*
inflated duckweed - *Lemna gibba*
inkberry - *Ilex glabra, Ilex, Randia aculeata*
inkwood - *Exothea paniculata*
inland
inland bluegrass - *Poa interior*
inland sedge - *Carex interior*
inland serviceberry - *Amelanchier interior*
innocence - *Collinsia verna, Hedyotis caerulea*
insect
Dalmatian insect-flower - *Chrysanthemum cinerariifolium*
Persian insect-flower - *Chrysanthemum coccineum*
inside-out-flower - *Vancouveria planipetala*
interior live oak - *Quercus wislizenii*
intermediate
intermediate oat grass - *Danthonia intermedia*
intermediate wheat grass - *Elytrigia intermedia*
interrupted
interrupted club-moss - *Lycopodium annotinum*
interrupted fern - *Osmunda claytoniana*
ipecac
American ipecac - *Gillenia stipulata*
Carolina ipecac - *Euphorbia ipecacuanhae*
false ipecac - *Asclepias curassavica*
ipecac spurge - *Euphorbia ipecacuanhae*
milk ipecac - *Apocynum androsaemifolium*
wild ipecac - *Euphorbia ipecacuanhae*
ipil
ipil - *Leucaena leucocephala*
ipil-ipil - *Leucaena leucocephala*
ipomoea, star ipomoea - *Ipomoea coccinea*
Ireland, bells-of-Ireland - *Moluccella laevis*
iris
African iris - *Dietes iridioides*
blue flag iris - *Iris versicolor*
copper iris - *Iris fulva*
dwarf-crested iris - *Iris cristata*
dwarf iris - *Iris verna*
English iris - *Iris latifolia*
Japanese iris - *Iris ensata*
Japanese water iris - *Iris ensata*
red iris - *Iris fulva*
Rocky Mountain iris - *Iris missouriensis*
Spanish iris - *Iris xiphium*
vernal iris - *Iris verna*
violet iris - *Iris verna*
wild iris - *Iris versicolor*
yellow flag iris - *Iris pseudacorus*
Irish
Irish moss - *Soleirolia soleirolii*
Irish potato - *Solanum tuberosum*
Irish shamrock - *Oxalis acetosella, Trifolium dubium*
iron
cast-iron-plant - *Aspidistra elatior*
iron fern - *Rumohra adiantiformis*
iron-tree - *Metrosideros*
ironbark
ironbark - *Eucalyptus*
red ironbark - *Eucalyptus sideroxylon*

ironweed
  desert ironweed - *Olneya tesota*
  ironweed - *Galeopsis tetrahit, Vernonia noveboracensis,
    Vernonia*
  little ironweed - *Vernonia cinerea*
  New York ironweed - *Vernonia noveboracensis*
  tall ironweed - *Vernonia gigantea*
  western ironweed - *Vernonia baldwinii*
  yellow-ironweed - *Verbesina alternifolia*
ironwood
  Catalina ironwood - *Lyonothamnus*
  ironwood - *Carpinus caroliniana, Carpinus, Cliftonia
    monophylla, Cyrilla racemiflora, Olneya tesota, Ostrya
    virginiana, Sideroxylon lycioides*
  South Sea ironwood - *Casuarina equisetifolia*
island
  Canary Island date palm - *Phoenix canariensis*
  Canary Island pine - *Pinus canariensis*
  Catalina Island cherry - *Prunus ilicifolia* subsp. *lyonii*
  island live oak - *Quercus tomentella*
  island oak - *Quercus tomentella*
  Norfolk Island pine - *Araucaria heterophylla*
  Rhode Island bent grass - *Agrostis capillaris*
  Sea Island cotton - *Gossypium barbadense*
  Solomon Island ivy - *Epipremnum aureum*
islay - *Prunus ilicifolia*
Italian
  Italian broccoli - *Brassica oleracea* var. *italica*
  Italian bugloss - *Anchusa azurea*
  Italian clover - *Trifolium incarnatum*
  Italian corn-salad - *Valerianella eriocarpa*
  Italian cypress - *Cupressus sempervirens*
  Italian millet - *Setaria italica*
  Italian rye grass - *Lolium multiflorum*
  Italian stone pine - *Pinus pinea*
  Italian thistle - *Carduus pycnocephalus*
  Italian woodbine - *Lonicera caprifolium*
itch
  cow-itch - *Campsis radicans, Toxicodendron radicans*
  itch grass - *Rottboellia cochinchinensis*
  itch-weed - *Veratrum viride*
ivy
  American ivy - *Parthenocissus quinquefolia*
  aralia-ivy - *Fatshedera lizei*
  Boston ivy - *Parthenocissus tricuspidata*
  Cape ivy - *Senecio mikanioides*
  coliseum-ivy - *Cymbalaria muralis*
  devil's-ivy - *Epipremnum aureum*
  English ivy - *Hedera helix*
  five-leaved ivy - *Parthenocissus quinquefolia*
  German ivy - *Senecio mikanioides*
  grape ivy - *Cissus rhombifolia, Cissus*
  ground-ivy - *Glechoma hederacea*
  ivy - *Cissus, Hedera, Kalmia latifolia*
  ivy-arum - *Epipremnum aureum*
  ivy-bush - *Kalmia latifolia*
  ivy geranium - *Pelargonium peltatum*
  ivy-leaf morning-glory - *Ipomoea hederacea*
  ivy-leaf red morning-glory - *Ipomoea hederifolia*
  ivy-leaf speedwell - *Veronica hederifolia*
  Japanese ivy - *Parthenocissus tricuspidata*
  Kenilworth ivy - *Cymbalaria muralis*

  marine ivy - *Cissus incisa*
  parlor-ivy - *Senecio mikanioides*
  poison-ivy - *Toxicodendron pubescens, T. radicans*
  red-flame-ivy - *Hemigraphis alternata*
  red-ivy - *Hemigraphis alternata*
  redwood-ivy - *Vancouveria planipetala*
  Solomon Island ivy - *Epipremnum aureum*
  spider-ivy - *Chlorophytum comosum*
  Swedish ivy - *Plectranthus*
  switch-ivy - *Leucothoe fontanesiana*
  tree-ivy - *Fatshedera lizei*
  water-ivy - *Senecio mikanioides*
  western poison-ivy - *Toxicodendron rydbergii, T. vernix*
ixora, scarlet ixora - *Ixora coccinea*
izote - *Yucca elephantipes*
jaboncillo - *Sapindus saponaria, S. saponaria* var.
    drummondii
jaboticaba - *Myrciaria cauliflora*
jack
  Alabama supple-jack - *Berchemia scandens*
  jack bean - *Canavalia ensiformis*
  Jack-go-to-bed-at-noon - *Tragopogon pratensis*
  Jack-in-the-pulpit - *Arisaema triphyllum*
  jack oak - *Quercus ellipsoidalis, Q. imbricaria,
    Q. marilandica*
  jack pine - *Pinus banksiana*
  sand-jack - *Quercus incana*
  small Jack-in-the-pulpit - *Arisaema triphyllum*
  supple-jack - *Berchemia scandens*
  swamp Jack-in-the-pulpit - *Arisaema triphyllum*
  three-cornered-jack - *Emex australis*
jackfruit - *Artocarpus heterophyllus*
Jackson's brier - *Smilax smallii*
Jacob's
  blue Jacob's-ladder - *Polemonium caeruleum*
  Jacob's-coat - *Acalypha wilkesiana*
  Jacob's-ladder - *Polemonium caeruleum* subsp. *villosum,
    Polemonium*
  Jacob's-staff - *Fouquieria splendens*
  spreading Jacob's-ladder - *Polemonium reptans*
jade
  Chinese jade-plant - *Crassula arborescens*
  jade-tree - *Crassula ovata*
  jade-vine - *Strongylodon macrobotrys*
  ragged-jade - *Lychnis flos-cuculi*
  silver jade-plant - *Crassula arborescens*
jagged chickweed - *Holosteum umbellatum*
jaggery palm - *Caryota urens*
jalap, wild jalap - *Podophyllum peltatum*
Jamaican
  Jamaican caper - *Capparis cynophallophora*
  Jamaican caper-tree - *Capparis cynophallophora*
  Jamaican cherry - *Muntingia calabura*
  Jamaican dogwood - *Piscidia piscipula*
  Jamaican fever-plant - *Tribulus cistoides*
  Jamaican gold fern - *Pityrogramma sulphurea*
  Jamaican honeysuckle - *Passiflora laurifolia*
  Jamaican nectandra - *Ocotea coriacea*
  Jamaican nutmeg - *Monodora myristica*

Jamaican sorrel - *Hibiscus sabdariffa*
Jamaican vervain - *Stachytarpheta jamaicensis*
jambolan
    jambolan - *Syzygium cumini*
    jambolan-plum - *Syzygium cumini*
jambool - *Syzygium cumini*
jambos
    jambos - *Syzygium jambos*
    pomerac jambos - *Syzygium malaccense*
jambu - *Syzygium cumini*
Jamestown-weed - *Datura stramonium*
jang
    cat-jang - *Vigna unguiculata* subsp. *cylindrica*
    cat-jang pea - *Cajanus cajan*
Japanese
    Japanese alder - *Alnus japonica*
    Japanese andromeda - *Pieris japonica*
    Japanese anemone - *Anemone hupehensis* var. *japonica*
    Japanese aucuba - *Aucuba japonica*
    Japanese azalea - *Rhododendron japonicum*
    Japanese barberry - *Berberis thunbergii*
    Japanese bead-tree - *Melia azedarach*
    Japanese bindweed - *Calystegia hederacea*
    Japanese black pine - *Pinus thunbergiana*
    Japanese boxwood - *Buxus microphylla* var. *japonica*
    Japanese brome - *Bromus japonicus*
    Japanese bunching onion - *Allium fistulosum*
    Japanese bush-clover - *Lespedeza striata*
    Japanese carpet grass - *Zoysia matrella*
    Japanese cedar - *Cryptomeria japonica*
    Japanese chess - *Bromus japonicus*
    Japanese chestnut - *Castanea crenata*
    Japanese climbing fern - *Lygodium japonicum*
    Japanese clover - *Lespedeza striata*
    Japanese elm - *Ulmus japonica*
    Japanese false brome grass - *Brachypodium pinnatum*
    Japanese false cypress - *Chamaecyparis obtusa*
    Japanese fatsia - *Fatsia japonica*
    Japanese fern-palm - *Cycas revoluta*
    Japanese flowering cherry - *Prunus serrulata*
    Japanese flowering crabapple - *Malus floribunda*
    Japanese hedge-parsley - *Torilis japonica*
    Japanese holly - *Ilex crenata*
    Japanese holly fern - *Cyrtomium falcata*
    Japanese honey locust - *Gleditsia japonica*
    Japanese honeysuckle - *Lonicera japonica*
    Japanese hop - *Humulus japonicus*
    Japanese horse-chestnut - *Aesculus turbinata*
    Japanese iris - *Iris ensata*
    Japanese ivy - *Parthenocissus tricuspidata*
    Japanese knotweed - *Polygonum cuspidatum*
    Japanese-lantern - *Physalis alkekengi*
    Japanese larch - *Larix kaempferi*
    Japanese laurel - *Aucuba japonica*
    Japanese lawn grass - *Zoysia japonica*
    Japanese lespedeza - *Lespedeza striata*
    Japanese maple - *Acer palmatum*
    Japanese mat rush - *Juncus effusus*
    Japanese medlar - *Eriobotrya japonica*
    Japanese millet - *Echinochloa frumentacea, Setaria italica*
    Japanese mint - *Mentha arvensis*
    Japanese morning-glory - *Ipomoea nil*

    Japanese moss - *Soleirolia soleirolii*
    Japanese pachysandra - *Pachysandra terminalis*
    Japanese pagoda-tree - *Sophora japonica*
    Japanese pear - *Pyrus pyrifolia*
    Japanese persimmon - *Diospyros kaki*
    Japanese photinia - *Photinia glabra*
    Japanese pittosporum - *Pittosporum tobira*
    Japanese plum - *Eriobotrya japonica, Prunus salicina*
    Japanese plum-yew - *Cephalotaxus harringtonia* var. *drupacea*
    Japanese plume grass - *Miscanthus sinensis*
    Japanese poinsettia - *Euphorbia heterophylla, Pedilanthus tithymaloides*
    Japanese privet - *Ligustrum japonicum*
    Japanese quince - *Chaenomeles speciosa*
    Japanese raisin-tree - *Hovenia dulcis*
    Japanese red pine - *Pinus densiflora*
    Japanese rose - *Kerria japonica, Rosa rugosa*
    Japanese rubber-plant - *Crassula ovata*
    Japanese sago-palm - *Cycas revoluta*
    Japanese spindle-tree - *Euonymus japonicus*
    Japanese spiraea - *Spiraea japonica*
    Japanese-spurge - *Pachysandra terminalis*
    Japanese timber bamboo - *Phyllostachys bambusoides*
    Japanese Turk's-cap lily - *Lilium hansonii*
    Japanese umbrella-pine - *Sciadopitys verticillata*
    Japanese varnish-tree - *Firmiana simplex*
    Japanese walnut - *Juglans ailantifolia*
    Japanese water iris - *Iris ensata*
    Japanese wisteria - *Wisteria floribunda*
    Japanese yew - *Taxus cuspidata*
    Japanese zelkova - *Zelkova serrata*
    northern Japanese hemlock - *Tsuga diversifolia*
    showy Japanese lily - *Lilium speciosum*
jaragua grass - *Hyparrhenia rufa*
jarrah - *Eucalyptus marginata*
jarvil, sweet jarvil - *Osmorhiza claytonii*
jasmine
    angel-wing jasmine - *Jasminum nitidum*
    Arabian jasmine - *Jasminum sambac*
    blue-jasmine - *Clematis crispa*
    Cape jasmine - *Gardenia jasminoides*
    Carolina jasmine - *Gelsemium sempervirens*
    cinnamon-jasmine - *Hedychium coronarium*
    Confederate-jasmine - *Trachelospermum jasminoides*
    crape-jasmine - *Tabernaemontana divaricata*
    jasmine - *Jasminum*
    jasmine tobacco - *Nicotiana alata*
    orange-jasmine - *Murraya paniculata*
    rock-jasmine - *Androsace*
    star jasmine - *Jasminum multiflorum, J. nitidum*
    windmill jasmine - *Jasminum nitidum*
    winter jasmine - *Jasminum nudiflorum*
jaundice-berry - *Berberis vulgaris*
Java
    Java fig - *Ficus benjamina*
    Java plum - *Syzygium cumini*
javillo - *Hura crepitans*
Jeffrey's pine - *Pinus jeffreyi*
jelly, South American jelly palm - *Butia capitata*
Jennie, creeping-Jennie - *Convolvulus arvensis, Lysimachia nummularia*

jequirity bean - *Abrus precatorius*
Jersey
   Jersey pine - *Pinus virginiana*
   Jersey tea ceanothus - *Ceanothus americanus*
   New Jersey blueberry - *Vaccinium caesariense*
   New Jersey tea - *Ceanothus americanus*
Jerusalem
   asp-of-Jerusalem - *Isatis tinctoria*
   false Jerusalem cherry - *Solanum capsicastrum*
   Jerusalem artichoke - *Helianthus tuberosus*
   Jerusalem cherry - *Solanum pseudocapsicum*
   Jerusalem-cross - *Lychnis chalcedonica*
   Jerusalem oak goosefoot - *Chenopodium botrys*
   Jerusalem pea - *Vigna unguiculata* subsp. *cylindrica*
   Jerusalem pine - *Pinus halepensis*
   Jerusalem sage - *Phlomis tuberosa*
   Jerusalem tea - *Chenopodium ambrosioides*
   Jerusalem thorn - *Parkinsonia aculeata*
jessamine
   day jessamine - *Cestrum diurnum*
   night jessamine - *Cestrum nocturnum*
   orange jessamine - *Murraya paniculata*
   yellow jessamine - *Gelsemium sempervirens*
Jesuit-nut - *Trapa natans*
Jesus, heart-of-Jesus - *Caladium bicolor*
jet
   black jet-bead - *Rhodotypos scandens*
   jet-bead - *Rhodotypos*
Jew
   green wandering-Jew - *Tradescantia albiflora*
   wandering-Jew - *Tradescantia fluminensis, T. zebrina*
Jew's, East Indian Jew's mallow - *Corchorus aestuans*
jewels-of-Opar - *Talinum paniculatum*
jewelweed - *Impatiens capensis, I. pallida, Impatiens*
jicama - *Pachyrhizus erosus*
Jim Hill's mustard - *Sisymbrium altissimum*
jimmy-weed - *Haplopappus heterophyllus*
jimson-weed - *Datura stramonium*
Job's-tears - *Coix lacryma-jobi*
jocote - *Spondias purpurea*
Joe
   green-stemmed Joe-Pye-weed - *Eupatorium purpureum*
   hollow Joe-Pye-weed - *Eupatorium fistulosum*
   hollow-stemmed Joe-Pye-weed - *Eupatorium fistulosum*
   Joe-Pye-weed - *Eupatorium maculatum, E. purpureum*
   poor-Joe - *Diodia teres*
   purple-node Joe-Pye-weed - *Eupatorium purpureum*
   spotted Joe-Pye-weed - *Eupatorium maculatum*
   sweet Joe-Pye-weed - *Eupatorium purpureum*
Johnny-jump-up - *Viola tricolor*
Johnson grass - *Sorghum halepense*
joint
   American joint-vetch - *Aeschynomene americana*
   blue-joint - *Calamagrostis canadensis*
   Indian joint-vetch - *Aeschynomene indica*
   joint-fir - *Ephedra*
   joint-head arthraxon - *Arthraxon hispidus*
   joint-vetch - *Aeschynomene americana*
   joint-weed - *Polygonella polygama*
   joint-wood - *Cassia javanica* var. *indochinensis*
   northern joint-vetch - *Aeschynomene virginica*

jointed
   jointed charlock - *Raphanus raphanistrum*
   jointed flat sedge - *Cyperus articulatus*
   jointed goat grass - *Aegilops cylindrica*
jojoba - *Simmondsia chinensis*
jonquil - *Narcissus jonquilla*
Joseph's-coat - *Amaranthus tricolor*
Joshua-tree - *Yucca brevifolia*
Jove's-fruit - *Lindera melissaefolium*
joy
   joy-weed - *Alternanthera ficoidea* var. *amoena*
   sessile joy-weed - *Alternanthera sessilis*
   simpler's-joy - *Verbena hastata*
Jubas-bush - *Iresine diffusa*
Judge Daly's sunflower - *Helianthus maximilianii*
jujube
   Chinese jujube - *Ziziphus jujuba*
   cottony jujube - *Ziziphus mauritiana*
   Indian jujube - *Ziziphus mauritiana*
   jujube - *Ziziphus jujuba*
jumbie-bean - *Leucaena leucocephala*
jump, Johnny-jump-up - *Viola tricolor*
jumping
   jumping cactus - *Opuntia fulgida*
   jumping cholla - *Opuntia fulgida*
June grass - *Danthonia spicata, Koeleria pyrimidata, Poa pratensis*
Juneberry
   Juneberry - *Amelanchier*
   mountain Juneberry - *Amelanchier bartramiana*
jungle
   jungle-rice - *Echinochloa colona*
   jungle rice grass - *Echinochloa colona, E. crus-galli*
juniper
   alligator juniper - *Juniperus deppeana*
   Ashe's juniper - *Juniperus ashei*
   California juniper - *Juniperus californica, J. occidentalis*
   cherry-stone juniper - *Juniperus monosperma*
   Chinese juniper - *Juniperus chinensis*
   creeping juniper - *Juniperus horizontalis*
   creeping savin juniper - *Juniperus horizontalis*
   Greek juniper - *Juniperus excelsa*
   juniper - *Juniperus communis, Juniperus*
   low juniper - *Juniperus communis* subsp. *depressa*
   Mexican drooping juniper - *Juniperus flaccida*
   mountain juniper - *Juniperus communis*
   one-seeded juniper - *Juniperus monosperma*
   Pinchot's juniper - *Juniperus pinchotii*
   prickly juniper - *Juniperus oxycedrus*
   red-berry juniper - *Juniperus pinchotii*
   Rocky Mountain juniper - *Juniperus scopulorum*
   Sargent's juniper - *Juniperus chinensis* var. *sargentii*
   shore juniper - *Juniperus conferta*
   sierra juniper - *Juniperus occidentalis*
   sweet-fruited juniper - *Juniperus deppeana*
   Utah juniper - *Juniperus osteosperma*
   western juniper - *Juniperus occidentalis*
junta-de-cobra-pintada - *Justicia brandegeana*
Jupiter's-beard - *Centranthus ruber*
jute
   bastard-jute - *Hibiscus cannabinus*

bimli-jute - *Hibiscus cannabinus*
China jute - *Abutilon theophrasti*
Congo jute - *Urena lobata*
Cuban jute - *Sida rhombifolia*
kaffir
  kaffir - *Sorghum bicolor*
  kaffir-corn - *Sorghum bicolor*
Kahili ginger - *Hedychium gardnerianum*
kai
  kai-apple - *Dovyalis caffra*
  kai-tsoi - *Brassica juncea*
kaki
  kaki - *Diospyros kaki*
  kaki persimmon - *Diospyros kaki*
kalanchoe
  Christmas-tree kalanchoe - *Kalanchoe laciniata*
  fig-tree kalanchoe - *Kalanchoe laciniata*
kale
  kale - *Brassica oleracea, B. oleracea* var. *acephala*
  sea kale - *Crambe maritima, Crambe*
Kalm's St. John's-wort - *Hypericum kalmianum*
kalmia, bog kalmia - *Kalmia polifolia*
kalo - *Colocasia esculenta*
Kamchatka lily - *Fritillaria camschatcensis*
kangkong - *Ipomoea aquatica*
kanniedood aloe - *Aloe variegata*
Kansas
  Kansas hawthorn - *Crataegus coccinioides*
  Kansas sunflower - *Helianthus petiolaris*
kapok - *Ceiba pentandra*
kapundung - *Baccaurea racemosa*
karanda - *Carissa carandas*
karashina - *Brassica juncea*
karee - *Rhus lancea*
kariba-weed - *Salvinia molesta*
karo - *Pittosporum crassifolium*
karoo-tree - *Rhus lancea*
karri-tree - *Paulownia tomentosa*
karum-tree - *Pongamia pinnata*
Kashgar tamarisk - *Tamarix hispida*
Kashmir boquet - *Clerodendrum*
kat
  kat - *Catha edulis*
  kat-sola - *Aeschynomene indica*
katsura-tree - *Cercidiphyllum japonicum*
kau-apple - *Dovyalis caffra*
kawa'ii - *Ilex anomala* f. *sandwicensis*
kee, gow-kee - *Lycium barbarum*
keeled, sickle-keeled lupine - *Lupinus albicaulis*
keg-fig - *Diospyros kaki*
kei-apple - *Dovyalis caffra*
Kellerman's sunflower - *Helianthus kellermanii*
Kellogg's oak - *Quercus kelloggii*
kenaf - *Hibiscus cannabinus*
Kenai birch - *Betula papyrifera* var. *kenaica*
Kenilworth ivy - *Cymbalaria muralis*
kentia palm - *Howeia forsteriana*

Kentucky
  Kentucky bluegrass - *Poa pratensis*
  Kentucky coffee-tree - *Gymnocladus dioica*
Kenyan
  Kenyan clover - *Trifolium semipilosum*
  Kenyan wild white clover - *Trifolium semipilosum*
kepau - *Bougainvillea umbellifera*
kerria
  kerria - *Kerria japonica*
  white kerria - *Rhodotypos*
ketmia, bladder ketmia - *Hibiscus trionum*
key lime - *Citrus aurantiifolia*
khaki-weed - *Alternanthera caracasana*
khat - *Catha edulis*
Khesari - *Lathyrus sativus*
khus-khus - *Vetiveria zizanioides*
kidney
  kidney bean - *Phaseolus vulgaris*
  kidney-leaf - *Ranunculus abortivus*
  kidney-leaf violet - *Viola renifolia*
  kidney-vetch - *Anthyllis vulneraria*
  kidney-vine - *Galium asprellum*
  kidney-wood - *Eysenhardtia polystachya*
  kidney-wort - *Baccharis pilularis*
  kidney-wort baccharis - *Baccharis pilularis* var. *consanguinea*
Kikuyu grass - *Pennisetum clandestinum*
kill, calf-kill - *Kalmia angustifolia*
king
  glaucous king-devil - *Hieracium piloselloides*
  good-King-Henry - *Chenopodium bonus-henricus*
  king-devil - *Hieracium aurantiacum, H. floribundum*
  king-devil hawkweed - *Hieracium piloselloides*
  king-of-the-meadow - *Thalictrum pubescens*
  king orange - *Citrus reticulata*
  king protea - *Protea cynaroides*
  King Ranch blue-stem - *Bothriochloa ischaemum, B. ischaemum* var. *songarica*
  yellow king-devil - *Hieracium caespitosum*
king's
  king's-crown - *Justicia carnea*
  king's-cup - *Caltha palustris*
  king's-head - *Ambrosia trifida*
  king's lupine - *Lupinus kingii*
  king's-nut - *Carya laciniosa*
  king's palm - *Archontophoenix alexandrae*
  king's-spear - *Eremurus robustus*
kingfisher daisy - *Felicia bergeriana*
kings, tree-of-kings - *Cordyline terminalis*
kinnikinick - *Arctostaphylos uva-ursi*
kino - *Coccoloba uvifera*
kiri-tree - *Paulownia tomentosa*
kiss-me-over-the-garden-gate - *Polygonum orientale*
kitten's-tail - *Besseya rubra, Buchnera rubra*
kittul-tree - *Caryota urens*
kiwi
  kiwi - *Actinidia deliciosa*
  kiwi-fruit - *Actinidia deliciosa*
Klamath
  Klamath plum - *Prunus subcordata*

Klamath-weed - *Hypericum perforatum*
Kleberg's
  Kleberg's blue-stem - *Dichanthium annulatum*
  Kleberg's grass - *Dichanthium annulatum*
Klein's grass - *Panicum coloratum*
knapweed
  American knapweed - *Centaurea americana*
  big-head knapweed - *Centaurea macrocephala*
  black knapweed - *Centaurea nigra*
  brown knapweed - *Centaurea jacea*
  creeping knapweed - *Centaurea repens*
  diffuse knapweed - *Centaurea diffusa*
  feather-head knapweed - *Centaurea trichocephala*
  knapweed - *Centaurea*
  meadow knapweed - *Centaurea pratensis*
  Russian knapweed - *Acroptilon repens, Centaurea repens*
  spotted knapweed - *Centaurea maculosa*
  spreading knapweed - *Centaurea diffusa*
  squarrose knapweed - *Centaurea squarrosa*
  tumble knapweed - *Centaurea diffusa*
  Vochin knapweed - *Centaurea nigrescens*
knawel - *Scleranthus annuus*
knesheneka - *Rubus stellatus*
knife, Dutch-case-knife bean - *Phaseolus coccineus*
knob
  knob celery - *Apium graveolens* var. *rapaceum*
  knob-cone pine - *Pinus attenuata*
  knob sedge - *Cyperus pseudovegetus*
  knob-styled dogwood - *Cornus amomum*
knot
  knot-root bristle grass - *Setaria geniculata*
  knot-root fox-tail - *Setaria geniculata*
knotgrass
  German knotgrass - *Scleranthus annuus*
  knotgrass - *Elytrigia repens, Paspalum distichum*
knotted
  knotted clover - *Trifolium striatum*
  knotted hedge-parsley - *Torilis nodosa*
knotweed
  black-fringe knotweed - *Polygonum cilinode*
  bushy knotweed - *Polygonum ramosissimum*
  Douglas's knotweed - *Polygonum douglasii*
  erect knotweed - *Polygonum erectum*
  giant knotweed - *Polygonum sachalinense*
  Japanese knotweed - *Polygonum cuspidatum*
  knotweed - *Polygonum lapathifolium, Polygonum*
  prostrate knotweed - *Polygonum aviculare*
  sea beach knotweed - *Polygonum glaucum*
  silver-sheath knotweed - *Polygonum argyrocoleon*
  striate knotweed - *Polygonum achoreum*
  tufted knotweed - *Polygonum caespitosum* var *longisetum*
Knowlton's hop-hornbeam - *Ostrya knowltonii*
koa - *Acacia koa*
kochia - *Bassia scoparia*
kohlrabi - *Brassica oleracea, B. oleracea* var. *gongylodes*
kokio - *Kokia drynarioides*
Kong, Hong Kong kumquat - *Fortunella hindsii*
kopiko-tea - *Psychotria kaduana*
korakan - *Eleusine coracana*
korari - *Phormium tenax*

Korean
  Korean boxwood - *Buxus microphylla* var. *koreana*
  Korean bush-clover - *Kummerowia stipulacea*
  Korean clover - *Kummerowia stipulacea*
  Korean grass - *Zoysia japonica*
  Korean lawn grass - *Zoysia japonica*
  Korean lespedeza - *Kummerowia stipulacea*
kudzu
  kudzu - *Pueraria lobata*
  kudzu-vine - *Pueraria lobata*
  tropical kudzu - *Pueraria phaseoloides*
kui ts'ai - *Allium tuberosum*
kumquat
  Hong Kong kumquat - *Fortunella hindsii*
  kumquat - *Fortunella*
  oval kumquat - *Fortunella margarita*
kura
  kura - *Trifolium ambiguum*
  kura clover - *Trifolium ambiguum*
kurumaba-zakuro-so - *Mollugo verticillata*
kwa
  la-kwa - *Momordica charantia*
  sing-kwa - *Luffa acutangula*
  zit-kwa - *Benincasa hispida*
kyasuma grass - *Pennisetum pedicellatum*
kyllinga, green kyllinga - *Cyperus brevifolius*
lablab bean - *Lablab purpureus*
Labrador
  glandular Labrador tea - *Ledum glandulosum*
  Labrador pine - *Pinus banksiana*
  Labrador tea - *Ledum groenlandicum, Ledum*
  western Labrador tea - *Ledum glandulosum*
labriform milkweed - *Asclepias labriformis*
laburnum, Indian laburnum - *Cassia fistula*
lace
  blue lace-flower - *Trachymene coerulea*
  lace fern - *Sphenomeris chinensis*
  lace grass - *Eragrostis capillaris*
  lace lip fern - *Cheilanthes gracillima*
  lace-pod - *Thysanocarpus curvipes*
  Queen Anne's-lace - *Daucus carota*
  queen's-lace - *Daucus carota*
  St. Catharine's-lace - *Eriogonum giganteum*
ladder
  blue Jacob's-ladder - *Polemonium caeruleum*
  Jacob's-ladder - *Polemonium caeruleum* subsp. *villosum, Polemonium*
  spreading Jacob's-ladder - *Polemonium reptans*
ladies, Quaker-ladies - *Hedyotis caerulea*
ladino clover - *Trifolium repens*
lady
  lady fern - *Athyrium felix-femina*
  lady-of-Mexico cactus - *Mammillaria hahniana*
  lady-of-the-night - *Brunfelsia americana*
  lady palm - *Rhapis excelsa, Rhapis*
  lady-rue - *Thalictrum clavatum*
  Lady Washington's geranium - *Pelargonium domesticum*
  naked-lady-lily - *Amaryllis belladonna*
  northern lady fern - *Athyrium felix-femina* var. *michauxii*
  slender lady palm - *Rhapis excelsa*

lady's
    alpine lady's-mantle - *Alchemilla alpina*
    creeping lady's-sorrel - *Oxalis corniculata*
    lady's bedstraw - *Galium verum*
    lady's-eardrops - *Brunnichia ovata, Fuchsia triphylla*
    lady's-earrings - *Impatiens capensis*
    lady's-finger - *Abelmoschus esculentus*
    lady's-fingers - *Anthyllis vulneraria*
    lady's leek - *Allium cernuum*
    lady's-mantle - *Alchemilla xanthochlora, Alchemilla*
    lady's-slipper - *Cypripedium*
    lady's-smock - *Cardamine pratensis*
    lady's-sorrel - *Oxalis*
    lady's-thumb - *Polygonum persicaria*
    lady's-tobacco - *Antennaria plantaginifolia, Antennaria*
    large lady's-slipper - *Cypripedium calceolus* var. *pubescens*
    northern slender lady's-tresses - *Spiranthes lacera*
    pink lady's-slipper - *Cypripedium acaule*
    showy lady's-slipper - *Cypripedium reginae*
    slender lady's-tresses - *Spiranthes lacera*
    small white lady's-slipper - *Cypripedium candidum*
    two-leaf lady's-slipper - *Cypripedium acaule*
    western lady's-mantle - *Aphanes occidentalis*
    white lady's-slipper - *Cypripedium candidum*
lake hygrophila - *Hygrophila lacustris*
Lakes, Great Lakes gentian - *Gentiana rubricaulis*
lamb
    lamb mint - *Mentha piperita*
    lamb-succory - *Arnoseris minima*
lamb's
    lamb's-lettuce - *Valerianella locusta, Valerianella*
    lamb's-quarters - *Chenopodium album*
    lamb's-tongue - *Plantago media*
    narrow-leaf lamb's-quarters - *Chenopodium desiccatum*
    net-seed lamb's-quarters - *Chenopodium berlandieri*
    slim-leaf lamb's-quarters - *Chenopodium leptophyllum*
Lambert's
    Lambert's crazyweed - *Oxytropis lambertii*
    Lambert's loco - *Oxytropis lambertii*
lambkill - *Kalmia angustifolia*
lanate lip fern - *Cheilanthes lanosa*
lance
    lance-leaf arrowhead - *Sagittaria lancifolia*
    lance-leaf buckthorn - *Rhamnus lanceolata*
    lance-leaf coreopsis - *Coreopsis lanceolata*
    lance-leaf crotalaria - *Crotalaria lanceolata*
    lance-leaf grape fern - *Botrychium lanceolatum*
    lance-leaf greenbrier - *Smilax smallii*
    lance-leaf ground-cherry - *Physalis angulata*
    lance-leaf loosestrife - *Lysimachia lanceolata, Lythrum alatum* var. *lanceolatum*
    lance-leaf pickerelweed - *Pontederia cordata* var. *lancifolia*
    lance-leaf ragweed - *Ambrosia bidentata*
    lance-leaf sage - *Salvia reflexa*
    lance-leaf sagittaria - *Sagittaria lancifolia*
    lance-leaf violet - *Viola lanceolata*
    lance-leaf water-plantain - *Alisma lanceolatum*
lanceolate milkweed - *Asclepias lanceolata*
lancewood - *Pseudopanax crassifolia*
land
    dry-land bitter-cress - *Cardamine parviflora*
    pine-land three-awn - *Aristida stricta*

langsat - *Lansium domesticum*
lantana
    lantana - *Lantana camara*
    large-leaf lantana - *Lantana camara*
    trailing lantana - *Lantana montevidensis*
    weeping lantana - *Lantana montevidensis*
lantern
    Chinese lantern-plant - *Physalis alkekengi*
    Japanese-lantern - *Physalis alkekengi*
Lapland
    Lapland rhododendron - *Rhododendron lapponicum*
    Lapland rose-bay - *Rhododendron lapponicum*
larch
    American larch - *Larix laricina*
    black larch - *Larix laricina*
    European larch - *Larix decidua*
    golden larch - *Pseudolarix amabilis*
    Japanese larch - *Larix kaempferi*
    larch - *Larix*
    subalpine larch - *Larix lyallii*
    western larch - *Larix occidentalis*
large
    large beard-tongue - *Penstemon grandiflorus*
    large-bracted tick-trefoil - *Desmodium cuspidatum*
    large cane - *Arundinaria gigantea*
    large coralroot - *Corallorrhiza maculata*
    large crabgrass - *Digitaria sanguinalis*
    large cranberry - *Vaccinium macrocarpon*
    large-flowered beard-tongue - *Penstemon grandiflorus*
    large-flowered bellwort - *Uvularia grandiflora*
    large-flowered leaf-cup - *Polymnia uvedalia*
    large-flowered trillium - *Trillium grandiflorum*
    large-flowered water-plantain - *Alisma plantago-aquatica*
    large-fruit dodder - *Cuscuta umbrosa*
    large-fruit rose-apple - *Syzygium malaccense*
    large gallberry - *Ilex coriacea*
    large hop clover - *Trifolium campestre*
    large houstonia - *Hedyotis purpurea*
    large lady's-slipper - *Cypripedium calceolus* var. *pubescens*
    large-leaf aster - *Aster macrophyllus*
    large-leaf avens - *Geum macrophyllum*
    large-leaf cucumber-tree - *Magnolia macrophylla*
    large-leaf holly - *Ilex ambigua* var. *montana*
    large-leaf lantana - *Lantana camara*
    large-leaf linden - *Tilia platyphyllos*
    large-leaf lupine - *Lupinus polyphyllus*
    large-leaf pondweed - *Potamogeton amplifolius*
    large-leaf water-leaf - *Hydrophyllum macrophyllum*
    large-leaf white violet - *Viola incognita*
    large purple aster - *Aster patens*
    large pussy willow - *Salix discolor*
    large Russian vetch - *Vicia villosa*
    large sand rocket - *Diplotaxis tenuifolia*
    large-seeded dodder - *Cuscuta indecora, C. pentagona*
    large-seeded false flax - *Camelina sativa*
    large-tooth aspen - *Populus grandidentata*
    large tupelo - *Nyssa aquatica*
    large twayblade - *Liparis liliifolia*
    large vetch - *Vicia gigantea*
    large water-lily - *Nymphaea ampla*
    large white petunia - *Petunia axillaris*
larger blue flag - *Iris versicolor*

larkspur
  bouquet larkspur - *Delphinium grandiflorum*
  candle larkspur - *Delphinium elatum*
  cardinal larkspur - *Delphinium cardinale*
  desert larkspur - *Delphinium andersonii*
  dunce-cap larkspur - *Delphinium occidentale*
  dwarf larkspur - *Delphinium nuttallianum, D. tricorne*
  field larkspur - *Delphinium consolida, D. menziesii*
  forking larkspur - *Delphinium consolida*
  Geyer's larkspur - *Delphinium geyeri*
  larkspur - *Delphinium*
  larkspur violet - *Viola palmata* var. *pedatifida*
  low larkspur - *Delphinium menziesii, D. nuttallianum*
  meadow larkspur - *Delphinium nuttallianum*
  mountain larkspur - *Delphinium glaucum*
  pale larkspur - *Delphinium glaucum*
  rocket larkspur - *Consolida ajacis*
  scarlet larkspur - *Delphinium cardinale*
  spring larkspur - *Delphinium tricorne*
  tall larkspur - *Delphinium barbeyi, D. exaltatum,
    D. trolliifolium*
  wood larkspur - *Delphinium trolliifolium*
lasiandra - *Tibouchina urvilleana*
late
  late boneset - *Eupatorium serotinum*
  late coralroot - *Corallorrhiza odontorhiza*
  late eupatorium - *Eupatorium serotinum*
  late-flowering goosefoot - *Chenopodium strictum* var.
    glaucophyllum
  late-flowering thoroughwort - *Eupatorium serotinum*
  late goldenrod - *Solidago gigantea*
  late milk-vetch - *Astragalus miser* var. *serotinus*
  late sweet blueberry - *Vaccinium angustifolium*
  late water grass - *Echinochloa phyllopogon*
laurel
  Alexandrian-laurel - *Calophyllum inophyllum*
  alpine-laurel - *Kalmia polifolia*
  American laurel - *Kalmia*
  Australian laurel - *Pittosporum tobira*
  beaver-tree-laurel - *Magnolia virginiana*
  black-laurel - *Gordonia lasianthus*
  bog-laurel - *Kalmia polifolia*
  California laurel - *Umbellularia californica*
  Carolina cherry-laurel - *Prunus caroliniana*
  cherry-laurel - *Prunus laurocerasus*
  Chinese laurel - *Antidesma bunius*
  drooping-laurel - *Leucothoe fontanesiana*
  dwarf-laurel - *Kalmia angustifolia*
  English cherry-laurel - *Prunus laurocerasus*
  great-laurel - *Rhododendron maximum*
  Grecian laurel - *Laurus nobilis*
  Indian laurel - *Calophyllum inophyllum*
  Indian laurel fig - *Ficus microcarpa*
  Japanese laurel - *Aucuba japonica*
  laurel - *Kalmia, Laurus nobilis*
  laurel-leaved greenbrier - *Smilax laurifolia*
  laurel-leaved oak - *Quercus laurifolia*
  laurel magnolia - *Magnolia virginiana*
  laurel oak - *Quercus imbricaria, Q. laurifolia*
  laurel sumac - *Malosma laurina*
  laurel-tree - *Persea borbonia*
  laurel willow - *Salix pentandra*

  laurel-wood - *Calophyllum inophyllum*
  mountain dog-laurel - *Leucothoe fontanesiana*
  mountain-laurel - *Kalmia latifolia*
  narrow-leaf-laurel - *Kalmia angustifolia*
  pale-laurel - *Kalmia polifolia*
  pig-laurel - *Kalmia angustifolia*
  poison-laurel - *Kalmia latifolia*
  Portuguese cherry-laurel - *Prunus lusitanica*
  purple laurel - *Rhododendron catawbiense*
  sheep-laurel - *Kalmia angustifolia*
  swamp dog-laurel - *Leucothoe axillaris*
  swamp-laurel - *Kalmia polifolia*
  Texas mountain-laurel - *Sophora secundiflora*
  tropic-laurel - *Ficus benjamina*
  variegated-laurel - *Codiaeum*
  white-laurel - *Rhododendron maximum*
laurustinus - *Viburnum tinus*
lavender
  desert lavender - *Hyptis emoryi*
  English lavender - *Lavandula angustifolia*
  lavender - *Lavandula angustifolia*
  lavender-cotton - *Santolina chamaecyparissus*
  lavender giant hyssop - *Agastache foeniculum*
  lavender-scallops - *Kalanchoe fedtschenkoi*
  sea-lavender - *Limonium*
law's
  mother-in-law's-tongue - *Sansevieria trifasciata*
  mother-in-law's-tongue-plant - *Dieffenbachia*
lawn
  Japanese lawn grass - *Zoysia japonica*
  Korean lawn grass - *Zoysia japonica*
  lawn burweed - *Soliva sessilis*
  lawn-leaf - *Dichondra carolinensis*
  lawn pennywort - *Hydrocotyle sibthorpioides*
Lawson's cypress - *Chamaecyparis lawsoniana*
lax water-milfoil - *Myriophyllum laxum*
lazy-man's grass - *Eremochloa ophiuroides*
lead
  birch-lead mountain-mahogany - *Cercocarpus betuloides*
  great lead-tree - *Leucaena pulverulenta*
  lead-grass - *Salicornia virginica*
  lead-plant - *Amorpha canescens*
  lead-tree - *Leucaena leucocephala*
  lead-wood - *Krugiodendron ferreum*
  little-leaf lead-tree - *Leucaena retusa*
  smooth lead-plant - *Amorpha nana*
leadwort, Cape leadwort - *Plumbago auriculata*
leafless cynanchum - *Cynanchum scoparium*
leafy
  leafy bladderwort - *Utricularia foliosa*
  leafy bog aster - *Aster nemoralis*
  leafy pondweed - *Potamogeton foliosus*
  leafy spurge - *Euphorbia esula*
  leafy-stemmed plantain - *Plantago psyllium*
  leafy thistle - *Cirsium foliosum*
least willow - *Salix rotundifolia*
leather
  leather fern - *Acrostichum aureum, Rumohra adiantiformis*
  leather-flower - *Clematis glaucophylla, C. versicolor,
    C. viorna, C. virginiana, Clematis*
  leather-leaf - *Chamaedaphne calyculata*

leather-leaf elm - *Ulmus parvifolia*
leather oak - *Quercus durata*
leather-root - *Psoralea macrostachya*
leather wood fern - *Dryopteris marginalis*
leatherwood
    leatherwood - *Cyrilla racemiflora, Dirca palustris*
    southern leatherwood - *Cyrilla racemiflora*
Leavenworth's goldenrod - *Solidago leavenworthii*
leaves, little five-leaves - *Isotria medeoloides*
lebbeck - *Albizia lebbeck*
lebbek-tree - *Albizia lebbeck*
Leconte's
    Leconte's sedge - *Cyperus lecontei*
    Leconte's violet - *Viola affinis*
leechee - *Litchi chinensis*
leek
    house-leek - *Sempervivum tectorum*
    lady's leek - *Allium cernuum*
    leek - *Allium porrum*
    meadow leek - *Allium canadense*
    rose leek - *Allium canadense*
    stone leek - *Allium fistulosum*
    wild leek - *Allium tricoccum*
Lehmann's
    Lehmann's gum - *Eucalyptus lehmannii*
    Lehmann's love grass - *Eragrostis lehmanniana*
lehua, 'ohi'a lehua - *Metrosideros polymorpha*
Leiberg's bluegrass - *Poa leibergii*
lemandarin - *Citrus limonia*
Lemmon's needle grass - *Stipa lemmonii*
lemon
    lemon - *Citrus limon*
    lemon balm - *Melissa officinalis*
    lemon bee balm - *Monarda citriodora*
    lemon-lily - *Hemerocallis lilioasphodelus*
    lemon mint - *Mentha piperita* var. *citrata, Monarda citriodora*
    lemon-scented gum - *Eucalyptus citriodora*
    lemon scurf pea - *Psoralidium lanceolatum*
    lemon sumac - *Rhus aromatica*
    lemon thyme - *Thymus serpyllum*
    rough lemon - *Citrus jambhiri*
    water-lemon - *Passiflora laurifolia*
    wild lemon - *Podophyllum peltatum*
    wild water-lemon - *Passiflora foetida*
lemonade
    lemonade-berry - *Rhus integrifolia*
    lemonade sumac - *Rhus integrifolia*
lemongrass
    East Indian lemongrass - *Cymbopogon flexuosus*
    lemongrass - *Cymbopogon citratus*
    West Indian lemongrass - *Cymbopogon citratus*
lens-podded white-top - *Cardaria chalepensis*
lenticular sedge - *Carex lenticularis*
lentil
    lentil - *Lens culinaris*
    lentil tare - *Vicia tetrasperma*
lentisco - *Rhus virens*
leopard
    leopard-bane - *Doronicum plantagineum*
    leopard-flower - *Belamcanda chinensis*

leopard lily - *Lilium catesbaei, L. pardalinum*
leopard palm - *Amorphophallus rivieri*
lepyrodiclis - *Lepyrodiclis holosteoides*
les, tous-les-mois - *Canna indica*
lespedeza
    annual lespedeza - *Lespedeza striata*
    bicolored lespedeza - *Lespedeza bicolor*
    Chinese lespedeza - *Lespedeza cuneata*
    downy trailing lespedeza - *Lespedeza procumbens*
    Japanese lespedeza - *Lespedeza striata*
    Korean lespedeza - *Kummerowia stipulacea*
    lespedeza - *Lespedeza striata*
    lespedeza dodder - *Cuscuta pentagona*
    perennial lespedeza - *Lespedeza cuneata*
    round-head lespedeza - *Lespedeza capitata*
    Sericea lespedeza - *Lespedeza cuneata*
    shrub lespedeza - *Lespedeza bicolor*
    slender lespedeza - *Lespedeza violacea, L. virginica*
    smooth trailing lespedeza - *Lespedeza repens*
    striate lespedeza - *Lespedeza striata*
    Thunberg's lespedeza - *Lespedeza thunbergii*
    velvety lespedeza - *Lespedeza stuevei*
    violet lespedeza - *Lespedeza violacea*
    Virginia lespedeza - *Lespedeza virginica*
    wand lespedeza - *Lespedeza intermedia*
lesser
    lesser broomrape - *Orobanche minor*
    lesser burdock - *Arctium minus*
    lesser celandine - *Ranunculus ficaria*
    lesser duckweed - *Lemna minor*
    lesser periwinkle - *Vinca minor*
    lesser prickly sedge - *Carex muricata*
    lesser pyrola - *Pyrola minor*
    lesser snapdragon - *Antirrhinum orontium*
    lesser starwort - *Stellaria graminea*
    lesser stitch-wort - *Stellaria graminea*
    lesser toadflax - *Chaenorhinum minus*
    lesser wintergreen - *Pyrola minor*
Letterman's needle grass - *Stipa lettermanii*
lettuce
    biennial lettuce - *Lactuca biennis*
    blue lettuce - *Lactuca tatarica* subsp. *pulchella*
    Canadian wild lettuce - *Lactuca canadensis*
    Cos lettuce - *Lactuca sativa* var. *longifolia*
    desert wire-lettuce - *Stephanomeria pauciflora*
    devil's-lettuce - *Amsinckia tessellata*
    Florida wild lettuce - *Lactuca floridana*
    garden lettuce - *Lactuca sativa*
    glaucous white-lettuce - *Prenanthes racemosa*
    grass-leaf lettuce - *Lactuca graminifolia*
    head lettuce - *Lactuca sativa* var. *capitata*
    Indian lettuce - *Montia linearis*
    lamb's-lettuce - *Valerianella locusta, Valerianella*
    lettuce - *Lactuca sativa, Lactuca*
    lettuce saxifrage - *Saxifraga micranthidifolia*
    miner's-lettuce - *Claytonia perfoliata, Montia*
    mountain-lettuce - *Saxifraga micranthidifolia*
    narrow-leaf miner's-lettuce - *Montia linearis*
    prickly lettuce - *Lactuca serriola*
    Romaine lettuce - *Lactuca sativa* var. *longifolia*
    slender wire-lettuce - *Stephanomeria tenuifolia*
    smooth white-lettuce - *Prenanthes racemosa*

tall lettuce - *Lactuca canadensis*
tall white-lettuce - *Prenanthes altissima*
wall lettuce - *Lactuca muralis*
water-lettuce - *Pistia stratiotes*
white-lettuce - *Prenanthes alba*
wild lettuce - *Lactuca biennis*
willow-leaf lettuce - *Lactuca saligna*
wire-lettuce - *Stephanomeria pauciflora, Stephanomeria*
woodland lettuce - *Lactuca floridana*
leucaena - *Leucaena leucocephala*
Levant cotton - *Gossypium herbaceum*
leverwood - *Ostrya virginiana*
liane, pomme-de-liane - *Passiflora laurifolia*
licaria, gulf licaria - *Licaria triandra*
lice
  beggar's-lice - *Cynoglossum, Hackelia*
  harvest-lice - *Agrimonia*
  nits-and-lice - *Hypericum drummondii*
licorice
  American licorice - *Glycyrrhiza lepidota*
  Chinook licorice - *Lupinus littoralis*
  Indian licorice - *Abrus precatorius*
  licorice - *Glycyrrhiza glabra*
  licorice fern - *Polypodium glycyrrhiza*
  licorice goldenrod - *Solidago odora*
  licorice milk-vetch - *Astragalus glycyphyllos*
  licorice-root - *Glycyrrhiza glabra*
  licorice-vine - *Abrus precatorius*
  wild licorice - *Abrus precatorius, Galium circaezans,*
    G. lanceolatum, Glycyrrhiza lepidota
  wild white licorice - *Galium circaezans*
  yellow wild licorice - *Galium lanceolatum*
lid, pale lid-flower - *Calyptranthes pallens*
lies, love-lies-bleeding - *Amaranthus caudatus*
life
  life-everlasting - *Anaphalis*
  life-of-man - *Aralia racemosa*
  life-plant - *Kalanchoe pinnata*
lightening, scarlet-lightening - *Lychnis chalcedonica*
lignum
  holy-wood lignum-vitae - *Guajacum sanctum*
  lignum-vitae - *Guajacum sanctum*
lilac
  Amur lilac - *Syringa amurensis*
  Catalina mountain-lilac - *Ceanothus arboreus*
  Chinese lilac - *Syringa chinensis*
  Hungarian lilac - *Syringa josikaea*
  Indian lilac - *Melia azedarach*
  lilac - *Syringa vulgaris, Syringa*
  lilac chaste-tree - *Vitex agnus-castus*
  Persian lilac - *Melia azedarach, Syringa persica*
  summer-lilac - *Buddleja davidii*
  wild lilac - *Ceanothus sanguineus*
Lilian, blonde-Lilian - *Erythronium albidum*
lily
  African blood-lily - *Haemanthus coccineus*
  African lily - *Agapanthus africanus*
  Amazon lily - *Eucharis grandiflora*
  Amazon water-lily - *Victoria amazonica*
  American Turk's-cap lily - *Lilium superbum*
  American water-lily - *Nymphaea odorata*

arum-lily - *Zantedeschia aethiopica*
atamasco-lily - *Zephyranthes atamasco*
avalanche-lily - *Erythronium grandiflorum*
banana water-lily - *Nymphaea mexicana*
Barbados lily - *Hippeastrum puniceum*
belladonna-lily - *Amaryllis belladonna*
Bermuda Easter lily - *Lilium longiflorum* var. *eximium*
big blue lily-turf - *Liriope muscari*
blackberry-lily - *Belamcanda chinensis*
blood-lily - *Haemanthus coccineus*
blue-bead-lily - *Clintonia borealis*
blue plantain-lily - *Hosta ventricosa*
blue water-lily - *Nymphaea elegans*
boat-lily - *Tradescantia spathacea*
butterfly-lily - *Hedychium coronarium*
calla-lily - *Zantedeschia aethiopica, Zantedeschia*
Canadian lily - *Lilium canadense*
candlestick lily - *Lilium dauricum, L. hollandicum*
Cape blue water-lily - *Nymphaea capensis*
Caucasian lily - *Lilium monadelphum*
checker-lily - *Fritillaria affines*
cobra-lily - *Darlingtonia californica*
Columbia lily - *Lilium columbianum*
corn-lily - *Clintonia borealis*
cow-lily - *Nuphar*
crane-lily - *Strelitzia reginae*
crinum-lily - *Crinum*
Easter lily - *Lilium longiflorum* var. *eximium*
Eucharist-lily - *Eucharis grandiflora*
European white water-lily - *Nymphaea alba*
fairy water-lily - *Nymphoides aquatica*
false lily-of-the-valley - *Maianthemum canadense,*
  Maianthemum
fawn-lily - *Erythronium*
Fenzel's water-lily - *Nymphaea glandulifera*
fire-lily - *Cyrtanthus angusitfolius*
flamingo-lily - *Anthurium andraeanum*
fragrant plantain-lily - *Hosta plantaginea*
fragrant water-lily - *Nymphaea odorata*
garland-lily - *Hedychium*
George's-lily - *Cyrtanthus elatus*
giant water-lily - *Victoria*
ginger-lily - *Alpinia, Hedychium coronarium, Hedychium*
glacier-lily - *Erythronium grandiflorum*
gloriosa-lily - *Gloriosa superba*
gold-banded lily - *Lilium auratum*
golden calla-lily - *Zantedeschia elliottiana*
golden-rayed lily - *Lilium auratum*
Guernsey lily - *Nerine sarniensis*
Japanese Turk's-cap lily - *Lilium hansonii*
Kamchatka lily - *Fritillaria camschatcensis*
large water-lily - *Nymphaea ampla*
lemon-lily - *Hemerocallis lilioasphodelus*
leopard lily - *Lilium catesbaei, L. pardalinum*
lily-of-the-Amazon - *Eucharis grandiflora*
lily-of-the-field - *Anemone*
lily-of-the-Nile - *Agapanthus africanus*
lily-of-the-valley - *Convallaria majalis*
lily-of-the-valley-bush - *Pieris japonica*
lily-royal - *Lilium superbum*
lily-turf - *Liriope*
Madonna-lily - *Eucharis grandiflora, Lilium candidum*
magic-lily - *Lycoris squamigera*

magnolia water-lily - *Nymphaea tuberosa*
mariposa-lily - *Calochortus*
Martagon lily - *Lilium martagon*
Mexican water-lily - *Nymphaea mexicana*
Michigan lily - *Lilium michiganense*
mountain lily - *Lilium auratum*
naked-lady-lily - *Amaryllis belladonna*
Nankeen lily - *Lilium testaceum*
narrow-leaf plantain-lily - *Hosta lancifolia*
orange-cup lily - *Lilium philadelphicum*
orange lily - *Lilium bulbiferum*
Oregon lily - *Lilium columbianum*
palm-lily - *Yucca gloriosa*
panther lily - *Lilium pardalinum*
pig-lily - *Zantedeschia aethiopica*
pine lily - *Lilium catesbaei*
pink calla-lily - *Zantedeschia rehmannii*
pond-lily - *Nymphaea odorata*
queen-lily - *Curcuma petiolata*
red calla-lily - *Zantedeschia rehmannii*
red spider-lily - *Lycoris radiata*
regal lily - *Lilium regale*
resurrection-lily - *Lycoris squamigera*
royal lily - *Lilium regale*
royal water-lily - *Victoria amazonica*
sago-lily - *Calochortus*
sand-lily - *Leucocrinum montanum, Leucocrinum*
Scarborough-lily - *Cyrtanthus elatus*
sego-lily - *Calochortus nuttallii*
showy Japanese lily - *Lilium speciosum*
showy lily - *Lilium speciosum*
Siberian lily - *Ixiolirion tataricum*
snowdon-lily - *Lloydia serotina*
southern red lily - *Lilium catesbaei*
spice-lily - *Manfreda maculosa*
spider-lily - *Crinum, Hymenocallis*
star-flowered lily-of-the-valley - *Smilacina stellata*
star-lily - *Leucocrinum montanum, Leucocrinum*
swamp lily - *Lilium superbum*
tartar-lily - *Ixiolirion tataricum*
tiger lily - *Lilium lancifolium*
torch-lily - *Kniphofia uvaria, Kniphofia*
triplet-lily - *Triteleia laxa*
trout-lily - *Erythronium americanum, Erythronium*
trumpet lily - *Lilium longiflorum*
tuberous water-lily - *Nymphaea tuberosa*
turban lily - *Lilium martagon*
Turk's-cap lily - *Lilium martagon, L. superbum*
Washington's lily - *Lilium washingtonianum*
water-lily - *Nymphaea*
western orange-cup lily - *Lilium philadelphicum* var. *andinum*
white calla-lily - *Zantedeschia aethiopica*
white-trumpet lily - *Lilium longiflorum*
white water-lily - *Nymphaea odorata*
wild lily-of-the-valley - *Maianthemum canadense, Pyrola elliptica, P. rotundifolia*
wild orange-red lily - *Lilium philadelphicum*
wild yellow lily - *Lilium canadense*
wood lily - *Lilium philadelphicum*
yellow bell lily - *Lilium canadense*
yellow calla-lily - *Zantedeschia elliottiana*
yellow pond-lily - *Nuphar*
yellow water-lily - *Nuphar lutea, Nymphaea mexicana*

zephyr-lily - *Zephyranthes candida*
lima bean - *Phaseolus lunatus*
limber
  limber honeysuckle - *Lonicera dioica*
  limber pine - *Pinus flexilis*
limbo, gumbo-limbo - *Bursera simaruba*
lime
  key lime - *Citrus aurantiifolia*
  lime - *Citrus aurantiifolia*
  lime-berry - *Triphasia trifolia*
  lime prickly-ash - *Zanthoxylum fagara*
  Mandarin lime - *Citrus limonia*
  Mexican lime - *Citrus aurantiifolia*
  Ogeechee lime tupelo - *Nyssa ogeche*
  pendant white-lime - *Tilia petiolaris*
  Rangpur lime - *Citrus limonia*
  Spanish lime - *Melicoccus bijugatus*
  Tahitian lime - *Citrus latifolia*
  weeping-lime - *Tilia petiolaris*
  West Indian lime - *Citrus aurantiifolia*
  wild lime - *Zanthoxylum fagara*
limnophila - *Limnophila sessiliflora*
limpo grass - *Hemarthria altissima*
linda-tarde - *Gaura coccinea*
linden
  American linden - *Tilia americana*
  big-leaf linden - *Tilia platyphyllos*
  broad-leaf linden - *Tilia platyphyllos*
  European linden - *Tilia cordata*
  large-leaf linden - *Tilia platyphyllos*
  little-leaf linden - *Tilia cordata*
  pendant silver linden - *Tilia petiolaris*
Linden's viburnum - *Viburnum dilatatum*
Lindheimer's prickly-pear - *Opuntia lindheimeri*
lined bulrush - *Scirpus pendulus*
ling
  ling - *Trapa natans*
  ling-berry - *Vaccinium vitis-idaea* var. *minus*
lingon - *Vaccinium vitis-idaea* var. *minus*
lingonberry - *Vaccinium vitis-idaea, V. vitis-idaea* var. *minus*
link-leaf - *Schlumbergera truncata*
linseed - *Linum usitatissimum*
lion's
  lion's-beard - *Anemone patens*
  lion's-ear - *Leonotis nepetifolia, Leonotis, Leonurus cardiaca*
  lion's-foot - *Prenanthes serpentaria*
  lion's-tail - *Leonotis, Leonurus cardiaca*
  lion's-tooth - *Leontodon autumnalis*
lip
  Fee's lip fern - *Cheilanthes feei*
  lace lip fern - *Cheilanthes gracillima*
  lanate lip fern - *Cheilanthes lanosa*
  lip fern - *Cheilanthes*
  slender lip fern - *Cheilanthes feei*
  woolly lip fern - *Cheilanthes lanosa*
lippia, mat lippia - *Phyla nodiflora*
lips, blue-lips - *Collinsia grandiflora*
lipstick
  lipstick-plant - *Aeschynanthus pulcher*
  lipstick-tree - *Bixa orellana*

lipstick-vine - *Aeschynanthus pulcher*
lis, fleur-de-lis - *Iris germanica, Iris*
litchi - *Litchi chinensis*
little
    little barley - *Hordeum pusillum*
    little bellwort - *Uvularia sessilifolia*
    little bitter-cress - *Cardamine oligosperma*
    little blue-stem - *Schizachyrium scoparium*
    little bur-clover - *Medicago minima*
    little club-moss - *Selaginella*
    little-fantasy peperomia - *Peperomia caperata*
    little five-leaves - *Isotria medeoloides*
    little grape fern - *Botrychium simplex*
    little-hip hawthorn - *Crataegus spathulata*
    little ironweed - *Vernonia cinerea*
    little-leaf boxwood - *Buxus microphylla*
    little-leaf horse-brush - *Tetradymia glabrata*
    little-leaf huckleberry - *Vaccinium scoparium*
    little-leaf lead-tree - *Leucaena retusa*
    little-leaf linden - *Tilia cordata*
    little-leaf sensitive-brier - *Mimosa quadrivalvis* var. *angustata*
    little-leaf spurge - *Euphorbia micromera*
    little-leaf tick-trefoil - *Desmodium ciliare*
    little love grass - *Eragrostis poaeoides*
    little mallow - *Malva parviflora*
    little prickly sedge - *Carex sterilis*
    little quaking grass - *Briza minor*
    little rice grass - *Oryzopsis exigua*
    little-seed canary grass - *Phalaris minor*
    little-seed rice grass - *Oryzopsis micrantha*
    little silver-bell - *Halesia tetraptera*
    little skullcap - *Scutellaria parvula*
    little starwort - *Stellaria graminea*
    little sundrops - *Oenothera perennis*
    little-tree willow - *Salix arbusculoides*
    little wale - *Lithospermum officinale*
    little walnut - *Juglans microcarpa*
    little whorled pogonia - *Isotria medeoloides*
live
    California live oak - *Quercus agrifolia*
    canyon live oak - *Quercus chrysolepis*
    coast live oak - *Quercus agrifolia*
    dwarf live oak - *Quercus minima*
    interior live oak - *Quercus wislizenii*
    island live oak - *Quercus tomentella*
    live-forever - *Sedum telephium, Sempervivum tectorum*
    live-forever stonecrop - *Sedum telephium*
    live oak - *Quercus virginiana*
    shrub live oak - *Quercus turbinella*
    southern live oak - *Quercus virginiana*
liver
    liver-leaf - *Hepatica*
    liver-seed grass - *Urochloa panicoides*
liverwort, noble-liverwort - *Hepatica*
livid amaranth - *Amaranthus lividus*
living
    living-stone - *Lithops hookeri*
    living-vase - *Aechmea*
lizard-plant - *Tetrastigma voinieranum*
lizard's-tail - *Gaura parviflora, Saururus cernuus*
Lizzy, busy-Lizzy - *Impatiens wallerana*

lobed
    lobed ground-cherry - *Physalis lobata*
    round-lobed hepatica - *Hepatica americana*
    three-lobed coneflower - *Rudbeckia triloba*
    three-lobed morning-glory - *Ipomoea triloba*
lobelia
    blue lobelia - *Lobelia siphilitica*
    brook lobelia - *Lobelia kalmii*
    downy lobelia - *Lobelia puberula*
    edging lobelia - *Lobelia erinus*
    great lobelia - *Lobelia siphilitica*
    lobelia - *Lobelia inflata*
    pale-spike lobelia - *Lobelia spicata*
    spiked lobelia - *Lobelia spicata*
    water lobelia - *Lobelia dortmanna*
loblolly
    loblolly-bay - *Gordonia lasianthus*
    loblolly pine - *Pinus taeda*
lobster-plant - *Euphorbia pulcherrima*
loco
    Big Bend loco - *Astragalus earlei*
    half-moon loco - *Astragalus allochrous*
    Lambert's loco - *Oxytropis lambertii*
    purple loco - *Astragalus agrestis, Oxytropis lambertii*
    spotted loco - *Astragalus lentiginosus*
    white loco - *Oxytropis lambertii*
    woolly loco - *Astragalus mollissimus*
    Wooton's loco - *Astragalus wootonii*
locoweed
    locoweed - *Astragalus, Oxytropis lambertii, Oxytropis*
    showy locoweed - *Oxytropis splendens*
    spotted locoweed - *Astragalus lentiginosus*
    woolly locoweed - *Astragalus mollissimus*
locust
    black locust - *Robinia pseudoacacia*
    bristly locust - *Robinia hispida*
    clammy locust - *Robinia viscosa*
    desert locust - *Robinia neomexicana*
    honey locust - *Gleditsia triacanthos*
    Japanese honey locust - *Gleditsia japonica*
    locust - *Robinia*
    locust bean - *Ceratonia siliqua*
    moss locust - *Robinia hispida*
    New Mexico locust - *Robinia neomexicana*
    sweet locust - *Gleditsia triacanthos*
    thornless honey locust - *Gleditsia triacanthos* var. *inermis*
    water locust - *Gleditsia aquatica*
    yellow locust - *Robinia pseudoacacia*
lodgepole pine - *Pinus contorta, P. contorta subsp.*
    murrayana
lofty fig - *Ficus altissima*
loganberry - *Rubus loganobaccus*
lomatium, slender-fruit lomatium - *Lomatium bicolor* var.
    leptocarpum
Lombardy poplar - *Populus nigra*
London
    London plane-tree - *Platanus acerifolia*
    London-pride - *Lychnis chalcedonica*
    London rocket - *Sisymbrium irio*
long
    Arizona long-leaf pine - *Pinus engelmannii*
    extra-long staple cotton - *Gossypium barbadense*

long-awn arctic sedge - *Carex podocarpa*
long-beaked willow - *Salix bebbiana*
long-flowered cat's-claw - *Acacia greggii*
long-fruited primrose-willow - *Ludwigia octovalvis*
long-headed anemone - *Anemone cylindrica*
long-headed thimbleweed - *Anemone cylindrica*
long-leaf chickweed - *Stellaria longifolia*
long-leaf dock - *Rumex longifolius*
long-leaf ephedra - *Ephedra trifurca*
long-leaf ground-cherry - *Physalis longifolia*
long-leaf pine - *Pinus palustris*
long-leaf pondweed - *Potamogeton nodosus*
long-leaf speedwell - *Veronica longifolia*
long-leaf starwort - *Stellaria longifolia*
long-leaf wild buckwheat - *Eriogonum longifolium*
long-plumed purple avens - *Geum triflorum*
long-root - *Arenaria caroliniana*
long-sheath elodea - *Elodea longivaginata*
long-spine sandbur - *Cenchrus longispinus*
long-spurred violet - *Viola rostrata*
long-stalked aster - *Aster dumosus*
long-stalked crane's-bill - *Geranium columbinum*
long-stalked sedge - *Carex pedunculata*
long-styled anise-root - *Osmorhiza longistylis*
long-tag pine - *Pinus echinata*
long-tongue mutton grass - *Poa fendleriana*
yard-long bean - *Vigna unguiculata, V. unguiculata* subsp.
  sesquipedalis
loofah
  loofah - *Luffa aegyptiaca*
  smooth loofah - *Luffa aegyptiaca*
looking
  looking-glass-plant - *Coprosma repens*
  small Venus's-looking-glass - *Triodanis biflora*
  Venus's-looking-glass - *Triodanis perfoliata*
loose
  loose-flowered alpine sedge - *Carex rariflora*
  loose-flowered milk-vetch - *Astragalus tenellus*
  loose-flowered sedge - *Carex laxiflora*
loosestrife
  creeping loosestrife - *Lysimachia nummularia*
  dotted loosestrife - *Lysimachia punctata*
  false loosestrife - *Ludwigia*
  fringed loosestrife - *Lysimachia ciliata*
  garden loosestrife - *Lysimachia punctata, L. vulgaris*
  hyssop-leaf loosestrife - *Lythrum hyssopifolium*
  hyssop loosestrife - *Lythrum hyssopifolium*
  lance-leaf loosestrife - *Lysimachia lanceolata, Lythrum
    alatum* var. *lanceolatum*
  loosestrife - *Lysimachia, Lythrum*
  many-fruited false loosestrife - *Ludwigia polycarpa*
  narrow-leaf loosestrife - *Lythrum lineare*
  prairie loosestrife - *Lysimachia quadrifolia*
  purple loosestrife - *Lythrum salicaria*
  smooth loosestrife - *Lysimachia quadrifolia*
  spiked loosestrife - *Lythrum salicaria*
  spotted loosestrife - *Lysimachia punctata*
  swamp loosestrife - *Decodon verticillatus, Lysimachia
    terrestris, L. thyrsiflora*
  tufted loosestrife - *Lysimachia thyrsiflora*
  whorled loosestrife - *Lysimachia quadrifolia*
  wing-angled loosestrife - *Lythrum alatum*

  winged loosestrife - *Lythrum alatum*
  yellow loosestrife - *Lysimachia terrestris*
lop-seed - *Phryma leptostachya*
loquat - *Eriobotrya japonica*
Lord's
  Lord's-candlestick - *Yucca gloriosa*
  Our-Lord's-candle - *Yucca whipplei*
lotus
  American lotus - *Nelumbo lutea*
  East Indian lotus - *Nelumbo nucifera*
  Hindu lotus - *Nelumbo nucifera*
  lotus milk-vetch - *Astragalus lotiflorus*
  Oriental lotus - *Nelumbo nucifera*
  sacred lotus - *Nelumbo nucifera*
  water lotus - *Nelumbo*
Louisiana
  Louisiana broomrape - *Orobanche ludoviciana*
  Louisiana sagebrush - *Artemisia ludoviciana*
  Louisiana sedge - *Carex louisianica*
  Louisiana wormwood - *Artemisia ludoviciana*
lousewort
  forest lousewort - *Pedicularis canadensis*
  lousewort - *Pedicularis canadensis, Pedicularis*
  small lousewort - *Pedicularis sylvatica*
  swamp lousewort - *Pedicularis lanceolata, P. palustris*
lovage
  lovage - *Levisticum officinale*
  Scotch lovage - *Ligusticum scothicum*
  sea-lovage - *Ligusticum scothicum*
love
  Bahia love grass - *Eragrostis bahiensis*
  big-top love grass - *Eragrostis hirsuta*
  Boer's love grass - *Eragrostis curvula*
  Elliott's love grass - *Eragrostis elliottii*
  flowers-of-love - *Tabernaemontana divaricata*
  gopher-tail love grass - *Eragrostis ciliaris*
  Indian love grass - *Eragrostis pilosa*
  Lehmann's love grass - *Eragrostis lehmanniana*
  little love grass - *Eragrostis poaeoides*
  love-apple - *Lycopersicon esculentum*
  love grass - *Eragrostis*
  love grass petticoat-climber - *Eragrostis tenella*
  love-in-a-mist - *Passiflora foetida*
  love-lies-bleeding - *Amaranthus caudatus*
  love pea - *Abrus precatorius*
  love-vine - *Cuscuta pentagona*
  Mediterranean love grass - *Eragrostis barrelieri*
  Mexican love grass - *Eragrostis mexicana*
  Mexican love-plant - *Kalanchoe pinnata*
  plains love grass - *Eragrostis intermedia*
  purple love grass - *Eragrostis spectabilis*
  red love grass - *Eragrostis secundiflora*
  sand love grass - *Eragrostis trichodes*
  tufted love grass - *Eragrostis pectinacea*
  weeping love grass - *Eragrostis curvula*
lover, mountain-lover - *Paxistima canbyi*
lowland
  lowland fir - *Abies grandis*
  lowland white fir - *Abies grandis*
Lowrie's aster - *Aster lowrieanus*
lucerne
  Brazilian lucerne - *Stylosanthes guianensis*

lucerne - *Medicago sativa*
yellow lucerne - *Medicago sativa* subsp. *falcata*
Lucia, Santa Lucia fir - *Abies bracteata*
Lucie, St. Lucie cherry - *Prunus mahaleb*
luck
    good-luck-leaf - *Kalanchoe pinnata*
    good-luck palm - *Chamaedorea elegans*
    good-luck-plant - *Cordyline terminalis*
lucky-nut - *Thevetia peruviana*
Lucy, patient-Lucy - *Impatiens wallerana*
luffa
    angled luffa - *Luffa acutangula*
    luffa - *Luffa aegyptiaca*
lungwort
    golden lungwort - *Hieracium murorum*
    lungwort - *Mertensia*
    sea lungwort - *Mertensia maritima*
lupine
    arctic lupine - *Lupinus arcticus*
    Bentham's annual lupine - *Lupinus benthamii*
    bicolored lupine - *Lupinus bicolor*
    blue lupine - *Lupinus angustifolius, L. pilosus*
    Egyptian lupine - *Lupinus albus*
    European blue lupine - *Lupinus angustifolius*
    European yellow lupine - *Lupinus luteus*
    false lupine - *Thermopsis*
    field lupine - *Lupinus albus*
    king's lupine - *Lupinus kingii*
    large-leaf lupine - *Lupinus polyphyllus*
    low lupine - *Lupinus pusillus*
    lupine - *Lupinus*
    miniature lupine - *Lupinus bicolor*
    Nootka lupine - *Lupinus nootkatensis*
    pearl lupine - *Lupinus mutabilis*
    perennial lupine - *Lupinus perennis*
    river bank lupine - *Lupinus rivularis*
    seashore lupine - *Lupinus littoralis*
    sickle-keeled lupine - *Lupinus albicaulis*
    silky lupine - *Lupinus sericeus*
    silver lupine - *Lupinus argenteus*
    spider lupine - *Lupinus benthamii*
    sundial lupine - *Lupinus perennis*
    tail-cup lupine - *Lupinus argenteus* var. *heteranthus*
    Texas lupine - *Lupinus subcarnosus*
    tree lupine - *Lupinus arboreus*
    velvet lupine - *Lupinus leucophyllus*
    Washington's lupine - *Lupinus polyphyllus*
    white lupine - *Lupinus albus*
    wild lupine - *Lupinus perennis*
    woolly-leaf lupine - *Lupinus leucophyllus*
    yellow bush lupine - *Lupinus arboreus*
    yellow lupine - *Lupinus luteus*
Lyall's nettle - *Urtica dioica* subsp. *gracilis*
lychee - *Litchi chinensis*
lychnis
    evening lychnis - *Silene latiflora*
    scarlet lychnis - *Lychnis chalcedonica*
lycopod, dwarf lycopod - *Selaginella rupestris*
Lyme
    Lyme grass - *Elymus, Leymus arenarius*
    Lyme rye grass - *Lolium perenne*

Lyon
    Lyon bean - *Mucuna pruriens* var. *utilis*
    Lyon-tree - *Lyonothamnus floribundus*
lyonia, tree lyonia - *Lyonia ferruginea*
lythrum, purple lythrum - *Lythrum salicaria*
maackia, Amur maackia - *Maackia amurensis*
macadamia-nut - *Macadamia integrifolia*
MacArthur's feather palm - *Ptychosperma macarthurii*
Macartney's rose - *Rosa bracteata*
MacDonald's oak - *Quercus macdonaldii*
macranthum azalea - *Rhododendron indicum*
mad
    mad-apple - *Datura stramonium, Solanum melongena*
    mad-dog skullcap - *Scutellaria lateriflora*
Madagascar
    Madagascar palm - *Chrysalidocarpus lutescens*
    Madagascar periwinkle - *Catharanthus roseus*
    Madagascar plum - *Flacourtia indica*
madake - *Phyllostachys bambusoides*
madar - *Calotropis gigantea, C. procera*
madder
    blue field madder - *Sherardia arvensis*
    field madder - *Sherardia arvensis*
    madder - *Rubia tinctorum*
    wild madder - *Galium mollugo*
Madeira
    Madeira-nut - *Juglans regia*
    Madeira redwood - *Swietenia mahagoni*
madia
    madia - *Madia elegans*
    madia-oil-plant - *Madia sativa*
madnip - *Pastinaca sativa*
Madonna-lily - *Eucharis grandiflora, Lilium candidum*
Madras
    Madras hemp - *Crotalaria juncea*
    Madras thorn - *Pithecellobium dulce*
Madrid brome - *Bromus madritensis*
madrone
    madrone - *Arbutus menziesii*
    Pacific madrone - *Arbutus menziesii*
madwort
    golden-tuft madwort - *Alyssum saxatile*
    madwort - *Alyssum, Asperugo*
    rock madwort - *Alyssum saxatile*
magic-lily - *Lycoris squamigera*
magnolia
    big-leaf magnolia - *Magnolia macrophylla*
    Chinese magnolia - *Magnolia soulangiana*
    Fraser's magnolia - *Magnolia fraseri*
    great-leaf magnolia - *Magnolia macrophylla*
    laurel magnolia - *Magnolia virginiana*
    magnolia water-lily - *Nymphaea tuberosa*
    mountain magnolia - *Magnolia fraseri*
    saucer magnolia - *Magnolia soulangiana*
    small magnolia - *Magnolia virginiana*
    southern magnolia - *Magnolia grandiflora*
    star magnolia - *Magnolia stellata*
    sweet-bay magnolia - *Magnolia virginiana*
    umbrella magnolia - *Magnolia tripetala*
maguey - *Agave americana, Furcraea selloa*

Mahaleb cherry - *Prunus mahaleb*
mahoe - *Hibiscus tiliaceus*
mahogany
  birch-lead mountain-mahogany - *Cercocarpus betuloides*
  Florida mahogany - *Persea borbonia*
  mahogany birch - *Betula lenta*
  mountain-mahogany - *Betula lenta, Cercocarpus montanus*
  Spanish mahogany - *Swietenia mahagoni*
  swamp-mahogany - *Eucalyptus robusta*
  West Indian mahogany - *Swietenia mahagoni*
mahonia
  Cascades mahonia - *Berberis nervosa*
  cluster mahonia - *Berberis pinnata*
  Fremont's mahonia - *Berberis fremontii*
  holly mahonia - *Berberis aquifolium*
maiden
  maiden-cane - *Panicum hemitomon*
  maiden pink - *Dianthus deltoides*
maidenhair
  American maidenhair fern - *Adiantum pedatum*
  brittle maidenhair fern - *Adiantum tenerum*
  delta maidenhair fern - *Adiantum raddianum* cv. 'decoru'
  fan maidenhair fern - *Adiantum tenerum*
  maidenhair-berry - *Gaultheria hispidula*
  maidenhair fern - *Adiantum*
  maidenhair spleenwort - *Asplenium trichomanes*
  maidenhair tree - *Ginkgo biloba*
  northern maidenhair fern - *Adiantum pedatum*
maids, red-maids - *Calandrinia ciliata*
maid's, old-maid's-pink - *Agrostemma githago, Gypsophila paniculata*
maize
  maize - *Zea mays*
  water-maize - *Victoria amazonica*
majagua - *Hibiscus tiliaceus*
Malabar
  Malabar cardamom - *Elettaria cardamomum*
  Malabar grass - *Cymbopogon flexuosus*
  Malabar nightshade - *Basella*
  Malabar plum - *Syzygium jambos*
  Malabar spinach - *Basella alba*
malachra - *Malachra alceifolia*
malanga - *Xanthosoma sagittifolium*
Malay apple - *Syzygium malaccense*
malcolm stock - *Malcolmia africana*
male
  male-berry - *Lyonia ligustrina*
  male-blueberry - *Lyonia ligustrina*
malka - *Rubus chamaemorus*
mallee, gooseberry mallee - *Eucalyptus calycogona*
mallow
  alkali mallow - *Malvella leprosa*
  apricot mallow - *Sphaeralcea ambigua*
  bristly mallow - *Modiola caroliniana*
  bull mallow - *Malva nicaeensis*
  bur mallow - *Urena lobata*
  checker mallow - *Sidalcea malvaeflora, Sidalcea*
  crimson-eyed rose mallow - *Hibiscus moscheutos* subsp. *palustris*
  curled mallow - *Malva verticillata* var. *crispa*
  desert mallow - *Sphaeralcea ambigua*

Drummond's wax mallow - *Malvaviscus arboreus* var. *drummondii*
East Indian Jew's mallow - *Corchorus aestuans*
European mallow - *Malva alcea*
false mallow - *Malvastrum, Sphaeralcea*
giant mallow - *Hibiscus*
glade mallow - *Napaea dioica*
globe mallow - *Sphaeralcea*
great rose mallow - *Hibiscus grandiflorus*
halberd-leaf mallow - *Hibiscus laevis*
hemp-leaved mallow - *Althaea cannabina*
high mallow - *Malva sylvestris*
hollyhock mallow - *Malva alcea*
Indian mallow - *Abutilon incanum, A. theophrasti, Abutilon*
little mallow - *Malva parviflora*
mallow - *Malva neglecta, Malva*
mallow-rose - *Hibiscus moscheutos, H. moscheutos* subsp. palustris
marsh mallow - *Althaea officinalis, Hibiscus moscheutos* subsp. *palustris*
musk mallow - *Malva moschata, Malva*
narrow-leaf globe mallow - *Sphaeralcea angustifolia*
pale poppy mallow - *Callirhoe alcaeoides*
plains poppy mallow - *Callirhoe alcaeoides*
poppy mallow - *Callirhoe*
prairie mallow - *Sphaeralcea coccinea*
prickly mallow - *Sida spinosa*
purple poppy mallow - *Callirhoe involucrata*
red false mallow - *Sphaeralcea coccinea*
rose mallow - *Hibiscus lasiocarpus, H. moscheutos, Hibiscus*
scarlet globe mallow - *Sphaeralcea coccinea*
scarlet mallow - *Sphaeralcea coccinea*
sleepy mallow - *Malvaviscus*
smooth rose mallow - *Hibiscus laevis*
soldier rose mallow - *Hibiscus laevis*
swamp rose mallow - *Hibiscus moscheutos, H. moscheutos* subsp. *palustris*
Texas mallow - *Malvaviscus arboreus* var. *drummondii*
tree mallow - *Lavatera arborea*
Venice mallow - *Hibiscus trionum*
verrain mallow - *Malva alcea*
Virginia mallow - *Sida hermaphrodita*
wax mallow - *Malvaviscus arboreus*
white mallow - *Malvella leprosa*
woolly rose mallow - *Hibiscus lasiocarpus*
yellow false mallow - *Malvastrum hispidum*
malojilla - *Eriochloa polystachya*
Malta star-thistle - *Centaurea melitensis*
Maltese-cross - *Lychnis chalcedonica*
mamane - *Sophora chrysophylla*
mamey-colorado - *Pouteria sapota*
mammee sapote - *Pouteria sapota*
mammoth wild rye - *Leymus racemosus*
mamoncillo - *Melicoccus bijugatus*
man
  coast man-root - *Marah oreganus*
  life-of-man - *Aralia racemosa*
  man-in-a-boat - *Tradescantia spathacea*
  man-of-the-earth - *Ipomoea leptophylla, I. pandurata*
  man-root - *Ipomoea leptophylla, I. pandurata, Marah*
  man tree-fern - *Cibotium splendens*
  new old-man cactus - *Espostoa lanata*

old-man - *Artemisia abrotanum*
old-man-in-the-ground - *Marah oreganus*
Peruvian old-man cactus - *Espostoa lanata*
man's
    lazy-man's grass - *Eremochloa ophiuroides*
    old-man's-beard - *Chionanthus virginicus*
    old-man's-whiskers - *Geum triflorum*
    poor-man's-pepper - *Lepidium virginicum*
    white-man's-foot - *Plantago major*
manî - *Arachis hypogaea*
Manchurian wild rice - *Zizania latifolia*
mandarin
    Mandarin lime - *Citrus limonia*
    Mandarin orange - *Citrus reticulata*
    nodding mandarin - *Disporum maculatum*
    rose mandarin - *Streptopus roseus*
    white mandarin - *Streptopus amplexifolius*
    yellow mandarin - *Disporum lanuginosum*
mandrake - Mandragora officinarum, Podophyllum
    peltatum
Manetti's rose - *Rosa noisettiana* cv. 'Manetti'
mangel - *Beta vulgaris*
mango
    mango - *Mangifera indica*
    mango pepper - *Capsicum annuum*
mangold - *Beta vulgaris*
mangosteen - *Garcinia mangostana*
mangrove
    American mangrove - *Rhizophora mangle*
    black mangrove - *Avicennia germinans*
    button mangrove - *Conocarpus erectus*
    mangrove - *Rhizophora mangle*
    red mangrove - *Rhizophora mangle*
    silver button mangrove - *Conocarpus erectus* var. *sericeus*
    white mangrove - *Laguncularia racemosa*
manila
    manila - *Pithecellobium dulce*
    Manila grass - *Zoysia matrella*
    Manila palm - *Veitchia merrillii*
    Manila tamarind - *Pithecellobium dulce*
manioc - *Manihot esculenta*
manna
    American manna grass - *Glyceria grandis*
    eastern manna grass - *Glyceria septentrionalis*
    floating manna grass - *Glyceria fluitans, G. septentrionalis*
    fowl manna grass - *Glyceria elata, G. striata*
    manna gum - *Eucalyptus viminalis*
    northern manna grass - *Glyceria borealis*
    rattlesnake manna grass - *Glyceria canadensis*
    small floating manna grass - *Glyceria borealis*
    tall manna grass - *Glyceria elata*
    tufted manna grass - *Glyceria septentrionalis*
    water manna grass - *Glyceria fluitans*
mansa, yerba-mansa - *Anemopsis californica*
mantle
    alpine lady's-mantle - *Alchemilla alpina*
    lady's-mantle - *Alchemilla xanthochlora, Alchemilla*
    western lady's-mantle - *Aphanes occidentalis*
manuka - *Leptospermum scoparium*
many
    many-fruited false loosestrife - *Ludwigia polycarpa*

many-seeded goosefoot - *Chenopodium polyspermum*
many-stemmed evax - *Filago verna*
manzanita
    Bragg's manzanita - *Arctostaphylos nummularia*
    dune manzanita - *Arctostaphylos pumila*
    Eastwood manzanita - *Arctostaphylos glandulosa*
    green-leaf manzanita - *Arctostaphylos patula*
    hairy manzanita - *Arctostaphylos columbiana*
    manzanita - *Arctostaphylos*
    Mexican manzanita - *Arctostaphylos pungens*
    Monterey manzanita - *Arctostaphylos hookeri*
    Parry's manzanita - *Arctostaphylos manzanita*
    pine-mat manzanita - *Arctostaphylos nevadensis*
    point-leaf manzanita - *Arctostaphylos pungens*
    sand-mat manzanita - *Arctostaphylos pumila*
    white-leaf manzanita - *Arctostaphylos viscida*
    woolly manzanita - *Arctostaphylos tomentosa*
maple
    Amur maple - *Acer ginnala*
    ash-leaf maple - *Acer negundo.*
    big-leaf maple - *Acer macrophyllum*
    big-tooth maple - *Acer grandidentatum*
    black maple - *Acer nigrum*
    black sugar maple - *Acer nigrum*
    canyon maple - *Acer macrophyllum*
    chalk maple - *Acer leucoderme*
    Douglas's maple - *Acer glabrum* subsp. *douglasii*
    dwarf maple - *Acer glabrum*
    field maple - *Acer campestre*
    Florida maple - *Acer barbatum*
    flowering-maple - *Abutilon pictum, Abutilon, Viburnum acerifolium*
    hard maple - *Acer saccharum*
    hedge maple - *Acer campestre*
    Japanese maple - *Acer palmatum*
    maple - *Acer*
    maple-leaf goosefoot - *Chenopodium hybridum, C. simplex*
    maple-leaf viburnum - *Viburnum acerifolium*
    maple-leaf water-leaf - *Hydrophyllum canadense*
    mono maple - *Acer mono*
    moose maple - *Acer spicatum*
    mountain maple - *Acer glabrum* subsp. *douglasii, A. spicatum*
    Norway maple - *Acer platanoides*
    Oregon maple - *Acer macrophyllum*
    parlor-maple - *Abutilon*
    Pennsylvania maple - *Acer pensylvanicum*
    red maple - *Acer rubrum*
    river maple - *Acer saccharinum*
    rock maple - *Acer nigrum, A. saccharum*
    Rocky Mountain maple - *Acer glabrum*
    scarlet maple - *Acer rubrum*
    Shantung maple - *Acer truncatum*
    silver maple - *Acer saccharinum*
    soft maple - *Acer rubrum, A. saccharinum*
    southern sugar maple - *Acer barbatum*
    striped maple - *Acer pensylvanicum*
    sugar maple - *Acer grandidentatum, A. saccharum*
    swamp maple - *Acer rubrum*
    sycamore maple - *Acer pseudoplatanus*
    trident maple - *Acer rubrum* var. *trilobum*
    vine maple - *Acer circinatum*
    white-bark maple - *Acer leucoderme*
    white maple - *Acer saccharinum*

maranon - *Anacardium occidentale*
marble
  marble pea - *Vigna unguiculata* subsp. *cylindrica*
  marble-seed - *Onosmodium molle*
mare's-tail - *Conyza canadensis, Hippuris vulgaris*
margin, white-margin spurge - *Euphorbia albomarginata*
marginal
  marginal shield fern - *Dryopteris marginalis*
  marginal wood fern - *Dryopteris marginalis*
marguerite
  marguerite - *Chrysanthemum frutescens, Leucanthemum vulgare*
  white marguerite - *Chrysanthemum frutescens*
Maria, Santa Maria - *Parthenium hysterophorus*
marigold
  African marigold - *Tagetes erecta*
  Aztec marigold - *Tagetes erecta*
  big marigold - *Tagetes erecta*
  blue-eyed Cape-marigold - *Dimorphotheca sinuata*
  bur-marigold - *Bidens pilosa, Bidens*
  Cape-marigold - *Dimorphotheca*
  corn-marigold - *Chrysanthemum segetum*
  desert-marigold - *Baileya multiradiacata*
  fetid-marigold - *Dyssodia papposa, Dyssodia*
  French marigold - *Tagetes erecta*
  marigold - *Tagetes*
  marsh-marigold - *Caltha palustris, Caltha*
  nodding bur-marigold - *Bidens cernua*
  pot-marigold - *Calendula officinalis*
  smaller bur-marigold - *Bidens cernua*
  stinking marigold - *Dyssodia papposa*
  water-marigold - *Bidens, Megalodonta beckii*
  wild marigold - *Tagetes minuta*
marijuana - *Cannabis sativa*
marine
  marine ivy - *Cissus incisa*
  marine-vine - *Cissus incisa*
mariposa
  desert mariposa - *Calochortus kennedyi*
  green-banded mariposa - *Calochortus macrocarpus*
  mariposa-lily - *Calochortus*
maritime pine - *Pinus pinaster*
marjoram
  annual marjoram - *Origanum majorana*
  marjoram - *Origanum vulgare*
  pot marjoram - *Origanum vulgare*
  sweet marjoram - *Origanum majorana*
  wild marjoram - *Origanum vulgare*
markry - *Toxicodendron radicans*
marlberry - *Ardisia escallonioides*
marlock, white-leaf marlock - *Eucalyptus tetragona*
marmalade
  marmalade - *Genipa americana*
  marmalade-box - *Genipa americana*
  marmalade-fruit - *Pouteria sapota*
  marmalade-plum - *Pouteria sapota*
marquerite, golden marquerite - *Anthemis tinctoria*
marram grass - *Ammophila arenaria*
marri - *Eucalyptus calophylla*
marrow
  marrow - *Cucurbita maxima, C. pepo*

wild marrow - *Cucurbita texana*
marrube - *Marrubium vulgare*
marsh
  alpine marsh violet - *Viola palustris*
  annual marsh-elder - *Iva annua*
  annual salt marsh aster - *Aster subulatus*
  big marsh-elder - *Iva xanthifolia*
  blue marsh violet - *Viola cucullata*
  marsh arrow grass - *Triglochin palustris*
  marsh beggar-ticks - *Bidens mitis*
  marsh bellflower - *Campanula aparinoides*
  marsh blue violet - *Viola cucullata*
  marsh cinquefoil - *Potentilla palustris*
  marsh club-moss - *Lycopodium inundatum*
  marsh-collard - *Nuphar*
  marsh-cress - *Rorippa palustris*
  marsh cyperus - *Cyperus pseudovegetus*
  marsh-dayflower - *Murdannia keisak*
  marsh-elder - *Iva xanthifolia, Iva*
  marsh eryngo - *Eryngium aquaticum*
  marsh fern - *Thelypteris palustris*
  marsh fire-finger - *Potentilla palustris*
  marsh fleabane - *Pluchea camphorata, Senecio congestus*
  marsh foxtail - *Alopecurus geniculatus*
  marsh grass - *Scolochloa festucacea, Spartina*
  marsh horsetail - *Equisetum palustre*
  marsh mallow - *Althaea officinalis, Hibiscus moscheutos* subsp. *palustris*
  marsh-marigold - *Caltha palustris, Caltha*
  marsh mermaid-weed - *Proserpinaca palustris*
  marsh pea-vine - *Lathyrus palustris*
  marsh pennywort - *Hydrocotyle americana*
  marsh-pepper smartweed - *Polygonum hydropiper*
  marsh-pink - *Sabatia angularis*
  marsh-rosemary - *Limonium*
  marsh skullcap - *Scutellaria galericulata*
  marsh skullwort - *Scutellaria galericulata*
  marsh sow-thistle - *Sonchus arvensis* spp. *uliginosus*
  marsh speedwell - *Veronica scutellata*
  marsh St. John's-wort - *Hypericum tubulosum, H. virginicum*
  marsh straw sedge - *Carex tenera*
  marsh thistle - *Cirsium palustre*
  marsh-trefoil - *Menyanthes trifoliata*
  marsh vetchling - *Lathyrus palustris*
  marsh violet - *Viola palustris*
  perennial salt marsh aster - *Aster tenuifolius*
  rough marsh-elder - *Iva annua*
  salt marsh cord grass - *Spartina alterniflora*
  salt marsh foxtail - *Setaria magna*
  western marsh-rosemary - *Limonium californicum*
Martagon lily - *Lilium martagon*
Martha Washington's geranium - *Pelargonium domesticum*
marvel
  marvel - *Marrubium vulgare*
  marvel-of-Peru - *Mirabilis jalapa*
Mary
  blue-eyed-Mary - *Collinsia verna*
  eastern blue-eyed-Mary - *Collinsia verna*
Mary's
  Mary's grass - *Microstegium vimineum* var. *imberbe*
  St. Mary's thistle - *Silybum marianum*

Maryland
  Maryland golden aster - *Chrysopsis mariana*
  Maryland meadow-beauty - *Rhexia mariana*
Mascarene grass - *Zoysia tenuifolia*
master, rattlesnake-master - *Eryngium aquaticum,*
    E. yuccifolium, Manfreda virginica
masterwort - *Angelica atropurpurea, Heracleum*
    sphondylium subsp. *montanum*
mastic
  false mastic - *Sideroxylon foetidissimum*
  Peruvian mastic-tree - *Schinus molle*
mat
  bog-mat - *Wolffiella gladiata*
  flat bog-mat - *Wolffiella floridana*
  Japanese mat rush - *Juncus effusus*
  mat amaranth - *Amaranthus graecizans*
  mat chaff-flower - *Alternanthera caracasana*
  mat grama - *Bouteloua simplex*
  mat-grass - *Phyla nodiflora*
  mat lippia - *Phyla nodiflora*
  mat rush - *Scirpus lacustris*
  mat sandpur - *Cenchrus longispinus*
  pine-mat manzanita - *Arctostaphylos nevadensis*
  sand-mat manzanita - *Arctostaphylos pumila*
match
  match-brush - *Gutierrezia*
  match-weed - *Gutierrezia*
mathers - *Vaccinium ovalifolium*
matricary - *Matricaria*
matrimony
  barbary matrimony-vine - *Lycium barbarum*
  Chinese matrimony-vine - *Lycium barbarum*
  matrimony-vine - *Lycium*
maul oak - *Quercus chrysolepis*
Maule's quince - *Chaenomeles japonica*
Mauritius
  Mauritius raspberry - *Rubus rosifolius*
  Mauritius velvet bean - *Mucuna pruriens* var. *utilis*
mauve sleekwort - *Liparis liliifolia*
Max's chrysanthemum - *Leucanthemum maximum*
Maximilian's sunflower - *Helianthus maximilianii*
may
  May-apple - *Podophyllum peltatum*
  May-blob - *Caltha palustris*
  May grass - *Phalaris caroliniana*
  May hawthorn - *Crataegus aestivalis*
  may-pop passionflower - *Passiflora incarnata*
  may-pops - *Passiflora incarnata*
mayflower
  Canadian mayflower - *Maianthemum canadense*
  mayflower - *Epigaea repens*
mayweed
  false mayweed - *Dyssodia papposa*
  mayweed - *Ambrosia artemisiifolia, Anthemis cotula*
  mayweed-chamomile - *Anthemis cotula*
mazus, Asian mazus - *Mazus pumilus*
mazzard cherry - *Prunus avium*
McNab's cypress - *Cupressus macnabiana*

meadow
  alpine meadow-rue - *Thalictrum alpinum*
  dull meadow-pitchers - *Rhexia mariana*
  dwarf meadow grass - *Poa annua*
  early meadow-rue - *Thalictrum dioicum*
  fowl meadow grass - *Poa palustris*
  king-of-the-meadow - *Thalictrum pubescens*
  Maryland meadow-beauty - *Rhexia mariana*
  meadow anemone - *Anemone canadensis*
  meadow barley - *Hordeum brachyantherum*
  meadow-berry - *Rhexia*
  meadow-bright - *Caltha palustris*
  meadow brome - *Bromus erectus*
  meadow buttercup - *Ranunculus acris*
  meadow campion - *Lychnis flos-cuculi*
  meadow chickweed - *Cerastium arvense*
  meadow closed gentian - *Gentiana clausa*
  meadow death camass - *Zigadenus venenosus*
  meadow drop-seed - *Sporobolus asper* var. *hookeri*
  meadow fern - *Thelypteris palustris*
  meadow fescue - *Festuca pratensis*
  meadow foxtail - *Alopecurus pratensis*
  meadow garlic - *Allium canadense*
  meadow geranium - *Geranium pratense*
  meadow-hyacinth - *Camassia scilloides*
  meadow knapweed - *Centaurea pratensis*
  meadow larkspur - *Delphinium nuttallianum*
  meadow leek - *Allium canadense*
  meadow parsnip - *Thaspium trifoliatum*
  meadow pea - *Lathyrus pratensis*
  meadow pea-vine - *Lathyrus pratensis*
  meadow phlox - *Phlox maculata*
  meadow-pine - *Equisetum arvense*
  meadow-pink - *Lychnis flos-cuculi*
  meadow-rue - *Thalictrum*
  meadow-saffron - *Colchicum autumnale*
  meadow sage - *Salvia pratensis*
  meadow salsify - *Tragopogon pratensis*
  meadow vetch - *Lathyrus pratensis*
  meadow-weed - *Ruellia tuberosa*
  meadow willow - *Salix petiolaris*
  mountain meadow-rue - *Thalictrum clavatum*
  northern meadow groundsel - *Senecio pauperculus*
  purple meadow-rue - *Thalictrum dasycarpum, T. revolutum*
  queen-of-the-meadow - *Filipendula ulmaria*
  salt meadow cord grass - *Spartina patens*
  skunk meadow-rue - *Thalictrum revolutum*
  smooth meadow parsnip - *Thaspium trifoliatum*
  tall meadow-rue - *Thalictrum pubescens*
  wax-leaf meadow-rue - *Thalictrum revolutum*
meadowsweet - *Filipendula, Spiraea alba, S. alba*
    var. *latifolia*
meal
  duck-meal - *Lemna perpusilla*
  meal-berry - *Arctostaphylos uva-ursi*
  meal-weed - *Chenopodium album*
  pimpled water-meal - *Wolffia papulifera*
  spotted water-meal - *Wolffia punctata*
  water-meal - *Wolffia columbiana*
mealy
  mealy-cup sage - *Salvia farinacea*
  mealy goosefoot - *Chenopodium incanum*

Mearns' sumac - *Rhus choriophylla*
medic
  black medic - *Medicago lupulina*
  medic - *Medicago*
  sickle medic - *Medicago sativa* subsp. *falcata*
  spotted medic - *Medicago arabica*
  toothed medic - *Medicago polymorpha*
medicinal
  medicinal agrimony - *Agrimonia eupatoria*
  medicinal aloe - *Aloe vera*
Mediterranean
  Mediterranean barley - *Hordeum geniculatum*
  Mediterranean grass - *Schismus barbatus*
  Mediterranean love grass - *Eragrostis barrelieri*
  Mediterranean sage - *Salvia aethiopis*
  Mediterranean saltwort - *Salsola vermiculata*
  Mediterranean wheat - *Triticum turgidum*
medlar
  Japanese medlar - *Eriobotrya japonica*
  medlar - *Mespilus germanica, Mimusops elengi*
Medusa's-head - *Taeniatherum caput-medusae*
meeting-houses - *Aquilegia canadensis*
melastoma, Bank's melastoma - *Melastoma malabathricum*
melic
  awned melic - *Melica smithii*
  California melic - *Melica imperfecta*
  false melic - *Schizachne purpurascens*
  Harford's melic - *Melica harfordii*
  melic - *Melica*
  Porter's melic - *Melica porteri*
  Smith's melic - *Melica smithii*
  three-flowered melic - *Melica nitens*
  two-flowered melic - *Melica mutica*
melilot
  melilot - *Melilotus alba, Melilotus*
  white melilot - *Melilotus alba*
  yellow melilot - *Melilotus officinalis*
melist - *Melilotus officinalis*
melon
  casaba melon - *Cucumis melo* var. *inodorus*
  Chinese preserving melon - *Benincasa hispida*
  Chinese winter melon - *Benincasa hispida*
  citron-melon - *Citrullus lanatus* var. *citroides*
  honeydew melon - *Cucumis melo* var. *inodorus*
  melon - *Cucumis melo*
  melon-leaf nightshade - *Solanum citrullifolium*
  melon-tree - *Carica papaya*
  netted melon - *Cucumis melo* var. *reticulatus*
  nutmeg melon - *Cucumis melo* var. *reticulatus*
  Persian melon - *Cucumis melo* var. *reticulatus*
  preserving melon - *Citrullus lanatus* var. *citroides*
  stock melon - *Citrullus lanatus* var. *citroides*
  winter melon - *Cucumis melo* var. *inodorus*
melongene - *Solanum melongena*
memorial rose - *Rosa wichuraiana*
men, three-men-in-a-boat - *Tradescantia spathacea*
Mendocino
  Mendocino cypress - *Cupressus pygmaea*
  Mendocino gentian - *Gentiana setigera*
menow-weed - *Ruellia tuberosa*

menteng - *Baccaurea racemosa*
Menzies tansy mustard - *Descurainia pinnata* subsp. *menziesii*
mercury
  mercury - *Chenopodium bonus-henricus, Toxicodendron radicans*
  mercury-weed - *Acalypha virginica*
  three-seeded-mercury - *Acalypha virginica*
mermaid
  cut-leaf mermaid-weed - *Proserpinaca pectinata*
  marsh mermaid-weed - *Proserpinaca palustris*
  mermaid-weed - *Proserpinaca palustris, Proserpinaca*
Merrill's palm - *Veitchia merrillii*
merry-bells - *Uvularia*
mertensia, sea mertensia - *Mertensia maritima*
mesa oak - *Quercus engelmannii*
mescal
  mescal - *Agave parryi*
  mescal bean - *Sophora secundiflora*
mescat acacia - *Acacia constricta*
mesquite
  curly-mesquite grass - *Hilaria belangeri*
  false mesquite - *Calliandra*
  glandular mesquite - *Prosopis glandulosa*
  honey mesquite - *Prosopis glandulosa*
  mesquite - *Prosopis chilensis, P. glandulosa, P. glandulosa* var. *torreyana, P. juliflora*
  mesquite grass - *Bouteloua*
  mock-mesquite - *Calliandra eriophylla*
  Syrian mesquite - *Prosopis farcta*
  velvet mesquite - *Prosopis velutina*
  vine mesquite - *Panicum obtusum*
  vine mesquite grass - *Panicum obtusum*
  western honey mesquite - *Prosopis glandulosa* var. *torreyana*
mesquitilla - *Calliandra eriophylla*
messmate
  messmate - *Eucalyptus obliqua*
  messmate stringy-bark - *Eucalyptus obliqua*
metel - *Datura metel*
Mexican
  Mexican apple - *Casimiroa edulis*
  Mexican bamboo - *Polygonum cuspidatum*
  Mexican blue oak - *Quercus oblongifolia*
  Mexican blue-wood - *Condalia mexicana*
  Mexican buckeye - *Ungnadia speciosa*
  Mexican clover - *Richardia scabra*
  Mexican condalia - *Condalia mexicana*
  Mexican cypress - *Cupressus lusitanica*
  Mexican drooping juniper - *Juniperus flaccida*
  Mexican elder - *Sambucus mexicana*
  Mexican fire-plant - *Euphorbia heterophylla*
  Mexican flame-leaf - *Euphorbia pulcherrima*
  Mexican flame-vine - *Senecio chenopodioides*
  Mexican lime - *Citrus aurantiifolia*
  Mexican love grass - *Eragrostis mexicana*
  Mexican love-plant - *Kalanchoe pinnata*
  Mexican manzanita - *Arctostaphylos pungens*
  Mexican mint - *Plectranthus amboinicus*
  Mexican muhly - *Muhlenbergia mexicana*
  Mexican palmetto - *Sabal mexicana*
  Mexican palo-verde - *Parkinsonia aculeata*
  Mexican persimmon - *Diospyros texana*

Mexican pinyon - *Pinus cembroides*
Mexican plum - *Prunus mexicana*
Mexican poppy - *Argemone mexicana*
Mexican prickly poppy - *Argemone mexicana*
Mexican sprangle-top - *Leptochloa uninervia*
Mexican stone pine - *Pinus cembroides*
Mexican sunflower - *Tithonia rotundifolia*
Mexican tea - *Chenopodium ambrosioides*
Mexican Washington palm - *Washingtonia robusta*
Mexican water-lily - *Nymphaea mexicana*
Mexican-weed - *Caperonia castaniifolia*
Mexican whorled milkweed - *Asclepias fascicularis*
Mexican yellow pine - *Pinus patula*
rough-bark Mexican pine - *Pinus montezumae*

Mexico
lady-of-Mexico cactus - *Mammillaria hahniana*
New Mexico alder - *Alnus oblongifolia*
New Mexico evergreen sumac - *Rhus choriophylla*
New Mexico feather grass - *Stipa neomexicana*
New Mexico locust - *Robinia neomexicana*
New Mexico muhly - *Muhlenbergia pauciflora*
New Mexico raspberry - *Rubus neomexicanus*

mezereum - *Daphne mezereum*

Miami-mist - *Phacelia purshii*

Michigan lily - *Lilium michiganense*

midwestern
midwestern blue heart-leaf aster - *Aster shortii*
midwestern tickseed-sunflower - *Bidens aristosa*

mignonette
cut-leaf mignonette - *Reseda lutea*
garden mignonette - *Reseda odorata*
mignonette - *Reseda odorata*
white mignonette - *Reseda alba*
yellow mignonette - *Reseda lutea*

mild
mild smartweed - *Polygonum hydropiperoides*
mild water-pepper - *Polygonum hydropiperoides*

mile
African mile-a-minute - *Mikania cordata*
gray-mile - *Lithospermum officinale*
mile-a-minute - *Mikania micrantha*
mile-tree - *Casuarina equisetifolia*

milfoil
eastern water-milfoil - *Myriophyllum pinnatum*
Eurasian milfoil - *Myriophyllum spicatum*
Eurasian water-milfoil - *Myriophyllum spicatum*
lax water-milfoil - *Myriophyllum laxum*
milfoil - *Achillea millefolium*
northern water-milfoil - *Myriophyllum exalbescens*
variable water-milfoil - *Myriophyllum heterophyllum*
whorled water-milfoil - *Myriophyllum verticillatum*

milk
alpine milk-vetch - *Astragalus alpinus*
American milk-bush - *Synadenium grantii*
blessed milk-thistle - *Silybum marianum*
cat-milk - *Euphorbia helioscopia*
cicer milk-vetch - *Astragalus cicer*
Columbia milk-vetch - *Astragalus miser* var. *serotinus*
field milk-vetch - *Astragalus agrestis*
fleckled milk-vetch - *Astragalus lentiginosus*
flexible milk-vetch - *Astragalus flexuosus*
hairy milk pea - *Galactia volubilis*

Indian milk-vetch - *Astragalus aboriginorum*
late milk-vetch - *Astragalus miser* var. *serotinus*
licorice milk-vetch - *Astragalus glycyphyllos*
loose-flowered milk-vetch - *Astragalus tenellus*
lotus milk-vetch - *Astragalus lotiflorus*
milk-grass - *Valerianella locusta*
milk ipecac - *Apocynum androsaemifolium*
milk pea - *Galactia*
milk-purslane - *Euphorbia maculata*
milk-thistle - *Lactuca serriola, Silybum marianum*
milk-vetch - *Astragalus*
Missouri milk-vetch - *Astragalus missouriensis*
narrow-leaf milk-vetch - *Astragalus pectinatus*
Nuttall's milk-vetch - *Astragalus nuttallianus, A. nuttallii*
pliant milk-vetch - *Astragalus flexuosus*
purse milk-vetch - *Astragalus tenellus*
Russian sickle milk-vetch - *Astragalus falcatus*
small-flowered milk-vetch - *Astragalus nuttallianus*
standing milk-vetch - *Astragalus adsurgens*
timber milk-vetch - *Astragalus miser*
tine-leaf milk-vetch - *Astragalus pectinatus*
two-grooved milk-vetch - *Astragalus bisulcatus*
Wasatch milk-vetch - *Astragalus miser* var. *oblongifolius*
weedy milk-vetch - *Astragalus miser*
wolf's-milk - *Euphorbia esula*
Yellowstone milk-vetch - *Astragalus miser* var. *hylophilus*

milkweed
blood-flower milkweed - *Asclepias curassavica*
blunt-leaf milkweed - *Asclepias amplexicaulis*
broad-leaf milkweed - *Asclepias latifolia*
butterfly milkweed - *Asclepias tuberosa*
clasping milkweed - *Asclepias amplexicaulis*
climbing milkweed - *Cynanchum louiseae, Sarcostemma cynanchoides*
eastern whorled milkweed - *Asclepias verticillata*
few-flowered milkweed - *Asclepias lanceolata*
four-leaved milkweed - *Asclepias quadrifolia*
giant milkweed - *Calotropis gigantea*
Greek milkweed - *Asclepias speciosa*
green milkweed - *Asclepias viridiflora*
horsetail milkweed - *Asclepias verticillata*
labriform milkweed - *Asclepias labriformis*
lanceolate milkweed - *Asclepias lanceolata*
Mexican whorled milkweed - *Asclepias fascicularis*
milkweed - *Asclepias syriaca, Asclepias*
narrow-leaf milkweed - *Asclepias fascicularis*
orange milkweed - *Asclepias tuberosa*
poison milkweed - *Asclepias subverticillata, Euphorbia corollata*
poke milkweed - *Asclepias exaltata*
purple milkweed - *Asclepias purpurascens*
red milkweed - *Asclepias rubra*
showy milkweed - *Asclepias speciosa*
smooth milkweed - *Asclepias sullivantii*
Sullivant's milkweed - *Asclepias sullivantii*
swamp milkweed - *Asclepias incarnata*
tall milkweed - *Asclepias exaltata*
wandering milkweed - *Apocynum androsaemifolium*
western whorled milkweed - *Asclepias subverticillata*
white-flowered milkweed - *Euphorbia corollata*
white milkweed - *Asclepias variegata*
whorled milkweed - *Asclepias subverticillata, A. verticillata*
woolly-pod milkweed - *Asclepias eriocarpa*

milkwort
  blood milkwort - *Polygala sanguinea*
  cross-leaf milkwort - *Polygala cruciata*
  field milkwort - *Polygala sanguinea*
  milkwort - *Polygala*
  orange milkwort - *Polygala lutea*
  sea-milkwort - *Glaux maritima*
  whorled milkwort - *Polygala verticillata*
  yellow milkwort - *Polygala lutea*
miller, dusty-miller - *Centaurea cineraria, Lychnis coronaria*
millet
  adlay millet - *Coix lacryma-jobi*
  African millet - *Eleusine coracana*
  broom corn millet - *Panicum miliaceum*
  broom millet - *Panicum miliaceum*
  brown-top millet - *Brachiaria fasciculata, B. ramosa*
  bulrush millet - *Pennisetum glaucum*
  cat-tail millet - *Pennisetum glaucum*
  Egyptian millet - *Sorghum halepense*
  finger millet - *Eleusine coracana*
  foxtail millet - *Setaria italica*
  German millet - *Setaria italica*
  hog millet - *Panicum miliaceum*
  Hungarian millet - *Setaria italica*
  Italian millet - *Setaria italica*
  Japanese millet - *Echinochloa frumentacea, Setaria italica*
  millet - *Panicum miliaceum*
  millet wood rush - *Luzula parviflora*
  pearl millet - *Pennisetum glaucum*
  Polish millet - *Digitaria sanguinalis*
  proso millet - *Panicum miliaceum*
  Sanwa millet - *Echinochloa frumentacea*
  spring millet - *Milium vernale*
  Texas millet - *Panicum texanum*
  wild proso millet - *Panicum miliaceum*
milo - *Sorghum bicolor*
mimosa
  cat-claw mimosa - *Mimosa pigra*
  mimosa - *Albizia julibrissin*
  mimosa-tree - *Albizia julibrissin*
  prairie-mimosa - *Desmanthus illinoensis*
  Texas mimosa - *Acacia greggii*
mind-your-own-business-plant - *Soleirolia soleirolii*
miner's
  miner's-lettuce - *Claytonia perfoliata, Montia*
  narrow-leaf miner's-lettuce - *Montia linearis*
Ming aralia - *Polyscias fruticosa*
miniature
  miniature date palm - *Phoenix roebelenii*
  miniature fan palm - *Rhapis excelsa*
  miniature holly - *Malpighia coccigera*
  miniature lupine - *Lupinus bicolor*
  miniature pansy - *Viola tricolor*
minnie-bush - *Menziesia pilosa*
mint
  apple mint - *Mentha rotundifolia, M. suaveolens*
  bergamot mint - *Mentha piperita* var. *citrata*
  bitter mint - *Hyptis mutabilis*
  brandy mint - *Mentha piperita*
  cat-mint - *Nepeta cataria*
  corn mint - *Mentha arvensis*
  coyote mint - *Monardella villosa*

  curled mint - *Mentha crispa*
  dotted mint - *Monarda punctata*
  field mint - *Mentha arvensis*
  hairy mountain mint - *Pycnanthemum pilosum*
  hairy wood mint - *Blephilia hirsuta*
  Indian mint - *Plectranthus amboinicus*
  Japanese mint - *Mentha arvensis*
  lamb mint - *Mentha piperita*
  lemon mint - *Mentha piperita* var. *citrata, Monarda citriodora*
  Mexican mint - *Plectranthus amboinicus*
  mint - *Mentha*
  mint-geranium - *Chrysanthemum balsamita*
  mint perilla - *Perilla frutescens*
  mint vervain - *Verbena menthifolia*
  mountain mint - *Monardella odoratissima, Pycnanthemum*
  perilla-mint - *Perilla frutescens*
  red mint - *Mentha gentilis*
  round-leaf mint - *Mentha rotundifolia*
  Scotch mint - *Mentha gentilis*
  soup mint - *Plectranthus amboinicus*
  squaw-mint - *Hedeoma pulegioides*
  stone mint - *Cunila origanoides*
  Virginia mountain mint - *Pycnanthemum virginianum*
  water mint - *Mentha aquatica*
  wild mint - *Mentha arvensis*
  wood mint - *Blephilia hirsuta*
minute
  African mile-a-minute - *Mikania cordata*
  mile-a-minute - *Mikania micrantha*
  minute bladderwort - *Utricularia olivacea*
  minute duckweed - *Lemna minuta*
miracle-leaf - *Kalanchoe pinnata*
miraculous-berry - *Synsepalum dulcificum*
mirasol - *Helianthus annuus*
mirror-plant - *Coprosma repens*
misery, mountain-misery - *Chamaebatia foliolosa*
missey-moosey - *Sorbus americana*
mission grass - *Pennisetum polystachyon*
Mississippi hackberry - *Celtis laevigata*
Missouri
  Missouri currant - *Ribes aureum, R. odoratum*
  Missouri goldenrod - *Solidago missouriensis*
  Missouri gooseberry - *Ribes missouriense, R. setosum*
  Missouri goosefoot - *Chenopodium missouriense*
  Missouri gourd - *Cucurbita foetidissima*
  Missouri milk-vetch - *Astragalus missouriensis*
  Missouri violet - *Viola missouriensis*
  Missouri willow - *Salix eriocephala*
mist
  love-in-a-mist - *Passiflora foetida*
  Miami-mist - *Phacelia purshii*
  Scotch-mist - *Galium sylvaticum*
mistflower - *Conoclinium coelestinum*
mistletoe
  American Christmas mistletoe - *Phoradendron serotinum*
  American mistletoe - *Phoradendron serotinum*
  dwarf mistletoe - *Arceuthobium pusillum*
  eastern dwarf mistletoe - *Arceuthobium pusillum*
  false mistletoe - *Phoradendron*
  mistletoe - *Phoradendron*

mistletoe ficus - *Ficus deltoidea*
mistletoe rubber-plant - *Ficus deltoidea*
miterwort
  false miterwort - *Tiarella*
  miterwort - *Mitella diphylla, Mitella*
  naked miterwort - *Mitella nuda*
mithridate mustard - *Thlaspi arvense*
moccasin-flower - *Cypripedium acaule, Cypripedium*
mock
  Atlantic mock bishop's-weed - *Ptilimnium capillaceum*
  house blooming mock-orange - *Pittosporum tobira*
  Indian mock-strawberry - *Duchesnea indica*
  mock-apple - *Echinocystis lobata*
  mock azalea - *Menziesia ferruginea*
  mock bishop's-weed - *Ptilimnium capillaceum*
  mock cucumber - *Echinocystis lobata*
  mock-mesquite - *Calliandra eriophylla*
  mock-orange - *Prunus caroliniana, Sideroxylon lycioides, Styrax americanus*
  mock pennyroyal - *Hedeoma pulegioides, Hedeoma*
  mock-plane - *Acer pseudoplatanus*
  mock-strawberry - *Duchesnea indica*
  sweet mock-orange - *Philadelphus coronarius*
mocker-nut hickory - *Carya tomentosa*
modesty - *Hibiscus trionum, Whipplea modesta*
modoc cypress - *Cupressus bakeri*
Mohr's oak - *Quercus mohriana*
mois, tous-les-mois - *Canna indica*
molasses grass - *Melinis minutiflora*
mole-plant - *Euphorbia heterophylla, E. lathyris*
Moline's elm - *Ulmus americana* cv. 'molin'
molle - *Schinus molle*
molly, green-molly - *Kochia americana*
Molucca
  Molucca balm - *Moluccella laevis*
  Molucca raspberry - *Rubus moluccanus*
mombin
  purple mombin - *Spondias purpurea*
  red mombin - *Spondias purpurea*
money
  money-plant - *Lunaria annua*
  money-tree - *Eucalyptus pulverulenta*
moneywort - *Lysimachia nummularia*
Mongolian oak - *Quercus mongolica*
monk's
  azure monk's-hood - *Aconitum carmichaelii*
  garden monk's-hood - *Aconitum napellus*
  monk's-hood - *Aconitum, Astrophytum myriostigma*
  monk's-hood-vine - *Ampelopsis aconitifolia*
  monk's-pepper-tree - *Vitex agnus-castus*
  monk's rhubarb - *Rumex alpinus, R. patientia*
  ornamental monk's-hood - *Astrophytum ornatum*
  southern monk's-hood - *Aconitum uncinatum*
  wild monk's-hood - *Aconitum uncinatum*
monkey
  Allegheny monkey-flower - *Mimulus ringens*
  monkey-flower - *Mimulus guttatus, Mimulus*
  monkey-nut - *Arachis hypogaea*
  monkey-pistol - *Hura crepitans*
  monkey-plant - *Ruellia makoyana*
  monkey-puzzle-tree - *Araucaria araucana*

  scarlet monkey-flower - *Mimulus cardinalis*
  sharp-winged monkey-flower - *Mimulus alatus*
  square-stemmed monkey-flower - *Mimulus ringens*
monkey's-dinner-bell - *Hura crepitans*
mono maple - *Acer mono*
monochoria
  arrow-leaf monochoria - *Monochoria hastata*
  monochoria - *Monochoria vaginalis*
monox - *Empetrum nigrum*
monstera - *Monstera deliciosa*
Montana blue-eyed-grass - *Sisyrinchium montanum*
Monterey
  Monterey ceanothus - *Ceanothus cuneatus* var. *rigidus*
  Monterey cypress - *Cupressus macrocarpa*
  Monterey Indian paintbrush - *Castilleja latifolia*
  Monterey manzanita - *Arctostaphylos hookeri*
  Monterey pine - *Pinus radiata*
moon
  half-moon loco - *Astragalus allochrous*
  white-moon petunia - *Petunia axillaris*
moonbeam - *Tabernaemontana divaricata*
moonflower
  bush moonflower - *Ipomoea leptophylla*
  moonflower - *Ipomoea alba*
  purple moonflower - *Ipomoea turbinata*
moonseed
  Canadian moonseed - *Menispermum canadense*
  Carolina moonseed - *Cocculus carolinus*
  red-berry moonseed - *Cocculus carolinus*
  red moonseed - *Cocculus carolinus*
moonshine - *Anaphalis margaritacea*
moonwort - *Botrychium lunaria, Botrychium, Lunaria annua*
moor
  moor-berry - *Vaccinium uliginosum*
  moor grass - *Molinia caerulea*
moort, round-leaf moort - *Eucalyptus platypus*
moose
  moose-berry - *Viburnum edule*
  moose maple - *Acer spicatum*
  moosewood - *Acer pensylvanicum, Dirca palustris, Viburnum alnifolium*
moosey, missey-moosey - *Sorbus americana*
morel, petty-morel - *Aralia racemosa*
Morello's cherry - *Prunus cerasus* var. *austera*
Moreton Bay chestnut - *Castanospermum australe*
Mormon tea - *Ephedra viridis*
morning
  beach morning-glory - *Ipomoea pes-caprae*
  big-root morning-glory - *Ipomoea pandurata*
  blue morning-glory - *Ipomoea indica*
  bush morning-glory - *Ipomoea leptophylla*
  Cairo morning-glory - *Ipomoea cairica*
  cotton morning-glory - *Ipomoea cordatotriloba* var. *torreyana*
  cypress-vine morning-glory - *Ipomoea quamoclit*
  ivy-leaf morning-glory - *Ipomoea hederacea*
  ivy-leaf red morning-glory - *Ipomoea hederifolia*
  Japanese morning-glory - *Ipomoea nil*
  morning campion - *Silene dioica*
  morning-glory - *Ipomoea purpurea, Ipomoea*
  multicolored morning-glory - *Ipomoea tricolor*
  orchard morning-glory - *Convolvulus arvensis*

palm-leaf morning-glory - *Ipomoea wrightii*
pitted morning-glory - *Ipomoea lacunosa*
red morning-glory - *Ipomoea coccinea*
sharp-pod morning-glory - *Ipomoea cordatotriloba*
small-flowered morning-glory - *Convolvulus arvensis,*
   Jacquemontia tamnifolia
small red morning-glory - *Ipomoea coccinea*
small white morning-glory - *Ipomoea lacunosa*
swamp morning-glory - *Ipomoea aquatica*
tall morning-glory - *Ipomoea purpurea*
three-lobed morning-glory - *Ipomoea triloba*
white-edge morning-glory - *Ipomoea nil*
white morning-glory - *Ipomoea lacunosa*
wild morning-glory - *Calystegia sepium, Convolvulus*
   arvensis
woolly morning-glory - *Argyreia nervosa*

Morrow's honeysuckle - *Lonicera morrowii*
mosaic-plant - *Fittonia verschaffeltii*
moschatel - *Adoxa moschatellina*
Moses
   Moses-in-a-boat - *Tradescantia spathacea*
   Moses-in-the-bulrushes - *Tradescantia spathacea*
   Moses-on-a-raft - *Tradescantia spathacea*
mosquito
   Carolina mosquito-fern - *Azolla caroliniana*
   mosquito-bills - *Dodecatheon hendersonii*
   mosquito-plant - *Hedeoma pulegioides*
   pinnate mosquito-fern - *Azolla pinnata*
moss
   ball-moss - *Tillandsia recurvata*
   bog club-moss - *Lycopodium inundatum*
   bog-moss - *Mayaca fluviatilis*
   bristly club-moss - *Lycopodium annotinum*
   bunch-moss - *Tillandsia recurvata*
   club-moss - *Lycopodium clavatum, Lycopodium*
   elk-moss - *Lycopodium clavatum*
   fir club-moss - *Lycopodium selago*
   gold-moss - *Sedum acre*
   interrupted club-moss - *Lycopodium annotinum*
   Irish moss - *Soleirolia soleirolii*
   Japanese moss - *Soleirolia soleirolii*
   little club-moss - *Selaginella*
   marsh club-moss - *Lycopodium inundatum*
   moss campion - *Silene acaulis*
   moss locust - *Robinia hispida*
   moss-pink - *Phlox subulata*
   moss-rose - *Portulaca grandiflora*
   moss vervain - *Verbena tenuisecta*
   Oregon spike-moss - *Selaginella oregana*
   rock-moss - *Sedum pulchellum*
   rock spike-moss - *Selaginella rupestris*
   rose-moss - *Portulaca grandiflora*
   running club-moss - *Lycopodium clavatum*
   shining club-moss - *Lycopodium lucidulum*
   Spanish-moss - *Tillandsia usneoides*
   spike-moss - *Selaginella*
   stiff club-moss - *Lycopodium annotinum*
   tree club-moss - *Lycopodium obscurum*
   Wallace's spike-moss - *Selaginella wallacei*
   Watson's spike-moss - *Selaginella watsonii*
mossy
   mossy-cup oak - *Quercus macrocarpa*

mossy stonecrop - *Sedum acre*
moth
   moth mullein - *Verbascum blattaria*
   white moth mullein - *Verbascum blattaria* var. *albiflora*
mother
   mother-in-law-plant - *Dieffenbachia seguine*
   mother-in-law's-tongue - *Sansevieria trifasciata*
   mother-in-law's-tongue-plant - *Dieffenbachia*
   mother-of-the-evening - *Hesperis matronalis*
   mother-of-thousands - *Saxifraga stolonifera*
   mother-of-thyme - *Acinos arvensis, Thymus serpyllum*
mothers, thousand-mothers - *Tolmiea menziesii*
motherwort
   horehound motherwort - *Leonurus marrubiastrum*
   motherwort - *Leonurus cardiaca*
   Siberian motherwort - *Leonurus sibiricus*
motie, water-motie - *Baccharis salicifolia*
mountain
   American mountain-ash - *Sorbus americana*
   birch-lead mountain-mahogany - *Cercocarpus betuloides*
   black mountain-ash - *Eucalyptus sieber*
   Catalina mountain-lilac - *Ceanothus arboreus*
   European mountain-ash - *Sorbus aucuparia*
   evergreen mountain fetterbush - *Pieris floribunda*
   fire-on-the-mountain - *Euphorbia cyathophora*
   hairy mountain mint - *Pycnanthemum pilosum*
   mountain alder - *Alnus tenuifolia, A. viridis* subsp. *crispa*
   mountain andromeda - *Pieris floribunda*
   mountain angelica - *Angelica triquinata*
   mountain-ash - *Sorbus*
   mountain-avens - *Dryas octopetala, D. octopetala* subsp.
    alaskensis
   mountain bilberry - *Vaccinium membranaceum*
   mountain birch - *Betula occidentalis*
   mountain bladder fern - *Cystopteris montana*
   mountain blueberry - *Vaccinium membranaceum*
   mountain bluet - *Centaurea montana*
   mountain-box - *Arctostaphylos uva-ursi*
   mountain brome - *Bromus carinatus*
   mountain bugbane - *Cimicifuga americana*
   mountain bunch grass - *Festuca viridula*
   mountain camellia - *Stewartia ovata*
   mountain cliff fern - *Woodsia scopulina*
   mountain cranberry - *Vaccinium erythrocarpum, V. vitis-*
    idaea, V. vitis-idaea var. *minus*
   mountain currant - *Ribes alpinum*
   mountain daisy - *Arenaria groenlandica*
   mountain death camass - *Zigadenus elegans*
   mountain dog-laurel - *Leucothoe fontanesiana*
   mountain dogwood - *Cornus nuttallii*
   mountain-ebony - *Bauhinia hookeri, B. variegata*
   mountain fly honeysuckle - *Lonicera villosa*
   mountain-glory - *Holodiscus dumosus*
   mountain grape - *Vitis rupestris*
   mountain grass - *Oryzopsis*
   mountain hair grass - *Deschampsia atropurpurea*
   mountain-heather - *Phyllodoce breweri, Phyllodoce*
   mountain hemlock - *Tsuga mertensiana*
   mountain holly - *Ilex ambigua* var. *montana, Nemopanthus,*
    Prunus ilicifolia
   mountain holly fern - *Polystichum lonchitis*
   mountain Juneberry - *Amelanchier bartramiana*

mountain juniper - *Juniperus communis*
mountain larkspur - *Delphinium glaucum*
mountain-laurel - *Kalmia latifolia*
mountain-lettuce - *Saxifraga micranthidifolia*
mountain lily - *Lilium auratum*
mountain-lover - *Paxistima canbyi*
mountain magnolia - *Magnolia fraseri*
mountain-mahogany - *Betula lenta, Cercocarpus montanus*
mountain maple - *Acer glabrum* subsp. *douglasii, A. spicatum*
mountain meadow-rue - *Thalictrum clavatum*
mountain mint - *Monardella odoratissima, Pycnanthemum*
mountain-misery - *Chamaebatia foliolosa*
mountain muhly - *Muhlenbergia montana*
mountain parsley fern - *Cryptogramma crispa*
mountain phlox - *Phlox ovata*
mountain pink currant - *Ribes nevadense*
mountain-pride - *Penstemon newberryi*
mountain rhubarb - *Rumex alpinus*
mountain rose-bay - *Rhododendron catawbiense*
mountain rye - *Secale montanum*
mountain sandwort - *Arenaria groenlandica*
mountain serviceberry - *Amelanchier bartramiana*
mountain-sorrel - *Oxyria digyna*
mountain St. John's-wort - *Hypericum mitchellianum*
mountain stewartia - *Stewartia ovata*
mountain strawberry - *Fragaria virginiana*
mountain sumac - *Rhus copallina*
mountain-sweet - *Ceanothus americanus*
mountain tarweed - *Madia glomerata*
mountain-tea - *Gaultheria procumbens*
mountain timothy - *Phleum pratense*
mountain white-alder - *Clethra acuminata*
mountain white potentilla - *Potentilla tridentata*
mountain whitethorn - *Ceanothus cordulatus*
mountain willow - *Salix scouleriana*
mountain winter-cress - *Cardamine rotundifolia*
mountain winterberry - *Ilex ambigua* var. *montana*
mountain wood fern - *Dryopteris campyloptera*
Pacific mountain-ash - *Sorbus sitchensis*
purple mountain saxifrage - *Saxifraga oppositifolia*
Rocky Mountain bee-plant - *Cleome serrulata*
Rocky Mountain Douglas fir - *Pseudotsuga menziesii* var. *glauca*
Rocky Mountain iris - *Iris missouriensis*
Rocky Mountain juniper - *Juniperus scopulorum*
Rocky Mountain maple - *Acer glabrum*
Rocky Mountain raspberry - *Rubus deliciosus*
Rocky Mountain scrub oak - *Quercus undulata*
Rocky Mountain spatter-dock - *Nuphar lutea* subsp. *polysepala*
Rocky Mountain woodsia - *Woodsia scopulina*
Rocky Mountain yellow pine - *Pinus ponderosa* var. *scopulorum*
showy mountain-ash - *Sorbus decora*
silver-leaf mountain gum - *Eucalyptus pulverulenta*
snow-on-the-mountain - *Euphorbia marginata*
Swiss mountain pine - *Pinus mugo*
table mountain pine - *Pinus pungens*
Texas mountain-laurel - *Sophora secundiflora*
Virginia mountain mint - *Pycnanthemum virginianum*
yellow mountain saxifrage - *Saxifraga aizoides*

Mountains, White Mountains dogwood - *Viburnum alnifolium*
mourning-bride - *Scabiosa atropurpurea*
mouse
church-mouse three-awn - *Aristida dichotoma*
mouse barley - *Hordeum murinum*
mouse-bloodwort - *Hieracium pilosella*
mouse-ear chickweed - *Cerastium glomeratum, C. vulgatum, Cerastium*
mouse-ear-cress - *Arabidopsis thaliana*
mouse-ear hawkweed - *Hieracium pilosella*
mouse foxtail - *Alopecurus myosuroides*
mouse's
mouse's-ears - *Antennaria plantaginifolia*
mouse's-tail - *Myosurus minimus*
moxie
moxie - *Gaultheria hispidula*
moxie-plum - *Gaultheria hispidula*
mu
mu - *Vernicia montana*
mu-oil-tree - *Vernicia montana*
mu-tree - *Vernicia montana*
muck, curly muck-weed - *Potamogeton crispus*
mud
mud sedge - *Carex limosa*
round-leaf mud-plantain - *Heteranthera reniformis*
Texas mud-baby - *Echinodorus cordifolius*
mudar
mudar - *Calotropis gigantea*
small mudar - *Calotropis procera*
muguet - *Maianthemum canadense*
mugwort
California mugwort - *Artemisia douglasiana*
mugwort - *Artemisia douglasiana, A. vulgaris, Artemisia*
western mugwort - *Artemisia ludoviciana*
muhly
alkali muhly - *Muhlenbergia asperifolia*
bush muhly - *Muhlenbergia porteri*
creeping muhly - *Muhlenbergia repens*
few-flowered muhly - *Muhlenbergia pauciflora*
foxtail muhly - *Muhlenbergia andina*
hairy-awn muhly - *Muhlenbergia capillaris*
Mexican muhly - *Muhlenbergia mexicana*
mountain muhly - *Muhlenbergia montana*
muhly - *Muhlenbergia*
New Mexico muhly - *Muhlenbergia pauciflora*
plains muhly - *Muhlenbergia cuspidata*
pull-up muhly - *Muhlenbergia filiformis*
slim-stem muhly - *Muhlenbergia filiculmis*
spike muhly - *Muhlenbergia wrightii*
wire-stem muhly - *Muhlenbergia frondosa*
mulberry
American mulberry - *Morus rubra*
black mulberry - *Morus nigra*
blite-mulberry - *Chenopodium capitatum*
French mulberry - *Callicarpa americana*
Indian mulberry - *Morinda citrifolia*
mulberry - *Morus*
paper-mulberry - *Broussonetia papyrifera*
red mulberry - *Morus rubra*
silkworm mulberry - *Morus alba* var. *multicaulis*

Texas mulberry - *Morus microphylla*
white mulberry - *Morus alba*
mule's-fat - *Baccharis viminea*
mullein
  clasping mullein - *Verbascum phlomoides*
  corn-mullein - *Agrostemma githago*
  moth mullein - *Verbascum blattaria*
  mullein - *Verbascum phlomoides, V. thapsus, Verbascum*
  mullein-foxglove - *Dasistoma macrophylla*
  mullein nightshade - *Solanum erianthum*
  mullein-pink - *Lychnis coronaria*
  purple-stamen mullein - *Verbascum virgatum*
  turkey-mullein - *Eremocarpus setigerus*
  wand mullein - *Verbascum virgatum*
  white moth mullein - *Verbascum blattaria* var. *albiflora*
  white mullein - *Verbascum lychnitis*
  woolly mullein - *Verbascum thapsus*
mullet, Indian mullet - *Oryzopsis hymenoides*
multicolored morning-glory - *Ipomoea tricolor*
multiflora rose - *Rosa multiflora*
multiplier onion - *Allium cepa*
mum - *Chrysanthemum morifolium*
mung bean - *Vigna radiata*
muraski-giboshi - *Hosta ventricosa*
Murray's red gum - *Eucalyptus camaldulensis*
muscadine grape - *Vitis rotundifolia*
muscle-wood - *Carpinus caroliniana*
musk
  musk - *Malva moschata*
  musk-clover - *Erodium moschatum*
  musk-flower - *Mimulus moschatus*
  musk mallow - *Malva moschata, Malva*
  musk-plant - *Malva moschata, Mimulus moschatus*
  musk-root - *Adoxa moschatellina*
  musk thistle - *Carduus nutans*
  wild musk - *Erodium cicutarium*
muskmelon - *Cucumis melo* var. *reticulatus*
muskrat-weed - *Cicuta maculata, Thalictrum pubescens*
musquash-root - *Cicuta maculata*
mustang grape - *Vitis mustangensis*
mustard
  African mustard - *Brassica tournefortii*
  ball mustard - *Neslia paniculata*
  bead-podded mustard - *Chorispora tenella*
  bird's-rape mustard - *Brassica rapa*
  black mustard - *Brassica nigra*
  blue mustard - *Chorispora tenella*
  brown mustard - *Brassica juncea, B. nigra*
  Chinese mustard - *Brassica juncea*
  dog-mustard - *Erucastrum gallicum*
  field mustard - *Brassica rapa*
  garlic mustard - *Alliaria petiolata*
  green tansy mustard - *Descurainia pinnata* var. *brachycarpa*
  hare-ear mustard - *Conringia orientalis*
  hedge-mustard - *Sisymbrium altissimum, S. officinale*
  hill-mustard - *Bunias orientalis*
  Indian mustard - *Brassica juncea*
  Jim Hill's mustard - *Sisymbrium altissimum*
  leaf mustard - *Brassica juncea*
  Menzies tansy mustard - *Descurainia pinnata* subsp. *menziesii*
  mithridate mustard - *Thlaspi arvense*

  mustard cabbage - *Brassica juncea*
  mustard-greens - *Brassica juncea*
  mustard-tree - *Nicotiana glauca*
  Oriental mustard - *Sisymbrium orientale*
  pinnate tansy mustard - *Descurainia pinnata*
  Richardson's tansy mustard - *Descurainia richardsonii*
  short-pod mustard - *Hirschfeldia incana*
  Siberian mustard - *Cardaria pubescens*
  spreading mustard - *Erysimum repandum*
  Swatow mustard - *Brassica juncea*
  Syrian mustard - *Euclidium syriacum*
  tall hedge mustard - *Sisymbrium altissimum*
  tall worm-seed mustard - *Erysimum hieraciifolium*
  tansy mustard - *Descurainia pinnata, D. sophia*
  tower mustard - *Arabis glabra*
  treacle mustard - *Erysimum cheiranthoides, E. repandum, Erysimum*
  tumble mustard - *Sisymbrium altissimum*
  wallflower mustard - *Erysimum cheiranthoides*
  water mustard - *Barbarea vulgaris*
  western tansy mustard - *Descurainia incisa*
  white mustard - *Sinapis alba*
  worm-seed mustard - *Erysimum cheiranthoides*
mutton
  long-tongue mutton grass - *Poa fendleriana*
  mutton bluegrass - *Poa fendleriana, Setaria faberi*
Myrobalan plum - *Prunus cerasifera*
myrtle
  bog myrtle - *Myrica gale*
  box sand-myrtle - *Leiophyllum buxifolium*
  California wax myrtle - *Myrica californica*
  Cape myrtle - *Myrsine africana*
  crape-myrtle - *Lagerstroemia indica*
  downy rose-myrtle - *Rhodomyrtus tomentosus*
  dwarf wax myrtle - *Myrica pusilla*
  Greek myrtle - *Myrtus communis*
  ground-myrtle - *Vinca*
  honey-myrtle - *Melaleuca*
  myrtle - *Cyrilla racemiflora, Myrtus communis*
  myrtle box-leaf - *Paxistima myrsinites*
  myrtle dahoon - *Ilex myrtifolia*
  myrtle oak - *Quercus myrtifolia*
  myrtle spurge - *Euphorbia lathyris*
  Oregon myrtle - *Umbellularia californica*
  Pacific wax myrtle - *Myrica californica*
  running-myrtle - *Vinca minor*
  sea-myrtle - *Baccharis halimifolia*
  southern wax myrtle - *Myrica cerifera*
  Swedish myrtle - *Myrtus communis*
  wax myrtle - *Myrica californica, M. cerifera, M. heterophylla*
  yellow-myrtle - *Lysimachia nummularia*
mysteria - *Colchicum autumnale*
nagoon-berry - *Rubus stellatus*
naiad
  brittle-leaf naiad - *Najas minor*
  grassy naiad - *Najas graminea*
  holly-leaf naiad - *Najas marina*
  slender naiad - *Najas flexilis*
  southern naiad - *Najas guadalupensis*
nail
  nail-rod - *Typha latifolia*
  nail-wort - *Paronychia*

naio - *Myoporum sandwicense*
naked
    naked broomrape - *Orobanche uniflora*
    naked-flowered tick-trefoil - *Desmodium nudiflorum*
    naked-lady-lily - *Amaryllis belladonna*
    naked miterwort - *Mitella nuda*
    naked oat - *Avena nuda*
    naked-spiked ragweed - *Ambrosia psilostachya*
    naked-stemmed sunflower - *Helianthus occidentalis*
    naked tick-trefoil - *Desmodium nudiflorum*
    naked viburnum - *Viburnum nudum*
nameless water-hyssop - *Bacopa innominata*
nance - *Byrsonima crassifolia*
Nankeen lily - *Lilium testaceum*
nannyberry - *Viburnum lentago, V. prunifolium*
nap-at-noon - *Ornithogalum umbellatum*
napier grass - *Pennisetum purpureum*
narcissus
    narcissus - *Narcissus poeticus, Narcissus*
    poet's narcissus - *Narcissus poeticus*
    polyanthus narcissus - *Narcissus tazetta*
    trumpet narcissus - *Narcissus pseudonarcissus*
narra - *Pterocarpus indicus*
narrow
    narrow-leaf blue-eyed-grass - *Sisyrinchium angustifolium*
    narrow-leaf bur-reed - *Sparganium emersum*
    narrow-leaf cat-tail - *Typha angustifolia*
    narrow-leaf cottonwood - *Populus angustifolia*
    narrow-leaf cudweed - *Gnaphalium purpureum* var. *falcatum*
    narrow-leaf dock - *Rumex stenophyllus*
    narrow-leaf fire-thorn - *Pyracantha angustifolia*
    narrow-leaf fritillary - *Fritillaria affines*
    narrow-leaf gentian - *Gentiana linearis*
    narrow-leaf globe mallow - *Sphaeralcea angustifolia*
    narrow-leaf hawk's-beard - *Crepis tectorum*
    narrow-leaf hawkweed - *Hieracium umbellatum*
    narrow-leaf lamb's-quarters - *Chenopodium desiccatum*
    narrow-leaf-laurel - *Kalmia angustifolia*
    narrow-leaf loosestrife - *Lythrum lineare*
    narrow-leaf milk-vetch - *Astragalus pectinatus*
    narrow-leaf milkweed - *Asclepias fascicularis*
    narrow-leaf miner's-lettuce - *Montia linearis*
    narrow-leaf pepper-weed - *Lepidium ruderale*
    narrow-leaf plantain - *Plantago lanceolata*
    narrow-leaf plantain-lily - *Hosta lancifolia*
    narrow-leaf puccoon - *Lithospermum incisum*
    narrow-leaf signal grass - *Brachiaria piligera*
    narrow-leaf spring-beauty - *Claytonia virginica*
    narrow-leaf sunflower - *Helianthus angustifolius*
    narrow-leaf trefoil - *Lotus tenuis*
    narrow-leaf vervain - *Verbena simplex*
    narrow-leaf water-plantain - *Alisma gramineum*
    narrow-leaf willow - *Salix exigua*
    narrow reed grass - *Calamagrostis neglecta*
naseberry - *Manilkara zapota*
nasturtium
    garden nasturtium - *Tropaeolum majus*
    nasturtium bauhinia - *Bauhinia galpinii*
    tall nasturtium - *Tropaeolum majus*
Natal
    Natal grass - *Rhynchelytrum repens*

Natal-plum - *Carissa macrocarpa*
navelwort - *Hydrocotyle*
navy bean - *Phaseolus vulgaris*
Nebraska
    Nebraska fern - *Conium maculatum*
    Nebraska sedge - *Carex nebrascensis*
neck
    coast fiddle-neck - *Amsinckia lycopsoides*
    crook-neck squash - *Cucurbita moschata*
    neck-weed - *Cannabis sativa, Veronica peregrina*
    rigid fiddle-neck - *Amsinckia retrorsa*
    small-flowered fiddle-neck - *Amsinckia menziesii*
    tarweed fiddle-neck - *Amsinckia lycopsoides*
    western fiddle-neck - *Amsinckia tessellata*
necklace
    necklace poplar - *Populus deltoides*
    necklace-weed - *Actaea*
necks, fiddle-necks - *Amsinckia lycopsoides*
nectandra, Jamaican nectandra - *Ocotea coriacea*
nectarine - *Prunus persica* var. *nucipersica*
needle
    Adam's-needle - *Yucca filamentosa*
    black-seed needle grass - *Stipa avenacea*
    California needle grass - *Stipa cernua, S. pulchra*
    Columbia needle grass - *Stipa columbiana*
    desert needle grass - *Stipa speciosa*
    devil's-darning-needle - *Clematis virginiana*
    foothill needle grass - *Stipa lepida*
    green needle grass - *Stipa viridula*
    Lemmon's needle grass - *Stipa lemmonii*
    Letterman's needle grass - *Stipa lettermanii*
    needle-and-thread - *Stipa comata*
    needle-and-thread grass - *Stipa comata*
    needle grama - *Bouteloua aristidoides*
    needle grass - *Stipa sibirica, Stipa*
    needle-palm - *Yucca filamentosa*
    needle rush - *Juncus roemerianus*
    needle spike rush - *Eleocharis acicularis*
    needle-weed - *Navarretia intertexta*
    Pringle's needle grass - *Stipa pringlei*
    purple needle grass - *Stipa pulchra*
    Richardson's needle grass - *Stipa richardsonii*
    Scribner's needle grass - *Stipa scribneri*
    shepherd's-needle - *Scandix pecten-veneris*
    six-weeks needle grama - *Bouteloua aristidoides*
    Texas needle grass - *Stipa leucotricha*
    Thurber's needle grass - *Stipa thurberiana*
    Venus's-needle - *Scandix pecten-veneris*
    western needle grass - *Stipa occidentalis*
needles, Spanish-needles - *Bidens bipinnata, B. pilosa, Bidens*
negi, ezo-negi - *Allium schoenoprasum*
Negro coffee - *Senna occidentalis*
nelumbo, yellow nelumbo - *Nelumbo lutea*
Nepal
    Nepal barley - *Hordeum vulgare*
    Nepal privet - *Ligustrum lucidum*
nephthytis - *Syngonium podophyllum*
nerve
    nerve-plant - *Fittonia verschaffeltii*
    nerve-root - *Cypripedium acaule, C. calceolus* var. *pubescens*
    silver nerve-plant - *Fittonia verschaffeltii*

nerved
- nerved waxy sedge - *Carex verrucosa*
- three-nerved duckweed - *Lemna trinervis*

nest
- bird's-nest - *Daucus carota, Pastinaca sativa*
- bird's-nest fern - *Asplenium nidus*
- climbing bird's nest fern - *Microsorium punctatum*
- giant bird's-nest - *Pterospora andromedea*

net
- net-leaf hackberry - *Celtis reticulata*
- net-leaf pawpaw - *Asimina reticulata*
- net-leaf willow - *Salix reticulata*
- net-seed lamb's-quarters - *Chenopodium berlandieri*
- net-seed spurge - *Euphorbia spathulata*
- silver-net-plant - *Fittonia verschaffeltii*

netbush
- blood-red netbush - *Calothamnus sanguineus*
- netbush - *Calothamnus sanguineus*

netted melon - *Cucumis melo* var. *reticulatus*

nettle
- bee-nettle - *Galeopsis tetrahit*
- bull-nettle - *Cnidoscolus stimulosus, Solanum elaeagnifolium*
- burning nettle - *Urtica urens*
- Carolina horse-nettle - *Solanum carolinense*
- Carolina nettle - *Solanum carolinense*
- cut-leaf nettle - *Solanum triflorum*
- dead-nettle - *Lamium*
- dog nettle - *Urtica urens*
- dwarf nettle - *Urtica urens*
- European nettle - *Urtica dioica*
- false nettle - *Boehmeria cylindrica*
- field-nettle betony - *Stachys arvensis*
- flowering-nettle - *Galeopsis tetrahit*
- hedge-nettle - *Stachys*
- hedge-nettle betony - *Stachys annua*
- hemp-nettle - *Galeopsis tetrahit*
- horse-nettle - *Solanum carolinense*
- low hedge-nettle - *Stachys arvensis*
- low nettle - *Stachys arvensis*
- Lyall's nettle - *Urtica dioica* subsp. *gracilis*
- nettle - *Urtica*
- nettle chain fern - *Woodwardia areolata*
- nettle-leaf bellflower - *Campanula trachelium*
- nettle-leaf giant hyssop - *Agastache urticifolia*
- nettle-leaf goosefoot - *Chenopodium murale*
- nettle-tree - *Celtis occidentalis, Celtis*
- purple dead-nettle - *Lamium purpureum*
- red dead-nettle - *Lamium purpureum*
- red hemp-nettle - *Galeopsis ladanum*
- robust horse-nettle - *Solanum dimidiatum*
- silver horse-nettle - *Solanum elaeagnifolium*
- silver-leaf-nettle - *Solanum elaeagnifolium*
- slender nettle - *Urtica dioica*
- smooth hedge-nettle - *Stachys tenuifolia*
- southern nettle - *Urtica chamaedryoides*
- spotted dead-nettle - *Lamium maculatum*
- spurge-nettle - *Cnidoscolus stimulosus*
- stinging nettle - *Urtica dioica*
- tall nettle - *Urtica dioica*
- Texas bull-nettle - *Cnidoscolus texanus*
- three-flowered-nettle - *Solanum triflorum*
- white dead-nettle - *Lamium album*
- white-horse-nettle - *Solanum elaeagnifolium*
- woolly hedge-nettle - *Stachys olympica*

Nevada
- Nevada bluegrass - *Poa secunda* subsp. *nevadensis*
- Nevada ephedra - *Ephedra nevadensis*

New
- New Caledonia pine - *Araucaria columnaris*
- New England aster - *Aster novae-angliae*
- New England grape - *Vitis novae-angliae*
- New England serviceberry - *Amelanchier sanguinea*
- New Guinea creeper - *Mucuna novaguineensis*
- New Jersey blueberry - *Vaccinium caesariense*
- New Jersey tea - *Ceanothus americanus*
- New Mexico alder - *Alnus oblongifolia*
- New Mexico evergreen sumac - *Rhus choriophylla*
- New Mexico feather grass - *Stipa neomexicana*
- New Mexico locust - *Robinia neomexicana*
- New Mexico muhly - *Muhlenbergia pauciflora*
- New Mexico raspberry - *Rubus neomexicanus*
- New York aster - *Aster novi-belgii*
- New York fern - *Thelypteris noveboracensis*
- New York ironweed - *Vernonia noveboracensis*
- New Zealand cliff brake - *Pellaea rotundifolia*
- New Zealand fireweed - *Erechtites glomerata*
- New Zealand flax - *Phormium tenax*
- New Zealand hemp - *Phormium tenax*
- New Zealand spinach - *Tetragonia tetragonioides*
- New Zealand tea-tree - *Leptospermum scoparium*

new old-man cactus - *Espostoa lanata*

nickers
- gray nickers - *Caesalpinia bonduc*
- nickers-tree - *Gymnocladus dioica*
- yellow nickers - *Caesalpinia bonduc*

Niger-seed - *Guizotia abyssinica*

Nigerian stylo - *Stylosanthes erecta*

night
- beauty-of-the-night - *Mirabilis jalapa*
- lady-of-the-night - *Brunfelsia americana*
- night-blooming cereus - *Hylocereus undatus*
- night-flowering catchfly - *Silene noctiflora*
- night jessamine - *Cestrum nocturnum*
- night-scented stock - *Matthiola longipetala*
- queen-of-the-night - *Hylocereus undatus*

nightshade
- American black nightshade - *Solanum americanum*
- ball nightshade - *Solanum carolinense*
- bittersweet nightshade - *Solanum dulcamara*
- black nightshade - *Solanum nigrum*
- cut-leaf nightshade - *Solanum triflorum*
- deadly nightshade - *Atropa belladonna, Solanum dulcamara*
- enchanter's-nightshade - *Circaea quadrisulcata, Circaea*
- hairy nightshade - *Solanum sarrachoides*
- Malabar nightshade - *Basella*
- melon-leaf nightshade - *Solanum citrullifolium*
- mullein nightshade - *Solanum erianthum*
- nightshade - *Solanum nigrum*
- poisonous nightshade - *Solanum dulcamara*
- purple nightshade - *Solanum xanti*
- silver-leaf nightshade - *Solanum elaeagnifolium*
- small enchanter's-nightshade - *Circaea alpina*
- soda-apple nightshade - *Solanum aculeatissimum*
- sticky nightshade - *Solanum sisymbriifolium*

stinking nightshade - *Hyoscyamus niger*
Torrey's nightshade - *Solanum dimidiatum*
Nile, lily-of-the-Nile - *Agapanthus africanus*
nimble-will - *Muhlenbergia schreberi*
ninebark - *Physocarpus opulifolius, Physocarpus*
nipplewort
   dwarf nipplewort - *Arnoseris minima*
   nipplewort - *Lapsana communis*
nira - *Allium tuberosum*
niruri - *Phyllanthus niruri*
nispero - *Manilkara zapota*
nits-and-lice - *Hypericum drummondii*
Noah's-ark - *Cypripedium calceolus* var. *pubescens*
noble
   noble fir - *Abies procera*
   noble-liverwort - *Hepatica*
nodding
   nodding beak rush - *Rhynchospora inexpansa*
   nodding beggar-ticks - *Bidens cernua*
   nodding brome - *Bromus anomalus*
   nodding broomrape - *Orobanche cernua*
   nodding bur-marigold - *Bidens cernua*
   nodding catchfly - *Silene pendula*
   nodding chickweed - *Cerastium nutans*
   nodding fescue - *Festuca obtusa*
   nodding foxtail - *Setaria faberi*
   nodding mandarin - *Disporum maculatum*
   nodding onion - *Allium cernuum*
   nodding semaphore grass - *Pleuropogon refractus*
   nodding spurge - *Euphorbia maculata, E. nutans*
   nodding star-of-Bethlehem - *Ornithogalum nutans*
   nodding thistle - *Carduus nutans*
   nodding trillium - *Trillium cernuum*
   nodding trisetum - *Trisetum cernuum*
   nodding wild onion - *Allium cernuum*
node, purple-node Joe-Pye-weed - *Eupatorium purpureum*
nogal - *Juglans major, J. microcarpa*
noires, gueules-noires - *Gaylussacia baccata*
none-so-pretty - *Silene armeria*
nonesuch clover - *Medicago lupulina*
noog - *Guizotia abyssinica*
noon
   Jack-go-to-bed-at-noon - *Tragopogon pratensis*
   John-go-to-bed-at-noon - *Tragopogon pratensis*
   nap-at-noon - *Ornithogalum umbellatum*
Nootka
   Nootka lupine - *Lupinus nootkatensis*
   Nootka rose - *Rosa nutkana*
   Nootka yellow-cedar - *Chamaecyparis nootkatensis*
Norfolk Island pine - *Araucaria heterophylla*
northeastern beard-tongue - *Penstemon hirsutus*
northern
   cultivated northern wild rice - *Zizania palustris*
   low northern sedge - *Carex concinnoides*
   northern adder's-tongue fern - *Ophioglossum pusillum*
   northern Bangalow palm - *Archontophoenix alexandrae*
   northern bayberry - *Myrica pensylvanica*
   northern bedstraw - *Galium boreale*
   northern black currant - *Ribes hudsonianum*

   northern blue flag - *Iris versicolor*
   northern bluebells - *Mertensia paniculata*
   northern bog goldenrod - *Solidago uliginosa*
   northern bog violet - *Viola sororia* subsp. *affinis*
   northern bugleweed - *Lycopus uniflorus*
   northern bush-honeysuckle - *Diervilla lonicera*
   northern California black walnut - *Juglans hindsii*
   northern catalpa - *Catalpa speciosa*
   northern clustered sedge - *Carex arcta*
   northern comandra - *Geocaulon lividum*
   northern coralroot - *Corallorrhiza trifida*
   northern dewberry - *Rubus flagellaris*
   northern drop-seed - *Sporobolus heterolepis*
   northern estuarine beggar-ticks - *Bidens hyperborea*
   northern fly honeysuckle - *Lonicera villosa*
   northern fog-fruit - *Phyla lanceolata*
   northern gooseberry - *Ribes oxyacanthoides*
   northern green orchid - *Habenaria hyperborea*
   northern hackberry - *Celtis occidentalis*
   northern hawkweed - *Hieracium umbellatum*
   northern holly fern - *Polystichum lonchitis*
   northern horse-balm - *Collinsonia canadensis*
   northern Japanese hemlock - *Tsuga diversifolia*
   northern joint-vetch - *Aeschynomene virginica*
   northern lady fern - *Athyrium felix-femina* var. *michauxii*
   northern maidenhair fern - *Adiantum pedatum*
   northern manna grass - *Glyceria borealis*
   northern meadow groundsel - *Senecio pauperculus*
   northern nut-grass - *Cyperus esculentus*
   northern pin oak - *Quercus ellipsoidalis*
   northern prickly-ash - *Zanthoxylum americanum*
   northern red currant - *Ribes rubrum*
   northern red oak - *Quercus rubra*
   northern reed grass - *Calamagrostis neglecta*
   northern sedge - *Carex deflexa*
   northern slender lady's-tresses - *Spiranthes lacera*
   northern swamp dogwood - *Cornus racemosa*
   northern swamp groundsel - *Senecio congestus*
   northern tickseed - *Bidens coronata*
   northern water-milfoil - *Myriophyllum exalbescens*
   northern water-nymph - *Najas flexilis*
   northern white-cedar - *Thuja occidentalis*
   northern white violet - *Viola macloskeyi* var. *pallens, V. renifolia*
   northern wild comfrey - *Cynoglossum boreale*
   northern wood-sorrel - *Oxalis acetosella*
northwest willow - *Salix sessilifolia*
Norway
   Norway maple - *Acer platanoides*
   Norway pine - *Pinus resinosa*
   Norway spruce - *Picea abies*
nose-bleed - *Achillea millefolium*
nosegay - *Plumeria rubra*
notch-seeded buckwheat - *Fagopyrum esculentum*
nug - *Guizotia abyssinica*
num, big num-num - *Carissa macrocarpa*
nut
   American bladder-nut - *Staphylea trifolia*
   Australian-nut - *Macadamia integrifolia*
   Barbados-nut - *Jatropha curcas*
   bitter-nut hickory - *Carya cordiformis*
   bladder-nut - *Staphylea*

bur-nut - *Tribulus terrestris*
Chinese wing-nut - *Pterocarya stenoptera*
deer-nut - *Simmondsia chinensis*
earth-nut - *Arachis hypogaea*
false nut sedge - *Cyperus strigosus*
goat-nut - *Simmondsia chinensis*
grass-nut - *Arachis hypogaea, Triteleia laxa*
heart-nut - *Juglans cordiformis*
Jesuit-nut - *Trapa natans*
king's-nut - *Carya laciniosa*
lucky-nut - *Thevetia peruviana*
macadamia-nut - *Macadamia integrifolia*
Madeira-nut - *Juglans regia*
mocker-nut hickory - *Carya tomentosa*
monkey-nut - *Arachis hypogaea*
northern nut-grass - *Cyperus esculentus*
nut-grass - *Cyperus rotundus*
nut pine - *Pinus monophylla*
nut rush - *Scleria*
nut sedge - *Cyperus esculentus, C. rotundus*
physic-nut - *Jatropha curcas*
pig-nut - *Carya cordiformis, Hoffmannseggia glauca,*
    Simmondsia chinensis
pig-nut hickory - *Carya glabra*
pistachio-nut - *Pistacia vera*
purple nut sedge - *Cyperus rotundus*
Queensland-nut - *Macadamia integrifolia*
rush-nut - *Cyperus esculentus*
sierra bladder-nut - *Staphylea bolanderi*
small-fruited Queensland-nut - *Macadamia tetraphylla*
square-nut - *Carya tomentosa*
sweet pig-nut - *Carya ovalis*
tiger-nut - *Cyperus esculentus*
trapa-nut - *Trapa natans*
two-leaf nut pine - *Pinus edulis*
water-nut - *Trapa natans*
wing-nut - *Pterocarya*
yellow nut-grass - *Cyperus esculentus*
yellow nut sedge - *Cyperus esculentus*
nutgall-tree - *Rhus chinensis*
Nutka rose - *Rosa nutkana*
nutmeg
African nutmeg - *Monodora myristica*
calabash nutmeg - *Monodora myristica*
California nutmeg - *Torreya californica*
Jamaican nutmeg - *Monodora myristica*
nutmeg hickory - *Carya myristiciformis*
nutmeg melon - *Cucumis melo* var. *reticulatus*
nuts, pond-nuts - *Nelumbo lutea*
Nuttall's
Nuttall's alkali grass - *Puccinellia nuttalliana*
Nuttall's death camass - *Zigadenus nuttallii*
Nuttall's milk-vetch - *Astragalus nuttallianus, A. nuttallii*
Nuttall's oak - *Quercus nuttallii*
Nuttall's pondweed - *Potamogeton epihydrus*
Nuttall's poverty-weed - *Monolepis nuttalliana*
Nuttall's scrub oak - *Quercus dumosa*
Nuttall's sunflower - *Helianthus nuttallii*
Nuttall's waterweed - *Elodea nuttallii*
nymph
alkaline water-nymph - *Najas marina*
eutrophic water-nymph - *Najas minor*

northern water-nymph - *Najas flexilis*
southern water-nymph - *Najas guadalupensis*
wood-nymph - *Moneses uniflora*
o'clock
Colorado four-o'clock - *Mirabilis multiflora*
eleven-o'clock - *Portulaca grandiflora*
four-o'clock - *Mirabilis jalapa*
trailing four-o'clock - *Allionia incarnata*
wild four-o'clock - *Mirabilis nyctaginea*
oak
Ajo oak - *Quercus ajoensis*
Arizona white oak - *Quercus arizonica*
Arkansas oak - *Quercus arkansana*
basket oak - *Quercus michauxii, Q. prinus*
bastard oak - *Quercus austrina*
bear oak - *Quercus ilicifolia*
black oak - *Quercus velutina*
blackjack oak - *Quercus marilandica*
blue oak - *Quercus douglasii*
bluejack oak - *Quercus incana*
bluff oak - *Quercus austrina*
Brazilian oak - *Casuarina glauca*
bur oak - *Quercus macrocarpa*
California black oak - *Quercus kelloggii*
California field oak - *Quercus agrifolia*
California live oak - *Quercus agrifolia*
California scrub oak - *Quercus dumosa*
California white oak - *Quercus lobata*
canyon live oak - *Quercus chrysolepis*
canyon oak - *Quercus chrysolepis*
Catesby's oak - *Quercus laevis*
Chapman's oak - *Quercus chapmanii*
cherry-bark oak - *Quercus falcata* var. *pagodifolia*
chestnut oak - *Quercus prinus*
chinkapin oak - *Quercus muehlenbergii*
chinquapin oak - *Quercus prinoides*
coast live oak - *Quercus agrifolia*
cork oak - *Quercus suber*
cow oak - *Quercus michauxii*
Darlington's oak - *Quercus hemisphaerica*
deer oak - *Quercus sadleriana*
drooping she-oak - *Casuarina stricta*
Durand's oak - *Quercus durandii*
dwarf chestnut oak - *Quercus prinoides*
dwarf chinquapin oak - *Quercus prinoides*
dwarf live oak - *Quercus minima*
eastern poison-oak - *Toxicodendron pubescens*
Emory's oak - *Quercus emoryi*
Engelmann's oak - *Quercus engelmannii*
English oak - *Quercus robur*
European turkey oak - *Quercus cerris*
Gambel's oak - *Quercus gambelii*
Graves's oak - *Quercus gravesii*
gray oak - *Quercus grisea*
Harvard's shin oak - *Quercus harvardii*
high-ground willow oak - *Quercus incana*
holly oak - *Quercus ilex*
Holm's oak - *Quercus ilex*
huckleberry oak - *Quercus vaccinifolia*
interior live oak - *Quercus wislizenii*
island live oak - *Quercus tomentella*
island oak - *Quercus tomentella*

jack oak - *Quercus ellipsoidalis, Q. imbricaria,*
  Q. marilandica
Jerusalem oak goosefoot - *Chenopodium botrys*
Kellogg's oak - *Quercus kelloggii*
laurel-leaved oak - *Quercus laurifolia*
laurel oak - *Quercus imbricaria, Q. laurifolia*
leather oak - *Quercus durata*
live oak - *Quercus virginiana*
MacDonald's oak - *Quercus macdonaldii*
maul oak - *Quercus chrysolepis*
mesa oak - *Quercus engelmannii*
Mexican blue oak - *Quercus oblongifolia*
Mohr's oak - *Quercus mohriana*
Mongolian oak - *Quercus mongolica*
mossy-cup oak - *Quercus macrocarpa*
myrtle oak - *Quercus myrtifolia*
northern pin oak - *Quercus ellipsoidalis*
northern red oak - *Quercus rubra*
Nuttall's oak - *Quercus nuttallii*
Nuttall's scrub oak - *Quercus dumosa*
oak - *Quercus*
oak fern - *Gymnocarpium dryopteris*
oak-leaf goosefoot - *Chenopodium glaucum*
Oglethorpe's oak - *Quercus oglethorpensis*
Oregon oak - *Quercus garryana*
Oregon white oak - *Quercus garryana*
over-cup oak - *Quercus lyrata*
Pacific poison-oak - *Rhus diversiloba*
pin oak - *Quercus palustris*
plateau oak - *Quercus virginiana var. fusiformis*
poison-oak - *Toxicodendron pubescens*
possum oak - *Quercus nigra*
post oak - *Quercus stellata*
post-oak grape - *Vitis lincecumii*
river she-oak - *Casuarina cunninghamiana*
rock chestnut oak - *Quercus prinus*
Rocky Mountain scrub oak - *Quercus undulata*
running oak - *Quercus pumila*
sandpaper oak - *Quercus pungens*
saw-tooth oak - *Quercus acutissima*
scarlet oak - *Quercus coccinea*
scrub oak - *Quercus ilicifolia*
she-oak - *Casuarina*
shingle oak - *Quercus imbricaria*
shinnery oak - *Quercus harvardii*
shrub live oak - *Quercus turbinella*
Shumard's red oak - *Quercus shumardii*
silk-oak - *Grevillea robusta*
silver-leaf oak - *Quercus hypoleucoides*
southern live oak - *Quercus virginiana*
southern red oak - *Quercus falcata*
Spanish oak - *Quercus palustris*
Spanish red oak - *Quercus falcata*
swamp chestnut oak - *Quercus michauxii, Q. prinus*
swamp post oak - *Quercus lyrata*
swamp red oak - *Quercus falcata var. pagodifolia*
swamp white oak - *Quercus bicolor, Q. michauxii*
tan-bark oak - *Lithocarpus densiflora*
tan oak - *Lithocarpus densiflora*
Texas red oak - *Quercus texana*
Toumey's oak - *Quercus toumeyi*
truffle oak - *Quercus robur*
turbinella oak - *Quercus turbinella*

turkey oak - *Quercus incana, Q. laevis*
valley oak - *Quercus lobata*
water oak - *Quercus nigra*
wavy-leaf oak - *Quercus undulata*
western oak - *Quercus garryana*
western poison-oak - *Rhus diversiloba*
white oak - *Quercus alba*
willow oak - *Quercus phellos*
yellow-bark oak - *Quercus velutina*
yellow chestnut oak - *Quercus muehlenbergii*
yellow oak - *Quercus muehlenbergii*
oat
  animated oat - *Avena sterilis*
  black oat grass - *Stipa avenacea*
  bristle oat - *Avena strigosa*
  California oat grass - *Danthonia californica*
  cultivated oat - *Avena sativa*
  downy oat grass - *Danthonia sericea*
  golden oat grass - *Trisetum flavescens*
  intermediate oat grass - *Danthonia intermedia*
  naked oat - *Avena nuda*
  oat - *Avena sativa*
  oat grass - *Danthonia*
  one-spike oat grass - *Danthonia unispicata*
  Parry's oat grass - *Danthonia parryi*
  poverty oat grass - *Danthonia spicata*
  sand oat - *Avena strigosa*
  slender oat - *Avena barbata*
  slender wild oat - *Avena barbata*
  small oat - *Avena strigosa*
  sterile oat - *Avena sterilis*
  tall oat grass - *Arrhenatherum elatius*
  Tartarian oat - *Avena fatua*
  timber oat grass - *Danthonia intermedia*
  tuber oat grass - *Arrhenatherum elatius*
  white oat grass - *Danthonia spicata*
  winter wild oat - *Avena ludoviciana*
  yellow oat grass - *Trisetum flavescens*
oats
  oats - *Avena*
  potato oats - *Avena fatua*
  red oats - *Avena fatua*
  sea-oats - *Uniola paniculata*
  side-oats grama - *Bouteloua curtipendula*
  swamp-oats - *Trisetum pensylvanicum*
  water-oats - *Zizania*
  wild oats - *Chasmanthium latifolium, Uvularia sessilifolia*
obedience-plant - *Maranta arundinacea*
obscure duckweed - *Lemna obscura*
ocean-spray - *Holodiscus discolor*
Oconee bells - *Shortia galacifolia*
ocotillo - *Fouquieria splendens*
octopus-tree - *Schefflera actinophylla*
ocumo - *Xanthosoma sagittifolium*
odorless bayberry - *Myrica inodora*
Ogeche tupelo - *Nyssa ogeche*
Ogeechee lime tupelo - *Nyssa ogeche*
Oglethorpe's oak - *Quercus oglethorpensis*
'ohi'a lehua - *Metrosideros polymorpha*
Ohio
  Ohio buckeye - *Aesculus glabra*

Ohio goldenrod - *Solidago ohioensis*
oil
  castor-oil-plant - *Ricinus communis*
  China-wood oil-tree - *Vernicia fordii*
  madia-oil-plant - *Madia sativa*
  mu-oil-tree - *Vernicia montana*
  oil-cloth-flower - *Anthurium andraeanum*
  poonga oil-tree - *Pongamia pinnata*
  Siberian oil-seed - *Camelina microcarpa*
  tung oil-tree - *Vernicia fordii*
oiticica - *Licania rigida*
okra - *Abelmoschus esculentus*
old
  new old-man cactus - *Espostoa lanata*
  old-field-balsam - *Gnaphalium obtusifolium*
  old-field birch - *Betula populifolia*
  old-field cinquefoil - *Potentilla simplex*
  old-field clover - *Trifolium arvense*
  old-field pine - *Pinus taeda*
  old-field toadflax - *Linaria canadensis*
  old-fog - *Danthonia spicata*
  old-maid's-pink - *Agrostemma githago, Gypsophila paniculata*
  old-man - *Artemisia abrotanum*
  old-man-in-the-ground - *Marah oreganus*
  old-man's-beard - *Chionanthus virginicus*
  old-man's-whiskers - *Geum triflorum*
  old-plainsman - *Hymenopappus scabiosaeus* var. *corymbosus*
  old-witch grass - *Panicum capillare*
  Old World arrowhead - *Sagittaria sagittifolia*
  Old World diamond-flower - *Hedyotis corymbosa*
  Peruvian old-man cactus - *Espostoa lanata*
oleander
  oleander - *Nerium oleander*
  water-oleander - *Decodon verticillatus*
  yellow oleander - *Thevetia peruviana*
oleaster - *Elaeagnus angustifolia*
olive
  American olive - *Osmanthus americanus*
  autumn-olive - *Elaeagnus umbellata*
  black-olive - *Bucida buceras*
  California olive - *Umbellularia californica*
  Chinese olive - *Osmanthus heterophyllus*
  desert-olive - *Forestiera pubescens*
  desert-olive forestiera - *Forestiera phillyreoides*
  fragrant-olive - *Osmanthus fragrans*
  holly-olive - *Osmanthus heterophyllus*
  olive - *Olea europaea*
  Russian olive - *Elaeagnus angustifolia*
  sweet-olive - *Osmanthus fragrans*
  tea-olive - *Osmanthus fragrans*
  wild olive - *Elaeagnus angustifolia, Halesia tetraptera, Nyssa aquatica, Osmanthus americanus*
one
  one-flowered cancer-root - *Orobanche uniflora*
  one-flowered gentian - *Gentiana autumnalis*
  one-flowered hydrolea - *Hydrolea uniflora*
  one-flowered-pyrola - *Moneses uniflora*
  one-flowered shin-leaf - *Moneses uniflora*
  one-flowered wintergreen - *Moneses uniflora*
  one-leaf-clover - *Alysicarpus vaginalis, Macroptilium lathyroides*

  one-leaf orchis - *Orchis rotundifolia*
  one-seeded croton - *Croton monanthogynus*
  one-seeded hawthorn - *Crataegus monogyna*
  one-seeded juniper - *Juniperus monosperma*
  one-sided bottlebrush - *Calothamnus sanguineus*
  one-sided pyrola - *Orthilia secunda*
  one-sided-wintergreen - *Orthilia secunda*
  one-spike oat grass - *Danthonia unispicata*
onigurumi - *Juglans ailantifolia*
onion
  Alaska onion grass - *Melica subulata*
  bracted green-onion - *Habenaria viridis* var. *bracteata*
  bunching onion - *Allium fistulosum*
  Egyptian onion - *Allium cepa*
  Geyer's onion grass - *Melica geyeri*
  Japanese bunching onion - *Allium fistulosum*
  multiplier onion - *Allium cepa*
  nodding onion - *Allium cernuum*
  nodding wild onion - *Allium cernuum*
  onion - *Allium cepa, Allium*
  onion dodder - *Cuscuta gronovii*
  onion grass - *Melica bulbosa*
  onion-weed - *Asphodelus fistulosus*
  potato onion - *Allium cepa*
  prairie onion - *Allium stellatum*
  purple onion grass - *Melica spectabilis*
  Spanish onion - *Allium fistulosum*
  spring onion - *Allium fistulosum*
  swamp onion - *Allium validum*
  top onion - *Allium cepa* var. *viviparum*
  tree onion - *Allium cepa*
  two-bladed onion - *Allium fistulosum*
  Welsh onion - *Allium fistulosum*
  wild onion - *Allium amplectens, A. canadense, A. cernuum, A. drummondii, A. stellatum, A. textile, A. vineale*
Opar, jewels-of-Opar - *Talinum paniculatum*
opium
  opium poppy - *Papaver somniferum*
  wild opium - *Lactuca serriola*
opiuma - *Pithecellobium dulce*
opossum-wood - *Halesia tetraptera*
orache
  garden orache - *Atriplex hortensis*
  halberd-leaf orache - *Atriplex patula* subsp. *hastata*
  orache - *Atriplex*
  red orache - *Atriplex rosea*
  sea beach orache - *Atriplex arenaria*
  spreading orache - *Atriplex patula*
  thorn orache - *Bassia hyssopifolia*
oracle
  spear oracle - *Atriplex patula*
  tumbling oracle - *Atriplex rosea*
orange
  bitter orange - *Citrus aurantium*
  Chinese box-orange - *Severinia buxifolia*
  hardy-orange - *Poncirus trifoliata*
  house blooming mock-orange - *Pittosporum tobira*
  king orange - *Citrus reticulata*
  Mandarin orange - *Citrus reticulata*
  mock-orange - *Prunus caroliniana, Sideroxylon lycioides, Styrax americanus*
  orange - *Citrus sinensis*

orange cosmos - *Cosmos sulphureus*
orange-cup lily - *Lilium philadelphicum*
orange daylily - *Hemerocallis fulva*
orange dwarf-dandelion - *Krigia biflora*
orange-eyed buddleja - *Buddleja davidii*
orange-glow vine - *Senecio chenopodioides*
orange-grass - *Hypericum gentianoides*
orange hawkweed - *Hieracium aurantiacum*
orange-jasmine - *Murraya paniculata*
orange jessamine - *Murraya paniculata*
orange lily - *Lilium bulbiferum*
orange milkweed - *Asclepias tuberosa*
orange milkwort - *Polygala lutea*
orange-paintbrush - *Hieracium aurantiacum*
orange-root - *Asclepias tuberosa, Hydrastis canadensis*
orange sneezeweed - *Helenium hoopesii*
orange touch-me-not - *Impatiens capensis*
osage-orange - *Maclura pomifera*
Panama orange - *Citrofortunella mitis*
Satsuma orange - *Citrus reticulata*
Seville orange - *Citrus aurantium*
sour orange - *Citrus aurantium*
sweet mock-orange - *Philadelphus coronarius*
sweet orange - *Citrus sinensis*
trifoliate-orange - *Poncirus trifoliata*
western orange-cup lily - *Lilium philadelphicum* var. *andinum*
wild orange - *Prunus caroliniana*
wild orange-red lily - *Lilium philadelphicum*
orchard
orchard grass - *Dactylis glomerata*
orchard morning-glory - *Convolvulus arvensis*
orchid
fringe orchid - *Habenaria*
leaf white orchid - *Habenaria dilatata*
northern green orchid - *Habenaria hyperborea*
orchid-tree - *Bauhinia variegata*
orchid-vine - *Stigmaphyllon ciliatum*
rein orchid - *Habenaria*
tall white bog orchid - *Habenaria dilatata*
orchis
one-leaf orchis - *Orchis rotundifolia*
showy orchis - *Orchis spectabilis*
small round-leaf orchis - *Orchis rotundifolia*
oregano - *Origanum vulgare*
Oregon
Oregon alder - *Alnus rubra*
Oregon ash - *Fraxinus latifolia*
Oregon bent grass - *Agrostis oregonensis*
Oregon boxwood - *Paxistima myrsinites*
Oregon cherry - *Prunus emarginata*
Oregon crabapple - *Malus fusca*
Oregon fleabane - *Erigeron speciosus*
Oregon grape - *Berberis aquifolium*
Oregon holly - *Ilex aquifolium*
Oregon lily - *Lilium columbianum*
Oregon maple - *Acer macrophyllum*
Oregon myrtle - *Umbellularia californica*
Oregon oak - *Quercus garryana*
Oregon spike-moss - *Selaginella oregana*
Oregon sunflower - *Balsamorhiza sagittata*
Oregon white oak - *Quercus garryana*
Oregon woodsia - *Woodsia oregana*

Orford, Port Orford cedar - *Chamaecyparis lawsoniana*
organ-pipe cactus - *Stenocereus thurberi*
organy - *Origanum vulgare*
Oriental
Oriental arborvitae - *Platycladus orientalis*
Oriental bittersweet - *Celastrus orbiculatus*
Oriental cherry - *Prunus serrulata*
Oriental garlic - *Allium tuberosum*
Oriental lotus - *Nelumbo nucifera*
Oriental mustard - *Sisymbrium orientale*
Oriental pear - *Pyrus pyrifolia*
Oriental plane-tree - *Platanus orientalis*
Oriental poppy - *Papaver orientale*
origano - *Origanum vulgare*
ornamental monk's-hood - *Astrophytum ornatum*
orpine
garden orpine - *Sedum alboroseum*
orpine - *Sedum telephium, Sedum*
osage-orange - *Maclura pomifera*
osier
green osier - *Cornus alternifolia*
osier - *Salix viminalis, Salix*
purple osier - *Salix purpurea*
red osier dogwood - *Cornus sericea*
silky osier - *Salix viminalis*
western osier - *Cornus occidentalis*
osmanthus
holly osmanthus - *Osmanthus heterophyllus*
sweet osmanthus - *Osmanthus fragrans*
osoberry - *Oemleria cerasiformis*
ostrich
ostrich fern - *Matteuccia struthiopteris*
ostrich-plume - *Brassica juncea*
Oswego tea - *Monarda didyma*
Otaheite-walnut - *Aleurites moluccana*
oudo - *Aralia cordata*
Our-Lord's-candle - *Yucca whipplei*
out, inside-out-flower - *Vancouveria planipetala*
oval
oval-headed sedge - *Carex cephalophora*
oval kumquat - *Fortunella margarita*
oval-leaf peperomia - *Peperomia obtusifolia*
oval-leaf willow - *Salix ovalifolia, S. stolonifera*
over
gill-over-the-ground - *Glechoma hederacea*
kiss-me-over-the-garden-gate - *Polygonum orientale*
over-cup oak - *Quercus lyrata*
ovoid spike rush - *Eleocharis ovata*
owl-clover - *Orthocarpus luteus*
own, mind-your-own-business-plant - *Soleirolia soleirolii*
ox
hawkweed ox-tongue - *Picris hieracioides*
ox-eye - *Heliopsis helianthoides, Heliopsis*
ox-eye daisy - *Leucanthemum vulgare*
ox-horn bucida - *Bucida buceras*
ox-tongue - *Picris hieracioides*
ox's, bristly ox's-tongue - *Picris echioides*
oxalis, buttercup oxalis - *Oxalis pes-caprae*
oxlip - *Primula elatior*

oyster
  black oyster-plant - *Scorzonera hispanica*
  oyster-leaf - *Mertensia maritima*
  oyster-plant - *Tradescantia spathacea, Tragopogon porrifolius*
  oyster-wood - *Gymnanthes lucida*
  Spanish oyster-plant - *Scorzonera hispanica*
  vegetable-oyster - *Tragopogon porrifolius*

Ozark
  Ozark hickory - *Carya texana*
  Ozark tickseed - *Bidens polylepis*
  Ozark white-cedar - *Juniperus ashei*

pacaya - *Chamaedorea elegans*

pachysandra
  Allegheny pachysandra - *Pachysandra procumbens*
  Japanese pachysandra - *Pachysandra terminalis*

Pacific
  Pacific bayberry - *Myrica californica*
  Pacific blackberry - *Rubus ursinus*
  Pacific bluegrass - *Poa secunda*
  Pacific Christmas fern - *Polystichum munitum*
  Pacific Coast red elder - *Sambucus callicarpa*
  Pacific dewberry - *Rubus ursinus*
  Pacific dogwood - *Cornus nuttallii*
  Pacific fescue - *Vulpia microstachys* var. *pauciflora*
  Pacific madrone - *Arbutus menziesii*
  Pacific mountain-ash - *Sorbus sitchensis*
  Pacific plum - *Prunus subcordata*
  Pacific poison-oak - *Rhus diversiloba*
  Pacific red elder - *Sambucus callicarpa*
  Pacific reed grass - *Calamagrostis nutkaensis*
  Pacific rhododendron - *Rhododendron macrophyllum*
  Pacific silver fir - *Abies amabilis*
  Pacific silver-weed - *Potentilla pacifica*
  Pacific wax myrtle - *Myrica californica*
  Pacific willow - *Salix lucida* subsp. *lasiandra*
  Pacific yew - *Taxus brevifolia*

padauk - *Pterocarpus indicus*

padded, heart-padded hoary-cress - *Cardaria draba*

padouk - *Pterocarpus indicus*

pagoda
  Japanese pagoda-tree - *Sophora japonica*
  pagoda dogwood - *Cornus alternifolia*
  pagoda-tree - *Plumeria rubra* f. *acutifolia, Sophora japonica*

paint
  Indian-paint - *Chenopodium capitatum, Lithospermum canescens*
  paint-leaf - *Euphorbia heterophylla*
  paint-root - *Lachnanthes caroliniana*

paintbrush
  devil's-paintbrush - *Hieracium aurantiacum*
  downy paintbrush - *Castilleja sessiliflora*
  Flora's paintbrush - *Emilia javanica*
  Indian paintbrush - *Asclepias tuberosa, Castilleja affinis, C. coccinea, Castilleja*
  Monterey Indian paintbrush - *Castilleja latifolia*
  orange-paintbrush - *Hieracium aurantiacum*
  scarlet paintbrush - *Castilleja coccinea*
  woolly Indian paintbrush - *Castilleja foliolosa*
  yellow-paintbrush - *Hieracium caespitosum*

painted
  painted buckeye - *Aesculus sylvatica*
  painted copper-leaf - *Acalypha wilkesiana*
  painted-cup - *Castilleja coccinea, Castilleja*
  painted daisy - *Chrysanthemum coccineum*
  painted euphorbia - *Euphorbia prunifolia*
  painted-leaf - *Euphorbia pulcherrima*
  painted poinsettia - *Euphorbia cyathophora*
  painted spurge - *Euphorbia heterophylla*
  painted-tongue - *Salpiglossis sinuata*
  painted trillium - *Trillium undulatum*
  seaside painted-cup - *Castilleja latifolia*
  woolly painted-cup - *Castilleja foliolosa*

pak-choi - *Brassica chinensis*

pakai - *Amaranthus viridis*

pale
  pale coralroot - *Corallorrhiza trifida*
  pale corydalis - *Corydalis sempervirens*
  pale dock - *Rumex altissimus*
  pale echinacea - *Echinacea pallida*
  pale-flowered leaf-cup - *Polymnia canadensis*
  pale hickory - *Carya pallida*
  pale Indian plantain - *Cacalia atriplicifolia*
  pale larkspur - *Delphinium glaucum*
  pale-laurel - *Kalmia polifolia*
  pale-leaf sunflower - *Helianthus strumosus*
  pale lid-flower - *Calyptranthes pallens*
  pale persicaria - *Polygonum lapathifolium*
  pale poppy mallow - *Callirhoe alcaeoides*
  pale purple coneflower - *Echinacea pallida*
  pale sedge - *Carex pallescens*
  pale-seed plantain - *Plantago virginica*
  pale smartweed - *Polygonum lapathifolium*
  pale-snapdragon - *Impatiens pallida*
  pale-spike lobelia - *Lobelia spicata*
  pale touch-me-not - *Impatiens pallida*
  pale vetch - *Vicia caroliniana*
  pale violet - *Viola striata*

palm
  Alexandra's palm - *Archontophoenix alexandrae*
  areca palm - *Chrysalidocarpus lutescens*
  areng palm - *Arenga pinnata*
  assai palm - *Euterpe edulis, E. oleracea*
  bamboo palm - *Chamaedorea erumpens, Rhapis excelsa*
  Belmore's palm - *Howeia belmoreana*
  Belmore's sentry palm - *Howeia belmoreana*
  black fiber palm - *Arenga pinnata*
  bottle-palm - *Nolina recurvata*
  bread-palm - *Cycas*
  Burmese fishtail palm - *Caryota mitis*
  butterfly palm - *Chrysalidocarpus lutescens*
  California fan palm - *Washingtonia filifera*
  California palm - *Washingtonia filifera*
  California Washington palm - *Washingtonia filifera*
  Canary Island date palm - *Phoenix canariensis*
  cane palm - *Chrysalidocarpus lutescens*
  Chinese windmill palm - *Trachycarpus fortunei*
  Christmas palm - *Veitchia merrillii*
  cluster palm - *Ptychosperma macarthurii*
  clustered fishtail palm - *Caryota mitis*
  coconut palm - *Cocos nucifera*
  curly palm - *Howeia belmoreana*
  date palm - *Phoenix dactylifera, Phoenix*
  desert fan palm - *Washingtonia filifera*

desert palm - *Washingtonia robusta*
doum palm - *Hyphaene thebaica*
dwarf date palm - *Phoenix roebelenii*
Everglades palm - *Acoelorraphe wrightii*
fan palm - *Coccothrinax, Livistona chinensis, Livistona*
fern-palm - *Cycas circinalis*
fishtail palm - *Caryota urens, Caryota*
Florida cabbage palm - *Sabal palmetto*
Florida royal palm - *Roystonea elata*
Florida silver palm - *Coccothrinax argentata*
Forster's palm - *Howeia forsteriana*
Forster's sentry palm - *Howeia forsteriana*
funeral-palm - *Cycas*
gingerbread palm - *Hyphaene thebaica*
golden cane palm - *Chrysalidocarpus lutescens*
golden feather palm - *Chrysalidocarpus lutescens*
Gomuti palm - *Arenga pinnata*
good-luck palm - *Chamaedorea elegans*
hemp palm - *Trachycarpus fortunei*
hurricane palm - *Ptychosperma macarthurii*
jaggery palm - *Caryota urens*
Japanese fern-palm - *Cycas revoluta*
Japanese sago-palm - *Cycas revoluta*
kentia palm - *Howeia forsteriana*
king's palm - *Archontophoenix alexandrae*
lady palm - *Rhapis excelsa, Rhapis*
leopard palm - *Amorphophallus rivieri*
MacArthur's feather palm - *Ptychosperma macarthurii*
Madagascar palm - *Chrysalidocarpus lutescens*
Manila palm - *Veitchia merrillii*
Merrill's palm - *Veitchia merrillii*
Mexican Washington palm - *Washingtonia robusta*
miniature date palm - *Phoenix roebelenii*
miniature fan palm - *Rhapis excelsa*
needle-palm - *Yucca filamentosa*
northern Bangalow palm - *Archontophoenix alexandrae*
palm grass - *Setaria palmifolia*
palm-leaf morning-glory - *Ipomoea wrightii*
palm-lily - *Yucca gloriosa*
pandanus-palm - *Pandanus tectorius*
parlor palm - *Chamaedorea elegans*
Paurotis's palm - *Acoelorraphe wrightii*
peach palm - *Bactris gasipaes*
petticoat palm - *Washingtonia filifera*
pupunha palm - *Syagrus inajai*
pygmy date palm - *Phoenix roebelenii*
queen palm - *Syagrus romanzoffianum*
rattan palm - *Rhapis excelsa*
Roebelin's palm - *Phoenix roebelenii*
royal palm - *Roystonea*
sago palm - *Caryota urens*
saw cabbage palm - *Acoelorraphe wrightii*
Senegal date palm - *Phoenix reclinata*
sentry palm - *Howeia forsteriana, Howeia*
silver saw palm - *Acoelorraphe wrightii*
slender lady palm - *Rhapis excelsa*
snake-palm - *Amorphophallus rivieri, Amorphophallus*
South American jelly palm - *Butia capitata*
spiny-club palm - *Bactris*
stinking-palm - *Hedeoma pulegioides*
sugar palm - *Arenga pinnata*
thatch-leaf palm - *Howeia forsteriana*
thatch palm - *Howeia forsteriana*

thread palm - *Washingtonia robusta*
ti-palm - *Cordyline terminalis*
toddy palm - *Caryota urens*
traveler's-palm - *Ravenala madagascariensis*
tufted fishtail palm - *Caryota mitis*
umbrella-palm - *Cyperus alternifolius*
Washington palm - *Washingtonia*
wild date palm - *Phoenix reclinata*
windmill palm - *Trachycarpus fortunei*
wine palm - *Caryota urens*
yellow butterfly palm - *Chrysalidocarpus lutescens*
yellow palm - *Chrysalidocarpus lutescens*
Palm Beach bells - *Kalanchoe*
palma-christi - *Ricinus communis*
palmate hop clover - *Trifolium aureum*
palmella - *Yucca elata*
Palmer's amaranth - *Amaranthus palmeri*
palmetto
   blue-stem palmetto - *Sabal minor*
   bush palmetto - *Sabal minor*
   cabbage palmetto - *Sabal palmetto*
   dwarf palmetto - *Sabal minor*
   Etonia palmetto - *Sabal etonia*
   Mexican palmetto - *Sabal mexicana*
   palmetto - *Sabal*
   Rio Grande palmetto - *Sabal mexicana*
   saw palmetto - *Serenoa repens*
   scrub palmetto - *Sabal etonia, S. minor, Serenoa repens*
   Texas palmetto - *Sabal mexicana*
   Victoria palmetto - *Sabal mexicana*
palo
   blue palo-verde - *Parkinsonia florida*
   Mexican palo-verde - *Parkinsonia aculeata*
   palo-verde - *Parkinsonia florida*
palouse
   palouse tarweed - *Amsinckia retrorsa*
   palouse thistle - *Cirsium brevifolium*
palta - *Persea americana*
pampas grass - *Cortaderia selloana*
pamplemousse - *Citrus paradisi*
Panama
   Panama-berry - *Muntingia calabura*
   Panama orange - *Citrofortunella mitis*
panamica - *Pilea involucrata*
panda-bear-plant - *Kalanchoe tomentosa*
pandang - *Pandanus odoratissimus*
pandanus-palm - *Pandanus tectorius*
pangola grass - *Digitaria eriantha* subsp. *pentzii*
panic
   bulb panic grass - *Panicum bulbosum*
   fall panic grass - *Panicum dichotomiflorum*
   panic grass - *Echinochloa crus-galli, Panicum anceps,*
      P. maximum, Panicum
panicled
   panicled aster - *Aster lanceolatus*
   panicled dogwood - *Cornus racemosa*
   panicled hydrangea - *Hydrangea paniculata*
   panicled tick-trefoil - *Desmodium paniculatum*
panicum
   broad-leaf panicum - *Panicum adspersum*
   brown-top panicum - *Brachiaria fasciculata*

cock-foot panicum - *Echinochloa crus-galli*
fall panicum - *Panicum dichotomiflorum*
fringed panicum - *Panicum ciliatum*
red-top panicum - *Panicum rigidulum*
sprawling-panicum - *Urochloa reptans*
Texas panicum - *Panicum texanum*

pansy
bedding pansy - *Viola cornuta*
European field pansy - *Viola arvensis*
European wild pansy - *Viola tricolor*
field pansy - *Viola rafinesquii, V. tricolor*
miniature pansy - *Viola tricolor*
pansy - *Viola tricolor*
pansy-flowered geranium - *Pelargonium domesticum*
tufted pansy - *Viola cornuta*

panther lily - *Lilium pardalinum*
papala - *Bougainvillea umbellifera*
papaya - *Carica papaya, Carica*

paper
Chinese rice-paper-plant - *Tetrapanax papyrifer*
glossy-leaved paper-plant - *Fatsia japonica*
paper-bark-tree - *Melaleuca quinquenervia*
paper birch - *Betula papyrifera*
paper-flower - *Bougainvillea glabra*
paper-mulberry - *Broussonetia papyrifera*
paper-plant - *Fatsia japonica*
rice-paper-plant - *Tetrapanax papyrifer*

papinac, white papinac - *Leucaena leucocephala*
papoose-root - *Caulophyllum thalictroides*
paprika pepper - *Capsicum annuum*

Pará
Pará grass - *Brachiaria mutica, Panicum purpurascens*
Pará rubber-tree - *Hevea brasiliensis*

para-para - *Pisonia umbellifera*

paradise
bird-of-paradise - *Caesalpinia gilliesii, Strelitzia reginae*
paradise apple - *Malus pumila, M. sylvestris*
paradise-flower - *Caesalpinia pulcherrima*
paradise-plant - *Justicia carnea*
paradise poinciana - *Caesalpinia gilliesii*
paradise-tree - *Melia azedarach, Simarouba glauca*
queen's bird-of-paradise - *Strelitzia reginae*

Paraguay
Paraguay bur - *Acanthospermum australe*
Paraguay star-bur - *Acanthospermum australe*

Paraguayan trumpet-tree - *Tabebuia argentea*
parakeet-flower - *Heliconia psittacorum*
Paraná pine - *Araucaria angustifolia*
parasol, Chinese parasol-tree - *Firmiana simplex*
parentucellia, yellow parentucellia - *Parentucellia viscosa*
parilla, yellow parilla - *Menispermum canadense*
Paris daisy - *Chrysanthemum frutescens*
park willow - *Salix monticola*

parlor
parlor-ivy - *Senecio mikanioides*
parlor-maple - *Abutilon*
parlor palm - *Chamaedorea elegans*

Parnassus
American grass-of-Parnassus - *Parnassia glauca*
grass-of-Parnassus - *Parnassia californica, Parnassia*

parrot
parrot-flower - *Heliconia psittacorum*
parrot-leaf - *Alternanthera ficoidea* var. *amoena*
parrot-plantain - *Heliconia psittacorum*

parrot's-feather - *Myriophyllum aquaticum*

Parry's
Parry's manzanita - *Arctostaphylos manzanita*
Parry's oat grass - *Danthonia parryi*
Parry's pinyon pine - *Pinus quadrifolia*
Parry's rabbit-brush - *Chrysothamnus parryi*

parsley
Chinese parsley - *Coriandrum sativum*
European parsley fern - *Cryptogramma crispa*
fool's-parsley - *Aethusa cynapium*
hedge-parsley - *Torilis arvensis*
hemlock-parsley - *Conioselinum pacificum*
ibapah spring parsley - *Cymopterus ibapensis*
Japanese hedge-parsley - *Torilis japonica*
knotted hedge-parsley - *Torilis nodosa*
mountain parsley fern - *Cryptogramma crispa*
parsley - *Coriandrum sativum, Petroselinum crispum*
parsley fern - *Cryptogramma crispa, Cryptogramma*
parsley hawthorn - *Crataegus marshallii*
parsley-leaf blackberry - *Rubus laciniatus*
parsley-piert - *Alchemilla arvensis, A. microcarpa*
poison-parsley - *Conium maculatum*
prairie-parsley - *Polytaenia nuttallii*
slender parsley-piert - *Alchemilla microcarpa*
spring parsley - *Cymopterus ibapensis*

parsnip
American cow-parsnip - *Heracleum sphondylium* subsp. *montanum*
bladder-parsnip - *Lomatium utriculatum*
buck-parsnip - *Lomatium triternatum*
cow-parsnip - *Heracleum sphondylium* subsp. *montanum, Heracleum*
meadow parsnip - *Thaspium trifoliatum*
parsnip - *Pastinaca sativa*
pestle parsnip - *Lomatium nudicaule*
smooth meadow parsnip - *Thaspium trifoliatum*
water-parsnip - *Sium suave*
wild parsnip - *Pastinaca sativa*

parthenium
eastern parthenium - *Parthenium integrifolium*
ragweed parthenium - *Parthenium hysterophorus*

partridge
partridge-breast - *Aloe variegata*
partridge pea - *Chamaecrista fasciculata*
sensitive partridge pea - *Chamaecrista nictitans*

partridgeberry - Mitchella repens, Vaccinium vitis-idaea

paspalum
bare-stem paspalum - *Paspalum setaceum* var. *longipedunculatum*
brown-seed paspalum - *Paspalum plicatulum*
bull paspalum - *Paspalum boscianum*
Comb's paspalum - *Paspalum almum*
field paspalum - *Paspalum laeve*
Florida paspalum - *Paspalum floridanum*
fringe-leaf paspalum - *Paspalum ciliatifolium, P. setaceum* var. *ciliatifolium*
seashore paspalum - *Paspalum vaginatum*

sour paspalum - *Paspalum conjugatum*
supine paspalum - *Paspalum setaceum* var. *supinum*
thin paspalum - *Paspalum setaceum*
water paspalum - *Paspalum fluitans*
pasqueflower
American pasqueflower - *Anemone patens*
eastern pasqueflower - *Anemone patens*
pasqueflower - *Anemone patens*
passion
passion-fruit - *Passiflora edulis*
purple-passion-vine - *Gynura aurantiaca*
passionflower
blue passionflower - *Passiflora caerulea*
corky-stem passionflower - *Passiflora suberosa*
may-pop passionflower - *Passiflora incarnata*
passionflower - *Passiflora*
red-fruit passionflower - *Passiflora foetida*
wild passionflower - *Passiflora incarnata*
pasture
pasture brake - *Pteridium aquilinum*
pasture gooseberry - *Ribes cynosbati*
pasture hawthorn - *Crataegus spathulata*
pasture rose - *Rosa carolina*
pasture thistle - *Cirsium pumilim, C. pumilum*
path rush - *Juncus tenuis*
pathfinder - *Adenocaulon bicolor*
patience
herb-patience - *Rumex patientia*
patience dock - *Rumex patientia*
patience-plant - *Impatiens wallerana*
patient-Lucy - *Impatiens wallerana*
paulownia, royal paulownia - *Paulownia tomentosa*
Paurotis's palm - *Acoelorraphe wrightii*
paw, bear's-paw fern - *Aglaomorpha meyeniana*
pawpaw
American pawpaw - *Asimina triloba*
dwarf pawpaw - *Asimina parviflora*
flag pawpaw - *Asimina incana*
net-leaf pawpaw - *Asimina reticulata*
pawpaw - *Asimina triloba, Carica papaya*
paws, pussy-paws - *Calyptridium umbellatum*
pe-tsai - *Brassica pekinensis*
pea
Angola pea - *Cajanus cajan*
asparagus pea - *Psophocarpus tetragonolobus*
Austrian field pea - *Sphaerophysa salsula*
Austrian pea-weed - *Sphaerophysa salsula*
Austrian winter pea - *Lathyrus hirsutus*
beach pea - *Lathyrus japonicus, L. littoralis*
black-eyed pea - *Vigna unguiculata*
blue pea - *Clitoria ternatea, Lupinus perennis*
bush pea - *Thermopsis mollis*
butterfly pea - *Centrosema pubescens, Centrosema, Clitoria
    mariana, C. ternatea*
Caley's pea - *Lathyrus hirsutus*
Canadian pea - *Vicia cracca*
cat-jang pea - *Cajanus cajan*
chickling pea - *Lathyrus sativus*
Congo pea - *Cajanus cajan*
cow pea - *Vigna unguiculata*
cow pea witch-weed - *Striga gesnerioides*

craw pea - *Lathyrus pratensis*
crowder pea - *Vigna unguiculata*
dog-tooth pea - *Lathyrus sativus*
earthnut pea - *Lathyrus tuberosus*
edible-podded pea - *Pisum sativum* var. *macrocarpon*
Egyptian pea - *Cicer arietinum*
English pea - *Pisum sativum*
everlasting pea - *Lathyrus latifolius, L. sylvestris*
everlasting pea-vine - *Lathyrus latifolius*
field pea - *Pisum sativum, P. sativum* var. *arvense*
flat pea - *Lathyrus sylvestris*
flat pea-vine - *Lathyrus sylvestris*
flat-pod pea-vine - *Lathyrus cicera*
garden pea - *Pisum sativum*
grass pea-vine - *Lathyrus sativus*
gray scurf-pea - *Psoralidium tenuiflorum*
groundnut pea-vine - *Lathyrus tuberosus*
hairy milk pea - *Galactia volubilis*
heath pea - *Lathyrus japonicus*
hoary pea - *Tephrosia*
Indian pea - *Lathyrus sativus*
Jerusalem pea - *Vigna unguiculata* subsp. *cylindrica*
lemon scurf pea - *Psoralidium lanceolatum*
love pea - *Abrus precatorius*
marble pea - *Vigna unguiculata* subsp. *cylindrica*
marsh pea-vine - *Lathyrus palustris*
meadow pea - *Lathyrus pratensis*
meadow pea-vine - *Lathyrus pratensis*
milk pea - *Galactia*
partridge pea - *Chamaecrista fasciculata*
pea - *Pisum sativum, Pisum*
pea-tree - *Caragana, Sesbania exaltata*
perennial pea - *Lathyrus latifolius, L. sylvestris*
perennial sweet pea - *Lathyrus latifolius*
pigeon pea - *Cajanus cajan, Cajanus*
rabbit pea - *Tephrosia virginiana*
riga pea - *Lathyrus sativus*
rosary pea - *Abrus precatorius*
rough pea-vine - *Lathyrus hirsutus*
scurfy pea - *Psoralea*
sea pea - *Lathyrus japonicus*
seaside pea - *Lathyrus japonicus*
sensitive partridge pea - *Chamaecrista nictitans*
Siberian pea-tree - *Caragana arborescens*
singletary pea - *Lathyrus hirsutus*
snap pea - *Pisum sativum*
snow pea - *Pisum sativum* var. *macrocarpon*
southern pea - *Vigna unguiculata*
spurred butterfly pea - *Centrosema virginianum*
sugar pea - *Pisum sativum, P. sativum* var. *macrocarpon*
sweet pea - *Lathyrus odoratus*
Tangier pea - *Lathyrus tingitanus*
white pea - *Lathyrus ochroleucus*
wild pea - *Crotalaria sagittalis, Lathyrus palustris, Lathyrus*
wild sweet pea - *Lathyrus latifolius*
wing-stemmed wild pea-vine - *Lathyrus palustris*
peace-in-the-home - *Soleirolia soleirolii*
peach
Chinese wild peach - *Prunus davidiana*
David's peach - *Prunus davidiana*
desert peach - *Prunus andersonii*
peach - *Prunus persica*
peach-bells - *Campanula persicifolia*

peach-leaved willow - *Salix amygdaloides*
peach palm - *Bactris gasipaes*
peacock-flower - *Delonix regia*
peanut
    hog-peanut - *Amphicarpaea bracteata*
    peanut - *Arachis hypogaea*
pear
    alligator-pear - *Persea americana*
    Asian pear - *Pyrus pyrifolia*
    balsam-pear - *Momordica charantia*
    Bradford's pear - *Pyrus calleryana*
    brittle prickly-pear - *Opuntia fragilis*
    Callery's pear - *Pyrus calleryana*
    Chinese pear - *Pyrus pyrifolia*
    eastern prickly-pear - *Opuntia austrina, O. compressa*
    Japanese pear - *Pyrus pyrifolia*
    Lindheimer's prickly-pear - *Opuntia lindheimeri*
    Oriental pear - *Pyrus pyrifolia*
    pear - *Pyrus communis, Pyrus*
    pear hawthorn - *Crataegus calpodendron*
    pear-thorn - *Crataegus calpodendron, C. laevigata*
    pear tomato - *Lycopersicon esculentum* var. *pyriforme*
    plains prickly-pear - *Opuntia polyacantha*
    prickly-pear - *Opuntia compressa, Opuntia*
    prickly-pear cactus - *Opuntia stricta* var. *dillenii*
    sand pear - *Pyrus pyrifolia*
    spreading prickly-pear - *Opuntia compressa*
    Texas prickly-pear - *Opuntia lindheimeri*
    vinegar-pear - *Passiflora laurifolia*
pearl
    arctic pearl-wort - *Sagina procumbens*
    bird's-eye pearl-wort - *Sagina procumbens*
    pearl-bush - *Exochorda racemosa*
    pearl gromwell - *Lithospermum officinale*
    pearl lupine - *Lupinus mutabilis*
    pearl millet - *Pennisetum glaucum*
    pearl-plant - *Lithospermum officinale*
    pearl-wort - *Sagina*
pearly everlasting - *Anaphalis margaritacea*
pecan
    bitter pecan - *Carya aquatica*
    pecan - *Carya illinoensis*
peco - *Delphinium menziesii*
peg-wood - *Cornus sanguinea*
peganum, harmel peganum - *Peganum harmala*
pejibeye - *Bactris gasipaes*
pejivalle - *Bactris gasipaes*
pellionia, satin pellionia - *Pellionia pulchra*
pellitory
    Florida pellitory - *Parietaria floridana*
    pellitory - *Parietaria*
    Pennsylvania pellitory - *Parietaria pensylvanica*
pen-wiper - *Kalanchoe marmorata*
pencil-cedar - *Juniperus virginiana*
pendant
    pendant silver linden - *Tilia petiolaris*
    pendant white-lime - *Tilia petiolaris*
pennisetum, Chinese pennisetum - *Pennisetum alopecuroides*
Pennsylvania
    Pennsylvania bitter-cress - *Cardamine pensylvanica*

Pennsylvania maple - *Acer pensylvanicum*
Pennsylvania pellitory - *Parietaria pensylvanica*
Pennsylvania smartweed - *Polygonum pensylvanicum*
penny
    penny-flower - *Lunaria annua*
    penny-grass - *Thlaspi*
pennycress
    field pennycress - *Thlaspi arvense*
    pennycress - *Thlaspi*
    thoroughwort pennycress - *Thlaspi perfoliatum*
pennyroyal
    American false pennyroyal - *Hedeoma pulegioides*
    American pennyroyal - *Hedeoma pulegioides*
    bastard-pennyroyal - *Trichostema dichotomum*
    false-pennyroyal - *Trichostema brachiatum*
    mock pennyroyal - *Hedeoma pulegioides, Hedeoma*
    rough false pennyroyal - *Hedeoma hispidum*
pennywort
    Asiatic pennywort - *Centella asiatica*
    coastal plain pennywort - *Hydrocotyle bonariensis*
    lawn pennywort - *Hydrocotyle sibthorpioides*
    marsh pennywort - *Hydrocotyle americana*
    pennywort - *Cymbalaria muralis, Hydrocotyle, Obolaria virginica*
    water pennywort - *Hydrocotyle americana, H. umbellata, Hydrocotyle*
    whorled pennywort - *Hydrocotyle verticillata*
Pensacola Bahia grass - *Paspalum notatum* var. *saurae*
penstemon, Eaton's penstemon - *Penstemon eatonii*
peony
    Chinese peony - *Paeonia lactiflora*
    garden peony - *Paeonia lactiflora*
    peony - *Paeonia officinalis, Paeonia*
    tree peony - *Paeonia suffruticosa*
peperomia
    emerald-ripple peperomia - *Peperomia caperata*
    green-ripple peperomia - *Peperomia caperata*
    little-fantasy peperomia - *Peperomia caperata*
    oval-leaf peperomia - *Peperomia obtusifolia*
pepper
    alder-leaf pepper-bush - *Clethra alnifolia*
    Australian pepper - *Schinus molle*
    baby-pepper - *Rivina humilis*
    bell pepper - *Capsicum annuum, C. frutescens* cv. 'grossu'
    bird pepper - *Capsicum frutescens*
    black pepper - *Piper nigrum*
    Brazilian pepper-tree - *Schinus terebinthifolius*
    California pepper-tree - *Schinus molle*
    Cayenne pepper - *Capsicum annuum, C. frutescens*
    Celebes pepper - *Piper ornatum*
    chili pepper - *Capsicum annuum, C. frutescens*
    clasping pepper-weed - *Lepidium perfoliatum*
    field pepper-weed - *Lepidium campestre*
    garden pepper - *Capsicum annuum*
    green-flowered pepper-weed - *Lepidium densiflorum*
    green pepper - *Capsicum annuum*
    mango pepper - *Capsicum annuum*
    marsh-pepper smartweed - *Polygonum hydropiper*
    mild water-pepper - *Polygonum hydropiperoides*
    monk's-pepper-tree - *Vitex agnus-castus*
    narrow-leaf pepper-weed - *Lepidium ruderale*
    paprika pepper - *Capsicum annuum*
    pepper - *Piper*

pepper-and-salt - *Erigenia bulbosa*
pepper-face - *Peperomia obtusifolia*
pepper-plant - *Piper nigrum*
pepper-vine - *Ampelopsis arborea*
pepper-wood - *Umbellularia californica, Zanthoxylum clava-herculis*
perennial pepper-weed - *Lepidium latifolium*
Peruvian pepper-tree - *Schinus molle*
poor-man's-pepper - *Lepidium virginicum*
stinking pepper-weed - *Lepidium ruderale*
sweet pepper - *Capsicum frutescens* cv. 'grossu'
sweet pepper-bush - *Clethra acuminata, C. alnifolia*
tabasco pepper - *Capsicum frutescens*
tongue pepper-weed - *Lepidium nitidum*
Virginia pepper-weed - *Lepidium virginicum*
water-pepper - *Polygonum hydropiper*
white pepper - *Piper nigrum*
wild pepper - *Vitex agnus-castus*

peppergrass
clasping-leaf peppergrass - *Lepidium perfoliatum*
downy peppergrass - *Lepidium campestre*
field peppergrass - *Lepidium campestre*
green-flowered peppergrass - *Lepidium densiflorum*
peppergrass - *Lepidium virginicum, Lepidium*
perennial peppergrass - *Cardaria draba, Lepidium latifolium*
roadside peppergrass - *Lepidium ruderale*

pepperidge - *Nyssa sylvatica*

peppermint - *Mentha piperita*

pepperwort
European pepperwort - *Marsilea quadrifolia*
hairy pepperwort - *Marsilea vestita*
hoary pepperwort - *Cardaria draba*
pepperwort - *Cardamine diphylla, Lepidium, Marsilea*

perennial
perennial flax - *Linum perenne*
perennial glasswort - *Salicornia virginica*
perennial goosefoot - *Chenopodium bonus-henricus*
perennial ground-cherry - *Physalis longifolia* var. *subglabrata*
perennial lespedeza - *Lespedeza cuneata*
perennial lupine - *Lupinus perennis*
perennial pea - *Lathyrus latifolius, L. sylvestris*
perennial pepper-weed - *Lepidium latifolium*
perennial peppergrass - *Cardaria draba, Lepidium latifolium*
perennial phlox - *Phlox paniculata*
perennial quaking grass - *Briza media*
perennial ragweed - *Ambrosia psilostachya, A. psilostachya* var. *coronopifolia*
perennial ray grass - *Lolium perenne*
perennial rye grass - *Lolium perenne*
perennial salt marsh aster - *Aster tenuifolius*
perennial sow-thistle - *Sonchus arvensis*
perennial spiraea - *Astilbe*
perennial sundrops - *Oenothera perennis*
perennial sweet pea - *Lathyrus latifolius*
perennial teosinte - *Zea perennis*
perennial thistle - *Cirsium arvense*
perennial wild bean - *Strophostyles umbellata*
summer perennial phlox - *Phlox paniculata*

perfoliate
perfoliate bellwort - *Uvularia perfoliata*
perfoliate horse-gentian - *Triosteum perfoliatum*
perfoliate pondweed - *Potamogeton perfoliatus*

perfumed cherry - *Prunus mahaleb*

perilla
mint perilla - *Perilla frutescens*
perilla - *Perilla frutescens*
perilla-mint - *Perilla frutescens*

periquito-saracura - *Alternanthera philoxeroides*

periwinkle
big-leaf periwinkle - *Vinca major*
big periwinkle - *Vinca major*
greater periwinkle - *Vinca major*
lesser periwinkle - *Vinca minor*
Madagascar periwinkle - *Catharanthus roseus*
periwinkle - *Vinca minor, Vinca*
rose periwinkle - *Catharanthus roseus*

perpetua - *Gomphrena globosa*

Persian
Persian buttercup - *Ranunculus asiaticus*
Persian clover - *Trifolium resupinatum*
Persian darnel - *Lolium persicum*
Persian insect-flower - *Chrysanthemum coccineum*
Persian lilac - *Melia azedarach, Syringa persica*
Persian melon - *Cucumis melo* var. *reticulatus*
Persian purslane - *Veronica persica*
Persian speedwell - *Veronica persica*
Persian violet - *Exacum affine*
Persian walnut - *Juglans regia*

persicaria, pale persicaria - *Polygonum lapathifolium*

persicary
glandular persicary - *Polygonum pensylvanicum*
swamp persicary - *Polygonum pensylvanicum*

persimmon
American persimmon - *Diospyros virginiana*
black persimmon - *Diospyros texana*
Japanese persimmon - *Diospyros kaki*
kaki persimmon - *Diospyros kaki*
Mexican persimmon - *Diospyros texana*
persimmon - *Diospyros virginiana, Diospyros*
Texas persimmon - *Diospyros texana*

pert, golden-pert - *Gratiola aurea*

Peru
apple-of-Peru - *Nicandra physalodes*
marvel-of-Peru - *Mirabilis jalapa*

perunkila - *Carissa carandas*

Peruvian
Peruvian apple cactus - *Cereus uruguayanus*
Peruvian cherry - *Physalis peruviana*
Peruvian daffodil - *Hymenocallis narcissiflora*
Peruvian ground-cherry - *Physalis peruviana*
Peruvian mastic-tree - *Schinus molle*
Peruvian old-man cactus - *Espostoa lanata*
Peruvian pepper-tree - *Schinus molle*
Peruvian snowball cactus - *Espostoa lanata*

pestle parsnip - *Lomatium nudicaule*

petal
ten-petal stick-leaf - *Mentzelia decapetala*
two-petal ash - *Fraxinus dipetala*

petals, ten-petals sunflower - *Helianthus decapetalus*

Peter's, St. Peter's-wort - *Hypericum crux-andreae*

petioled sunflower - *Helianthus petiolaris*

petticoat
love grass petticoat-climber - *Eragrostis tenella*

petticoat palm - *Washingtonia filifera*
petty
    petty-morel - *Aralia racemosa*
    petty spurge - *Euphorbia peplus*
    rose petty - *Erigeron pulchellus*
petunia
    garden petunia - *Petunia hybrida*
    large white petunia - *Petunia axillaris*
    petunia - *Petunia hybrida*
    seaside petunia - *Petunia parviflora*
    white-moon petunia - *Petunia axillaris*
    wild petunia - *Petunia parviflora, Ruellia caroliniensis*
phasey bean - *Macroptilium lathyroides*
pheasant-eye adonis - *Adonis annua*
pheasant's
    pheasant's-eye - *Adonis annua, Narcissus poeticus*
    pheasant's-wings - *Aloe variegata*
Philadelphia fleabane - *Erigeron philadelphicus*
Philippine violet - *Barleria cristata*
philodendron
    cut-leaf philodendron - *Monstera deliciosa*
    heart-leaf philodendron - *Philodendron cordatum,*
      P. scandens
    variegated philodendron - *Epipremnum aureum*
    velvet-leaf philodendron - *Philodendron scandens* f. *micans*
phlox
    Allegheny phlox - *Phlox ovata*
    annual phlox - *Phlox drummondii*
    blue phlox - *Phlox divaricata*
    broad-leaf phlox - *Phlox amplifolia*
    Buckley's phlox - *Phlox buckleyi*
    crawling phlox - *Phlox stolonifera*
    creeping phlox - *Phlox stolonifera*
    downy phlox - *Phlox pilosa*
    Drummond's phlox - *Phlox drummondii*
    fall phlox - *Phlox paniculata*
    forest phlox - *Phlox divaricata*
    meadow phlox - *Phlox maculata*
    mountain phlox - *Phlox ovata*
    perennial phlox - *Phlox paniculata*
    prairie phlox - *Phlox pilosa*
    prickly phlox - *Leptodactylon californicum*
    shale-barren phlox - *Phlox buckleyi*
    smooth phlox - *Phlox glaberrima*
    spotted phlox - *Phlox nivalis*
    summer perennial phlox - *Phlox paniculata*
    sword-leaf phlox - *Phlox buckleyi*
    thick-leaf phlox - *Phlox carolina*
    trailing phlox - *Phlox nivalis*
    wide-leaf phlox - *Phlox amplifolia*
Phoenix-tree - *Firmiana simplex*
photinia, Japanese photinia - *Photinia glabra*
physic
    Indian physic - *Apocynum cannabinum, Gillenia trifoliata*
    physic-nut - *Jatropha curcas*
pichi
    pichi - *Fabiana imbricata*
    pichi-pichi - *Fabiana imbricata*
pick-a-back-plant - *Tolmiea menziesii*
pickerelweed
    lance-leaf pickerelweed - *Pontederia cordata* var. *lancifolia*

pickerelweed - *Pontederia cordata*
pie
    cherry-pie - *Heliotropium arborescens*
    pie cherry - *Prunus cerasus*
    pie-plant - *Rheum rhabarbarum*
Piedmont
    Piedmont azalea - *Rhododendron canescens*
    Piedmont buckbean - *Thermopsis mollis*
    Piedmont rhododendron - *Rhododendron minus*
piemacker - *Abutilon theophrasti*
piert
    parsley-piert - *Alchemilla arvensis, A. microcarpa*
    slender parsley-piert - *Alchemilla microcarpa*
pig
    pig-laurel - *Kalmia angustifolia*
    pig-lily - *Zantedeschia aethiopica*
    pig-nut - *Carya cordiformis, Hoffmannseggia glauca,*
      Simmondsia chinensis
    pig-nut hickory - *Carya glabra*
    sweet pig-nut - *Carya ovalis*
pigeon
    Asian pigeon-wings - *Clitoria ternatea*
    pigeon-berry - *Duranta erecta, Phytolacca americana*
    pigeon grape - *Vitis aestivalis, V. cinerea*
    pigeon grass - *Digitaria sanguinalis, Setaria verticillata*
    pigeon pea - *Cajanus cajan, Cajanus*
    pigeon-plum - *Coccoloba diversifolia*
    pigeon-weed - *Lithospermum arvense*
pigeon's-foot - *Salicornia europaea*
piggyback-plant - *Tolmiea menziesii*
pigweed
    amaranth pigweed - *Amaranthus hybridus, A. retroflexus*
    fringed pigweed - *Amaranthus fimbriatus*
    pigweed - *Amaranthus hybridus, A. retroflexus, Chenopodium*
      album, Chenopodium
    prostrate pigweed - *Amaranthus blitoides, A. graecizans*
    red-root pigweed - *Amaranthus retroflexus*
    rough pigweed - *Amaranthus hybridus, A. retroflexus*
    Russian pigweed - *Axyris amaranthoides*
    smooth pigweed - *Amaranthus hybridus*
    spiny pigweed - *Amaranthus spinosus*
    spreading pigweed - *Amaranthus graecizans*
    strawberry pigweed - *Chenopodium capitatum*
    strong-scented pigweed - *Chenopodium ambrosioides*
    tumble pigweed - *Amaranthus albus, A. graecizans*
    tumbling pigweed - *Amaranthus albus*
    white pigweed - *Amaranthus albus*
    winged pigweed - *Cycloloma atriplicifolium*
pilea, watermelon pilea - *Pilea cadierei*
pilewort - *Erechtites hieracifolia, Ranunculus ficaria,*
    Scrophularia marilandica
pilipiliula - *Chrysopogon aciculatus*
pimbina - *Viburnum trilobum*
pimento - *Capsicum annuum, Pimenta dioica*
pimiento - *Capsicum annuum*
pimpernel
    false pimpernel - *Lindernia anagallidea, L. procumbens*
    low false pimpernel - *Lindernia dubia*
    round-leaf false pimpernel - *Lindernia grandiflora*
    scarlet pimpernel - *Anagallis arvensis*
    water pimpernel - *Samolus parviflorus*

yellow pimpernel - *Taenidia integerrima*
pimpled water-meal - *Wolffia papulifera*
pin
   northern pin oak - *Quercus ellipsoidalis*
   pin cherry - *Prunus pensylvanica*
   pin-clover - *Erodium cicutarium*
   pin grass - *Erodium cicutarium*
   pin oak - *Quercus palustris*
Pinchot's juniper - *Juniperus pinchotii*
pincushion
   pincushion - *Mammillaria, Scabiosa atropurpurea*
   pincushion-flower - *Scabiosa*
pindar - *Arachis hypogaea*
pine
   air-pine - *Aechmea*
   Aleppo pine - *Pinus halepensis*
   Apache pine - *Pinus engelmannii*
   Arizona long-leaf pine - *Pinus engelmannii*
   Australian pine - *Casuarina equisetifolia, Casuarina*
   Austrian pine - *Pinus nigra*
   beach pine - *Pinus contorta*
   Bhutan pine - *Pinus wallichiana*
   big-cone pine - *Pinus coulteri*
   bishop's pine - *Pinus muricata*
   black pine - *Pinus nigra*
   blue pine - *Pinus wallichiana*
   Brazilian pine - *Araucaria angustifolia*
   Canadian pine - *Pinus resinosa*
   Canary Island pine - *Pinus canariensis*
   Caribbean pine - *Pinus caribaea*
   cedar pine - *Pinus glabra*
   Chihuahuan pine - *Pinus leiophylla*
   Chilean pine - *Araucaria araucana*
   cluster pine - *Pinus pinaster*
   Colorado pine - *Pinus edulis*
   Corsican pine - *Pinus nigra*
   Coulter's pine - *Pinus coulteri*
   cow's-tail-pine - *Cephalotaxus harringtonia* var. *drupacea*
   Cuban pine - *Pinus caribaea*
   digger pine - *Pinus sabiniana*
   eastern white pine - *Pinus strobus*
   flat-branch ground-pine - *Lycopodium obscurum*
   foothill pine - *Pinus sabiniana*
   foxtail pine - *Pinus balfouriana*
   frankincense pine - *Pinus taeda*
   Georgia pine - *Pinus palustris*
   giant pine - *Pinus lambertiana*
   gray pine - *Pinus banksiana, P. sabiniana*
   Great Basin bristle-cone pine - *Pinus longaeva*
   ground-pine - *Lycopodium clavatum, L. tristachyum*
   hickory pine - *Pinus longaeva, P. pungens*
   Himalayan white pine - *Pinus wallichiana*
   house-pine - *Araucaria heterophylla*
   Italian stone pine - *Pinus pinea*
   jack pine - *Pinus banksiana*
   Japanese black pine - *Pinus thunbergiana*
   Japanese red pine - *Pinus densiflora*
   Japanese umbrella-pine - *Sciadopitys verticillata*
   Jeffrey's pine - *Pinus jeffreyi*
   Jersey pine - *Pinus virginiana*
   Jerusalem pine - *Pinus halepensis*
   knob-cone pine - *Pinus attenuata*

Labrador pine - *Pinus banksiana*
limber pine - *Pinus flexilis*
loblolly pine - *Pinus taeda*
lodgepole pine - *Pinus contorta, P. contorta* subsp. *murrayana*
long-leaf pine - *Pinus palustris*
long-tag pine - *Pinus echinata*
maritime pine - *Pinus pinaster*
meadow-pine - *Equisetum arvense*
Mexican stone pine - *Pinus cembroides*
Mexican yellow pine - *Pinus patula*
Monterey pine - *Pinus radiata*
New Caledonia pine - *Araucaria columnaris*
Norfolk Island pine - *Araucaria heterophylla*
Norway pine - *Pinus resinosa*
nut pine - *Pinus monophylla*
old-field pine - *Pinus taeda*
Paraná pine - *Araucaria angustifolia*
Parry's pinyon pine - *Pinus quadrifolia*
pine - *Pinus*
pine-barren cyperus - *Cyperus cylindricus*
pine-barren gentian - *Gentiana autumnalis*
pine-barren goldenrod - *Solidago fistulosa*
pine-barren sandwort - *Arenaria caroliniana*
pine-barren tick-trefoil - *Desmodium strictum*
pine bluegrass - *Poa scabrella, P. secunda*
pine grass - *Calamagrostis rubescens*
pine-hill beak rush - *Rhynchospora globularis*
pine-land three-awn - *Aristida stricta*
pine lily - *Lilium catesbaei*
pine-mat manzanita - *Arctostaphylos nevadensis*
pine-sap - *Monotropa uniflora*
pine violet - *Viola lobata*
pine-weed - *Hypericum gentianoides*
pinyon pine - *Pinus edulis*
pitch pine - *Pinus rigida*
Pocosin pine - *Pinus serotina*
pond pine - *Pinus serotina*
ponderosa pine - *Pinus ponderosa*
poverty pine - *Pinus virginiana*
prickly pine - *Pinus pungens*
princess-pine - *Lycopodium obscurum*
red pine - *Pinus resinosa*
Rocky Mountain yellow pine - *Pinus ponderosa* var. *scopulorum*
rough-bark Mexican pine - *Pinus montezumae*
running-pine - *Lycopodium clavatum*
sand pine - *Pinus clausa*
Scot's pine - *Pinus sylvestris*
Scotch pine - *Pinus sylvestris*
screw-pine - *Pandanus utilis, Pandanus*
scrub pine - *Pinus banksiana, P. virginiana*
shore pine - *Pinus contorta*
short-leaf pine - *Pinus echinata*
single-leaf pinyon pine - *Pinus monophylla*
slash pine - *Pinus elliottii*
soledad pine - *Pinus torreyana*
southern pine - *Pinus palustris*
southern yellow pine - *Pinus palustris*
spruce pine - *Pinus echinata, P. glabra, P. virginiana*
stone pine - *Pinus monophylla*
sugar pine - *Pinus lambertiana*
Swiss mountain pine - *Pinus mugo*

Swiss stone pine - *Pinus cembra*
table mountain pine - *Pinus pungens*
thatch screw-pine - *Pandanus tectorius*
torch pine - *Pinus rigida*
Torrey's pine - *Pinus torreyana*
two-leaf nut pine - *Pinus edulis*
umbrella pine - *Pinus pinea*
Veitch's screw-pine - *Pandanus veitchii*
Virginia pine - *Pinus virginiana*
western bristle-cone pine - *Pinus longaeva*
western white pine - *Pinus monticola*
western yellow pine - *Pinus ponderosa*
white-bark pine - *Pinus albicaulis*
white pine - *Pinus strobus*
yellow pine - *Pinus echinata*

pineapple
    pineapple - *Ananas comosus*
    pineapple-guava - *Acca sellowiana*
    pineapple-shrub - *Calycanthus floridus*
    pineapple-weed - *Matricaria matricarioides*
    wild pineapple - *Tillandsia fasciculata*

pinedrops - *Pterospora andromedea*

pinesap
    pinesap - *Monotropa hypopithys*
    sweet pinesap - *Monotropsis odorata*

piney
    hairy piney-woods goldenrod - *Solidago flexicaulis*
    piney-woods drop-seed - *Sporobolus junceus*

pingue - *Hymenoxys richardsonii* var. *floribunda*

pink
    California Indian pink - *Silene californica*
    childing pink - *Dianthus prolifer*
    clove pink - *Dianthus caryophyllus*
    cottage pink - *Dianthus plumarius*
    cushion-pink - *Silene acaulis*
    Deptford pink - *Dianthus armeria*
    election-pink - *Rhododendron austrinum*
    fire-pink - *Silene virginica*
    garden pink - *Dianthus plumarius*
    grass pink - *Dianthus armeria, D. plumarius*
    ground-pink - *Phlox subulata*
    Indian pink - *Lobelia cardinalis, Lychnis flos-cuculi, Spigelia marilandica*
    maiden pink - *Dianthus deltoides*
    marsh-pink - *Sabatia angularis*
    meadow-pink - *Lychnis flos-cuculi*
    moss-pink - *Phlox subulata*
    mountain pink currant - *Ribes nevadense*
    mullein-pink - *Lychnis coronaria*
    old-maid's-pink - *Agrostemma githago, Gypsophila paniculata*
    pink - *Dianthus*
    pink ammannia - *Ammannia latifolia*
    pink-and-white-shower - *Cassia javanica* var. *indochinensis*
    pink calla-lily - *Zantedeschia rehmannii*
    pink cockle - *Vaccaria hispanica*
    pink cryptanthus - *Cryptanthus bromelioides*
    pink honeysuckle - *Lonicera hispidula*
    pink lady's-slipper - *Cypripedium acaule*
    pink purslane - *Portulaca pilosa*
    pink pyrola - *Pyrola asarifolia*
    pink-root - *Spigelia marilandica*
    pink-shell azalea - *Rhododendron vaseyi*
    pink sundew - *Drosera capillaris*
    pink-weed - *Polygonum pensylvanicum*
    pink weigela - *Weigela florida*
    pink wild bean - *Strophostyles umbellata*
    pink wintergreen - *Pyrola asarifolia*
    rainbow pink - *Dianthus chinensis*
    rose-pink - *Sabatia angularis*
    rush-pink - *Lygodesmia juncea*
    sea-pink - *Armeria*
    swamp-pink - *Calopogonium*
    wild pink - *Silene caroliniana, S. regia*
    windmill pink - *Silene gallica*

pinnate
    pinnate mosquito-fern - *Azolla pinnata*
    pinnate poppy - *Papaver argemone*
    pinnate tansy mustard - *Descurainia pinnata*

pintada, junta-de-cobra-pintada - *Justicia brandegeana*

pinwheel-flower - *Tabernaemontana divaricata*

pinxter
    Florida pinxter - *Rhododendron canescens*
    pinxter-bloom - *Rhododendron periclymenoides*

pinyon
    Colorado pinyon - *Pinus edulis*
    Mexican pinyon - *Pinus cembroides*
    Parry's pinyon pine - *Pinus quadrifolia*
    pinyon pine - *Pinus edulis*
    pinyon rice grass - *Piptochaetium fimbriatum*
    single-leaf pinyon pine - *Pinus monophylla*

pipe
    Dutchman's-pipe - *Aristolochia macrophyllum*
    Indian-pipe - *Monotropa uniflora*
    organ-pipe cactus - *Stenocereus thurberi*
    pipe-plant - *Aeschynanthus pulcher*
    pipe-vine - *Aristolochia macrophyllum*
    prince's-pipe - *Chimaphila umbellata, C. umbellata* var. *cisatlantica*
    pudding-pipe-tree - *Cassia fistula*

pipes
    horse-pipes - *Equisetum arvense*
    pygmy-pipes - *Monotropsis odorata*

piprage - *Berberis vulgaris*
    pipsissewa - *Chimaphila umbellata, C. umbellata* var. *cisatlantica*

pirul - *Schinus molle*

pistachio
    Chinese pistachio - *Pistacia chinensis*
    pistachio - *Pistacia*
    pistachio-nut - *Pistacia vera*

pistol, monkey-pistol - *Hura crepitans*

pit
    pit-scale grass - *Hackelochloa granularis*
    pit-seed goosefoot - *Chenopodium berlandieri*

pitanga - *Eugenia uniflora*

pitaya - *Echinocereus viridiflorus*

pitch pine - *Pinus rigida*

pitcher
    California pitcher-plant - *Darlingtonia californica*
    pitcher-plant - *Sarracenia purpurea, Sarracenia*
    pitcher sage - *Salvia spathacea*
    southern pitcher-plant - *Sarracenia purpurea*

sweet pitcher-plant - *Sarracenia purpurea*
yellow pitcher-plant - *Sarracenia flava*
pitcher's sage - *Salvia azurea* var. *grandiflora*
pitchers, dull meadow-pitchers - *Rhexia mariana*
pitchforks - *Bidens cernua, Bidens*
pitted
  pitted beard grass - *Bothriochloa pertusa*
  pitted blue-stem - *Bothriochloa pertusa*
  pitted morning-glory - *Ipomoea lacunosa*
pittosporum, Japanese pittosporum - *Pittosporum tobira*
plague, devil's-plague - *Daucus carota*
plain
  coastal plain dewberry - *Rubus trivialis*
  coastal plain gentian - *Gentiana catesbaei*
  coastal plain pennywort - *Hydrocotyle bonariensis*
  coastal plain tickseed - *Bidens mitis*
  coastal plain willow - *Salix caroliniana*
Plains, Great Plains cottonwood - *Populus deltoides* var. *occidentalis*
plains
  plains blazing-star - *Liatris squarrosa*
  plains blue-stem - *Bothriochloa ischaemum*
  plains bluegrass - *Poa arida*
  plains bristle grass - *Setaria macrostachya*
  plains coreopsis - *Coreopsis tinctoria*
  plains cottonwood - *Populus deltoides* var. *occidentalis*
  plains love grass - *Eragrostis intermedia*
  plains muhly - *Muhlenbergia cuspidata*
  plains poppy mallow - *Callirhoe alcaeoides*
  plains prickly-pear - *Opuntia polyacantha*
  plains puccoon - *Lithospermum caroliniense*
  plains reed grass - *Calamagrostis montanensis*
plainsman, old-plainsman - *Hymenopappus scabiosaeus* var. *corymbosus*
plaintain, western rattlesnake plaintain - *Goodyera oblongifolia*
plane
  London plane-tree - *Platanus acerifolia*
  mock-plane - *Acer pseudoplatanus*
  Oriental plane-tree - *Platanus orientalis*
  plane-tree - *Platanus occidentalis*
planer-tree - *Planera aquatica*
plant
  air-plant - *Kalanchoe pinnata*
  aluminum-plant - *Pilea cadierei*
  American century-plant - *Agave americana*
  American radiator-plant - *Peperomia obtusifolia*
  artillery-plant - *Pilea microphylla*
  baby-rubber-plant - *Peperomia obtusifolia*
  banana-plant - *Nymphoides aquatica*
  band-plant - *Vinca major*
  barroom-plant - *Aspidistra elatior*
  beef-plant - *Iresine herbstii*
  beefsteak-plant - *Acalypha wilkesiana, Iresine herbstii*
  black oyster-plant - *Scorzonera hispanica*
  blackening-plant - *Hibiscus rosa-sinensis*
  button-plant - *Spermacoce assurgens*
  calico-plant - *Alternanthera bettzichiana*
  California pitcher-plant - *Darlingtonia californica*
  cart-track-plant - *Plantago major*
  cast-iron-plant - *Aspidistra elatior*

castor-oil-plant - *Ricinus communis*
centipede-plant - *Homalocladium platycladum*
century-plant - *Agave americana*
cheese-plant - *Malva neglecta*
Chinese jade-plant - *Crassula arborescens*
Chinese lantern-plant - *Physalis alkekengi*
Chinese rice-paper-plant - *Tetrapanax papyrifer*
Chinese rubber-plant - *Crassula ovata*
Chinese silk-plant - *Boehmeria nivea*
Christ-plant - *Euphorbia milii*
cochineal-plant - *Opuntia cochenillifera*
compass-plant - *Lactuca serriola, Silphium laciniatum*
coral-bead-plant - *Abrus precatorius*
coral-plant - *Russelia equisetiformis*
corn-plant - *Dracaena fragrans* cv. 'massangean'
corpse-plant - *Monotropa uniflora*
Corsican carpet-plant - *Soleirolia soleirolii*
creeping rubber-plant - *Ficus pumila*
crown-plant - *Calotropis gigantea*
cup-plant - *Silphium perfoliatum*
curtain-plant - *Kalanchoe pinnata*
dagger-plant - *Yucca aloifolia*
dollar-plant - *Crassula ovata*
dwarf rubber-plant - *Crassula ovata*
elephant-ear-plant - *Alocasia, Colocasia esculenta*
empress-candle-plant - *Senna alata*
false hop-plant - *Justicia brandegeana*
fern-leaf inch-plant - *Tripogandra multiflora*
fever-plant - *Oenothera biennis*
firecracker-plant - *Cuphea ignea*
flamingo-plant - *Justicia carnea*
flannel-plant - *Verbascum thapsus*
fountain-plant - *Russelia equisetiformis*
friendship-plant - *Pilea involucrata*
fruit-salad-plant - *Monstera deliciosa*
gas-plant - *Dictamnus albus*
giant sensitive-plant - *Mimosa invisa*
glossy-leaved paper-plant - *Fatsia japonica*
good-luck-plant - *Cordyline terminalis*
gopher-plant - *Euphorbia lathyris*
gum-plant - *Grindelia squarrosa*
Hartshorn's-plant - *Anemone patens*
hedge-plant - *Ligustrum*
hemp-plant - *Agave sisalana*
honesty-plant - *Lunaria annua*
honey-plant - *Hoya carnosa*
hurricane-plant - *Monstera deliciosa*
ice-plant - *Mesembryanthemum crystallinum*
inch-plant - *Tradescantia zebrina*
Indian cup-plant - *Sarracenia purpurea*
Jamaican fever-plant - *Tribulus cistoides*
Japanese rubber-plant - *Crassula ovata*
lead-plant - *Amorpha canescens*
life-plant - *Kalanchoe pinnata*
lipstick-plant - *Aeschynanthus pulcher*
lizard-plant - *Tetrastigma voinieranum*
lobster-plant - *Euphorbia pulcherrima*
looking-glass-plant - *Coprosma repens*
madia-oil-plant - *Madia sativa*
Mexican fire-plant - *Euphorbia heterophylla*
Mexican love-plant - *Kalanchoe pinnata*
mind-your-own-business-plant - *Soleirolia soleirolii*
mirror-plant - *Coprosma repens*

mistletoe rubber-plant - *Ficus deltoidea*
mole-plant - *Euphorbia heterophylla, E. lathyris*
money-plant - *Lunaria annua*
monkey-plant - *Ruellia makoyana*
mosaic-plant - *Fittonia verschaffeltii*
mosquito-plant - *Hedeoma pulegioides*
mother-in-law-plant - *Dieffenbachia seguine*
mother-in-law's-tongue-plant - *Dieffenbachia*
musk-plant - *Malva moschata, Mimulus moschatus*
nerve-plant - *Fittonia verschaffeltii*
obedience-plant - *Maranta arundinacea*
oyster-plant - *Tradescantia spathacea, Tragopogon porrifolius*
panda-bear-plant - *Kalanchoe tomentosa*
paper-plant - *Fatsia japonica*
paradise-plant - *Justicia carnea*
patience-plant - *Impatiens wallerana*
pearl-plant - *Lithospermum officinale*
pepper-plant - *Piper nigrum*
pick-a-back-plant - *Tolmiea menziesii*
pie-plant - *Rheum rhabarbarum*
piggyback-plant - *Tolmiea menziesii*
pipe-plant - *Aeschynanthus pulcher*
pitcher-plant - *Sarracenia purpurea, Sarracenia*
plume-plant - *Justicia carnea*
plush-plant - *Kalanchoe tomentosa*
poker-plant - *Kniphofia uvaria, Kniphofia*
prayer-plant - *Maranta leuconeura*
radiator-plant - *Peperomia*
red-bead-plant - *Abrus precatorius*
ribbon-plant - *Chlorophytum comosum*
rice-paper-plant - *Tetrapanax papyrifer*
Rocky Mountain bee-plant - *Cleome serrulata*
rouge-plant - *Rivina humilis*
royal-vine-plant - *Gynura aurantiaca*
rubber-plant - *Ficus elastica*
sensitive-plant - *Mimosa pudica*
shoofly-plant - *Nicandra physalodes*
shrimp-plant - *Justicia brandegeana*
silver jade-plant - *Crassula arborescens*
silver nerve-plant - *Fittonia verschaffeltii*
silver-net-plant - *Fittonia verschaffeltii*
slipper-plant - *Pedilanthus tithymaloides*
small-leaf rubber-plant - *Ficus benjamina*
smooth lead-plant - *Amorpha nana*
snake-plant - *Sansevieria trifasciata*
soap-plant - *Chlorogalum pomeridianum, Zigadenus venenosus*
South American air-plant - *Kalanchoe fedtschenkoi*
southern pitcher-plant - *Sarracenia purpurea*
Spanish oyster-plant - *Scorzonera hispanica*
spider-plant - *Chlorophytum comosum, Cleome*
starfish-plant - *Cryptanthus acaulis*
stepladder-plant - *Costus malortieanus*
sun-plant - *Portulaca grandiflora*
sweating-plant - *Eupatorium perfoliatum*
sweet pitcher-plant - *Sarracenia purpurea*
Swiss-cheese-plant - *Monstera deliciosa*
tall umbrella-plant - *Cyperus eragrostis*
tapeworm-plant - *Homalocladium platycladum*
tapioca-plant - *Manihot esculenta*
tea-plant - *Camellia sinensis, Viburnum lentago*
teddy-bear-plant - *Cyanotis kewensis*

telegraph-plant - *Heterotheca grandiflora*
telegraph-plant-weed - *Heterotheca grandiflora*
trail-plant - *Adenocaulon bicolor*
trailing velvet-plant - *Ruellia makoyana*
tuna-plant - *Opuntia*
umbrella-plant - *Cyperus alternifolius, Eriogonum*
unicorn-plant - *Proboscidea louisianica*
urn-plant - *Aechmea fasciata*
velvet-plant - *Gynura aurantiaca, Verbascum thapsus*
wax-plant - *Hoya carnosa*
weather-plant - *Abrus precatorius*
wild sensitive-plant - *Chamaecrista nictitans*
wine-plant - *Rheum rhabarbarum*
wire-plant - *Muhlenbergia*
yellow bee-plant - *Cleome lutea*
yellow pitcher-plant - *Sarracenia flava*
zebra-plant - *Aphelandra squarrosa*
plantain
American plantain - *Plantago rugelii*
black-seed plantain - *Plantago rugelii*
blue plantain-lily - *Hosta ventricosa*
bracted plantain - *Plantago aristata*
broad-leaf plantain - *Plantago major*
buck-horn plantain - *Plantago lanceolata*
English plantain - *Plantago lanceolata*
fragrant plantain-lily - *Hosta plantaginea*
great Indian plantain - *Cacalia muhlenbergii*
great plantain - *Plantago major*
hastate Indian plantain - *Cacalia suaveolens*
hoary plantain - *Plantago media*
lance-leaf water-plantain - *Alisma lanceolatum*
large-flowered water-plantain - *Alisma plantago-aquatica*
leafy-stemmed plantain - *Plantago psyllium*
narrow-leaf plantain - *Plantago lanceolata*
narrow-leaf plantain-lily - *Hosta lancifolia*
narrow-leaf water-plantain - *Alisma gramineum*
pale Indian plantain - *Cacalia atriplicifolia*
pale-seed plantain - *Plantago virginica*
parrot-plantain - *Heliconia psittacorum*
plantain - *Musa acuminata, M. paradisiaca, Plantago major, Plantago*
plantain-leaf everlasting - *Antennaria plantaginifolia*
plantain-leaf pussy-toes - *Antennaria plantaginifolia*
plantain-leaved sedge - *Carex plantaginea*
poor-Robin's-plantain - *Hieracium venosum*
round-leaf mud-plantain - *Heteranthera reniformis*
Rugel's plantain - *Plantago rugelii*
seaside plantain - *Plantago maritima* var. *juncoides*
slender plantain - *Plantago elongata*
sweet-scented Indian plantain - *Cacalia suaveolens*
water-plantain - *Alisma plantago-aquatica, Alisma*
white-plantain - *Antennaria plantaginifolia*
woolly plantain - *Plantago patagonica*
plateau oak - *Quercus virginiana* var. *fusiformis*
Platte groundsel - *Senecio plattensis*
platter
Amazon water-platter - *Victoria amazonica*
platter-leaf - *Coccoloba uvifera*
water-platter - *Victoria*
plenty, horn-of-plenty - *Datura metel*
pleroma - *Tibouchina urvilleana*
pleurisy-root - *Asclepias tuberosa*

pliant milk-vetch - *Astragalus flexuosus*
plum
  Alleghany plum - *Prunus alleghaniensis*
  American plum - *Prunus americana*
  apricot plum - *Prunus simonii*
  August plum - *Prunus americana*
  Batoko-plum - *Flacourtia indica*
  beach plum - *Prunus maritima*
  big tree plum - *Prunus mexicana*
  black-plum - *Syzygium cumini*
  bullace plum - *Prunus domestica* subsp. *insititia*
  Canadian plum - *Prunus nigra*
  cherry plum - *Prunus cerasifera*
  Chickasaw plum - *Prunus angustifolia*
  coco-plum - *Chrysobalanus icaco*
  Damson's plum - *Prunus domestica* subsp. *insititia*
  date-plum - *Diospyros kaki, D. virginiana*
  dove-plum - *Coccoloba diversifolia*
  European plum - *Prunus domestica*
  flat-woods plum - *Prunus umbellata*
  garden plum - *Prunus domestica*
  goose plum - *Prunus americana*
  governor's-plum - *Flacourtia indica*
  Harrington's plum-yew - *Cephalotaxus harringtonia*
  hog plum - *Prunus americana, P. umbellata*
  hortulan plum - *Prunus hortulana*
  jambolan-plum - *Syzygium cumini*
  Japanese plum - *Eriobotrya japonica, Prunus salicina*
  Japanese plum-yew - *Cephalotaxus harringtonia* var.
    drupacea
  Java plum - *Syzygium cumini*
  Klamath plum - *Prunus subcordata*
  Madagascar plum - *Flacourtia indica*
  Malabar plum - *Syzygium jambos*
  marmalade-plum - *Pouteria sapota*
  Mexican plum - *Prunus mexicana*
  moxie-plum - *Gaultheria hispidula*
  Myrobalan plum - *Prunus cerasifera*
  Natal-plum - *Carissa macrocarpa*
  Pacific plum - *Prunus subcordata*
  pigeon-plum - *Coccoloba diversifolia*
  plum - *Prunus domestica*
  plum-fruited yew - *Cephalotaxus harringtonia* var. *drupacea*
  plum-leaf apple - *Malus prunifolia*
  plum-leaf azalea - *Rhododendron prunifolium*
  plum-yew - *Cephalotaxus*
  prune plum - *Prunus domestica*
  sand plum - *Prunus angustifolia*
  shore plum - *Prunus maritima*
  sierra plum - *Prunus subcordata*
  Simon's plum - *Prunus simonii*
  sloe plum - *Prunus umbellata*
  Spanish plum - *Spondias purpurea*
  wild goose plum - *Prunus hortulana, P. munsoniana*
  wild plum - *Prunus americana*
plume
  bent-awn plume grass - *Saccharum contortum*
  Brazilian-plume - *Justicia carnea*
  brown plume grass - *Saccharum brevibarbe*
  desert-plume - *Stanleya pinnata*
  feather-plume - *Dalea formosa*
  green-plume rabbit-brush - *Chrysothamnus nauseosus* subsp.
    graveolens

  Japanese plume grass - *Miscanthus sinensis*
  ostrich-plume - *Brassica juncea*
  plume-flower - *Justicia carnea*
  plume-plant - *Justicia carnea*
  plume thistle - *Cirsium virginianum, Cirsium*
  prince's-plume - *Stanleya*
  silver plume grass - *Saccharum alopecuroideum*
  sugar-cane plume grass - *Saccharum giganteum*
plumed, long-plumed purple avens - *Geum triflorum*
plumeless thistle - *Carduus acanthoides, C. nutans*
plush-plant - *Kalanchoe tomentosa*
pocan - *Phytolacca americana*
pocketbook-flower - *Calceolaria mexicana*
Pocosin pine - *Pinus serotina*
pod
  angle-pod - *Cynanchum laeve*
  bag-pod sesbania - *Sesbania vesicaria*
  black-pod vetch - *Vicia sativa* subsp. *nigra*
  bladder-pod - *Lesquerella, Lobelia inflata*
  flat-pod pea-vine - *Lathyrus cicera*
  fringe-pod - *Thysanocarpus curvipes*
  Gordon's bladder-pod - *Lesquerella gordonii*
  lace-pod - *Thysanocarpus curvipes*
  pod corn - *Zea mays*
  rough-pod copper-leaf - *Acalypha ostryifolia*
  sharp-pod morning-glory - *Ipomoea cordatotriloba*
  short-pod mustard - *Hirschfeldia incana*
  sickle-pod - *Arabis canadensis, Senna obtusifolia, S. tora*
  square-pod water-primrose - *Ludwigia alternifolia*
  water-pod - *Ellisia nyctelea*
  woolly-pod milkweed - *Asclepias eriocarpa*
podded
  bead-podded mustard - *Chorispora tenella*
  edible-podded pea - *Pisum sativum* var. *macrocarpon*
  globe-podded hoary-cress - *Cardaria pubescens*
  lens-podded white-top - *Cardaria chalepensis*
  top-podded water-primrose - *Ludwigia polycarpa*
podocarpus, yew podocarpus - *Podocarpus macrophyllus*
poet's narcissus - *Narcissus poeticus*
pogonia
  little whorled pogonia - *Isotria medeoloides*
  rose pogonia - *Pogonia ophioglossoides*
  small whorled pogonia - *Isotria medeoloides*
poha - *Physalis peruviana*
poinciana
  dwarf poinciana - *Caesalpinia pulcherrima*
  flower-fence poinciana - *Caesalpinia pulcherrima*
  paradise poinciana - *Caesalpinia gilliesii*
  royal poinciana - *Delonix regia*
poinsettia
  annual poinsettia - *Euphorbia heterophylla*
  Japanese poinsettia - *Euphorbia heterophylla, Pedilanthus*
    tithymaloides
  painted poinsettia - *Euphorbia cyathophora*
  poinsettia - *Euphorbia pulcherrima*
  wild poinsettia - *Euphorbia heterophylla*
point
  point-leaf manzanita - *Arctostaphylos pungens*
  Point Reyes ceanothus - *Ceanothus gloriosus*
  Point Reyes creeper - *Ceanothus gloriosus*
  sharp-point fluvellin - *Kickxia elatine*

Yankee Point ceanothus - *Ceanothus griseus* var. *horizontalis*
pointed
    pointed broom-sedge - *Carex scoparia*
    pointed-leaf tick-trefoil - *Desmodium glutinosum*
poison
    beaver-poison - *Cicuta maculata*
    bushman's-poison - *Acokanthera oppositifolia, Acokanthera*
    cow-poison - *Delphinium menziesii, D. trolliifolium*
    crow-poison - *Zigadenus densus*
    eastern poison-oak - *Toxicodendron pubescens*
    fish-poison-tree - *Piscidia piscipula*
    Pacific poison-oak - *Rhus diversiloba*
    poison-berry - *Solanum nigrum*
    poison-bush - *Acokanthera*
    poison camass - *Zigadenus nuttallii*
    poison-chickweed - *Anagallis arvensis*
    poison darnel - *Lolium temulentum*
    poison-dogwood - *Toxicodendron vernix*
    poison-elder - *Toxicodendron vernix*
    poison flag - *Iris versicolor*
    poison-hemlock - *Conium maculatum*
    poison-ivy - *Toxicodendron pubescens, T. radicans*
    poison-laurel - *Kalmia latifolia*
    poison milkweed - *Asclepias subverticillata, Euphorbia
      corollata*
    poison-oak - *Toxicodendron pubescens*
    poison-parsley - *Conium maculatum*
    poison primrose - *Primula obconica*
    poison rye grass - *Lolium temulentum*
    poison stinkweed - *Conium maculatum*
    poison sumac - *Toxicodendron vernix*
    poison-tree - *Acokanthera*
    poison-vetch - *Astragalus*
    poison-weed - *Delphinium menziesii*
    poison-wood - *Gymnanthes lucida*
    sheep-poison - *Kalmia angustifolia*
    two-grooved poison-vetch - *Astragalus bisulcatus*
    western poison-ivy - *Toxicodendron rydbergii, T. vernix*
    western poison-oak - *Rhus diversiloba*
poisonous nightshade - *Solanum dulcamara*
poke
    Indian poke - *Veratrum viride*
    poke milkweed - *Asclepias exaltata*
    Virginia poke - *Phytolacca americana*
pokeberry - *Phytolacca americana*
poker
    poker-plant - *Kniphofia uvaria, Kniphofia*
    red-hot-poker - *Kniphofia*
pokeweed - *Phytolacca americana*
polar grass - *Arctagrostis latifolia*
pole
    fish-pole bamboo - *Phyllostachys aurea*
    pole-cat-geranium - *Lantana montevidensis*
polecat
    polecat-bush - *Rhus aromatica*
    polecat-weed - *Symplocarpus foetidus*
polemonium, annual polemonium - *Polemonium
    micranthum*
Polish
    Polish millet - *Digitaria sanguinalis*
    Polish wheat - *Triticum polonicum*

polished willow - *Salix laevigata*
Pollyana-vine - *Soleirolia soleirolii*
polyanthus
    polyanthus - *Primula polyantha*
    polyanthus narcissus - *Narcissus tazetta*
polygala, fringed polygala - *Polygala paucifolia*
polygonum dodder - *Cuscuta polygonorum*
polypody
    California polypody - *Polypodium californicum*
    polypody - *Polypodium virginianum*
    rock polypody - *Polypodium virginianum*
    western polypody - *Polypodium hesperium*
polypogon
    ditch polypogon - *Polypogon interruptus*
    rabbit's-foot polypogon - *Polypogon monspeliensis*
pomegranate - *Punica granatum*
pomelo - *Citrus paradisi*
pomerac jambos - *Syzygium malaccense*
pomme-de-liane - *Passiflora laurifolia*
pond
    pond-apple - *Annona glabra*
    pond cypress - *Taxodium distichum* var. *imbricarium*
    pond-lily - *Nymphaea odorata*
    pond-nuts - *Nelumbo lutea*
    pond pine - *Pinus serotina*
    pond-spice - *Litsea aestivalis*
    yellow pond-lily - *Nuphar*
ponderosa pine - *Pinus ponderosa*
pondweed
    American pondweed - *Potamogeton nodosus*
    big-leaf pondweed - *Potamogeton amplifolius*
    broad-leaf pondweed - *Potamogeton amplifolius*
    clasping-leaf pondweed - *Potamogeton perfoliatus*
    crispate-leaf pondweed - *Potamogeton crispus*
    curly-leaf pondweed - *Potamogeton crispus*
    curly pondweed - *Potamogeton crispus*
    diverse-leaf pondweed - *Potamogeton diversifolius*
    eel-grass pondweed - *Potamogeton zosteriformis*
    fennel-leaf pondweed - *Potamogeton pectinatus*
    fine-leaf pondweed - *Potamogeton filiformis*
    flat-stem pondweed - *Potamogeton zosteriformis*
    floating-leaf pondweed - *Potamogeton natans*
    floating pondweed - *Potamogeton natans*
    Fries's pondweed - *Potamogeton friesii*
    grass-leaf pondweed - *Potamogeton gramineus*
    horned-pondweed - *Zannichellia palustris*
    Illinois pondweed - *Potamogeton illinoensis*
    large-leaf pondweed - *Potamogeton amplifolius*
    leafy pondweed - *Potamogeton foliosus*
    long-leaf pondweed - *Potamogeton nodosus*
    Nuttall's pondweed - *Potamogeton epihydrus*
    perfoliate pondweed - *Potamogeton perfoliatus*
    pondweed - *Potamogeton*
    ribbon-leaf pondweed - *Potamogeton epihydrus*
    sago pondweed - *Potamogeton pectinatus*
    shining pondweed - *Potamogeton illinoensis*
    slender pondweed - *Potamogeton pusillus*
    small pondweed - *Potamogeton pusillus*
    snail-seed pondweed - *Potamogeton diversifolius*
    thread-leaf pondweed - *Potamogeton filiformis*
    variable pondweed - *Potamogeton gramineus*

water-leaf pondweed - *Potamogeton diversifolius*
white-stem pondweed - *Potamogeton praelongus*
pongam - *Pongamia pinnata*
pontic azalea - *Rhododendron luteum*
ponytail - *Nolina recurvata*
poonga oil-tree - *Pongamia pinnata*
poor
  poor-Joe - *Diodia teres*
  poor-man's-pepper - *Lepidium virginicum*
  poor-Robin's fleabane - *Erigeron pulchellus*
  poor-Robin's-plantain - *Hieracium venosum*
  whip-poor-will-shoe - *Cypripedium calceolus* var. *pubescens*
poorland-flower - *Leucanthemum vulgare*
poorman's-weatherglass - *Anagallis arvensis*
pop
  may-pop passionflower - *Passiflora incarnata*
  pop ash - *Fraxinus caroliniana*
  running-pop - *Passiflora foetida*
popcorn
  popcorn - *Zea mays*
  popcorn-flower - *Plagiobothrys figuratus, Plagiobothrys*
popinac - *Acacia farnesiana*
poplar
  balsam poplar - *Populus balsamifera*
  black poplar - *Populus nigra*
  Chinese white poplar - *Populus tomentosa*
  downy poplar - *Populus heterophylla*
  gray poplar - *Populus canescens*
  Lombardy poplar - *Populus nigra*
  necklace poplar - *Populus deltoides*
  poplar - *Populus*
  silver-leaf poplar - *Populus alba*
  Simon's poplar - *Populus simonii*
  tulip-poplar - *Liriodendron tulipifera*
  western balsam poplar - *Populus trichocarpa*
  white poplar - *Populus alba*
  yellow-poplar - *Liriodendron tulipifera*
poppy
  annual prickly poppy - *Argemone polyanthemos*
  arctic poppy - *Papaver nudicaule*
  blue poppy - *Meconopsis betonicifolia*
  blue-stem prickle poppy - *Argemone albiflora*
  bubble-poppy - *Silene csereii*
  bush poppy - *Dendromecon rigida*
  California poppy - *Eschscholzia californica*
  celandine poppy - *Stylophorum diphyllum*
  corn poppy - *Papaver rhoeas*
  crested prickle poppy - *Argemone platyceras*
  field poppy - *Papaver dubium, P. rhoeas*
  Flander's poppy - *Papaver rhoeas*
  Iceland poppy - *Papaver nudicaule*
  Mexican poppy - *Argemone mexicana*
  Mexican prickly poppy - *Argemone mexicana*
  opium poppy - *Papaver somniferum*
  Oriental poppy - *Papaver orientale*
  pale poppy mallow - *Callirhoe alcaeoides*
  pinnate poppy - *Papaver argemone*
  plains poppy mallow - *Callirhoe alcaeoides*
  poppy - *Papaver*
  poppy anemone - *Anemone coronaria*
  poppy mallow - *Callirhoe*
  prickle poppy - *Argemone*

  prickly poppy - *Argemone mexicana*
  purple poppy mallow - *Callirhoe involucrata*
  red poppy - *Papaver rhoeas*
  Roemer's poppy - *Roemeria refracta*
  Shirley's poppy - *Papaver rhoeas*
  tree poppy - *Dendromecon rigida*
  white prickle poppy - *Argemone albiflora*
  wood poppy - *Stylophorum diphyllum*
pops, may-pops - *Passiflora incarnata*
porch-vine - *Wedelia trilobata*
porcupine
  porcupine grass - *Stipa spartea*
  porcupine sedge - *Carex hystricina*
porlieria, Texas porlieria - *Porlieria angustifolia*
Port Orford cedar - *Chamaecyparis lawsoniana*
Porter's melic - *Melica porteri*
portia-tree - *Thespesia populnea*
Portuguese
  Portuguese cherry-laurel - *Prunus lusitanica*
  Portuguese cypress - *Cupressus lusitanica*
portulaca, wild portulaca - *Portulaca oleracea*
possum
  possum-apple - *Diospyros virginiana*
  possum grape - *Cissus verticillata, Vitis baileyana*
  possum-haw - *Ilex decidua, Viburnum acerifolium, V. nudun*
  possum oak - *Quercus nigra*
  possum-wood - *Diospyros virginiana*
post
  post oak - *Quercus stellata*
  post-oak grape - *Vitis lincecumii*
  swamp post oak - *Quercus lyrata*
pot
  pot-herb - *Barbarea vulgaris*
  pot-marigold - *Calendula officinalis*
  pot marjoram - *Origanum vulgare*
potato
  air-potato - *Dioscorea bulbifera*
  American potato bean - *Apios americana*
  duck-potato - *Sagittaria latifolia*
  hog-potato - *Hoffmannseggia glauca*
  Irish potato - *Solanum tuberosum*
  potato - *Solanum tuberosum*
  potato oats - *Avena fatua*
  potato onion - *Allium cepa*
  potato yam - *Dioscorea bulbifera*
  prairie-potato - *Psoralea esculenta*
  swamp-potato - *Sagittaria*
  swan-potato - *Sagittaria sagittifolia*
  sweet-potato - *Ipomoea batatas*
  sweet-potato-tree - *Manihot esculenta*
  white potato - *Solanum tuberosum*
  wild potato - *Chlorogalum pomeridianum, Ipomoea pandurata, Solanum jamesii*
  wild sweet-potato-vine - *Ipomoea pandurata*
potentilla
  diffuse potentilla - *Potentilla paradoxa*
  mountain white potentilla - *Potentilla tridentata*
  strawberry potentilla - *Potentilla sterilis*
  tall potentilla - *Potentilla arguta*
pothos
  golden pothos - *Epipremnum aureum*

pothos - *Epipremnum aureum*
pouch-flower - *Calceolaria mexicana*
poulard wheat - *Triticum turgidum*
poverty
    Nuttall's poverty-weed - *Monolepis nuttalliana*
    poverty brome - *Bromus sterilis*
    poverty drop-seed - *Sporobolus vaginiflorus*
    poverty grass - *Aristida dichotoma, Danthonia spicata,*
      Sporobolus vaginiflorus
    poverty oat grass - *Danthonia spicata*
    poverty pine - *Pinus virginiana*
    poverty rush - *Juncus tenuis*
    poverty sump-weed - *Iva axillaris*
    poverty three-awn - *Aristida divaricata*
    poverty-weed - *Ambrosia confertiflora, Anaphalis*
      margaritacea, Iva axillaris
    silver-leaf poverty-weed - *Ambrosia tomentosa*
    woolly-leaf poverty-weed - *Ambrosia grayi*
powder
    powder-puff - *Calliandra*
    red powder-puff - *Calliandra haematocephala*
Powell's amaranth - *Amaranthus powellii*
prairie
    climbing prairie rose - *Rosa setigera*
    columnar prairie coneflower - *Ratibida columnifera*
    downy prairie-clover - *Dalea villosa*
    dwarf prairie rose - *Rosa arkansana*
    prairie acacia - *Acacia angustissima*
    prairie bayonet-grass - *Scirpus maritimus*
    prairie beard grass - *Schizachyrium scoparium*
    prairie blazing-star - *Liatris pycnostachya*
    prairie bottle gentian - *Gentiana andrewsii*
    prairie broomrape - *Orobanche ludoviciana*
    prairie buttercup - *Ranunculus rhomboideus*
    prairie cinquefoil - *Potentilla pensylvanica*
    prairie coneflower - *Ratibida*
    prairie cord grass - *Spartina pectinata*
    prairie crabapple - *Malus ioensis*
    prairie crowfoot - *Ranunculus rhomboideus*
    prairie cup grass - *Eriochloa contracta*
    prairie-dock - *Silphium terebinthinaceum*
    prairie dogbane - *Apocynum sibiricum*
    prairie drop-seed - *Sporobolus heterolepis*
    prairie evening-primrose - *Oenothera albicaulis*
    prairie flax - *Linum perenne* subsp. *lewisii*
    prairie-gentian - *Eustoma russellianum*
    prairie golden aster - *Chrysopsis camporum*
    prairie grass - *Bromus catharticus*
    prairie groundsel - *Senecio plattensis*
    prairie loosestrife - *Lysimachia quadrifolia*
    prairie mallow - *Sphaeralcea coccinea*
    prairie-mimosa - *Desmanthus illinoensis*
    prairie onion - *Allium stellatum*
    prairie-parsley - *Polytaenia nuttallii*
    prairie phlox - *Phlox pilosa*
    prairie-potato - *Psoralea esculenta*
    prairie rocket - *Erysimum asperum*
    prairie rose - *Rosa arkansana, R. setigera*
    prairie rosinweed - *Silphium integrifolium*
    prairie sage-wort - *Artemisia frigida*
    prairie sand-reed - *Calamovilfa longifolia*
    prairie-smoke - *Anemone patens*

    prairie spurge - *Euphorbia spathulata*
    prairie sumac - *Rhus lanceolata*
    prairie sunflower - *Helianthus petiolaris*
    prairie-tea - *Croton monanthogynus*
    prairie-tea croton - *Croton monanthogynus*
    prairie thistle - *Cirsium flodmanii*
    prairie three-awn - *Aristida adscensionis*
    prairie trefoil - *Lotus unifoliatus*
    prairie trillium - *Trillium recurvatum*
    prairie-turnip - *Psoralea esculenta*
    prairie wedge grass - *Sphenopholis obtusata*
    prairie wild rose - *Rosa arkansana*
    prairie willow - *Salix humilis*
    purple prairie-clover - *Dalea purpurea*
    purple prairie violet - *Viola palmata* var. *pedatifida*
    queen-of-the-prairie - *Filipendula rubra*
    shaggy prairie-turnip - *Psoralea esculenta*
    silky prairie-clover - *Dalea villosa*
    upright prairie coneflower - *Ratibida columnifera*
    white prairie-clover - *Dalea candida*
    yellow prairie violet - *Viola nuttallii*
prayer
    prayer-beads - *Abrus precatorius*
    prayer-plant - *Maranta leuconeura*
precatory bean - *Abrus precatorius*
preserving
    Chinese preserving melon - *Benincasa hispida*
    preserving melon - *Citrullus lanatus* var. *citroides*
pretty
    none-so-pretty - *Silene armeria*
    pretty-face - *Triteleia ixioides*
prickle
    blue-stem prickle poppy - *Argemone albiflora*
    crested prickle poppy - *Argemone platyceras*
    prickle poppy - *Argemone*
    white prickle poppy - *Argemone albiflora*
prickleweed - *Desmanthus illinoensis*
prickly
    annual prickly poppy - *Argemone polyanthemos*
    brittle prickly-pear - *Opuntia fragilis*
    eastern prickly-pear - *Opuntia austrina, O. compressa*
    lesser prickly sedge - *Carex muricata*
    lime prickly-ash - *Zanthoxylum fagara*
    Lindheimer's prickly-pear - *Opuntia lindheimeri*
    little prickly sedge - *Carex sterilis*
    Mexican prickly poppy - *Argemone mexicana*
    northern prickly-ash - *Zanthoxylum americanum*
    plains prickly-pear - *Opuntia polyacantha*
    prickly-ash - *Aralia spinosa, Zanthoxylum americanum*
    prickly comfrey - *Symphytum asperum*
    prickly cucumber - *Echinocystis lobata*
    prickly gooseberry - *Ribes cynosbati*
    prickly juniper - *Juniperus oxycedrus*
    prickly lettuce - *Lactuca serriola*
    prickly mallow - *Sida spinosa*
    prickly-pear - *Opuntia compressa, Opuntia*
    prickly-pear cactus - *Opuntia stricta* var. *dillenii*
    prickly phlox - *Leptodactylon californicum*
    prickly pine - *Pinus pungens*
    prickly poppy - *Argemone mexicana*
    prickly rose - *Rosa acicularis*
    prickly sida - *Sida spinosa*

prickly smartweed - *Polygonum bungeanum*
prickly sow-thistle - *Sonchus asper*
southern prickly-ash - *Zanthoxylum clava-herculis*
spreading prickly-pear - *Opuntia compressa*
Texas prickly-pear - *Opuntia lindheimeri*

pride
London-pride - *Lychnis chalcedonica*
mountain-pride - *Penstemon newberryi*
pride-of-Barbados - *Caesalpinia pulcherrima*
pride-of-Bolivia - *Tipuana tipu*
pride-of-China - *Melia azedarach*
pride-of-India - *Koelreuteria paniculata, Melia azedarach*

prim - *Ligustrum vulgare*

primrose
baby primrose - *Primula malacoides*
bird's-eye primrose - *Primula laurentiana*
bushy water-primrose - *Ludwigia alternifolia, L. decurrens*
Cape primrose - *Streptocarpus rexii*
Chinese primrose - *Primula sinensis*
creeping water-primrose - *Ludwigia peploides*
cut-leaf evening-primrose - *Oenothera laciniata*
desert evening-primrose - *Oenothera deltoides*
English primrose - *Primula vulgaris*
evening-primrose - *Oenothera biennis, Oenothera*
fairy primrose - *Primula malacoides*
field-primrose - *Oenothera biennis*
floating water-primrose - *Ludwigia repens*
German primrose - *Primula obconica*
Greenland primrose - *Primula egaliksensis*
long-fruited primrose-willow - *Ludwigia octovalvis*
poison primrose - *Primula obconica*
prairie evening-primrose - *Oenothera albicaulis*
primrose - *Primula*
primrose-leaf violet - *Viola primulifolia*
primrose-willow - *Ludwigia octovalvis*
sea beach evening-primrose - *Oenothera humifusa*
showy evening-primrose - *Oenothera speciosa*
showy water-primrose - *Ludwigia uruguayensis*
small-flowered evening-primrose - *Oenothera nuttallii, O. parviflora*
spike-primrose - *Boisduvalia densiflora*
spreading evening-primrose - *Oenothera humifusa*
square-pod water-primrose - *Ludwigia alternifolia*
top-podded water-primrose - *Ludwigia polycarpa*
tree-primrose - *Oenothera biennis*
Uruguay water-primrose - *Ludwigia uruguayensis*
white evening-primrose - *Oenothera speciosa*
white-stem evening-primrose - *Oenothera nuttallii*
wing-stemmed water-primrose - *Ludwigia decurrens*
winged water-primrose - *Ludwigia decurrens*

prince's
prince's-feather - *Polygonum orientale*
prince's-pipe - *Chimaphila umbellata, C. umbellata* var. *cisatlantica*
prince's-plume - *Stanleya*

princess
princess-feather - *Polygonum orientale*
princess-flower - *Tibouchina urvilleana*
princess-pine - *Lycopodium obscurum*
princess-tree - *Paulownia tomentosa*
princess-vine - *Cissus sicyoides*

Pringle's needle grass - *Stipa pringlei*

print, butter-print - *Abutilon theophrasti*

privet
Amur privet - *Ligustrum amurense*
California privet - *Ligustrum obtusifolium, L. ovalifolium*
Chinese privet - *Ligustrum lucidum, L. sinense*
glossy privet - *Ligustrum lucidum*
Japanese privet - *Ligustrum japonicum*
Nepal privet - *Ligustrum lucidum*
privet - *Ligustrum vulgare, Ligustrum*
swamp-privet - *Forestiera acuminata*
wax-leaf privet - *Ligustrum japonicum, L. lucidum*

proboscis-flower - *Proboscidea louisianica*

proso
proso millet - *Panicum miliaceum*
wild proso millet - *Panicum miliaceum*

prostrate
prostrate amaranth - *Amaranthus albus, A. blitoides, A. graecizans*
prostrate-coleus - *Plectranthus*
prostrate knotweed - *Polygonum aviculare*
prostrate pigweed - *Amaranthus blitoides, A. graecizans*
prostrate spurge - *Euphorbia humistrata*
prostrate tick-trefoil - *Desmodium rotundifolium*
prostrate vervain - *Verbena bracteata*

protea
giant protea - *Protea cynaroides*
king protea - *Protea cynaroides*

provision-tree - *Pachira aquatica*

prune plum - *Prunus domestica*

psoralea, scurfy psoralea - *Psoralidium tenuiflorum*

psyllium - *Plantago psyllium*

pubescent wheat grass - *Elytrigia intermedia*

puccoon
hoary puccoon - *Lithospermum canescens*
narrow-leaf puccoon - *Lithospermum incisum*
plains puccoon - *Lithospermum caroliniense*
puccoon - *Lithospermum arvense, L. caroliniense, Lithospermum*
red-puccoon - *Sanguinaria canadensis*

pudding
pudding-berry - *Cornus canadensis*
pudding-grass - *Hedeoma pulegioides*
pudding-pipe-tree - *Cassia fistula*

puero - *Pueraria phaseoloides*

puff
powder-puff - *Calliandra*
red powder-puff - *Calliandra haematocephala*

puke-weed - *Lobelia inflata*

pull-up muhly - *Muhlenbergia filiformis*

pulpit
Jack-in-the-pulpit - *Arisaema triphyllum*
small Jack-in-the-pulpit - *Arisaema triphyllum*
swamp Jack-in-the-pulpit - *Arisaema triphyllum*

pummelo - *Citrus maxima*

pumpkin
bush pumpkin - *Cucurbita pepo* var. *melopepo*
Canadian pumpkin - *Cucurbita moschata*
fetid wild pumpkin - *Cucurbita foetidissima*
pumpkin - *Cucurbita maxima, C. moschata, C. pepo, Cucurbita*
white pumpkin - *Benincasa hispida*

wild pumpkin - *Cucurbita foetidissima*
puncture-vine - *Tribulus terrestris*
punk-tree - *Melaleuca quinquenervia*
pupuna - *Bactris gasipaes*
pupunha palm - *Syagrus inajai*
purging
  purging cassia - *Cassia fistula*
  purging fistula - *Cassia fistula*
purple
  large purple aster - *Aster patens*
  long-plumed purple avens - *Geum triflorum*
  pale purple coneflower - *Echinacea pallida*
  purple alpine saxifrage - *Saxifraga oppositifolia*
  purple amaranth - *Amaranthus cruentus*
  purple ammannia - *Ammannia coccinea*
  purple anise - *Illicium floridanum*
  purple avens - *Geum rivale*
  purple bean - *Macroptilium atropurpureum*
  purple beauty-berry - *Callicarpa dichotoma*
  purple bell - *Campanula rapunculoides*
  purple bladderwort - *Utricularia purpurea*
  purple boneset - *Eupatorium maculatum, E. perfoliatum*
  purple chokeberry - *Aronia prunifolia, Malus floribunda*
  purple coneflower - *Echinacea purpurea, Echinacea*
  purple crabgrass - *Digitaria sanguinalis*
  purple-cress - *Cardamine douglasii*
  purple crowberry - *Empetrum atropurpureum*
  purple cudweed - *Gnaphalium purpureum*
  purple dead-nettle - *Lamium purpureum*
  purple elephant's-foot - *Elephantopus nudatus*
  purple-eyed-grass - *Sisyrinchium douglasii*
  purple fanwort - *Cabomba caroliniana*
  purple-flowered salsify - *Tragopogon porrifolius*
  purple-flowering raspberry - *Rubus odoratus*
  purple giant hyssop - *Agastache scrophulariaefolia*
  purple glory-tree - *Tibouchina urvilleana*
  purple grama - *Bouteloua radicosa*
  purple granadilla - *Passiflora edulis*
  purple ground-cherry - *Physalis lobata*
  purple-head - *Polygonum pensylvanicum*
  purple-head sneezeweed - *Helenium flexuosum*
  purple laurel - *Rhododendron catawbiense*
  purple-leaf willow-herb - *Epilobium coloratum*
  purple-leaved spiderwort - *Tradescantia spathacea*
  purple loco - *Astragalus agrestis, Oxytropis lambertii*
  purple loosestrife - *Lythrum salicaria*
  purple love grass - *Eragrostis spectabilis*
  purple lythrum - *Lythrum salicaria*
  purple meadow-rue - *Thalictrum dasycarpum, T. revolutum*
  purple milkweed - *Asclepias purpurascens*
  purple mombin - *Spondias purpurea*
  purple moonflower - *Ipomoea turbinata*
  purple mountain saxifrage - *Saxifraga oppositifolia*
  purple needle grass - *Stipa pulchra*
  purple nightshade - *Solanum xanti*
  purple-node Joe-Pye-weed - *Eupatorium purpureum*
  purple nut sedge - *Cyperus rotundus*
  purple onion grass - *Melica spectabilis*
  purple osier - *Salix purpurea*
  purple-passion-vine - *Gynura aurantiaca*
  purple poppy mallow - *Callirhoe involucrata*
  purple prairie-clover - *Dalea purpurea*

purple prairie violet - *Viola palmata* var. *pedatifida*
purple reed grass - *Calamagrostis purpurascens*
purple sand grass - *Triplasis purpurea*
purple sanicle - *Sanicula bipinnatifida*
purple-stamen mullein - *Verbascum virgatum*
purple star-thistle - *Centaurea calcitrapa*
purple-stem angelica - *Angelica atropurpurea*
purple-stem aster - *Aster puniceus*
purple-stem beggar-ticks - *Bidens connata*
purple strawberry guava - *Psidium littorale* var. *longipes*
purple three-awn - *Aristida purpurea*
purple-top - *Tridens flavus*
purple trillium - *Trillium erectum*
purple vetch - *Vicia americana, V. benghalensis*
purple wan-dock - *Brasenia schreberi*
purple willow - *Salix purpurea*
robust purple foxtail - *Setaria viridis* var. *robusta-purpurea*
purret - *Allium porrum*
purse
  purse milk-vetch - *Astragalus tenellus*
  shepherd's-purse - *Capsella bursa-pastoris*
purslane
  horse-purslane - *Trianthema portulacastrum*
  milk-purslane - *Euphorbia maculata*
  Persian purslane - *Veronica persica*
  pink purslane - *Portulaca pilosa*
  purslane - *Portulaca oleracea, Portulaca*
  purslane speedwell - *Veronica peregrina, V. peregrina* subsp. *xalapensis*
  rock-purslane - *Calandrinia*
  sea-purslane - *Honckenya peploides, Sesuvium portulacastrum*
  Siberian purslane - *Claytonia sibirica*
  water-purslane - *Didiplis diandra, Ludwigia palustris*
  western purslane speedwell - *Veronica peregrina* subsp. *xalapensis*
  winter-purslane - *Claytonia perfoliata*
pusley
  Brazilian pusley - *Richardia brasiliensis*
  Florida pusley - *Richardia scabra*
  pusley - *Portulaca oleracea*
pussy
  Canadian pussy-toes - *Antennaria neglecta* var. *canadensis*
  field pussy-toes - *Antennaria neglecta*
  large pussy willow - *Salix discolor*
  plantain-leaf pussy-toes - *Antennaria plantaginifolia*
  pussy-paws - *Calyptridium umbellatum*
  pussy-toes - *Antennaria plantaginifolia, Antennaria*
  pussy willow - *Salix discolor*
  small pussy willow - *Salix humilis*
  solitary pussy-toes - *Antennaria solitaria*
pussy's
  pussy's-ears - *Calochortus tolmiei, Kalanchoe tomentosa*
  pussy's-foot - *Ageratum*
puttyroot - *Aplectrum hyemale*
puzzle, monkey-puzzle-tree - *Araucaria araucana*
Pye
  green-stemmed Joe-Pye-weed - *Eupatorium purpureum*
  hollow Joe-Pye-weed - *Eupatorium fistulosum*
  hollow-stemmed Joe-Pye-weed - *Eupatorium fistulosum*
  Joe-Pye-weed - *Eupatorium maculatum, E. purpureum*
  purple-node Joe-Pye-weed - *Eupatorium purpureum*
  spotted Joe-Pye-weed - *Eupatorium maculatum*

sweet Joe-Pye-weed - *Eupatorium purpureum*
pygmy
    pygmy date palm - *Phoenix roebelenii*
    pygmy-pipes - *Monotropsis odorata*
pyrethrum
    Dalmatian pyrethrum - *Chrysanthemum cinerariifolium*
    pyrethrum - *Chrysanthemum cinerariifolium, C. coccineum*
pyrola
    arctic pyrola - *Pyrola glandiflora*
    greenish-flowered pyrola - *Pyrola chlorantha*
    lesser pyrola - *Pyrola minor*
    one-flowered-pyrola - *Moneses uniflora*
    one-sided pyrola - *Orthilia secunda*
    pink pyrola - *Pyrola asarifolia*
    pyrola - *Pyrola*
    round-leaf pyrola - *Pyrola rotundifolia*
qat - *Catha edulis*
quack salvers-grass - *Euphorbia cyparissias*
quackgrass - *Elytrigia repens*
quakegrass - *Bromus briziformis*
Quaker
    Quaker comfrey - *Symphytum uplandicum*
    Quaker-ladies - *Hedyotis caerulea*
Quaker's-bonnet - *Lupinus perennis*
quaking
    little quaking grass - *Briza minor*
    perennial quaking grass - *Briza media*
    quaking aspen - *Populus tremuloides*
    quaking grass - *Briza media*
quamash - *Camassia quamash*
queen
    Honolulu-queen - *Hylocereus undatus*
    queen-lily - *Curcuma petiolata*
    queen-of-the-meadow - *Filipendula ulmaria*
    queen-of-the-night - *Hylocereus undatus*
    queen-of-the-prairie - *Filipendula rubra*
    queen palm - *Syagrus romanzoffianum*
    queen sago - *Cycas circinalis*
Queen Anne's-lace - *Daucus carota*
queen's
    queen's bird-of-paradise - *Strelitzia reginae*
    queen's-cup - *Clintonia uniflora*
    queen's-delight - *Stillingia sylvatica*
    queen's-lace - *Daucus carota*
    queen's-root - *Stillingia sylvatica*
Queensland
    Queensland arrowroot - *Canna indica*
    Queensland-nut - *Macadamia integrifolia*
    Queensland umbrella-tree - *Schefflera actinophylla*
    small-fruited Queensland-nut - *Macadamia tetraphylla*
quercitron - *Quercus velutina*
quick
    quick-beam - *Sorbus aucuparia*
    quick grass - *Elytrigia repens*
    quick-set-thorn - *Crataegus laevigata*
    quick-silver-weed - *Thalictrum dioicum*
    quick-weed - *Galinsoga parviflora*
quilete - *Amaranthus caudatus*
quince
    Chinese quince - *Pseudocydonia sinensis*
    flowering quince - *Chaenomeles japonica*

Japanese quince - *Chaenomeles speciosa*
Maule's quince - *Chaenomeles japonica*
quince - *Cydonia oblonga*
quinine
    quinine - *Cinchona officinalis*
    quinine-weed - *Parthenium hysterophorus*
    wild quinine - *Parthenium integrifolium*
quinoa - *Chenopodium quinoa*
quinua - *Chenopodium quinoa*
quitch grass - *Elytrigia repens*
quiver-leaf - *Populus tremuloides*
rabbit
    desert rabbit-brush - *Chrysothamnus paniculatus*
    Douglas's rabbit-brush - *Chrysothamnus viscidiflorus*
    gray rabbit-brush - *Chrysothamnus nauseosus*
    green-plume rabbit-brush - *Chrysothamnus nauseosus* subsp.
      graveolens
    Parry's rabbit-brush - *Chrysothamnus parryi*
    rabbit-bells - *Crotalaria rotundifolia*
    rabbit-berry - *Shepherdia canadensis*
    rabbit-brush - *Chrysothamnus*
    rabbit-foot clover - *Trifolium arvense*
    rabbit pea - *Tephrosia virginiana*
    rabbit-tail grass - *Lagurus ovatus*
    rabbit-tobacco - *Gnaphalium obtusifolium*
    rabbit-tracks - *Maranta leuconeura* var. *kerchoveana*
    rubber rabbit-brush - *Chrysothamnus nauseosus*
    southwestern rabbit-brush - *Chrysothamnus pulchellus*
    yellow rabbit-brush - *Chrysothamnus viscidiflorus*
rabbit's
    rabbit's-eye blueberry - *Vaccinium ashei*
    rabbit's-foot - *Maranta leuconeura* var. *kerchoveana*
    rabbit's-foot grass - *Polypogon monspeliensis*
    rabbit's-foot polypogon - *Polypogon monspeliensis*
raccoon-berry - *Podophyllum peltatum*
racoon-grape - *Ampelopsis cordata*
radiate finger grass - *Chloris radiata*
radiator
    American radiator-plant - *Peperomia obtusifolia*
    radiator-plant - *Peperomia*
radish
    Chinese radish - *Raphanus sativus* cv. 'longipinnatu'
    radish - *Raphanus sativus*
    wild radish - *Raphanus raphanistrum*
raft, Moses-on-a-raft - *Tradescantia spathacea*
ragged
    ragged-jade - *Lychnis flos-cuculi*
    ragged-robin - *Lychnis flos-cuculi*
ragi - *Eleusine coracana*
ragweed
    blood ragweed - *Ambrosia trifida* var. *texana*
    bur ragweed - *Ambrosia tomentosa*
    false ragweed - *Ambrosia tenuifolia*
    giant ragweed - *Ambrosia trifida*
    great ragweed - *Ambrosia trifida*
    lance-leaf ragweed - *Ambrosia bidentata*
    naked-spiked ragweed - *Ambrosia psilostachya*
    perennial ragweed - *Ambrosia psilostachya, A. psilostachya*
      var. *coronopifolia*
    ragweed - *Ambrosia artemisiifolia, Ambrosia*
    ragweed parthenium - *Parthenium hysterophorus*

western ragweed - *Ambrosia psilostachya, A. psilostachya* var. *coronopifolia*

ragwort
  golden ragwort - *Senecio aureus*
  ragwort - *Senecio jacobaea, Senecio*
  tansy ragwort - *Senecio jacobaea*

railroad-vine - *Ipomoea pes-caprae*

rain
  golden-rain - *Cassia fistula*
  golden-rain-tree - *Koelreuteria paniculata, Koelreuteria*
  rain-tree - *Albizia saman*

rainbow
  rainbow pink - *Dianthus chinensis*
  rainbow-star - *Cryptanthus bromelioides* var. *tricolor*
  rainbow-vine - *Pellionia pulchra*

raisin
  Japanese raisin-tree - *Hovenia dulcis*
  wild raisin - *Viburnum cassinoides*

ram's-horn - *Proboscidea louisianica*

rambutan - *Nephelium lappaceum*

ramie - *Boehmeria nivea*

ramona - *Salvia*

ramontchi - *Flacourtia indica*

ramp - *Allium tricoccum*

rampion, German rampion - *Oenothera biennis*

ramsted - *Linaria vulgaris*

ramtilla - *Guizotia abyssinica*

Ranch, King Ranch blue-stem - *Bothriochloa ischaemum, B. ischaemum* var. *songarica*

rancher's fire-weed - *Amsinckia menziesii*

Rangoon creeper - *Quisqualis indica*

Rangpur lime - *Citrus limonia*

rapanea, Guianan rapanea - *Myrsine guianensis*

rape
  bird's-rape - *Brassica rapa*
  bird's-rape mustard - *Brassica rapa*
  California rape - *Sinapis arvensis*
  rape - *Brassica napus*
  swede rape - *Brassica napus*

rapeseed - *Brassica napus*

raspberry
  American red raspberry - *Rubus strigosus*
  black raspberry - *Rubus occidentalis*
  dwarf raspberry - *Rubus pubescens*
  dwarf red raspberry - *Rubus pubescens*
  European red raspberry - *Rubus idaeus*
  flowering raspberry - *Rubus odoratus*
  Mauritius raspberry - *Rubus rosifolius*
  Molucca raspberry - *Rubus moluccanus*
  New Mexico raspberry - *Rubus neomexicanus*
  purple-flowering raspberry - *Rubus odoratus*
  red raspberry - *Rubus idaeus*
  Rocky Mountain raspberry - *Rubus deliciosus*

rat-tail fescue - *Vulpia myuros*

rat's-tail - *Vulpia bromoides*

rattan
  dwarf ground rattan - *Rhapis excelsa*
  rattan palm - *Rhapis excelsa*
  rattan-vine - *Berchemia scandens*

rattle
  Drummond's rattle-bush - *Sesbania drummondii*
  low rattle-box - *Crotalaria rotundifolia*
  rattle-box - *Crotalaria sagittalis, Crotalaria, Ludwigia alternifolia, Rhinanthus crista-galli*
  rattle-bush - *Sesbania punicea*
  rattle-top - *Cimicifuga*
  rattle-weed - *Astragalus lentiginosus, Crotalaria sagittalis*
  showy rattle-box - *Crotalaria spectabilis*
  water-leaf rattle-box - *Crotalaria retusa*
  wedge-leaf rattle-box - *Crotalaria retusa*
  weedy rattle-box - *Crotalaria sagittalis*
  yellow-rattle - *Rhinanthus minor, Rhinanthus*

rattlesnake
  rattlesnake brome - *Bromus briziformis*
  rattlesnake chess - *Bromus briziformis*
  rattlesnake fern - *Botrychium virginianum*
  rattlesnake manna grass - *Glyceria canadensis*
  rattlesnake-master - *Eryngium aquaticum, E. yuccifolium, Manfreda virginica*
  rattlesnake-root - *Prenanthes alba, Prenanthes*
  rattlesnake-weed - *Daucus pusillus, Hieracium venosum*
  slender rattlesnake-root - *Prenanthes autumnalis*
  western rattlesnake plaintain - *Goodyera oblongifolia*

Ravenna's grass - *Saccharum ravennae*

ray, perennial ray grass - *Lolium perenne*

rayed, golden-rayed lily - *Lilium auratum*

rayless
  rayless dog-fennel - *Matricaria matricarioides*
  rayless goldenrod - *Haplopappus heterophyllus*

razor-sedge - *Scleria*

redbud
  California redbud - *Cercis occidentalis*
  Chinese redbud - *Cercis chinensis*
  eastern redbud - *Cercis canadensis*
  western redbud - *Cercis occidentalis*

redtop
  Idaho redtop - *Agrostis idahoensis*
  redtop - *Agrostis gigantea*
  Ross's redtop - *Agrostis rossiae*
  spike redtop - *Agrostis exarata*
  Thurber's redtop - *Agrostis thurberiana*

redwood
  coastal redwood - *Sequoia sempervirens*
  dawn redwood - *Metasequoia glyptostroboides*
  giant redwood - *Sequoiadendron giganteum*
  Madeira redwood - *Swietenia mahagoni*
  redwood - *Sequoia sempervirens*
  redwood-ivy - *Vancouveria planipetala*

reed
  branched bur-reed - *Sparganium erectum*
  branching bur-reed - *Sparganium androcladum*
  broad-fruited bur-reed - *Sparganium eurycarpum*
  bur-reed - *Sparganium*
  Burma reed - *Neyraudia reynaudiana*
  drooping wood-reed - *Cinna latifolia*
  giant bur-reed - *Sparganium eurycarpum*
  giant reed - *Arundo donax*
  green-fruited bur-reed - *Sparganium chlorocarpum*
  narrow-leaf bur-reed - *Sparganium emersum*
  narrow reed grass - *Calamagrostis neglecta*
  northern reed grass - *Calamagrostis neglecta*

Pacific reed grass - *Calamagrostis nutkaensis*
plains reed grass - *Calamagrostis montanensis*
prairie sand-reed - *Calamovilfa longifolia*
purple reed grass - *Calamagrostis purpurascens*
reed - *Phragmites australis*
reed canary grass - *Phalaris arundinacea*
reed foxtail - *Alopecurus arundinaceus*
reed grass - *Calamagrostis*
sand-reed - *Calamovilfa longifolia*
Scribner's reed grass - *Calamagrostis scribneri*
stout wood-reed - *Cinna arundinacea*
three-square bur-reed - *Sparganium americanum*
water bur-reed - *Sparganium fluctuans*
wood-reed - *Cinna arundinacea*
regal
regal geranium - *Pelargonium domesticum*
regal lily - *Lilium regale*
rein orchid - *Habenaria*
reina, zapatica-de-la-reina - *Clitoria ternatea*
Rendle's foxtail - *Alopecurus rendlei*
rennet, cheese-rennet - *Galium verum*
rescue grass - *Bromus catharticus*
resin
resin birch - *Betula glandulosa*
resin-bush - *Viguiera stenoloba*
resin-weed - *Gutierrezia*
resurrection
resurrection fern - *Polypodium polypodioides*
resurrection-lily - *Lycoris squamigera*
retaima - *Parkinsonia aculeata*
retrorse sedge - *Carex retrorsa*
Reverchon's three-awn - *Aristida purpurea*
reversed clover - *Trifolium resupinatum*
rex begonia - *Begonia rex-cultorum*
Reyes
Point Reyes ceanothus - *Ceanothus gloriosus*
Point Reyes creeper - *Ceanothus gloriosus*
rhapis, fern rhapis - *Rhapis excelsa*
rheumatism-weed - *Apocynum cannabinum*
Rhode Island bent grass - *Agrostis capillaris*
Rhodes's grass - *Chloris gayana*
rhododendron
Canadian rhododendron - *Rhododendron canadense*
Carolina rhododendron - *Rhododendron carolinianum*
catawba rhododendron - *Rhododendron catawbiense*
great rhododendron - *Rhododendron maximum*
honey-bell rhododendron - *Rhododendron campylocarpum*
Indian rhododendron - *Melastoma malabathricum*
Lapland rhododendron - *Rhododendron lapponicum*
Pacific rhododendron - *Rhododendron macrophyllum*
Piedmont rhododendron - *Rhododendron minus*
rhododendron - *Rhododendron*
rose-bay rhododendron - *Rhododendron maximum*
silvery rhododendron - *Rhododendron grande*
Smirnow's rhododendron - *Rhododendron smirnowii*
tree rhododendron - *Rhododendron arboreum*
West Coast rhododendron - *Rhododendron macrophyllum*
rhodora - *Rhododendron canadense*
rhombic copper-leaf - *Acalypha rhomboidea*

rhubarb
Chinese rhubarb - *Rheum officinale*
monk's rhubarb - *Rumex alpinus, R. patientia*
mountain rhubarb - *Rumex alpinus*
rhubarb - *Rheum rhabarbarum*
wild rhubarb - *Rumex hymenosepalus*
rhus, willow rhus - *Rhus lancea*
rib, wing-rib sumac - *Rhus copallina*
ribbon
ribbon-bush - *Homalocladium platycladum*
ribbon-cactus - *Pedilanthus tithymaloides*
ribbon fern - *Polypodium phyllitidis*
ribbon-leaf pondweed - *Potamogeton epihydrus*
ribbon-plant - *Chlorophytum comosum*
ribgrass - *Plantago lanceolata*
ribwort - *Plantago*
rice
annual wild rice - *Zizania aquatica*
Chinese rice-paper-plant - *Tetrapanax papyrifer*
Chinese wild rice - *Leymus chinensis*
cultivated northern wild rice - *Zizania palustris*
eastern wild rice - *Zizania aquatica*
Formosan rice-tree - *Fatsia japonica*
hungry-rice - *Digitaria exilis*
Indian rice - *Oryzopsis hymenoides, Zizania aquatica*
jungle-rice - *Echinochloa colona*
jungle rice grass - *Echinochloa colona, E. crus-galli*
little rice grass - *Oryzopsis exigua*
little-seed rice grass - *Oryzopsis micrantha*
Manchurian wild rice - *Zizania latifolia*
pinyon rice grass - *Piptochaetium fimbriatum*
red rice - *Oryza sativa*
rice - *Oryza sativa*
rice cut grass - *Leersia oryzoides*
rice-field bulrush - *Scirpus mucronatus*
rice flat sedge - *Cyperus iria*
rice-grain fritillary - *Fritillaria affines*
rice grass - *Oryzopsis*
rice-paper-plant - *Tetrapanax papyrifer*
southern wild rice - *Zizaniopsis miliacea*
upland rice - *Oryza sativa*
wild rice - *Zizania aquatica, Zizania*
Richardson's
Richardson's needle grass - *Stipa richardsonii*
Richardson's tansy mustard - *Descurainia richardsonii*
Richardson's willow - *Salix lanata* subsp. *richardsonii*
richweed - *Ageratina altissima, Collinsonia canadensis, Pilea pumila*
Riddell's groundsel - *Senecio riddellii*
Ridge, Blue Ridge St. John's-wort - *Hypericum mitchellianum*
ridge-seed spurge - *Euphorbia glyptosperma*
riga pea - *Lathyrus sativus*
rigid
rigid fiddle-neck - *Amsinckia retrorsa*
rigid goldenrod - *Solidago rigida*
rigid rye grass - *Lolium rigidum*
ringed beard grass - *Dichanthium annulatum*
ringworm
ringworm-bush - *Senna alata*
ringworm senna - *Senna alata*

ringworm-shrub - *Senna alata*
Rio Grande palmetto - *Sabal mexicana*
rip
  rip-gut brome - *Bromus diandrus, B. rigidus*
  rip-gut brome grass - *Bromus rigidus*
  rip-gut grass - *Bromus rigidus*
  rip-gut sedge - *Carex lacustris*
ripple
  emerald-ripple peperomia - *Peperomia caperata*
  green-ripple peperomia - *Peperomia caperata*
  ripple-grass - *Plantago lanceolata*
River
  Colorado River hemp - *Sesbania exaltata*
  Vanstadens River daisy - *Osteospermum ecklonis*
river
  river bank grape - *Vitis riparia*
  river bank lupine - *Lupinus rivularis*
  river bank sedge - *Carex riparia*
  river-beauty - *Epilobium latifolium*
  river birch - *Betula nigra*
  river bulrush - *Scirpus fluviatilis*
  river hawthorn - *Crataegus rivularis*
  river maple - *Acer saccharinum*
  river red gum - *Eucalyptus camaldulensis*
  river she-oak - *Casuarina cunninghamiana*
  river tea-tree - *Melaleuca leucadendra*
  river walnut - *Juglans microcarpa*
  river willow - *Salix fluviatilis*
rivet wheat - *Triticum turgidum*
roadside
  roadside agrimony - *Agrimonia striata*
  roadside peppergrass - *Lepidium ruderale*
Roanoke bells - *Mertensia virginica*
robe, hunter's-robe - *Epipremnum aureum*
Robert, herb-Robert - *Geranium robertianum*
Robert's geranium - *Geranium robertianum*
robin
  ragged-robin - *Lychnis flos-cuculi*
  red-robin - *Geranium robertianum*
  robin-run-away - *Dalibarda repens*
  runaway-robin - *Glechoma hederacea*
  Virginia wake-robin - *Peltandra virginica*
  wake-robin - *Trillium erectum, Trillium*
Robin's
  poor-Robin's fleabane - *Erigeron pulchellus*
  poor-Robin's-plantain - *Hieracium venosum*
roble - *Quercus lobata*
robust
  robust horse-nettle - *Solanum dimidiatum*
  robust purple foxtail - *Setaria viridis* var. *robusta-purpurea*
  robust white foxtail - *Setaria viridis* var. *robusta-alba*
robusta
  robusta coffee - *Coffea canephora*
  wild robusta coffee - *Coffea canephora*
rock
  green rock-cress - *Arabis missouriensis*
  hairy rock-cress - *Arabis hirsuta*
  rock aster - *Aster alpinus*
  rock brake - *Cryptogramma crispa, Cryptogramma*
  rock chestnut oak - *Quercus prinus*
  rock cranberry - *Vaccinium vitis-idaea* var. *minus*

  rock-cress - *Arabis, Sibara virginica*
  rock elm - *Ulmus thomasii*
  rock finger grass - *Chloris petraea*
  rock-foil - *Saxifraga*
  rock-geranium - *Heuchera americana*
  rock grape - *Vitis rupestris*
  rock harlequin - *Corydalis sempervirens*
  rock-jasmine - *Androsace*
  rock madwort - *Alyssum saxatile*
  rock maple - *Acer nigrum, A. saccharum*
  rock-moss - *Sedum pulchellum*
  rock polypody - *Polypodium virginianum*
  rock-purslane - *Calandrinia*
  rock-rose - *Cistus, Helianthemum*
  rock sandwort - *Arenaria stricta*
  rock spike-moss - *Selaginella rupestris*
  rock-spiraea - *Holodiscus discolor, H. dumosus*
  smooth rock-cress - *Arabis laevigata*
  Steller's rock brake - *Cryptogramma stelleri*
rockberry - *Empetrum eamesii*
rocket
  American sea-rocket - *Cakile edentula*
  dame's-rocket - *Hesperis matronalis*
  dyer's-rocket - *Reseda lutea*
  garden rocket - *Eruca vesicaria* subsp. *sativa*
  large sand rocket - *Diplotaxis tenuifolia*
  London rocket - *Sisymbrium irio*
  prairie rocket - *Erysimum asperum*
  rocket - *Hesperis*
  rocket candytuft - *Iberis amara*
  rocket-cress - *Barbarea vulgaris*
  rocket larkspur - *Consolida ajacis*
  salad rocket - *Eruca vesicaria* subsp. *sativa*
  sand rocket - *Diplotaxis muralis*
  sea-rocket - *Cakile maritima*
  sweet-rocket - *Hesperis matronalis*
  wall-rocket - *Diplotaxis tenuifolia*
  yellow-rocket - *Barbarea vulgaris*
Rocky
  Rocky Mountain bee-plant - *Cleome serrulata*
  Rocky Mountain Douglas fir - *Pseudotsuga menziesii* var. *glauca*
  Rocky Mountain iris - *Iris missouriensis*
  Rocky Mountain juniper - *Juniperus scopulorum*
  Rocky Mountain maple - *Acer glabrum*
  Rocky Mountain raspberry - *Rubus deliciosus*
  Rocky Mountain scrub oak - *Quercus undulata*
  Rocky Mountain spatter-dock - *Nuphar lutea* subsp. *polysepala*
  Rocky Mountain woodsia - *Woodsia scopulina*
  Rocky Mountain yellow pine - *Pinus ponderosa* var. *scopulorum*
rod
  nail-rod - *Typha latifolia*
  silver-rod - *Solidago bicolor*
  smooth withe-rod - *Viburnum nudum*
  withe-rod - *Viburnum cassinoides*
Roebelin's palm - *Phoenix roebelenii*
Roemer's poppy - *Roemeria refracta*
Romaine lettuce - *Lactuca sativa* var. *longifolia*
Roman
  Roman-candle - *Yucca gloriosa*

Roman chamomile - *Chamaemelum nobile*
Roman wormwood - *Ambrosia artemisiifolia, Artemisia pontica, Corydalis sempervirens*
root
  ague-root - *Aletris farinosa*
  anise-root - *Osmorhiza longistylis*
  balsam-root - *Balsamorhiza*
  big-root - *Marah*
  big-root morning-glory - *Ipomoea pandurata*
  biscuit-root - *Lomatium bicolor, Lomatium*
  bitter-root - *Lewisia rediviva*
  black-root - *Veronicastrum virginicum*
  bowman's-root - *Apocynum cannabinum, Gillenia trifoliata, Veronicastrum virginicum*
  cancer-root - *Epifagus virginiana*
  canker-root - *Coptis groenlandica, C. trifolia*
  China-root - *Smilax tamnoides*
  chocolate-root - *Geum rivale*
  Choctaw-root - *Apocynum cannabinum*
  coast man-root - *Marah oreganus*
  colic-root - *Aletris, Dioscorea villosa*
  convulsion-root - *Monotropa uniflora*
  crinkle-root - *Cardamine diphylla*
  Culver's-root - *Veronicastrum virginicum*
  dragon-root - *Arisaema triphyllum*
  dye-root - *Lachnanthes caroliniana*
  fit's-root - *Astragalus glycyphyllos, Monotropa uniflora*
  flag-root - *Acorus calamus*
  gag-root - *Lobelia inflata*
  ginger-root - *Tussilago farfara*
  Indian cucumber-root - *Medeola virginica*
  knot-root bristle grass - *Setaria geniculata*
  knot-root fox-tail - *Setaria geniculata*
  leather-root - *Psoralea macrostachya*
  licorice-root - *Glycyrrhiza glabra*
  long-root - *Arenaria caroliniana*
  long-styled anise-root - *Osmorhiza longistylis*
  man-root - *Ipomoea leptophylla, I. pandurata, Marah*
  musk-root - *Adoxa moschatellina*
  musquash-root - *Cicuta maculata*
  nerve-root - *Cypripedium acaule, C. calceolus* var. *pubescens*
  one-flowered cancer-root - *Orobanche uniflora*
  orange-root - *Asclepias tuberosa, Hydrastis canadensis*
  paint-root - *Lachnanthes caroliniana*
  papoose-root - *Caulophyllum thalictroides*
  pink-root - *Spigelia marilandica*
  pleurisy-root - *Asclepias tuberosa*
  queen's-root - *Stillingia sylvatica*
  rattlesnake-root - *Prenanthes alba, Prenanthes*
  red-root - *Amaranthus retroflexus, Cannabis sativa, Ceanothus americanus, Ceanothus, Lachnanthes caroliniana, Lithospermum arvense*
  red-root flat sedge - *Cyperus erythrorhizos*
  red-root pigweed - *Amaranthus retroflexus*
  rose-root - *Sedum rosea*
  shrub yellow-root - *Xanthorhiza simplicissima*
  slender rattlesnake-root - *Prenanthes autumnalis*
  spreading sweet-root - *Osmorhiza chilensis*
  stone-root - *Collinsonia canadensis*
  sweet-root - *Glycyrrhiza lepidota*
  tapering sweet-root - *Osmorhiza chilensis*
  tobacco-root - *Valeriana edulis*
  tuber-root - *Asclepias tuberosa*
  tuft-root - *Dieffenbachia*
  umbil-root - *Cypripedium calceolus* var. *pubescens*
  unicorn-root - *Aletris farinosa*
  white-root - *Asclepias tuberosa*
  yellow colic-root - *Aletris aurea*
rooted
  red-rooted cyperus - *Cyperus erythrorhizos*
  tap-rooted valerian - *Valeriana edulis*
  turnip-rooted celery - *Apium graveolens* var. *rapaceum*
rope-bark - *Dirca palustris*
rosary pea - *Abrus precatorius*
rose
  Arkansas rose - *Rosa arkansana*
  Austrian yellow rose - *Rosa foetida*
  baby rose - *Rosa multiflora*
  Bengal rose - *Rosa chinensis*
  bristly rose - *Rosa acicularis*
  broad-leaf rose-bay - *Tabernaemontana divaricata*
  Burnet's rose - *Rosa spinosissima*
  cabbage rose - *Rosa centifolia*
  California rose - *Rosa californica*
  California rose-bay - *Rhododendron macrophyllum*
  California wild rose - *Rosa californica*
  Cherokee rose - *Rosa laevigata*
  China rose - *Hibiscus rosa-sinensis, Rosa chinensis*
  Christmas-rose - *Helleborus niger*
  cinnamon rose - *Rosa majalis*
  cliff-rose - *Cowania mexicana*
  climbing prairie rose - *Rosa setigera*
  climbing rose - *Rosa setigera*
  cluster rose - *Rosa pisocarpa*
  corn-rose - *Agrostemma githago*
  cotton-rose - *Filago*
  crimson-eyed rose mallow - *Hibiscus moscheutos* subsp. *palustris*
  desert-rose - *Adenium obesum*
  dog rose - *Rosa canina*
  downy rose-myrtle - *Rhodomyrtus tomentosus*
  dwarf prairie rose - *Rosa arkansana*
  East Indian rose-bay - *Tabernaemontana divaricata*
  French rose - *Rosa gallica*
  great rose mallow - *Hibiscus grandiflorus*
  guelder-rose - *Viburnum opulus*
  Japanese rose - *Kerria japonica, Rosa rugosa*
  Lapland rose-bay - *Rhododendron lapponicum*
  large-fruit rose-apple - *Syzygium malaccense*
  Macartney's rose - *Rosa bracteata*
  mallow-rose - *Hibiscus moscheutos, H. moscheutos* subsp. *palustris*
  Manetti's rose - *Rosa noisettiana* cv. 'Manetti'
  memorial rose - *Rosa wichuraiana*
  moss-rose - *Portulaca grandiflora*
  mountain rose-bay - *Rhododendron catawbiense*
  multiflora rose - *Rosa multiflora*
  Nootka rose - *Rosa nutkana*
  Nutka rose - *Rosa nutkana*
  pasture rose - *Rosa carolina*
  prairie rose - *Rosa arkansana, R. setigera*
  prairie wild rose - *Rosa arkansana*
  prickly rose - *Rosa acicularis*
  rock-rose - *Cistus, Helianthemum*
  rose - *Rosa*

rose-acacia - *Robinia hispida, R. viscosa*
rose-apple - *Syzygium jambos, S. malaccense*
rose-balsam - *Impatiens balsamina*
rose-bay - *Nerium oleander*
rose-bay rhododendron - *Rhododendron maximum*
rose campion - *Lychnis coronaria*
rose clover - *Trifolium hirtum*
rose daphne - *Daphne cneorum*
rose-gentian - *Sabatia*
rose geranium - *Pelargonium graveolens*
rose gum - *Eucalyptus grandis*
rose leek - *Allium canadense*
rose mallow - *Hibiscus lasiocarpus, H. moscheutos, Hibiscus*
rose mandarin - *Streptopus roseus*
rose-moss - *Portulaca grandiflora*
rose-of-China - *Hibiscus rosa-sinensis*
rose-of-Sharon - *Hibiscus syriacus, Hypericum calycinum*
rose periwinkle - *Catharanthus roseus*
rose petty - *Erigeron pulchellus*
rose-pink - *Sabatia angularis*
rose pogonia - *Pogonia ophioglossoides*
rose-ring blanket-flower - *Gaillardia pulchella*
rose-ring gaillardia - *Gaillardia pulchella*
rose-root - *Sedum rosea*
rose twisted-stalk - *Streptopus roseus*
rose verbena - *Verbena canadensis*
rose vervain - *Verbena canadensis*
rosin-rose - *Hypericum perforatum*
rugosa rose - *Rosa rugosa*
Scotch rose - *Rosa pimpinellifolia, R. spinosissima*
smooth rose - *Rosa blanda*
smooth rose mallow - *Hibiscus laevis*
soldier rose mallow - *Hibiscus laevis*
Sturt's desert-rose - *Gossypium sturtianum*
sun-rose - *Helianthemum nummularium, Helianthemum*
swamp rose - *Rosa palustris*
swamp rose mallow - *Hibiscus moscheutos, H. moscheutos* subsp. *palustris*
tea rose - *Rosa odorata*
Turkestan rose - *Rosa rugosa*
Virginia rose - *Rosa virginiana*
western rose - *Rosa woodsii*
wood rose - *Rosa gymnocarpa*
Wood's rose - *Rosa woodsii*
woolly rose mallow - *Hibiscus lasiocarpus*
roselle - *Hibiscus sabdariffa*
rosemary
  bog-rosemary - *Andromeda glaucophylla, Andromeda*
  marsh-rosemary - *Limonium*
  rosemary - *Rosmarinus officinalis*
  western marsh-rosemary - *Limonium californicum*
  wild rosemary - *Ledum palustre*
rosewood
  Burmese rosewood - *Pterocarpus indicus*
  Indian rosewood - *Dalbergia sissoo*
  rosewood - *Tipuana tipu*
rosin
  rosin-brush - *Baccharis sarothroides*
  rosin-rose - *Hypericum perforatum*
rosinweed
  basal-leaved rosinweed - *Silphium terebinthinaceum*
  cup rosinweed - *Silphium perfoliatum*

prairie rosinweed - *Silphium integrifolium*
rosinweed - *Grindelia squarrosa, Silphium integrifolium, Silphium*
showy rosinweed - *Silphium radula*
whole-leaf rosinweed - *Silphium integrifolium*
whorled rosinweed - *Silphium trifoliatum*
Ross's redtop - *Agrostis rossiae*
Rothrock's grama - *Bouteloua rothrockii*
rouge-plant - *Rivina humilis*
rough
  rough avens - *Geum virginianum*
  rough-bark Arizona cypress - *Cupressus arizonica*
  rough-bark Mexican pine - *Pinus montezumae*
  rough bedstraw - *Galium asprellum*
  rough bent grass - *Agrostis scabra*
  rough bluegrass - *Poa trivialis*
  rough bugleweed - *Lycopus asper*
  rough button-weed - *Diodia teres*
  rough cat's-ear - *Hypochaeris radicata*
  rough cinquefoil - *Potentilla norvegica*
  rough daisy fleabane - *Erigeron strigosus*
  rough dog's-tail - *Cynosurus echinatus*
  rough false pennyroyal - *Hedeoma hispidum*
  rough fescue - *Festuca scabrella*
  rough fleabane - *Erigeron strigosus*
  rough-fruited cinquefoil - *Potentilla recta*
  rough hawk's-beard - *Crepis biennis*
  rough hawk's-bit - *Leontodon hirtus, L. taraxacoides*
  rough hawkweed - *Hieracium scabrum*
  rough-leaf dogwood - *Cornus drummondii*
  rough-leaf goldenrod - *Solidago patula*
  rough-leaf sunflower - *Helianthus strumosus*
  rough-leaf velvet-seed - *Guettarda scabra*
  rough lemon - *Citrus jambhiri*
  rough marsh-elder - *Iva annua*
  rough pea-vine - *Lathyrus hirsutus*
  rough pigweed - *Amaranthus hybridus, A. retroflexus*
  rough-pod copper-leaf - *Acalypha ostryifolia*
  rough sedge - *Carex scabrata, C. senta*
  rough-seed buttercup - *Ranunculus muricatus*
  rough-seed clammy-weed - *Polanisia dodecandra*
  rough senta - *Carex senta*
  rough-stalk bluegrass - *Poa palustris, P. trivialis*
  rough sump-weed - *Iva annua*
  rough sunflower - *Helianthus hirsutus*
  rough tree-fern - *Cyathea australis*
  small rough-leaf cornel - *Cornus asperifolia*
round
  round-head bush-clover - *Lespedeza capitata*
  round-head lespedeza - *Lespedeza capitata*
  round-leaf catchfly - *Silene rotundifolia*
  round-leaf dogwood - *Cornus rugosa*
  round-leaf eucalypt - *Eucalyptus polyanthemos*
  round-leaf eupatorium - *Eupatorium rotundifolium*
  round-leaf false pimpernel - *Lindernia grandiflora*
  round-leaf greenbrier - *Smilax rotundifolia*
  round-leaf hawthorn - *Crataegus chrysocarpa*
  round-leaf mint - *Mentha rotundifolia*
  round-leaf moort - *Eucalyptus platypus*
  round-leaf mud-plantain - *Heteranthera reniformis*
  round-leaf pyrola - *Pyrola rotundifolia*
  round-leaf serviceberry - *Amelanchier sanguinea*

round-leaf spurge - *Euphorbia serpens*
round-leaf sundew - *Drosera rotundifolia*
round-leaf thoroughwort - *Eupatorium rotundifolium*
round-leaf tick-trefoil - *Desmodium rotundifolium*
round-leaf violet - *Viola rotundifolia*
round-leaf yellow violet - *Viola rotundifolia*
round-lobed hepatica - *Hepatica americana*
round-wood - *Sorbus americana*
small round-leaf orchis - *Orchis rotundifolia*
western round-leaf violet - *Viola orbiculata*
rounded shin-leaf - *Pyrola rotundifolia*
rover bellflower - *Campanula rapunculoides*
row, two-row stonecrop - *Sedum spurium*
rowan - *Sorbus aucuparia*
royal
   Florida royal palm - *Roystonea elata*
   lily-royal - *Lilium superbum*
   royal catchfly - *Silene regia*
   royal fern - *Osmunda regalis*
   royal lily - *Lilium regale*
   royal palm - *Roystonea*
   royal paulownia - *Paulownia tomentosa*
   royal poinciana - *Delonix regia*
   royal-red-bugler - *Aeschynanthus pulcher*
   royal-vine-plant - *Gynura aurantiaca*
   royal water-lily - *Victoria amazonica*
rubber
   Assam rubber - *Ficus elastica*
   baby-rubber-plant - *Peperomia obtusifolia*
   bitter rubber-weed - *Hymenoxys odorata*
   Chinese rubber-plant - *Crassula ovata*
   creeping rubber-plant - *Ficus pumila*
   dwarf rubber-plant - *Crassula ovata*
   Indian rubber fig - *Ficus elastica*
   Indian rubber-tree - *Ficus elastica*
   Japanese rubber-plant - *Crassula ovata*
   mistletoe rubber-plant - *Ficus deltoidea*
   Pará rubber-tree - *Hevea brasiliensis*
   rubber-plant - *Ficus elastica*
   rubber rabbit-brush - *Chrysothamnus nauseosus*
   rubber-tree - *Hevea brasiliensis, Schefflera actinophylla*
   small-leaf rubber-plant - *Ficus benjamina*
rue
   African rue - *Peganum harmala*
   alpine meadow-rue - *Thalictrum alpinum*
   early meadow-rue - *Thalictrum dioicum*
   goat's-rue - *Galega officinalis*
   lady-rue - *Thalictrum clavatum*
   meadow-rue - *Thalictrum*
   mountain meadow-rue - *Thalictrum clavatum*
   purple meadow-rue - *Thalictrum dasycarpum, T. revolutum*
   rue - *Ruta graveolens*
   rue anemone - *Thalictrum thalictroides*
   skunk meadow-rue - *Thalictrum revolutum*
   tall meadow-rue - *Thalictrum pubescens*
   tassel-rue - *Trautvetteria carolinensis*
   wax-leaf meadow-rue - *Thalictrum revolutum*
ruellia, smooth ruellia - *Ruellia strepens*
Rugel's plantain - *Plantago rugelii*
rugosa rose - *Rosa rugosa*
rum cherry - *Prunus serotina*
runaway-robin - *Glechoma hederacea*

runner
   runner bean - *Phaseolus vulgaris*
   scarlet runner bean - *Phaseolus coccineus*
running
   running blackberry - *Rubus flagellaris, R. hispidus*
   running buffalo clover - *Trifolium stoloniferum*
   running club-moss - *Lycopodium clavatum*
   running-evergreen - *Lycopodium complanatum*
   running five-fingers - *Potentilla canadensis*
   running-myrtle - *Vinca minor*
   running oak - *Quercus pumila*
   running-pine - *Lycopodium clavatum*
   running-pop - *Passiflora foetida*
   running strawberry-bush - *Euonymus obovatus*
rush
   annual fringe-rush - *Fimbristylis annua*
   Baltic rush - *Juncus balticus*
   beak rush - *Rhynchospora*
   beaked spike rush - *Eleocharis rostellata*
   blunt spike rush - *Eleocharis ovata*
   bog rush - *Juncus effusus, Juncus*
   brown beak rush - *Rhynchospora fusca*
   bull-rush - *Typha latifolia*
   chairmaker's rush - *Scirpus americanus*
   creeping spike rush - *Eleocharis palustris*
   dwarf spike rush - *Eleocharis parvula*
   field rush - *Juncus tenuis*
   field wood rush - *Luzula campestris*
   flowering-rush - *Butomus umbellatus*
   forked fringe-rush - *Fimbristylis dichotoma*
   foxtail-rush - *Equisetum arvense*
   globe fringe-rush - *Fimbristylis littoralis, F. miliacea*
   grassy-rush - *Butomus umbellatus*
   Gulf Coast spike rush - *Eleocharis cellulosa*
   hog rush - *Juncus gerardii*
   horned rush - *Rhynchospora corniculata*
   Japanese mat rush - *Juncus effusus*
   mat rush - *Scirpus lacustris*
   millet wood rush - *Luzula parviflora*
   needle rush - *Juncus roemerianus*
   needle spike rush - *Eleocharis acicularis*
   nodding beak rush - *Rhynchospora inexpansa*
   nut rush - *Scleria*
   ovoid spike rush - *Eleocharis ovata*
   path rush - *Juncus tenuis*
   pine-hill beak rush - *Rhynchospora globularis*
   poverty rush - *Juncus tenuis*
   rush - *Juncus*
   rush fuirena - *Fuirena scirpoidea*
   rush grass - *Sporobolus*
   rush-nut - *Cyperus esculentus*
   rush-pink - *Lygodesmia juncea*
   rush skeleton-weed - *Chondrilla juncea*
   salt rush - *Juncus lesueurii*
   scouring-rush - *Equisetum arvense, E. hyemale, Equisetum*
   slender fringe-rush - *Fimbristylis autumnalis*
   slender rush - *Juncus tenuis*
   slender spike rush - *Eleocharis acicularis, E. baldwinii*
   slender yard rush - *Juncus tenuis*
   small spike rush - *Eleocharis parvula*
   smooth scouring-rush - *Equisetum laevigatum*
   soft rush - *Juncus effusus*
   spike rush - *Eleocharis*

sprouting spike rush - *Eleocharis vivipara*
square-stem spike rush - *Eleocharis quadrangulata*
stone-rush - *Scleria*
swaying rush - *Scirpus subterminalis*
toad rush - *Juncus bufonius*
tufted club rush - *Scirpus caespitosus*
tufted rush - *Juncus acuminatus*
variegated scouring-rush - *Equisetum variegatum*
white beak rush - *Rhynchospora alba*
wire rush - *Juncus balticus*
wood rush - *Luzula campestris, Luzula*

russet
russet buffalo-berry - *Shepherdia canadensis*
russet cotton grass - *Eriophorum russeolum*
russet sedge - *Carex saxatilis*

Russian
barb-wire Russian thistle - *Salsola paulsenii*
dwarf Russian almond - *Prunus tenella*
large Russian vetch - *Vicia villosa*
Russian cedar - *Pinus cembra*
Russian comfrey - *Symphytum uplandicum*
Russian dandelion - *Taraxacum kok-saghyz*
Russian knapweed - *Acroptilon repens, Centaurea repens*
Russian olive - *Elaeagnus angustifolia*
Russian pigweed - *Axyris amaranthoides*
Russian sickle milk-vetch - *Astragalus falcatus*
Russian thistle - *Salsola australis*
Russian wild rye - *Elymus junceus*

rust-weed - *Polypremum procumbens*

rusty
rusty black-haw - *Viburnum rufidulum*
rusty-leaf - *Menziesia ferruginea*

rutabaga - *Brassica napus, B. napus* var. *napobrassica*

rye
Altai wild rye - *Leymus angustus*
annual rye grass - *Lolium multiflorum*
Australian rye grass - *Lolium multiflorum*
basin wild rye - *Leymus cinereus*
beardless wild rye - *Leymus triticoides*
blue wild rye - *Elymus glaucus*
Canadian wild rye - *Elymus canadensis*
downy wild rye - *Elymus villosus*
English rye grass - *Lolium perenne*
giant wild rye - *Leymus cinereus*
hairy wild rye - *Elymus villosus*
Italian rye grass - *Lolium multiflorum*
Lyme rye grass - *Lolium perenne*
mammoth wild rye - *Leymus racemosus*
mountain rye - *Secale montanum*
perennial rye grass - *Lolium perenne*
poison rye grass - *Lolium temulentum*
rigid rye grass - *Lolium rigidum*
Russian wild rye - *Elymus junceus*
rye - *Secale cereale*
rye brome - *Bromus secalinus*
rye grass - *Lolium*
Salina wild rye - *Leymus salinus*
Siberian wild rye - *Elymus sibiricus*
Virginia wild rye - *Elymus virginicus*
wild rye - *Elymus*
Wimmera rye grass - *Lolium rigidum*

sabicu - *Lysiloma latisiliqua*

sacahuista - *Nolina microcarpa*
sacaton, alkali sacaton - *Sporobolus airoides*
sachaline - *Polygonum sachalinense*
Sacramento bur - *Triumfetta semitriloba*

sacred
sacred-bamboo - *Nandina domestica*
sacred-bean - *Nelumbo*
sacred datura - *Datura innoxia*
sacred lotus - *Nelumbo nucifera*

saddle, side-saddle-flower - *Sarracenia purpurea*

safflower - *Carthamus tinctorius*

saffron
bastard-saffron - *Carthamus tinctorius*
false saffron - *Carthamus tinctorius*
meadow-saffron - *Colchicum autumnale*
saffron-spike - *Aphelandra squarrosa*

sage
annual bur-sage - *Ambrosia acanthicarpa*
autumn sage - *Salvia greggii*
black sage - *Salvia mellifera*
blue sage - *Salvia azurea*
creeping sage - *Salvia sonomensis*
field sage-wort - *Artemisia campestris* subsp. *caudata*
garden sage - *Salvia officinalis*
germander-sage - *Teucrium scorodonia*
gray-ball sage - *Salvia dorrii*
Jerusalem sage - *Phlomis tuberosa*
lance-leaf sage - *Salvia reflexa*
meadow sage - *Salvia pratensis*
mealy-cup sage - *Salvia farinacea*
Mediterranean sage - *Salvia aethiopis*
pitcher sage - *Salvia spathacea*
pitcher's sage - *Salvia azurea* var. *grandiflora*
prairie sage-wort - *Artemisia frigida*
sage - *Salvia officinalis, Salvia*
sage-leaf willow - *Salix candida*
sage-tree - *Vitex agnus-castus*
sage-wort - *Artemisia campestris*
scarlet sage - *Salvia coccinea, S. splendens*
skeleton-leaf bur-sage - *Ambrosia tomentosa*
slim-leaf bur-sage - *Ambrosia confertiflora*
Texas sage - *Salvia coccinea*
white sage - *Artemisia ludoviciana, Salvia apiana*
whorled sage - *Salvia verticillata*
wood-sage - *Teucrium canadense, T. scorodonia*
woolly-leaf bur-sage - *Ambrosia grayi*
yellow-sage - *Lantana camara*

sagebrush
basin sagebrush - *Artemisia tridentata*
big sagebrush - *Artemisia tridentata*
California sagebrush - *Artemisia californica*
fringed sagebrush - *Artemisia frigida*
Louisiana sagebrush - *Artemisia ludoviciana*
low sagebrush - *Artemisia arbuscula*
sagebrush - *Artemisia tridentata, Artemisia*
sand sagebrush - *Artemisia filifolia*
silver sagebrush - *Artemisia cana*

sagittaria
grass-leaved sagittaria - *Sagittaria graminea*
Hudson sagittaria - *Sagittaria subulata*
lance-leaf sagittaria - *Sagittaria lancifolia*

sago
  Japanese sago-palm - *Cycas revoluta*
  queen sago - *Cycas circinalis*
  sago conehead - *Cycas*
  sago cycas - *Zamia pumila*
  sago-lily - *Calochortus*
  sago palm - *Caryota urens*
  sago pondweed - *Potamogeton pectinatus*
sagrada, cascara sagrada - *Rhamnus purshiana*
saguaro
  giant saguaro - *Carnegiea gigantea*
  saguaro - *Carnegiea gigantea*
sahuaro - *Carnegiea gigantea*
sailor
  creeping-sailor - *Saxifraga stolonifera*
  sailor-caps - *Dodecatheon hendersonii*
sailors, blue-sailors - *Cichorium intybus*
sainfoil - *Desmodium canadense*
sainfoin - *Onobrychis viciifolia*
salad
  beaked corn-salad - *Valerianella radiata*
  corn-salad - *Valerianella locusta, Valerianella*
  duck-salad - *Heteranthera limosa*
  fruit-salad-plant - *Monstera deliciosa*
  Indian-salad - *Hydrophyllum virginianum*
  Italian corn-salad - *Valerianella eriocarpa*
  salad bean - *Phaseolus vulgaris*
  salad burnet - *Sanguisorba minor*
  salad rocket - *Eruca vesicaria* subsp. *sativa*
  Shawnee-salad - *Hydrophyllum virginianum*
salal - *Gaultheria shallon*
salamander-tree - *Antidesma bunius*
saligot - *Trapa natans*
Salina wild rye - *Leymus salinus*
sallow sedge - *Carex lurida*
salmonberry - *Rubus chamaemorus, R. parviflorus, R. spectabilis*
salsify
  black salsify - *Scorzonera hispanica*
  meadow salsify - *Tragopogon pratensis*
  purple-flowered salsify - *Tragopogon porrifolius*
  salsify - *Tragopogon porrifolius*
  Spanish salsify - *Scorzonera hispanica*
  western salsify - *Tragopogon dubius*
  yellow salsify - *Tragopogon dubius*
salt
  annual salt marsh aster - *Aster subulatus*
  desert salt grass - *Distichlis stricta*
  pepper-and-salt - *Erigenia bulbosa*
  perennial salt marsh aster - *Aster tenuifolius*
  salt-bush - *Salsola australis*
  salt-cedar - *Tamarix ramosissima*
  salt grass - *Distichlis spicata, Distichlis*
  salt heliotrope - *Heliotropium curassavicum*
  salt marsh cord grass - *Spartina alterniflora*
  salt marsh foxtail - *Setaria magna*
  salt meadow cord grass - *Spartina patens*
  salt rush - *Juncus lesueurii*
  salt-water cord grass - *Spartina alterniflora*
  seashore salt grass - *Distichlis spicata*

saltbush
  Australian saltbush - *Atriplex semibaccata*
  berry saltbush - *Atriplex semibaccata*
  desert saltbush - *Atriplex polycarpa*
  four-wing saltbush - *Atriplex canescens*
  saltbush - *Atriplex*
  Salton's saltbush - *Atriplex elegans* var. *fasciculata*
  silver-scale saltbush - *Atriplex argentea*
  spiny saltbush - *Atriplex confertifolia*
  wheel-scale saltbush - *Atriplex elegans*
Salton's saltbush - *Atriplex elegans* var. *fasciculata*
saltwort
  dwarf saltwort - *Salicornia bigelovii*
  Mediterranean saltwort - *Salsola vermiculata*
salvers
  quack salvers-grass - *Euphorbia cyparissias*
  salvers spurge - *Euphorbia cyparissias*
salvinia
  giant salvinia - *Salvinia auriculata, S. molesta*
  salvinia - *Salvinia auriculata*
saman - *Albizia saman*
samphire - *Salicornia europaea, Salicornia*
Sampson's snakeroot - *Gentiana villosa*
sand
  box sand-myrtle - *Leiophyllum buxifolium*
  large sand rocket - *Diplotaxis tenuifolia*
  prairie sand-reed - *Calamovilfa longifolia*
  purple sand grass - *Triplasis purpurea*
  sand-berry - *Arctostaphylos uva-ursi*
  sand blackberry - *Rubus cuneifolius*
  sand blue-stem - *Andropogon hallii*
  sand brome - *Bromus arenarius*
  sand catchfly - *Silene conica*
  sand cherry - *Prunus pumila, P. susquehanae*
  sand drop-seed - *Sporobolus cryptandrus*
  sand-dune willow - *Salix syrticola*
  sand grape - *Vitis rupestris*
  sand grass - *Triplasis purpurea*
  sand hickory - *Carya pallida*
  sand-hills amaranth - *Amaranthus arenicola*
  sand-jack - *Quercus incana*
  sand-lily - *Leucocrinum montanum, Leucocrinum*
  sand love grass - *Eragrostis trichodes*
  sand-mat manzanita - *Arctostaphylos pumila*
  sand oat - *Avena strigosa*
  sand pear - *Pyrus pyrifolia*
  sand pine - *Pinus clausa*
  sand plum - *Prunus angustifolia*
  sand-reed - *Calamovilfa longifolia*
  sand rocket - *Diplotaxis muralis*
  sand sagebrush - *Artemisia filifolia*
  sand-verbena - *Abronia*
  sand vetch - *Vicia acutifolia*
  sand-vine - *Cynanchum laeve*
  western sand cherry - *Prunus pumila* var. *besseyi*
  yellow sand-verbena - *Abronia latifolia*
sandbar willow - *Salix exigua, S. sessilifolia*
Sandberg's
  Sandberg's birch - *Betula sandbergii*
  Sandberg's bluegrass - *Poa secunda*
sandbox-tree - *Hura crepitans*

sandbur
  coast sandbur - *Cenchrus incertus*
  dune sandbur - *Cenchrus tribuloides*
  field sandbur - *Cenchrus incertus*
  long-spine sandbur - *Cenchrus longispinus*
  sandbur - *Cenchrus longispinus, Cenchrus*
  southern sandbur - *Cenchrus echinatus*
sandle, bastard-sandle-wood - *Myoporum sandwicense*
sandpaper oak - *Quercus pungens*
sandpur, mat sandpur - *Cenchrus longispinus*
sands, red-sands spurry - *Spergularia rubra*
Sandwich's dodder - *Cuscuta sandwichiana*
sandwort
  grove sandwort - *Moehringia lateriflora*
  mountain sandwort - *Arenaria groenlandica*
  pine-barren sandwort - *Arenaria caroliniana*
  rock sandwort - *Arenaria stricta*
  sandwort - *Arenaria*
  sea beach sandwort - *Honckenya peploides*
  thyme-leaf sandwort - *Arenaria serpyllifolia*
sanfoin - *Onobrychis viciifolia*
sang - *Panax quinquefolius*
sanguinary - *Achillea millefolium*
sanicle
  Indian sanicle - *Ageratina altissima*
  purple sanicle - *Sanicula bipinnatifida*
  sanicle - *Sanicula*
  white sanicle - *Ageratina altissima*
Santa
  Santa Lucia fir - *Abies bracteata*
  Santa Maria - *Parthenium hysterophorus*
santa, yerba-santa - *Eriodictyon californicum*
santolina, green santolina - *Santolina rosmarinifolia*
Sanwa millet - *Echinochloa frumentacea*
sap, pine-sap - *Monotropa uniflora*
sapodilla
  sapodilla - *Manilkara zapota*
  wild sapodilla - *Manilkara jaimiqui* subsp. *emarginata*
sapota - *Pouteria sapota*
sapote
  mammee sapote - *Pouteria sapota*
  sapote - *Pouteria sapota*
  sapote-amarillo - *Pouteria campechiana*
  sapote-borracho - *Pouteria campechiana*
  white sapote - *Casimiroa edulis*
sapotilla - *Manilkara zapota*
sapphire-berry sweet-leaf - *Symplocos tinctoria*
saracura, periquito-saracura - *Alternanthera philoxeroides*
Saramolla grass - *Ischaemum rugosum*
Sargent's
  Sargent's cypress - *Cupressus sargentii*
  Sargent's juniper - *Juniperus chinensis* var. *sargentii*
sarsaparilla
  bristly sarsaparilla - *Aralia hispida*
  sarsaparilla-vine - *Smilax pumila*
  wild sarsaparilla - *Aralia nudicaulis*
Sarvis's holly - *Ilex amelanchier*
Sasanqua camellia - *Camellia sasanqua*
Saskatoon serviceberry - *Amelanchier alnifolia*

sassafras
  sassafras - *Sassafras albidum*
  white sassafras - *Sassafras albidum*
satin
  Brazilian satin-tail - *Imperata brasiliensis*
  satin pellionia - *Pellionia pulchra*
satinflower - *Clarkia amoena*
satinwood - *Murraya paniculata*
Satsuma orange - *Citrus reticulata*
saucer magnolia - *Magnolia soulangiana*
sausage-tree - *Kigelia africana*
savanna grass - *Axonopus compressus*
savannah fern - *Dicranopteris linearis*
savin
  creeping savin juniper - *Juniperus horizontalis*
  savin - *Juniperus sabina*
savory
  savory - *Satureja*
  savory-leaf aster - *Aster linariifolius*
  summer savory - *Satureja hortensis*
savoy
  savoy - *Brassica oleracea* var. *capitata*
  savoy cabbage - *Brassica oleracea* var. *capitata*
saw
  saw-brier - *Smilax glauca*
  saw cabbage palm - *Acoelorraphe wrightii*
  saw-leaf zelkova - *Zelkova serrata*
  saw palmetto - *Serenoa repens*
  saw-tooth fog-fruit - *Phyla nodiflora* var. *incisa*
  saw-tooth oak - *Quercus acutissima*
  saw-tooth spurge - *Euphorbia serrata*
  saw-tooth sunflower - *Helianthus grosseserratus*
  silver saw palm - *Acoelorraphe wrightii*
Sawara cypress - *Chamaecyparis pisifera*
sawgrass - *Cladium jamaicense*
saxifrage
  alpine-brook saxifrage - *Saxifraga rivularis*
  brook saxifrage - *Boykinia aconitifolia*
  burnet-saxifrage - *Pimpinella saxifraga*
  early saxifrage - *Saxifraga virginiensis*
  lettuce saxifrage - *Saxifraga micranthidifolia*
  purple alpine saxifrage - *Saxifraga oppositifolia*
  purple mountain saxifrage - *Saxifraga oppositifolia*
  swamp saxifrage - *Saxifraga pensylvanica*
  yellow alpine saxifrage - *Saxifraga aizoides*
  yellow mountain saxifrage - *Saxifraga aizoides*
scab-wort - *Inula helenium*
scabious
  field scabious - *Knautia arvensis*
  scabious - *Scabiosa*
  southern-scabious - *Succisa australis*
  sweet scabious - *Scabiosa atropurpurea*
scale
  all-scale - *Atriplex polycarpa*
  pit-scale grass - *Hackelochloa granularis*
  red-scale - *Atriplex rosea*
  shad-scale - *Atriplex confertifolia*
  silver-scale - *Atriplex argentea*
  silver-scale saltbush - *Atriplex argentea*
  spreading scale-seed - *Spermolepis divaricata*
  western scale-seed - *Spermolepis divaricata*

wheel-scale saltbush - *Atriplex elegans*
scaled, hollow-scaled bulrush - *Scirpus koilolepis*
scallops, lavender-scallops - *Kalanchoe fedtschenkoi*
scaly
  scaly-bark beefwood - *Casuarina glauca*
  scaly blazing-star - *Liatris squarrosa*
Scarborough-lily - *Cyrtanthus elatus*
scarlet
  Chinese scarlet eggplant - *Solanum integrifolium*
  scarlet basket-vine - *Aeschynanthus pulcher*
  scarlet bugler - *Penstemon centranthifolius*
  scarlet-bush - *Hamelia patens*
  scarlet eggplant - *Solanum integrifolium*
  scarlet fire-thorn - *Pyracantha coccinea*
  scarlet gaura - *Gaura coccinea*
  scarlet globe mallow - *Sphaeralcea coccinea*
  scarlet hawthorn - *Crataegus coccinea*
  scarlet ixora - *Ixora coccinea*
  scarlet larkspur - *Delphinium cardinale*
  scarlet-lightening - *Lychnis chalcedonica*
  scarlet lychnis - *Lychnis chalcedonica*
  scarlet mallow - *Sphaeralcea coccinea*
  scarlet maple - *Acer rubrum*
  scarlet monkey-flower - *Mimulus cardinalis*
  scarlet oak - *Quercus coccinea*
  scarlet paintbrush - *Castilleja coccinea*
  scarlet pimpernel - *Anagallis arvensis*
  scarlet runner bean - *Phaseolus coccineus*
  scarlet sage - *Salvia coccinea, S. splendens*
  scarlet sumac - *Rhus glabra*
  scarlet wisteria-tree - *Sesbania punicea*
scentless-chamomile - *Matricaria maritima, M. perforata*
schnittlauch - *Allium schoenoprasum*
Schott's yucca - *Yucca schottii*
Schrader's brome - *Bromus catharticus*
schrankia, cat-claw schrankia - *Mimosa quadrivalvis* var.
  nuttallii
Schreber's aster - *Aster schreberi*
scoke - *Phytolacca americana*
scoop, sugar-scoop - *Tiarella unifoliata*
scorpion
  field scorpion-grass - *Myosotis arvensis*
  scorpion-grass - *Myosotis, Plagiobothrys figuratus*
  scorpion-weed - *Phacelia purshii, Phacelia*
Scot's pine - *Pinus sylvestris*
Scotch
  Scotch-attorney - *Clusia rosea*
  Scotch broom - *Cytisus scoparius*
  Scotch elm - *Ulmus glabra*
  Scotch heath - *Erica cinerea*
  Scotch heather - *Calluna vulgaris*
  Scotch lovage - *Ligusticum scothicum*
  Scotch mint - *Mentha gentilis*
  Scotch-mist - *Galium sylvaticum*
  Scotch pine - *Pinus sylvestris*
  Scotch rose - *Rosa pimpinellifolia, R. spinosissima*
  Scotch spearmint - *Mentha gentilis*
  Scotch thistle - *Onopordum acanthium*
Scouler's willow - *Salix scouleriana*
scouring
  scouring-rush - *Equisetum arvense, E. hyemale, Equisetum*

smooth scouring-rush - *Equisetum laevigatum*
variegated scouring-rush - *Equisetum variegatum*
scratch grass - *Muhlenbergia asperifolia*
screw
  screw-pine - *Pandanus utilis, Pandanus*
  screw-stem - *Bartonia paniculata*
  thatch screw-pine - *Pandanus tectorius*
  Veitch's screw-pine - *Pandanus veitchii*
Scribner's
  Scribner's needle grass - *Stipa scribneri*
  Scribner's reed grass - *Calamagrostis scribneri*
scrub
  California scrub oak - *Quercus dumosa*
  Nuttall's scrub oak - *Quercus dumosa*
  Rocky Mountain scrub oak - *Quercus undulata*
  scrub hickory - *Carya floridana*
  scrub oak - *Quercus ilicifolia*
  scrub palmetto - *Sabal etonia, S. minor, Serenoa repens*
  scrub pine - *Pinus banksiana, P. virginiana*
scuppernong grape - *Vitis rotundifolia*
scurf
  gray scurf-pea - *Psoralidium tenuiflorum*
  lemon scurf pea - *Psoralidium lanceolatum*
scurfy
  scurfy pea - *Psoralea*
  scurfy psoralea - *Psoralidium tenuiflorum*
scurvy
  scurvy - *Brassica nigra*
  scurvy-grass - *Barbarea verna, Cochlearia officinalis,*
    Crambe maritima
scutch grass - *Cynodon dactylon, Elytrigia repens*
sea
  American sea-rocket - *Cakile edentula*
  sea-ash - *Zanthoxylum clava-herculis*
  sea beach dock - *Rumex pallidus*
  sea beach evening-primrose - *Oenothera humifusa*
  sea beach knotweed - *Polygonum glaucum*
  sea beach orache - *Atriplex arenaria*
  sea beach sandwort - *Honckenya peploides*
  sea beet - *Beta vulgaris*
  sea-chickweed - *Honckenya peploides*
  sea-daffodil - *Hymenocallis*
  sea-fig - *Carpobrotus chilensis*
  sea-grape - *Coccoloba uvifera*
  sea hibiscus - *Hibiscus tiliaceus*
  sea hollyhock - *Hibiscus moscheutos* subsp. *palustris*
  Sea Island cotton - *Gossypium barbadense*
  sea kale - *Crambe maritima, Crambe*
  sea-lavender - *Limonium*
  sea-lovage - *Ligusticum scothicum*
  sea lungwort - *Mertensia maritima*
  sea mertensia - *Mertensia maritima*
  sea-milkwort - *Glaux maritima*
  sea-myrtle - *Baccharis halimifolia*
  sea-oats - *Uniola paniculata*
  sea pea - *Lathyrus japonicus*
  sea-pink - *Armeria*
  sea-purslane - *Honckenya peploides, Sesuvium portulacastrum*
  sea-rocket - *Cakile maritima*
  South Sea ironwood - *Casuarina equisetifolia*

seal

    false Solomon's-seal - *Smilacina racemosa*

    great Solomon's-seal - *Polygonatum biflorum, P. commutatum*

    Solomon's-seal - *Polygonatum biflorum, Polygonatum*

    starry false Solomon's-seal - *Smilacina stellata, S. trifolia*

    thousand-seal - *Achillea millefolium*

    three-leaf false Solomon's seal - *Smilacina trifolia*

    two-leaf Solomon's-seal - *Maianthemum canadense*

seashore

    seashore drop-seed - *Sporobolus virginicus*

    seashore lupine - *Lupinus littoralis*

    seashore paspalum - *Paspalum vaginatum*

    seashore salt grass - *Distichlis spicata*

    seashore vervain - *Verbena litoralis*

seaside

    seaside alder - *Alnus maritima*

    seaside arrow grass - *Triglochin maritima*

    seaside bluebells - *Mertensia maritima*

    seaside clover - *Trifolium wormskioldii*

    seaside crowfoot - *Ranunculus cymbalaria*

    seaside dock - *Rumex pallidus*

    seaside goldenrod - *Solidago sempervirens*

    seaside heliotrope - *Heliotropium curassavicum, H. curassavicum var. obovatum*

    seaside painted-cup - *Castilleja latifolia*

    seaside pea - *Lathyrus japonicus*

    seaside petunia - *Petunia parviflora*

    seaside plantain - *Plantago maritima var. juncoides*

    seaside spurge - *Euphorbia polygonifolia*

sedge

    annual sedge - *Cyperus compressus*

    awn-fruited sedge - *Carex stipata*

    beaked sedge - *Carex rostrata*

    black-edged sedge - *Carex nigromarginata* var. *elliptica*

    black sedge - *Carex atratiformis*

    bladder sedge - *Carex intumescens*

    blunt-broom sedge - *Carex tribuloides*

    bottlebrush sedge - *Carex hystricina*

    bristle-stalk sedge - *Carex leptalea*

    bristly sedge - *Carex comosa*

    broom-sedge - *Andropogon virginicus*

    broom sedge blue-stem - *Andropogon virginicus*

    browned sedge - *Carex adusta*

    brownish sedge - *Carex brunnescens*

    coco sedge - *Cyperus esculentus, C. rotundus*

    creeping sedge - *Carex chordorrhiza*

    crested sedge - *Carex cristatella*

    cylindric sedge - *Cyperus retrorsus*

    drooping wood sedge - *Carex arctata*

    dry-spiked sedge - *Carex siccata*

    elk sedge - *Carex geyeri*

    false nut sedge - *Cyperus strigosus*

    few-flowered sedge - *Carex pauciflora*

    few-seeded sedge - *Carex oligosperma*

    field sedge - *Carex conoidea*

    flaccid sedge - *Carex leptalea*

    flat sedge - *Cyperus odoratus, Cyperus*

    fox sedge - *Carex vulpinoidea*

    foxtail sedge - *Carex alopecoidea*

    Frank's sedge - *Carex frankii*

    Fraser's sedge - *Cymophyllus fraseri*

    fringe sedge - *Carex crinita*

    giant white-top sedge - *Rhynchospora latifolia*

    globe sedge - *Cyperus globulosus*

    golden-fruited sedge - *Carex aurea*

    graceful sedge - *Carex gracillima*

    green sedge - *Carex viridula, Cyperus virens*

    hay sedge - *Carex foenea*

    heavy sedge - *Carex gravida*

    hillside sedge - *Carex siccata*

    hop-like sedge - *Carex lupuliformis*

    hop sedge - *Carex lupulina*

    inland sedge - *Carex interior*

    jointed flat sedge - *Cyperus articulatus*

    knob sedge - *Cyperus pseudovegetus*

    Leconte's sedge - *Cyperus lecontei*

    lenticular sedge - *Carex lenticularis*

    lesser prickly sedge - *Carex muricata*

    little prickly sedge - *Carex sterilis*

    long-awn arctic sedge - *Carex podocarpa*

    long-stalked sedge - *Carex pedunculata*

    loose-flowered alpine sedge - *Carex rariflora*

    loose-flowered sedge - *Carex laxiflora*

    Louisiana sedge - *Carex louisianica*

    low northern sedge - *Carex concinnoides*

    marsh straw sedge - *Carex tenera*

    mud sedge - *Carex limosa*

    Nebraska sedge - *Carex nebrascensis*

    nerved waxy sedge - *Carex verrucosa*

    northern clustered sedge - *Carex arcta*

    northern sedge - *Carex deflexa*

    nut sedge - *Cyperus esculentus, C. rotundus*

    oval-headed sedge - *Carex cephalophora*

    pale sedge - *Carex pallescens*

    plantain-leaved sedge - *Carex plantaginea*

    pointed broom-sedge - *Carex scoparia*

    porcupine sedge - *Carex hystricina*

    purple nut sedge - *Cyperus rotundus*

    razor-sedge - *Scleria*

    red-root flat sedge - *Cyperus erythrorhizos*

    retrorse sedge - *Carex retrorsa*

    rice flat sedge - *Cyperus iria*

    rip-gut sedge - *Carex lacustris*

    river bank sedge - *Carex riparia*

    rough sedge - *Carex scabrata, C. senta*

    russet sedge - *Carex saxatilis*

    sallow sedge - *Carex lurida*

    sedge - *Carex*

    sedge galingale - *Cyperus diandrus*

    sedge grass - *Andropogon virginicus*

    silvery sedge - *Carex canescens*

    slender sedge - *Carex lasiocarpa*

    small-flowered umbrella sedge - *Cyperus difformis*

    stellate sedge - *Carex rosea*

    straw sedge - *Carex straminea*

    sugar-grass sedge - *Carex atherodes*

    summer sedge - *Carex aestivalis*

    Surinam sedge - *Cyperus surinamensis*

    thicket sedge - *Carex platyphylla*

    thin-leaf sedge - *Carex cephaloidea*

    thread-leaf sedge - *Carex filifolia*

    three-fruited sedge - *Carex trisperma*

    tussock sedge - *Carex stricta*

    twisted sedge - *Carex torta*

umbrella sedge - *Cyperus alternifolius, Cyperus*
velvet sedge - *Carex vestita*
water sedge - *Carex aquatilis*
waxy sedge - *Carex glaucescens*
weak arctic sedge - *Carex supina*
white bear sedge - *Carex albursina*
white-top sedge - *Rhynchospora colorata*
wide-fruit sedge - *Carex angustata*
wolf-tail sedge - *Carex cherokeensis*
wood sedge - *Carex tetanica*
wool-fruit sedge - *Carex lasiocarpa*
woolly sedge - *Carex lanuginosa*
yellow nut sedge - *Cyperus esculentus*
yellow-white sedge - *Carex albolutescens*

seep
seep-willow - *Baccharis salicifolia*
western seep-weed - *Suaeda calceoliformis*

sego-lily - *Calochortus nuttallii*
self-heal - *Prunella vulgaris*
selva, yerba-de-selva - *Whipplea modesta*
semaphore, nodding semaphore grass - *Pleuropogon refractus*

Seminole
Seminole balsam - *Psychotria nervosa*
Seminole bread - *Zamia pumila*

Seneca snakeroot - *Polygala senega*
senecio, bush senecio - *Senecio douglasii*
Senegal date palm - *Phoenix reclinata*
senita - *Pachycereus schottii*
senji - *Melilotus indica*

senna
Alexandrian senna - *Senna alexandrina*
bladder-senna - *Colutea arborescens*
candlestick senna - *Senna alata*
coffee senna - *Senna occidentalis*
desert senna - *Senna covesii*
Indian senna - *Senna alexandrina*
ringworm senna - *Senna alata*
senna - *Cassia*
sickle senna - *Senna tora*
southern wild senna - *Senna marilandica*
Tinnevelly senna - *Senna alexandrina*
wild senna - *Senna hebecarpa, S. marilandica*
wormwood senna - *Cassia artemisioides*

sensitive
cat-claw sensitive-brier - *Mimosa quadrivalvis* var. *angustata*
giant sensitive-plant - *Mimosa invisa*
little-leaf sensitive-brier - *Mimosa quadrivalvis* var. *angustata*
sensitive-brier - *Mimosa quadrivalvis* var. *nuttallii*
sensitive fern - *Onoclea sensibilis*
sensitive partridge pea - *Chamaecrista nictitans*
sensitive-plant - *Mimosa pudica*
wild sensitive-plant - *Chamaecrista nictitans*

senta, rough senta - *Carex senta*
sentry
Belmore's sentry palm - *Howeia belmoreana*
Forster's sentry palm - *Howeia forsteriana*
sentry palm - *Howeia forsteriana, Howeia*

senvil - *Brassica nigra*
September elm - *Ulmus serotina*
sequoia, giant-sequoia - *Sequoiadendron giganteum*

Sericea lespedeza - *Lespedeza cuneata*
serpent
serpent-cucumber - *Trichosanthes anguina*
serpent-grass - *Polygonum viviparum*

serradella - *Ornithopus sativus*
serrated tussock - *Nassella trichotoma*
service-tree - *Sorbus domestica*
serviceberry
Allegheny serviceberry - *Amelanchier laevis*
cluster serviceberry - *Amelanchier alnifolia* var. *pumila*
Cusick's serviceberry - *Amelanchier alnifolia* var. *cusickii*
downy serviceberry - *Amelanchier arborea, A. canadensis*
eastern serviceberry - *Amelanchier canadensis*
inland serviceberry - *Amelanchier interior*
mountain serviceberry - *Amelanchier bartramiana*
New England serviceberry - *Amelanchier sanguinea*
round-leaf serviceberry - *Amelanchier sanguinea*
Saskatoon serviceberry - *Amelanchier alnifolia*
serviceberry - *Amelanchier*
serviceberry willow - *Salix monticola*
smooth serviceberry - *Amelanchier laevis*
Utah serviceberry - *Amelanchier utahensis*
western serviceberry - *Amelanchier alnifolia*

sesame - *Sesamum indicum*
sesban - *Sesbania exaltata*
sesbania, bag-pod sesbania - *Sesbania vesicaria*
sessile
sessile bellwort - *Uvularia sessilifolia*
sessile blazing-star - *Liatris spicata*
sessile joy-weed - *Alternanthera sessilis*
sessile-leaf tick-trefoil - *Desmodium sessilifolium*
sessile tick-clover - *Desmodium sessilifolium*
sessile tick-trefoil - *Desmodium sessilifolium*

set, quick-set-thorn - *Crataegus laevigata*
seven-bark - *Hydrangea arborescens*
Seville orange - *Citrus aurantium*
sewee bean - *Phaseolus lunatus*
seymeria
smooth seymeria - *Seymeria cassioides*
sticky seymeria - *Seymeria pectinata*

shad
shad - *Amelanchier*
shad-scale - *Atriplex confertifolia*

shadbush - *Amelanchier canadensis, Amelanchier*
shaddock - *Citrus maxima*
shade
shade fescue - *Festuca rubra* var. *heterophylla*
shade horsetail - *Equisetum palustre*

shagbark
false shagbark - *Carya ovalis*
shagbark hickory - *Carya ovata*

shaggy
shaggy golden aster - *Chrysopsis mariana*
shaggy prairie-turnip - *Psoralea esculenta*

shale-barren phlox - *Phlox buckleyi*
shallon - *Gaultheria shallon*
shallot - *Allium ascalonicum, A. cepa*
shamrock
Irish shamrock - *Oxalis acetosella, Trifolium dubium*
wood shamrock - *Oxalis montana*

Shantung
  Shantung cabbage - *Brassica pekinensis*
  Shantung maple - *Acer truncatum*
  Sharon, rose-of-Sharon - *Hibiscus syriacus, Hypericum calycinum*
sharp
  sharp-pod morning-glory - *Ipomoea cordatotriloba*
  sharp-point fluvellin - *Kickxia elatine*
  sharp-winged monkey-flower - *Mimulus alatus*
Shasta daisy - *Leucanthemum maximum*
shatter-cane - *Sorghum bicolor*
shaving
  Cupid's-shaving-brush - *Emilia fosbergii*
  shaving-brush-tree - *Pseudobombax ellipticum*
Shawnee-salad - *Hydrophyllum virginianum*
she
  drooping she-oak - *Casuarina stricta*
  river she-oak - *Casuarina cunninghamiana*
  she-balsam - *Abies fraseri*
  she-oak - *Casuarina*
sheath
  long-sheath elodea - *Elodea longivaginata*
  silver-sheath knotweed - *Polygonum argyrocoleon*
sheathed cotton grass - *Eriophorum vaginatum*
sheep
  sheep fescue - *Festuca ovina*
  sheep-laurel - *Kalmia angustifolia*
  sheep-poison - *Kalmia angustifolia*
  sheep sorrel - *Rumex acetosella*
sheep's
  sheep's-bit - *Jasione montana*
  sheep's-fat - *Atriplex confertifolia*
sheepberry - *Viburnum lentago, V. prunifolium*
shell
  pink-shell azalea - *Rhododendron vaseyi*
  shell-flower - *Moluccella laevis*
  shell-ginger - *Alpinia zerumbet*
shellbark hickory - *Carya laciniosa, C. ovata*
shellflower - *Alpinia zerumbet*
Shelly's grass - *Elytrigia repens*
shepherd's
  shepherd's-clock - *Anagallis arvensis*
  shepherd's-cress - *Teesdalia nudicaulis*
  shepherd's-needle - *Scandix pecten-veneris*
  shepherd's-purse - *Capsella bursa-pastoris*
shield
  coastal shield fern - *Dryopteris arguta*
  marginal shield fern - *Dryopteris marginalis*
  shield fern - *Dryopteris, Polystichum*
  water-shield - *Brasenia schreberi*
shin
  Harvard's shin oak - *Quercus harvardii*
  one-flowered shin-leaf - *Moneses uniflora*
  rounded shin-leaf - *Pyrola rotundifolia*
  shin-leaf - *Pyrola elliptica, P. rotundifolia, Pyrola*
  white-veined shin-leaf - *Pyrola picta*
shingle oak - *Quercus imbricaria*
shining
  shining club-moss - *Lycopodium lucidulum*
  shining cyperus - *Cyperus rivularis*
  shining pondweed - *Potamogeton illinoensis*

shining spurge - *Euphorbia lucida*
  shining sumac - *Rhus copallina*
  shining willow - *Salix lucida*
shinnery oak - *Quercus harvardii*
Shirley's poppy - *Papaver rhoeas*
shittim-wood - *Halesia tetraptera, Sideroxylon lanuginosa*
shoe
  shoe-buttons - *Sanicula bipinnatifida*
  Venus's-shoe - *Cypripedium calceolus var. pubescens*
  whip-poor-will-shoe - *Cypripedium calceolus var. pubescens*
  yellow Indian-shoe - *Cypripedium calceolus var. pubescens*
shoestring, devil's-shoestring - *Lygodesmia juncea, Viburnum alnifolium*
shoofly
  shoofly - *Alternanthera ficoidea var. amoena, Hibiscus trionum*
  shoofly-plant - *Nicandra physalodes*
shooting
  alpine shooting-star - *Dodecatheon alpinum*
  eastern shooting-star - *Dodecatheon meadia*
  shooting-star - *Dodecatheon conjugens, D. meadia, Dodecatheon*
shore
  shore aster - *Aster tradescantii*
  shore buttercup - *Ranunculus cymbalaria*
  shore juniper - *Juniperus conferta*
  shore pine - *Pinus contorta*
  shore plum - *Prunus maritima*
short
  short-awn barley - *Hordeum brevisubulatum*
  short-awn foxtail - *Alopecurus aequalis*
  short-bristled umbrella-grass - *Fuirena breviseta*
  short-leaf fig - *Ficus citrifolia*
  short-leaf pine - *Pinus echinata*
  short-pod mustard - *Hirschfeldia incana*
  short-stalk copper-leaf - *Acalypha gracilens*
Short's aster - *Aster shortii*
shot
  Indian-shot - *Canna indica*
  shot huckleberry - *Vaccinium ovatum*
show geranium - *Pelargonium domesticum*
shower
  golden-shower - *Cassia fistula, Pyrostegia venusta*
  pink-and-white-shower - *Cassia javanica var. indochinensis*
  shower-tree - *Cassia artemisioides, Cassia*
showy
  showy crabapple - *Malus floribunda*
  showy crazyweed - *Oxytropis splendens*
  showy crotalaria - *Crotalaria spectabilis*
  showy evening-primrose - *Oenothera speciosa*
  showy goldenrod - *Solidago speciosa*
  showy Japanese lily - *Lilium speciosum*
  showy lady's-slipper - *Cypripedium reginae*
  showy lily - *Lilium speciosum*
  showy locoweed - *Oxytropis splendens*
  showy milkweed - *Asclepias speciosa*
  showy mountain-ash - *Sorbus decora*
  showy orchis - *Orchis spectabilis*
  showy rattle-box - *Crotalaria spectabilis*
  showy rosinweed - *Silphium radula*
  showy speedwell - *Hebe speciosa*
  showy sunflower - *Helianthus laetiflorus*

showy sunray - *Helipterum splendidum*
showy tarweed - *Madia elegans*
showy tick-trefoil - *Desmodium canadense*
showy water-primrose - *Ludwigia uruguayensis*
shrimp
  shrimp-bush - *Justicia brandegeana*
  shrimp-plant - *Justicia brandegeana*
shrub
  banana-shrub - *Michelia figo*
  California sweet-shrub - *Calycanthus occidentalis*
  pineapple-shrub - *Calycanthus floridus*
  ringworm-shrub - *Senna alata*
  shrub althea - *Hibiscus syriacus*
  shrub bush-clover - *Lespedeza bicolor*
  shrub lespedeza - *Lespedeza bicolor*
  shrub live oak - *Quercus turbinella*
  shrub-verbena - *Lantana*
  shrub yellow-root - *Xanthorhiza simplicissima*
  strawberry-shrub - *Calycanthus floridus*
  sweet-shrub - *Calycanthus occidentalis, Calycanthus*
shrubby
  shrubby bittersweet - *Celastrus scandens, Celastrus*
  shrubby cinquefoil - *Potentilla fruticosa*
  shrubby-fern - *Comptonia peregrina*
  shrubby red-cedar - *Juniperus horizontalis*
  shrubby St. John's-wort - *Hypericum prolificum*
  shrubby trefoil - *Ptelea*
shu, feng-hsiang-shu - *Liquidambar formosana*
shuck, honey-shuck - *Gleditsia triacanthos*
Shumard's red oak - *Quercus shumardii*
sibara - *Sibara virginica*
Siberian
  Siberian crabapple - *Malus baccata*
  Siberian crane's-bill - *Geranium sibiricum*
  Siberian elm - *Ulmus pumila*
  Siberian lily - *Ixiolirion tataricum*
  Siberian motherwort - *Leonurus sibiricus*
  Siberian mustard - *Cardaria pubescens*
  Siberian oil-seed - *Camelina microcarpa*
  Siberian pea-tree - *Caragana arborescens*
  Siberian purslane - *Claytonia sibirica*
  Siberian squill - *Scilla sibirica*
  Siberian wheat grass - *Agropyron fragile* subsp. *sibiricum*
  Siberian wild rye - *Elymus sibiricus*
sickle
  Russian sickle milk-vetch - *Astragalus falcatus*
  sickle alfalfa - *Medicago sativa* subsp. *falcata*
  sickle grass - *Parapholis incurva*
  sickle-keeled lupine - *Lupinus albicaulis*
  sickle-leaf golden aster - *Chrysopsis falcata*
  sickle medic - *Medicago sativa* subsp. *falcata*
  sickle-pod - *Arabis canadensis, Senna obtusifolia, S. tora*
  sickle senna - *Senna tora*
  sickle-weed - *Falcaria vulgaris*
sida
  alkali-sida - *Malvella leprosa*
  arrow-leaf sida - *Sida rhombifolia*
  heart-leaf sida - *Sida cordifolia*
  prickly sida - *Sida spinosa*
  southern sida - *Sida acuta*
side
  side-oats grama - *Bouteloua curtipendula*

side-saddle-flower - *Sarracenia purpurea*
sided
  one-sided bottlebrush - *Calothamnus sanguineus*
  one-sided pyrola - *Orthilia secunda*
  one-sided-wintergreen - *Orthilia secunda*
sierra
  sierra alder - *Alnus rhombifolia*
  sierra bladder-nut - *Staphylea bolanderi*
  sierra chinquapin - *Castanopsis sempervirens*
  sierra clover - *Trifolium wormskioldii*
  sierra currant - *Ribes nevadense*
  sierra gooseberry - *Ribes roezlii*
  sierra juniper - *Juniperus occidentalis*
  sierra plum - *Prunus subcordata*
  sierra star-tulip - *Calochortus nudus*
sieva bean - *Phaseolus lunatus*
signal
  broad-leaf signal grass - *Brachiaria platyphylla*
  fringed signal grass - *Brachiaria ciliatissima*
  narrow-leaf signal grass - *Brachiaria piligera*
silk
  Chinese silk-plant - *Boehmeria nivea*
  floss-silk-tree - *Chorisia speciosa*
  red silk cotton-tree - *Bombax ceiba*
  silk cotton-tree - *Ceiba pentandra*
  silk grass - *Oryzopsis hymenoides*
  silk-oak - *Grevillea robusta*
  silk-tassel-bush - *Garrya*
  silk-tree - *Albizia julibrissin*
  silk-tree albizia - *Albizia julibrissin*
  silk-vine - *Periploca graeca*
  wavy-leaf silk-tassel - *Garrya elliptica*
  white silk cotton-tree - *Ceiba pentandra*
silkweed - *Asclepias syriaca*
silkworm mulberry - *Morus alba* var. *multicaulis*
silky
  silky-camellia - *Stewartia melachodendron*
  silky crazyweed - *Oxytropis sericea*
  silky dogwood - *Cornus amomum, C. obliqua*
  silky lupine - *Lupinus sericeus*
  silky osier - *Salix viminalis*
  silky prairie-clover - *Dalea villosa*
  silky sophora - *Sophora nuttalliana, S. sericea*
  silky stewartia - *Stewartia melachodendron*
  silky-white everlasting - *Helipterum splendidum*
  silky willow - *Salix sericea*
silver
  coastal silver-weed - *Potentilla pacifica*
  Florida silver palm - *Coccothrinax argentata*
  gold-and-silver-flower - *Lonicera japonica*
  little silver-bell - *Halesia tetraptera*
  Pacific silver fir - *Abies amabilis*
  Pacific silver-weed - *Potentilla pacifica*
  pendant silver linden - *Tilia petiolaris*
  quick-silver-weed - *Thalictrum dioicum*
  silver beard grass - *Bothriochloa saccharoides*
  silver-bell - *Halesia*
  silver blue-stem - *Bothriochloa saccharoides*
  silver buffalo-berry - *Shepherdia argentea*
  silver-bush - *Convolvulus cneorum, Sophora tomentosa*
  silver button - *Anaphalis margaritacea*
  silver button mangrove - *Conocarpus erectus* var. *sericeus*

silver buttonwood - *Conocarpus erectus* var. *sericeus*
silver-dollar - *Crassula arborescens, Lunaria annua*
silver-dollar gum - *Eucalyptus polyanthemos*
silver-dollar-tree - *Eucalyptus cinerea*
silver fern - *Pityrogramma calomelanos*
silver fir - *Abies alba, A. amabilis, A. grandis*
silver fittonia - *Fittonia verschaffeltii*
silver hair grass - *Aira caryophyllea*
silver horse-nettle - *Solanum elaeagnifolium*
silver jade-plant - *Crassula arborescens*
silver-leaf - *Anaphalis margaritacea*
silver-leaf desmodium - *Desmodium uncinatum*
silver-leaf grape - *Vitis aestivalis* var. *argentifolia*
silver-leaf gum - *Eucalyptus pulverulenta*
silver-leaf mountain gum - *Eucalyptus pulverulenta*
silver-leaf-nettle - *Solanum elaeagnifolium*
silver-leaf nightshade - *Solanum elaeagnifolium*
silver-leaf oak - *Quercus hypoleucoides*
silver-leaf poplar - *Populus alba*
silver-leaf poverty-weed - *Ambrosia tomentosa*
silver-leaf sunflower - *Helianthus argophyllus*
silver lupine - *Lupinus argenteus*
silver maple - *Acer saccharinum*
silver nerve-plant - *Fittonia verschaffeltii*
silver-net-plant - *Fittonia verschaffeltii*
silver plume grass - *Saccharum alopecuroideum*
silver-rod - *Solidago bicolor*
silver sagebrush - *Artemisia cana*
silver saw palm - *Acoelorraphe wrightii*
silver-scale - *Atriplex argentea*
silver-scale saltbush - *Atriplex argentea*
silver-sheath knotweed - *Polygonum argyrocoleon*
silver-threads - *Fittonia verschaffeltii*
silver-tree - *Conocarpus erectus* var. *sericeus*
silver trumpet-tree - *Tabebuia argentea*
silver-weed - *Potentilla anserina*
silver-weed cinquefoil - *Potentilla anserina*
silver willow - *Salix candida*
silver wormwood - *Artemisia ludoviciana*
two-winged silver-bell - *Halesia diptera*
silverberry - *Elaeagnus angustifolia, E. commutata,*
    Shepherdia argentea
silverling - *Baccharis halimifolia*
silvertop-ash - *Eucalyptus sieber*
silvery
silvery cinquefoil - *Potentilla argentea*
silvery rhododendron - *Rhododendron grande*
silvery sedge - *Carex canescens*
silvery spleenwort - *Athyrium thelypteroides*
Simon's
Simon's plum - *Prunus simonii*
Simon's poplar - *Populus simonii*
simpler's-joy - *Verbena hastata*
sing-kwa - *Luffa acutangula*
Singapore holly - *Malpighia coccigera*
single
single-leaf ash - *Fraxinus anomala*
single-leaf pinyon pine - *Pinus monophylla*
singletary pea - *Lathyrus hirsutus*
siratro - *Macroptilium atropurpureum*
siris-tree - *Albizia lebbeck*

sisal
sisal - *Agave sisalana*
sisal-hemp - *Agave sisalana*
sissoo - *Dalbergia sissoo*
sisu - *Dalbergia sissoo*
Sitka
Sitka alder - *Alnus sinuata*
Sitka burnet - *Sanguisorba stipulata*
Sitka spruce - *Picea sitchensis*
Sitka vetch - *Vicia gigantea*
Sitka willow - *Salix sitchensis*
six
six-weeks grama - *Bouteloua barbata*
six-weeks grass - *Poa annua*
six-weeks needle grama - *Bouteloua aristidoides*
six-weeks three-awn - *Aristida adscensionis*
skeels - *Sophora davidii*
skeleton
rush skeleton-weed - *Chondrilla juncea*
skeleton-leaf bur - *Ambrosia tomentosa*
skeleton-leaf bur-sage - *Ambrosia tomentosa*
skeleton-leaf willow - *Salix petiolaris*
skeleton-weed - *Chondrilla juncea, Lygodesmia juncea*
skevish - *Erigeron philadelphicus*
skullcap
blue skullcap - *Scutellaria lateriflora*
downy skullcap - *Scutellaria incana*
hairy skullcap - *Scutellaria elliptica*
hoary skullcap - *Scutellaria incana*
hyssop skullcap - *Scutellaria integrifolia*
little skullcap - *Scutellaria parvula*
mad-dog skullcap - *Scutellaria lateriflora*
marsh skullcap - *Scutellaria galericulata*
skullcap - *Scutellaria*
smaller skullcap - *Scutellaria parvula*
skullwort, marsh skullwort - *Scutellaria galericulata*
skunk
skunk-brush - *Rhus trilobata*
skunk-bush - *Rhus trilobata*
skunk-cabbage - *Symplocarpus foetidus*
skunk currant - *Ribes glandulosum*
skunk grape - *Vitis labrusca*
skunk meadow-rue - *Thalictrum revolutum*
skunk-tail grass - *Hordeum jubatum*
skunk-vine - *Paederia foetida*
skunk-weed - *Croton texensis, Navarretia squarrosa*
skunk-weed-gilia - *Navarretia squarrosa*
western skunk-cabbage - *Lysichiton americanum*
yellow skunk-cabbage - *Lysichiton americanum*
sky
Brazilian sky-flower - *Duranta erecta*
sky-flower - *Duranta erecta*
skyline bluegrass - *Poa epilis*
slash pine - *Pinus elliottii*
sleekwort, mauve sleekwort - *Liparis liliifolia*
sleepy
sleepy catchfly - *Silene antirrhina*
sleepy grass - *Stipa robusta*
sleepy mallow - *Malvaviscus*
slender
northern slender lady's-tresses - *Spiranthes lacera*

slender amaranth - *Amaranthus viridis*
slender arrowhead - *Sagittaria graminea*
slender bluegrass - *Poa secunda*
slender bugleweed - *Lycopus uniflorus*
slender bulrush - *Scirpus heterochaetus*
slender bush-clover - *Lespedeza virginica*
slender chess - *Bromus tectorum*
slender cliff brake - *Cryptogramma stelleri*
slender copper-leaf - *Acalypha gracilens*
slender corydalis - *Corydalis micrantha*
slender crabgrass - *Digitaria filiformis*
slender cyperus - *Cyperus filiculmis*
slender false brome grass - *Brachypodium sylvaticum*
slender fescue - *Festuca tenuifolia*
slender-flower thistle - *Carduus tenuiflorus*
slender fringe-rush - *Fimbristylis autumnalis*
slender-fruit lomatium - *Lomatium bicolor* var. *leptocarpum*
slender fume-wort - *Corydalis micrantha*
slender glasswort - *Salicornia europaea*
slender goldenrod - *Solidago erecta*
slender hair grass - *Deschampsia elongata*
slender lady palm - *Rhapis excelsa*
slender lady's-tresses - *Spiranthes lacera*
slender-leaf crotalaria - *Crotalaria brevidens*
slender lespedeza - *Lespedeza violacea, L. virginica*
slender lip fern - *Cheilanthes feei*
slender naiad - *Najas flexilis*
slender nettle - *Urtica dioica*
slender oat - *Avena barbata*
slender parsley-piert - *Alchemilla microcarpa*
slender plantain - *Plantago elongata*
slender pondweed - *Potamogeton pusillus*
slender rattlesnake-root - *Prenanthes autumnalis*
slender rush - *Juncus tenuis*
slender sedge - *Carex lasiocarpa*
slender speedwell - *Veronica filiformis*
slender spike rush - *Eleocharis acicularis, E. baldwinii*
slender-stem androsace - *Androsace filiformis*
slender-stem bladderwort - *Utricularia gibba*
slender trefoil - *Lotus tenuis*
slender vetch - *Vicia tetrasperma*
slender wheat grass - *Elymus trachycaulus*
slender wild oat - *Avena barbata*
slender wire-lettuce - *Stephanomeria tenuifolia*
slender yard rush - *Juncus tenuis*

slim
slim-leaf bur-sage - *Ambrosia confertiflora*
slim-leaf lamb's-quarters - *Chenopodium leptophyllum*
slim-spike foxtail - *Alopecurus myosuroides*
slim-spike three-awn - *Aristida longespica*
slim-stem muhly - *Muhlenbergia filiculmis*

slipper
fairy-slipper - *Calypso bulbosa*
golden-slipper - *Cypripedium calceolus* var. *pubescens*
lady's-slipper - *Cypripedium*
large lady's-slipper - *Cypripedium calceolus* var. *pubescens*
pink lady's-slipper - *Cypripedium acaule*
showy lady's-slipper - *Cypripedium reginae*
slipper-flower - *Calceolaria mexicana, Pedilanthus tithymaloides*
slipper-plant - *Pedilanthus tithymaloides*
small white lady's-slipper - *Cypripedium candidum*
two-leaf lady's-slipper - *Cypripedium acaule*

violet-slipper gloxinia - *Sinningia speciosa*
white lady's-slipper - *Cypripedium candidum*
slipperwort - *Calceolaria mexicana*
slippery elm - *Ulmus rubra*
slobber-weed - *Euphorbia maculata*

sloe
sloe - *Prunus alleghaniensis, P. americana, P. spinosa*
sloe plum - *Prunus umbellata*

slough
American slough grass - *Beckmannia syzigachne*
slough grass - *Spartina pectinata*

small
small bindweed - *Convolvulus arvensis*
small bugloss - *Anchusa arvensis*
small-bulrush - *Typha angustifolia*
small celandine - *Ranunculus ficaria*
small crabgrass - *Digitaria ischaemum*
small cranberry - *Vaccinium oxycoccos*
small crown-flower - *Calotropis procera*
small datura - *Datura discolor*
small duckweed - *Lemna valdiviana*
small enchanter's-nightshade - *Circaea alpina*
small floating manna grass - *Glyceria borealis*
small-flowered Alexander grass - *Brachiaria subquadripara*
small-flowered bird's-foot trefoil - *Lotus micranthus*
small-flowered bitter-cress - *Cardamine parviflora*
small-flowered buttercup - *Ranunculus abortivus*
small-flowered crane's-bill - *Geranium pusillum*
small-flowered crowfoot - *Ranunculus abortivus*
small-flowered evening-primrose - *Oenothera nuttallii, O. parviflora*
small-flowered fiddle-neck - *Amsinckia menziesii*
small-flowered galinsoga - *Galinsoga parviflora*
small-flowered geranium - *Geranium pusillum*
small-flowered leaf-cup - *Polymnia canadensis*
small-flowered milk-vetch - *Astragalus nuttallianus*
small-flowered morning-glory - *Convolvulus arvensis, Jacquemontia tamnifolia*
small-flowered tamarisk - *Tamarix parviflora*
small-flowered thistle - *Cirsium arvense*
small-flowered umbrella sedge - *Cyperus difformis*
small-flowered wild bean - *Strophostyles leiosperma*
small-fruited hickory - *Carya glabra*
small-fruited Queensland-nut - *Macadamia tetraphylla*
small-headed goldenrod - *Solidago minor*
small-headed sunflower - *Helianthus microcephalus*
small hop clover - *Trifolium dubium*
small Jack-in-the-pulpit - *Arisaema triphyllum*
small-leaf rubber-plant - *Ficus benjamina*
small-leaf tick-trefoil - *Desmodium ciliare*
small lousewort - *Pedicularis sylvatica*
small magnolia - *Magnolia virginiana*
small mudar - *Calotropis procera*
small oat - *Avena strigosa*
small pondweed - *Potamogeton pusillus*
small pussy willow - *Salix humilis*
small red morning-glory - *Ipomoea coccinea*
small rough-leaf cornel - *Cornus asperifolia*
small round-leaf orchis - *Orchis rotundifolia*
small-seeded alfalfa dodder - *Cuscuta approximata*
small-seeded dodder - *Cuscuta planiflora*
small-seeded false flax - *Camelina microcarpa*

small snapdragon - *Chaenorhinum minus*
small spike rush - *Eleocharis parvula*
small Venus's-looking-glass - *Triodanis biflora*
small water-wort - *Elatine minima*
small white aster - *Aster laterifolius*
small white lady's-slipper - *Cypripedium candidum*
small white morning-glory - *Ipomoea lacunosa*
small whorled pogonia - *Isotria medeoloides*
small-wood sunflower - *Helianthus microcephalus*
small yellow water-crowfoot - *Ranunculus gmelinii*
Small's groundsel - *Senecio anonymus*
smaller
  smaller bur-marigold - *Bidens cernua*
  smaller burdock - *Arctium minus*
  smaller forget-me-not - *Myosotis laxa*
  smaller fringed gentian - *Gentianopsis procera*
  smaller skullcap - *Scutellaria parvula*
smartweed
  alpine smartweed - *Polygonum viviparum*
  dotted smartweed - *Polygonum punctatum*
  hedge smartweed - *Polygonum scandens*
  marsh-pepper smartweed - *Polygonum hydropiper*
  mild smartweed - *Polygonum hydropiperoides*
  pale smartweed - *Polygonum lapathifolium*
  Pennsylvania smartweed - *Polygonum pensylvanicum*
  prickly smartweed - *Polygonum bungeanum*
  smartweed - *Polygonum hydropiper, Polygonum*
  smartweed dodder - *Cuscuta polygonorum*
  swamp smartweed - *Polygonum coccineum*
  water smartweed - *Polygonum amphibium, P. punctatum*
smilax asparagus - *Asparagus asparagoides*
smilo grass - *Oryzopsis miliacea*
Smirnow's rhododendron - *Rhododendron smirnowii*
Smith's melic - *Melica smithii*
smock, lady's-smock - *Cardamine pratensis*
smoke
  American smoke-tree - *Cotinus obovatus*
  desert smoke-tree - *Psorothamnus spinosa*
  earth-smoke - *Fumaria officinalis*
  prairie-smoke - *Anemone patens*
  smoke-bush - *Cotinus coggygria*
  smoke-tree - *Cotinus coggygria*
  smoke-weed - *Eupatorium maculatum*
smooth
  smooth alder - *Alnus rugosa, A. serrulata*
  smooth arrow-wood - *Viburnum recognitum*
  smooth aster - *Aster laevis*
  smooth azalea - *Rhododendron arborescens*
  smooth bedstraw - *Galium mollugo*
  smooth blackberry - *Rubus canadensis*
  smooth bramble - *Rubus canadensis*
  smooth brome - *Bromus inermis*
  smooth brome grass - *Bromus commutatus*
  smooth button-weed - *Spermacoce glabra*
  smooth cat's-ear - *Hypochaeris glabra*
  smooth cliff brake - *Pellaea glabella*
  smooth cliff fern - *Woodsia glabella*
  smooth cord grass - *Spartina alterniflora*
  smooth crabgrass - *Digitaria ischaemum*
  smooth crotalaria - *Crotalaria pallida*
  smooth dock - *Rumex altissimus*
  smooth false dandelion - *Hypochaeris glabra*

smooth ground-cherry - *Physalis longifolia* var. *subglabrata*
smooth hawk's-beard - *Crepis capillaris*
smooth hawkweed - *Hieracium floribundum*
smooth hedge-nettle - *Stachys tenuifolia*
smooth hydrangea - *Hydrangea arborescens*
smooth lead-plant - *Amorpha nana*
smooth-leaved elm - *Ulmus minor*
smooth loofah - *Luffa aegyptiaca*
smooth loosestrife - *Lysimachia quadrifolia*
smooth meadow parsnip - *Thaspium trifoliatum*
smooth milkweed - *Asclepias sullivantii*
smooth phlox - *Phlox glaberrima*
smooth pigweed - *Amaranthus hybridus*
smooth rock-cress - *Arabis laevigata*
smooth rose - *Rosa blanda*
smooth rose mallow - *Hibiscus laevis*
smooth ruellia - *Ruellia strepens*
smooth scouring-rush - *Equisetum laevigatum*
smooth-seeded wild bean - *Strophostyles leiosperma*
smooth serviceberry - *Amelanchier laevis*
smooth seymeria - *Seymeria cassioides*
smooth spiderwort - *Tradescantia ohiensis*
smooth sumac - *Rhus glabra*
smooth tare - *Vicia tetrasperma*
smooth trailing lespedeza - *Lespedeza repens*
smooth white-lettuce - *Prenanthes racemosa*
smooth winterberry - *Ilex laevigata*
smooth withe-rod - *Viburnum nudum*
smooth woodsia - *Woodsia glabella*
smooth yellow violet - *Viola pubescens*
snail
  snail-seed - *Coccoloba diversifolia, Cocculus carolinus*
  snail-seed pondweed - *Potamogeton diversifolius*
snake
  snake-flower - *Echium vulgare*
  snake gourd - *Trichosanthes anguina*
  snake-grass - *Equisetum arvense*
  snake-palm - *Amorphophallus rivieri, Amorphophallus*
  snake-plant - *Sansevieria trifasciata*
  snake-weed - *Conium maculatum*
snakeberry - *Actaea rubra*
snakehead - *Chelone glabra, Chelone*
snakemouth - *Pogonia ophioglossoides*
snakeroot
  black snakeroot - *Cimicifuga racemosa, Sanicula marilandica, Sanicula, Zigadenus densus*
  broom snakeroot - *Gutierrezia sarothrae*
  button snakeroot - *Eryngium aquaticum, E. yuccifolium, Liatris*
  Sampson's snakeroot - *Gentiana villosa*
  Seneca snakeroot - *Polygala senega*
  snakeroot - *Asarum canadense*
  Virginia snakeroot - *Aristolochia serpentaria*
  white snakeroot - *Ageratina altissima*
snakeweed
  broom snakeweed - *Gutierrezia sarothrae*
  snakeweed - *Gutierrezia, Polygonum bistortoides*
  sticky snakeweed - *Gutierrezia microcephala*
  thread-leaf snakeweed - *Gutierrezia microcephala*
snap
  snap bean - *Phaseolus vulgaris*
  snap pea - *Pisum sativum*

snap-weed - *Impatiens*
spotted snap-weed - *Impatiens capensis*
snapdragon
  dwarf snapdragon - *Chaenorhinum minus, Chaenorhinum*
  garden snapdragon - *Antirrhinum majus*
  lesser snapdragon - *Antirrhinum orontium*
  pale-snapdragon - *Impatiens pallida*
  small snapdragon - *Chaenorhinum minus*
  snapdragon - *Antirrhinum majus, Antirrhinum*
  spurred snapdragon - *Linaria*
  wild snapdragon - *Linaria vulgaris*
sneezeweed
  bitter sneezeweed - *Helenium amarum*
  orange sneezeweed - *Helenium hoopesii*
  purple-head sneezeweed - *Helenium flexuosum*
  sneezeweed - *Achillea ptarmica, Helenium autumnale,*
    H. microcephalum, Helenium
  sneezeweed yarrow - *Achillea ptarmica*
  southern sneezeweed - *Helenium flexuosum*
sneezewort
  sneezewort - *Achillea ptarmica*
  sneezewort yarrow - *Achillea ptarmica*
snow
  hills-of-snow - *Hydrangea arborescens*
  snow azalea - *Rhododendron mucronatum*
  snow gum - *Eucalyptus niphophila, E. pauciflora*
  snow-heather - *Erica carnea*
  snow-in-summer - *Cerastium tomentosum*
  snow-on-the-mountain - *Euphorbia marginata*
  snow pea - *Pisum sativum* var. *macrocarpon*
  snow trillium - *Trillium nivale*
snowball
  Peruvian snowball cactus - *Espostoa lanata*
  snowball-bush - *Viburnum opulus, V. opulus* var. *roseum*
  wild snowball - *Ceanothus americanus*
snowbell
  American snowbell - *Styrax americanus*
  big-leaf snowbell - *Styrax grandifolus*
  snowbell - *Styrax*
snowberry
  creeping snowberry - *Gaultheria hispidula*
  snowberry - *Chiococca alba, Symphoricarpos albus, S. albus*
    var. *laevigatus, Symphoricarpos*
  western snowberry - *Symphoricarpos occidentalis*
snowbush - *Ceanothus cordulatus*
snowdon-lily - *Lloydia serotina*
snowdrop
  snowdrop - *Galanthus nivalis*
  snowdrop-tree - *Halesia*
snowflake
  snowflake - *Lamium album, Leucojum aestivum*
  spring-snowflake - *Leucojum vernum*
  summer-snowflake - *Leucojum aestivum, Ornithogalum*
    umbellatum
snowy
  snowy campion - *Silene nivea*
  snowy cockle - *Silene noctiflora*
snuffbox fern - *Thelypteris palustris*
soap
  Hawaiian soap-tree - *Sapindus oahuensis*
  soap-bush - *Clidemia hirta*

soap-plant - *Chlorogalum pomeridianum, Zigadenus*
  venenosus
soap-tree - *Yucca elata*
soap-tree yucca - *Yucca elata*
soap-weed - *Yucca elata, Y. glauca*
soap-well - *Yucca glauca*
soapberry
  soapberry - *Sapindus, Shepherdia canadensis*
  southern soapberry - *Sapindus saponaria*
  western soapberry - *Sapindus saponaria* var. *drummondii*
  wing-leaf soapberry - *Sapindus saponaria*
soapwort
  cow soapwort - *Vaccaria hispanica*
  soapwort - *Saponaria officinalis*
  soapwort gentian - *Gentiana saponaria*
society-garlic - *Tulbaghia violacea*
soda-apple nightshade - *Solanum aculeatissimum*
Sodom, vine-of-Sodom - *Citrullus colocynthis*
soft
  creeping soft grass - *Holcus mollis*
  soft brome - *Bromus hordeaceus*
  soft chess - *Bromus hordeaceus*
  soft-flag - *Typha angustifolia*
  soft hemp - *Cannabis sativa*
  soft maple - *Acer rubrum, A. saccharinum*
  soft rush - *Juncus effusus*
  soft-stem bulrush - *Scirpus validus*
  soft sunflower - *Helianthus mollis*
softly, tread-softly - *Cnidoscolus stimulosus, C. texanus*
sojabean - *Glycine max*
sola, kat-sola - *Aeschynomene indica*
soldier
  soldier rose mallow - *Hibiscus laevis*
  soldier-weed - *Amaranthus spinosus, Colubrina elliptica*
soldier's-cap - *Aconitum napellus*
soledad pine - *Pinus torreyana*
solitary pussy-toes - *Antennaria solitaria*
Solomon
  fat-Solomon - *Smilacina racemosa*
  Solomon Island ivy - *Epipremnum aureum*
Solomon's
  false Solomon's-seal - *Smilacina racemosa*
  great Solomon's-seal - *Polygonatum biflorum, P.*
    commutatum
  Solomon's-seal - *Polygonatum biflorum, Polygonatum*
  Solomon's-zigzag - *Smilacina racemosa*
  starry false Solomon's-seal - *Smilacina stellata, S. trifolia*
  three-leaf false Solomon's seal - *Smilacina trifolia*
  two-leaf Solomon's-seal - *Maianthemum canadense*
Somali tea - *Catha edulis*
Sophia, herb-Sophia - *Descurainia sophia*
sophora
  silky sophora - *Sophora nuttalliana, S. sericea*
  Texas sophora - *Sophora affinis*
  vetch-leaf sophora - *Sophora davidii*
sorghum
  grain sorghum - *Sorghum bicolor*
  sorghum - *Sorghum bicolor*
  sorghum-almum - *Sorghum almum*
  sweet sorghum - *Sorghum bicolor*

sorrel
  creeping lady's-sorrel - *Oxalis corniculata*
  creeping wood-sorrel - *Oxalis corniculata*
  creeping yellow wood-sorrel - *Oxalis corniculata*
  dock sorrel - *Rumex*
  European wood-sorrel - *Oxalis acetosella, O. europaea*
  Florida yellow wood-sorrel - *Oxalis dillenii*
  garden sorrel - *Rumex acetosa*
  green sorrel - *Rumex acetosa*
  Indian sorrel - *Hibiscus sabdariffa*
  Jamaican sorrel - *Hibiscus sabdariffa*
  lady's-sorrel - *Oxalis*
  mountain-sorrel - *Oxyria digyna*
  northern wood-sorrel - *Oxalis acetosella*
  red sorrel - *Rumex acetosella*
  red wood-sorrel - *Oxalis oregana*
  sheep sorrel - *Rumex acetosella*
  sorrel - *Rumex acetosella, Rumex*
  sorrel-tree - *Oxydendrum arboreum*
  southern yellow wood-sorrel - *Oxalis dillenii*
  upright yellow sorrel - *Oxalis europaea*
  violet wood-sorrel - *Oxalis corymbosa, O. violacea*
  wood-sorrel - *Oxalis*
  yellow wood-sorrel - *Oxalis europaea, O. stricta*
sotol - *Dasylirion*
soup mint - *Plectranthus amboinicus*
sour
  sour-berry - *Rhus integrifolia*
  sour cherry - *Prunus cerasus*
  sour-clover - *Melilotus indica*
  sour dock - *Rumex acetosa, R. acetosella*
  sour grass - *Digitaria insularis, Paspalum conjugatum*
  sour-greens - *Rumex venosus*
  sour-gum - *Nyssa sylvatica*
  sour orange - *Citrus aurantium*
  sour paspalum - *Paspalum conjugatum*
  sour-top blueberry - *Vaccinium myrtilloides*
soursop - *Annona reticulata*
sourwood - *Oxydendrum arboreum*
South
  South American air-plant - *Kalanchoe fedtschenkoi*
  South American jelly palm - *Butia capitata*
  South American waterweed - *Egeria densa*
  South Sea ironwood - *Casuarina equisetifolia*
southern
  southern aster - *Aster hemisphericus*
  southern bayberry - *Myrica cerifera, M. heterophylla*
  southern black-haw - *Viburnum rufidulum*
  southern blackberry - *Rubus argutus*
  southern blue flag - *Iris virginica* var. *schrevei*
  southern brass-buttons - *Cotula australis*
  southern buckthorn - *Sideroxylon lycioides*
  southern bulrush - *Scirpus californicus*
  southern California walnut - *Juglans californica*
  southern cane - *Arundinaria gigantea*
  southern cat-tail - *Typha domingensis*
  southern catalpa - *Catalpa bignonioides*
  southern chervil - *Chaerophyllum tainturieri*
  southern cole - *Brassica juncea*
  southern cottonwood - *Populus deltoides*
  southern crabapple - *Malus angustifolia*
  southern crabgrass - *Digitaria ciliaris*

  southern cut grass - *Leersia hexandra*
  southern dewberry - *Rubus trivialis*
  southern dodder - *Cuscuta obtusiflora*
  southern flat-seed-sunflower - *Verbesina occidentalis*
  southern fox grape - *Vitis rotundifolia*
  southern hackberry - *Celtis laevigata*
  southern harebell - *Campanula divaricata*
  southern high-bush blueberry - *Vaccinium elliottii*
  southern leatherwood - *Cyrilla racemiflora*
  southern live oak - *Quercus virginiana*
  southern magnolia - *Magnolia grandiflora*
  southern monk's-hood - *Aconitum uncinatum*
  southern naiad - *Najas guadalupensis*
  southern nettle - *Urtica chamaedryoides*
  southern pea - *Vigna unguiculata*
  southern pine - *Pinus palustris*
  southern pitcher-plant - *Sarracenia purpurea*
  southern prickly-ash - *Zanthoxylum clava-herculis*
  southern red-cedar - *Juniperus silicicola*
  southern red lily - *Lilium catesbaei*
  southern red oak - *Quercus falcata*
  southern sandbur - *Cenchrus echinatus*
  southern-scabious - *Succisa australis*
  southern sida - *Sida acuta*
  southern sneezeweed - *Helenium flexuosum*
  southern soapberry - *Sapindus saponaria*
  southern sugar maple - *Acer barbatum*
  southern sundrops - *Oenothera fruticosa*
  southern swamp crinum - *Crinum americanum*
  southern swamp dogwood - *Cornus stricta*
  southern water grass - *Luziola fluitans*
  southern water-nymph - *Najas guadalupensis*
  southern wax myrtle - *Myrica cerifera*
  southern white-cedar - *Chamaecyparis thyoides*
  southern wild crabapple - *Malus angustifolia*
  southern wild rice - *Zizaniopsis miliacea*
  southern wild senna - *Senna marilandica*
  southern wormwood - *Artemisia abrotanum*
  southern yellow pine - *Pinus palustris*
  southern yellow wood-sorrel - *Oxalis dillenii*
southernwood - *Artemisia abrotanum*
southwestern
  southwestern carrot - *Daucus pusillus*
  southwestern chokecherry - *Prunus serotina* subsp. *virens*
  southwestern coral bean - *Erythrina flabelliformis*
  southwestern rabbit-brush - *Chrysothamnus pulchellus*
sow
  annual sow-thistle - *Sonchus oleraceus*
  creeping sow-thistle - *Sonchus arvensis*
  field sow-thistle - *Sonchus arvensis*
  marsh sow-thistle - *Sonchus arvensis* spp. *uliginosus*
  perennial sow-thistle - *Sonchus arvensis*
  prickly sow-thistle - *Sonchus asper*
  sow-bane - *Chenopodium hybridum, C. murale*
  sow-teat blackberry - *Rubus allegheniensis*
  sow-teat strawberry - *Fragaria vesca*
  sow-thistle - *Sonchus oleraceus, Sonchus*
  spiny-leaved sow-thistle - *Sonchus asper*
  spiny sow-thistle - *Sonchus asper*
soybean - *Glycine max*
spaghetti squash - *Cucurbita pepo*

Spanish
  Spanish arrowroot - *Canna indica*
  Spanish bayonet - *Yucca aloifolia, Y. baccata*
  Spanish brome - *Bromus madritensis*
  Spanish brome grass - *Bromus madritensis*
  Spanish broom - *Spartium junceum*
  Spanish cherry - *Mimusops elengi*
  Spanish chestnut - *Castanea sativa*
  Spanish-dagger - *Yucca gloriosa*
  Spanish grape - *Vitis berlandieri*
  Spanish iris - *Iris xiphium*
  Spanish lime - *Melicoccus bijugatus*
  Spanish mahogany - *Swietenia mahagoni*
  Spanish-moss - *Tillandsia usneoides*
  Spanish-needles - *Bidens bipinnata, B. pilosa, Bidens*
  Spanish oak - *Quercus palustris*
  Spanish onion - *Allium fistulosum*
  Spanish oyster-plant - *Scorzonera hispanica*
  Spanish plum - *Spondias purpurea*
  Spanish red oak - *Quercus falcata*
  Spanish salsify - *Scorzonera hispanica*
  Spanish-tea - *Chenopodium ambrosioides*
  Spanish thyme - *Lippia micromera, Plectranthus amboinicus*
  Spanish tick-clover - *Desmodium uncinatum*
sparkleberry - *Vaccinium arboreum*
spatter
  Rocky Mountain spatter-dock - *Nuphar lutea* subsp. *polysepala*
  spatter-dock - *Nuphar lutea* subsp. *advena, Nuphar*
spatulate-leaved heliotrope - *Heliotropium curassavicum* var. *obovatum*
spear
  cardinal-spear - *Erythrina herbacea*
  king's-spear - *Eremurus robustus*
  low spear grass - *Poa annua*
  spear grass - *Poa pratensis, Stipa*
  spear-leaf violet - *Viola hastata*
  spear oracle - *Atriplex patula*
  spear thistle - *Cirsium virginianum*
  tufted spear grass - *Eragrostis pectinacea*
spearmint
  Scotch spearmint - *Mentha gentilis*
  spearmint - *Mentha spicata*
spearwort
  creeping spearwort - *Ranunculus reptans*
  spearwort - *Ranunculus flammula*
speckled alder - *Alnus incana, A. rugosa*
speedwell
  beach speedwell - *Veronica longifolia*
  corn speedwell - *Veronica arvensis*
  creeping speedwell - *Veronica filiformis, V. serpyllifolia* subsp. *humifusa*
  field speedwell - *Veronica agrestis*
  germander speedwell - *Veronica chamaedrys*
  ivy-leaf speedwell - *Veronica hederifolia*
  long-leaf speedwell - *Veronica longifolia*
  marsh speedwell - *Veronica scutellata*
  Persian speedwell - *Veronica persica*
  purslane speedwell - *Veronica peregrina, V. peregrina* subsp. *xalapensis*
  showy speedwell - *Hebe speciosa*
  slender speedwell - *Veronica filiformis*

  speedwell - *Veronica officinalis, Veronica*
  spike speedwell - *Veronica spicata*
  thyme-leaved speedwell - *Veronica serpyllifolia* subsp. *humifusa*
  water speedwell - *Veronica anagallis-aquatica*
  wayside speedwell - *Veronica polita*
  western purslane speedwell - *Veronica peregrina* subsp. *xalapensis*
  winter speedwell - *Veronica persica*
spelt - *Triticum spelta*
spicate indigo - *Indigofera spicata*
spice
  Indian spice - *Vitex agnus-castus*
  pond-spice - *Litsea aestivalis*
  spice-lily - *Manfreda maculosa*
spiceberry - *Ardisia crenata*
spicebush - *Calycanthus occidentalis, Lindera benzoin*
spider
  red spider-lily - *Lycoris radiata*
  spider-flower - *Cleome hassleriana, Grevillea*
  spider grass - *Aristida ternipes*
  spider-ivy - *Chlorophytum comosum*
  spider-lily - *Crinum, Hymenocallis*
  spider lupine - *Lupinus benthamii*
  spider-plant - *Chlorophytum comosum, Cleome*
  spiny spider-flower - *Cleome spinosa*
spiderling
  Coulter's spiderling - *Boerhavia coulteri*
  erect spiderling - *Boerhavia erecta*
  red spiderling - *Boerhavia diffusa*
  spiderling - *Boerhavia*
spiderwort
  purple-leaved spiderwort - *Tradescantia spathacea*
  smooth spiderwort - *Tradescantia ohiensis*
  spiderwort - *Tradescantia ohiensis, T. virginiana*
  tropical spiderwort - *Commelina benghalensis*
  Virginia spiderwort - *Tradescantia virginiana*
  white-flowered spiderwort - *Tradescantia fluminensis*
spike
  beaked spike rush - *Eleocharis rostellata*
  blunt spike rush - *Eleocharis ovata*
  creeping spike rush - *Eleocharis palustris*
  dwarf spike rush - *Eleocharis parvula*
  Gulf Coast spike rush - *Eleocharis cellulosa*
  needle spike rush - *Eleocharis acicularis*
  one-spike oat grass - *Danthonia unispicata*
  Oregon spike-moss - *Selaginella oregana*
  ovoid spike rush - *Eleocharis ovata*
  pale-spike lobelia - *Lobelia spicata*
  rock spike-moss - *Selaginella rupestris*
  saffron-spike - *Aphelandra squarrosa*
  slender spike rush - *Eleocharis acicularis, E. baldwinii*
  slim-spike foxtail - *Alopecurus myosuroides*
  slim-spike three-awn - *Aristida longespica*
  small spike rush - *Eleocharis parvula*
  spike drop-seed - *Sporobolus contractus*
  spike-moss - *Selaginella*
  spike muhly - *Muhlenbergia wrightii*
  spike-primrose - *Boisduvalia densiflora*
  spike redtop - *Agrostis exarata*
  spike rush - *Eleocharis*
  spike speedwell - *Veronica spicata*

spike trisetum - *Trisetum spicatum*
spike-weed - *Hemizonia pungens*
sprouting spike rush - *Eleocharis vivipara*
square-stem spike rush - *Eleocharis quadrangulata*
thick-spike blazing-star - *Liatris pycnostachya*
thick-spike wheat grass - *Elymus lanceolatus*
Wallace's spike-moss - *Selaginella wallacei*
Watson's spike-moss - *Selaginella watsonii*
spiked
    dry-spiked sedge - *Carex siccata*
    naked-spiked ragweed - *Ambrosia psilostachya*
    spiked lobelia - *Lobelia spicata*
    spiked loosestrife - *Lythrum salicaria*
spikenard
    American spikenard - *Aralia racemosa*
    false spikenard - *Smilacina racemosa*
    spikenard - *Aralia cordata, A. racemosa*
spinach
    country-spinach - *Basella alba*
    Indian spinach - *Basella alba*
    Malabar spinach - *Basella alba*
    New Zealand spinach - *Tetragonia tetragonioides*
    spinach - *Spinacia oleracea*
    spinach beet - *Beta vulgaris* subsp. *cicla*
    spinach dock - *Rumex patientia*
    strawberry-spinach - *Chenopodium capitatum*
    vine-spinach - *Basella alba*
    water-spinach - *Ipomoea aquatica*
    wild spinach - *Chenopodium bonus-henricus*
spindle
    European spindle-tree - *Euonymus europaeus*
    Japanese spindle-tree - *Euonymus japonicus*
    spindle-tree - *Euonymus*
    winged spindle-tree - *Euonymus alatus*
spine
    long-spine sandbur - *Cenchrus longispinus*
    yellow-spine thistle - *Cirsium ochrocentrum*
spineless
    spineless cactus - *Opuntia ficus-indica*
    spineless horse-brush - *Tetradymia canescens*
    spineless yucca - *Yucca elephantipes*
spinulose wood fern - *Dryopteris carthusiana*
spiny
    spiny amaranth - *Amaranthus spinosus*
    spiny aster - *Chloranthus spinosa*
    spiny ceanothus - *Ceanothus spinosus*
    spiny cholla - *Opuntia spinosior*
    spiny-club palm - *Bactris*
    spiny cocklebur - *Xanthium spinosum*
    spiny emex - *Emex spinosa*
    spiny-fruit buttercup - *Ranunculus muricatus*
    spiny hackberry - *Celtis pallida*
    spiny-leaved sow-thistle - *Sonchus asper*
    spiny pigweed - *Amaranthus spinosus*
    spiny red-berry - *Rhamnus crocea*
    spiny saltbush - *Atriplex confertifolia*
    spiny sow-thistle - *Sonchus asper*
    spiny spider-flower - *Cleome spinosa*
spiraea
    Japanese spiraea - *Spiraea japonica*
    perennial spiraea - *Astilbe*
    rock-spiraea - *Holodiscus discolor, H. dumosus*

spiral
    spiral eucalyptus - *Eucalyptus cinerea*
    spiral-flag - *Costus malortieanus*
    spiral-ginger - *Costus malortieanus*
spire, sweet-spire - *Itea virginica*
spleen amaranthus - *Amaranthus hybridus*
spleenwort
    maidenhair spleenwort - *Asplenium trichomanes*
    silvery spleenwort - *Athyrium thelypteroides*
    spleenwort - *Asplenium*
    spleenwort-bush - *Comptonia peregrina*
splendid everlasting - *Helipterum splendidum*
sponge
    sponge gourd - *Luffa aegyptiaca*
    sponge-tree - *Acacia farnesiana*
    vegetable-sponge - *Luffa aegyptiaca*
spoon-leaf yucca - *Yucca filamentosa*
spoonflower - *Dasylirion wheeleri, Peltandra sagittifolia*
spoonwood - *Kalmia latifolia*
spotted
    spotted bee balm - *Monarda punctata*
    spotted bladderwort - *Utricularia purpurea*
    spotted bur-clover - *Medicago arabica*
    spotted cat's-ear - *Hypochaeris radicata*
    spotted coralroot - *Corallorrhiza maculata*
    spotted cowbane - *Cicuta maculata*
    spotted crane's-bill - *Geranium maculatum*
    spotted dead-nettle - *Lamium maculatum*
    spotted dracaena - *Dracaena surculosa*
    spotted dumb-cane - *Dieffenbachia maculata*
    spotted geranium - *Geranium maculatum*
    spotted-hemlock - *Conium maculatum*
    spotted Joe-Pye-weed - *Eupatorium maculatum*
    spotted knapweed - *Centaurea maculosa*
    spotted loco - *Astragalus lentiginosus*
    spotted locoweed - *Astragalus lentiginosus*
    spotted loosestrife - *Lysimachia punctata*
    spotted medic - *Medicago arabica*
    spotted phlox - *Phlox nivalis*
    spotted snap-weed - *Impatiens capensis*
    spotted spurge - *Euphorbia maculata*
    spotted St. John's-wort - *Hypericum punctatum*
    spotted touch-me-not - *Impatiens capensis*
    spotted water-hemlock - *Cicuta maculata*
    spotted water-meal - *Wolffia punctata*
    spotted wintergreen - *Chimaphila maculata*
sprangle
    Amazon sprangle-top - *Leptochloa panicoides*
    bearded sprangle-top - *Leptochloa fascicularis*
    Chinese sprangle-top - *Leptochloa chinensis*
    green sprangle-top - *Leptochloa dubia*
    Mexican sprangle-top - *Leptochloa uninervia*
    red sprangle-top - *Leptochloa filiformis*
    sprangle-top - *Leptochloa, Scolochloa festucacea*
sprawling
    sprawling horseweed - *Calyptocarpus vialis*
    sprawling-panicum - *Urochloa reptans*
spray, ocean-spray - *Holodiscus discolor*
spreading
    spreading dayflower - *Commelina diffusa*
    spreading dogbane - *Apocynum androsaemifolium*

spreading evening-primrose - *Oenothera humifusa*
spreading globeflower - *Trollius laxus*
spreading Jacob's-ladder - *Polemonium reptans*
spreading knapweed - *Centaurea diffusa*
spreading mustard - *Erysimum repandum*
spreading orache - *Atriplex patula*
spreading pigweed - *Amaranthus graecizans*
spreading prickly-pear - *Opuntia compressa*
spreading scale-seed - *Spermolepis divaricata*
spreading sweet-root - *Osmorhiza chilensis*
spreading witch grass - *Panicum dichotomiflorum*
spreading wood fern - *Dryopteris campyloptera*
spreading yellow-cress - *Rorippa sinuata*
Sprenger's asparagus - *Asparagus densiflorus*
spring
spring blue spring daisy - *Erigeron pulchellus*
broad-leaf spring-beauty - *Claytonia caroliniana*
Carolina spring-beauty - *Claytonia caroliniana*
farewell-to-spring - *Clarkia amoena, Clarkia*
harbinger-of-spring - *Erigenia bulbosa*
ibapah spring parsley - *Cymopterus ibapensis*
narrow-leaf spring-beauty - *Claytonia virginica*
spring adonis - *Adonis vernalis*
spring-beauty - *Claytonia megarrhiza, C. virginica, Claytonia*
spring cockle - *Vaccaria hispanica*
spring-cress - *Cardamine bulbosa*
spring forget-me-not - *Myosotis verna*
spring-gold - *Lomatium utriculatum*
spring heath - *Erica carnea*
spring larkspur - *Delphinium tricorne*
spring millet - *Milium vernale*
spring onion - *Allium fistulosum*
spring parsley - *Cymopterus ibapensis*
spring-snowflake - *Leucojum vernum*
spring-tape - *Sagittaria subulata*
spring-vetch - *Vicia sativa*
spring whitlow-grass - *Draba verna*
sprite, water-sprite - *Ceratopteris thalictroides*
sprouting
sprouting broccoli - *Brassica oleracea* var. *italica*
sprouting-crabgrass - *Panicum dichotomiflorum*
sprouting-leaf - *Kalanchoe pinnata*
sprouting spike rush - *Eleocharis vivipara*
sprouts, Brussels sprouts - *Brassica oleracea* var.
 gemmifera
spruce
big-cone-spruce - *Pseudotsuga macrocarpa*
black spruce - *Picea mariana*
bog spruce - *Picea mariana*
Brewer's spruce - *Picea breweriana*
cat spruce - *Picea glauca*
Colorado blue spruce - *Picea pungens*
Colorado spruce - *Picea pungens*
double spruce - *Picea mariana*
Engelmann's spruce - *Picea engelmannii*
hemlock-spruce - *Tsuga*
Norway spruce - *Picea abies*
red spruce - *Picea rubens*
Sitka spruce - *Picea sitchensis*
spruce - *Picea*
spruce pine - *Pinus echinata, P. glabra, P. virginiana*
weeping spruce - *Picea breweriana*
white spruce - *Picea glauca*

spur
hook-spur violet - *Viola adunca*
spur-flower - *Plectranthus*
spurge
Allegheny-spurge - *Pachysandra procumbens*
broad-leaf spurge - *Euphorbia lucida*
caper spurge - *Euphorbia lathyris*
creeping spurge - *Euphorbia serpens*
cypress spurge - *Euphorbia cyparissias*
flowering spurge - *Euphorbia corollata*
garden spurge - *Euphorbia hirta*
gopher spurge - *Euphorbia lathyris*
ground spurge - *Euphorbia prostrata*
hairy spurge - *Euphorbia vermiculata*
hyssop spurge - *Euphorbia hyssopifolia*
ipecac spurge - *Euphorbia ipecacuanhae*
Japanese-spurge - *Pachysandra terminalis*
leafy spurge - *Euphorbia esula*
little-leaf spurge - *Euphorbia micromera*
myrtle spurge - *Euphorbia lathyris*
net-seed spurge - *Euphorbia spathulata*
nodding spurge - *Euphorbia maculata, E. nutans*
painted spurge - *Euphorbia heterophylla*
petty spurge - *Euphorbia peplus*
prairie spurge - *Euphorbia spathulata*
prostrate spurge - *Euphorbia humistrata*
ridge-seed spurge - *Euphorbia glyptosperma*
round-leaf spurge - *Euphorbia serpens*
salvers spurge - *Euphorbia cyparissias*
saw-tooth spurge - *Euphorbia serrata*
seaside spurge - *Euphorbia polygonifolia*
shining spurge - *Euphorbia lucida*
spotted spurge - *Euphorbia maculata*
spurge - *Euphorbia*
spurge-nettle - *Cnidoscolus stimulosus*
stubble spurge - *Euphorbia maculata*
sun spurge - *Euphorbia helioscopia*
thyme-leaved spurge - *Euphorbia serpyllifolia*
tinted spurge - *Euphorbia commutata*
toothed spurge - *Euphorbia dentata, E. serrata*
tramp's spurge - *Euphorbia corollata*
wart spurge - *Euphorbia helioscopia*
white-margin spurge - *Euphorbia albomarginata*
wood spurge - *Euphorbia commutata*
spurred
great-spurred violet - *Viola selkirkii*
long-spurred violet - *Viola rostrata*
spurred anoda - *Anoda cristata*
spurred butterfly pea - *Centrosema virginianum*
spurred-gentian - *Halenia elliptica*
spurred snapdragon - *Linaria*
spurry
corn spurry - *Spergula arvensis*
red-sands spurry - *Spergularia rubra*
spurry - *Spergula*
umbrella-spurry - *Holosteum umbellatum*
square
carpenter's-square - *Scrophularia marilandica*
square-nut - *Carya tomentosa*
square-pod water-primrose - *Ludwigia alternifolia*
square-stem spike rush - *Eleocharis quadrangulata*
square-stemmed monkey-flower - *Mimulus ringens*

three-square - *Scirpus americanus*
three-square bur-reed - *Sparganium americanum*
squarrose
    squarrose knapweed - *Centaurea squarrosa*
    squarrose umbrella-grass - *Fuirena squarrosa*
    squarrose white aster - *Aster ericoides*
squash
    autumn squash - *Cucurbita maxima*
    bush squash - *Cucurbita pepo* var. *melopepo*
    crook-neck squash - *Cucurbita moschata*
    spaghetti squash - *Cucurbita pepo*
    squash - *Cucurbita maxima, Cucurbita*
    squash-berry - *Viburnum edule*
    summer squash - *Cucurbita pepo*
    winter squash - *Cucurbita maxima, C. moschata, C. pepo*
squaw
    squaw-apple - *Peraphyllum ramosissimum*
    squaw-berry - *Mitchella repens*
    squaw-bush - *Condalia spathulata, Rhus aromatica*
    squaw currant - *Ribes cereum*
    squaw-grass - *Xerophyllum tenax*
    squaw huckleberry - *Vaccinium caesium, V. stamineum*
    squaw-mint - *Hedeoma pulegioides*
    squaw-weed - *Ageratina altissima, Senecio aureus, Senecio*
squawroot - *Perideridia gairdneri, Trillium erectum*
squill
    blue-flower squill - *Hyacinthoides hispanica*
    indigo-squill - *Camassia scilloides*
    Siberian squill - *Scilla sibirica*
    squill - *Scilla*
squirrel
    squirrel-corn - *Dicentra canadensis*
    squirrel-foot fern - *Davallia trichomanoides*
    squirrel-tail barley - *Hordeum jubatum*
    squirrel-tail fescue - *Vulpia bromoides*
squirrel's-tail - *Elymus elymoides*
St.
    Blue Ridge St. John's-wort - *Hypericum mitchellianum*
    Canadian St. John's-wort - *Hypericum canadense*
    creeping St. John's-wort - *Hypericum adpressum,*
      H. calycinum
    dwarf St. John's-wort - *Hypericum mutilum*
    golden St. John's-wort - *Hypericum frondosum*
    great St. John's-wort - *Hypericum pyramidatum*
    Kalm's St. John's-wort - *Hypericum kalmianum*
    marsh St. John's-wort - *Hypericum tubulosum, H. virginicum*
    mountain St. John's-wort - *Hypericum mitchellianum*
    shrubby St. John's-wort - *Hypericum prolificum*
    spotted St. John's-wort - *Hypericum punctatum*
    St. Andrew's-cross - *Hypericum hypericoides*
    St. Augustine's grass - *Stenotaphrum secundatum*
    St. Barbara's cress - *Barbarea vulgaris*
    St. Catharine's-lace - *Eriogonum giganteum*
    St. John's-bread - *Ceratonia siliqua*
    St. John's-wort - *Hypericum perforatum, Hypericum*
    St. Lucie cherry - *Prunus mahaleb*
    St. Mary's thistle - *Silybum marianum*
    St. Peter's-wort - *Hypericum crux-andreae*
    St. Vincent's cistus - *Cistus palhinhai*
staff
    Jacob's-staff - *Fouquieria splendens*
    staff-vine - *Celastrus scandens*

stag-bush - *Viburnum prunifolium*
stag's garlic - *Allium vineale*
stagger
    stagger-bush - *Lyonia ferruginea, L. mariana*
    stagger-weed - *Delphinium menziesii*
    stagger-wort - *Helenium autumnale*
staghorn
    staghorn cholla - *Opuntia versicolor*
    staghorn-evergreen - *Lycopodium clavatum*
    staghorn fern - *Platycerium bifurcatum*
    staghorn sumac - *Rhus hirta*
stalk
    bristle-stalk sedge - *Carex leptalea*
    rose twisted-stalk - *Streptopus roseus*
    rough-stalk bluegrass - *Poa palustris, P. trivialis*
    short-stalk copper-leaf - *Acalypha gracilens*
    twisted-stalk - *Streptopus amplexifolius, Streptopus*
stalked
    long-stalked aster - *Aster dumosus*
    long-stalked crane's-bill - *Geranium columbinum*
    long-stalked sedge - *Carex pedunculata*
stamen
    five-stamen tamarisk - *Tamarix chinensis*
    purple-stamen mullein - *Verbascum virgatum*
stanch, blood-stanch - *Conyza canadensis*
standard crested wheat grass - *Agropyron desertorum*
standing
    standing bugle - *Ajuga genevensis*
    standing milk-vetch - *Astragalus adsurgens*
staple, extra-long staple cotton - *Gossypium barbadense*
star
    alpine shooting-star - *Dodecatheon alpinum*
    blazing-star - *Liatris spicata, Liatris, Mentzelia laevicaulis,*
      Mentzelia
    blue-star - *Amsonia tabernaemontana, Amsonia*
    bristly star-bur - *Acanthospermum hispidum*
    Christmas-star - *Euphorbia pulcherrima*
    cylindric blazing-star - *Liatris cylindracea*
    dense blazing-star - *Liatris spicata*
    dotted blazing-star - *Liatris punctata*
    earth-star - *Cryptanthus*
    eastern shooting-star - *Dodecatheon meadia*
    Egyptian star-cluster - *Pentas lanceolata*
    few-headed blazing-star - *Liatris cylindracea*
    Iberian star-thistle - *Centaurea iberica*
    Malta star-thistle - *Centaurea melitensis*
    nodding star-of-Bethlehem - *Ornithogalum nutans*
    Paraguay star-bur - *Acanthospermum australe*
    plains blazing-star - *Liatris squarrosa*
    prairie blazing-star - *Liatris pycnostachya*
    purple star-thistle - *Centaurea calcitrapa*
    rainbow-star - *Cryptanthus bromelioides* var. *tricolor*
    scaly blazing-star - *Liatris squarrosa*
    sessile blazing-star - *Liatris spicata*
    shooting-star - *Dodecatheon conjugens, D. meadia,*
      Dodecatheon
    sierra star-tulip - *Calochortus nudus*
    star cactus - *Astrophytum ornatum, Astrophytum*
    star chickweed - *Stellaria pubera*
    star-cluster - *Pentas lanceolata*
    star-cucumber - *Sicyos angulatus*
    star duckweed - *Lemna trisulca*

star-flowered lily-of-the-valley - *Smilacina stellata*
star-fruit - *Averrhoa carambola*
star-glory - *Ipomoea quamoclit*
star grass - *Aletris*
star ipomoea - *Ipomoea coccinea*
star jasmine - *Jasminum multiflorum, J. nitidum*
star-leaf - *Schefflera actinophylla*
star-lily - *Leucocrinum montanum, Leucocrinum*
star magnolia - *Magnolia stellata*
star-of-Bethlehem - *Ornithogalum umbellatum*
star-tulip - *Calochortus*
thick-spike blazing-star - *Liatris pycnostachya*
water star grass - *Heteranthera dubia*
woodland star - *Lithophragma affine*
yellow-star - *Helenium autumnale*
yellow star-thistle - *Centaurea solstitialis*
starfish-plant - *Cryptanthus acaulis*
starflower - *Smilacina stellata, Trientalis borealis, Trientalis*
starry
starry campion - *Silene stellata*
starry false Solomon's-seal - *Smilacina stellata, S. trifolia*
starry glasswort - *Cerastium arvense*
stars, bog-stars - *Parnassia californica, Parnassia*
starved aster - *Aster lateriflorus*
starwort
Easter-bell starwort - *Stellaria holostea*
European water-starwort - *Callitriche stagnalis*
lesser starwort - *Stellaria graminea*
little starwort - *Stellaria graminea*
long-leaf starwort - *Stellaria longifolia*
starwort - *Aster, Spergula arvensis, Stellaria longipes, Stellaria*
water-starwort - *Callitriche verna, Callitriche*
yellow starwort - *Inula helenium*
statice - *Limonium*
staves-acre - *Delphinium menziesii*
steeplebush - *Spiraea tomentosa*
stellate sedge - *Carex rosea*
Steller's rock brake - *Cryptogramma stelleri*
stepladder-plant - *Costus malortieanus*
steppe-cabbage - *Rapistrum perenne*
sterile
sterile brome - *Bromus sterilis*
sterile oat - *Avena sterilis*
stevia - *Piqueria trinervia*
stewartia
mountain stewartia - *Stewartia ovata*
silky stewartia - *Stewartia melachodendron*
Virginia stewartia - *Stewartia melachodendron*
stick
desert stick-leaf - *Mentzelia multiflora*
devil's-walking-stick - *Aralia spinosa*
stick-leaf - *Mentzelia oligosperma*
stick-tights - *Bidens cernua, Bidens*
ten-petal stick-leaf - *Mentzelia decapetala*
walking-stick cholla - *Opuntia imbricata*
white-stem stick-leaf - *Mentzelia albicaulis*
stickseed
stickseed - *Hackelia floribunda, Hackelia, Lappula redowskii*
western stickseed - *Hackelia floribunda*

sticktight
sticktight - *Bidens frondosa*
western sticktight - *Lappula occidentalis*
stickweed - *Agrimonia gryposepala, Hackelia diffusa*
stickwort - *Spergula arvensis*
sticky
sticky baccharis - *Baccharis salicifolia*
sticky chickweed - *Cerastium glomeratum*
sticky cockle - *Silene noctiflora*
sticky currant - *Ribes viscosissimum*
sticky groundsel - *Senecio viscosus*
sticky hawkweed - *Hieracium scabrum*
sticky nightshade - *Solanum sisymbriifolium*
sticky seymeria - *Seymeria pectinata*
sticky snakeweed - *Gutierrezia microcephala*
stiff
stiff aster - *Aster linariifolius*
stiff club-moss - *Lycopodium annotinum*
stiff coreopsis - *Coreopsis palmata*
stiff-cornel dogwood - *Cornus stricta*
stiff dogwood - *Cornus stricta*
stiff gentian - *Gentianella quinquefolia*
stiff goldenrod - *Solidago rigida*
stiff-haired sunflower - *Helianthus hirsutus*
stiff sunflower - *Helianthus pauciflorus*
stiff tick-trefoil - *Desmodium strictum*
still, be-still-tree - *Thevetia peruviana*
stinging nettle - *Urtica dioica*
stink
stink currant - *Ribes bracteosum*
stink grass - *Eragrostis cilianensis*
stink-wort - *Datura stramonium*
stinking
stinking-ash - *Ptelea trifoliata*
stinking-Benjamin - *Trillium erectum*
stinking-cedar - *Torreya taxifolia*
stinking-chamomile - *Anthemis cotula*
stinking-clover - *Cleome serrulata*
stinking-daisy - *Anthemis cotula*
stinking fleabane - *Pluchea foetida*
stinking goosefoot - *Chenopodium vulvaria*
stinking marigold - *Dyssodia papposa*
stinking nightshade - *Hyoscyamus niger*
stinking-palm - *Hedeoma pulegioides*
stinking pepper-weed - *Lepidium ruderale*
stinking tarweed - *Madia glomerata*
stinking-Willie - *Senecio jacobaea*
stinkweed
poison stinkweed - *Conium maculatum*
stinkweed - *Anthemis cotula, Dyssodia papposa, Pluchea camphorata, Thlaspi arvense*
stipa, western stipa - *Stipa comata*
stitch
greater stitch-wort - *Stellaria holostea*
lesser stitch-wort - *Stellaria graminea*
stitch-wort - *Stellaria media*
stock
Brampton's stock - *Matthiola incana*
evening-scented stock - *Matthiola longipetala*
evening stock - *Matthiola longipetala*
giant stock bean - *Canavalia ensiformis*

imperial stock - *Matthiola incana*
malcolm stock - *Malcolmia africana*
night-scented stock - *Matthiola longipetala*
stock - *Matthiola incana, Matthiola*
stock melon - *Citrullus lanatus* var. *citroides*
ten-weeks stock - *Matthiola incana* var. *annua*
Stokes's aster - *Stokesia laevis*
stone
  cherry-stone juniper - *Juniperus monosperma*
  Italian stone pine - *Pinus pinea*
  living-stone - *Lithops hookeri*
  Mexican stone pine - *Pinus cembroides*
  stone clover - *Trifolium arvense*
  stone leek - *Allium fistulosum*
  stone mint - *Cunila origanoides*
  stone pine - *Pinus monophylla*
  stone-root - *Collinsonia canadensis*
  stone-rush - *Scleria*
  stone-seed - *Lithospermum arvense*
  Swiss stone pine - *Pinus cembra*
stonecrop
  ditch-stonecrop - *Penthorum sedoides*
  live-forever stonecrop - *Sedum telephium*
  mossy stonecrop - *Sedum acre*
  stonecrop - *Sedum*
  two-row stonecrop - *Sedum spurium*
stonewort, cliff stonewort - *Sedum glaucophyllum*
storax - *Styrax*
stork's-bill - *Erodium, Pelargonium*
stout wood-reed - *Cinna arundinacea*
stramonium thorn-apple - *Datura stramonium*
strand-wheat - *Lolium perenne*
strangler
  Florida strangler fig - *Ficus aurea*
  strangler-vine - *Morrenia odorata*
strap
  strap fern - *Polypodium phyllitidis*
  strap-leaf violet - *Viola lanceolata*
straw
  marsh straw sedge - *Carex tenera*
  straw-colored cyperus - *Cyperus strigosus*
  straw-flower - *Helichrysum bracteatum, Helipterum*
  straw foxglove - *Digitalis lutea*
  straw sedge - *Carex straminea*
strawberry
  alpine strawberry - *Fragaria vesca*
  barren-strawberry - *Waldsteinia fragarioides*
  Chilean strawberry - *Fragaria chiloensis*
  cultivated strawberry - *Fragaria ananassa*
  European strawberry - *Fragaria vesca*
  false strawberry - *Duchesnea indica*
  garden strawberry - *Fragaria ananassa*
  Indian mock-strawberry - *Duchesnea indica*
  Indian strawberry - *Duchesnea indica*
  mock-strawberry - *Duchesnea indica*
  mountain strawberry - *Fragaria virginiana*
  purple strawberry guava - *Psidium littorale* var. *longipes*
  running strawberry-bush - *Euonymus obovatus*
  sow-teat strawberry - *Fragaria vesca*
  strawberry - *Fragaria*
  strawberry-begonia - *Saxifraga stolonifera*

strawberry-blite - *Chenopodium capitatum*
strawberry-bush - *Euonymus americanus*
strawberry cactus - *Mammillaria*
strawberry clover - *Trifolium fragiferum*
strawberry-geranium - *Saxifraga stolonifera*
strawberry ground-cherry - *Physalis alkekengi*
strawberry-leaf cinquefoil - *Potentilla sterilis*
strawberry pigweed - *Chenopodium capitatum*
strawberry potentilla - *Potentilla sterilis*
strawberry-shrub - *Calycanthus floridus*
strawberry-spinach - *Chenopodium capitatum*
strawberry tomato - *Physalis peruviana, P. pubescens,*
  P. pubescens var. *grisea*
strawberry-tree - *Arbutus unedo*
Virginia strawberry - *Fragaria virginiana*
wild strawberry - *Fragaria virginiana*
wood strawberry - *Fragaria vesca*
woodland strawberry - *Fragaria vesca, F. vesca* subsp.
  americana
stream violet - *Viola glabella*
striate
  striate clover - *Trifolium striatum*
  striate knotweed - *Polygonum achoreum*
  striate lespedeza - *Lespedeza striata*
string bean - *Phaseolus vulgaris*
stringy
  mealy stringy-bark - *Eucalyptus cinerea*
  messmate stringy-bark - *Eucalyptus obliqua*
  stringy-bark - *Eucalyptus*
striped
  striped crotalaria - *Crotalaria pallida*
  striped gentian - *Gentiana villosa*
  striped maple - *Acer pensylvanicum*
  striped toadflax - *Linaria repens*
  striped violet - *Viola striata*
strong-scented pigweed - *Chenopodium ambrosioides*
stubble spurge - *Euphorbia maculata*
Stueve's bush-clover - *Lespedeza stuevei*
stunt grape - *Vitis labrusca*
Sturt's desert-rose - *Gossypium sturtianum*
styled
  knob-styled dogwood - *Cornus amomum*
  long-styled anise-root - *Osmorhiza longistylis*
stylo
  Brazilian stylo - *Stylosanthes guianensis*
  Caribbean stylo - *Stylosanthes hamata*
  Nigerian stylo - *Stylosanthes erecta*
  Townsville stylo - *Stylosanthes humilis*
styptic-weed - *Senna occidentalis*
subalpine
  subalpine fir - *Abies lasiocarpa*
  subalpine larch - *Larix lyallii*
subterranean clover - *Trifolium subterraneum*
succory
  gum succory - *Chondrilla juncea*
  lamb-succory - *Arnoseris minima*
  succory - *Cichorium intybus*
  succory-dock - *Lapsana communis*
Sudan grass - *Sorghum drummondii*
sugar
  black sugar maple - *Acer nigrum*

horse-sugar - *Symplocos tinctoria*
southern sugar maple - *Acer barbatum*
sugar-apple - *Annona squamosa*
sugar beet - *Beta vulgaris*
sugar-bush sumac - *Rhus ovata*
sugar-cane - *Saccharum officinarum*
sugar-cane plume grass - *Saccharum giganteum*
sugar grape - *Vitis rupestris*
sugar-grass sedge - *Carex atherodes*
sugar maple - *Acer grandidentatum, A. saccharum*
sugar palm - *Arenga pinnata*
sugar pea - *Pisum sativum, P. sativum* var. *macrocarpon*
sugar pine - *Pinus lambertiana*
sugar-scoop - *Tiarella unifoliata*
sugar sumac - *Rhus ovata*
sugar-tree - *Acer barbatum*
wild sugar-cane - *Saccharum spontaneum*
sugarberry - *Celtis laevigata, C. occidentalis, Celtis*
sugarplum - *Amelanchier*
Sullivant's milkweed - *Asclepias sullivantii*
sulphur
sulphur cinquefoil - *Potentilla recta*
sulphur-flower - *Eriogonum umbellatum*
sultana - *Impatiens wallerana*
sumac
Chinese sumac - *Rhus chinensis*
dwarf sumac - *Rhus copallina*
evergreen sumac - *Rhus sempervirens, R. virens*
flame-tree sumac - *Rhus copallina*
fragrant sumac - *Rhus aromatica*
laurel sumac - *Malosma laurina*
lemon sumac - *Rhus aromatica*
lemonade sumac - *Rhus integrifolia*
Mearns' sumac - *Rhus choriophylla*
mountain sumac - *Rhus copallina*
New Mexico evergreen sumac - *Rhus choriophylla*
poison sumac - *Toxicodendron vernix*
prairie sumac - *Rhus lanceolata*
scarlet sumac - *Rhus glabra*
shining sumac - *Rhus copallina*
smooth sumac - *Rhus glabra*
staghorn sumac - *Rhus hirta*
sugar-bush sumac - *Rhus ovata*
sugar sumac - *Rhus ovata*
swamp sumac - *Toxicodendron vernix*
sweet-scented sumac - *Rhus aromatica*
tobacco sumac - *Rhus virens*
velvet sumac - *Rhus hirta*
Venetian sumac - *Cotinus coggygria*
Virginia sumac - *Rhus hirta*
wing-rib sumac - *Rhus copallina*
winged sumac - *Rhus copallina*
summer
snow-in-summer - *Cerastium tomentosum*
summer-azalea - *Pelargonium domesticum*
summer-berry - *Viburnum trilobum*
summer cohosh - *Cimicifuga americana*
summer-cypress - *Bassia scoparia*
summer grape - *Vitis aestivalis, V. aestivalis* var. *argentifolia*
summer-haw - *Crataegus flava*
summer-lilac - *Buddleja davidii*
summer perennial phlox - *Phlox paniculata*

summer savory - *Satureja hortensis*
summer sedge - *Carex aestivalis*
summer-snowflake - *Leucojum aestivum, Ornithogalum umbellatum*
summer squash - *Cucurbita pepo*
summer-sweet - *Clethra alnifolia*
sump
poverty sump-weed - *Iva axillaris*
rough sump-weed - *Iva annua*
sun
California sun-cup - *Camissonia bistorta*
glory-of-the-sun - *Leucocoryne ixioides*
sun-choke - *Helianthus tuberosus*
sun-cup - *Camissonia bistorta, C. subacaulis*
sun-plant - *Portulaca grandiflora*
sun-rose - *Helianthemum nummularium, Helianthemum*
sun spurge - *Euphorbia helioscopia*
sundew
pink sundew - *Drosera capillaris*
round-leaf sundew - *Drosera rotundifolia*
sundial
sundial - *Lupinus perennis*
sundial lupine - *Lupinus perennis*
sundrops
little sundrops - *Oenothera perennis*
perennial sundrops - *Oenothera perennis*
southern sundrops - *Oenothera fruticosa*
sundrops - *Oenothera fruticosa, Oenothera*
sunflower
Appalachian sunflower - *Helianthus atrorubens*
ashy sunflower - *Helianthus mollis*
beach sunflower - *Helianthus debilis*
bright sunflower - *Helianthus laetiflorus*
cheerful sunflower - *Helianthus laetiflorus*
dark-eye sunflower - *Helianthus atrorubens*
dark-red sunflower - *Helianthus atrorubens*
different-leaf sunflower - *Helianthus heterophyllus*
divaricate sunflower - *Helianthus divaricatus*
divergent sunflower - *Helianthus divaricatus*
forest sunflower - *Helianthus decapetalus*
giant sunflower - *Helianthus decapetalus, H. giganteus*
hairy sunflower - *Helianthus heterophyllus, H. hirsutus, H. mollis*
hairy-wood sunflower - *Helianthus atrorubens*
Hopi sunflower - *Helianthus annuus*
Judge Daly's sunflower - *Helianthus maximilianii*
Kansas sunflower - *Helianthus petiolaris*
Kellerman's sunflower - *Helianthus kellermanii*
Maximilian's sunflower - *Helianthus maximilianii*
Mexican sunflower - *Tithonia rotundifolia*
midwestern tickseed-sunflower - *Bidens aristosa*
naked-stemmed sunflower - *Helianthus occidentalis*
narrow-leaf sunflower - *Helianthus angustifolius*
Nuttall's sunflower - *Helianthus nuttallii*
Oregon sunflower - *Balsamorhiza sagittata*
pale-leaf sunflower - *Helianthus strumosus*
petioled sunflower - *Helianthus petiolaris*
prairie sunflower - *Helianthus petiolaris*
rough-leaf sunflower - *Helianthus strumosus*
rough sunflower - *Helianthus hirsutus*
saw-tooth sunflower - *Helianthus grosseserratus*
showy sunflower - *Helianthus laetiflorus*

silver-leaf sunflower - *Helianthus argophyllus*
small-headed sunflower - *Helianthus microcephalus*
small-wood sunflower - *Helianthus microcephalus*
soft sunflower - *Helianthus mollis*
southern flat-seed-sunflower - *Verbesina occidentalis*
stiff-haired sunflower - *Helianthus hirsutus*
stiff sunflower - *Helianthus pauciflorus*
sunflower - *Balsamorhiza, Helianthus annuus, Helianthus*
sunflower-everlasting - *Heliopsis helianthoides*
swamp sunflower - *Helenium autumnale, Helianthus angustifolius, H. giganteus*
swollen sunflower - *Helianthus strumosus*
tall sunflower - *Helianthus giganteus*
ten-petals sunflower - *Helianthus decapetalus*
thin-leaf sunflower - *Helianthus decapetalus*
tickseed-sunflower - *Bidens aristosa, B. coronata*
weak sunflower - *Helianthus debilis*
western sunflower - *Helianthus occidentalis*
wild sunflower - *Helianthus annuus*
woodland sunflower - *Helianthus decapetalus, H. divaricatus, H. strumosus, H. tuberosus*

sunn hemp - *Crotalaria juncea*
sunray, showy sunray - *Helipterum splendidum*
supine paspalum - *Paspalum setaceum* var. *supinum*
supple
    Alabama supple-jack - *Berchemia scandens*
    supple-jack - *Berchemia scandens*
Surinam
    Surinam cherry - *Eugenia uniflora*
    Surinam sedge - *Cyperus surinamensis*
Susan
    black-eyed-Susan - *Rudbeckia hirta, R. hirta* var. *pulcherrima*
    brown-eyed-Susan - *Rudbeckia triloba*
swallow
    black swallow-wort - *Cynanchum louiseae*
    swallow-wort - *Chelidonium majus, Vincetoxicum hirundinaria*
    white swallow-wort - *Vincetoxicum hirundinaria*
swamp
    coastal swamp goldenrod - *Solidago elliottii*
    downy swamp huckleberry - *Vaccinium atrococcum*
    evergreen swamp fetterbush - *Lyonia lucida*
    northern swamp dogwood - *Cornus racemosa*
    northern swamp groundsel - *Senecio congestus*
    southern swamp crinum - *Crinum americanum*
    southern swamp dogwood - *Cornus stricta*
    swamp azalea - *Rhododendron viscosum*
    swamp-bay - *Magnolia virginiana, Persea palustris*
    swamp birch - *Betula pumila*
    swamp blackberry - *Rubus hispidus*
    swamp blackgum - *Nyssa sylvatica* var. *biflora*
    swamp blueberry - *Vaccinium corymbosum*
    swamp buttercup - *Ranunculus hispidus*
    swamp-candles - *Lysimachia terrestris*
    swamp chestnut oak - *Quercus michauxii, Q. prinus*
    swamp cottonwood - *Populus heterophylla*
    swamp currant - *Ribes lacustre*
    swamp dewberry - *Rubus hispidus*
    swamp dock - *Rumex verticillatus*
    swamp dodder - *Cuscuta gronovii*
    swamp dog-laurel - *Leucothoe axillaris*
    swamp fern - *Acrostichum*

swamp fly honeysuckle - *Lonicera oblongifolia*
swamp gooseberry - *Ribes lacustre*
swamp haw - *Viburnum cassinoides*
swamp hickory - *Carya cordiformis*
swamp-honeysuckle - *Rhododendron viscosum*
swamp Jack-in-the-pulpit - *Arisaema triphyllum*
swamp-laurel - *Kalmia polifolia*
swamp lily - *Lilium superbum*
swamp loosestrife - *Decodon verticillatus, Lysimachia terrestris, L. thyrsiflora*
swamp lousewort - *Pedicularis lanceolata, P. palustris*
swamp-mahogany - *Eucalyptus robusta*
swamp maple - *Acer rubrum*
swamp milkweed - *Asclepias incarnata*
swamp morning-glory - *Ipomoea aquatica*
swamp-oats - *Trisetum pensylvanicum*
swamp onion - *Allium validum*
swamp persicary - *Polygonum pensylvanicum*
swamp-pink - *Calopogonium*
swamp post oak - *Quercus lyrata*
swamp-potato - *Sagittaria*
swamp-privet - *Forestiera acuminata*
swamp red-bay - *Persea borbonia*
swamp red currant - *Ribes triste*
swamp red oak - *Quercus falcata* var. *pagodifolia*
swamp rose - *Rosa palustris*
swamp rose mallow - *Hibiscus moscheutos, H. moscheutos* subsp. *palustris*
swamp saxifrage - *Saxifraga pensylvanica*
swamp smartweed - *Polygonum coccineum*
swamp sumac - *Toxicodendron vernix*
swamp sunflower - *Helenium autumnale, Helianthus angustifolius, H. giganteus*
swamp tea-tree - *Melaleuca quinquenervia*
swamp thistle - *Cirsium muticum*
swamp tupelo - *Nyssa sylvatica* var. *biflora*
swamp white-cedar - *Chamaecyparis thyoides*
swamp white oak - *Quercus bicolor, Q. michauxii*
white swamp azalea - *Rhododendron viscosum*

swan-potato - *Sagittaria sagittifolia*
Swatow mustard - *Brassica juncea*
swaying rush - *Scirpus subterminalis*
sweating-plant - *Eupatorium perfoliatum*
swede
    swede - *Brassica napus* var. *napobrassica*
    swede rape - *Brassica napus*
Swedish
    Swedish begonia - *Plectranthus*
    Swedish ivy - *Plectranthus*
    Swedish myrtle - *Myrtus communis*
    Swedish turnip - *Brassica napus* var. *napobrassica*
sweet
    American sweet-gum - *Liquidambar styraciflua*
    bland sweet cicely - *Osmorhiza claytonii*
    California sweet grass - *Hierochloe occidentalis*
    California sweet-shrub - *Calycanthus occidentalis*
    early sweet bilberry - *Vaccinium vacillans*
    floating sweet grass - *Glyceria fluitans*
    hairy sweet cicely - *Osmorhiza claytonii*
    Indian sweet-clover - *Melilotus indica*
    late sweet blueberry - *Vaccinium angustifolium*
    low sweet blueberry - *Vaccinium angustifolium*

mountain-sweet - *Ceanothus americanus*
perennial sweet pea - *Lathyrus latifolius*
sapphire-berry sweet-leaf - *Symplocos tinctoria*
spreading sweet-root - *Osmorhiza chilensis*
summer-sweet - *Clethra alnifolia*
sweet acacia - *Acacia farnesiana*
sweet-alyssum - *Lobularia maritima*
sweet Annie - *Artemisia annua*
sweet azalea - *Rhododendron arborescens*
sweet balm - *Melissa officinalis*
sweet basil - *Ocimum basilicum*
sweet-bay - *Laurus nobilis, Magnolia virginiana, Persea borbonia*
sweet-bay magnolia - *Magnolia virginiana*
sweet-bells - *Leucothoe racemosa*
sweet-berry honeysuckle - *Lonicera caerulea*
sweet birch - *Betula lenta*
sweet blueberry - *Vaccinium vacillans*
sweet buckeye - *Aesculus flava*
sweet cherry - *Prunus avium*
sweet-clover - *Melilotus*
sweet-colt's-foot - *Petasites*
sweet corn - *Zea mays*
sweet crabapple - *Malus coronaria*
sweet elder - *Sambucus canadensis*
sweet everlasting - *Gnaphalium obtusifolium*
sweet-fern - *Comptonia asplenifolia, C. peregrina*
sweet-flag - *Acorus calamus*
sweet-fruited juniper - *Juniperus deppeana*
sweet gale - *Myrica gale*
sweet gallberry - *Ilex coriacea*
sweet goldenrod - *Solidago odora*
sweet grass - *Hierochloe odorata*
sweet-gum - *Liquidambar styraciflua*
sweet-haw - *Viburnum prunifolium*
sweet horsemint - *Cunila origanoides*
sweet-hurts - *Vaccinium angustifolium*
sweet jarvil - *Osmorhiza claytonii*
sweet Joe-Pye-weed - *Eupatorium purpureum*
sweet-leaf - *Symplocos tinctoria*
sweet locust - *Gleditsia triacanthos*
sweet marjoram - *Origanum majorana*
sweet mock-orange - *Philadelphus coronarius*
sweet-olive - *Osmanthus fragrans*
sweet orange - *Citrus sinensis*
sweet osmanthus - *Osmanthus fragrans*
sweet pea - *Lathyrus odoratus*
sweet pepper - *Capsicum frutescens* cv. 'grossu'
sweet pepper-bush - *Clethra acuminata, C. alnifolia*
sweet pig-nut - *Carya ovalis*
sweet pinesap - *Monotropsis odorata*
sweet pitcher-plant - *Sarracenia purpurea*
sweet-potato - *Ipomoea batatas*
sweet-potato-tree - *Manihot esculenta*
sweet-rocket - *Hesperis matronalis*
sweet-root - *Glycyrrhiza lepidota*
sweet scabious - *Scabiosa atropurpurea*
sweet-scented bedstraw - *Galium triflorum*
sweet-scented crabapple - *Malus coronaria*
sweet-scented geranium - *Pelargonium graveolens*
sweet-scented Indian plantain - *Cacalia suaveolens*
sweet-scented sumac - *Rhus aromatica*
sweet-shrub - *Calycanthus occidentalis, Calycanthus*

sweet sorghum - *Sorghum bicolor*
sweet-spire - *Itea virginica*
sweet tangle-head - *Heteropogon melanocarpus*
sweet vernal grass - *Anthoxanthum odoratum*
sweet viburnum - *Viburnum lentago, V. odoratissimum*
sweet violet - *Viola odorata*
sweet white violet - *Viola blanda*
sweet-William - *Dianthus barbatus*
sweet-William catchfly - *Silene armeria*
sweet woodruff - *Galium odoratum*
sweet wormwood - *Artemisia annua*
tall yellow sweet-clover - *Melilotus altissima*
tapering sweet-root - *Osmorhiza chilensis*
western sweet white violet - *Viola macloskeyi*
white sweet-clover - *Melilotus alba*
wild sweet crabapple - *Malus coronaria*
wild sweet pea - *Lathyrus latifolius*
wild sweet-potato-vine - *Ipomoea pandurata*
wild sweet-William - *Phlox divaricata, P. maculata*
winter-sweet - *Chimonanthus praecox*
woolly sweet cicely - *Osmorhiza claytonii*
yellow sweet-clover - *Melilotus officinalis*
sweetbrier - *Rosa eglanteria*
sweetsop - *Annona squamosa*
swine
swine-bane - *Chenopodium murale*
swine-cress - *Coronopus didymus*
Swiss
Swiss-chard - *Beta vulgaris* subsp. *cicla*
Swiss-cheese-plant - *Monstera deliciosa*
Swiss mountain pine - *Pinus mugo*
Swiss stone pine - *Pinus cembra*
switch
switch-cane - *Arundinaria gigantea* subsp. *tecta*
switch grass - *Panicum virgatum*
switch-ivy - *Leucothoe fontanesiana*
swollen
swollen bladderwort - *Utricularia radiata*
swollen sunflower - *Helianthus strumosus*
sword
sword bean - *Canavalia ensiformis, C. gladiata*
sword brake - *Pteris ensiformis*
sword fern - *Nephrolepis biserrata*
sword-leaf phlox - *Phlox buckleyi*
western sword fern - *Polystichum munitum*
sycamore
American sycamore - *Platanus occidentalis*
Arizona sycamore - *Platanus wrightii*
California sycamore - *Platanus racemosa*
eastern sycamore - *Platanus occidentalis*
sycamore maple - *Acer pseudoplatanus*
western sycamore - *Platanus racemosa*
Sydney blue gum - *Eucalyptus saligna*
Sylvan's horsetail - *Equisetum sylvaticum*
Syrian
Syrian bead-tree - *Melia azedarach*
Syrian bean-caper - *Zygophyllum fabago*
Syrian mesquite - *Prosopis farcta*
Syrian mustard - *Euclidium syriacum*
tabasco pepper - *Capsicum frutescens*
table mountain pine - *Pinus pungens*

tacamahac - *Populus balsamifera*
tag, long-tag pine - *Pinus echinata*
Tahitian lime - *Citrus latifolia*
Tahoka daisy - *Machaeranthera tanacetifolia*
tail
  annual-cat-tail - *Rostraria cristata*
  blue cat-tail - *Typha glauca*
  Brazilian satin-tail - *Imperata brasiliensis*
  broad-leaf cat-tail - *Typha latifolia*
  cat-tail - *Typha latifolia*
  cat-tail grass - *Setaria pallide-fusca*
  cat-tail millet - *Pennisetum glaucum*
  coon's-tail - *Ceratophyllum demersum*
  cow's-tail-pine - *Cephalotaxus harringtonia* var. *drupacea*
  crested dog-tail grass - *Cynosurus cristatus*
  crested dog's-tail - *Cynosurus cristatus*
  flicker-tail grass - *Hordeum glaucum*
  gopher-tail love grass - *Eragrostis ciliaris*
  hare's-tail - *Lagurus ovatus*
  hedgehog dog-tail grass - *Cynosurus echinatus*
  kitten's-tail - *Besseya rubra, Buchnera rubra*
  knot-root fox-tail - *Setaria geniculata*
  lion's-tail - *Leonotis, Leonurus cardiaca*
  lizard's-tail - *Gaura parviflora, Saururus cernuus*
  mare's-tail - *Conyza canadensis, Hippuris vulgaris*
  mouse's-tail - *Myosurus minimus*
  narrow-leaf cat-tail - *Typha angustifolia*
  rabbit-tail grass - *Lagurus ovatus*
  rat-tail fescue - *Vulpia myuros*
  rat's-tail - *Vulpia bromoides*
  rough dog's-tail - *Cynosurus echinatus*
  skunk-tail grass - *Hordeum jubatum*
  southern cat-tail - *Typha domingensis*
  squirrel-tail barley - *Hordeum jubatum*
  squirrel-tail fescue - *Vulpia bromoides*
  squirrel's-tail - *Elymus elymoides*
  tail-cup lupine - *Lupinus argenteus* var. *heteranthus*
  wolf-tail sedge - *Carex cherokeensis*
tale-wort - *Borago officinalis*
tall
  tall ambrosia - *Ambrosia trifida*
  tall anemone - *Anemone virginiana*
  tall beggar-ticks - *Bidens vulgata*
  tall bellflower - *Campanula americana*
  tall bilberry - *Vaccinium ovalifolium*
  tall blueberry willow - *Salix novaeangliae*
  tall boneset - *Eupatorium maculatum*
  tall buttercup - *Ranunculus acris*
  tall cinquefoil - *Potentilla arguta*
  tall coreopsis - *Coreopsis tripteris*
  tall fescue - *Festuca arundinacea*
  tall five-finger - *Potentilla norvegica*
  tall flat-top white aster - *Aster umbellatus*
  tall fleabane - *Conyza floribunda*
  tall goldenrod - *Solidago canadensis, S. canadensis* var. *scabra*
  tall hawkweed - *Hieracium praealtum* var. *decipiens*
  tall hedge mustard - *Sisymbrium altissimum*
  tall ironweed - *Vernonia gigantea*
  tall larkspur - *Delphinium barbeyi, D. exaltatum, D. trolliifolium*
  tall lettuce - *Lactuca canadensis*
  tall manna grass - *Glyceria elata*
  tall meadow-rue - *Thalictrum pubescens*
  tall milkweed - *Asclepias exaltata*
  tall morning-glory - *Ipomoea purpurea*
  tall nasturtium - *Tropaeolum majus*
  tall nettle - *Urtica dioica*
  tall oat grass - *Arrhenatherum elatius*
  tall potentilla - *Potentilla arguta*
  tall sunflower - *Helianthus giganteus*
  tall thistle - *Cirsium altissimum*
  tall tickseed - *Coreopsis tripteris*
  tall trisetum - *Trisetum canescens*
  tall umbrella-plant - *Cyperus eragrostis*
  tall vervain - *Verbena bonariensis*
  tall water-hemp - *Amaranthus tuberculatus*
  tall wheat grass - *Elytrigia elongata*
  tall white bog orchid - *Habenaria dilatata*
  tall white-lettuce - *Prenanthes altissima*
  tall white violet - *Viola canadensis*
  tall worm-seed mustard - *Erysimum hieraciifolium*
  tall worm-seed wallflower - *Erysimum hieraciifolium*
  tall yellow sweet-clover - *Melilotus altissima*
tallerack - *Eucalyptus tetragona*
tallow
  Chinese tallow-tree - *Sapium sebiferum*
  tallow-tree - *Sapium sebiferum*
  tallow-wood - *Ximenia americana*
tamanu - *Calophyllum inophyllum*
tamarack - *Larix laricina*
tamarind
  Manila tamarind - *Pithecellobium dulce*
  wild tamarind - *Lysiloma latisiliqua*
tamarisk
  Athel tamarisk - *Tamarix aphylla*
  Chinese tamarisk - *Tamarix chinensis*
  five-stamen tamarisk - *Tamarix chinensis*
  French tamarisk - *Tamarix gallica*
  Kashgar tamarisk - *Tamarix hispida*
  small-flowered tamarisk - *Tamarix parviflora*
tamono - *Calophyllum inophyllum*
tampala - *Amaranthus tricolor*
tan
  tan-bark oak - *Lithocarpus densiflora*
  tan oak - *Lithocarpus densiflora*
tangel-foot - *Viburnum alnifolium*
tangelo - *Citrus tangelo*
tangerine - *Citrus reticulata*
Tangier pea - *Lathyrus tingitanus*
tangle
  blue-tangle - *Gaylussacia frondosa*
  sweet tangle-head - *Heteropogon melanocarpus*
  tangle-head - *Heteropogon contortus*
  turkey-tangle - *Phyla nodiflora*
tangled bladderwort - *Utricularia biflora*
tankard, cool-tankard - *Borago officinalis*
tanner's dock - *Rumex hymenosepalus*
tannia - *Xanthosoma sagittifolium*
tansy
  false tansy - *Artemisia biennis*
  garden tansy - *Tanacetum vulgare*
  goose-tansy - *Potentilla anserina*

green tansy mustard - *Descurainia pinnata* var. *brachycarpa*
Menzies tansy mustard - *Descurainia pinnata* subsp. *menziesii*
pinnate tansy mustard - *Descurainia pinnata*
Richardson's tansy mustard - *Descurainia richardsonii*
tansy - *Tanacetum vulgare, Tanacetum*
tansy mustard - *Descurainia pinnata, D. sophia*
tansy ragwort - *Senecio jacobaea*
western tansy mustard - *Descurainia incisa*
wild tansy - *Ambrosia artemisiifolia*
tanyah - *Xanthosoma sagittifolium*
tap-rooted valerian - *Valeriana edulis*
tapa-cloth-tree - *Broussonetia papyrifera*
tape
spring-tape - *Sagittaria subulata*
tape-weed - *Vallisneria spiralis*
tapering sweet-root - *Osmorhiza chilensis*
tapeworm-plant - *Homalocladium platycladum*
tapioca-plant - *Manihot esculenta*
tar-fitch - *Lathyrus pratensis*
tarde, linda-tarde - *Gaura coccinea*
tare
lentil tare - *Vicia tetrasperma*
smooth tare - *Vicia tetrasperma*
tare - *Vicia sativa, Vicia*
yellow tare - *Lathyrus pratensis*
tarhui - *Lupinus mutabilis*
taro
Chinese taro - *Alocasia cucullata*
taro - *Colocasia esculenta*
taro-vine - *Epipremnum aureum*
wild taro - *Colocasia esculenta*
tarragon - *Artemisia dracunculus*
tarry cockle - *Silene antirrhina*
tartar-lily - *Ixiolirion tataricum*
Tartarian oat - *Avena fatua*
tarweed
Chilean tarweed - *Madia sativa*
cluster tarweed - *Madia glomerata*
coast tarweed - *Madia sativa*
hayfield tarweed - *Hemizonia congesta*
mountain tarweed - *Madia glomerata*
palouse tarweed - *Amsinckia retrorsa*
showy tarweed - *Madia elegans*
stinking tarweed - *Madia glomerata*
tarweed - *Amsinckia intermedia, Cuphea viscosissima,*
Grindelia squarrosa, Madia elegans, Madia
tarweed cuphea - *Cuphea carthagenensis*
tarweed fiddle-neck - *Amsinckia lycopsoides*
virgate tarweed - *Holocarpha virgata*
tarwi - *Lupinus mutabilis*
tasajillo - *Opuntia leptocaulis*
tasi - *Morrenia odorata*
Tasmanian blue gum - *Eucalyptus globulus*
tassel
red tassel-flower - *Emilia sonchifolia*
silk-tassel-bush - *Garrya*
tassel-flower - *Amaranthus caudatus, Brickellia grandiflora,*
Emilia javanica
tassel-hyacinth - *Muscari comosum*
tassel-rue - *Trautvetteria carolinensis*

tassel-white - *Itea virginica*
wavy-leaf silk-tassel - *Garrya elliptica*
Tatarian
Tatarian buckwheat - *Fagopyrum tataricum*
Tatarian honeysuckle - *Lonicera tatarica*
tawny daylily - *Hemerocallis fulva*
tea
Abyssinian tea - *Catha edulis*
Appalachian tea - *Ilex glabra, Viburnum cassinoides*
Arabian tea - *Catha edulis*
broad-leaf tea-tree - *Melaleuca quinquenervia*
crystal-tea - *Ledum palustre*
glandular Labrador tea - *Ledum glandulosum*
Jersey tea ceanothus - *Ceanothus americanus*
Jerusalem tea - *Chenopodium ambrosioides*
kopiko-tea - *Psychotria kaduana*
Labrador tea - *Ledum groenlandicum, Ledum*
Mexican tea - *Chenopodium ambrosioides*
Mormon tea - *Ephedra viridis*
mountain-tea - *Gaultheria procumbens*
New Jersey tea - *Ceanothus americanus*
New Zealand tea-tree - *Leptospermum scoparium*
Oswego tea - *Monarda didyma*
prairie-tea - *Croton monanthogynus*
prairie-tea croton - *Croton monanthogynus*
river tea-tree - *Melaleuca leucadendra*
Somali tea - *Catha edulis*
Spanish-tea - *Chenopodium ambrosioides*
swamp tea-tree - *Melaleuca quinquenervia*
tea - *Camellia sinensis*
tea-leaf willow - *Salix planifolia*
tea-olive - *Osmanthus fragrans*
tea-plant - *Camellia sinensis, Viburnum lentago*
tea rose - *Rosa odorata*
trapper's-tea - *Ledum glandulosum*
weeping tea-tree - *Melaleuca leucadendra*
western Labrador tea - *Ledum glandulosum*
teaberry - *Gaultheria procumbens, Viburnum cassinoides*
teak - *Tectona grandis*
tear
arrow-leaf tear-thumb - *Polygonum sagittatum*
halberd-leaf tear-thumb - *Polygonum arifolium*
tears
angel's-tears - *Soleirolia soleirolii*
baby's-tears - *Soleirolia soleirolii*
Job's-tears - *Coix lacryma-jobi*
widow's-tears - *Tradescantia virginiana*
teasel
cut-leaf teasel - *Dipsacus laciniatus*
Fuller's teasel - *Dipsacus sativus*
teasel - *Dipsacus fullonum*
teasel gourd - *Cucumis dipsaceus*
wild teasel - *Dipsacus fullonum*
teat
sow-teat blackberry - *Rubus allegheniensis*
sow-teat strawberry - *Fragaria vesca*
tecate cypress - *Cupressus guadalupensis*
teddy
teddy-bear-plant - *Cyanotis kewensis*
teddy-bear-vine - *Cyanotis kewensis*
telegraph
telegraph-plant - *Heterotheca grandiflora*

telegraph-plant-weed - *Heterotheca grandiflora*
temple-tree - *Plumeria*
ten
    ten-commandments - *Maranta leuconeura*
    ten-petal stick-leaf - *Mentzelia decapetala*
    ten-petals sunflower - *Helianthus decapetalus*
    ten-weeks stock - *Matthiola incana* var. *annua*
teosinte
    perennial teosinte - *Zea perennis*
    teosinte - *Zea mays* subsp. *mexicana*
Tepary bean - *Phaseolus acutifolius*
tepeguaje - *Leucaena pulverulenta*
terete yellow-cress - *Rorippa teres*
terongan - *Solanum torvum*
Terrell's grass - *Lolium perenne*
tesota - *Olneya tesota*
testiculate buttercup - *Ranunculus testiculatus*
tetrazygia, Florida tetrazygia - *Tetrazygia bicolor*
tetterfush - *Lyonia lucida*
Texas
    Texas black walnut - *Juglans microcarpa*
    Texas blue-weed - *Helianthus ciliaris*
    Texas bluebonnet - *Lupinus subcarnosus, L. texensis*
    Texas bluegrass - *Poa arachnifera*
    Texas buckeye - *Aesculus glabra* var. *arguta*
    Texas bull-nettle - *Cnidoscolus texanus*
    Texas croton - *Croton texensis*
    Texas ebony - *Pithecellobium flexicaule*
    Texas filaree - *Erodium texanum*
    Texas gourd - *Cucurbita texana*
    Texas Hercules's-club - *Zanthoxylum hirsutum*
    Texas lupine - *Lupinus subcarnosus*
    Texas mallow - *Malvaviscus arboreus* var. *drummondii*
    Texas millet - *Panicum texanum*
    Texas mimosa - *Acacia greggii*
    Texas mountain-laurel - *Sophora secundiflora*
    Texas mud-baby - *Echinodorus cordifolius*
    Texas mulberry - *Morus microphylla*
    Texas needle grass - *Stipa leucotricha*
    Texas palmetto - *Sabal mexicana*
    Texas panicum - *Panicum texanum*
    Texas persimmon - *Diospyros texana*
    Texas porlieria - *Porlieria angustifolia*
    Texas prickly-pear - *Opuntia lindheimeri*
    Texas red oak - *Quercus texana*
    Texas sage - *Salvia coccinea*
    Texas sophora - *Sophora affinis*
    Texas toadflax - *Linaria texana*
    Texas walnut - *Juglans microcarpa*
    Texas-weed - *Caperonia palustris*
    Texas white-brush - *Aloysia gratissima*
thale-cress - *Arabidopsis thaliana*
Thanksgiving cactus - *Schlumbergera truncata*
thatch
    thatch-leaf palm - *Howeia forsteriana*
    thatch palm - *Howeia forsteriana*
    thatch screw-pine - *Pandanus tectorius*
thick
    thick-leaf phlox - *Phlox carolina*
    thick-spike blazing-star - *Liatris pycnostachya*
    thick-spike wheat grass - *Elymus lanceolatus*

thicket
    thicket bean - *Phaseolus polystachios*
    thicket sedge - *Carex platyphylla*
thief, wheat-thief - *Bromus secalinus, Lithospermum arvense*
thimbleberry
    thimbleberry - *Rubus occidentalis, R. odoratus, R. parviflorus*
    western thimbleberry - *Rubus parviflorus*
thimbleweed
    long-headed thimbleweed - *Anemone cylindrica*
    thimbleweed - *Anemone cylindrica, A. virginiana*
thin
    thin grass - *Agrostis pallens*
    thin-leaf alder - *Alnus tenuifolia*
    thin-leaf coneflower - *Rudbeckia triloba*
    thin-leaf huckleberry - *Vaccinium membranaceum*
    thin-leaf sedge - *Carex cephaloidea*
    thin-leaf sunflower - *Helianthus decapetalus*
    thin paspalum - *Paspalum setaceum*
thistle
    annual sow-thistle - *Sonchus oleraceus*
    barb-wire Russian thistle - *Salsola paulsenii*
    blessed milk-thistle - *Silybum marianum*
    blessed thistle - *Cnicus benedictus, Silybum marianum*
    blue-thistle - *Echium vulgare*
    bull thistle - *Cirsium pumilim, C. vulgare*
    Canadian thistle - *Cirsium arvense*
    cotton thistle - *Onopordum acanthium*
    creeping sow-thistle - *Sonchus arvensis*
    creeping thistle - *Cirsium arvense*
    distaff thistle - *Carthamus lanatus*
    field sow-thistle - *Sonchus arvensis*
    field thistle - *Cirsium discolor*
    Flodman's thistle - *Cirsium flodmanii*
    fragrant thistle - *Cirsium pumilim*
    globe thistle - *Echinops*
    great globe thistle - *Echinops sphaerocephalus*
    green thistle - *Cirsium arvense*
    holy thistle - *Silybum marianum*
    horse-thistle - *Lactuca serriola*
    Iberian star-thistle - *Centaurea iberica*
    Indian thistle - *Cirsium edule*
    Italian thistle - *Carduus pycnocephalus*
    leafy thistle - *Cirsium foliosum*
    Malta star-thistle - *Centaurea melitensis*
    marsh sow-thistle - *Sonchus arvensis* spp. *uliginosus*
    marsh thistle - *Cirsium palustre*
    milk-thistle - *Lactuca serriola, Silybum marianum*
    musk thistle - *Carduus nutans*
    nodding thistle - *Carduus nutans*
    palouse thistle - *Cirsium brevifolium*
    pasture thistle - *Cirsium pumilim, C. pumilum*
    perennial sow-thistle - *Sonchus arvensis*
    perennial thistle - *Cirsium arvense*
    plume thistle - *Cirsium virginianum, Cirsium*
    plumeless thistle - *Carduus acanthoides, C. nutans*
    prairie thistle - *Cirsium flodmanii*
    prickly sow-thistle - *Sonchus asper*
    purple star-thistle - *Centaurea calcitrapa*
    Russian thistle - *Salsola australis*
    Scotch thistle - *Onopordum acanthium*
    slender-flower thistle - *Carduus tenuiflorus*
    small-flowered thistle - *Cirsium arvense*

sow-thistle - *Sonchus oleraceus, Sonchus*
spear thistle - *Cirsium virginianum*
spiny-leaved sow-thistle - *Sonchus asper*
spiny sow-thistle - *Sonchus asper*
St. Mary's thistle - *Silybum marianum*
swamp thistle - *Cirsium muticum*
tall thistle - *Cirsium altissimum*
thistle - *Cirsium*
thornless thistle - *Centaurea americana*
Turkestan thistle - *Centaurea repens*
Virginia thistle - *Cirsium virginianum*
wavy-leaf thistle - *Cirsium undulatum*
welted thistle - *Carduus crispus*
western thistle - *Cirsium occidentale*
woolly distaff thistle - *Carthamus lanatus*
yellow-spine thistle - *Cirsium ochrocentrum*
yellow star-thistle - *Centaurea solstitialis*
yellow thistle - *Cirsium horridulum*

thorn
black-thorn - *Crataegus calpodendron, Prunus spinosa*
camel-thorn - *Alhagi maurorum, A. pseudalhagi*
Christ-thorn - *Euphorbia milii*
cockspur-thorn - *Crataegus crus-galli*
desert-thorn - *Lycium pallidum*
downy thorn-apple - *Datura innoxia, D. metel*
fire-thorn - *Pyracantha*
Jerusalem thorn - *Parkinsonia aculeata*
Madras thorn - *Pithecellobium dulce*
narrow-leaf fire-thorn - *Pyracantha angustifolia*
pear-thorn - *Crataegus calpodendron, C. laevigata*
quick-set-thorn - *Crataegus laevigata*
scarlet fire-thorn - *Pyracantha coccinea*
stramonium thorn-apple - *Datura stramonium*
thorn-apple - *Crataegus, Datura stramonium, Datura*
thorn orache - *Bassia hyssopifolia*
Washington's-thorn - *Crataegus phaenopyrum*
West Indian black-thorn - *Acacia farnesiana*
white-thorn acacia - *Acacia constricta*
yellow-fruited-thorn - *Crataegus flava*

thornless
thornless honey locust - *Gleditsia triacanthos* var. *inermis*
thornless thistle - *Centaurea americana*

thorns, crown-of-thorns - *Euphorbia milii*

thorny
thorny amaranth - *Amaranthus spinosus*
thorny elaeagnus - *Elaeagnus pungens*

thorough-wax - *Bupleurum rotundi-folium, Bupleurum*

thoroughwort
hyssop-leaf thoroughwort - *Eupatorium hyssopifolium*
late-flowering thoroughwort - *Eupatorium serotinum*
round-leaf thoroughwort - *Eupatorium rotundifolium*
thoroughwort - *Eupatorium perfoliatum, Eupatorium*
thoroughwort pennycress - *Thlaspi perfoliatum*
white thoroughwort - *Eupatorium album*

thousand
thousand-leaf - *Achillea millefolium*
thousand-mothers - *Tolmiea menziesii*
thousand-seal - *Achillea millefolium*

thousands, mother-of-thousands - *Saxifraga stolonifera*

thread
dew-thread - *Drosera filiformis*
gold-thread - *Coptis groenlandica, Coptis*

gold-thread-vine - *Cuscuta gronovii*
needle-and-thread - *Stipa comata*
needle-and-thread grass - *Stipa comata*
thread-leaf pondweed - *Potamogeton filiformis*
thread-leaf sedge - *Carex filifolia*
thread-leaf snakeweed - *Gutierrezia microcephala*
thread-leaf tickseed - *Coreopsis verticillata*
thread palm - *Washingtonia robusta*

threads, silver-threads - *Fittonia verschaffeltii*

three
Arizona three-awn grass - *Aristida arizonica*
arrow-feather three-awn - *Aristida purpurascens*
church-mouse three-awn - *Aristida dichotoma*
pine-land three-awn - *Aristida stricta*
poverty three-awn - *Aristida divaricata*
prairie three-awn - *Aristida adscensionis*
purple three-awn - *Aristida purpurea*
red three-awn - *Aristida purpurea*
Reverchon's three-awn - *Aristida purpurea*
six-weeks three-awn - *Aristida adscensionis*
slim-spike three-awn - *Aristida longespica*
three-awn - *Aristida*
three-cornered-jack - *Emex australis*
three-flowered beggar-weed - *Desmodium triflorum*
three-flowered melic - *Melica nitens*
three-flowered-nettle - *Solanum triflorum*
three-fruited sedge - *Carex trisperma*
three-horn bedstraw - *Galium tricornutum*
three-leaf false Solomon's seal - *Smilacina trifolia*
three-leaf groundsel - *Senecio longilobus*
three-lobed coneflower - *Rudbeckia triloba*
three-lobed morning-glory - *Ipomoea triloba*
three-men-in-a-boat - *Tradescantia spathacea*
three-nerved duckweed - *Lemna trinervis*
three-seeded croton - *Croton lindheimerianus*
three-seeded-mercury - *Acalypha virginica*
three-square - *Scirpus americanus*
three-square bur-reed - *Sparganium americanum*
three-tooth cinquefoil - *Potentilla tridentata*
Wooton's three-awn - *Aristida pansa*

thrift - *Armeria*

throat-wort - *Campanula trachelium*

thumb
arrow-leaf tear-thumb - *Polygonum sagittatum*
halberd-leaf tear-thumb - *Polygonum arifolium*
lady's-thumb - *Polygonum persicaria*

Thunberg's lespedeza - *Lespedeza thunbergii*

Thurber's
Thurber's fescue - *Festuca thurberi*
Thurber's needle grass - *Stipa thurberiana*
Thurber's redtop - *Agrostis thurberiana*

thyme
creeping thyme - *Thymus serpyllum*
French thyme - *Plectranthus amboinicus*
garden thyme - *Thymus vulgaris*
lemon thyme - *Thymus serpyllum*
mother-of-thyme - *Acinos arvensis, Thymus serpyllum*
Spanish thyme - *Lippia micromera, Plectranthus amboinicus*
thyme - *Thymus vulgaris*
thyme dodder - *Cuscuta epithymum*
thyme-leaf sandwort - *Arenaria serpyllifolia*

thyme-leaved speedwell - *Veronica serpyllifolia* subsp.
  humifusa
thyme-leaved spurge - *Euphorbia serpyllifolia*
wild thyme - *Thymus serpyllum*
ti
  ti - *Cordyline terminalis*
  ti-es - *Pouteria campechiana*
  ti-palm - *Cordyline terminalis*
tibisee - *Lasiacis divaricata*
tick
  beggar-tick grass - *Aristida orcuttiana*
  big tick-trefoil - *Desmodium cuspidatum*
  Canadian tick-trefoil - *Desmodium canadense*
  cluster-leaf tick-trefoil - *Desmodium glutinosum*
  hoary tick-clover - *Desmodium canescens*
  hoary tick-trefoil - *Desmodium canescens*
  large-bracted tick-trefoil - *Desmodium cuspidatum*
  little-leaf tick-trefoil - *Desmodium ciliare*
  naked-flowered tick-trefoil - *Desmodium nudiflorum*
  naked tick-trefoil - *Desmodium nudiflorum*
  panicled tick-trefoil - *Desmodium paniculatum*
  pine-barren tick-trefoil - *Desmodium strictum*
  pointed-leaf tick-trefoil - *Desmodium glutinosum*
  prostrate tick-trefoil - *Desmodium rotundifolium*
  round-leaf tick-trefoil - *Desmodium rotundifolium*
  sessile-leaf tick-trefoil - *Desmodium sessilifolium*
  sessile tick-clover - *Desmodium sessilifolium*
  sessile tick-trefoil - *Desmodium sessilifolium*
  showy tick-trefoil - *Desmodium canadense*
  small-leaf tick-trefoil - *Desmodium ciliare*
  Spanish tick-clover - *Desmodium uncinatum*
  stiff tick-trefoil - *Desmodium strictum*
  tick-trefoil - *Desmodium canadense, Desmodium*
  tick-weed - *Verbesina virginica*
  velvety tick-trefoil - *Desmodium viridiflorum*
  tickle grass - *Agrostis hyemalis, A. scabra, Hordeum glaucum,*
    Panicum capillare
ticks
  bearded beggar-ticks - *Bidens aristosa*
  beggar-ticks - *Agrimonia gryposepala, Bidens bipinnata,*
    B. pilosa, Bidens, Desmodium
  Bigelow's beggar-ticks - *Bidens bigelovii*
  connate beggar-ticks - *Bidens connata*
  coreopsis beggar-ticks - *Bidens polylepis*
  devil's beggar-ticks - *Bidens frondosa*
  estuary beggar-ticks - *Bidens hyperborea*
  hairy beggar-ticks - *Bidens pilosa*
  marsh beggar-ticks - *Bidens mitis*
  nodding beggar-ticks - *Bidens cernua*
  northern estuarine beggar-ticks - *Bidens hyperborea*
  purple-stem beggar-ticks - *Bidens connata*
  tall beggar-ticks - *Bidens vulgata*
  trifid beggar-ticks - *Bidens tripartita*
  western beggar-ticks - *Bidens vulgata*
tickseed
  coastal plain tickseed - *Bidens mitis*
  finger tickseed - *Coreopsis palmata*
  hyssop-leaf tickseed - *Corispermum hyssopifolium*
  midwestern tickseed-sunflower - *Bidens aristosa*
  northern tickseed - *Bidens coronata*
  Ozark tickseed - *Bidens polylepis*
  tall tickseed - *Coreopsis tripteris*

thread-leaf tickseed - *Coreopsis verticillata*
tickseed - *Bidens, Coreopsis lanceolata, Coreopsis*
tickseed-sunflower - *Bidens aristosa, B. coronata*
tidestromia, woolly tidestromia - *Tidestromia lanuginosa*
tiger
  tiger aloe - *Aloe variegata*
  tiger-flower - *Tigridia pavonia*
  tiger lily - *Lilium lancifolium*
  tiger-nut - *Cyperus esculentus*
tights, stick-tights - *Bidens cernua, Bidens*
timber
  hardy timber bamboo - *Phyllostachys bambusoides*
  Japanese timber bamboo - *Phyllostachys bambusoides*
  timber milk-vetch - *Astragalus miser*
  timber oat grass - *Danthonia intermedia*
timberline bluegrass - *Poa rupicola*
timothy
  alpine cat timothy - *Phleum alpinum*
  alpine timothy - *Phleum alpinum*
  cultivated timothy - *Phleum pratense*
  mountain timothy - *Phleum pratense*
  timothy - *Phleum pratense, Phleum*
  timothy canary grass - *Phalaris angusta*
  water timothy - *Alopecurus geniculatus*
tine-leaf milk-vetch - *Astragalus pectinatus*
tinker's-weed - *Triosteum perfoliatum*
Tinnevelly senna - *Senna alexandrina*
tinted spurge - *Euphorbia commutata*
tiny vetch - *Vicia hirsuta*
tip, white-tip clover - *Trifolium variegatum*
tipa - *Tipuana tipu*
Tipton's-weed - *Hypericum perforatum*
tipu-tree - *Tipuana tipu*
tiss-wood - *Persea borbonia*
titi
  black titi - *Cliftonia monophylla, Cyrilla racemiflora*
  red titi - *Cyrilla racemiflora*
  titi - *Oxydendrum arboreum*
  titi-tree - *Cliftonia monophylla*
toad rush - *Juncus bufonius*
toadflax
  bastard-toadflax - *Comandra umbellata, Comandra*
  blue toadflax - *Linaria canadensis*
  broad-leaf toadflax - *Linaria genistifolia*
  Dalmatian toadflax - *Linaria genistifolia* subsp. *dalmatica*
  Florida toadflax - *Linaria floridana*
  lesser toadflax - *Chaenorhinum minus*
  old-field toadflax - *Linaria canadensis*
  striped toadflax - *Linaria repens*
  Texas toadflax - *Linaria texana*
  toadflax - *Linaria*
  yellow toadflax - *Linaria vulgaris*
tobacco
  Aztec tobacco - *Nicotiana rustica*
  desert tobacco - *Nicotiana trigonophylla*
  fiddle-leaf tobacco - *Nicotiana repanda*
  flowering tobacco - *Nicotiana alata*
  Indian tobacco - *Lobelia inflata*
  jasmine tobacco - *Nicotiana alata*
  lady's-tobacco - *Antennaria plantaginifolia, Antennaria*
  rabbit-tobacco - *Gnaphalium obtusifolium*

tobacco - *Nicotiana tabacum*
tobacco-root - *Valeriana edulis*
tobacco sumac - *Rhus virens*
tobacco-weed - *Elephantopus tomentosus*
tobacco witch-weed - *Striga gesnerioides*
tree tobacco - *Nicotiana glauca*
wild tobacco - *Nicotiana repanda*
winged tobacco - *Nicotiana alata*
tobasco cimarron - *Nicotiana repanda*
tobosa grass - *Hilaria mutica*
tocalote - *Centaurea melitensis*
toddy palm - *Caryota urens*
toes
  Canadian pussy-toes - *Antennaria neglecta* var. *canadensis*
  field pussy-toes - *Antennaria neglecta*
  plantain-leaf pussy-toes - *Antennaria plantaginifolia*
  pussy-toes - *Antennaria plantaginifolia, Antennaria*
  solitary pussy-toes - *Antennaria solitaria*
tolguacha - *Datura innoxia*
tomatillo - *Physalis philadelphica*
tomato
  cherry tomato - *Lycopersicon esculentum* var. *cerasiforme,*
    Physalis peruviana
  currant tomato - *Lycopersicon pimpinellifolium*
  gooseberry-tomato - *Physalis peruviana*
  husk-tomato - *Physalis pubescens, Physalis*
  pear tomato - *Lycopersicon esculentum* var. *pyriforme*
  strawberry tomato - *Physalis peruviana, P. pubescens,*
    P. pubescens var. *grisea*
  tomato - *Lycopersicon esculentum, Lycopersicon*
  tomato-fruited eggplant - *Solanum integrifolium*
  tomato-tree - *Cyphomandra betacea*
  tree-tomato - *Cyphomandra betacea*
  wild tomato - *Solanum triflorum*
tongue
  adder's-tongue - *Erythronium*
  adder's-tongue fern - *Ophioglossum*
  Appalachian beard-tongue - *Penstemon canescens*
  beard-tongue - *Penstemon*
  bristly ox's-tongue - *Picris echioides*
  deer's-tongue - *Panicum clandestinum*
  devil's-tongue - *Amorphophallus rivieri, Amorphophallus*
  gray beard-tongue - *Penstemon canescens*
  hairy beard-tongue - *Penstemon hirsutus*
  hawkweed ox-tongue - *Picris hieracioides*
  hound's-tongue - *Cynoglossum officinale, Cynoglossum*
  lamb's-tongue - *Plantago media*
  large beard-tongue - *Penstemon grandiflorus*
  large-flowered beard-tongue - *Penstemon grandiflorus*
  long-tongue mutton grass - *Poa fendleriana*
  mother-in-law's-tongue - *Sansevieria trifasciata*
  mother-in-law's-tongue-plant - *Dieffenbachia*
  northeastern beard-tongue - *Penstemon hirsutus*
  northern adder's-tongue fern - *Ophioglossum pusillum*
  ox-tongue - *Picris hieracioides*
  painted-tongue - *Salpiglossis sinuata*
  tongue-grass - *Lepidium virginicum, Lepidium*
  tongue pepper-weed - *Lepidium nitidum*
  whip-tongue - *Galium mollugo*
  wild tongue-grass - *Lepidium densiflorum*
  yellow adder's-tongue - *Erythronium americanum*
toog - *Bischofia javanica*

tooth
  big-tooth aspen - *Populus grandidentata*
  big-tooth maple - *Acer grandidentatum*
  dog-tooth grass - *Cynodon dactylon*
  dog-tooth pea - *Lathyrus sativus*
  dog-tooth-violet - *Erythronium*
  Indian tooth-cup - *Rotala indica*
  large-tooth aspen - *Populus grandidentata*
  lion's-tooth - *Leontodon autumnalis*
  saw-tooth fog-fruit - *Phyla nodiflora* var. *incisa*
  saw-tooth oak - *Quercus acutissima*
  saw-tooth spurge - *Euphorbia serrata*
  saw-tooth sunflower - *Helianthus grosseserratus*
  three-tooth cinquefoil - *Potentilla tridentata*
  tooth-cup - *Rotala ramosior*
  tooth-leaved croton - *Croton glandulosa*
  white dog-tooth-violet - *Erythronium albidum*
toothache
  toothache grass - *Ctenium aromaticum*
  toothache-tree - *Zanthoxylum americanum*
toothed
  toothed bur-clover - *Medicago polymorpha*
  toothed medic - *Medicago polymorpha*
  toothed spurge - *Euphorbia dentata, E. serrata*
  toothed wood fern - *Dryopteris carthusiana*
toothpick ammi - *Ammi visnaga*
toothwort
  cut-leaf toothwort - *Cardamine concatenata*
  toothwort - *Cardamine diphylla*
top
  Amazon sprangle-top - *Leptochloa panicoides*
  Arizona cotton-top - *Digitaria californica*
  bearded sprangle-top - *Leptochloa fascicularis*
  big-top love grass - *Eragrostis hirsuta*
  brown-top - *Agrostis capillaris*
  brown-top millet - *Brachiaria fasciculata, B. ramosa*
  brown-top panicum - *Brachiaria fasciculata*
  Chinese sprangle-top - *Leptochloa chinensis*
  flat-top white aster - *Aster umbellatus*
  giant white-top sedge - *Rhynchospora latifolia*
  golden-top - *Lamarckia aurea*
  green sprangle-top - *Leptochloa dubia*
  hairy white-top - *Cardaria pubescens*
  lens-podded white-top - *Cardaria chalepensis*
  Mexican sprangle-top - *Leptochloa uninervia*
  purple-top - *Tridens flavus*
  rattle-top - *Cimicifuga*
  red sprangle-top - *Leptochloa filiformis*
  red-top panicum - *Panicum rigidulum*
  sour-top blueberry - *Vaccinium myrtilloides*
  sprangle-top - *Leptochloa, Scolochloa festucacea*
  tall flat-top white aster - *Aster umbellatus*
  top onion - *Allium cepa* var. *viviparum*
  top-podded water-primrose - *Ludwigia polycarpa*
  white-top - *Cardaria draba, C. pubescens, Erigeron annuus,*
    E. strigosus
  white-top sedge - *Rhynchospora colorata*
torch
  torch azalea - *Rhododendron kaempferi*
  torch cactus - *Echinopsis spachiana*
  torch-lily - *Kniphofia uvaria, Kniphofia*
  torch pine - *Pinus rigida*

white torch cactus - *Echinopsis spachiana*
Toringo crabapple - *Malus sieboldii*
torpedo grass - *Panicum repens*
Torrey's
    Torrey's amaranth - *Amaranthus bigelovii*
    Torrey's nightshade - *Solanum dimidiatum*
    Torrey's pine - *Pinus torreyana*
    Torrey's vauquelinia - *Vauquelinia californica*
torreya
    California torreya - *Torreya californica*
    Florida torreya - *Torreya taxifolia*
touch
    orange touch-me-not - *Impatiens capensis*
    pale touch-me-not - *Impatiens pallida*
    spotted touch-me-not - *Impatiens capensis*
    touch-me-not - *Impatiens noli-tangere, Impatiens*
    yellow touch-me-not - *Impatiens pallida*
tough buckthorn - *Sideroxylon tenax*
Toumey's oak - *Quercus toumeyi*
tous-les-mois - *Canna indica*
towel gourd - *Luffa acutangula*
tower mustard - *Arabis glabra*
townhall's-clock - *Adoxa moschatellina*
Townsville stylo - *Stylosanthes humilis*
track, cart-track-plant - *Plantago major*
tracks, rabbit-tracks - *Maranta leuconeura* var.
    kerchoveana
Tracy's willow - *Salix lasiolepis*
Tradescant's aster - *Aster tradescantii*
trail-plant - *Adenocaulon bicolor*
trailing
    downy trailing lespedeza - *Lespedeza procumbens*
    smooth trailing lespedeza - *Lespedeza repens*
    trailing arbutus - *Epigaea repens*
    trailing bitter-cress - *Cardamine rotundifolia*
    trailing bramble - *Rubus flagellaris*
    trailing bush-clover - *Lespedeza procumbens*
    trailing crown-vetch - *Coronilla varia*
    trailing-evergreen - *Lycopodium complanatum*
    trailing four-o'clock - *Allionia incarnata*
    trailing lantana - *Lantana montevidensis*
    trailing phlox - *Phlox nivalis*
    trailing velvet-plant - *Ruellia makoyana*
    trailing watermelon-begonia - *Pellionia repens*
    trailing wedelia - *Wedelia trilobata*
    trailing wild bean - *Strophostyles helvola*
    trailing wolf-bane - *Aconitum reclinatum*
tramp's
    tramp's spurge - *Euphorbia corollata*
    tramp's-trouble - *Smilax bona-nox*
Transvaal daisy - *Gerbera jamesonii*
trapa-nut - *Trapa natans*
trapper's-tea - *Ledum glandulosum*
traveler's-palm - *Ravenala madagascariensis*
treacle
    treacle-berry - *Smilacina racemosa*
    treacle mustard - *Erysimum cheiranthoides, E. repandum,*
    Erysimum
tread-softly - *Cnidoscolus stimulosus, C. texanus*
treasure-flower - *Gazania rigens*

tree
    African tulip-tree - *Spathodea campanulata*
    American smoke-tree - *Cotinus obovatus*
    American wayfaring-tree - *Viburnum alnifolium*
    angelica-tree - *Aralia spinosa*
    anise-tree - *Illicium*
    auricula-tree - *Calotropis procera*
    Australian flame-tree - *Brachychiton acerifolius*
    Australian tree-fern - *Cyathea australis, C. cooperi*
    Australian umbrella-tree - *Schefflera actinophylla*
    be-still-tree - *Thevetia peruviana*
    bead-tree - *Melia*
    bean-tree - *Laburnum*
    beaver-tree-laurel - *Magnolia virginiana*
    beefwood-tree - *Casuarina equisetifolia, Casuarina*
    Benjamin's-tree - *Ficus benjamina*
    big-tree - *Sequoiadendron giganteum*
    big tree plum - *Prunus mexicana*
    bird-catcher-tree - *Pisonia umbellifera*
    blond tree-fern - *Cibotium splendens*
    Brazilian grape tree - *Myrciaria cauliflora*
    Brazilian pepper-tree - *Schinus terebinthifolius*
    broad-leaf tea-tree - *Melaleuca quinquenervia*
    buckwheat-tree - *Cliftonia monophylla*
    cabbage-tree - *Cordyline, Cussonia spicata*
    cajeput-tree - *Melaleuca quinquenervia*
    calabash-tree - *Crescentia cujete*
    California hop-tree - *Ptelea crenulata*
    California pepper-tree - *Schinus molle*
    camphor-tree - *Cinnamomum camphora*
    candleberry-tree - *Aleurites moluccana*
    candlenut-tree - *Aleurites moluccana*
    caouthchouc-tree - *Hevea brasiliensis*
    catawba-tree - *Catalpa speciosa*
    chaste-tree - *Vitex agnus-castus*
    China-tree - *Koelreuteria paniculata, Melia azedarach*
    China-wood oil-tree - *Vernicia fordii*
    Chinese parasol-tree - *Firmiana simplex*
    Chinese tallow-tree - *Sapium sebiferum*
    chocolate tree - *Theobroma cacao*
    Christmas-berry-tree - *Schinus terebinthifolius*
    Christmas-tree kalanchoe - *Kalanchoe laciniata*
    cigar-tree - *Catalpa speciosa*
    cockspur coral-tree - *Erythrina crista-galli*
    coffee-tree - *Polyscias guilfoylei*
    copa-tree - *Ailanthus altissima*
    coral-tree - *Erythrina*
    cosmetic-bark-tree - *Murraya paniculata*
    council-tree - *Ficus altissima*
    cry-baby-tree - *Erythrina crista-galli*
    cucumber-tree - *Magnolia acuminata*
    desert smoke-tree - *Psorothamnus spinosa*
    dragon-tree - *Dracaena draco*
    ear-leaved umbrella-tree - *Magnolia fraseri*
    East Indian fig-tree - *Ficus benghalensis*
    egg-fruit-tree - *Pouteria campechiana*
    elephant-foot-tree - *Nolina recurvata*
    empress-tree - *Paulownia tomentosa*
    European spindle-tree - *Euonymus europaeus*
    fever-tree - *Pinckneya pubens*
    fig-tree kalanchoe - *Kalanchoe laciniata*
    fire-wheel-tree - *Stenocarpus sinuatus*
    fish-poison-tree - *Piscidia piscipula*

flamboyant-tree - *Delonix regia*
flame bottle-tree - *Brachychiton acerifolius*
flame-tree - *Delonix regia*
flame-tree sumac - *Rhus copallina*
Florida anise-tree - *Illicium floridanum*
floss-silk-tree - *Chorisia speciosa*
Formosan rice-tree - *Fatsia japonica*
fringe-tree - *Chionanthus virginicus*
ginberbread-tree - *Hyphaene thebaica*
glory-tree - *Clerodendrum thompsoniae*
golden-rain-tree - *Koelreuteria paniculata, Koelreuteria*
great lead-tree - *Leucaena pulverulenta*
groundsel-tree - *Baccharis halimifolia*
Hawaiian soap-tree - *Sapindus oahuensis*
Hawaiian tree-fern - *Cibotium chamissoi, C. splendens*
hemp-tree - *Vitex agnus-castus*
hop-tree - *Ptelea trifoliata, Ptelea*
horsetail-tree - *Casuarina equisetifolia*
iigiri-tree - *Idesia polycarpa*
Indian rubber-tree - *Ficus elastica*
iron-tree - *Metrosideros*
jade-tree - *Crassula ovata*
Jamaican caper-tree - *Capparis cynophallophora*
Japanese bead-tree - *Melia azedarach*
Japanese pagoda-tree - *Sophora japonica*
Japanese raisin-tree - *Hovenia dulcis*
Japanese spindle-tree - *Euonymus japonicus*
Japanese varnish-tree - *Firmiana simplex*
Joshua-tree - *Yucca brevifolia*
karoo-tree - *Rhus lancea*
karri-tree - *Paulownia tomentosa*
karum-tree - *Pongamia pinnata*
katsura-tree - *Cercidiphyllum japonicum*
Kentucky coffee-tree - *Gymnocladus dioica*
kiri-tree - *Paulownia tomentosa*
kittul-tree - *Caryota urens*
large-leaf cucumber-tree - *Magnolia macrophylla*
laurel-tree - *Persea borbonia*
lead-tree - *Leucaena leucocephala*
lebbek-tree - *Albizia lebbeck*
lilac chaste-tree - *Vitex agnus-castus*
lipstick-tree - *Bixa orellana*
little-leaf lead-tree - *Leucaena retusa*
little-tree willow - *Salix arbusculoides*
London plane-tree - *Platanus acerifolia*
Lyon-tree - *Lyonothamnus floribundus*
maidenhair tree - *Ginkgo biloba*
man tree-fern - *Cibotium splendens*
melon-tree - *Carica papaya*
mile-tree - *Casuarina equisetifolia*
mimosa-tree - *Albizia julibrissin*
money-tree - *Eucalyptus pulverulenta*
monk's-pepper-tree - *Vitex agnus-castus*
monkey-puzzle-tree - *Araucaria araucana*
mu-oil-tree - *Vernicia montana*
mu-tree - *Vernicia montana*
mustard-tree - *Nicotiana glauca*
nettle-tree - *Celtis occidentalis, Celtis*
New Zealand tea-tree - *Leptospermum scoparium*
nickers-tree - *Gymnocladus dioica*
nutgall-tree - *Rhus chinensis*
octopus-tree - *Schefflera actinophylla*
orchid-tree - *Bauhinia variegata*

Oriental plane-tree - *Platanus orientalis*
pagoda-tree - *Plumeria rubra* f. *acutifolia, Sophora japonica*
paper-bark-tree - *Melaleuca quinquenervia*
Pará rubber-tree - *Hevea brasiliensis*
paradise-tree - *Melia azedarach, Simarouba glauca*
Paraguayan trumpet-tree - *Tabebuia argentea*
pea-tree - *Caragana, Sesbania exaltata*
Peruvian mastic-tree - *Schinus molle*
Peruvian pepper-tree - *Schinus molle*
Phoenix-tree - *Firmiana simplex*
plane-tree - *Platanus occidentalis*
planer-tree - *Planera aquatica*
poison-tree - *Acokanthera*
poonga oil-tree - *Pongamia pinnata*
portia-tree - *Thespesia populnea*
princess-tree - *Paulownia tomentosa*
provision-tree - *Pachira aquatica*
pudding-pipe-tree - *Cassia fistula*
punk-tree - *Melaleuca quinquenervia*
purple glory-tree - *Tibouchina urvilleana*
Queensland umbrella-tree - *Schefflera actinophylla*
rain-tree - *Albizia saman*
red silk cotton-tree - *Bombax ceiba*
river tea-tree - *Melaleuca leucadendra*
rough tree-fern - *Cyathea australis*
rubber-tree - *Hevea brasiliensis, Schefflera actinophylla*
sage-tree - *Vitex agnus-castus*
salamander-tree - *Antidesma bunius*
sandbox-tree - *Hura crepitans*
sausage-tree - *Kigelia africana*
scarlet wisteria-tree - *Sesbania punicea*
service-tree - *Sorbus domestica*
shaving-brush-tree - *Pseudobombax ellipticum*
shower-tree - *Cassia artemisioides, Cassia*
Siberian pea-tree - *Caragana arborescens*
silk cotton-tree - *Ceiba pentandra*
silk-tree - *Albizia julibrissin*
silk-tree albizia - *Albizia julibrissin*
silver-dollar-tree - *Eucalyptus cinerea*
silver-tree - *Conocarpus erectus* var. *sericeus*
silver trumpet-tree - *Tabebuia argentea*
siris-tree - *Albizia lebbeck*
smoke-tree - *Cotinus coggygria*
snowdrop-tree - *Halesia*
soap-tree - *Yucca elata*
soap-tree yucca - *Yucca elata*
sorrel-tree - *Oxydendrum arboreum*
spindle-tree - *Euonymus*
sponge-tree - *Acacia farnesiana*
strawberry-tree - *Arbutus unedo*
sugar-tree - *Acer barbatum*
swamp tea-tree - *Melaleuca quinquenervia*
sweet-potato-tree - *Manihot esculenta*
Syrian bead-tree - *Melia azedarach*
tallow-tree - *Sapium sebiferum*
tapa-cloth-tree - *Broussonetia papyrifera*
temple-tree - *Plumeria*
tipu-tree - *Tipuana tipu*
titi-tree - *Cliftonia monophylla*
tomato-tree - *Cyphomandra betacea*
toothache-tree - *Zanthoxylum americanum*
tree-bine - *Cissus*
tree-clover - *Melilotus alba*

tree club-moss - *Lycopodium obscurum*
tree cotton - *Gossypium arboreum, G. barbadense*
tree-fern - *Cyathea*
tree huckleberry - *Rhododendron arboreum*
tree-ivy - *Fatshedera lizei*
tree lupine - *Lupinus arboreus*
tree lyonia - *Lyonia ferruginea*
tree mallow - *Lavatera arborea*
tree-of-gold - *Tabebuia argentea*
tree-of-heaven - *Ailanthus altissima*
tree-of-kings - *Cordyline terminalis*
tree onion - *Allium cepa*
tree peony - *Paeonia suffruticosa*
tree poppy - *Dendromecon rigida*
tree-primrose - *Oenothera biennis*
tree rhododendron - *Rhododendron arboreum*
tree tobacco - *Nicotiana glauca*
tree-tomato - *Cyphomandra betacea*
trumpet-tree - *Tabebuia*
tulip-tree - *Liriodendron tulipifera*
tung oil-tree - *Vernicia fordii*
umbrella-tree - *Magnolia macrophylla, M. tripetala*
varnish-tree - *Ailanthus altissima, Aleurites moluccana,*
   Koelreuteria paniculata
Venezuelan tree-bine - *Cissus rhombifolia*
vinegar-tree - *Rhus glabra*
wayfaring-tree - *Viburnum lantana*
weeping tea-tree - *Melaleuca leucadendra*
West Indian tree-fern - *Cyathea arborea*
white silk cotton-tree - *Ceiba pentandra*
white wax-tree - *Ligustrum lucidum*
whitten-tree - *Viburnum opulus*
wig-tree - *Cotinus coggygria*
winged spindle-tree - *Euonymus alatus*
wonder-tree - *Ricinus communis*
yellow anise-tree - *Illicium parviflorum*
yellow cucumber-tree - *Magnolia acuminata* var. *subcordata*
trefoil
   big tick-trefoil - *Desmodium cuspidatum*
   big trefoil - *Lotus uliginosus*
   bird's-foot trefoil - *Lotus corniculatus*
   Canadian tick-trefoil - *Desmodium canadense*
   cluster-leaf tick-trefoil - *Desmodium glutinosum*
   greater bird's-foot trefoil - *Lotus pedunculatus*
   hoary tick-trefoil - *Desmodium canescens*
   large-bracted tick-trefoil - *Desmodium cuspidatum*
   little-leaf tick-trefoil - *Desmodium ciliare*
   marsh-trefoil - *Menyanthes trifoliata*
   naked-flowered tick-trefoil - *Desmodium nudiflorum*
   naked tick-trefoil - *Desmodium nudiflorum*
   narrow-leaf trefoil - *Lotus tenuis*
   panicled tick-trefoil - *Desmodium paniculatum*
   pine-barren tick-trefoil - *Desmodium strictum*
   pointed-leaf tick-trefoil - *Desmodium glutinosum*
   prairie trefoil - *Lotus unifoliatus*
   prostrate tick-trefoil - *Desmodium rotundifolium*
   round-leaf tick-trefoil - *Desmodium rotundifolium*
   sessile-leaf tick-trefoil - *Desmodium sessilifolium*
   sessile tick-trefoil - *Desmodium sessilifolium*
   showy tick-trefoil - *Desmodium canadense*
   shrubby trefoil - *Ptelea*
   slender trefoil - *Lotus tenuis*
   small-flowered bird's-foot trefoil - *Lotus micranthus*

   small-leaf tick-trefoil - *Desmodium ciliare*
   stiff tick-trefoil - *Desmodium strictum*
   tick-trefoil - *Desmodium canadense, Desmodium*
   trefoil - *Trifolium*
   velvety tick-trefoil - *Desmodium viridiflorum*
   yellow-trefoil - *Medicago lupulina*
trema
   Florida trema - *Trema micrantha*
   West Indian trema - *Trema lamarckiana*
trembling aspen - *Populus tremuloides*
tresses
   northern slender lady's-tresses - *Spiranthes lacera*
   slender lady's-tresses - *Spiranthes lacera*
trident maple - *Acer rubrum* var. *trilobum*
trifid beggar-ticks - *Bidens tripartita*
trifoliate-orange - *Poncirus trifoliata*
trillium
   bent trillium - *Trillium flexipes*
   big white trillium - *Trillium grandiflorum*
   drooping trillium - *Trillium flexipes*
   dwarf white trillium - *Trillium nivale*
   large-flowered trillium - *Trillium grandiflorum*
   nodding trillium - *Trillium cernuum*
   painted trillium - *Trillium undulatum*
   prairie trillium - *Trillium recurvatum*
   purple trillium - *Trillium erectum*
   red trillium - *Trillium erectum*
   snow trillium - *Trillium nivale*
   trillium - *Trifolium striatum, Trillium*
   white trillium - *Trillium grandiflorum*
triplet-lily - *Triteleia laxa*
trisetum
   nodding trisetum - *Trisetum cernuum*
   spike trisetum - *Trisetum spicatum*
   tall trisetum - *Trisetum canescens*
   Wolf's trisetum - *Trisetum wolfii*
tritoma - *Kniphofia*
trompillo - *Solanum elaeagnifolium*
tropic
   tropic ageratum - *Ageratum conyzoides*
   tropic-laurel - *Ficus benjamina*
tropical
   tropical almond - *Terminalia cattapa*
   tropical bleeding-heart - *Clerodendrum thompsoniae*
   tropical carpet grass - *Axonopus compressus*
   tropical crabgrass - *Digitaria bicornis*
   tropical croton - *Croton glandulosus* var. *septentrionalis*
   tropical kudzu - *Pueraria phaseoloides*
   tropical spiderwort - *Commelina benghalensis*
trouble, tramp's-trouble - *Smilax bona-nox*
trout-lily - *Erythronium americanum, Erythronium*
true
   true aloe - *Aloe vera*
   true forget-me-not - *Myosotis scorpioides*
   true ginger - *Zingiber officinale*
   true hemp - *Cannabis sativa*
truffle oak - *Quercus robur*
trumpet
   angel's-trumpet - *Datura innoxia*
   cat-claw trumpet - *Macfadyena unguis-cati*
   evening trumpet-flower - *Gelsemium sempervirens*

flaming-trumpet - *Pyrostegia venusta*
golden-trumpet - *Allamanda cathartica*
Paraguayan trumpet-tree - *Tabebuia argentea*
silver trumpet-tree - *Tabebuia argentea*
trumpet-creeper - *Campsis radicans*
trumpet-flower - *Tecoma stans*
trumpet-honeysuckle - *Campsis radicans, Lonicera sempervirens*
trumpet-leaf - *Sarracenia flava*
trumpet lily - *Lilium longiflorum*
trumpet narcissus - *Narcissus pseudonarcissus*
trumpet-tree - *Tabebuia*
trumpet-vine - *Campsis radicans*
trumpet-weed - *Eupatorium fistulosum, E. maculatum*
white-trumpet lily - *Lilium longiflorum*
trumpets
trumpets - *Sarracenia flava*
umbrella-trumpets - *Sarracenia flava*
yellow trumpets - *Sarracenia alata*
ts'ai, kui ts'ai - *Allium tuberosum*
tsai, pe-tsai - *Brassica pekinensis*
tsoi, kai-tsoi - *Brassica juncea*
tsuru-na - *Tetragonia tetragonioides*
tube-flower - *Clerodendrum*
tuber
tuber oat grass - *Arrhenatherum elatius*
tuber-root - *Asclepias tuberosa*
tuberose
tuberose - *Polianthes tuberosa*
wild tuberose - *Manfreda maculosa*
tuberous
hybrid tuberous begonia - *Begonia tuberhybrida*
tuberous vetch - *Lathyrus tuberosus*
tuberous vetchling - *Lathyrus tuberosus*
tuberous water-lily - *Nymphaea tuberosa*
tuckahoe - *Peltandra virginica*
tuft
golden-tuft alyssum - *Alyssum saxatile*
golden-tuft madwort - *Alyssum saxatile*
tuft-root - *Dieffenbachia*
tufted
tufted club rush - *Scirpus caespitosus*
tufted fishtail palm - *Caryota mitis*
tufted hair grass - *Deschampsia caespitosa*
tufted hard grass - *Sclerochloa dura*
tufted knotweed - *Polygonum caespitosum* var *longisetum*
tufted loosestrife - *Lysimachia thyrsiflora*
tufted love grass - *Eragrostis pectinacea*
tufted manna grass - *Glyceria septentrionalis*
tufted pansy - *Viola cornuta*
tufted rush - *Juncus acuminatus*
tufted spear grass - *Eragrostis pectinacea*
tufted vetch - *Vicia cracca*
tule - *Scirpus acutus* var. *occidentalis*
tulip
African tulip-tree - *Spathodea campanulata*
butterfly-tulip - *Calochortus*
globe-tulip - *Calochortus*
sierra star-tulip - *Calochortus nudus*
star-tulip - *Calochortus*
tulip-poplar - *Liriodendron tulipifera*

tulip-tree - *Liriodendron tulipifera*
tumble
feather tumble grass - *Eragrostis tenella*
tumble grass - *Schedonnardus paniculatus*
tumble knapweed - *Centaurea diffusa*
tumble mustard - *Sisymbrium altissimum*
tumble pigweed - *Amaranthus albus, A. graecizans*
tumble windmill grass - *Chloris verticillata*
tumbleweed
tumbleweed - *Amaranthus albus, A. graecizans, Cycloloma atriplicifolium, Salsola kali*
tumbleweed grass - *Panicum capillare*
tumbling
tumbling atriplex - *Atriplex rosea*
tumbling oracle - *Atriplex rosea*
tumbling pigweed - *Amaranthus albus*
tumeric - *Hydrastis canadensis*
tuna-plant - *Opuntia*
tung
tung - *Vernicia montana*
tung oil-tree - *Vernicia fordii*
tunka - *Benincasa hispida*
tupelo
black tupelo - *Nyssa sylvatica*
large tupelo - *Nyssa aquatica*
Ogeche tupelo - *Nyssa ogeche*
Ogeechee lime tupelo - *Nyssa ogeche*
swamp tupelo - *Nyssa sylvatica* var. *biflora*
upland tupelo - *Nyssa sylvatica*
water tupelo - *Nyssa aquatica*
turban
turban buttercup - *Ranunculus asiaticus*
turban lily - *Lilium martagon*
turbinella oak - *Quercus turbinella*
turf
big blue lily-turf - *Liriope muscari*
lily-turf - *Liriope*
Turk's
American Turk's-cap lily - *Lilium superbum*
Japanese Turk's-cap lily - *Lilium hansonii*
Turk's-cap - *Aconitum napellus, Malvaviscus arboreus, M. arboreus* var. *drummondii*
Turk's-cap lily - *Lilium martagon, L. superbum*
Turkestan
Turkestan blue-stem - *Bothriochloa ischaemum*
Turkestan rose - *Rosa rugosa*
Turkestan thistle - *Centaurea repens*
turkey
European turkey oak - *Quercus cerris*
turkey-beard - *Xerophyllum asphodeloides*
turkey-berry - *Solanum torvum*
turkey-foot - *Andropogon gerardii, A. hallii*
turkey-mullein - *Eremocarpus setigerus*
turkey oak - *Quercus incana, Q. laevis*
turkey-tangle - *Phyla nodiflora*
Turkish boxwood - *Buxus sempervirens*
turmeric - *Curcuma longa*
turnip
Indian turnip - *Arisaema triphyllum, Psoralea esculenta*
prairie-turnip - *Psoralea esculenta*
shaggy prairie-turnip - *Psoralea esculenta*

Swedish turnip - *Brassica napus* var. *napobrassica*
turnip - *Brassica napus, B. rapa*
turnip-rooted celery - *Apium graveolens* var. *rapaceum*
turnip-weed - *Rapistrum rugosum*
wild turnip - *Rapistrum rugosum*
turnpike goosefoot - *Chenopodium botrys*
turnsole - *Heliotropium indicum, Heliotropium*
turpentine-weed - *Gutierrezia*
turtle
red turtle-head - *Chelone lyonii*
turtle-head - *Chelone glabra, Chelone*
white turtle-head - *Chelone glabra*
tussock
serrated tussock - *Nassella trichotoma*
tussock bellflower - *Campanula carpatica*
tussock cotton grass - *Eriophorum vaginatum*
tussock sedge - *Carex stricta*
twayblade, large twayblade - *Liparis liliifolia*
twig, blood-twig dogwood - *Cornus sanguinea*
twinberry - *Lonicera involucrata*
twinflower - *Linnaea borealis*
twinleaf - *Jeffersonia diphylla*
twist-wood - *Viburnum lantana*
twisted
rose twisted-stalk - *Streptopus roseus*
twisted acacia - *Acacia tortuosa*
twisted heath - *Erica cinerea*
twisted-leaf yucca - *Yucca rupicola*
twisted sedge - *Carex torta*
twisted-stalk - *Streptopus amplexifolius, Streptopus*
two
two-bladed onion - *Allium fistulosum*
two-eyed violet - *Viola ocellata*
two-flowered cynthia - *Krigia biflora*
two-flowered melic - *Melica mutica*
two-grooved milk-vetch - *Astragalus bisulcatus*
two-grooved poison-vetch - *Astragalus bisulcatus*
two-leaf lady's-slipper - *Cypripedium acaule*
two-leaf nut pine - *Pinus edulis*
two-leaf Solomon's-seal - *Maianthemum canadense*
two-petal ash - *Fraxinus dipetala*
two-row stonecrop - *Sedum spurium*
two-winged silver-bell - *Halesia diptera*
udo - *Aralia cordata*
umbil-root - *Cypripedium calceolus* var. *pubescens*
umbrella
Australian umbrella-tree - *Schefflera actinophylla*
dwarf umbrella-grass - *Fuirena pumila*
ear-leaved umbrella-tree - *Magnolia fraseri*
Japanese umbrella-pine - *Sciadopitys verticillata*
Queensland umbrella-tree - *Schefflera actinophylla*
short-bristled umbrella-grass - *Fuirena breviseta*
small-flowered umbrella sedge - *Cyperus difformis*
squarrose umbrella-grass - *Fuirena squarrosa*
tall umbrella-plant - *Cyperus eragrostis*
umbrella arum - *Amorphophallus rivieri*
umbrella dodder - *Cuscuta umbellata*
umbrella magnolia - *Magnolia tripetala*
umbrella-palm - *Cyperus alternifolius*
umbrella pine - *Pinus pinea*
umbrella-plant - *Cyperus alternifolius, Eriogonum*

umbrella sedge - *Cyperus alternifolius, Cyperus*
umbrella-spurry - *Holosteum umbellatum*
umbrella-tree - *Magnolia macrophylla, M. tripetala*
umbrella-trumpets - *Sarracenia flava*
umbrella-wort - *Mirabilis nyctaginea, Mirabilis*
unbranched umbrella-grass - *Fuirena simplex*
umkokolo - *Dovyalis caffra*
unbranched umbrella-grass - *Fuirena simplex*
under-green willow - *Salix commutata*
Unguentine-cactus - *Aloe vera*
unicorn
unicorn-flower - *Proboscidea louisianica*
unicorn-plant - *Proboscidea louisianica*
unicorn-root - *Aletris farinosa*
union-leaf - *Gaultheria shallon*
up
Johnny-jump-up - *Viola tricolor*
pull-up muhly - *Muhlenbergia filiformis*
upland
upland bent grass - *Agrostis perennans*
upland boneset - *Eupatorium sessilifolium*
upland cotton - *Gossypium hirsutum*
upland-cress - *Barbarea vulgaris*
upland rice - *Oryza sativa*
upland tupelo - *Nyssa sylvatica*
upland white aster - *Solidago ptarmicoides*
upland willow - *Salix humilis*
white upland aster - *Solidago ptarmicoides*
upright
upright brome - *Bromus erectus*
upright cinquefoil - *Potentilla recta*
upright prairie coneflower - *Ratibida columnifera*
upright yellow sorrel - *Oxalis europaea*
urd - *Vigna mungo*
urn-plant - *Aechmea fasciata*
Uruguay water-primrose - *Ludwigia uruguayensis*
usambava violet - *Saintpaulia ionantha*
Utah
Utah juniper - *Juniperus osteosperma*
Utah serviceberry - *Amelanchier utahensis*
valerian
American valerian - *Cypripedium calceolus* var. *pubescens*
edible valerian - *Valeriana edulis*
Greek valerian - *Polemonium caeruleum* subsp. *villosum, P. reptans, Polemonium*
red valerian - *Centranthus ruber*
tap-rooted valerian - *Valeriana edulis*
valerian - *Valeriana officinalis*
valle, cardo-del-valle - *Centaurea americana*
valley
false lily-of-the-valley - *Maianthemum canadense, Maianthemum*
lily-of-the-valley - *Convallaria majalis*
lily-of-the-valley-bush - *Pieris japonica*
star-flowered lily-of-the-valley - *Smilacina stellata*
valley oak - *Quercus lobata*
wild lily-of-the-valley - *Maianthemum canadense, Pyrola elliptica, P. rotundifolia*
vamp, wood-vamp - *Decumaria barbara*
vanilla
vanilla - *Vanilla planifolia*

vanilla grass - *Hierochloe odorata*
vanilla-leaf - *Achlys triphylla*
Vanstadens River daisy - *Osteospermum ecklonis*
variable
  variable pondweed - *Potamogeton gramineus*
  variable water-milfoil - *Myriophyllum heterophyllum*
variegated
  variegated horsetail - *Equisetum variegatum*
  variegated-laurel - *Codiaeum*
  variegated philodendron - *Epipremnum aureum*
  variegated scouring-rush - *Equisetum variegatum*
varnish
  Japanese varnish-tree - *Firmiana simplex*
  varnish-tree - *Ailanthus altissima, Aleurites moluccana,*
    Koelreuteria paniculata
vase
  living-vase - *Aechmea*
  vase-vine - *Clematis viorna, Clematis*
Vasey's grass - *Paspalum urvillei*
  vauquelinia, Torrey's vauquelinia - *Vauquelinia californica*
vegetable
  vegetable-oyster - *Tragopogon porrifolius*
  vegetable-sponge - *Luffa aegyptiaca*
veined
  red-veined dock - *Rumex obtusifolius*
  white-veined shin-leaf - *Pyrola picta*
  white-veined wintergreen - *Pyrola picta*
veiny
  veiny dock - *Rumex venosus*
  veiny hawkweed - *Hieracium venosum*
Veitch's screw-pine - *Pandanus veitchii*
veldt daisy - *Gerbera jamesonii*
veludo-branco - *Celosia argentea*
velvet
  Bengal velvet bean - *Mucuna pruriens* var. *utilis*
  cowage velvet bean - *Mucuna pruriens* var. *utilis*
  creeping velvet grass - *Holcus mollis*
  Florida velvet bean - *Mucuna pruriens* var. *utilis*
  German velvet grass - *Holcus mollis*
  Mauritius velvet bean - *Mucuna pruriens* var. *utilis*
  rough-leaf velvet-seed - *Guettarda scabra*
  trailing velvet-plant - *Ruellia makoyana*
  velvet ash - *Fraxinus velutina*
  velvet bean - *Mucuna pruriens* var. *utilis*
  velvet-bells - *Bartsia alpina*
  velvet bent grass - *Agrostis canina*
  velvet bundle-flower - *Desmanthus velutinus*
  velvet finger grass - *Digitaria velutina*
  velvet grass - *Holcus lanatus*
  velvet-leaf - *Abutilon theophrasti*
  velvet-leaf blueberry - *Vaccinium myrtilloides*
  velvet-leaf philodendron - *Philodendron scandens* f. *micans*
  velvet lupine - *Lupinus leucophyllus*
  velvet mesquite - *Prosopis velutina*
  velvet-plant - *Gynura aurantiaca, Verbascum thapsus*
  velvet sedge - *Carex vestita*
  velvet sumac - *Rhus hirta*
  velvet-weed - *Abutilon theophrasti, Gaura parviflora*
  Yokohama velvet bean - *Mucuna pruriens* var. *utilis*
velvety
  velvety lespedeza - *Lespedeza stuevei*

velvety tick-trefoil - *Desmodium viridiflorum*
Venetian sumac - *Cotinus coggygria*
Venezuelan tree-bine - *Cissus rhombifolia*
Venice mallow - *Hibiscus trionum*
ventenata - *Ventenata dubia*
Venus's
  small Venus's-looking-glass - *Triodanis biflora*
  Venus's-comb - *Scandix pecten-veneris*
  Venus's-flytrap - *Dionaea muscipula*
  Venus's-hair fern - *Adiantum capillaris-veneris*
  Venus's-looking-glass - *Triodanis perfoliata*
  Venus's-needle - *Scandix pecten-veneris*
  Venus's-shoe - *Cypripedium calceolus* var. *pubescens*
verbena
  blue verbena - *Verbena hastata*
  clump verbena - *Verbena canadensis*
  garden verbena - *Verbena hybrida*
  rose verbena - *Verbena canadensis*
  sand-verbena - *Abronia*
  shrub-verbena - *Lantana*
  white verbena - *Verbena urticifolia*
  wild verbena - *Verbena bracteata*
  woolly verbena - *Verbena stricta*
  yellow sand-verbena - *Abronia latifolia*
verde
  blue palo-verde - *Parkinsonia florida*
  Mexican palo-verde - *Parkinsonia aculeata*
  palo-verde - *Parkinsonia florida*
vernal
  sweet vernal grass - *Anthoxanthum odoratum*
  vernal iris - *Iris verna*
verrain mallow - *Malva alcea*
vervain
  blue vervain - *Verbena hastata*
  bracted vervain - *Verbena bracteata*
  Brazilian vervain - *Verbena brasiliensis*
  creeping vervain - *Verbena canadensis*
  Dakota vervain - *Verbena bipinnatifida*
  European vervain - *Verbena officinalis*
  hoary vervain - *Verbena stricta*
  Jamaican vervain - *Stachytarpheta jamaicensis*
  mint vervain - *Verbena menthifolia*
  moss vervain - *Verbena tenuisecta*
  narrow-leaf vervain - *Verbena simplex*
  prostrate vervain - *Verbena bracteata*
  rose vervain - *Verbena canadensis*
  seashore vervain - *Verbena litoralis*
  tall vervain - *Verbena bonariensis*
  white vervain - *Verbena urticifolia*
vetch
  alpine milk-vetch - *Astragalus alpinus*
  American joint-vetch - *Aeschynomene americana*
  American vetch - *Vicia americana*
  bird vetch - *Vicia cracca*
  black-pod vetch - *Vicia sativa* subsp. *nigra*
  boreal vetch - *Vicia cracca*
  bramble vetch - *Vicia tenuifolia*
  Carolina vetch - *Vicia caroliniana*
  cicer milk-vetch - *Astragalus cicer*
  Columbia milk-vetch - *Astragalus miser* var. *serotinus*
  cow vetch - *Vicia cracca*
  crown-vetch - *Coronilla varia*

deer-vetch - *Lotus unifoliatus*
field milk-vetch - *Astragalus agrestis*
fleckled milk-vetch - *Astragalus lentiginosus*
flexible milk-vetch - *Astragalus flexuosus*
four-seeded vetch - *Vicia tetrasperma*
hairy vetch - *Vicia villosa*
hedge vetch - *Vicia sepium*
Hungarian vetch - *Vicia pannonica*
Indian joint-vetch - *Aeschynomene indica*
Indian milk-vetch - *Astragalus aboriginorum*
joint-vetch - *Aeschynomene americana*
kidney-vetch - *Anthyllis vulneraria*
large Russian vetch - *Vicia villosa*
large vetch - *Vicia gigantea*
late milk-vetch - *Astragalus miser* var. *serotinus*
licorice milk-vetch - *Astragalus glycyphyllos*
loose-flowered milk-vetch - *Astragalus tenellus*
lotus milk-vetch - *Astragalus lotiflorus*
meadow vetch - *Lathyrus pratensis*
milk-vetch - *Astragalus*
Missouri milk-vetch - *Astragalus missouriensis*
narrow-leaf milk-vetch - *Astragalus pectinatus*
northern joint-vetch - *Aeschynomene virginica*
Nuttall's milk-vetch - *Astragalus nuttallianus, A. nuttallii*
pale vetch - *Vicia caroliniana*
pliant milk-vetch - *Astragalus flexuosus*
poison-vetch - *Astragalus*
purple vetch - *Vicia americana, V. benghalensis*
purse milk-vetch - *Astragalus tenellus*
Russian sickle milk-vetch - *Astragalus falcatus*
sand vetch - *Vicia acutifolia*
Sitka vetch - *Vicia gigantea*
slender vetch - *Vicia tetrasperma*
small-flowered milk-vetch - *Astragalus nuttallianus*
spring-vetch - *Vicia sativa*
standing milk-vetch - *Astragalus adsurgens*
timber milk-vetch - *Astragalus miser*
tine-leaf milk-vetch - *Astragalus pectinatus*
tiny vetch - *Vicia hirsuta*
trailing crown-vetch - *Coronilla varia*
tuberous vetch - *Lathyrus tuberosus*
tufted vetch - *Vicia cracca*
two-grooved milk-vetch - *Astragalus bisulcatus*
two-grooved poison-vetch - *Astragalus bisulcatus*
vetch - *Vicia sativa, V. sativa* subsp. *nigra, Vicia*
vetch-leaf sophora - *Sophora davidii*
Wasatch milk-vetch - *Astragalus miser* var. *oblongifolius*
weedy milk-vetch - *Astragalus miser*
winter vetch - *Vicia villosa, V. villosa* subsp. *varia*
wood vetch - *Vicia caroliniana*
Yellowstone milk-vetch - *Astragalus miser* var. *hylophilus*
vetchling
chickling vetchling - *Lathyrus sativus*
marsh vetchling - *Lathyrus palustris*
tuberous vetchling - *Lathyrus tuberosus*
vetchling - *Lathyrus*
yellow vetchling - *Lathyrus ochroleucus, L. pratensis*
vetiver - *Vetiveria zizanioides*
viburnum
arrow-wood viburnum - *Viburnum dentatum*
black-haw viburnum - *Viburnum prunifolium*
double-file viburnum - *Viburnum plicatum* var. *tomentosum*
Linden's viburnum - *Viburnum dilatatum*

maple-leaf viburnum - *Viburnum acerifolium*
naked viburnum - *Viburnum nudum*
sweet viburnum - *Viburnum lentago, V. odoratissimum*
Walter's viburnum - *Viburnum obovatum*
Victoria palmetto - *Sabal mexicana*
Vincent's, St. Vincent's cistus - *Cistus palhinhai*
vine
Allegheny-vine - *Adlumia fungosa*
Amazon-vine - *Stigmaphyllon ciliatum*
apricot-vine - *Passiflora incarnata*
arrow-vine - *Polygonum sagittatum*
arrowhead-vine - *Syngonium podophyllum*
balloon-vine - *Cardiospermum halicababum*
bamboo-vine - *Smilax laurifolia*
barbary matrimony-vine - *Lycium barbarum*
bean-vine - *Phaseolus polystachios*
blaspheme-vine - *Smilax laurifolia*
blue-vine - *Cynanchum laeve*
Brazilian golden-vine - *Stigmaphyllon ciliatum*
breadfruit-vine - *Monstera deliciosa*
buckwheat-vine - *Brunnichia ovata*
butterfly-vine - *Stigmaphyllon ciliatum*
cat-claw-vine - *Macfadyena unguis-cati*
chestnut-vine - *Tetrastigma voinieranum*
Chinese matrimony-vine - *Lycium barbarum*
chocolate-vine - *Akebia quinata*
cinnamon-vine - *Dioscorea batatas*
cross-vine - *Bignonia capreolata*
cypress-vine - *Ipomoea quamoclit*
cypress-vine morning-glory - *Ipomoea quamoclit*
devil's-vine - *Calystegia sepium*
everlasting pea-vine - *Lathyrus latifolius*
flame-vine - *Pyrostegia venusta*
flat pea-vine - *Lathyrus sylvestris*
flat-pod pea-vine - *Lathyrus cicera*
garlic-vine - *Cydista aequinoctialis*
gold-thread-vine - *Cuscuta gronovii*
grass pea-vine - *Lathyrus sativus*
green-vine - *Convolvulus arvensis*
groundnut pea-vine - *Lathyrus tuberosus*
honey-vine - *Cynanchum laeve*
jade-vine - *Strongylodon macrobotrys*
kidney-vine - *Galium asprellum*
kudzu-vine - *Pueraria lobata*
licorice-vine - *Abrus precatorius*
lipstick-vine - *Aeschynanthus pulcher*
love-vine - *Cuscuta pentagona*
marine-vine - *Cissus incisa*
marsh pea-vine - *Lathyrus palustris*
matrimony-vine - *Lycium*
meadow pea-vine - *Lathyrus pratensis*
Mexican flame-vine - *Senecio chenopodioides*
monk's-hood-vine - *Ampelopsis aconitifolia*
orange-glow vine - *Senecio chenopodioides*
orchid-vine - *Stigmaphyllon ciliatum*
pepper-vine - *Ampelopsis arborea*
pipe-vine - *Aristolochia macrophyllum*
Pollyana-vine - *Soleirolia soleirolii*
porch-vine - *Wedelia trilobata*
princess-vine - *Cissus sicyoides*
puncture-vine - *Tribulus terrestris*
purple-passion-vine - *Gynura aurantiaca*
quarter-vine - *Bignonia capreolata*

railroad-vine - *Ipomoea pes-caprae*
rainbow-vine - *Pellionia pulchra*
rattan-vine - *Berchemia scandens*
red bugle-vine - *Aeschynanthus pulcher*
red-vine - *Brunnichia ovata*
rough pea-vine - *Lathyrus hirsutus*
royal-vine-plant - *Gynura aurantiaca*
sand-vine - *Cynanchum laeve*
sarsaparilla-vine - *Smilax pumila*
scarlet basket-vine - *Aeschynanthus pulcher*
silk-vine - *Periploca graeca*
skunk-vine - *Paederia foetida*
staff-vine - *Celastrus scandens*
strangler-vine - *Morrenia odorata*
taro-vine - *Epipremnum aureum*
teddy-bear-vine - *Cyanotis kewensis*
trumpet-vine - *Campsis radicans*
vase-vine - *Clematis viorna, Clematis*
vine cactus - *Fouquieria splendens*
vine maple - *Acer circinatum*
vine mesquite - *Panicum obtusum*
vine mesquite grass - *Panicum obtusum*
vine-of-Sodom - *Citrullus colocynthis*
vine-spinach - *Basella alba*
weather-vine - *Abrus precatorius*
wild bamboo-vine - *Smilax auriculata*
wild sweet-potato-vine - *Ipomoea pandurata*
wing-stemmed wild pea-vine - *Lathyrus palustris*

vinegar
vinegar-pear - *Passiflora laurifolia*
vinegar-tree - *Rhus glabra*
vinegar-weed - *Trichostema lanceolatum*

viola - *Viola cornuta*

violet
African violet - *Saintpaulia ionantha*
Alaska violet - *Viola langsdorfii*
alpine marsh violet - *Viola palustris*
American dog violet - *Viola conspersa*
arrow-leaf violet - *Viola sagittata*
arrowhead violet - *Viola sagittata*
bird's-foot violet - *Viola pedata*
blue marsh violet - *Viola cucullata*
blue violet - *Viola papilionacea*
bouquet-violet - *Lythrum salicaria*
Canadian violet - *Viola canadensis*
cream violet - *Viola striata*
creamy violet - *Viola striata*
dame's-violet - *Hesperis matronalis*
dog-tooth-violet - *Erythronium*
downy yellow violet - *Viola pubescens*
early blue violet - *Viola palmata*
early yellow violet - *Viola rotundifolia*
English violet - *Viola odorata*
field violet - *Viola arvensis*
florist's violet - *Viola odorata*
German violet - *Exacum affine*
great-spurred violet - *Viola selkirkii*
green-violet - *Hybanthus concolor*
halberd-leaf violet - *Viola hastata*
hook-spur violet - *Viola adunca*
horned violet - *Viola cornuta*
kidney-leaf violet - *Viola renifolia*
lance-leaf violet - *Viola lanceolata*

large-leaf white violet - *Viola incognita*
larkspur violet - *Viola palmata* var. *pedatifida*
Leconte's violet - *Viola affinis*
long-spurred violet - *Viola rostrata*
marsh blue violet - *Viola cucullata*
marsh violet - *Viola palustris*
Missouri violet - *Viola missouriensis*
northern bog violet - *Viola sororia* subsp. *affinis*
northern white violet - *Viola macloskeyi* var. *pallens,*
 V. renifolia
pale violet - *Viola striata*
Persian violet - *Exacum affine*
Philippine violet - *Barleria cristata*
pine violet - *Viola lobata*
primrose-leaf violet - *Viola primulifolia*
purple prairie violet - *Viola palmata* var. *pedatifida*
round-leaf violet - *Viola rotundifolia*
round-leaf yellow violet - *Viola rotundifolia*
smooth yellow violet - *Viola pubescens*
spear-leaf violet - *Viola hastata*
strap-leaf violet - *Viola lanceolata*
stream violet - *Viola glabella*
striped violet - *Viola striata*
sweet violet - *Viola odorata*
sweet white violet - *Viola blanda*
tall white violet - *Viola canadensis*
two-eyed violet - *Viola ocellata*
usambava violet - *Saintpaulia ionantha*
violet - *Viola*
violet crabgrass - *Digitaria violascens*
violet iris - *Iris verna*
violet lespedeza - *Lespedeza violacea*
violet-slipper gloxinia - *Sinningia speciosa*
violet wood-sorrel - *Oxalis corymbosa, O. violacea*
water violet - *Viola lanceolata*
western dog violet - *Viola adunca*
western round-leaf violet - *Viola orbiculata*
western sweet white violet - *Viola macloskeyi*
white dog-tooth-violet - *Erythronium albidum*
wild violet - *Viola tricolor*
wood violet - *Viola palmata*
woolly blue violet - *Viola sororia*
yellow prairie violet - *Viola nuttallii*
yellow wood violet - *Viola lobata*

violettas - *Anoda cristata*

viper-bugloss - *Echium vulgare*

viper's
viper's-gourd - *Trichosanthes anguina*
viper's-grass - *Scorzonera hispanica*

virgate tarweed - *Holocarpha virgata*

virgilia - *Cladrastis lutea*

virgin's-bower - *Clematis virginiana, Clematis*

Virginia
Virginia beard grass - *Andropogon virginicus*
Virginia bluebells - *Mertensia virginica*
Virginia button-weed - *Diodia virginiana*
Virginia chain fern - *Woodwardia virginica*
Virginia copper-leaf - *Acalypha virginica*
Virginia cotton grass - *Eriophorum virginicum*
Virginia cowslip - *Mertensia virginica*
Virginia creeper - *Parthenocissus quinquefolia*
Virginia dayflower - *Commelina virginica*

Virginia dwarf-dandelion - *Krigia virginica*
Virginia ground-cherry - *Physalis virginiana*
Virginia lespedeza - *Lespedeza virginica*
Virginia mallow - *Sida hermaphrodita*
Virginia mountain mint - *Pycnanthemum virginianum*
Virginia pepper-weed - *Lepidium virginicum*
Virginia pine - *Pinus virginiana*
Virginia poke - *Phytolacca americana*
Virginia rose - *Rosa virginiana*
Virginia snakeroot - *Aristolochia serpentaria*
Virginia spiderwort - *Tradescantia virginiana*
Virginia stewartia - *Stewartia melachodendron*
Virginia strawberry - *Fragaria virginiana*
Virginia sumac - *Rhus hirta*
Virginia thistle - *Cirsium virginianum*
Virginia wake-robin - *Peltandra virginica*
Virginia water-leaf - *Hydrophyllum virginianum*
Virginia wild rye - *Elymus virginicus*
Virginia willow - *Itea virginica*
Virginia yellow flax - *Linum virginianum*
vitae
    holy-wood lignum-vitae - *Guajacum sanctum*
    lignum-vitae - *Guajacum sanctum*
Vochin knapweed - *Centaurea nigrescens*
volunteer corn - *Zea mays*
wahoo
    eastern wahoo - *Euonymus atropurpureus*
    wahoo - *Euonymus atropurpureus*
    wahoo elm - *Ulmus alata*
    western wahoo - *Euonymus occidentalis*
wake
    Virginia wake-robin - *Peltandra virginica*
    wake-robin - *Trillium erectum, Trillium*
waldmeister - *Galium odoratum*
wale, little wale - *Lithospermum officinale*
walking
    devil's-walking-stick - *Aralia spinosa*
    walking anthericum - *Chlorophytum comosum*
    walking fern - *Asplenium rhizophyllum*
    walking-stick cholla - *Opuntia imbricata*
wall
    American wall fern - *Polypodium virginianum*
    wall barley - *Hordeum glaucum, H. murinum*
    wall hawkweed - *Hieracium murorum*
    wall lettuce - *Lactuca muralis*
    wall-rocket - *Diplotaxis tenuifolia*
Wallace's spike-moss - *Selaginella wallacei*
wallflower
    bushy wallflower - *Erysimum repandum*
    coast wallflower - *Erysimum capitatum*
    English wallflower - *Erysimum cheiri*
    tall worm-seed wallflower - *Erysimum hieraciifolium*
    wallflower - *Erysimum cheiri, Erysimum*
    wallflower mustard - *Erysimum cheiranthoides*
    western wallflower - *Erysimum asperum, E. capitatum*
walnut
    Arizona walnut - *Juglans major*
    black walnut - *Juglans nigra*
    California black walnut - *Juglans californica*
    California walnut - *Juglans californica*
    country-walnut - *Aleurites moluccana*
    English walnut - *Juglans regia*

    Hinds' walnut - *Juglans hindsii*
    Indian walnut - *Aleurites moluccana*
    Japanese walnut - *Juglans ailantifolia*
    little walnut - *Juglans microcarpa*
    northern California black walnut - *Juglans hindsii*
    Otaheite-walnut - *Aleurites moluccana*
    Persian walnut - *Juglans regia*
    river walnut - *Juglans microcarpa*
    southern California walnut - *Juglans californica*
    Texas black walnut - *Juglans microcarpa*
    Texas walnut - *Juglans microcarpa*
    walnut - *Juglans*
    white walnut - *Juglans cinerea*
Walter's viburnum - *Viburnum obovatum*
waltheria, Indian waltheria - *Waltheria indica*
wampi - *Clausena lansium*
wan, purple wan-dock - *Brasenia schreberi*
wand
    wand lespedeza - *Lespedeza intermedia*
    wand-like bush-clover - *Lespedeza intermedia*
    wand mullein - *Verbascum virgatum*
wandering
    green wandering-Jew - *Tradescantia albiflora*
    wandering cudweed - *Gnaphalium pensylvanicum*
    wandering-Jew - *Tradescantia fluminensis, T. zebrina*
    wandering milkweed - *Apocynum androsaemifolium*
wapato - *Sagittaria cuneata, S. latifolia*
Ward's willow - *Salix caroliniana*
warlock - *Brassica nigra*
Warrigal's cabbage - *Tetragonia tetragonioides*
warrior, Indian-warrior - *Pedicularis densiflora*
warron - *Cucurbita pepo*
wart
    wart-cress - *Coronopus*
    wart spurge - *Euphorbia helioscopia*
    wart-weed - *Chelidonium majus, Euphorbia helioscopia*
Wasatch milk-vetch - *Astragalus miser* var. *oblongifolius*
Washington
    California Washington palm - *Washingtonia filifera*
    Mexican Washington palm - *Washingtonia robusta*
    Washington palm - *Washingtonia*
Washington's
    Lady Washington's geranium - *Pelargonium domesticum*
    Martha Washington's geranium - *Pelargonium domesticum*
    Washington's hawthorn - *Crataegus phaenopyrum*
    Washington's lily - *Lilium washingtonianum*
    Washington's lupine - *Lupinus polyphyllus*
    Washington's-thorn - *Crataegus phaenopyrum*
watches - *Sarracenia flava*
water
    alkaline water-nymph - *Najas marina*
    Amazon water-lily - *Victoria amazonica*
    Amazon water-platter - *Victoria amazonica*
    American water-lily - *Nymphaea odorata*
    American water-willow - *Justicia americana*
    anchored water-hyacinth - *Eichhornia azurea*
    banana water-lily - *Nymphaea mexicana*
    blue water-lily - *Nymphaea elegans*
    broad-leaf water-leaf - *Hydrophyllum canadense*
    bulb-bearing water-hemlock - *Cicuta bulbifera*
    bushy water-primrose - *Ludwigia alternifolia, L. decurrens*

Cape blue water-lily - *Nymphaea capensis*
Carolina water-hyssop - *Bacopa caroliniana*
creeping water-primrose - *Ludwigia peploides*
cut-leaf water-horehound - *Lycopus americanus*
disc water-hyssop - *Bacopa rotundifolia*
early water grass - *Echinochloa oryzoides*
eastern water-leaf - *Hydrophyllum virginianum*
eastern water-milfoil - *Myriophyllum pinnatum*
Eisen's water-hyssop - *Bacopa eisenii*
Eurasian water-milfoil - *Myriophyllum spicatum*
European water-clover - *Marsilea quadrifolia*
European water horehound - *Lycopus europaeus*
European water-starwort - *Callitriche stagnalis*
European white water-lily - *Nymphaea alba*
eutrophic water-nymph - *Najas minor*
fairy water-lily - *Nymphoides aquatica*
Fenzel's water-lily - *Nymphaea glandulifera*
floating water fern - *Ceratopteris thalictroides*
floating water-primrose - *Ludwigia repens*
fragrant water-lily - *Nymphaea odorata*
giant water-lily - *Victoria*
hairy water-leaf - *Hydrophyllum canadense, H. macrophyllum*
Japanese water iris - *Iris ensata*
lance-leaf water-plantain - *Alisma lanceolatum*
large-flowered water-plantain - *Alisma plantago-aquatica*
large-leaf water-leaf - *Hydrophyllum macrophyllum*
large water-lily - *Nymphaea ampla*
late water grass - *Echinochloa phyllopogon*
lax water-milfoil - *Myriophyllum laxum*
magnolia water-lily - *Nymphaea tuberosa*
maple-leaf water-leaf - *Hydrophyllum canadense*
Mexican water-lily - *Nymphaea mexicana*
mild water-pepper - *Polygonum hydropiperoides*
nameless water-hyssop - *Bacopa innominata*
narrow-leaf water-plantain - *Alisma gramineum*
northern water-milfoil - *Myriophyllum exalbescens*
northern water-nymph - *Najas flexilis*
pimpled water-meal - *Wolffia papulifera*
royal water-lily - *Victoria amazonica*
salt-water cord grass - *Spartina alterniflora*
showy water-primrose - *Ludwigia uruguayensis*
small water-wort - *Elatine minima*
small yellow water-crowfoot - *Ranunculus gmelinii*
southern water grass - *Luziola fluitans*
southern water-nymph - *Najas guadalupensis*
spotted water-hemlock - *Cicuta maculata*
spotted water-meal - *Wolffia punctata*
square-pod water-primrose - *Ludwigia alternifolia*
tall water-hemp - *Amaranthus tuberculatus*
top-podded water-primrose - *Ludwigia polycarpa*
tuberous water-lily - *Nymphaea tuberosa*
Uruguay water-primrose - *Ludwigia uruguayensis*
variable water-milfoil - *Myriophyllum heterophyllum*
Virginia water-leaf - *Hydrophyllum virginianum*
water-arum - *Calla palustris*
water ash - *Fraxinus caroliniana*
water avens - *Geum rivale*
water-beech - *Carpinus caroliniana*
water-bent - *Polypogon semiverticillatus*
water-berry - *Lonicera caerulea*
water birch - *Betula occidentalis*
water bulrush - *Scirpus subterminalis*
water bur-reed - *Sparganium fluctuans*

water-celery - *Oenanthe sarmentosa, Vallisneria americana*
water-chestnut - *Pachira aquatica, Trapa natans*
water-chickweed - *Callitriche*
water-chinquapin - *Nelumbo lutea*
water-clover - *Marsilea*
water-collard - *Nuphar*
water dock - *Rumex altissimus, R. orbiculatus, R. verticillatus*
water-dragon - *Calla palustris, Saururus cernuus*
water-dropwort - *Oenanthe, Oxypolis rigidior*
water elm - *Ulmus americana*
water-feather - *Myriophyllum aquaticum*
water fern - *Ceratopteris thalictroides, Salvinia minima*
water-flaxseed - *Spirodela polyrhiza*
water foxtail - *Alopecurus geniculatus*
water fringe - *Nymphoides peltata*
water-gladiolus - *Butomus umbellatus*
water grass - *Echinochloa crus-galli*
water-hairbrush - *Catabrosa aquatica*
water-hemlock - *Cicuta maculata, Cicuta*
water-hemp - *Amaranthus rudis*
water hickory - *Carya aquatica*
water-horehound - *Lycopus americanus, Lycopus*
water horsetail - *Equisetum fluviatile*
water-hyacinth - *Eichhornia crassipes*
water-hyssop - *Bacopa monnieri*
water-ivy - *Senecio mikanioides*
water-leaf - *Hydrophyllum*
water-leaf pondweed - *Potamogeton diversifolius*
water-leaf rattle-box - *Crotalaria retusa*
water-lemon - *Passiflora laurifolia*
water-lettuce - *Pistia stratiotes*
water-lily - *Nymphaea*
water lobelia - *Lobelia dortmanna*
water locust - *Gleditsia aquatica*
water lotus - *Nelumbo*
water-maize - *Victoria amazonica*
water manna grass - *Glyceria fluitans*
water-marigold - *Bidens, Megalodonta beckii*
water-meal - *Wolffia columbiana*
water mint - *Mentha aquatica*
water-motie - *Baccharis salicifolia*
water mustard - *Barbarea vulgaris*
water-nut - *Trapa natans*
water oak - *Quercus nigra*
water-oats - *Zizania*
water-oleander - *Decodon verticillatus*
water-parsnip - *Sium suave*
water paspalum - *Paspalum fluitans*
water pennywort - *Hydrocotyle americana, H. umbellata, Hydrocotyle*
water-pepper - *Polygonum hydropiper*
water pimpernel - *Samolus parviflorus*
water-plantain - *Alisma plantago-aquatica, Alisma*
water-platter - *Victoria*
water-pod - *Ellisia nyctelea*
water-purslane - *Didiplis diandra, Ludwigia palustris*
water sedge - *Carex aquatilis*
water-shield - *Brasenia schreberi*
water smartweed - *Polygonum amphibium, P. punctatum*
water speedwell - *Veronica anagallis-aquatica*
water-spinach - *Ipomoea aquatica*
water-sprite - *Ceratopteris thalictroides*
water star grass - *Heteranthera dubia*

water-starwort - *Callitriche verna, Callitriche*
water timothy - *Alopecurus geniculatus*
water tupelo - *Nyssa aquatica*
water violet - *Viola lanceolata*
water-willow - *Baccharis salicifolia, Decodon verticillatus*
water-wisteria - *Hygrophila difformis*
water-wort - *Elatine triandra*
western water-hemlock - *Cicuta douglasii*
white water buttercup - *Ranunculus aquatilis* var. *capillaceus*
white water-crowfoot - *Ranunculus trichophyllus*
white water-lily - *Nymphaea odorata*
whorled water-milfoil - *Myriophyllum verticillatum*
wild water-lemon - *Passiflora foetida*
wing-stemmed water-primrose - *Ludwigia decurrens*
winged water-primrose - *Ludwigia decurrens*
yellow water buttercup - *Ranunculus flabellaris*
yellow water crowfoot - *Ranunculus flabellaris*
yellow water-lily - *Nuphar lutea, Nymphaea mexicana*
watercress - *Rorippa nasturtium-aquaticum*
watermelon
  Chinese watermelon - *Benincasa hispida*
  trailing watermelon-begonia - *Pellionia repens*
  watermelon - *Citrullus lanatus*
  watermelon pilea - *Pilea cadierei*
waterweed
  Brazilian waterweed - *Egeria densa*
  free-flowered waterweed - *Elodea nuttallii*
  Nuttall's waterweed - *Elodea nuttallii*
  South American waterweed - *Egeria densa*
  waterweed - *Elodea canadensis*
Watson's spike-moss - *Selaginella watsonii*
wattle, coastal wattle - *Acacia cyclopis*
wavy
  wavy-leaf aster - *Aster undulatus*
  wavy-leaf gaura - *Gaura sinuata*
  wavy-leaf oak - *Quercus undulata*
  wavy-leaf silk-tassel - *Garrya elliptica*
  wavy-leaf thistle - *Cirsium undulatum*
wax
  blue wax-weed - *Cuphea viscosissima*
  California wax myrtle - *Myrica californica*
  Drummond's wax mallow - *Malvaviscus arboreus* var. *drummondii*
  dwarf wax myrtle - *Myrica pusilla*
  Pacific wax myrtle - *Myrica californica*
  southern wax myrtle - *Myrica cerifera*
  thorough-wax - *Bupleurum rotundi-folium, Bupleurum*
  wax-balls - *Acalypha virginica*
  wax bean - *Phaseolus vulgaris*
  wax currant - *Ribes cereum*
  wax-flower - *Chimaphila, Tabernaemontana divaricata*
  wax gourd - *Benincasa hispida*
  wax-leaf meadow-rue - *Thalictrum revolutum*
  wax-leaf privet - *Ligustrum japonicum, L. lucidum*
  wax mallow - *Malvaviscus arboreus*
  wax myrtle - *Myrica californica, M. cerifera, M. heterophylla*
  wax-plant - *Hoya carnosa*
  wax-work - *Celastrus scandens*
  white wax-tree - *Ligustrum lucidum*
waxberry - *Myrica cerifera, Symphoricarpos albus, S. albus* var. *laevigatus*
waxen, woad-waxen - *Genista tinctoria*

waxy
  nerved waxy sedge - *Carex verrucosa*
  waxy cloak fern - *Notholaena sinuata*
  waxy sedge - *Carex glaucescens*
wayfaring
  American wayfaring-tree - *Viburnum alnifolium*
  wayfaring-tree - *Viburnum lantana*
wayside speedwell - *Veronica polita*
weak
  weak arctic sedge - *Carex supina*
  weak sunflower - *Helianthus debilis*
weather
  weather grass - *Stipa spartea*
  weather-plant - *Abrus precatorius*
  weather-vine - *Abrus precatorius*
  weatherglass, poorman's-weatherglass - *Anagallis arvensis*
weaver's-broom - *Spartium junceum*
wedelia
  trailing wedelia - *Wedelia trilobata*
  wedelia - *Wedelia trilobata*
wedge
  prairie wedge grass - *Sphenopholis obtusata*
  wedge grass - *Sphenopholis obtusata*
  wedge-leaf arrowhead - *Sagittaria cuneata*
  wedge-leaf fog-fruit - *Phyla cuneifolia*
  wedge-leaf rattle-box - *Crotalaria retusa*
weed
  ague-weed - *Eupatorium perfoliatum, Gentianella quinquefolia*
  alkali-weed - *Cressa truxillensis*
  alligator-weed - *Alternanthera philoxeroides*
  American burn-weed - *Erechtites hieracifolia*
  annual broom-weed - *Amphiachyris dracunculoides*
  arctic willow-weed - *Epilobium davuricum*
  artillery-weed - *Pilea microphylla*
  asthma-weed - *Lobelia inflata*
  Atlantic mock bishop's-weed - *Ptilimnium capillaceum*
  Australian burn-weed - *Erechtites minima*
  Austrian pea-weed - *Sphaerophysa salsula*
  bacon-weed - *Chenopodium album*
  beggar-weed - *Desmodium canadense, D. tortuosum*
  bishop's gout-weed - *Aegopodium podagraria*
  bishop's-weed - *Ptilimnium capillaceum*
  bitter rubber-weed - *Hymenoxys odorata*
  bitter-weed - *Ambrosia artemisiifolia, A. trifida, Artemisia biennis, Conyza canadensis, Helenium amarum*
  black-weed - *Ambrosia acanthicarpa*
  blue wax-weed - *Cuphea viscosissima*
  blue-weed - *Helianthus ciliaris*
  broad-leaf button-weed - *Spermacoce alata*
  broom-weed - *Amphiachyris dracunculoides, Gutierrezia, Sida acuta*
  buckthorn-weed - *Amsinckia intermedia*
  buffalo-weed - *Ambrosia trifida*
  burn-weed - *Erechtites hieracifolia*
  burnt-weed - *Epilobium angustifolium*
  burro-weed - *Haplopappus tenuisectus*
  butter-weed - *Abutilon theophrasti*
  butterfly-weed - *Asclepias tuberosa*
  button-weed - *Diodia teres, Spermacoce glabra*
  caesar-weed - *Urena lobata*
  candy-weed - *Polygala lutea*

Cape-weed - *Phyla nodiflora*
careless-weed - *Amaranthus retroflexus*
carpenter-weed - *Prunella vulgaris*
catch-weed - *Asperugo procumbens*
catch-weed bedstraw - *Galium aparine*
chafe-weed - *Gnaphalium obtusifolium*
cheese-weed - *Malva parviflora*
chicken-weed - *Galium boreale*
clasping pepper-weed - *Lepidium perfoliatum*
clay-weed - *Tussilago farfara*
coastal silver-weed - *Potentilla pacifica*
coffee-weed - *Cichorium intybus*
consumption-weed - *Baccharis halimifolia*
copper-weed - *Iva acerosa*
cotton weed - *Anaphalis margaritacea, Asclepias syriaca*
cow pea witch-weed - *Striga gesnerioides*
creeping beggar-weed - *Desmodium incanum*
crofton-weed - *Ageratina adenophora*
crown-weed - *Ambrosia trifida*
curly-cup gum-weed - *Grindelia squarrosa*
curly muck-weed - *Potamogeton crispus*
cut-leaf mermaid-weed - *Proserpinaca pectinata*
death-weed - *Iva axillaris*
devil-weed - *Osmanthus*
dill-weed - *Anthemis cotula*
dog-weed - *Dyssodia*
dove-weed - *Eremocarpus setigerus, Murdannia nudiflora*
Drummond's golden-weed - *Haplopappus drummondii*
emetic-weed - *Lobelia inflata*
eola-weed - *Hypericum perforatum*
false broom-weed - *Ericameria austrotexana*
fan-weed - *Thlaspi arvense*
field pepper-weed - *Lepidium campestre*
finger-weed - *Amsinckia intermedia*
flat-weed - *Hypochaeris radicata*
flax-weed - *Linaria vulgaris*
flix-weed - *Descurainia sophia*
Florida beggar-weed - *Desmodium tortuosum*
French-weed - *Galinsoga quadriradiata, Thlaspi arvense*
ghost-weed - *Euphorbia marginata*
goat-weed - *Hypericum perforatum*
golden-weed - *Haplopappus, Isocoma coronopifolia*
goose-weed - *Sphenoclea zeylandica*
gout-weed - *Aegopodium podagraria*
graveyard-weed - *Euphorbia cyparissias*
green-flowered pepper-weed - *Lepidium densiflorum*
green-stemmed Joe-Pye-weed - *Eupatorium purpureum*
gum-weed - *Grindelia squarrosa*
gut-weed - *Sonchus arvensis*
gypsy-weed - *Veronica officinalis*
hair-weed - *Cuscuta epilinum*
hairy willow-weed - *Epilobium hirsutum*
hay-fever-weed - *Ambrosia acanthicarpa*
hollow Joe-Pye-weed - *Eupatorium fistulosum*
hollow-stemmed Joe-Pye-weed - *Eupatorium fistulosum*
horse-weed - *Ambrosia trifida*
itch-weed - *Veratrum viride*
Jamestown-weed - *Datura stramonium*
jimmy-weed - *Haplopappus heterophyllus*
jimson-weed - *Datura stramonium*
Joe-Pye-weed - *Eupatorium maculatum, E. purpureum*
joint-weed - *Polygonella polygama*
joy-weed - *Alternanthera ficoidea* var. *amoena*

kariba-weed - *Salvinia molesta*
khaki-weed - *Alternanthera caracasana*
Klamath-weed - *Hypericum perforatum*
marsh mermaid-weed - *Proserpinaca palustris*
match-weed - *Gutierrezia*
meadow-weed - *Ruellia tuberosa*
meal-weed - *Chenopodium album*
menow-weed - *Ruellia tuberosa*
mercury-weed - *Acalypha virginica*
mermaid-weed - *Proserpinaca palustris, Proserpinaca*
Mexican-weed - *Caperonia castaniifolia*
mock bishop's-weed - *Ptilimnium capillaceum*
muskrat-weed - *Cicuta maculata, Thalictrum pubescens*
narrow-leaf pepper-weed - *Lepidium ruderale*
neck-weed - *Cannabis sativa, Veronica peregrina*
necklace-weed - *Actaea*
needle-weed - *Navarretia intertexta*
Nuttall's poverty-weed - *Monolepis nuttalliana*
onion-weed - *Asphodelus fistulosus*
Pacific silver-weed - *Potentilla pacifica*
perennial pepper-weed - *Lepidium latifolium*
pigeon-weed - *Lithospermum arvense*
pine-weed - *Hypericum gentianoides*
pineapple-weed - *Matricaria matricarioides*
pink-weed - *Polygonum pensylvanicum*
poison-weed - *Delphinium menziesii*
polecat-weed - *Symplocarpus foetidus*
poverty sump-weed - *Iva axillaris*
poverty-weed - *Ambrosia confertiflora, Anaphalis margaritacea, Iva axillaris*
puke-weed - *Lobelia inflata*
purple-node Joe-Pye-weed - *Eupatorium purpureum*
quick-silver-weed - *Thalictrum dioicum*
quick-weed - *Galinsoga parviflora*
quinine-weed - *Parthenium hysterophorus*
rancher's fire-weed - *Amsinckia menziesii*
rattle-weed - *Astragalus lentiginosus, Crotalaria sagittalis*
rattlesnake-weed - *Daucus pusillus, Hieracium venosum*
red-weed - *Melochia corchorifolia*
resin-weed - *Gutierrezia*
rheumatism-weed - *Apocynum cannabinum*
rough button-weed - *Diodia teres*
rough-seed clammy-weed - *Polanisia dodecandra*
rough sump-weed - *Iva annua*
rush skeleton-weed - *Chondrilla juncea*
rust-weed - *Polypremum procumbens*
scorpion-weed - *Phacelia purshii, Phacelia*
sessile joy-weed - *Alternanthera sessilis*
sickle-weed - *Falcaria vulgaris*
silver-leaf poverty-weed - *Ambrosia tomentosa*
silver-weed - *Potentilla anserina*
silver-weed cinquefoil - *Potentilla anserina*
skeleton-weed - *Chondrilla juncea, Lygodesmia juncea*
skunk-weed - *Croton texensis, Navarretia squarrosa*
skunk-weed-gilia - *Navarretia squarrosa*
slobber-weed - *Euphorbia maculata*
smoke-weed - *Eupatorium maculatum*
smooth button-weed - *Spermacoce glabra*
snake-weed - *Conium maculatum*
snap-weed - *Impatiens*
soap-weed - *Yucca elata, Y. glauca*
soldier-weed - *Amaranthus spinosus, Colubrina elliptica*
spike-weed - *Hemizonia pungens*

spotted Joe-Pye-weed - *Eupatorium maculatum*
spotted snap-weed - *Impatiens capensis*
squaw-weed - *Ageratina altissima, Senecio aureus, Senecio*
stagger-weed - *Delphinium menziesii*
stinking pepper-weed - *Lepidium ruderale*
styptic-weed - *Senna occidentalis*
sweet Joe-Pye-weed - *Eupatorium purpureum*
tape-weed - *Vallisneria spiralis*
telegraph-plant-weed - *Heterotheca grandiflora*
Texas blue-weed - *Helianthus ciliaris*
Texas-weed - *Caperonia palustris*
three-flowered beggar-weed - *Desmodium triflorum*
tick-weed - *Verbesina virginica*
tinker's-weed - *Triosteum perfoliatum*
Tipton's-weed - *Hypericum perforatum*
tobacco-weed - *Elephantopus tomentosus*
tobacco witch-weed - *Striga gesnerioides*
tongue pepper-weed - *Lepidium nitidum*
trumpet-weed - *Eupatorium fistulosum, E. maculatum*
turnip-weed - *Rapistrum rugosum*
turpentine-weed - *Gutierrezia*
velvet-weed - *Abutilon theophrasti, Gaura parviflora*
vinegar-weed - *Trichostema lanceolatum*
Virginia button-weed - *Diodia virginiana*
Virginia pepper-weed - *Lepidium virginicum*
wart-weed - *Chelidonium majus, Euphorbia helioscopia*
western clammy-weed - *Polanisia dodecandra* subsp. *trachysperma*
western seep-weed - *Suaeda calceoliformis*
white-weed - *Ageratum conyzoides, Ambrosia tomentosa, Chrysanthemum frutescens, Heliotropium curassavicum, Leucanthemum vulgare, Malvella leprosa*
willow-weed - *Polygonum lapathifolium*
witch-weed - *Striga asiatica*
woolly-leaf poverty-weed - *Ambrosia grayi*
woolly white drought-weed - *Eremocarpus setigerus*
Yankee-weed - *Eupatorium compositifolium*
yellow burn-weed - *Amsinckia lycopsoides*
yellow-weed - *Barbarea vulgaris*
weedy
weedy milk-vetch - *Astragalus miser*
weedy rattle-box - *Crotalaria sagittalis*
weeks
six-weeks grama - *Bouteloua barbata*
six-weeks grass - *Poa annua*
six-weeks needle grama - *Bouteloua aristidoides*
six-weeks three-awn - *Aristida adscensionis*
ten-weeks stock - *Matthiola incana* var. *annua*
weeping
weeping alkali grass - *Puccinellia distans*
weeping bottlebrush - *Callistemon viminalis*
weeping fig - *Ficus benjamina*
weeping lantana - *Lantana montevidensis*
weeping-lime - *Tilia petiolaris*
weeping love grass - *Eragrostis curvula*
weeping spruce - *Picea breweriana*
weeping tea-tree - *Melaleuca leucadendra*
weeping willow - *Salix babylonica*
weigela, pink weigela - *Weigela florida*
well, soap-well - *Yucca glauca*
Welsh onion - *Allium fistulosum*
welted thistle - *Carduus crispus*

West
West Coast rhododendron - *Rhododendron macrophyllum*
West Indian birch - *Bursera simaruba*
West Indian black-thorn - *Acacia farnesiana*
West Indian dogwood - *Piscidia piscipula*
West Indian gherkin - *Cucumis anguria*
West Indian holly - *Leea coccinea*
West Indian lemongrass - *Cymbopogon citratus*
West Indian lime - *Citrus aurantiifolia*
West Indian mahogany - *Swietenia mahagoni*
West Indian tree-fern - *Cyathea arborea*
West Indian trema - *Trema lamarckiana*
western
western androsace - *Androsace occidentalis*
western azalea - *Rhododendron occidentale*
western balsam poplar - *Populus trichocarpa*
western beggar-ticks - *Bidens vulgata*
western bistort - *Polygonum bistortoides*
western bleeding-heart - *Dicentra formosa*
western blue flag - *Iris missouriensis*
western blueberry - *Vaccinium occidentalis*
western bracken - *Pteridium aquilinum* var. *pubescens*
western bracken fern - *Pteridium aquilinum* var. *pubescens*
western bristle-cone pine - *Pinus longaeva*
western burning-bush - *Euonymus occidentalis*
western catalpa - *Catalpa speciosa*
western chokecherry - *Prunus virginiana* var. *demissa*
western clammy-weed - *Polanisia dodecandra* subsp. *trachysperma*
western cliff fern - *Woodsia oregana*
western coral bean - *Erythrina flabelliformis*
western cornel - *Cornus glabrata*
western daisy - *Astranthium integrifolium*
western dock - *Rumex occidentalis*
western dog violet - *Viola adunca*
western elderberry - *Sambucus cerulea*
western elodea - *Elodea nuttallii*
western false gromwell - *Onosmodium molle*
western false hellebore - *Veratrum californicum*
western fescue - *Festuca occidentalis*
western fiddle-neck - *Amsinckia tessellata*
western field buttercup - *Ranunculus occidentalis*
western flax - *Camelina microcarpa*
western flax dodder - *Cuscuta campestris*
western gromwell - *Lithospermum ruderale*
western hawk's-beard - *Crepis occidentalis*
western hazel - *Corylus cornuta* var. *californica*
western heart's-ease - *Viola ocellata*
western hemlock - *Tsuga heterophylla*
western honey mesquite - *Prosopis glandulosa* var. *torreyana*
western ironweed - *Vernonia baldwinii*
western juniper - *Juniperus occidentalis*
western Labrador tea - *Ledum glandulosum*
western lady's-mantle - *Aphanes occidentalis*
western larch - *Larix occidentalis*
western marsh-rosemary - *Limonium californicum*
western mugwort - *Artemisia ludoviciana*
western needle grass - *Stipa occidentalis*
western oak - *Quercus garryana*
western orange-cup lily - *Lilium philadelphicum* var. *andinum*
western osier - *Cornus occidentalis*
western poison-ivy - *Toxicodendron rydbergii, T. vernix*
western poison-oak - *Rhus diversiloba*

western polypody - *Polypodium hesperium*
western purslane speedwell - *Veronica peregrina* subsp. *xalapensis*
western ragweed - *Ambrosia psilostachya, A. psilostachya* var. *coronopifolia*
western rattlesnake plaintain - *Goodyera oblongifolia*
western red-cedar - *Thuja plicata*
western redbud - *Cercis occidentalis*
western rose - *Rosa woodsii*
western round-leaf violet - *Viola orbiculata*
western salsify - *Tragopogon dubius*
western sand cherry - *Prunus pumila* var. *besseyi*
western scale-seed - *Spermolepis divaricata*
western seep-weed - *Suaeda calceoliformis*
western serviceberry - *Amelanchier alnifolia*
western skunk-cabbage - *Lysichiton americanum*
western snowberry - *Symphoricarpos occidentalis*
western soapberry - *Sapindus saponaria* var. *drummondii*
western stickseed - *Hackelia floribunda*
western sticktight - *Lappula occidentalis*
western stipa - *Stipa comata*
western sunflower - *Helianthus occidentalis*
western sweet white violet - *Viola macloskeyi*
western sword fern - *Polystichum munitum*
western sycamore - *Platanus racemosa*
western tansy mustard - *Descurainia incisa*
western thimbleberry - *Rubus parviflorus*
western thistle - *Cirsium occidentale*
western wahoo - *Euonymus occidentalis*
western wallflower - *Erysimum asperum, E. capitatum*
western water-hemlock - *Cicuta douglasii*
western wheat grass - *Elymus smithii*
western white pine - *Pinus monticola*
western whorled milkweed - *Asclepias subverticillata*
western wild cucumber - *Marah oreganus*
western wild grape - *Vitis californica*
western yarrow - *Achillea millefolium*
western yellow pine - *Pinus ponderosa*
western yew - *Taxus brevifolia*

wheat
Alaska wheat - *Triticum turgidum*
beardless wheat grass - *Pseudoroegneria spicata*
blue-bunch wheat grass - *Pseudoroegneria spicata*
bread wheat - *Triticum aestivum*
cone wheat - *Triticum turgidum*
cow-wheat - *Melampyrum lineare*
crested wheat grass - *Agropyron cristatum*
desert crested wheat grass - *Agropyron desertorum*
durum wheat - *Triticum durum*
English wheat - *Triticum turgidum*
fairway crested wheat grass - *Agropyron cristatum*
false wheat grass - *Leymus chinensis*
Inca wheat - *Amaranthus caudatus*
Indian wheat - *Fagopyrum tataricum*
intermediate wheat grass - *Elytrigia intermedia*
Mediterranean wheat - *Triticum turgidum*
Polish wheat - *Triticum polonicum*
poulard wheat - *Triticum turgidum*
pubescent wheat grass - *Elytrigia intermedia*
rivet wheat - *Triticum turgidum*
Siberian wheat grass - *Agropyron fragile* subsp. *sibiricum*
slender wheat grass - *Elymus trachycaulus*
standard crested wheat grass - *Agropyron desertorum*

strand-wheat - *Lolium perenne*
tall wheat grass - *Elytrigia elongata*
thick-spike wheat grass - *Elymus lanceolatus*
western wheat grass - *Elymus smithii*
wheat - *Triticum aestivum, Triticum*
wheat grass - *Agropyron, Elytrigia repens*
wheat-thief - *Bromus secalinus, Lithospermum arvense*

wheel
fire-wheel - *Gaillardia pulchella*
fire-wheel-tree - *Stenocarpus sinuatus*
wheel-of-fire - *Stenocarpus sinuatus*
wheel-scale saltbush - *Atriplex elegans*

Wheeler's bluegrass - *Poa nervosa*
whin-berry - *Vaccinium myrtillus*
whip
coach-whip - *Fouquieria splendens*
whip-grass - *Scleria triglomerata*
whip-poor-will-shoe - *Cypripedium calceolus* var. *pubescens*
whip-tongue - *Galium mollugo*

whisk-fern - *Psilotum nudum*
whisker cactus - *Pachycereus schottii*
whiskers, old-man's-whiskers - *Geum triflorum*
whistle
whistle-wood - *Acer pensylvanicum*
wode-whistle - *Conium maculatum*

white
Arizona white oak - *Quercus arizonica*
Atlantic white-cedar - *Chamaecyparis thyoides*
big white trillium - *Trillium grandiflorum*
California white oak - *Quercus lobata*
Chinese white poplar - *Populus tomentosa*
coast white-alder - *Clethra alnifolia*
dwarf white birch - *Betula minor*
dwarf white trillium - *Trillium nivale*
eastern white pine - *Pinus strobus*
European white birch - *Betula pendula*
European white water-lily - *Nymphaea alba*
flat-top white aster - *Aster umbellatus*
giant white-top sedge - *Rhynchospora latifolia*
glaucous white-lettuce - *Prenanthes racemosa*
hairy white-top - *Cardaria pubescens*
Himalayan white pine - *Pinus wallichiana*
Kenyan wild white clover - *Trifolium semipilosum*
large-leaf white violet - *Viola incognita*
large white petunia - *Petunia axillaris*
leaf white orchid - *Habenaria dilatata*
lens-podded white-top - *Cardaria chalepensis*
lowland white fir - *Abies grandis*
mountain white-alder - *Clethra acuminata*
mountain white potentilla - *Potentilla tridentata*
northern white-cedar - *Thuja occidentalis*
northern white violet - *Viola macloskeyi* var. *pallens, V. renifolia*
Oregon white oak - *Quercus garryana*
Ozark white-cedar - *Juniperus ashei*
pendant white-lime - *Tilia petiolaris*
pink-and-white-shower - *Cassia javanica* var. *indochinensis*
red-white-and-blue-flower - *Cuphea ignea*
robust white foxtail - *Setaria viridis* var. *robusta-alba*
silky-white everlasting - *Helipterum splendidum*
small white aster - *Aster laterifolius*
small white lady's-slipper - *Cypripedium candidum*

small white morning-glory - *Ipomoea lacunosa*
smooth white-lettuce - *Prenanthes racemosa*
southern white-cedar - *Chamaecyparis thyoides*
squarrose white aster - *Aster ericoides*
swamp white-cedar - *Chamaecyparis thyoides*
swamp white oak - *Quercus bicolor, Q. michauxii*
sweet white violet - *Viola blanda*
tall flat-top white aster - *Aster umbellatus*
tall white bog orchid - *Habenaria dilatata*
tall white-lettuce - *Prenanthes altissima*
tall white violet - *Viola canadensis*
tassel-white - *Itea virginica*
Texas white-brush - *Aloysia gratissima*
upland white aster - *Solidago ptarmicoides*
western sweet white violet - *Viola macloskeyi*
western white pine - *Pinus monticola*
white alder - *Alnus incana, A. rhombifolia*
white ash - *Fraxinus americana*
white avens - *Geum canadense*
white baby-blue-eyes - *Nemophila menziesii* var. *atomaria*
white baneberry - *Actaea alba*
white-bark maple - *Acer leucoderme*
white-bark pine - *Pinus albicaulis*
white basswood - *Tilia heterophylla*
white beak rush - *Rhynchospora alba*
white bear sedge - *Carex albursina*
white birch - *Betula papyrifera, B. populifolia*
white-bottle - *Silene csereii*
white-bracted eupatorium - *Eupatorium album*
white-bracted hymenopappus - *Hymenopappus scabiosaeus*
white-brush - *Aloysia gratissima*
white bryony - *Bryonia alba*
white buttonwood - *Laguncularia racemosa*
white calla-lily - *Zantedeschia aethiopica*
white camass - *Zigadenus elegans*
white campion - *Silene latiflora*
white chervil - *Cryptotaenia canadensis*
white cinquefoil - *Potentilla arguta*
white clover - *Trifolium repens*
white cockle - *Silene latiflora*
white cohosh - *Actaea alba*
white currant - *Ribes rubrum*
white daisy - *Chrysanthemum frutescens, Leucanthemum vulgare*
white dead-nettle - *Lamium album*
white dock - *Rumex pallidus*
white dog-tooth-violet - *Erythronium albidum*
white dogwood - *Cornus florida*
white-edge morning-glory - *Ipomoea nil*
white elm - *Ulmus americana*
white evening-primrose - *Oenothera speciosa*
white field aster - *Aster lanceolatus*
white fir - *Abies amabilis, A. concolor*
white flax - *Linum catharticum*
white-flowered currant - *Ribes cereum*
white-flowered gourd - *Lagenaria siceraria*
white-flowered milkweed - *Euphorbia corollata*
white-flowered spiderwort - *Tradescantia fluminensis*
white forget-me-not - *Cryptantha intermedia*
white-ginger - *Hedychium coronarium*
white goldenrod - *Solidago bicolor*
white goosefoot - *Chenopodium album*
white gourd - *Benincasa hispida*

white grass - *Leersia*
white-heart hickory - *Carya tomentosa*
white heart-leaf aster - *Aster divaricatus*
white heath aster - *Aster pilosus*
white hedge bedstraw - *Galium mollugo*
white hellebore - *Veratrum viride*
white hickory - *Carya tomentosa*
white horehound - *Marrubium vulgare*
white-horse - *Danthonia spicata*
white-horse-nettle - *Solanum elaeagnifolium*
white kerria - *Rhodotypos*
white lady's-slipper - *Cypripedium candidum*
white-laurel - *Rhododendron maximum*
white-leaf fittonia - *Fittonia verschaffeltii*
white-leaf franseria - *Ambrosia tomentosa*
white-leaf manzanita - *Arctostaphylos viscida*
white-leaf marlock - *Eucalyptus tetragona*
white-lettuce - *Prenanthes alba*
white loco - *Oxytropis lambertii*
white lupine - *Lupinus albus*
white mallow - *Malvella leprosa*
white-man's-foot - *Plantago major*
white mandarin - *Streptopus amplexifolius*
white mangrove - *Laguncularia racemosa*
white maple - *Acer saccharinum*
white-margin spurge - *Euphorbia albomarginata*
white marguerite - *Chrysanthemum frutescens*
white melilot - *Melilotus alba*
white mignonette - *Reseda alba*
white milkweed - *Asclepias variegata*
white-moon petunia - *Petunia axillaris*
white morning-glory - *Ipomoea lacunosa*
white moth mullein - *Verbascum blattaria* var. *albiflora*
white mulberry - *Morus alba*
white mullein - *Verbascum lychnitis*
white mustard - *Sinapis alba*
white oak - *Quercus alba*
white oat grass - *Danthonia spicata*
white papinac - *Leucaena leucocephala*
white pea - *Lathyrus ochroleucus*
white pepper - *Piper nigrum*
white pigweed - *Amaranthus albus*
white pine - *Pinus strobus*
white-plantain - *Antennaria plantaginifolia*
white poplar - *Populus alba*
white potato - *Solanum tuberosum*
white prairie-clover - *Dalea candida*
white prickle poppy - *Argemone albiflora*
white pumpkin - *Benincasa hispida*
white-root - *Asclepias tuberosa*
white sage - *Artemisia ludoviciana, Salvia apiana*
white sanicle - *Ageratina altissima*
white sapote - *Casimiroa edulis*
white sassafras - *Sassafras albidum*
white silk cotton-tree - *Ceiba pentandra*
white snakeroot - *Ageratina altissima*
white spruce - *Picea glauca*
white-stem evening-primrose - *Oenothera nuttallii*
white-stem filaree - *Erodium moschatum*
white-stem gooseberry - *Ribes divaricatum*
white-stem pondweed - *Potamogeton praelongus*
white-stem stick-leaf - *Mentzelia albicaulis*
white swallow-wort - *Vincetoxicum hirundinaria*

white swamp azalea - *Rhododendron viscosum*
white sweet-clover - *Melilotus alba*
white-thorn acacia - *Acacia constricta*
white thoroughwort - *Eupatorium album*
white-tip clover - *Trifolium variegatum*
white-top - *Cardaria draba, C. pubescens, Erigeron annuus, E. strigosus*
white-top sedge - *Rhynchospora colorata*
white torch cactus - *Echinopsis spachiana*
white trillium - *Trillium grandiflorum*
white-trumpet lily - *Lilium longiflorum*
white turtle-head - *Chelone glabra*
white upland aster - *Solidago ptarmicoides*
white-veined shin-leaf - *Pyrola picta*
white-veined wintergreen - *Pyrola picta*
white verbena - *Verbena urticifolia*
white vervain - *Verbena urticifolia*
white walnut - *Juglans cinerea*
white water buttercup - *Ranunculus aquatilis* var. *capillaceus*
white water-crowfoot - *Ranunculus trichophyllus*
white water-lily - *Nymphaea odorata*
white wax-tree - *Ligustrum lucidum*
white-weed - *Ageratum conyzoides, Ambrosia tomentosa, Chrysanthemum frutescens, Heliotropium curassavicum, Leucanthemum vulgare, Malvella leprosa*
white willow - *Salix alba*
white-wood aster - *Aster divaricatus*
wild white licorice - *Galium circaezans*
woolly white drought-weed - *Eremocarpus setigerus*
yellow-white sedge - *Carex albolutescens*
White Mountains dogwood - *Viburnum alnifolium*
whitethorn, mountain whitethorn - *Ceanothus cordulatus*
whitewood - *Liriodendron tulipifera, Tabebuia rosea, Tilia americana*
whitlow
Carolina whitlow-grass - *Draba reptans*
spring whitlow-grass - *Draba verna*
whitlow-grass - *Draba verna*
whitlow-wort - *Paronychia*
wood whitlow-grass - *Draba nemorosa*
whitten-tree - *Viburnum opulus*
whole-leaf rosinweed - *Silphium integrifolium*
whorled
eastern whorled milkweed - *Asclepias verticillata*
little whorled pogonia - *Isotria medeoloides*
Mexican whorled milkweed - *Asclepias fascicularis*
small whorled pogonia - *Isotria medeoloides*
western whorled milkweed - *Asclepias subverticillata*
whorled aster - *Aster acuminatus*
whorled carpetweed - *Mollugo verticillata*
whorled coreopsis - *Coreopsis verticillata*
whorled loosestrife - *Lysimachia quadrifolia*
whorled milkweed - *Asclepias subverticillata, A. verticillata*
whorled milkwort - *Polygala verticillata*
whorled pennywort - *Hydrocotyle verticillata*
whorled rosinweed - *Silphium trifoliatum*
whorled sage - *Salvia verticillata*
whorled water-milfoil - *Myriophyllum verticillatum*
whorled wood aster - *Aster acuminatus*
whortleberry - *Vaccinium corymbosum, V. myrtillus*
wickup
Indian wickup - *Epilobium angustifolium*

wickup - *Epilobium angustifolium*
wicky - *Kalmia angustifolia*
wicopy - *Dirca palustris*
widdy - *Potentilla fruticosa*
wide
wide-fruit sedge - *Carex angustata*
wide-leaf phlox - *Phlox amplifolia*
widgeon-grass - *Ruppia maritima*
widow, grass-widow - *Sisyrinchium douglasii*
widow's
widow's-cross - *Sedum pulchellum*
widow's-frill - *Silene stellata*
widow's-tears - *Tradescantia virginiana*
wig-tree - *Cotinus coggygria*
wild
Altai wild rye - *Leymus angustus*
annual wild bean - *Strophostyles helvola*
annual wild rice - *Zizania aquatica*
Arizona wild cotton - *Gossypium thurberi*
basin wild rye - *Leymus cinereus*
beardless wild rye - *Leymus triticoides*
blue wild indigo - *Baptisia australis*
blue wild rye - *Elymus glaucus*
California wild grape - *Vitis californica*
California wild rose - *Rosa californica*
Canadian wild lettuce - *Lactuca canadensis*
Canadian wild rye - *Elymus canadensis*
Chinese wild peach - *Prunus davidiana*
Chinese wild rice - *Leymus chinensis*
cultivated northern wild rice - *Zizania palustris*
downy wild rye - *Elymus villosus*
dwarf wild indigo - *Amorpha nana*
eastern wild rice - *Zizania aquatica*
European wild pansy - *Viola tricolor*
fetid wild pumpkin - *Cucurbita foetidissima*
Florida wild lettuce - *Lactuca floridana*
giant wild rye - *Leymus cinereus*
hairy wild rye - *Elymus villosus*
Kenyan wild white clover - *Trifolium semipilosum*
long-leaf wild buckwheat - *Eriogonum longifolium*
mammoth wild rye - *Leymus racemosus*
Manchurian wild rice - *Zizania latifolia*
nodding wild onion - *Allium cernuum*
northern wild comfrey - *Cynoglossum boreale*
perennial wild bean - *Strophostyles umbellata*
pink wild bean - *Strophostyles umbellata*
prairie wild rose - *Rosa arkansana*
Russian wild rye - *Elymus junceus*
Salina wild rye - *Leymus salinus*
Siberian wild rye - *Elymus sibiricus*
slender wild oat - *Avena barbata*
small-flowered wild bean - *Strophostyles leiosperma*
smooth-seeded wild bean - *Strophostyles leiosperma*
southern wild crabapple - *Malus angustifolia*
southern wild rice - *Zizaniopsis miliacea*
southern wild senna - *Senna marilandica*
trailing wild bean - *Strophostyles helvola*
Virginia wild rye - *Elymus virginicus*
western wild cucumber - *Marah oreganus*
western wild grape - *Vitis californica*
wild allspice - *Lindera benzoin*
wild apple - *Malus sylvestris*

wild balsam-apple - *Echinocystis lobata*
wild bamboo-vine - *Smilax auriculata*
wild barley - *Hordeum jubatum, H. leporinum*
wild basil - *Clinopodium vulgare*
wild bean - *Apios americana, Phaseolus polystachios, Strophostyles helvola*
wild beet - *Amaranthus hybridus, A. retroflexus, Saxifraga pensylvanica*
wild begonia - *Rumex venosus*
wild bergamot - *Monarda fistulosa, Monarda*
wild black cherry - *Prunus serotina*
wild black currant - *Ribes americanum*
wild blue flax - *Linum grandiflorum, L. perenne*
wild buckwheat - *Eriogonum, Polygonum convolvulus*
wild cabbage - *Brassica oleracea*
wild calla - *Calla palustris*
wild carrot - *Daucus carota*
wild celery - *Ciclospermum leptophyllum*
wild chamomile - *Matricaria chamomilla*
wild cherry - *Prunus ilicifolia*
wild chervil - *Anthriscus sylvestris, Cryptotaenia canadensis*
wild cinnamon - *Canella winteriana*
wild cocoa - *Pachira aquatica*
wild coffee - *Colubrina arborescens, Polyscias guilfoylei, Triosteum aurantiacum, T. perfoliatum*
wild columbine - *Aquilegia canadensis*
wild comfrey - *Cynoglossum virginianum, Hackelia virginiana*
wild cotton - *Hibiscus moscheutos, H. moscheutos* subsp. *palustris*
wild crabapple - *Malus ioensis, Peraphyllum ramosissimum*
wild crocus - *Anemone patens*
wild cucumber - *Echinocystis lobata*
wild date palm - *Phoenix reclinata*
wild dilly - *Manilkara jaimiqui* subsp. *emarginata*
wild emmer - *Triticum dicoccoides*
wild four-o'clock - *Mirabilis nyctaginea*
wild garlic - *Allium canadense, A. vineale*
wild geranium - *Geranium maculatum*
wild ginger - *Asarum canadense, Asarum*
wild goose plum - *Prunus hortulana, P. munsoniana*
wild gourd - *Cucumis dipsaceus, Cucurbita foetidissima*
wild heliotrope - *Heliotropium curassavicum*
wild hemp - *Ambrosia trifida, Galeopsis tetrahit*
wild hippo - *Euphorbia corollata*
wild honeysuckle - *Gaura coccinea, Lonicera dioica*
wild hyacinth - *Camassia scilloides, Dichelostemma pulchellum, Scilla*
wild hydrangea - *Hydrangea arborescens, Rumex venosus*
wild ipecac - *Euphorbia ipecacuanhae*
wild iris - *Iris versicolor*
wild jalap - *Podophyllum peltatum*
wild leek - *Allium tricoccum*
wild lemon - *Podophyllum peltatum*
wild lettuce - *Lactuca biennis*
wild licorice - *Abrus precatorius, Galium circaezans, G. lanceolatum, Glycyrrhiza lepidota*
wild lilac - *Ceanothus sanguineus*
wild lily-of-the-valley - *Maianthemum canadense, Pyrola elliptica, P. rotundifolia*
wild lime - *Zanthoxylum fagara*
wild lupine - *Lupinus perennis*
wild madder - *Galium mollugo*

wild marigold - *Tagetes minuta*
wild marjoram - *Origanum vulgare*
wild marrow - *Cucurbita texana*
wild mint - *Mentha arvensis*
wild monk's-hood - *Aconitum uncinatum*
wild morning-glory - *Calystegia sepium, Convolvulus arvensis*
wild musk - *Erodium cicutarium*
wild oats - *Chasmanthium latifolium, Uvularia sessilifolia*
wild olive - *Elaeagnus angustifolia, Halesia tetraptera, Nyssa aquatica, Osmanthus americanus*
wild onion - *Allium amplectens, A. canadense, A. cernuum, A. drummondii, A. stellatum, A. textile, A. vineale*
wild opium - *Lactuca serriola*
wild orange - *Prunus caroliniana*
wild orange-red lily - *Lilium philadelphicum*
wild parsnip - *Pastinaca sativa*
wild passionflower - *Passiflora incarnata*
wild pea - *Crotalaria sagittalis, Lathyrus palustris, Lathyrus*
wild pepper - *Vitex agnus-castus*
wild petunia - *Petunia parviflora, Ruellia caroliniensis*
wild pineapple - *Tillandsia fasciculata*
wild pink - *Silene caroliniana, S. regia*
wild plum - *Prunus americana*
wild poinsettia - *Euphorbia heterophylla*
wild portulaca - *Portulaca oleracea*
wild potato - *Chlorogalum pomeridianum, Ipomoea pandurata, Solanum jamesii*
wild proso millet - *Panicum miliaceum*
wild pumpkin - *Cucurbita foetidissima*
wild quinine - *Parthenium integrifolium*
wild radish - *Raphanus raphanistrum*
wild raisin - *Viburnum cassinoides*
wild red cherry - *Prunus pensylvanica*
wild rhubarb - *Rumex hymenosepalus*
wild rice - *Zizania aquatica, Zizania*
wild robusta coffee - *Coffea canephora*
wild rosemary - *Ledum palustre*
wild rye - *Elymus*
wild sapodilla - *Manilkara jaimiqui* subsp. *emarginata*
wild sarsaparilla - *Aralia nudicaulis*
wild senna - *Senna hebecarpa, S. marilandica*
wild sensitive-plant - *Chamaecrista nictitans*
wild snapdragon - *Linaria vulgaris*
wild snowball - *Ceanothus americanus*
wild spinach - *Chenopodium bonus-henricus*
wild strawberry - *Fragaria virginiana*
wild sugar-cane - *Saccharum spontaneum*
wild sunflower - *Helianthus annuus*
wild sweet crabapple - *Malus coronaria*
wild sweet pea - *Lathyrus latifolius*
wild sweet-potato-vine - *Ipomoea pandurata*
wild sweet-William - *Phlox divaricata, P. maculata*
wild tamarind - *Lysiloma latisiliqua*
wild tansy - *Ambrosia artemisiifolia*
wild taro - *Colocasia esculenta*
wild teasel - *Dipsacus fullonum*
wild thyme - *Thymus serpyllum*
wild tobacco - *Nicotiana repanda*
wild tomato - *Solanum triflorum*
wild tongue-grass - *Lepidium densiflorum*
wild tuberose - *Manfreda maculosa*
wild turnip - *Rapistrum rugosum*

wild verbena - *Verbena bracteata*
wild violet - *Viola tricolor*
wild water-lemon - *Passiflora foetida*
wild white licorice - *Galium circaezans*
wild yam - *Dioscorea villosa*
wild yellow lily - *Lilium canadense*
wing-stemmed wild pea-vine - *Lathyrus palustris*
winter wild oat - *Avena ludoviciana*
yellow wild licorice - *Galium lanceolatum*

will
nimble-will - *Muhlenbergia schreberi*
whip-poor-will-shoe - *Cypripedium calceolus* var. *pubescens*

William
sweet-William - *Dianthus barbatus*
sweet-William catchfly - *Silene armeria*
wild sweet-William - *Phlox divaricata, P. maculata*

Willie, stinking-Willie - *Senecio jacobaea*

willow
Alaska bog willow - *Salix fuscescens*
American water-willow - *Justicia americana*
arctic willow - *Salix arctica*
arctic willow-weed - *Epilobium davuricum*
arroyo willow - *Salix lasiolepis*
autumn willow - *Salix serissima*
balsam willow - *Salix pyrifolia*
barren ground willow - *Salix brachycarpa* subsp. *niphoclada*
basket willow - *Salix purpurea, S. viminalis*
bastard-willow - *Rhus lancea*
bay-leaved willow - *Salix pentandra*
bay willow - *Salix pentandra*
Bebb's willow - *Salix bebbiana*
black willow - *Salix nigra*
brittle willow - *Salix fragilis*
Carolina willow - *Salix caroliniana*
Chamisso willow - *Salix chamissonis*
coastal plain willow - *Salix caroliniana*
coastal willow - *Salix hookeriana*
coyote willow - *Salix exigua*
crack willow - *Salix fragilis*
desert-willow - *Chilopsis linearis*
diamond-leaf willow - *Salix planifolia*
diamond willow - *Salix eriocephala*
ditch bank willow - *Salix exigua*
dune willow - *Salix cordata*
eastern willow-herb - *Epilobium coloratum*
felt-leaf willow - *Salix alaxensis*
Florida willow - *Salix floridana*
flowering-willow - *Chilopsis linearis, Epilobium angustifolium*
goat willow - *Salix caprea*
golden willow - *Salix alba* var. *vitellina*
Goodding's willow - *Salix gooddingii*
gray-leaf willow - *Salix glauca*
gray willow - *Salix bebbiana, S. cinerea, S. humilis*
great willow-herb - *Epilobium angustifolium*
hairy willow-herb - *Epilobium hirsutum*
hairy willow-weed - *Epilobium hirsutum*
heart-leaf willow - *Salix cordata, S. eriocephala*
high-ground willow oak - *Quercus incana*
hoary willow - *Salix caprea*
Hooker's willow - *Salix hookeriana*
large pussy willow - *Salix discolor*

laurel willow - *Salix pentandra*
least willow - *Salix rotundifolia*
little-tree willow - *Salix arbusculoides*
long-beaked willow - *Salix bebbiana*
long-fruited primrose-willow - *Ludwigia octovalis*
low blueberry willow - *Salix myrtillifolia*
meadow willow - *Salix petiolaris*
Missouri willow - *Salix eriocephala*
mountain willow - *Salix scouleriana*
narrow-leaf willow - *Salix exigua*
net-leaf willow - *Salix reticulata*
northwest willow - *Salix sessilifolia*
oval-leaf willow - *Salix ovalifolia, S. stolonifera*
Pacific willow - *Salix lucida* subsp. *lasiandra*
park willow - *Salix monticola*
peach-leaved willow - *Salix amygdaloides*
polished willow - *Salix laevigata*
prairie willow - *Salix humilis*
primrose-willow - *Ludwigia octovalvis*
purple-leaf willow-herb - *Epilobium coloratum*
purple willow - *Salix purpurea*
pussy willow - *Salix discolor*
red willow - *Salix laevigata*
Richardson's willow - *Salix lanata* subsp. *richardsonii*
river willow - *Salix fluviatilis*
sage-leaf willow - *Salix candida*
sand-dune willow - *Salix syrticola*
sandbar willow - *Salix exigua, S. sessilifolia*
Scouler's willow - *Salix scouleriana*
seep-willow - *Baccharis salicifolia*
serviceberry willow - *Salix monticola*
shining willow - *Salix lucida*
silky willow - *Salix sericea*
silver willow - *Salix candida*
Sitka willow - *Salix sitchensis*
skeleton-leaf willow - *Salix petiolaris*
small pussy willow - *Salix humilis*
tall blueberry willow - *Salix novaeangliae*
tea-leaf willow - *Salix planifolia*
Tracy's willow - *Salix lasiolepis*
under-green willow - *Salix commutata*
upland willow - *Salix humilis*
Virginia willow - *Itea virginica*
Ward's willow - *Salix caroliniana*
water-willow - *Baccharis salicifolia, Decodon verticillatus*
weeping willow - *Salix babylonica*
white willow - *Salix alba*
willow - *Salix*
willow amsonia - *Amsonia tabernaemontana*
willow baccharis - *Baccharis salicina*
willow bellflower - *Campanula persicifolia*
willow-herb - *Epilobium hirsutum, Epilobium*
willow-leaf lettuce - *Lactuca saligna*
willow oak - *Quercus phellos*
willow rhus - *Rhus lancea*
willow-weed - *Polygonum lapathifolium*
yew-leaf willow - *Salix taxifolia*

Wimmera rye grass - *Lolium rigidum*

wind-flower - *Anemone*

windmill
Chinese windmill palm - *Trachycarpus fortunei*
tumble windmill grass - *Chloris verticillata*
windmill grass - *Chloris verticillata, Chloris*

windmill jasmine - *Jasminum nitidum*
windmill palm - *Trachycarpus fortunei*
windmill pink - *Silene gallica*
windmills - *Allionia incarnata*
window-leaf - *Monstera*
Windsor bean - *Vicia faba*
wine
  wine-berry - *Rubus phoenicolasius*
  wine-flower - *Boerhavia*
  wine grape - *Vitis vinifera*
  wine palm - *Caryota urens*
  wine-plant - *Rheum rhabarbarum*
wing
  angel-wing jasmine - *Jasminum nitidum*
  bird-on-the-wing - *Polygala paucifolia*
  Chinese wing-nut - *Pterocarya stenoptera*
  four-wing saltbush - *Atriplex canescens*
  wing-angled loosestrife - *Lythrum alatum*
  wing-leaf soapberry - *Sapindus saponaria*
  wing-nut - *Pterocarya*
  wing-rib sumac - *Rhus copallina*
  wing-stem - *Verbesina alternifolia*
  wing-stemmed water-primrose - *Ludwigia decurrens*
  wing-stemmed wild pea-vine - *Lathyrus palustris*
winged
  sharp-winged monkey-flower - *Mimulus alatus*
  two-winged silver-bell - *Halesia diptera*
  winged bean - *Psophocarpus tetragonolobus*
  winged burning-bush - *Euonymus alatus*
  winged dock - *Rumex venosus*
  winged elm - *Ulmus alata*
  winged euonymus - *Euonymus alatus*
  winged loosestrife - *Lythrum alatum*
  winged pigweed - *Cycloloma atriplicifolium*
  winged spindle-tree - *Euonymus alatus*
  winged sumac - *Rhus copallina*
  winged tobacco - *Nicotiana alata*
  winged water-primrose - *Ludwigia decurrens*
wings
  Asian pigeon-wings - *Clitoria ternatea*
  pheasant's-wings - *Aloe variegata*
winter
  Austrian winter pea - *Lathyrus hirsutus*
  Chinese winter melon - *Benincasa hispida*
  early winter-cress - *Barbarea verna*
  mountain winter-cress - *Cardamine rotundifolia*
  winter bent grass - *Agrostis hyemalis*
  winter-cherry - *Physalis alkekengi, P. peruviana*
  winter-cress - *Barbarea orthoceras, B. vulgaris*
  winter daphne - *Daphne odora*
  winter-fat - *Krascheninnikovia lanata*
  winter-fern - *Conium maculatum*
  winter grape - *Vitis berlandieri, V. vulpina*
  winter heath - *Erica carnea*
  winter jasmine - *Jasminum nudiflorum*
  winter melon - *Cucumis melo* var. *inodorus*
  winter-purslane - *Claytonia perfoliata*
  winter speedwell - *Veronica persica*
  winter squash - *Cucurbita maxima, C. moschata, C. pepo*
  winter-sweet - *Chimonanthus praecox*
  winter vetch - *Vicia villosa, V. villosa* subsp. *varia*
  winter wild oat - *Avena ludoviciana*

winterberry
  mountain winterberry - *Ilex ambigua* var. *montana*
  smooth winterberry - *Ilex laevigata*
  winterberry - *Ilex glabra, I. verticillata, Ilex*
wintergreen
  alpine wintergreen - *Gaultheria humifusa*
  arctic wintergreen - *Pyrola glandiflora*
  flowering-wintergreen - *Polygala paucifolia*
  green-flowered wintergreen - *Pyrola chlorantha*
  lesser wintergreen - *Pyrola minor*
  one-flowered wintergreen - *Moneses uniflora*
  one-sided-wintergreen - *Orthilia secunda*
  pink wintergreen - *Pyrola asarifolia*
  spotted wintergreen - *Chimaphila maculata*
  white-veined wintergreen - *Pyrola picta*
  wintergreen - *Chimaphila, Gaultheria procumbens, Pyrola, Trientalis*
wiper, pen-wiper - *Kalanchoe marmorata*
wire
  barb-wire Russian thistle - *Salsola paulsenii*
  desert wire-lettuce - *Stephanomeria pauciflora*
  slender wire-lettuce - *Stephanomeria tenuifolia*
  wire-lettuce - *Stephanomeria pauciflora, Stephanomeria*
  wire-plant - *Muhlenbergia*
  wire rush - *Juncus balticus*
  wire-stem muhly - *Muhlenbergia frondosa*
wiregrass
  broom wiregrass - *Schizachyrium scoparium*
  drop-seed wiregrass - *Muhlenbergia schreberi*
  wiregrass - *Aristida dichotoma, Cynodon dactylon, Danthonia spicata, Eleusine indica, Juncus tenuis, Poacompressa*
wiss, blue wiss - *Teramnus labialis*
wisteria
  Chinese wisteria - *Wisteria sinensis*
  Japanese wisteria - *Wisteria floribunda*
  scarlet wisteria-tree - *Sesbania punicea*
  water-wisteria - *Hygrophila difformis*
witch
  cow pea witch-weed - *Striga gesnerioides*
  Gattinger's witch grass - *Panicum gattingeri*
  old-witch grass - *Panicum capillare*
  spreading witch grass - *Panicum dichotomiflorum*
  tobacco witch-weed - *Striga gesnerioides*
  witch grass - *Elytrigia repens, Panicum capillare*
  witch-hazel - *Hamamelis virginiana*
  witch-weed - *Striga asiatica*
witch's
  witch's-hair - *Panicum capillare*
  witch's-hobble - *Viburnum alnifolium*
withe
  smooth withe-rod - *Viburnum nudum*
  withe-rod - *Viburnum cassinoides*
witloof - *Cichorium intybus*
woad
  dyer's woad - *Isatis tinctoria*
  woad-waxen - *Genista tinctoria*
wode-whistle - *Conium maculatum*
wolf
  garden wolf-bane - *Aconitum napellus*
  trailing wolf-bane - *Aconitum reclinatum*
  wolf-bane - *Aconitum*

wolf bean - *Lupinus albus*
wolf-tail sedge - *Carex cherokeensis*
Wolf's trisetum - *Trisetum wolfii*
wolf's
  wolf's-claws - *Lycopodium clavatum*
  wolf's-milk - *Euphorbia esula*
wolfberry
  Berlandier's wolfberry - *Lycium berlandieri*
  wolfberry - *Symphoricarpos occidentalis*
wonder
  botanical-wonder - *Fatshedera lizei*
  wonder-apple - *Momordica balsamina*
  wonder bean - *Canavalia ensiformis*
  wonder-bulb - *Colchicum autumnale*
  wonder-tree - *Ricinus communis*
wonkapin - *Nelumbo lutea*
wood
  amboyna-wood - *Pterocarpus indicus*
  arrow-wood - *Viburnum acerifolium, V. dentatum, Viburnum*
  arrow-wood viburnum - *Viburnum dentatum*
  Australian black-wood - *Acacia melanoxylon*
  bastard-sandle-wood - *Myoporum sandwicense*
  beetle-wood - *Galax aphylla*
  bitter-wood - *Simarouba glauca*
  black-wood - *Acacia melanoxylon, Avicennia germinans*
  black-wood acacia - *Acacia melanoxylon*
  blue-wood aster - *Aster cordifolius*
  bow-wood - *Maclura pomifera*
  China-wood oil-tree - *Vernicia fordii*
  coastal wood fern - *Dryopteris arguta*
  crab-wood - *Gymnanthes lucida*
  creeping wood-sorrel - *Oxalis corniculata*
  creeping yellow wood-sorrel - *Oxalis corniculata*
  devil-wood - *Osmanthus americanus*
  downy arrow-wood - *Viburnum rafinesquianum*
  downy-leaf arrow-wood - *Viburnum rafinesquianum*
  drooping wood-reed - *Cinna latifolia*
  drooping wood sedge - *Carex arctata*
  European wood anemone - *Anemone nemorosa*
  European wood-sorrel - *Oxalis acetosella, O. europaea*
  field wood rush - *Luzula campestris*
  Florida yellow wood-sorrel - *Oxalis dillenii*
  fry-wood - *Albizia lebbeck*
  hairy wood mint - *Blephilia hirsuta*
  hairy-wood sunflower - *Helianthus atrorubens*
  holy-wood lignum-vitae - *Guajacum sanctum*
  Indian wood-apple - *Limonia acidissima*
  joint-wood - *Cassia javanica var. indochinensis*
  kidney-wood - *Eysenhardtia polystachya*
  laurel-wood - *Calophyllum inophyllum*
  lead-wood - *Krugiodendron ferreum*
  leather wood fern - *Dryopteris marginalis*
  marginal wood fern - *Dryopteris marginalis*
  Mexican blue-wood - *Condalia mexicana*
  millet wood rush - *Luzula parviflora*
  mountain wood fern - *Dryopteris campyloptera*
  muscle-wood - *Carpinus caroliniana*
  northern wood-sorrel - *Oxalis acetosella*
  opossum-wood - *Halesia tetraptera*
  oyster-wood - *Gymnanthes lucida*
  peg-wood - *Cornus sanguinea*

pepper-wood - *Umbellularia californica, Zanthoxylum clava-herculis*
poison-wood - *Gymnanthes lucida*
possum-wood - *Diospyros virginiana*
red wood-sorrel - *Oxalis oregana*
round-wood - *Sorbus americana*
shittim-wood - *Halesia tetraptera, Sideroxylon lanuginosa*
small-wood sunflower - *Helianthus microcephalus*
smooth arrow-wood - *Viburnum recognitum*
southern yellow wood-sorrel - *Oxalis dillenii*
spinulose wood fern - *Dryopteris carthusiana*
spreading wood fern - *Dryopteris campyloptera*
stout wood-reed - *Cinna arundinacea*
tallow-wood - *Ximenia americana*
tiss-wood - *Persea borbonia*
toothed wood fern - *Dryopteris carthusiana*
twist-wood - *Viburnum lantana*
violet wood-sorrel - *Oxalis corymbosa, O. violacea*
whistle-wood - *Acer pensylvanicum*
white-wood aster - *Aster divaricatus*
whorled wood aster - *Aster acuminatus*
wood anemone - *Anemone quinquefolia*
wood-apple - *Limonia acidissima*
wood-betony - *Pedicularis canadensis, Pedicularis*
wood bluegrass - *Poa nemoralis*
wood fern - *Dryopteris*
wood germander - *Teucrium scorodonia*
wood larkspur - *Delphinium trolliifolium*
wood lily - *Lilium philadelphicum*
wood mint - *Blephilia hirsuta*
wood-nymph - *Moneses uniflora*
wood poppy - *Stylophorum diphyllum*
wood-reed - *Cinna arundinacea*
wood rose - *Rosa gymnocarpa*
wood rush - *Luzula campestris, Luzula*
wood-sage - *Teucrium canadense, T. scorodonia*
wood sedge - *Carex tetanica*
wood shamrock - *Oxalis montana*
wood-sorrel - *Oxalis*
wood spurge - *Euphorbia commutata*
wood strawberry - *Fragaria vesca*
wood-vamp - *Decumaria barbara*
wood vetch - *Vicia caroliniana*
wood violet - *Viola palmata*
wood whitlow-grass - *Draba nemorosa*
yellow-wood - *Cladrastis lutea, Symplocos tinctoria*
yellow wood-sorrel - *Oxalis europaea, O. stricta*
yellow wood violet - *Viola lobata*
Wood's rose - *Rosa woodsii*
woodbine
  European woodbine - *Lonicera periclymenum*
  Italian woodbine - *Lonicera caprifolium*
  woodbine - *Clematis virginiana, Lonicera periclymenum, Parthenocissus quinquefolia*
woodland
  woodland flax - *Linum virginianum*
  woodland groundsel - *Senecio sylvaticus*
  woodland horsetail - *Equisetum sylvaticum*
  woodland lettuce - *Lactuca floridana*
  woodland star - *Lithophragma affine*
  woodland strawberry - *Fragaria vesca, F. vesca subsp. americana*

woodland sunflower - *Helianthus decapetalus, H. divaricatus, H. strumosus, H. tuberosus*

woodmint, downy woodmint - *Blephilia ciliata*

woodruff, sweet woodruff - *Galium odoratum*

woods

    flame-of-the-woods - *Ixora coccinea*

    flat-woods plum - *Prunus umbellata*

    hairy piney-woods goldenrod - *Solidago flexicaulis*

    piney-woods drop-seed - *Sporobolus junceus*

woodsia

    Oregon woodsia - *Woodsia oregana*

    Rocky Mountain woodsia - *Woodsia scopulina*

    smooth woodsia - *Woodsia glabella*

woody glasswort - *Salicornia virginica*

wool

    wool-flower - *Celosia argentea*

    wool-fruit sedge - *Carex lasiocarpa*

    wool grass - *Scirpus congdonii, S. cyperinus*

    wool grass bulrush - *Scirpus cyperinus*

woolen-breeches - *Hydrophyllum capitatum*

woolly

    woolly blue violet - *Viola sororia*

    woolly burdock - *Arctium tomentosum*

    woolly croton - *Croton capitatus*

    woolly cup grass - *Eriochloa villosa*

    woolly distaff thistle - *Carthamus lanatus*

    woolly finger grass - *Digitaria eriantha* subsp. *pentzii*

    woolly-gilia - *Navarretia intertexta*

    woolly hedge-nettle - *Stachys olympica*

    woolly Indian paintbrush - *Castilleja foliolosa*

    woolly-leaf bur-sage - *Ambrosia grayi*

    woolly-leaf lupine - *Lupinus leucophyllus*

    woolly-leaf poverty-weed - *Ambrosia grayi*

    woolly lip fern - *Cheilanthes lanosa*

    woolly loco - *Astragalus mollissimus*

    woolly locoweed - *Astragalus mollissimus*

    woolly manzanita - *Arctostaphylos tomentosa*

    woolly morning-glory - *Argyreia nervosa*

    woolly mullein - *Verbascum thapsus*

    woolly painted-cup - *Castilleja foliolosa*

    woolly plantain - *Plantago patagonica*

    woolly-pod milkweed - *Asclepias eriocarpa*

    woolly rose mallow - *Hibiscus lasiocarpus*

    woolly sedge - *Carex lanuginosa*

    woolly sweet cicely - *Osmorhiza claytonii*

    woolly tidestromia - *Tidestromia lanuginosa*

    woolly verbena - *Verbena stricta*

    woolly white drought-weed - *Eremocarpus setigerus*

Wooton's

    Wooton's loco - *Astragalus wootonii*

    Wooton's three-awn - *Aristida pansa*

World

    Old World arrowhead - *Sagittaria sagittifolia*

    Old World diamond-flower - *Hedyotis corymbosa*

worm

    American worm-seed - *Chenopodium ambrosioides*

    tall worm-seed mustard - *Erysimum hieraciifolium*

    tall worm-seed wallflower - *Erysimum hieraciifolium*

    worm-seed - *Chenopodium ambrosioides*

    worm-seed mustard - *Erysimum cheiranthoides*

wormgrass - *Spigelia marilandica*

wormwood

    absinth wormwood - *Artemisia absinthium*

    annual wormwood - *Artemisia annua*

    beach wormwood - *Artemisia stelleriana*

    biennial wormwood - *Artemisia biennis*

    Louisiana wormwood - *Artemisia ludoviciana*

    Roman wormwood - *Ambrosia artemisiifolia, Artemisia pontica, Corydalis sempervirens*

    silver wormwood - *Artemisia ludoviciana*

    southern wormwood - *Artemisia abrotanum*

    sweet wormwood - *Artemisia annua*

    wormwood - *Artemisia absinthium, A. vulgaris, Artemisia*

    wormwood cassia - *Cassia artemisioides*

    wormwood senna - *Cassia artemisioides*

wort

    arctic pearl-wort - *Sagina procumbens*

    back-wort - *Symphytum officinale*

    bird's-eye pearl-wort - *Sagina procumbens*

    black swallow-wort - *Cynanchum louiseae*

    Blue Ridge St. John's-wort - *Hypericum mitchellianum*

    bruise-wort - *Symphytum officinale*

    Canadian St. John's-wort - *Hypericum canadense*

    cool-wort - *Mitella diphylla, Pilea pumila*

    cough-wort - *Tussilago farfara*

    creeping St. John's-wort - *Hypericum adpressum, H. calycinum*

    curd-wort - *Galium verum*

    dane-wort - *Sambucus ebulus*

    deer-wort - *Ageratina altissima*

    dwarf St. John's-wort - *Hypericum mutilum*

    elf-wort - *Inula helenium*

    field sage-wort - *Artemisia campestris* subsp. *caudata*

    frost-wort - *Helianthemum canadense*

    golden St. John's-wort - *Hypericum frondosum*

    great St. John's-wort - *Hypericum pyramidatum*

    greater stitch-wort - *Stellaria holostea*

    gypsy-wort - *Lycopus*

    hog-wort - *Croton capitatus*

    Kalm's St. John's-wort - *Hypericum kalmianum*

    kidney-wort - *Baccharis pilularis*

    kidney-wort baccharis - *Baccharis pilularis* var. *consanguinea*

    lesser stitch-wort - *Stellaria graminea*

    marsh St. John's-wort - *Hypericum tubulosum, H. virginicum*

    mountain St. John's-wort - *Hypericum mitchellianum*

    nail-wort - *Paronychia*

    pearl-wort - *Sagina*

    prairie sage-wort - *Artemisia frigida*

    sage-wort - *Artemisia campestris*

    scab-wort - *Inula helenium*

    shrubby St. John's-wort - *Hypericum prolificum*

    slender fume-wort - *Corydalis micrantha*

    small water-wort - *Elatine minima*

    spotted St. John's-wort - *Hypericum punctatum*

    St. John's-wort - *Hypericum perforatum, Hypericum*

    St. Peter's-wort - *Hypericum crux-andreae*

    stagger-wort - *Helenium autumnale*

    stink-wort - *Datura stramonium*

    stitch-wort - *Stellaria media*

    swallow-wort - *Chelidonium majus, Vincetoxicum hirundinaria*

    tale-wort - *Borago officinalis*

    throat-wort - *Campanula trachelium*

    umbrella-wort - *Mirabilis nyctaginea, Mirabilis*

water-wort - *Elatine triandra*
white swallow-wort - *Vincetoxicum hirundinaria*
whitlow-wort - *Paronychia*
yellow fume-wort - *Corydalis flavula*
woundwort - *Anthyllis vulneraria, Stachys palustris, Stachys*

wreath
bridal-wreath - *Spiraea prunifolia, S. vanhouttei, Spiraea*
wreath goldenrod - *Solidago caesia*

Wych's elm - *Ulmus glabra*

yam
aerial yam - *Dioscorea bulbifera*
bulbil-bearing yam - *Dioscorea bulbifera*
Chinese yam - *Dioscorea batatas*
coco-yam - *Colocasia esculenta*
potato yam - *Dioscorea bulbifera*
wild yam - *Dioscorea villosa*
yam - *Dioscorea, Ipomoea batatas*
yam bean - *Pachyrhizus erosus*

Yankee
Yankee Point ceanothus - *Ceanothus griseus* var. *horizontalis*
Yankee-weed - *Eupatorium compositifolium*

yanquapin - *Nelumbo lutea*

yard
slender yard rush - *Juncus tenuis*
yard dock - *Rumex longifolius*
yard grass - *Eleusine indica*
yard-long bean - *Vigna unguiculata, V. unguiculata* subsp. *sesquipedalis*

yarrow
sneezeweed yarrow - *Achillea ptarmica*
sneezewort yarrow - *Achillea ptarmica*
western yarrow - *Achillea millefolium*
yarrow - *Achillea millefolium, Achillea*

yate, bushy yate - *Eucalyptus lehmannii*

yaupon - *Ilex cassine, I. vomitoria*

yautia - *Xanthosoma sagittifolium*

Yedda's-hawthorn - *Rhaphiolepis umbellata*

yellow
Alaska yellow-cedar - *Chamaecyparis nootkatensis*
Austrian yellow rose - *Rosa foetida*
creeping yellow-cress - *Rorippa sylvestris*
creeping yellow wood-sorrel - *Oxalis corniculata*
downy yellow violet - *Viola pubescens*
early yellow violet - *Viola rotundifolia*
European yellow lupine - *Lupinus luteus*
Florida yellow wood-sorrel - *Oxalis dillenii*
great yellow-cress - *Rorippa amphibia*
Mexican yellow pine - *Pinus patula*
Nootka yellow-cedar - *Chamaecyparis nootkatensis*
Rocky Mountain yellow pine - *Pinus ponderosa* var. *scopulorum*
round-leaf yellow violet - *Viola rotundifolia*
shrub yellow-root - *Xanthorhiza simplicissima*
small yellow water-crowfoot - *Ranunculus gmelinii*
smooth yellow violet - *Viola pubescens*
southern yellow pine - *Pinus palustris*
southern yellow wood-sorrel - *Oxalis dillenii*
spreading yellow-cress - *Rorippa sinuata*
tall yellow sweet-clover - *Melilotus altissima*
terete yellow-cress - *Rorippa teres*
upright yellow sorrel - *Oxalis europaea*
Virginia yellow flax - *Linum virginianum*

western yellow pine - *Pinus ponderosa*
wild yellow lily - *Lilium canadense*
yellow adder's-tongue - *Erythronium americanum*
yellow-alder - *Turnera ulmifolia*
yellow alfalfa - *Medicago sativa* subsp. *falcata*
yellow alpine saxifrage - *Saxifraga aizoides*
yellow alyssum - *Alyssum alyssoides*
yellow anise-tree - *Illicium parviflorum*
yellow azalea - *Rhododendron calendulaceum*
yellow bachelor's-button - *Polygala lutea*
yellow-bark oak - *Quercus velutina*
yellow bedstraw - *Galium verum*
yellow bee-plant - *Cleome lutea*
yellow bell lily - *Lilium canadense*
yellow-bells - *Tecoma stans*
yellow-berry - *Rubus chamaemorus*
yellow birch - *Betula alleghaniensis, B. pumila*
yellow blue-stem - *Bothriochloa ischaemum*
yellow bristle grass - *Setaria pumila*
yellow buckeye - *Aesculus flava*
yellow buckthorn - *Rhamnus caroliniana*
yellow bugleweed - *Ajuga chamaepitys*
yellow burn-weed - *Amsinckia lycopsoides*
yellow bush lupine - *Lupinus arboreus*
yellow butterfly palm - *Chrysalidocarpus lutescens*
yellow calla-lily - *Zantedeschia elliottiana*
yellow-chamomile - *Anthemis tinctoria*
yellow chestnut oak - *Quercus muehlenbergii*
yellow cleome - *Cleome lutea*
yellow clintonia - *Clintonia borealis*
yellow clover - *Trifolium dubium*
yellow colic-root - *Aletris aurea*
yellow corydalis - *Corydalis flavula*
yellow cosmos - *Cosmos sulphureus*
yellow-cress - *Rorippa*
yellow cucumber-tree - *Magnolia acuminata* var. *subcordata*
yellow daylily - *Hemerocallis lilioasphodelus*
yellow-devil - *Hieracium caespitosum*
yellow-devil hawkweed - *Hieracium floribundum*
yellow dock - *Rumex crispus*
yellow dog-fennel - *Helenium amarum*
yellow elder - *Tecoma stans*
yellow-eyed-grass - *Xyris caroliniana, Xyris*
yellow false mallow - *Malvastrum hispidum*
yellow field-cress - *Rorippa sylvestris*
yellow flag iris - *Iris pseudacorus*
yellow flax - *Linum virginianum*
yellow floating-heart - *Nymphoides peltata*
yellow-flowered alfalfa - *Medicago sativa* subsp. *falcata*
yellow-flowered gourd - *Cucurbita pepo* var. *ovifera*
yellow-flowered leaf-cup - *Polymnia uvedalia*
yellow forget-me-not - *Amsinckia lycopsoides*
yellow foxtail - *Setaria pumila*
yellow-fruited-thorn - *Crataegus flava*
yellow fume-wort - *Corydalis flavula*
yellow giant hyssop - *Agastache nepetoides*
yellow-gowan - *Ranunculus repens*
yellow gram - *Cicer arietinum*
yellow granadilla - *Passiflora laurifolia*
yellow guava - *Psidium guajava*
yellow harlequin - *Corydalis flavula*
yellow hawkweed - *Hieracium caespitosum*
yellow hawthorn - *Crataegus flava*

yellow hedge-hyssop - *Gratiola aurea*
yellow honeysuckle - *Lonicera flava*
yellow Indian-shoe - *Cypripedium calceolus* var. *pubescens*
yellow-ironweed - *Verbesina alternifolia*
yellow jessamine - *Gelsemium sempervirens*
yellow king-devil - *Hieracium caespitosum*
yellow locust - *Robinia pseudoacacia*
yellow loosestrife - *Lysimachia terrestris*
yellow lucerne - *Medicago sativa* subsp. *falcata*
yellow lupine - *Lupinus luteus*
yellow mandarin - *Disporum lanuginosum*
yellow melilot - *Melilotus officinalis*
yellow mignonette - *Reseda lutea*
yellow milkwort - *Polygala lutea*
yellow mountain saxifrage - *Saxifraga aizoides*
yellow-myrtle - *Lysimachia nummularia*
yellow nelumbo - *Nelumbo lutea*
yellow nickers - *Caesalpinia bonduc*
yellow nut-grass - *Cyperus esculentus*
yellow nut sedge - *Cyperus esculentus*
yellow oak - *Quercus muehlenbergii*
yellow oat grass - *Trisetum flavescens*
yellow oleander - *Thevetia peruviana*
yellow-paintbrush - *Hieracium caespitosum*
yellow palm - *Chrysalidocarpus lutescens*
yellow parentucellia - *Parentucellia viscosa*
yellow parilla - *Menispermum canadense*
yellow pimpernel - *Taenidia integerrima*
yellow pine - *Pinus echinata*
yellow pitcher-plant - *Sarracenia flava*
yellow pond-lily - *Nuphar*
yellow-poplar - *Liriodendron tulipifera*
yellow prairie violet - *Viola nuttallii*
yellow rabbit-brush - *Chrysothamnus viscidiflorus*
ycllow-rattle - *Rhinanthus minor, Rhinanthus*
yellow-rocket - *Barbarea vulgaris*
yellow-sage - *Lantana camara*
yellow salsify - *Tragopogon dubius*
yellow sand-verbena - *Abronia latifolia*
yellow skunk-cabbage - *Lysichiton americanum*
yellow-spine thistle - *Cirsium ochrocentrum*
yellow-star - *Helenium autumnale*
yellow star-thistle - *Centaurea solstitialis*
yellow starwort - *Inula helenium*
yellow sweet-clover - *Melilotus officinalis*
yellow tare - *Lathyrus pratensis*
yellow thistle - *Cirsium horridulum*
yellow toadflax - *Linaria vulgaris*
yellow touch-me-not - *Impatiens pallida*
yellow-trefoil - *Medicago lupulina*
yellow trumpets - *Sarracenia alata*
yellow vetchling - *Lathyrus ochroleucus, L. pratensis*
yellow water buttercup - *Ranunculus flabellaris*
yellow water crowfoot - *Ranunculus flabellaris*
yellow water-lily - *Nuphar lutea, Nymphaea mexicana*
yellow-weed - *Barbarea vulgaris*
yellow-white sedge - *Carex albolutescens*
yellow wild licorice - *Galium lanceolatum*
yellow-wood - *Cladrastis lutea, Symplocos tinctoria*
yellow wood-sorrel - *Oxalis europaea, O. stricta*
yellow wood violet - *Viola lobata*
Yellowstone milk-vetch - *Astragalus miser* var.
  hylophilus

yellowtop - *Senecio glabellus*
yerba
  yerba-buena - *Satureja douglasii*
  yerba-de-selva - *Whipplea modesta*
  yerba-mansa - *Anemopsis californica*
  yerba-santa - *Eriodictyon californicum*
yew
  American yew - *Taxus canadensis*
  Canadian yew - *Taxus canadensis*
  English yew - *Taxus baccata*
  Florida yew - *Taxus floridana*
  Harrington's plum-yew - *Cephalotaxus harringtonia*
  Japanese plum-yew - *Cephalotaxus harringtonia* var.
    drupacea
  Japanese yew - *Taxus cuspidata*
  Pacific yew - *Taxus brevifolia*
  plum-fruited yew - *Cephalotaxus harringtonia* var. *drupacea*
  plum-yew - *Cephalotaxus*
  western yew - *Taxus brevifolia*
  yew - *Taxus*
  yew-leaf willow - *Salix taxifolia*
  yew podocarpus - *Podocarpus macrophyllus*
yoke cactus - *Schlumbergera truncata*
Yokohama velvet bean - *Mucuna pruriens* var. *utilis*
Yorkshire fog - *Holcus lanatus*
youth-on-age - *Tolmiea menziesii*
yuca - *Manihot esculenta*
yucca
  aloe yucca - *Yucca aloifolia*
  banana yucca - *Yucca baccata*
  blue yucca - *Yucca baccata*
  Schott's yucca - *Yucca schottii*
  soap-tree yucca - *Yucca elata*
  spineless yucca - *Yucca elephantipes*
  spoon-leaf yucca - *Yucca filamentosa*
  twisted-leaf yucca - *Yucca rupicola*
zakuro, kurumaba-zakuro-so - *Mollugo verticillata*
Zanzibar balsam - *Impatiens wallerana*
zapatica-de-la-reina - *Clitoria ternatea*
zapote-blanco - *Casimiroa edulis*
Zealand
  New Zealand cliff brake - *Pellaea rotundifolia*
  New Zealand fireweed - *Erechtites glomerata*
  New Zealand flax - *Phormium tenax*
  New Zealand hemp - *Phormium tenax*
  New Zealand spinach - *Tetragonia tetragonioides*
  New Zealand tea-tree - *Leptospermum scoparium*
zebra-plant - *Aphelandra squarrosa*
zelkova
  elm-leaf zelkova - *Zelkova carpinifolia*
  Japanese zelkova - *Zelkova serrata*
  saw-leaf zelkova - *Zelkova serrata*
zephyr-lily - *Zephyranthes candida*
zerumbet ginger - *Zingiber zerumbet*
zigzag
  Solomon's-zigzag - *Smilacina racemosa*
  zigzag clover - *Trifolium medium*
  zigzag goldenrod - *Solidago flexicaulis*
zinnia - *Zinnia violacea*
zit-kwa - *Benincasa hispida*
zonal geranium - *Pelargonium hortorum*

zoysia
zoysia - *Zoysia japonica*
zoysia grass - *Zoysia matrella*

zucchini - *Cucurbita pepo*

Synonyms

Aberia caffra Hook.f. & Harv. - *Dovyalis caffra*
Abies arizonica Merriam - *Abies lasiocarpa* var. *arizonica*
  magnifica var. shastensis Lemm*on - Abies magnifica*
  nobilis (Douglas ex D. Don) Lin*dl. - Abies procera*
Abutilon avicennae Gaertn. - *Abutilon theophrasti*
Acer dasycarpum Ehrh. - *Acer saccharinum*
  douglasii Hook. - *Acer glabrum* subsp. *douglasii*
  floridanum (Chapm.) Pax - *Acer barbatum*
  negundo var. californicum (Torr. & A. Gray) Sarg. - *Acer negundo* subsp. *californicum*
  platanoides f. schwedleri K. Koch - *Acer platanoides*
  saccharophorum K. Koch - *Acer saccharum*
  saccharum subsp. grandidentatum (Nutt.) Desmarais - *Acer grandidentatum*
  saccharum subsp. leucoderme (Small) Desmarais - *Acer leucoderme*
  saccharum subsp. nigrum (F. Michx.) Desmarais - *Acer nigrum*
  saccharum var. nigrum (F. Michx.) Britton - *Acer nigrum*
  truncatum subsp. mono (Maxim.) A. E. Murray - *Acer mono*
Acerates viridiflora (Raf.) Eaton - *Asclepias viridiflora*
  viridiflora var. lanceolata (Ives) Torr. - *Asclepias viridiflora*
Achillea lanulosa Nutt. - *Achillea millefolium*
  millefolium subsp. lanulosa (Nutt.) Piper - *Achillea millefolium*
Achras zapota L. - *Manilkara zapota*
Acinos thymoides (L.) Moench - *Acinos arvensis*
Acnida tamariscina (Nutt.) A. Wood - *Amaranthus rudis*
  tuberculata Moq. - *Amaranthus tuberculatus*
Actaea arguta Nutt. ex Torr. & A. Gray - *Actaea rubra*
  eburnea Rydb. - *Actaea rubra*
  spicata var. arguta (Nutt. ex Torr. & A. Gray) E. Murray - *Actaea rubra*
Actinidia chinensis hort. non Planch. - *Actinidia deliciosa*
Actinomeris alternifolia (L.) DC. - *Verbesina alternifolia*
Adenium coetaneum Stapf - *Adenium obesum*
Adiantum decorum T. Moore - *Adiantum raddianum* cv. 'decorum'
Aegilops squarrosa L. - *Aegilops triuncialis*
Aesculus discolor Pursh - *Aesculus pavia*
  georgiana Sarg. - *Aesculus sylvatica*
  glabra var. leucodermis Sarg. - *Aesculus glabra*
  glabra var. sargentii Rehder - *Aesculus glabra*
  glaucescens Sarg. - *Aesculus sylvatica*
  neglecta Lindl. - *Aesculus sylvatica*
  neglecta var. georgiana (Sarg.) Sarg. - *Aesculus sylvatica*
  neglecta var. lanceolata (Sarg.) Sarg. - *Aesculus sylvatica*
  neglecta var. pubescens (Sarg.) Sarg. - *Aesculus sylvatica*
  neglecta var. tomentosa Sarg. - *Aesculus sylvatica*
  octandra Marsh. - *Aesculus flava*
  parviflora f. serotina Rehder - *Aesculus parviflora*
  splendens Sarg. - *Aesculus pavia*
Agapanthus umbellatus L'Hér. - *Agapanthus africanus*
Agave virginica L. - *Manfreda virginica*
Agropyron caninum (L.) P. Beauv. - *Elymus trachycaulus*
  dasystachyum (Hook.) Vasey - *Elymus lanceolatus*
  elongatum (Host) P. Beauv. - *Elytrigia elongata*

  inerme (J. G. Sm.) Rydb. - *Pseudoroegneria spicata*
  intermedium (Host) P. Beauv. - *Elytrigia intermedia*
  occidentale (Scribn.) Scribn. - *Elymus smithii*
  pauciflorum (Schwein.) Hitchc. - *Elymus trachycaulus*
  repens (L.) P. Beauv. - *Elytrigia repens*
  richardsonii Schrad. - *Elymus trachycaulus*
  rigidum (Schrad.) P. Beauv. - *Elytrigia elongata*
  riparium Scribn. & J. G. Sm. - *Elymus lanceolatus*
  sibiricum (Willd.) P. Beauv. - *Agropyron fragile* subsp. *sibiricum*
  smithii Rydb. - *Elymus smithii*
  smithii var. molle (Scribn. & J. G. Sm.) M. E. Jones - *Elymus smithii*
  spicatum Scribn. & J. G. Sm. - *Pseudoroegneria spicata*
  spicatum var. molle Scribn. & J. G. Sm - *Elymus smithii*
  subsecundum (Link) Hitchc. - *Elymus trachycaulus*
  tenerum Vasey - *Elymus trachycaulus*
  trachycaulum (Link) Malte - *Elymus trachycaulus*
  trichophorum (Link) K. Richt. - *Elytrigia intermedia*
Agrostis alba of American authors - *Agrostis stolonifera*
  diegoensis Vasey - *Agrostis pallens*
  exarata var. monolepis (Torr.) Hitchc. - *Agrostis exarata*
  geminata Trin. - *Agrostis scabra*
  latifolia Trevir. ex Goepp. - *Cinna latifolia*
  palustris Huds. - *Agrostis stolonifera*
  scabra var. geminata (Trin.) Swallen - *Agrostis scabra*
  tenuis Sibth. - *Agrostis capillaris*
  tenuis cv. 'Astoria' - *Agrostis capillaris*
Ailanthus glandulosa Dcsf. - *Ailanthus altissima*
Alchemilla occidentalis Nutt. - *Aphanes occidentalis*
Aleurites fordii Hemsl. - *Vernicia fordii*
  javanica Gand. - *Aleurites moluccana*
  montana (Lour.) E. H. Wilson - *Vernicia montana*
  triloba J. R. Forst. & G. Forst. - *Aleurites moluccana*
Alisma geyeri Torr. - *Alisma gramineum*
  plantago L. - *Alisma plantago-aquatica*
  subcordatum Raf. - *Alisma gramineum*
  triviale Pursh - *Alisma plantago-aquatica*
Allamanda neriifolia Hook. - *Allamanda schottii*
Allium ampeloprasum var. porrum (L.) Gay - *Allium porrum*
  attenuifolium Kellogg - *Allium amplectens*
  cepa var. aggregatum G. Don - *Allium cepa*
  cepa var. proliferum (Moench) Regel - *Allium cepa*
  cepa var. solaninum Alef. - *Allium cepa*
  mutabile Michx. - *Allium canadense*
  nuttallii S. Watson - *Allium drummondii*
  reticulatum Nutt. ex G. Don - *Allium textile*
  sibiricum L. - *Allium schoenoprasum*
Alnus crispa (Aiton) Pursh - *Alnus viridis* subsp. *crispa*
  oregona Nutt. - *Alnus rubra*
  vulgaris Pers. - *Alnus glutinosa*
Aloe barbadensis Mill. - *Aloe vera*
Aloysia lycioides Cham. - *Aloysia gratissima*
Alsophila australis R. Br. - *Cyathea australis*
Alternanthera amoena (Lem.) Voss - *Alternanthera ficoidea* var. *amoena*
  pungens Kunth - *Alternanthera caracasana*

Althaea ficifolia L. - *Alcea ficifolia*
  rosea (L.) Cav. - *Alcea rosea*
Amaranthus edulis Speg. - *Amaranthus caudatus*
  gangeticus L. - *Amaranthus tricolor*
  gracilis Desf. - *Amaranthus viridis*
Ambrosia aptera DC. - *Ambrosia trifida*
Amelanchier bakeri Greene - *Amelanchier utahensis*
  cusickii Fernald - *Amelanchier alnifolia* var. *cusickii*
  elliptica A. Nelson - *Amelanchier utahensis*
  florida Lindl. - *Amelanchier alnifolia*
  intermedia Spach - *Amelanchier canadensis*
  mormonica C. K. Schneid. - *Amelanchier utahensis*
  oblongifolia M. Roem. - *Amelanchier canadensis*
  oreophila A. Nelson - *Amelanchier utahensis*
  polycarpa Greene - *Amelanchier alnifolia* var. *pumila*
  prunifolia Greene - *Amelanchier utahensis*
  pumila Nutt. - *Amelanchier alnifolia* var. *pumila*
Ammocallis rosea (L.) Small - *Catharanthus roseus*
Ammodenia peploides (L.) Rupr. - *Honckenya peploides*
Ammophila longifolia Benth. ex Vasey - *Calamovilfa longifolia*
Amomum zingiber L. - *Zingiber officinale*
Ampelamus albidus (Nutt.) Britton - *Cynanchum laeve*
Ampelopsis quinquefolia (L.) Michx. - *Parthenocissus quinquefolia*
Amphicarpaea monoica (L.) Elliott - *Amphicarpaea bracteata*
Amygdalus communis L. - *Prunus dulcis*
  dulcis Mill. - *Prunus dulcis*
  persica L. - *Prunus persica*
Anaphalis margaritacea var. occidentalis Greene - *Anaphalis margaritacea*
Andromeda ferruginea Walter - *Lyonia ferruginea*
  mariana L. - *Lyonia mariana*
Andropogon annulatus Forssk. - *Dichanthium annulatum*
  barbinodis Lag. - *Bothriochloa barbinodis*
  citratus DC. ex Nees - *Cymbopogon citratus*
  furcatus Scribn. - *Andropogon gerardii*
  ischaemum L. - *Bothriochloa ischaemum*
  muricatus Retz. - *Vetiveria zizanioides*
  provincialis Scribn. - *Andropogon gerardii*
  saccharoides Sw. - *Bothriochloa saccharoides*
  scoparius Michx. - *Schizachyrium scoparium*
  ternarius Scribn. - *Andropogon gerardii*
Anemone japonica (Thunb.) Siebold & Zucc. - *Anemone hupehensis* var. *japonica*
  nuttalliana DC. - *Anemone patens*
  patens var. wolfgangiana (Besser) K. Koch - *Anemone patens*
  riparia Fernald - *Anemone virginiana*
  thalictroides L. - *Thalictrum thalictroides*
Anemonella thalictroides (L.) Spach - *Thalictrum thalictroides*
Anisum vulgare Gaertn. - *Pimpinella anisum*
Anoda hastata Cav. - *Anoda cristata*
Antennaria canadensis Greene - *Antennaria neglecta* var. *canadensis*
Antheropogon curtipendulus (Michx.) E. Fourn. - *Bouteloua curtipendula*
Anthriscus scandicina (Weber) Mansf. - *Anthriscus caucalis*
Anychia canadensis (L.) Britton, et al. - *Paronychia canadensis*
Apios tuberosa Moench - *Apios americana*
Apium leptophyllum (Pers.) F. Muell. ex Benth. - *Ciclospermum leptophyllum*
  petroselinum L. - *Petroselinum crispum*
Apocynum hypericifolium Aiton - *Apocynum sibiricum*
Arabis virginica (L.) Trel. - *Sibara virginica*

Aragallus lambertii Greene - *Oxytropis lambertii*
Aralia edulis Siebold & Zucc. - *Aralia cordata*
Araucaria brasiliana A. Rich. - *Araucaria angustifolia*
  excelsa (Lamb.) R. Br. - *Araucaria columnaris*
  imbricata Pav. - *Araucaria araucana*
Arctagrostis arundinacea (Trin.) Beal - *Arctagrostis latifolia*
  latifolia var. arundinacea (Trin.) Griseb. - *Arctagrostis latifolia*
Arctostaphylos uva-ursi var. coactilis Fernald & J. F. Macbr. - *Arctostaphylos uva-ursi*
Arctotheca calendula (L.) Levyns - *Arctostaphylos alpina*
Arctous alpina (L.) Nied. - *Arctostaphylos alpina*
Arecastrum romanzoffianum (Cham.) Becc. - *Syagrus romanzoffianum*
Arenaria peploides L. - *Honckenya peploides*
  peploides var. major Hook. - *Honckenya peploides*
Arenga saccharifera Labill. ex DC. - *Arenga pinnata*
Argemone alba F. Lestib. - *Argemone mexicana*
  intermedia Eastw. - *Argemone polyanthemos*
Argyreia speciosa (L.f.) Sweet - *Argyreia nervosa*
Arisaema atrorubens (Aiton) Blume - *Arisaema triphyllum*
Aristida fendleriana Steud. - *Poa fendleriana*
  glauca (Nees) Walp. - *Aristida purpurea*
  intermedia Scribn. & Ball - *Aristida longespica*
  longiseta Steud. - *Aristida purpurea*
  longiseta var. robusta Merr. - *Aristida purpurea*
  oligantha Michx. - *Aristida adscensionis*
  virgata Trin. - *Aristida purpurascens*
Aristolochia durior Hill - *Aristolochia macrophyllum*
Armeria labradorica Wallr. - *Armeria maritima* subsp. *sibirica*
Armoracia lapathifolia Gilib. ex Usteri - *Armoracia rusticana*
Aronia floribunda Spach - *Aronia prunifolia*
Arrhenatherum elatius var. bulbosum (Willd.) St.-Amans - *Arrhenatherum elatius*
Artemisia aromatica A. Nelson - *Artemisia dracunculus*
  caudata Michx. - *Artemisia campestris* subsp. *caudata*
  dracunculoides Pursh - *Artemisia dracunculus*
  glauca Pall. ex Willd. - *Artemisia dracunculus*
  gnaphalodes Nutt. - *Artemisia ludoviciana*
  heterophylla Nutt. - *Artemisia douglasiana*
  purshiana Besser - *Artemisia ludoviciana*
  vulgaris var. heterophylla (Nutt.) Jeps. - *Artemisia douglasiana*
Artocarpus communis J. R. Forst. & G. Forst. - *Artocarpus altilis*
  incisus L.f. - *Artocarpus altilis*
  integer of authors - *Artocarpus heterophyllus*
  integrifolius of authors - *Artocarpus heterophyllus*
Arum esculentum L. - *Colocasia esculenta*
Arundinaria macrosperma Michx. - *Arundinaria gigantea*
  tecta (Walter) Muhl. - *Arundinaria gigantea* subsp. *tecta*
Asclepias cornuti Decne. - *Asclepias syriaca*
  galioides Kunth - *Asclepias verticillata*
  phytolaccoides Pursh - *Asclepias exaltata*
Ascyrum hypericoides L. - *Hypericum hypericoides*
  stans Michx. - *Hypericum crux-andreae*
Asparagus myriocladus Baker - *Asparagus densiflorus*
  plumosus Baker - *Asparagus setaceus*
  sprengeri Regel - *Asparagus densiflorus*
Aspidium marginale (L.) Sw. - *Dryopteris marginalis*
  spinulosum (Wattull.) A. Gray - *Dryopteris carthusiana*
Aster azureus Lindl. - *Aster oolentangiensis*
  chinensis L. - *Callistephus chinensis*

exilis Elliott - *Aster subulatus*
longifolius Lam. - *Aster novi-belgii*
multiflorus Aiton - *Aster ericoides*
paniculatus var. simplex (Willd.) Burgess - *Aster lanceolatus*
ptarmicoides (Nees) Torr. & A. Gray - *Solidago ptarmicoides*
simplex Willd. - *Aster lanceolatus*
spinosus Benth. - *Chloranthus spinosa*
vimineus Lam. - *Aster laterifolius*
Astragalus goniatus Nutt. - *Astragalus agrestis*
menziesii A. Gray - *Astragalus nuttallii*
multiflorus A. Gray - *Astragalus tenellus*
nitidus Douglas ex Hook. - *Astragalus adsurgens*
serotinus A. Gray - *Astragalus miser* var. *serotinus*
Athyrium acrostichoides (Sw.) Diels - *Athyrium thelypteroides*
angustum (Willd.) J. Presl - *Athyrium felix-femina* var.
  michauxii
Atriplex hastata L. - *Atriplex patula* subsp. *hastata*
patula var. hastata (L.) A. Gray - *Atriplex patula* subsp.
  hastata
Audibertia humilis Benth. - *Salvia sonomensis*
stachyoides Benth. - *Salvia mellifera*
Avena byzantina K. Koch - *Avena sativa*
orientalis Schreb. - *Avena fatua*
Avicennia nitida Jacq. - *Avicennia germinans*
Azalea calendulacea Michx. - *Rhododendron calendulaceum*
canescens Michx. - *Rhododendron canescens*
nudiflora L. - *Rhododendron periclymenoides*
viscosa L. - *Rhododendron viscosum*
Baccharis glutinosa Pers. - *Baccharis salicifolia*
Baeria chrysostoma Fisch. & C. A. Mey. - *Lasthenia*
  chrysostoma
Basella rubra L. - *Basella alba*
Bauhinia punctata Bolle - *Bauhinia galpinii*
Beaucarnea recurvata Lem. - *Nolina recurvata*
Benincasa cerifera Savi - *Benincasa hispida*
Benzoin aestivale (L.) Nees - *Lindera benzoin*
Berchemia volubilis (L.f.) DC. - *Berchemia scandens*
Beta vulgaris var. cicla L. - *Beta vulgaris* subsp. *cicla*
Betula alba L. - *Betula pendula*
caerulea-grandia Blanch. - *Betula caerulea*
fontinalis Sarg. - *Betula occidentalis*
kenaica W. H. Evans - *Betula papyrifera* var. *kenaica*
lutea F. Michx. - *Betula pumila*
papyrifera var. occidentalis (Hook.) Sarg. - *Betula*
  occidentalis
rubra F. Michx. - *Betula nigra*
Bidens leucantha Willd. - *Bidens pilosa*
Bistorta bistortoides (Pursh) Small - *Polygonum bistortoides*
Bombax malabaricum DC. - *Bombax ceiba*
Borreria alata (Aubl.) DC. - *Spermacoce alata*
alata (Aubl.) DC. - *Spermococe alata*
Bouteloua oligostachya (Nutt.) Torr. ex A. Gray - *Bouteloua*
  gracilis
Brachiaria extensa Chase - *Brachiaria platyphylla*
Brassaia actinophylla Endl. - *Schefflera actinophylla*
Brassica campestris var. napobrassica (L.) DC. - *Brassica*
  napus var. *napobrassica*
alba (L.) Rabenh. - *Sinapis alba*
arvensis (L.) Rabenh. - *Sinapis arvensis*
campestris L. - *Brassica rapa*
caulorapa Pasq. - *Brassica oleracea*
geniculata (Desf.) Ball - *Hirschfeldia incana*
hirta Moench - *Sinapis alba*

kaber (DC.) L. C. Wheeler - *Sinapis arvensis*
napobrassica (L.) Mill. - *Brassica napus* var. *napobrassica*
oleracea var. chinensis (L.) Prain - *Brassica chinensis*
pe-tsai L. H. Bailey - *Brassica pekinensis*
rapa var. chinensis (L.) Kitam. - *Brassica chinensis*
Brauneria purpurea Britton - *Echinacea purpurea*
Brodiaea capitata Benth. - *Dichelostemma pulchellum*
ixioides Hook. - *Leucocoryne ixioides*
laxa (Benth.) S. Watson - *Triteleia laxa*
pulchella (Salisb.) Greene - *Dichelostemma pulchellum*
Bromelia comosa L. - *Ananas comosus*
Bromus aleutensis Rydb. - *Bromus carinatus*
breviaristatus Buckley - *Bromus carinatus*
commutatus var. aparicorum Simonk. - *Bromus commutatus*
gussonii Parl. - *Bromus rigidus*
marginatus Nees ex Steud. - *Bromus carinatus*
mollis L. - *Bromus hordeaceus*
polyanthus Scribn. ex Shear - *Bromus carinatus*
porteri (J. M. Coult.) Nash - *Bromus anomalus*
pumpellianus Scribn. - *Bromus inermis*
racemosus L. - *Bromus hordeaceus*
secalinus var. velutinus K. Koch - *Bromus secalinus*
villosus Forssk. - *Bromus rigidus*
Brunella vulgaris (L.) Greene - *Prunella vulgaris*
Brunnichia cirrhosa Gaertn. - *Brunnichia ovata*
Bumelia lanuginosa (Michx.) Pers. - *Sideroxylon lanuginosa*
lycioides (L.) Pers. - *Sideroxylon lycioides*
tenax (L.) Willd. - *Sideroxylon tenax*
Buphthalmum helianthoides L. - *Heliopsis helianthoides*
Cacalia reniformis Muhl. - *Cacalia muhlenbergii*
Calamagrostis aleutica Trin. - *Calamagrostis nutkaensis*
hyperborea Lange - *Calamagrostis neglecta*
inexpansa A. Gray - *Calamagrostis neglecta*
inexpansa var. brevior (Vasey) Stebbins - *Calamagrostis*
  neglecta
Calandrinia caulescens Kunth - *Calandrinia ciliata*
Callistemon lanceolatus (Sm.) DC. - *Callistemon citrinus*
Callistephus hortensis (Cass.) Cass. - *Callistephus chinensis*
Calocarpum sapota (Jacq.) Merr. - *Pouteria sapota*
Calonyction aculeatum (L.) House - *Ipomoea alba*
Caltha biflora DC. - *Caltha palustris*
howellii Greene - *Caltha palustris*
Camellia thea Link - *Camellia sinensis*
theifera Griff. - *Camellia sinensis*
Campanula petiolata A. DC. - *Campanula rotundifolia*
uliginosa Rydb. - *Campanula aparinoides*
Camptosorus rhizophyllus (L.) Link - *Asplenium rhizophyllum*
Canna edulis Ker Gawl. - *Canna indica*
Capparis jamaicensis Jacq. - *Capparis cynophallophora*
Carduus undulatus Nutt. - *Cirsium undulatum*
Carex angustior Mack. - *Carex muricata*
aquatilis var. altior (Rydb.) Fernald - *Carex aquatilis*
convoluta Mack. - *Carex rosea*
cristata Schwein. - *Carex tribuloides*
eurycarpa Holm - *Carex angustata*
fraseri Andr. - *Cymophyllus fraseri*
garberi Fernald - *Carex aurea*
inflata Huds. - *Carex rostrata*
kelloggii W. Boott - *Carex lenticularis*
oederi Retz. - *Carex viridula*
peckii Howe - *Carex nigromarginata* var. *elliptica*
physocarpa C. Presl - *Carex saxatilis*
rostrata var. utriculata Boott - *Carex rostrata*

stygia Fr. - *Carex rariflora*

Carissa grandiflora (E. Mey.) A. DC. - *Carissa macrocarpa*

Carum gairdneri (Hook. & Arn.) A. Gray - *Perideridia gairdneri*

  petroselinum Benth. & Hook. f. - *Petroselinum crispum*

  velenovskyi Rohlena - *Carum carvi*

Carya alba Nutt. - *Carya ovata*

  amara Raf. - *Carya cordiformis*

  oliviformis (Michx.) Nutt. - *Carya illinoensis*

  pecan (Marsh.) Engl. & Graebn. - *Carya illinoensis*

Cassandra calyculata (L.) D. Don - *Chamaedaphne calyculata*

Cassia acutifolia Delile - *Senna alexandrina*

  alata L. - *Senna alata*

  angustifolia Vahl - *Senna alexandrina*

  covesii A. Gray - *Senna covesii*

  fasciculata Michx. - *Chamaecrista fasciculata*

  hebecarpa Fernald - *Senna hebecarpa*

  marilandica L. - *Senna marilandica*

  nictitans L. - *Chamaecrista nictitans*

  nodosa Buch.-Ham. ex Roxb. - *Cassia javanica* var. *indochinensis*

  obtusifolia L. - *Senna obtusifolia*

  occidentalis L. - *Senna occidentalis*

  procumbens L. - *Chamaecrista nictitans*

  senna L. - *Senna alexandrina*

  tora L. - *Senna tora*

Castalia odorata Woodv. & Wood - *Nymphaea odorata*

Castanea bungeana Blume - *Castanea mollissima*

  formosana (Hayata) Hayata - *Castanea mollissima*

  japonica Blume - *Castanea crenata*

  pubinervis (Hassk.) C. K. Schneid. - *Castanea crenata*

  stricta Siebold & Zucc. - *Castanea crenata*

  vesca Gaertn. - *Castanea sativa*

Castilleja californica Abrams - *Castilleja affinis*

Catalpa cordifolia Moench - *Catalpa bignonioides*

Ceanothus rigidus Nutt. - *Ceanothus cuneatus* var. *rigidus*

  thyrsiflorus var. griseus Trel. - *Ceanothus griseus*

Cedrela sinensis A. Juss. - *Toona sinensis*

Celastrus edulis Vahl - *Catha edulis*

Celosia cristata L. - *Celosia argentea* var. *cristata*

Celtis douglasii Planch. - *Celtis reticulata*

Cenchrus carolinianus Walter - *Cenchrus incertus*

  pauciflorus Benth. - *Cenchrus incertus*

Centaurea picris Pall. - *Centaurea repens*

  virgata var. squarrosa (Willd.) Boiss. - *Centaurea squarrosa*

Centaurium minus of authors - *Centaurium erythraea*

  umbellatum Gilib. - *Centaurium erythraea*

Cephalotaxus drupacea Siebold & Zucc. - *Cephalotaxus harringtonia* var. *drupacea*

Cerastium arvense var. viscidulum J. F. Gmel. - *Cerastium arvense*

  campestre Greene - *Cerastium arvense*

  occidentale Greene - *Cerastium arvense*

  oreophilum Greene - *Cerastium arvense*

  scopulorum Greene - *Cerastium arvense*

  strictum L. - *Cerastium arvense*

Cerasus laurocerasus (L.) Loisel. - *Prunus laurocerasus*

  mahaleb (L.) Mill. - *Prunus mahaleb*

Ceratoides lanata (Pursh) J. T. Howell - *Krascheninnikovia lanata*

Cercidium floridum Benth. ex A. Gray - *Parkinsonia florida*

Cercocarpus betulaefolius Nutt. ex Hook. - *Cercocarpus montanus*

Cereus giganteus Engelm. - *Carnegiea gigantea*

  peruvianus (L.) Mill. - *Cereus uruguayanus*

  thurberi Engelm. - *Stenocereus thurberi*

Chaenomeles lagenaria (Loisel.) Koidz. - *Chaenomeles speciosa*

  sinensis (Thouin) Koehne - *Pseudocydonia sinensis*

Chaetochloa glauca (L.) Scribn. - *Setaria pumila*

  viridis (L.) Scribn. - *Setaria viridis*

Chamaenerion angustifolium (L.) Scop. - *Epilobium angustifolium*

Cheiranthus cheiri L. - *Erysimum cheiri*

Chenopodium ambrosioides var. anthelminticum (L.) A. Gray - *Chenopodium ambrosioides*

  anthelminticum L. - *Chenopodium ambrosioides*

  gigantospermum Aellen - *Chenopodium simplex*

Chimaphila corymbosa Pursh - *Chimaphila umbellata*

Chiogenes hispidula (L.) Torr. & A. Gray ex Torr. - *Gaultheria hispidula*

Chrysanthemum leucanthemum L. - *Leucanthemum vulgare*

  maximum Ramond - *Leucanthemum maximum*

  parthenium (L.) Bernh. - *Tanacetum parthenium*

  roseum Adams - *Chrysanthemum coccineum*

  vulgare (L.) Bernh. - *Tanacetum vulgare*

Chrysolepis chrysophylla (Douglas ex Hook.) Hjelmq. - *Castanopsis chrysophylla*

  sempervirens (Kellogg) Hjelmq. - *Castanopsis sempervirens*

Chrysopsis nervosa (Willd.) Fernald - *Pityopsis graminifolia*

Chrysothamnus baileyi Wooton & Standl. - *Chrysothamnus pulchellus*

  graveolens (Nutt.) Greene - *Chrysothamnus nauseosus* subsp. *graveolens*

Cibotium menziesii Hook. - *Cibotium chamissoi*

Cicer lens Willd. - *Lens culinaris*

Cinnamomum zeylanicum Blume - *Cinnamomum verum*

Circaea latifolia Hill - *Circaea quadrisulcata*

Cirsium lanceolatum Scop. - *Cirsium vulgare*

Citrullus vulgaris Schrad. - *Citrullus lanatus*

Citrus aurantium var. sinensis L. - *Citrus sinensis*

  decumana (L.) L. - *Citrus maxima*

  grandis (L.) Osbeck - *Citrus maxima*

  nobilis Andr. - *Citrus reticulata*

  nobilis var. deliciosa Swingle - *Citrus reticulata*

  reticulata var. deliciosa Swingle - *Citrus reticulata*

  trifoliata L. - *Poncirus trifoliata*

Cladothrix lanuginosa (Nutt.) Nutt. ex S. Watson - *Tidestromia lanuginosa*

Claytonia linearis Douglas ex Hook. - *Montia linearis*

Cnicus altissimus (L.) Willd. - *Cirsium altissimum*

  arvensis (L.) Roth - *Cirsium arvense*

  undulatus (Nutt.) A. Gray - *Cirsium undulatum*

Coccoloba floridana Meisn. - *Coccoloba diversifolia*

Cochlearia armoracia L. - *Armoracia rusticana*

Coeloglossum viride var. bracteatum (Muhl. ex Willd.) Hultén - *Habenaria viridis* var. *bracteata*

Coffea robusta L. Linden - *Coffea canephora*

Coix lacryma L. - *Coix lacryma-jobi*

Colocasia antiquorum Schott - *Colocasia esculenta*

Comandra livida Richardson - *Geocaulon lividum*

Commelina nudiflora Davidson - *Commelina diffusa*

Comptonia peregrina var. asplenifolia (L.) Fernald - *Comptonia asplenifolia*

Conioselinum benthamii (S. Watson) Fernald - *Conioselinum pacificum*

  chinense (L.) Britton, et al. - *Conioselinum pacificum*

Convolvulus japonicus Thunb. - *Calystegia hederacea*
  sepium L. - *Calystegia sepium*
  spithamaeus L. - *Calystegia spithamaea*
Corallorrhiza multiflora Nutt. - *Corallorrhiza maculata*
Cornus foemina Mill. - *Cornus stricta*
  paniculata L'Hér. - *Cornus racemosa*
  stolonifera Michx. - *Cornus sericea*
Corydalis glauca Pursh - *Corydalis sempervirens*
Corylus californica (A. DC.) Rose - *Corylus cornuta* var.
  californica
  rostrata Aiton - *Corylus cornuta*
  rostrata var. californica A. DC. - *Corylus cornuta* var.
  californica
Cotinus americanus Nutt. - *Cotinus obovatus*
Cotoneaster pyracantha (L.) Spach - *Pyracantha coccinea*
Cowania stansburiana Torr. - *Cowania mexicana*
Crassula argentea L.f. - *Crassula ovata*
Crataegus apiifolia (Marsh.) Michx. - *Crataegus marshallii*
  collina Chapm. - *Crataegus punctata*
  holmesiana Ashe - *Crataegus coccinea*
  michauxii Pers. - *Crataegus flava*
  oxycantha of authors, non L. - *Crataegus laevigata*
Crepis japonica (L.) Benth. - *Youngia japonica*
Crocanthemum canadense Britton - *Helianthemum canadense*
Crotalaria mucronata Desv. - *Crotalaria pallida*
  striata DC. - *Crotalaria pallida*
Cucurbita pepo var. condensa L. H. Bailey - *Cucurbita pepo*
  var. *melopepo*
Cuphea petiolata (L.) Koehne - *Cuphea viscosissima*
  platycentra Lem. - *Cuphea ignea*
Cupressus sempervirens var. stricta Aiton - *Cupressus*
  sempervirens
  thyoides L. - *Chamaecyparis thyoides*
Cyamopsis psoralioides DC. - *Cyamopsis tetragonoloba*
Cydonia japonica (Thunb.) Pers. - *Chaenomeles japonica*
  sinensis Thouin - *Pseudocydonia sinensis*
  vulgaris Pers. - *Cydonia oblonga*
Cymopterus watsonii (J. M. Coult. & Rose) M. E. Jones -
  Cymopterus ibapensis
Cynanchum nigrum (L.) Pers. - *Cynanchum louiseae*
  vincetoxicum (L.) Pers. - *Vincetoxicum hirundinaria*
Cyperus aristatus Rottb. - *Cyperus squarrosus*
  inflexus Muhl. - *Cyperus squarrosus*
Cyphomandra crassifolia (Ortega) Kuntze - *Cyphomandra*
  betacea
Cypripedium pubescens Willd. - *Cypripedium calceolus* var.
  pubescens
Cytisus monspessulanus L. - *Genista monspessulanus*
Dalea spinosa A. Gray - *Psorothamnus spinosa*
Danthonia americana Scribn. - *Danthonia californica*
  californica var. americana (Scribn.) Hitchc. - *Danthonia*
  californica
Datura meteloides DC. ex Dunal - *Datura innoxia*
Daubentonia tripetii F. T. Hubb. - *Sesbania punicea*
Daucus carota var. sativa Hoffm. - *Daucus carota* subsp. *sativus*
Delphinium ajacis L. - *Consolida ajacis*
  pauciflorum Nutt. ex Torr. & A. Gray - *Delphinium*
  nuttallianum
Dentaria diphylla Michx. - *Cardamine diphylla*
  laciniata Muhl. - *Cardamine concatenata*
Deringa canadensis (L.) Kuntze - *Cryptotaenia canadensis*
Descurainia richardsonii subsp. incisa (A. Gray) Detling -
  Descurainia incisa

Desmodium bracteosum Daniels - *Desmodium cuspidatum*
  bracteosum var. longifolium (Torr. & A. Gray) B. L. Rob. -
  Desmodium cuspidatum
  canum Schinz & Thell. - *Desmodium incanum*
  longifolium (Torr. & A. Gray) Daniels - *Desmodium*
  cuspidatum
  purpureum (Mill.) Fawc. & Rendle - *Desmodium tortuosum*
Dianthera americana L. - *Justicia americana*
Dichondra repens J. R. Forst. & G. Forst. - *Dichondra*
  carolinensis
  repens var. carolinensis (Michx.) Choisy - *Dichondra*
  carolinensis
Dichromena colorata (L.) Hitchc. - *Rhynchospora colorata*
  latifolia Baldwin ex Elliott - *Rhynchospora latifolia*
Dictamnus fraxinellus Pers. - *Dictamnus albus*
Dieffenbachia picta (Lodd.) Schott - *Dieffenbachia maculata*
  picta cv. 'rudolph' - *Dieffenbachia maculata*
Diervilla florida (Bunge) Siebold & Zucc. - *Weigela florida*
Dietes vegeta N. E. Br. - *Dietes iridioides*
Digitalis nevadensis Kunze - *Digitalis purpurea*
Digitaria decumbens Stent - *Digitaria eriantha* subsp. *pentzii*
  pentzii Stent - *Digitaria eriantha* subsp. *pentzii*
  valida Stent - *Digitaria eriantha* subsp. *pentzii*
Diospyros chinensis Blume - *Diospyros kaki*
Dipsacus sylvestris Huds., misapplied - *Dipsacus fullonum*
Distichlis maritima Raf. - *Distichlis spicata*
Dolichos angularis Willd. - *Vigna angularis*
  biflorus of authors - *Macrotyloma uniflorum*
  lablab L. - *Lablab purpureus*
  lobatus Willd. - *Pueraria lobata*
  uniflorus Lam. - *Macrotyloma uniflorum*
Doxantha unguis-cati (L.) Miers - *Macfadyena unguis-cati*
Draba caroliniana Walter - *Draba reptans*
Dracaena aurea H. Mann - *Pleomele aurea*
  godseffiana hort. ex Baker - *Dracaena surculosa*
Dryas alaskensis Porsild - *Dryas octopetala* subsp. *alaskensis*
Dryopteris austriaca (Jacq.) Woyn. - *Dryopteris campyloptera*
  austriaca var. spinulosa (O. Müll.) Fiori - *Dryopteris*
  carthusiana
  disjuncta (Ledeb.) C. V. Morton - *Gymnocarpium dryopteris*
  linnaeana C. Chr. - *Gymnocarpium dryopteris*
  spinulosa (O. Müll.) Watt - *Dryopteris carthusiana*
  thelypteris (L.) A. Gray - *Thelypteris palustris*
Dugaldia hoopesii (A. Gray) Rydb. - *Helenium hoopesii*
Duranta repens L. - *Duranta erecta*
Eatonia obtusata A. Gray - *Sphenopholis obtusata*
Echinochloa crus-galli var. frumentacea (Link) W. Wight -
  Echinochloa frumentacea
  pungens (Poir.) Rydb. - *Echinochloa crus-galli*
Echinocystis oregana (Torr. & A. Gray) Cogn. - *Marah*
  oreganus
Elaeagnus argentea Pursh - *Elaeagnus commutata*
  canadensis (L.) A. Nelson - *Shepherdia canadensis*
Eleocharis calva Torr. - *Eleocharis palustris*
  obtusa (Willd.) Schult. - *Eleocharis ovata*
  smallii Britton - *Eleocharis palustris*
Elodea densa (Planch.) Casp. - *Egeria densa*
Elymus angustus Trin. - *Leymus angustus*
  arenarius L. - *Leymus arenarius*
  caput-medusae L. - *Taeniatherum caput-medusae*
  chinensis (Trin.) Keng - *Leymus chinensis*

cinereus Scribn. & Merr. - *Leymus cinereus*
giganteus Vahl - *Leymus racemosus*
glaucus var. jepsonii Burtt Davy - *Elymus glaucus*
hirsutus Schreb. ex Roem. & Schult. - *Elymus villosus*
mollis Trin. - *Leymus mollis*
pseudoagropyrum (Trin. ex Griseb.) Turcz. - *Leymus chinensis*
salina M. E. Jones - *Leymus salinus*
striatus Willd. - *Elymus virginicus*
triticoides Buckley - *Leymus triticoides*
wiegandii Fisch. - *Elymus canadensis*
Elytrigia smithii (Rydb.) A. Löve - *Elymus smithii*
Emilia sagittata (Vahl) DC. - *Emilia javanica*
Equisetum kansanum J. H. Schaffn. - *Equisetum laevigatum*
Eragrostis diffusa Buckley - *Eragrostis pectinacea*
major Host - *Eragrostis cilianensis*
Eranthemum nervosum (Vahl) R. Br. ex Roem. & Schult. - *Eranthemum pulchellum*
Erechtites arguta (A. Rich.) DC. - *Erechtites glomerata*
Erianthus alopecuroides (L.) Elliott - *Saccharum alopecuroideum*
brevibarbis Michx. - *Saccharum brevibarbe*
contortus Baldwin ex Elliott - *Saccharum contortum*
giganteus (Walter) Michx. - *Saccharum giganteum*
ravennae (L.) P. Beauv. - *Saccharum ravennae*
Erigeron bonariensis L. - *Conyza bonariensis*
canadensis A. Br. - *Conyza canadensis*
ramosus (Walter) Britton, et al. - *Erigeron strigosus*
Eriochloa acuminata (C. Presl) Kunth - *Eriochloa polystachya*
Eriogonum polifolium Benth. - *Eriogonum fasciculatum*
stellatum Benth. - *Eriogonum umbellatum*
Ervatamia coronaria (Jacq.) Stapf - *Tabernaemontana divaricata*
divaricata (L.) Burkill - *Tabernaemontana divaricata*
Ervum lens L. - *Lens culinaris*
Erysimum arkansanum Nutt. - *Erysimum capitatum*
Erythraea centaurium of authors - *Centaurium erythraea*
Erythronium parviflorum Goodd. - *Erythronium grandiflorum*
Eucalyptus longirostris F. Muell. - *Eucalyptus camaldulensis*
rostrata Schltdl, - *Eucalyptus camaldulensis*
sideroxylon var. rosea Ingham - *Eucalyptus sideroxylon*
sieberana of authors, non F. Muell. - *Eucalyptus sieber*
Euchlaena mexicana Schrad. - *Zea mays* subsp. *mexicana*
perennis Hitchc. - *Zea perennis*
Eugenia jambos L. - *Syzygium jambos*
malaccensis L. - *Syzygium malaccense*
michelii Lam. - *Eugenia uniflora*
Eupatorium adenophorum Spreng. - *Ageratina adenophora*
ageratoides (L.) L.f. - *Ageratina altissima*
coelestinum L. - *Conoclinium coelestinum*
coronopifolium Willd. - *Eupatorium compositifolium*
rugosum Houtt. - *Ageratina altissima*
urticifolium Reichard - *Ageratina altissima*
Euphorbia dictyosperma Fisch. & C. A. Mey. - *Euphorbia spathulata*
lathyrus L. - *Euphorbia lathyris*
missouriensis Small - *Euphorbia spathulata*
preslii Guss. - *Euphorbia maculata*
Eurotia lanata (Pursh) Moq. - *Krascheninnikovia lanata*
Euryops athanasiae (L.f.) Less. ex Harv. - *Euryops speciosissimus*
Eustoma andrewsii A. Nelson - *Eustoma russellianum*
Evax multicaulis DC. - *Filago verna*

Faba vulgaris Moench - *Vicia faba*
Fagopyrum sagittatum Gilib. - *Fagopyrum esculentum*
vulgare T. Nees - *Fagopyrum esculentum*
Fagus americana Sweet - *Fagus grandifolia*
ferruginea Aiton - *Fagus grandifolia*
Falcata comosa Kuntze - *Amphicarpaea bracteata*
Fatsia horrida (Sm.) Benth. & Hook. f. ex Brewer & S. Watson - *Oplopanax horridum*
Feijoa sellowiana (O. Berg) O. Berg - *Acca sellowiana*
Feronia limonia (L.) Swingle - *Limonia acidissima*
Festuca brachyphylla Schult. - *Festuca ovina*
capillata Lam. - *Festuca tenuifolia*
dertonensis (All.) Asch. & Graebn. - *Vulpia bromoides*
elatior L. - *Festuca pratensis*
elatior var. arundinacea (Schreb.) Wimm. - *Festuca arundinacea*
megalura Nutt. - *Vulpia myuros*
myuros L. - *Vulpia myuros*
ovina var. brachyphylla (Schult.) Hitchc. - *Festuca ovina*
ovina var. duriuscula (L.) W. Koch - *Festuca brevipila*
rubra var. lanuginosa Mert. & W. Koch - *Festuca rubra*
rubra var. prolifera Piper - *Festuca rubra*
Ficus nitida Thunb. - *Ficus benjamina*
repens Rottler - *Ficus pumila*
retusa of authors - *Ficus microcarpa*
Firmiana platanifolia (L.f.) Marsili - *Firmiana simplex*
Fluminia festucacea (Willd.) Hitchc. - *Scolochloa festucacea*
Forestiera neomexicana A. Gray - *Forestiera pubescens*
Fragaria americana (Porter) Britton - *Fragaria vesca* subsp. *americana*
chiloensis var. ananassa (Duchesne) hort. ex L. H. Bailey - *Fragaria ananassa*
vesca var. americana Porter - *Fragaria vesca* subsp. *americana*
Frangula alnus Mill. - *Rhamnus frangula*
Franseria discolor Nutt. - *Ambrosia tomentosa*
Fraxinus lanceolata Borkh. - *Fraxinus pennsylvanica*
oregona Nutt. - *Fraxinus latifolia*
pennsylvanica var. lanceolata (Borkh.) Sarg. - *Fraxinus pennsylvanica*
pennsylvanica var. subintegerrima (Vahl) Fernald - *Fraxinus pennsylvanica*
platycarpa Michx. - *Fraxinus caroliniana*
profunda (Bush) Bush - *Fraxinus pennsylvanica*
tomentosa Michx. - *Fraxinus pennsylvanica*
viridis F. Michx. - *Fraxinus pennsylvanica*
Fremontia californica Torr. - *Fremontodendron californica*
Fritillaria lanceolata Pursh - *Fritillaria affines*
Funastrum cynanchoides (Decne.) Schltdl. - *Sarcostemma cynanchoides*
Galinsoga ciliata (Raf.) Blake - *Galinsoga quadriradiata*
Gardenia florida L. - *Gardenia jasminoides*
Gaura odorata Sessé ex Lag. - *Gaura coccinea*
Gaylussacia resinosa Torr. & A. Gray - *Gaylussacia baccata*
Gentian crinita Froel. - *Gentianopsis crinita*
procera T. Holm - *Gentianopsis procera*
Gentiana puberula Michx. - *Gentiana saponaria*
Gerardia pedicularia L. - *Aureolaria pedicularia*
quercifolia Britton, et al. - *Aureolaria virginica*
virginica (L.) Britton, et al. - *Aureolaria virginica*
Geum aleppicum var. strictum (Aiton) Fernald - *Geum aleppicum*
strictum Aiton - *Geum aleppicum*

Gleichenia dichotoma Burm. f. - *Dicranopteris linearis*
  linearis (Burm. f.) C. B. Clarke - *Dicranopteris linearis*
Glyceria canadensis var. laxa (Scribn.) Hitchc. - *Glyceria canadensis*
  nervata (Willd.) Trin. - *Glyceria striata*
Glycine gracilis Skvortsov - *Glycine max*
Gnaphalium chilense Spreng. - *Gnaphalium stramineum*
  decurrens Ives - *Gnaphalium viscosum*
  purpureum var. spathulatum (Lam.) Ahles - *Gnaphalium purpureum*
Godetia amoena (Lehm.) G. Don - *Clarkia amoena*
Gonolobus laevis Michx. - *Cynanchum laeve*
Gossypium mexicanum Tod. - *Gossypium hirsutum*
  nanking Meyen - *Gossypium arboreum*
  obtusifolium Roxb. - *Gossypium herbaceum*
  peruvianum Cav. - *Gossypium barbadense*
  sturtii F. Muell. - *Gossypium sturtianum*
  vitifolium Lam. - *Gossypium barbadense*
Grossularia reclinata (L.) Mill. - *Ribes uva-crispa*
Gutierrezia dracunculoides (DC.) Blake - *Amphiachyris dracunculoides*
Halesia carolina L. - *Halesia tetraptera*
  parviflora Michx. - *Halesia tetraptera*
Hamelia erecta Jacq. - *Hamelia patens*
Haplopappus pluriflorus (A. Gray) Hall - *Haplopappus heterophyllus*
Helenium tenuifolium Nutt. - *Helenium amarum*
Helianthus dalyi Britton - *Helianthus maximilianii*
  fascicularis Greene - *Helianthus nuttallii*
  lenticularis Douglas - *Helianthus annuus*
  rigidus (Cass.) Desf. - *Helianthus pauciflorus*
  scaberrimus Elliott - *Helianthus laetiflorus*
  tomentosus Michx. - *Helianthus tuberosus*
Heliotropium peruvianum L. - *Heliotropium arborescens*
Hemerocallis flava L. - *Hemerocallis lilioasphodelus*
Hemigraphis colorata (Blume) Hallier f. - *Hemigraphis alternata*
Hepatica triloba of authors - *Hepatica americana*
Heracleum lanatum Michx. - *Heracleum sphondylium* subsp. *montanum*
  maximum Bartram - *Heracleum sphondylium* subsp. *montanum*
Heyderia decurrens (Torr.) K. Koch - *Calocedrus decurrens*
Hibiscus esculentus L. - *Abelmoschus esculentus*
  militaris Cav. - *Hibiscus laevis*
  palustris L. - *Hibiscus moscheutos* subsp. *palustris*
Hicoria microcarpa (Wangenh.) Britton - *Carya ovalis*
  ovata (Mill.) Britton - *Carya ovata*
Hicorius alba Raf. - *Carya ovata*
Hieracium florentinum All. - *Hieracium piloselloides*
  pratense Tausch - *Hieracium caespitosum*
  vulgatum Fr. - *Hieracium lachenalii*
Hierochloe macrophylla Thurb. ex Bol. - *Hierochloe occidentalis*
Hilaria cenchroides Kunth - *Hilaria belangeri*
Holcus halepensis L. - *Sorghum halepense*
  sorghum L. - *Sorghum bicolor*
Holodiscus ariaefolius (Sm.) Greene - *Holodiscus discolor*
Hordeum aegiceras Royle ex Walp. - *Hordeum vulgare*
  boreale Scribn. & J. G. Sm. - *Hordeum brachyantherum*
  distichon L. - *Hordeum vulgare*
  gussoneanum Parl. - *Hordeum geniculatum*
  hexastichon L. - *Hordeum vulgare*

hystrix Roth - *Hordeum geniculatum*
irregulare Åberg & Wiebe - *Hordeum vulgare*
jubatum var. caespitosum (Scribn.) Hitchc. - *Hordeum jubatum*
nodosum Scribn. - *Hordeum brachyantherum*
nodosum var. boreale (Scribn. & J. G. Sm.) Hitchc. - *Hordeum brachyantherum*
sativum Pers. - *Hordeum vulgare*
stebbinsii Covas - *Hordeum glaucum*
vulgare var. trifurcatum (Schltdl.) Alef. - *Hordeum vulgare*
Hosackia parviflora Benth. - *Lotus micranthus*
Hosta japonica (Thunb.) Voss - *Hosta lancifolia*
Houstonia caerulea L. - *Hedyotis caerulea*
  purpurea L. - *Hedyotis purpurea*
Humulus americanus Nutt. - *Humulus lupulus*
Hydrangea petiolaris Siebold & Zucc. - *Hydrangea anomala* subsp. *petiolaris*
Hydrocotyle asiatica L. - *Centella asiatica*
Hymenocallis calathina (Ker Gawl.) G. Nicholson - *Hymenocallis narcissiflora*
Hymenopappus caroliniensis Porter - *Hymenopappus scabiosaeus*
  corymbosus Torr. & A. Gray - *Hymenopappus scabiosaeus* var. *corymbosus*
Hypericum spathulatum (Spach) Steud. - *Hypericum prolificum*
  stans (Michx.) Adams & E. Robson - *Hypericum crux-andreae*
Hystrix patula Moench - *Elymus hystrix*
Ilex heterophylla G. Don - *Osmanthus heterophyllus*
  montana A. Gray - *Ilex ambigua* var. *montana*
Impatiens biflora Walter - *Impatiens capensis*
  fulva Nutt. - *Impatiens capensis*
  sultanii Hook.f. - *Impatiens wallerana*
Ionidium concolor (T. F. Forst.) Benth. & Hook. f. ex S. Watson - *Hybanthus concolor*
Ipomoea bona-nox L. - *Ipomoea alba*
  congesta R. Br. - *Ipomoea indica*
  reptans of authors - *Ipomoea aquatica*
  speciosa L.f. - *Argyreia nervosa*
Iresine celosia L. - *Iresine diffusa*
Iris kaempferi Siebold ex Lem. - *Iris ensata*
  pelogonus Goodd. - *Iris missouriensis*
  schrevei Small - *Iris virginica* var. *schrevei*
  xiphioides Ehrh. - *Iris latifolia*
Isanthus brachiatus (L.) Britton, et al. - *Trichostema brachiatum*
Isocoma heterophylla (A. Gray) Greene - *Haplopappus heterophyllus*
Iva ciliata Willd. - *Iva annua*
Jacobinia carnea (Lindl.) G. Nicholson - *Justicia carnea*
Jatropha manihot L. - *Manihot esculenta*
Juglans alba L. - *Carya tomentosa*
  ailantifolia var. cordiformis (Maxim.) Rehder - *Juglans cordiformis*
  californica var. hindsii Jeps. - *Juglans hindsii*
  sieboldiana Maxim. - *Juglans ailantifolia*
Juncus balticus var. montanus Engelm. - *Juncus balticus*
  dudleyi Wiegand - *Juncus tenuis*
  interior Wiegand - *Juncus tenuis*
Juniperus californica var. utahensis Engelm. - *Juniperus osteosperma*
  communis var. montana Aiton - *Juniperus communis*
  communis var. saxatilis Pall. - *Juniperus communis*
  communis var. siberica Burgsd. - *Juniperus communis*

deppeana var. pachyphloea (Torr.) Martinez - *Juniperus deppeana*

glauca Salisb. - *Juniperus oxycedrus*

pachyphloea Torr. - *Juniperus deppeana*

prostrata Pers. - *Juniperus horizontalis*

siberica Burgsd. - *Juniperus communis*

utahensis (Engelm.) Lemmon - *Juniperus osteosperma*

Jussiaea californica (S. Watson) Jeps. - *Ludwigia peploides*

decurrens (Walter) DC. - *Ludwigia decurrens*

diffusa S. Watson - *Ludwigia peploides*

scabra Willd. - *Ludwigia octovalvis*

Kalanchoe somaliensis Baker - *Kalanchoe marmorata*

Kigelia pinnata (Jacq.) DC. - *Kigelia africana*

Kochia scoparia (L.) Schrad. - *Bassia scoparia*

Koeleria nitida Nutt. - *Koeleria pyrimidata*

Koelreuteria apiculata Rehder & E. H. Wilson - *Koelreuteria paniculata*

Kokia rockii Lewton - *Kokia drynarioides*

Kuhnia eupatorioides L. - *Brickellia eupatorioides*

glutinosa Elliott - *Brickellia eupatorioides*

Kuhnistera candida (Willd.) Kuntze - *Dalea candida*

Kyllinga brevifolia Rottb. - *Cyperus brevifolius*

Lablab niger Medik. - *Lablab purpureus*

vulgaris Savi - *Lablab purpureus*

Lactuca leucophaea A. Gray - *Lactuca biennis*

oblongifolia Nutt. - *Lactuca tatarica* subsp. *pulchella*

pulchella (Pursh) DC. - *Lactuca tatarica* subsp. *pulchella*

scariola L. - *Lactuca serriola*

spicata (Lam.) Hitchc. - *Lactuca biennis*

Lappa major Gaertn. - *Arctium lappa*

Lappula floribunda (Lehm.) Greene - *Hackelia floribunda*

texana (Scheele) Britton - *Lappula redowskii*

virginiana (L.) Greene - *Hackelia virginiana*

Larix europaea DC. - *Larix decidua*

leptolepis (Siebold & Zucc.) Gordon - *Larix kaempferi*

Larrea mexicana Moric. - *Larrea tridentata*

Laurocerasus caroliniana (Mill.) M. Roem. - *Prunus caroliniana*

officinalis M. Roem. - *Prunus laurocerasus*

Laurus benzoin L. - *Lindera benzoin*

Lavandula officinalis Chaix - *Lavandula angustifolia*

vera DC. - *Lavandula angustifolia*

Ledum palustre subsp. groenlandicum (Oeder) Hultén - *Ledum groenlandicum*

Lemaireocereus thurberi (Engelm.) Britton & Rose - *Stenocereus thurberi*

Lens esculenta Moench - *Lens culinaris*

Leontodon leysseri (Wallr.) Beck - *Leontodon taraxacoides*

nudicaulis (L.) Mérat - *Leontodon hirtus*

Lepachys columnaris (Sims) Torr. & A. Gray - *Ratibida columnifera*

Lepidium apetalum Willd. - *Lepidium densiflorum*

repens (Schrenk) Boiss. - *Cardaria draba*

Leptilon canadense (L.) Britton & A. Br. - *Conyza canadensis*

Leptodactylon pungens subsp. hookeri (Douglas ex Hook.) Wherry - *Leptodactylon pungens*

Lespedeza stipulacea Maxim. - *Kummerowia stipulacea*

Leucaena glauca (L.) Benth. - *Leucaena leucocephala*

Leucothoe axillaris var. editorum (Fernald & Schub.) Ahles - *Leucothoe fontanesiana*

editorum Fernald & Schub. - *Leucothoe fontanesiana*

Libocedrus decurrens Torr. - *Calocedrus decurrens*

Ligustrum ibota Siebold - *Ligustrum obtusifolium*

Lilium pensylvanicum Ker Gawl. - *Lilium dauricum*

tigrinum Ker Gawl. - *Lilium lancifolium*

umbellatum Pursh - *Lilium philadelphicum* var. *andinum*

Linaria cymbalaria (L.) Mill. - *Cymbalaria muralis*

Linum lewisii Pursh - *Linum perenne* subsp. *lewisii*

Lippia canescens Kunth - *Phyla nodiflora* var. *canescens*

cuneifolia Torr. - *Phyla cuneifolia*

lanceolata Michx. - *Phyla lanceolata*

lycioides (Cham.) Steud. - *Aloysia gratissima*

nodiflora (L.) Michaux - *Phyla nodiflora*

Lithospermum angustifolium Michx. - *Lithospermum incisum*

croceum Fernald - *Lithospermum caroliniense*

Lochnera rosea (L.) Rchb. - *Catharanthus roseus*

Lolium marschallii Steven - *Lolium perenne*

Lomatium leptocarpum (Torr. & A. Gray) J. M. Coult. & Rose - *Lomatium bicolor* var. *leptocarpum*

Lonicera dioica var. glaucescens (Rydb.) Butters - *Lonicera dioica*

Lophocereus schottii (Engelm.) Britton & Rose - *Pachycereus schottii*

Lophochloa cristata (L.) Hyl. - *Rostraria cristata*

Lotus americanus (Nutt.) Bisch. - *Lotus unifoliatus*

purshianus (Benth.) Clem. & E. G. Clem. - *Lotus unifoliatus*

Lucuma mammosa of authors - *Pouteria sapota*

nervosa A. DC. - *Pouteria campechiana*

salicifolia Kunth - *Pouteria campechiana*

Luffa cylindrica M. Roem. - *Luffa aegyptiaca*

Lupinus caudatus Kellogg - *Lupinus argenteus* var. *heteranthus*

hirsutus L. - *Lupinus pilosus*

leucopsis J. Agardh - *Lupinus sericeus*

lignipes A. Heller - *Lupinus rivularis*

termis Forssk. - *Lupinus albus*

Lychnis alba Mill. - *Silene latiflora*

dioica L. - *Silene dioica*

Lycium chinense Mill. - *Lycium barbarum*

halimifolium Mill. - *Lycium barbarum*

vulgare Dunal - *Lycium barbarum*

Lycopersicon cerasiforme Dunal - *Lycopersicon esculentum* var. *cerasiforme*

lycopersicum var. pyriforme (Dunal) Alef. - *Lycopersicon esculentum* var. *pyriforme*

pyriforme H. Karst. - *Lycopersicon esculentum* var. *pyriforme*

Lyonia ligustrina var. foliosiflora (Michx.) Fernald - *Lyonia ligustrina*

Madia capitata Nutt. - *Madia sativa*

Magnolia cordata Michx. - *Magnolia acuminata* var. *subcordata*

glauca L. - *Magnolia virginiana*

Mahonia aquifolium (Pursh) Nutt. - *Berberis aquifolium*

fremontii (Torr.) Fedde - *Berberis fremontii*

haematocarpa (Woot.) Fedde - *Berberis haematocarpa*

nervosa (Pursh) Nutt. - *Berberis nervosa*

pinnata (Lag.) Fedde - *Berberis pinnata*

repens (Lindl.) G. Don - *Berberis repens*

Malpighia punicifolia L. - *Malpighia glabra*

Malus diversifolia (Bong.) M. Roem. - *Malus fusca*

Malvastrum coccineum (Pursh) A. Gray - *Sphaeralcea coccinea*

Malvaviscus drummondii Torr. & A. Gray - *Malvaviscus arboreus* var. *drummondii*

Manihot aipi Pohl - *Manihot esculenta*

dulcis (J. F. Gmel.) Pax - *Manihot esculenta*

melanobasis Müll. Arg. - *Manihot esculenta*

utilissima Pohl - *Manihot esculenta*

Manilkara bahamensis (Baker) H. J. Lam & A. Meeuse - *Manilkara jaimiqui* subsp. *emarginata*
Marsilea minuta L. - *Marsilea vestita*
Martynia louisiana Wooton. - *Proboscidea louisianica*
proboscidea Wootonin - *Proboscidea louisianica*
Mastichodendron foetidissimum (Jacq.) Cronquist - *Sideroxylon foetidissimum*
Matthiola bicornis (Sibth. & Sm.) DC. - *Matthiola longipetala*
Medicago falcata L. - *Medicago sativa* subsp. *falcata*
hispida Gacrtn. - *Medicago polymorpha*
polymorpha var. vulgaris (Benth.) Shinners - *Medicago polymorpha*
Melaleuca leucadendron of authors - *Melaleuca quinquenervia*
Mentha cardiaca (A. Gray) Baker - *Mentha gentilis*
citrata Ehrh. - *Mentha piperita* var. *citrata*
penardii (Briq.) Rydb. - *Mentha arvensis*
viridis (L.) L. - *Mentha spicata*
Metrosideros collina subsp. polymorpha (Gaudich.) Rock - *Metrosideros polymorpha*
Michelia fuscata (Andr.) Blume - *Michelia figo*
Micromeria chamissonis (Benth.) Greene - *Satureja douglasii*
Mimosa farnesiana L. - *Acacia farnesiana*
lebeck L. - *Albizia lebbeck*
Monarda dispersa Small - *Monarda citriodora*
media Willd. - *Monarda fistulosa*
mollis L. - *Monarda fistulosa*
Monardella odoratissima subsp. pallida (A. Heller) Epling - *Monardella odoratissima*
Montia perfoliata (Donn ex Willd.) Howell - *Claytonia perfoliata*
sibirica (L.) J. T. Howell - *Claytonia sibirica*
Moraea iridioides L. - *Dietes iridioides*
Morus multicaulis Perr. - *Morus alba* var. *multicaulis*
Mucuna deeringiana (Bort) Merr. - *Mucuna pruriens* var. *utilis*
Muehlenbeckia platyclada (F. v. Muell.) Meisn. - *Homalocladium platycladum*
Musa cavendishii Lamb. ex Paxton - *Musa acuminata*
sapientum L. - *Musa paradisiaca*
Myosotis virginica Britton, et al. - *Myosotis verna*
Myrica asplenifolia J. M. Coult. - *Comptonia asplenifolia*
caroliniensis Mill. - *Myrica cerifera*
Myriophyllum brasiliense Cambess. - *Myriophyllum aquaticum*
Nabalus racemosus Hook. - *Prenanthes racemosa*
Nasturtium armoracia (L.) Fr. - *Armoracia rusticana*
nasturtium-aquaticum (L.) H. Karst. - *Rorippa nasturtium-aquaticum*
officinales R. Br. - *Rorippa nasturtium-aquaticum*
Nectandra coriacea (Sw.) Griseb. - *Ocotea coriacea*
Nelumbo pentapetala (Walter) Fernald - *Nelumbo lutea*
Neostyphonia integrifolia (Nutt.) Shafer - *Rhus integrifolia*
Nepeta hederacea (L.) Trevir. - *Glechoma hederacea*
Nephelium litchi Cambess. - *Litchi chinensis*
Nesaea verticillata (L.) Kunth - *Decodon verticillatus*
Nolina tuberculata hort. - *Nolina recurvata*
Nopalea cochenillifera (L.) Salm-Dyck - *Opuntia cochenillifera*
Notholcus lanatus (L.) Nash ex Hitchc. - *Holcus lanatus*
Nothopanax fruticosus (L.) Miq. - *Polyscias fruticosa*
Nuphar lutea subsp. macrophylla (Sm.) E. O. Beal - *Nuphar lutea* subsp. *advena*
Nuttallia cerasiformis Hook. & Arn. - *Oemleria cerasiformis*
Nymphaea advena Aiton - *Nuphar lutea* subsp. *advena*
blanda var. fenzliana (Lehm.) Casp. - *Nymphaea glandulifera*

Nyssa biflora Walter - *Nyssa sylvatica* var. *biflora*
multiflora Wangenh. - *Nyssa sylvatica*
Oakesia sessilifolia S. Watson - *Uvularia sessilifolia*
Oenanthe californica J. Presl - *Oenanthe sarmentosa*
sarmentosa var. californica J. M. Coult. & Rose - *Oenanthe sarmentosa*
Oenothera bistorta Nutt. ex Torr. & A. Gray - *Camissonia bistorta*
heterantha Nutt. - *Camissonia subacaulis*
tanacetifolia Torr. & A. Gray - *Oenothera nuttallii*
Olea aquifolium Siebold & Zucc. - *Osmanthus heterophyllus*
ilicifolia Hassk. - *Osmanthus heterophyllus*
Onoclea struthiopteris (L.) Hoffm. - *Matteuccia struthiopteris*
Opuntia dillenii (Ker Gawl.) Haw. - *Opuntia stricta* var. *dillenii*
engelmannii Salm-Dyck - *Opuntia ficus-indica*
humifusa (Raf.) Raf. - *Opuntia compressa*
humifusa var. austrina (Small) Dress - *Opuntia austrina*
megacantha Salm-Dyck - *Opuntia ficus-indica*
rafinesquii Engelm. - *Opuntia compressa*
Osmanthus ilicifolius (Hassk.) Carrière - *Osmanthus heterophyllus*
Osmaronia cerasiformis (Hook. & Arn.) Greene - *Oemleria cerasiformis*
Osmorhiza brevipes Suksd. - *Osmorhiza chilensis*
divaricata (Britton) Piper - *Osmorhiza chilensis*
intermedia (Rydb.) Blank. - *Osmorhiza chilensis*
nuda Torr. - *Osmorhiza chilensis*
Osterdamia tenuifolia (Trin.) Kuntze - *Zoysia tenuifolia*
Oxalis europea Jord. - *Oxalis corniculata*
florida Salisb. - *Oxalis dillenii*
Oxybaphus nyctagineus (Michx.) Sweet - *Mirabilis nyctaginea*
Oxygraphis cymbalaria (Pursh) Prantl - *Ranunculus cymbalaria*
Oxytenia acerosa Nutt. - *Iva acerosa*
Oxytropis albiflora (A. Nelson) K. Schum. - *Oxytropis sericea*
saximontana (A. Nelson) A. Nelson - *Oxytropis sericea*
Pachyrhizus angulatus Rich. ex DC. - *Pachyrhizus erosus*
bulbosus (L.) Kurz - *Pachyrhizus erosus*
Paeonia albiflora Pall. - *Paeonia lactiflora*
Pandanus odorifer Kuntze - *Pandanus odoratissimus*
Panicularia americana (Torr.) MacMill. - *Glyceria grandis*
elata Nash - *Glyceria elata*
Panicum agrostoides Spreng. - *Panicum rigidulum*
capillare var. occidentale Rydb. - *Panicum capillare*
crus-galli L. - *Echinochloa crus-galli*
fasciulatum Sw. - *Brachiaria fasciculata*
plenum Hitchc. & Chase - *Panicum bulbosum*
purpuascens Raddi - *Brachiaria mutica*
sanguinale L. - *Digitaria sanguinalis*
Papaya carica Gaertn. - *Carica papaya*
Pascopyrum smithii (Rydb.) A. Löve - *Elymus smithii*
Paspalum platycaule Willd. - *Axonopus compressus*
Pediomelum esculentum (Pursh) Rydb. - *Psoralea esculenta*
Pellaea gracilis Hook. - *Cryptogramma stelleri*
Pellionia daveauana (Carrière) N. E. Br. - *Pellionia repens*
Pennisetum americanum (L.) K. Schum. - *Pennisetum glaucum*
spicatum (L.) Körn. - *Pennisetum glaucum*
typhoides (Burm. f.) Stapf & F. T. Hubb. - *Pennisetum glaucum*
typhoideum Rich. - *Pennisetum glaucum*
Peplis diandra Nutt. - *Didiplis diandra*
Persea carolinensis Nees - *Persea borbonia*
pubescens (Pursh) Sarg. - *Persea borbonia*
Persica vulgaris Mill. - *Prunus persica*

Petalostemon candidum (Willd.) Michx. - *Dalea candida*
  oligophyllum (Torr.) Rydb. - *Dalea candida*
  purpureum (Vent.) Rydb. - *Dalea purpurea*
  villosum Nutt. - *Dalea villosa*
  violaceus Michx. - *Dalea purpurea*
Petroselinum hortense Hoffm. - *Petroselinum crispum*
  sativum Hoffm. - *Petroselinum crispum*
  vulgare Lag. - *Petroselinum crispum*
Phalaris arundinacea var. picta L. - *Phalaris arundinacea*
  tuberosa L. - *Phalaris aquatica*
  tuberosa var. stenoptera (Hack.) Hitchc. - *Phalaris stenoptera*
Phanerophlebia falcatum (L.f.) H. F. Copel. - *Cyrtomium falcata*
Pharbitis nil (L.) Choisy - *Ipomoea nil*
Phaseolus angularis (Willd.) W. Wight - *Vigna angularis*
  atropurpureus Moç. & Sessé ex DC. - *Macroptilium atropurpureum*
  aureus Roxb. - *Vigna radiata*
  inamoenus Macfad. - *Phaseolus lunatus*
  limensis Macfad. - *Phaseolus lunatus*
  max L. - *Glycine max*
  multiflorus Lam. - *Phaseolus coccineus*
  mungo L. - *Vigna mungo*
  nanus L. - *Phaseolus vulgaris*
  radiatus L. - *Vigna radiata*
  tunkinensis Macfad. - *Phaseolus lunatus*
Phegopteris dryopteris (L.) Fée - *Gymnocarpium dryopteris*
Philibertia cynanchoides (Schltdl.) A. Gray - *Sarcostemma cynanchoides*
Philodendron micans Klotzsch ex K. Koch - *Philodendron scandens* f. *micans*
  pertusum Kunth & Bouché - *Monstera deliciosa*
Phlox decussata Lyon ex Pursh - *Phlox paniculata*
Phoradendron flavescens (Pursh) Nutt. - *Phoradendron serotinum*
Photinia arbutifolia Lindl. - *Heteromeles arbutifolia*
Phragmites communis Trin. - *Phragmites australis*
Phyllostachys henonis Mitford - *Phyllostachys nigra*
  quilioi Rivière & C. Rivière - *Phyllostachys bambusoides*
Physalis edulis Sims - *Physalis peruviana*
  ixocarpa Hornem. - *Physalis philadelphica*
  lancifolia Nees - *Physalis angulata*
  pruinosa L. - *Physalis pubescens* var. *grisea*
  subglabrata Mack. & Bush - *Physalis longifolia*
  virginica A. Gray - *Physalis virginiana*
Phytolacca decandra L. - *Phytolacca americana*
Picea canadensis (Mill.) Britton, et al. - *Picea glauca*
  excelsa (Lam.) Link - *Picea abies*
  glauca var. albertiana (S. Br.) Sarg. - *Picea glauca*
  nigra (Aiton) Link - *Picea mariana*
  rubra Sarg. - *Picea rubens*
Pieris nitida (W. Bartram) Benth. & Hook. f. - *Lyonia lucida*
Pimenta officinalis Lindl. - *Pimenta dioica*
Pinus alba (Aiton) Link - *Picea glauca*
  aristata Engelm. - *Pinus longaeva*
  australis F. Michx. - *Pinus palustris*
  cembroides var. monophylla (Torr. & Frém.) Voss - *Pinus monophylla*
  concolor Engelm. ex Gordon - *Abies concolor*
  contorta var. latifolia Engelm. ex S. Watson - *Pinus contorta* subsp. *murrayana*
  excelsa Wall ex D. Don - *Pinus wallichiana*
  griffithii McClell. - *Pinus wallichiana*
  insignis Douglas ex Loudon - *Pinus radiata*

  montana Mill. - *Pinus mugo*
  murrayana Grev. & Balf. - *Pinus contorta* subsp. *murrayana*
  ponderosa var. jeffreyi (Grev. & Balf.) S. Watson - *Pinus jeffreyi*
  rigida var. serotina (Michx.) Loudon - *Pinus serotina*
  thunbergii Parl. - *Pinus thunbergiana*
Piscidia erythrina L. - *Piscidia piscipula*
Pisum arvense L. - *Pisum sativum* var. *arvense*
Pithecellobium guadalupense Pers. - *Pithecellobium keyense*
  saman (Jacq.) Benth. - *Albizia saman*
Plantago arenaria Waldst. & Kit. - *Plantago psyllium*
  asiatica L. - *Plantago major*
  oliganthos Roem. & Schult. - *Plantago maritima* var. *juncoides*
Plumeria acuminata Aiton - *Plumeria rubra* f. *acutifolia*
  acutifolia Poir. - *Plumeria rubra* f. *acutifolia*
Poa ampla Merr. - *Poa secunda* subsp. *nevadensis*
  canbyi (Scribn.) Howell - *Poa secunda*
  gracillima Scribn. - *Poa secunda*
  juncifolia Scribn. - *Poa secunda* subsp. *nevadensis*
  longiligula Scribn. - *Poa fendleriana*
  nevadensis Scribn. - *Poa secunda* subsp. *nevadensis*
  sandbergii Scribn. - *Poa secunda*
  serotina Ehrh. - *Poa palustris*
  wheeleri Vasey - *Poa nervosa*
Poinciana gilliesii Hook. - *Caesalpinia gilliesii*
  regia Bojer ex Hook. - *Delonix regia*
Poinsettia pulcherrima (Willd. ex Klotzsch) J. Graham - *Euphorbia pulcherrima*
Polycodium neglectum Small - *Vaccinium stamineum*
  stamineum (L.) Greene - *Vaccinium stamineum*
Polygonatum canaliculatum (Schult.f.) Pursh - *Polygonatum biflorum*
Polygonum acre Kunth - *Polygonum punctatum*
  bistorta L. - *Polygonum bistortoides*
  bistorta subsp. plumosum (Small) Hultén - *Polygonum bistortoides*
  buxiforme Small - *Polygonum aviculare*
  emersum S. Watson - *Polygonum coccineum*
  exsertum B. L. Rob. - *Polygonum ramosissimum*
  muhlenbergii (Meisn.) S. Watson - *Polygonum coccineum*
  natans f. hartwrightii (A. Gray) Stanford - *Polygonum amphibium*
  plumosum Small - *Polygonum bistortoides*
  prolificum (Small) B. L. Rob. - *Polygonum ramosissimum*
Polypodium vulgare var. occidentale Hook. - *Polypodium glycyrrhiza*
Polyscias balfouriana (Sander ex André) L. H. Bailey - *Polyscias scutellaria*
Polystichum adiantiforme (G. Forst.) Small - *Rumohra adiantiformis*
Pongamia glabra Vent. - *Pongamia pinnata*
Pontederia lanceolata Nutt. - *Pontederia cordata* var. *lancifolia*
Populus angulata Aiton - *Populus deltoides*
  candicans Aiton - *Populus balsamifera*
  deltoides subsp. monilifera (Aiton) Eckenw. - *Populus deltoides* var. *occidentalis*
  deltoides var. virginiana (Foug.) Sudw. - *Populus deltoides*
  monilifera Aiton - *Populus deltoides* var. *occidentalis*
  nigra var. italica Münchh. - *Populus nigra*
  occidentalis (Rydb.) Britton - *Populus deltoides* var. *occidentalis*
  sargentii Dode - *Populus deltoides* var. *occidentalis*

tacamahaca Mill. - *Populus balsamifera*
Potamogeton americanus Cham. & Schltdl. - *Potamogeton nodosus*
heterophyllus Schreb. - *Potamogeton gramineus*
Potentilla pensylvanica var. strigosa (Pall. ex Pursh) Pursh - *Potentilla pensylvanica*
pumila Poir. - *Potentilla canadensis*
strigosa Pall. ex Pursh - *Potentilla pensylvanica*
Poterium officinale (L.) A. Gray - *Sanguisorba officinalis*
Pothos aureus Lindl. & André - *Epipremnum aureum*
Pouteria dulcifica (Schumach. & Thonn.) Baehni - *Synsepalum dulcificum*
Prenanthes vulgaris Gueldenst. - *Lactuca muralis*
Prinos coriacea Pursh - *Ilex coriacea*
Proboscidea jussieui Wootonuthors - *Proboscidea louisianica*
louisiana (Mill.) Wooton & Standl. - *Proboscidea louisianica*
Prosopis juliflora var. torreyana L. D. Benson - *Prosopis glandulosa* var. *torreyana*
juliflora var. velutina (Woot.) Sarg. - *Prosopis velutina*
Prunus amygdalus Batsch - *Prunus dulcis*
besseyi L. H. Bailey - *Prunus pumila* var. *besseyi*
communis (L.) Arcang., non Huds. - *Prunus dulcis*
demissa (Nutt.) Walp. - *Prunus virginiana* var. *demissa*
depressa Pursh - *Prunus pumila*
lyonii (Eastw.) Sarg. - *Prunus ilicifolia* subsp. *lyonii*
melanocarpa (A. Nelson) Rydb. - *Prunus virginiana* var. *melanocarpa*
persica var. nectarina (Aiton) Maxim. - *Prunus persica* var. *nucipersica*
triflora Roxb. - *Prunus salicina*
Psedera quinquefolia (L.) Greene - *Parthenocissus quinquefolia*
tricuspidata (Siebold & Zucc.) Rehder - *Parthenocissus tricuspidata*
Pseudotsuga mucronata (Raf.) Sudw. - *Pseudotsuga menziesii*
taxifolia (Lamb.) Britton - *Pseudotsuga menziesii*
taxifolia var. glauca (Beissn.) Sudw. - *Pseudotsuga menziesii* var. *glauca*
Psidium cattleianum Sabine - *Psidium littorale* var. *longipes*
Psoralea floribunda Nutt. ex Torr. & A. Gray - *Psoralidium tenuiflorum*
lanceolata Pursh - *Psoralidium lanceolatum*
lanceolata var. scabra (Nutt.) Piper - *Psoralidium lanceolatum*
tenuiflora Pursh - *Psoralidium tenuiflorum*
tenuiflora var. floribunda (Nutt. ex Torr. & A. Gray) Rydb. - *Psoralidium tenuiflorum*
Pteridium aquilinum var. lanuginosum (Bong.) Fernald - *Pteridium aquilinum* var. *pubescens*
Pteris aquilina L. - *Pteridium aquilinum*
aquilina var. lanuginosa (Bong.) Kühn - *Pteridium aquilinum* var. *pubescens*
aquilina var. pubescens (Underw.) Clute - *Pteridium aquilinum* var. *pubescens*
Puccinellia airoides (Nutt.) S. Watson & J. M. Coult. - *Puccinellia nuttalliana*
Pueraria hirsuta (Thunb.) C. K. Schneid. - *Pueraria lobata*
thunbergiana (Siebold & Zucc.) Benth. - *Pueraria lobata*
Pulsatilla hirsutissima Britton - *Anemone patens*
Pyrethrum cinerariifolium Sch. Bip. - *Chrysanthemum cinerariifolium*
Pyrola secunda L. - *Orthilia secunda*
uniflora L. - *Moneses uniflora*
virens Schweigg. & Körte - *Pyrola chlorantha*

Pyrus angustifolia Aiton - *Malus angustifolia*
arbutifolia (L.) L.f. - *Aronia arbutifolia*
baccata L. - *Malus baccata*
coronaria L. - *Malus coronaria*
cydonia L. - *Cydonia oblonga*
fusca Raf. - *Malus fusca*
ioensis (A. Wood) L. H. Bailey - *Malus ioensis*
malus L. - *Malus domestica*
prunifolia Willd. - *Malus prunifolia*
rivularis Douglas ex Hook. - *Malus fusca*
serotina Rehder - *Pyrus pyrifolia*
sinensis Lindl. - *Pseudocydonia sinensis*
Quamoclit coccinea (L.) Moench - *Ipomoea coccinea*
pennata (Desr.) Bojer - *Ipomoea quamoclit*
vulgaris Choisy - *Ipomoea quamoclit*
Quercus aquatica Walter - *Quercus nigra*
borealis var. maxima (Marsh.) Sarg. - *Quercus rubra*
catesbaei Michx. - *Quercus laevis*
cinerea Michx. - *Quercus incana*
densiflora Hook. & Arn. - *Lithocarpus densiflora*
geminata Small - *Quercus virginiana*
gunnisonii Rydb. - *Quercus gambelii*
maxima (Marsh.) Sarg. - *Quercus rubra*
montana Willd. - *Quercus prinus*
shumardii var. texana (Buckley) Ashe - *Quercus texana*
tinctoria W. Bartram - *Quercus velutina*
utahensis (A. DC.) Rydb. - *Quercus gambelii*
virens Aiton - *Quercus virginiana*
Quincula lobata (Torr.) Raf. - *Physalis lobata*
Radicula armoracia (L.) B. L. Rob. - *Armoracia rusticana*
Ramona polystachya Greene - *Salvia apiana*
stachyoides Briq. - *Salvia mellifera*
Ranunculums septentrionalis Poir. - *Ranunculus hispidus*
Ranunculus apiophyllus St.-Lég. - *Ranunculus sceleratus*
bongardii Greene - *Ranunculus occidentalis*
calthaefolius Jord. - *Ranunculus ficaria*
delphinifolius Torr. - *Ranunculus flabellaris*
multifidus Pursh - *Ranunculus flabellaris*
Rapanea guianensis Aubl. - *Myrsine guianensis*
Ratibida columnaris (Sims) D. Don - *Ratibida columnifera*
Reynoutria japonica Houtt. - *Polygonum cuspidatum*
Rhacoma ilicifolia (Poir.) Trel. - *Crossopetalum ilicifolia*
Rhamnus crocea subsp. ilicifolia (Kellogg) C. B. Wolf - *Rhamnus ilicifolia*
Rhinanthus kyrollae Chabrey - *Rhinanthus crista-galli*
Rhipsalidopsis gaertneri (Regel) Moran - *Hatiora gaertneri*
Rhododendron californicum Hook. - *Rhododendron macrophyllum*
nudiflora (L.) Torr. - *Rhododendron periclymenoides*
obtusum var. kaempferi (Planch.) Wilson - *Rhododendron kaempferi*
roseum (Loisel.) Rehder - *Rhododendron austrinum*
Rhoeo discolor (L'Hér.) Hance - *Tradescantia spathacea*
spathacea (Sw.) Stearn - *Tradescantia spathacea*
Rhus cotinus L. - *Cotinus coggygria*
diversifolia Engl. - *Toxicodendron pubescens*
laurina Nutt. - *Malosma laurina*
radicans L. - *Toxicodendron radicans*
toxicodendron L. - *Toxicodendron pubescens*
typhina L. - *Rhus hirta*
vernix L. - *Toxicodendron vernix*
Rhynchelytrum roseum (Nees) Stapf & F. T. Hubb. ex Bews - *Rhynchelytrum repens*

Rhynchospora laxa Vahl - *Rhynchospora corniculata*
Ribes floridum L'Hér. - *Ribes americanum*
  grossularia L. - *Ribes uva-crispa*
  petiolare Douglas - *Ribes hudsonianum*
  prostratum L'Hér. - *Ribes glandulosum*
  sativum (Rchb.) Syme - *Ribes rubrum*
  saxosum Hook. - *Ribes oxyacanthoides*
  vulgare Lam. - *Ribes rubrum*
  wolfii Rothr. - *Ribes sanguineum*
Richardsonia scabra (L.) A. St.-Hil. - *Richardia scabra*
Rollinia deliciosa Saff. - *Rollinia pulchrinervis*
Rorippa armoracia (L.) Hitchc. - *Armoracia rusticana*
  islandica (Oeder ex Js. Murray) Borbás - *Rorippa palustris*
Rosa aciculata Cockerell - *Rosa nutkana*
  engelmannii S. Watson - *Rosa acicularis*
  fendleri Crép. - *Rosa woodsii*
  humilis Marsh. - *Rosa carolina*
  lucida Ehrh. - *Rosa virginiana*
  macounii Greene - *Rosa woodsii*
  manettii Bals.-Criv. ex Rivers - *Rosa noisettiana* cv. 'Manettii'
  neomexicana Cockerell - *Rosa woodsii*
  praticola Greene - *Rosa arkansana*
  rubiginosa L. - *Rosa eglanteria*
Rottboellia exaltata L.f. - *Rottboellia cochinchinensis*
Rubus floridus Tratt. - *Rubus argutus*
  idaeus var. strigosus (Michx.) Maxim. - *Rubus strigosus*
  nigrobaccus L. H. Bailey - *Rubus alleghceniensis*
  nutkanus Moç. ex Ser. - *Rubus parviflorus*
  procerus P. J. Müll. - *Rubus discolor*
  procumbens Muhl. - *Rubus canadensis*
  rosaefolius Sm. - *Rubus rosifolius*
  ursinus var. loganobaccus (L. H. Bailey) L. H. Bailey - *Rubus loganobaccus*
Rudbeckia amplexicaulis Vahl - *Dracopis amplexicaulis*
  serotina Nutt. - *Rudbeckia hirta* var. *pulcherrima*
Rumex pauciflorus Campd. - *Rumex alpinus*
Sabal serrulata (Michx.) Nutt. ex Schult. & Schult. f. - *Serenoa repens*
  texana (Cook) Becc. - *Sabal mexicana*
Sagittaria arifolia Nutt. - *Sagittaria sagittifolia*
  chinensis Pursh - *Sagittaria latifolia*
  variabilis Engelm. - *Sagittaria sagittifolia*
Salix adenophylla Hook. - *Salix cordata*
  anglorum Cham. - *Salix arctica*
  arbusculoides var. glabra Andersson - *Salix arbusculoides*
  bebbiana var. perrostrata (Rydb.) C. K. Schneid. - *Salix bebbiana*
  brachystachys Benth. - *Salix scouleriana*
  coulteri Andersson - *Salix sitchensis*
  flavescens Nutt. - *Salix scouleriana*
  glauca var. acutifolia (Hook.) C. K. Schneid. - *Salix glauca*
  glauca var. aliceae C. R. Ball - *Salix glauca*
  glauca var. glabrescens C. K. Schneid. - *Salix glauca*
  glaucops Andersson - *Salix glauca*
  hindsiana Benth. - *Salix exigua*
  interior Rowlee - *Salix exigua*
  lasiandra Benth. - *Salix lucida* subsp. *lasiandra*
  longifolia Muhl. - *Salix exigua*
  longipes Shuttlew. ex Andersson - *Salix caroliniana*
  microphylla Cham. & Schltdl. - *Salix taxifolia*
  nelsonii C. R. Ball - *Salix planifolia*
  nigra var. falcata (Pursh) Torr. - *Salix nigra*

  niphoclada Rydb. - *Salix brachycarpa* subsp. *niphoclada*
  nivalis Hook. - *Salix reticulata*
  nuttallii Sarg. - *Salix scouleriana*
  ovalifolia var. camdensis C. K. Schneid. - *Salix ovalifolia*
  padophylla Rydb. - *Salix monticola*
  perrostrata Rydb. - *Salix bebbiana*
  petrophila Rydb. - *Salix arctica*
  pseudocordata Andersson - *Salix novaeangliae*
  reticulata var. gigantifolia C. R. Ball - *Salix reticulata*
  richardsonii Hook. - *Salix lanata* subsp. *richardsonii*
  rigida Muhl. - *Salix eriocephala*
  rostrata Richardson - *Salix bebbiana*
  seemanii Rydb. - *Salix glauca*
  tracyi C. R. Ball - *Salix lasiolepis*
Salmalia malabarica (DC.) Schott & Endl. - *Bombax ceiba*
Salsola iberica A. Nelson & Pau - *Salsola australis*
  kali subsp. ruthenica (Iljin) Soó - *Salsola australis*
  kali var. tenuifolia Tausch - *Salsola australis*
  pestifer A. Nelson - *Salsola australis*
Salvia lanceolata Brouss. - *Salvia reflexa*
Samanea saman (Jacq.) Merr. - *Albizia saman*
Sambucus glauca Nutt. - *Sambucus cerulea*
  microbotrys Rydb. - *Sambucus racemosa* var. *microbotrys*
Santolina virens Mill. - *Santolina rosmarinifolia*
Sapindus drummondii Hook. & Arn. - *Sapindus saponaria* var. *drummondii*
Saponaria vaccaria L. - *Vaccaria hispanica*
Sarothamnus scoparius (L.) Wimm. ex W. Koch - *Cytisus scoparius*
Sassafras officinale Nees & Eberm. - *Sassafras albidum*
  variifolium (Salisb.) Kuntze - *Sassafras albidum*
Satureja vulgaris (L.) Fritsch - *Clinopodium vulgare*
Schrankia microphylla (Dryand.) J. F. Macbr. - *Mimosa quadrivalvis* var. *angustata*
  nuttallii (DC.) Standl. - *Mimosa quadrivalvis* var. *nuttallii*
  uncinata Willd. - *Mimosa quadrivalvis* var. *angustata*
Scilla hispanica Mill. - *Hyacinthoides hispanica*
Scindapsus aureus (Lindl. & André) Engl. - *Epipremnum aureum*
Scirpus caespitosus var. callosus Bigelow - *Scirpus caespitosus*
  campestris A. Nelson - *Scirpus maritimus*
  occidentalis S. Watson - *Scirpus acutus* var. *occidentalis*
  olneyi A. Gray - *Scirpus americanus*
  paludosus A. Nelson - *Scirpus maritimus*
  robustus A. Nelson - *Scirpus maritimus*
Scrophularia leporella C. Bicknell - *Scrophularia lanceolata*
  neglecta Rydb. - *Scrophularia marilandica*
Senecio balsamitae Muhl. ex Willd. - *Senecio pauperculus*
  confusus Britton - *Senecio chenopodioides*
  pauperculus var. balsamitae (Muhl. ex Willd.) Fernald - *Senecio pauperculus*
Sequoia gigantea (Lindl.) Buchholz - *Sequoiadendron giganteum*
Serenoa serrulata (Michx.) G. Nicholson - *Serenoa repens*
Serinia oppositifolia Kuntze - *Krigia virginica*
Sesamum orientale L. - *Sesamum indicum*
Sesbania macrocarpa Muhl. - *Sesbania exaltata*
Sesbanis tripetii (Poit.) hort. ex F. T. Hubb. - *Sesbania punicea*
Setaria glauca (L.) P. Beauv. - *Setaria pumila*
  lutescens (Weigel) F. T. Hubb. - *Setaria pumila*
Seymeria macrophyllum Nutt. - *Dasistoma macrophylla*

Sida hederacea (Douglas ex Hook.) Torr. ex A. Gray - *Malvella leprosa*

spinosa var. angustifolia (Lam.) Trin. & Planch. - *Sida spinosa*

Sidalcea delphinifolia Greene - *Sidalcea malvaeflora*

Sieversia ciliata G. Don - *Geum triflorum*

Silene alba (Mill.) E. H. L. Krause - *Silene latiflora*

cucubalus Wibel - *Silene vulgaris*

latifolia (Mill.) Britten & Rendle - *Silene vulgaris*

Silphium asperrimum Hook. - *Silphium radula*

Simmondsia californica Nutt. - *Simmondsia chinensis*

Siphonia brasiliensis Willd. ex Adr. Juss. - *Hevea brasiliensis*

Sisymbrium canescens var. brachycarpon S. Watson - *Descurainia pinnata* var. *brachycarpa*

thalianum (L.) Gay & Monnard - *Arabidopsis thaliana*

Sisyrinchium gramineum Lam. - *Sisyrinchium angustifolium*

graminoides C. Bicknell - *Sisyrinchium angustifolium*

Sitanion hystrix (Nutt.) J. G. Sm. - *Elymus elymoides*

Sium cicutifolium Schrank - *Sium suave*

Smilacina amplexicaulis Nutt. - *Smilacina racemosa*

racemosa var. amplexicaulis (Nutt.) S. Watson - *Smilacina racemosa*

sessilifolia (Baker) Nutt. ex S. Watson - *Smilacina stellata*

Smilax lanceolata L. - *Smilax smallii*

tamnoides var. hispida Muhl. - *Smilax hispida*

Soja hispida Moench - *Glycine max*

max (L.) Piper - *Glycine max*

Solanum ptycanthum Dunal - *Solanum nigrum*

Solidago altissima L. - *Solidago canadensis* var. *scabra*

conferta Mack. - *Solidago speciosa*

edisoniana Mack. - *Solidago elliottii*

gigantea var. leiophylla Fernald - *Solidago gigantea*

latifolia L. - *Solidago flexicaulis*

monticola Torr. & A. Gray - *Solidago hispida*

pitchcri Nutt. - *Solidago gigantea*

serotina Aiton - *Solidago gigantea*

Soliva pterosperma (Juss.) Less. - *Soliva sessilis*

Sorghum caffrorum (Retz.) P. Beauv. - *Sorghum bicolor*

dochna (Forssk.) Snowden - *Sorghum bicolor*

durra (Forssk.) Stapf - *Sorghum bicolor*

guineense Stapf - *Sorghum bicolor*

miliaceum (Roxb.) Snowden - *Sorghum halepense*

saccharatum (L.) Moench - *Sorghum bicolor*

sudanense (Piper) Stapf - *Sorghum drummondii*

vulgare Pers. - *Sorghum bicolor*

vulgare var. caffrorum (Retz.) F. T. Hubb. & Rehder - *Sorghum bicolor*

vulgare var. saccharatum (L.) Boerl. - *Sorghum bicolor*

vulgare var. sudanense (Piper) Hitchc. - *Sorghum drummondii*

vulgare var. technicum (K. D. Koernig) Jáv. - *Sorghum bicolor*

Spartina michauxiana Hitchc. - *Spartina pectinata*

Specularia biflora (Ruiz & Pav.) Fisch. & C. A. Mey. - *Triodanis biflora*

perfoliata (L.) A. DC. - *Triodanis perfoliata*

Sphaeralcea angusta (A. Gray) Fernald - *Malvastrum hispidum*

Sphaeropteris cooperi (F. v. Muell.) R. M. Tryon - *Cyathea cooperi*

Sphenomeris chusana (L.) H. F. Copel. - *Sphenomeris chinensis*

Sphenopholis intermedia Rydb. - *Sphenopholis obtusata*

Spiraea filipendula L. - *Filipendula vulgaris*

latifolia (Aiton) Borkh. - *Spiraea alba* var. *latifolia*

opulifolia L. - *Physocarpus opulifolius*

rubra Britton - *Filipendula rubra*

Sporobolus asperifolius Nees & C. A. Mey. - *Muhlenbergia asperifolia*

cuspidatus A. Wood - *Muhlenbergia cuspidata*

gracilis (Trin.) Merr. - *Sporobolus junceus*

vaginiflorus var. inaequalis Fernald - *Sporobolus vaginiflorus*

wrightii Munro ex Scribn. - *Sporobolus airoides*

Spraguea umbellata Torr. - *Calyptridium umbellatum*

Steironema ciliatum (L.) Raf. - *Lysimachia ciliata*

lanceolatum (Walter) A. Gray - *Lysimachia lanceolata*

Stenolobium stans (L.) Seem. - *Tecoma stans*

Sterculia platanifolia L.f. - *Firmiana simplex*

Stillingia sebifera (L.) Michx. - *Sapium sebiferum*

Stipa elmeri Scribn. - *Stipa occidentalis*

williamsii Scribn. - *Stipa columbiana*

Stizolobium deeringianum Bort - *Mucuna pruriens* var. *utilis*

Striga lutea Lour. - *Striga asiatica*

Strophostyles missouriensis Small - *Strophostyles helvola*

pauciflora (Benth.) S. Watson - *Strophostyles leiosperma*

Stylosanthes gracilis Kunth - *Stylosanthes guianensis*

guineensis Schumach. - *Stylosanthes erecta*

Suaeda occidentalis S. Watson - *Suaeda calceoliformis*

Symphoricarpos racemosus Michx. - *Symphoricarpos albus*

rivularis Suksd. - *Symphoricarpos albus* var. *laevigatus*

Symphytum peregrinum Ledeb. - *Symphytum uplandicum*

Tabacum nicotianum Bercht. & Opiz - *Nicotiana tabacum*

Tabernaemontana coronaria (Jacq.) Willd. - *Tabernaemontana divaricata*

Tagetes patula L. - *Tagetes erecta*

Tanacetum cinerariifolium (Trevir.) Sch. Bip. - *Chrysanthemum cinerariifolium*

Taraxacum vulgare Schrank - *Taraxacum officinale*

Taxodium ascendens Brongn. - *Taxodium distichum* var. *imbricarium*

distichum var. nutans of authors - *Taxodium distichum* var. *imbricarium*

Tecoma radicans (L.) Juss. - *Campsis radicans*

Tecomaria capensis (Thunb.) Spach - *Tecoma capensis*

Tetragonia expansa Thunb. ex Js. Murray - *Tetragonia tetragonioides*

Thalictrum polygamum Muhl. - *Thalictrum pubescens*

Thaspium aureum Nutt. - *Thaspium trifoliatum*

Thea bohea L. - *Camellia sinensis*

sinensis L. - *Camellia sinensis*

viridis L. - *Camellia sinensis*

Thelypteris thelypteris (L.) Nieuwl. - *Thelypteris palustris*

Thuja compacta Standish - *Thuja occidentalis*

occidentalis cv. 'compacta' - *Thuja occidentalis*

orientalis L. - *Platycladus orientalis*

Tibouchina semidecandra Cogn. - *Tibouchina urvilleana*

Tilia floridana Small - *Tilia caroliniana*

glabra Vent. - *Tilia americana*

Tithymalus missouriensis Small - *Euphorbia spathulata*

Toxicodendron diversilobum (Torr. & A. Gray) Greene - *Rhus diversiloba*

toxicarium (Salisb.) Gillis - *Toxicodendron pubescens*

Tragopogon major Jacq. - *Tragopogon dubius*

Trautvetteria grandis Nutt. ex Torr. & A. Gray - *Trautvetteria carolinensis*

Trichachne californica (Benth.) Chase - *Digitaria californica*

Trichocereus spachianus (Lem.) Riccob. - *Echinopsis spachiana*

Tricholaena repens (Willd.) Hitchc. - *Rhynchelytrum repens*

rosea Nees - *Rhynchelytrum repens*

Trichophorum caespitosum subsp. austriacum (Palla) Hegi - *Scirpus caespitosus*

Trientalis americana Pursh - *Trientalis borealis*

Trifolium agrarium L., misapplied - *Trifolium aureum*

  fendleri Greene - *Trifolium wormskioldii*

  involucratum Ortega - *Trifolium wormskioldii*

  involucratum var. fimbriatum (Lindl.) McDermott - *Trifolium wormskioldii*

  procumbens L. - *Trifolium dubium*

  willdenovii Spreng. - *Trifolium wormskioldii*

Triodia flava (L.) Smyth - *Tridens flavus*

Tristania conferta R. Br. - *Lophostemon conferta*

Triticum dicoccum Schrank - *Triticum dicoccon*

  persicum (Boiss.) Aitch. & Hemsl. - *Aegilops triuncialis*

  pyramidale Percival - *Triticum durum*

  repens P. Beauv. - *Elytrigia repens*

  sativum Lam. - *Triticum aestivum*

  timopheevii (Zhuk.) Zhuk. - *Triticum aestivum*

  vulgare Vill. - *Triticum aestivum*

Trollius albiflorus (A. Gray) Rydb. - *Trollius laxus*

Ulmus davidiana var. japonica (Rehder) Nakai - *Ulmus japonica*

  fulva Michx. - *Ulmus rubra*

  procera Salisb. - *Ulmus minor*

  pubescens Walter - *Ulmus rubra*

  racemosa D. Thomas - *Ulmus thomasii*

Uniola latifolia Michx. - *Chasmanthium latifolium*

Urtica lyallii S. Watson - *Urtica dioica* subsp. *gracilis*

  procera Muhl. ex Willd. - *Urtica dioica*

Utricularia fibrosa Walter - *Utricularia gibba*

Vaccaria pyramidata Medik. - *Vaccaria hispanica*

  segetalis Garcke ex Asch. - *Vaccaria hispanica*

Vaccinium amoenum Aiton - *Vaccinium corymbosum*

  australe Small - *Vaccinium corymbosum*

  canadense Kalm ex Richardson - *Vaccinium myrtilloides*

  pallidum Aiton - *Vaccinium vacillans*

Valeriana ceratophylla (Hook.) Piper - *Valeriana edulis*

  ciliata Torr. & A. Gray - *Valeriana edulis*

  furfurescens A. Nelson - *Valeriana edulis*

Vallota speciosa (L.f.) Durand & Schinz - *Cyrtanthus elatus*

Vancouveria parviflora Greene - *Vancouveria planipetala*

Vanilla fragrans (Salisb.) Ames - *Vanilla planifolia*

Verbena angustifolia Michx. - *Verbena simplex*

  bracteosa Michx. - *Verbena bracteata*

Verbesina texana Buckley - *Verbesina virginica*

Vernonia altissima Nutt. - *Vernonia gigantea*

  ovalifolia Torr. & A. Gray - *Vernonia noveboracensis*

Veronica humifusa Dicks. - *Veronica serpyllifolia* subsp. *humifusa*

  maritima L. - *Veronica longifolia*

  virginica L. - *Veronicastrum virginicum*

Viburnum pauciflorum Raf. - *Viburnum edule*

  pubescens (Aiton) Pursh - *Viburnum dentatum*

  tomentosum Thunb. - *Viburnum plicatum* var. *tomentosum*

Vicia americana var. truncata (Nutt.) Brewer - *Vicia americana*

  angustifolia L. - *Vicia sativa* subsp. *nigra*

  atropurpurea Desf. - *Vicia benghalensis*

  linearis Greene - *Vicia americana*

  sparsifolia Nutt. ex Torr. & A. Gray - *Vicia americana*

  trifida D. Dietr. - *Vicia americana*

Vigna cylindrica (L.) Skeels - *Vigna unguiculata* subsp. *cylindrica*

  sesquipedalis (L.) Fruwirth - *Vigna unguiculata* subsp. *sesquipedalis*

  sinensis (L.) Savi ex Hassk. - *Vigna unguiculata*

Vinca rosea L. - *Catharanthus roseus*

Viola epipsila Ledeb. - *Viola palustris*

  eriocarpa Schwein. - *Viola pubescens*

  kitaibeliana var. rafinesquii (Greene) Fernald - *Viola rafinesquii*

  linguaefolia Nutt. ex Torr. & A. Gray - *Viola nuttallii*

  nephrophylla Greene - *Viola sororia* subsp. *affinis*

  nuttallii var. linguaefolia (Nutt. ex Torr. & A. Gray) Piper - *Viola nuttallii*

  pallens (Banks) Brainerd - *Viola macloskeyi* var. *pallens*

  pedatifida G. Don - *Viola palmata* var. *pedatifida*

  pensylvanica Michx. - *Viola pubescens*

  priceana Pollard - *Viola sororia*

  tricolor var. hortensis DC. - *Viola tricolor*

  vallicola A. Nelson - *Viola nuttallii*

Vitis cordifolia Michx. - *Vitis vulpina*

  labruscana L. H. Bailey - *Vitis labrusca*

Vulpia megalura (Nutt.) Rydb. - *Vulpia myuros*

Wisteria chinensis DC. - *Wisteria sinensis*

Xanthium canadense Mill. - *Xanthium strumarium* var. *canadense*

  chinense Mill. - *Xanthium strumarium*

  commune Britton - *Xanthium strumarium* var. *canadense*

  echinatum A. Murray bis - *Xanthium strumarium* var. *canadense*

  italicum Moretti - *Xanthium strumarium* var. *canadense*

  orientale L. - *Xanthium strumarium* var. *canadense*

  oviforme Wallr. - *Xanthium strumarium* var. *canadense*

  pensylvanicum Wallr. - *Xanthium strumarium*

  speciosum Kearney - *Xanthium strumarium* var. *canadense*

Xolisma ferrugineum Nash - *Lyonia ferruginea*

Yucca arborescens (Torr.) Trel. - *Yucca brevifolia*

  draconis L. - *Dracaena draco*

  flaccida Haw. - *Yucca filamentosa*

  macrocarpa Engelm. - *Yucca schottii*

  radiosa Trel. - *Yucca elata*

  smalliana Fernald - *Yucca filamentosa*

Zamia floridana L.f. - *Zamia pumila*

  furfuracea L.f. - *Zamia pumila*

  umbrosa L.f. - *Zamia pumila*

Zauschneria californica C. Presl - *Epilobium canum*

Zea mays var. rugosa Bonaf. - *Zea mays*

  mays var. saccharata (Sturtev.) L. H. Bailey - *Zea mays*

  mexicana (Schrad.) Kuntze - *Zea mays* subsp. *mexicana*

Zebrina pendula Schnizl. - *Tradescantia zebrina*

  pendula Schnizl. - *Trandescantia zebrina*

Zigadenus coloradensis Rydb. - *Zigadenus elegans*

  gramineus Rydb. - *Zigadenus venenosus*

Zinnia elegans Jacq. - *Zinnia violacea*

Zizania caduciflora (Trin.) Hand.-Mazz. - *Zizania latifolia*

Zizia cordata K. Koch ex DC. - *Thaspium trifoliatum*

Zygocactus truncatus (Haw.) K. Schum. - *Schlumbergera truncata*

Families and Genera

Acanthaceae
Anisacanthus
Aphelandra
Barleria
Crossandra
Eranthemum
Fittonia
Hemigraphis
Hygrophila
Justicia
Ruellia

Aceraceae
Acer

Acoraceae
Acorus

Actinidiaceae
Actinidia

Adoxaceae
Adoxa

Agavaceae
Agave
Cordyline
Dasylirion
Dracaena
Furcraea
Manfreda
Nolina
Phormium
Pleomele
Polianthes
Sansevieria
Yucca

Aizoaceae
Carpobrotus
Galenia
Lampranthus
Lithops
Mesembryanthemum
Sesuvium
Tetragonia
Trianthema

Alismataceae
Alisma
Echinodorus
Sagittaria

Aloaceae
Aloe
Kniphofia

Amaranthaceae
Alternanthera
Amaranthus
Celosia
Froelichia

Gomphrena
Iresine
Tidestromia

Anacardiaceae
Anacardium
Cotinus
Malosma
Mangifera
Pistacia
Rhus
Schinus
Spondias
Toxicodendron

Annonaceae
Annona
Asimina
Monodora
Rollinia

Apiaceae
Aegopodium
Aethusa
Ammi
Anethum
Angelica
Anthriscus
Apium
Bupleurum
Carum
Centella
Chaerophyllum
Ciclospermum
Cicuta
Conioselinum
Conium
Coriandrum
Cryptotaenia
Cuminum
Cymopterus
Daucus
Erigenia
Eryngium
Falcaria
Foeniculum
Heracleum
Hydrocotyle
Levisticum
Ligusticum
Lomatium
Oenanthe
Osmorhiza
Oxypolis
Pastinaca
Perideridia
Petroselinum
Pimpinella
Polytaenia
Ptilimnium

Sanicula
Scandix
Sium
Spermolepis
Taenidia
Thaspium
Torilis
Trachymene
Zizia

Apocynaceae
Acokanthera
Adenium
Allamanda
Alstonia
Amsonia
Apocynum
Carissa
Catharanthus
Nerium
Plumeria
Tabernaemontana
Thevetia
Trachelospermum
Vinca

Aquifoliaceae
Ilex
Nemopanthus

Araceae
Aglaonema
Alocasia
Amorphophallus
Anthurium
Arisaema
Caladium
Calla
Colocasia
Dieffenbachia
Epipremnum
Lysichiton
Monstera
Orontium
Peltandra
Philodendron
Pistia
Symplocarpus
Syngonium
Xanthosoma
Zantedeschia

Araliaceae
Aralia
Cussonia
Fatshedera
Fatsia
Hedera
Oplopanax
Panax

Polyscias
Pseudopanax
Schefflera
Tetrapanax

Araucariaceae
Araucaria

Arecaceae
Acoelorraphe
Archontophoenix
Arenga
Bactris
Butia
Caryota
Chamaedorea
Chrysalidocarpus
Coccothrinax
Cocos
Euterpe
Howeia
Hyphaene
Livistona
Phoenix
Ptychosperma
Rhapis
Roystonea
Sabal
Serenoa
Syagrus
Trachycarpus
Veitchia
Washingtonia

Aristolochiaceae
Aristolochia
Asarum

Asclepiadaceae
Asclepias
Calotropis
Cynanchum
Hoya
Morrenia
Periploca
Sarcostemma
Vincetoxicum

Asteraceae
Acanthospermum
Achillea
Acroptilon
Adenocaulon
Ageratina
Ageratum
Ambrosia
Amphiachyris
Anaphalis
Antennaria
Anthemis
Arctium

Arctotis
Arnoseris
Artemisia
Aster
Astranthium
Baccharis
Baileya
Balsamorhiza
Bellis
Bidens
Brickellia
Cacalia
Calendula
Callistephus
Calyptocarpus
Carduus
Carthamus
Centaurea
Chamaemelum
Chondrilla
Chromolaena
Chrysanthemum
Chrysopsis
Chrysothamnus
Cichorium
Cirsium
Cnicus
Conoclinium
Conyza
Coreopsis
Cosmos
Cotula
Crepis
Crupina
Cynara
Dimorphotheca
Doronicum
Dracopis
Dyssodia
Echinacea
Echinops
Elephantopus
Emilia
Engelmannia
Erechtites
Ericameria
Erigeron
Eupatorium
Euryops
Felicia
Filago
Flaveria
Gaillardia
Galinsoga
Gazania
Gerbera
Gnaphalium
Grindelia
Guizotia
Gutierrezia
Gynura
Haplopappus
Helenium

Helianthus
Helichrysum
Heliopsis
Helipterum
Hemizonia
Heterotheca
Hieracium
Holocarpha
Hymenopappus
Hymenoxys
Hypochaeris
Inula
Isocoma
Iva
Krigia
Lactuca
Lapsana
Lasthenia
Leontodon
Leucanthemum
Liatris
Lygodesmia
Machaeranthera
Madia
Matricaria
Megalodonta
Mikania
Onopordum
Osteospermum
Parthenium
Petasites
Picris
Piqueria
Pityopsis
Pluchea
Polymnia
Prenanthes
Pulicaria
Pyrrhopappus
Ratibida
Rudbeckia
Santolina
Scorzonera
Senecio
Silphium
Silybum
Solidago
Soliva
Sonchus
Stephanomeria
Stokesia
Tagetes
Tanacetum
Taraxacum
Tetradymia
Tithonia
Tragopogon
Tridax
Tussilago
Verbesina
Vernonia
Viguiera
Wedelia

Xanthium
Youngia
Zinnia

Azollaceae
Azolla

Balsaminaceae
Impatiens

Basellaceae
Basella

Begoniaceae
Begonia

Berberidaceae
Achlys
Berberis
Caulophyllum
Jeffersonia
Nandina
Podophyllum
Vancouveria

Betulaceae
Alnus
Betula
Carpinus
Corylus
Ostrya

Bignoniaceae
Bignonia
Campsis
Catalpa
Chilopsis
Crescentia
Cydista
Kigelia
Macfadyena
Pyrostegia
Spathodea
Tabebuia
Tecoma

Bixaceae
Bixa

Bombacaceae
Bombax
Ceiba
Chorisia
Pachira
Pseudobombax

Boraginaceae
Amsinckia
Anchusa
Asperugo
Borago
Cordia
Cryptantha
Cynoglossum
Echium
Hackelia
Heliotropium

Lappula
Lithospermum
Mertensia
Myosotis
Onosmodium
Plagiobothrys
Symphytum

Brassicaceae
Alliaria
Alyssum
Arabidopsis
Arabis
Armoracia
Barbarea
Berteroa
Brassica
Bunias
Cakile
Camelina
Capsella
Cardamine
Cardaria
Chorispora
Cochlearia
Conringia
Coronopus
Crambe
Descurainia
Diplotaxis
Draba
Eruca
Erucastrum
Erysimum
Euclidium
Hesperis
Hirschfeldia
Iberis
Isatis
Lepidium
Lesquerella
Lobularia
Lunaria
Malcolmia
Matthiola
Neslia
Raphanus
Rapistrum
Rorippa
Sibara
Sinapis
Sisymbrium
Stanleya
Teesdalia
Thlaspi
Thysanocarpus

Bromeliaceae
Aechmea
Ananas
Cryptanthus
Tillandsia

Burseraceae
Bursera

Butomaceae
Butomus

Buxaceae
Buxus
Pachysandra

Cabombaceae
Brasenia
Cabomba

Cactaceae
Astrophytum
Carnegiea
Cereus
Echinocereus
Echinopsis
Espostoa
Hatiora
Hylocereus
Mammillaria
Opuntia
Pachycereus
Schlumbergera
Stenocereus

Callitrichaceae
Callitriche

Calycanthaceae
Calycanthus
Chimonanthus

Campanulaceae
Campanula
Jasione
Lobelia
Platycodon
Triodanis
Wahlenbergia

Canellaceae
Canella

Cannabaceae
Cannabis
Humulus

Cannaceae
Canna

Capparaceae
Capparis
Cleome
Polanisia

Caprifoliaceae
Abelia
Diervilla
Kolkwitzia
Linnaea
Lonicera
Sambucus
Symphoricarpos
Triosteum

Viburnum
Weigela

Caricaceae
Carica

Caryophyllaceae
Agrostemma
Arenaria
Cerastium
Dianthus
Drymaria
Gypsophila
Holosteum
Honckenya
Lepyrodiclis
Lychnis
Moehringia
Paronychia
Sagina
Saponaria
Scleranthus
Silene
Spergula
Spergularia
Stellaria
Vaccaria

Casuarinaceae
Casuarina

Celastraceae
Canotia
Catha
Celastrus
Crossopetalum
Euonymus
Paxistima

Cephalotaxaceae
Cephalotaxus

Ceratophyllaceae
Ceratophyllum

Cercidiphyllaceae
Cercidiphyllum

Chenopodiaceae
Atriplex
Axyris
Bassia
Beta
Chenopodium
Corispermum
Cycloloma
Halogeton
Kochia
Krascheninnikovia
Monolepis
Salicornia
Salsola
Sarcobatus
Spinacia
Suaeda

Chloranthaceae
Chloranthus

Chrysobalanaceae
Chrysobalanus
Licania

Cistaceae
Cistus
Helianthemum
Hudsonia

Clethraceae
Clethra

Clusiaceae
Calophyllum
Clusia
Garcinia
Hypericum

Combretaceae
Bucida
Conocarpus
Laguncularia
Quisqualis
Terminalia

Commelinaceae
Commelina
Cyanotis
Murdannia
Tradescantia
Tripogandra

Convolvulaceae
Argyreia
Calystegia
Convolvulus
Cressa
Dichondra
Ipomoea
Jacquemontia

Cornaceae
Aucuba
Cornus
Nyssa

Costaceae
Costus

Crassulaceae
Crassula
Kalanchoe
Sedum
Sempervivum

Cucurbitaceae
Benincasa
Bryonia
Citrullus
Cucumis
Cucurbita
Echinocystis
Lagenaria
Luffa
Marah

Melothria
Momordica
Sechium
Sicana
Sicyos
Trichosanthes

Cupressaceae
Calocedrus
Chamaecyparis
Cupressus
Juniperus
Platycladus
Thuja
Thujopsis

Cuscutaceae
Cuscuta

Cycadaceae
Cycas

Cyperaceae
Carex
Cladium
Cymophyllus
Cyperus
Eleocharis
Eriophorum
Fimbristylis
Fuirena
Rhynchospora
Scirpus
Scleria

Cyrillaceae
Cliftonia
Cyrilla

Diapensiaceae
Galax
Shortia

Dioscoreaceae
Dioscorea

Dipsacaceae
Dipsacus
Knautia
Scabiosa
Succisa

Droseraceae
Dionaea
Drosera

Ebenaceae
Diospyros

Elaeagnaceae
Elaeagnus
Shepherdia

Elatinaceae
Elatine

Empetraceae
Empetrum

Ephedraceae
Ephedra

Equisetaceae
Equisetum

Ericaceae
Andromeda
Arbutus
Arctostaphylos
Calluna
Chamaedaphne
Chimaphila
Epigaea
Erica
Gaultheria
Gaylussacia
Kalmia
Ledum
Leiophyllum
Leucothoe
Loiseleuria
Lyonia
Menziesia
Moneses
Monotropa
Monotropsis
Orthilia
Oxydendrum
Phyllodoce
Pieris
Pterospora
Pyrola
Rhododendron
Vaccinium

Euphorbiaceae
Acalypha
Aleurites
Antidesma
Baccaurea
Bischofia
Caperonia
Cnidoscolus
Codiaeum
Croton
Eremocarpus
Euphorbia
Gymnanthes
Hevea
Hura
Jatropha
Manihot
Pedilanthus
Phyllanthus
Ricinus
Sapium
Stillingia
Synadenium
Vernicia

Fabaceae
Abrus
Acacia
Aeschynomene

Albizia
Alhagi
Alysicarpus
Amorpha
Amphicarpaea
Anthyllis
Apios
Arachis
Astragalus
Baptisia
Bauhinia
Caesalpinia
Cajanus
Calliandra
Calopogonium
Canavalia
Caragana
Cassia
Castanospermum
Centrosema
Ceratonia
Cercis
Chamaecrista
Cicer
Cladrastis
Clitoria
Colutea
Coronilla
Crotalaria
Cyamopsis
Cytisus
Dalbergia
Dalea
Delonix
Desmanthus
Desmodium
Erythrina
Eysenhardtia
Galactia
Galega
Genista
Gleditsia
Glycine
Glycyrrhiza
Gymnocladus
Hoffmannseggia
Indigofera
Kummerowia
Lablab
Laburnum
Lathyrus
Lens
Lespedeza
Leucaena
Lotus
Lupinus
Lysiloma
Maackia
Macroptilium
Macrotyloma
Medicago
Melilotus
Mimosa

Mucuna
Olneya
Onobrychis
Ornithopus
Oxytropis
Pachyrhizus
Parkinsonia
Phaseolus
Piscidia
Pisum
Pithecellobium
Pongamia
Prosopis
Psophocarpus
Psoralea
Psoralidium
Psorothamnus
Pterocarpus
Pueraria
Robinia
Senna
Sesbania
Sophora
Spartium
Sphaerophysa
Strongylodon
Strophostyles
Stylosanthes
Tephrosia
Teramnus
Thermopsis
Tipuana
Trifolium
Trigonella
Ulex
Vicia
Vigna
Wisteria

Fagaceae
Castanea
Castanopsis
Fagus
Lithocarpus
Quercus

Flacourtiaceae
Azara
Dovyalis
Flacourtia
Idesia

Fouquieriaceae
Fouquieria

Garryaceae
Garrya

Gentianaceae
Bartonia
Centaurium
Eustoma
Exacum
Frasera
Gentiana

Gentianella
Gentianopsis
Halenia
Obolaria
Sabatia

Geraniaceae
Erodium
Geranium
Pelargonium

Gesneriaceae
Aeschynanthus
Saintpaulia
Sinningia
Streptocarpus

Ginkgoaceae
Ginkgo

Grossulariaceae
Itea
Ribes

Haemodoraceae
Lachnanthes

Haloragaceae
Myriophyllum
Proserpinaca

Hamamelidaceae
Hamamelis
Liquidambar

Heliconiaceae
Heliconia

Hippocastanaceae
Aesculus

Hippuridaceae
Hippuris

Hydrangeaceae
Decumaria
Hydrangea
Jamesia
Philadelphus
Whipplea

Hydrocharitaceae
Egeria
Elodea
Hydrilla
Limnobium
Stratiotes
Vallisneria

Hydrophyllaceae
Ellisia
Eriodictyon
Hydrolea
Hydrophyllum
Nemophila
Phacelia

Illiciaceae
Illicium

Iridaceae
  Belamcanda
  Dietes
  Gladiolus
  Iris
  Sisyrinchium
  Tigridia

Juglandaceae
  Carya
  Juglans
  Pterocarya

Juncaceae
  Juncus
  Luzula

Juncaginaceae
  Triglochin

Lamiaceae
  Acinos
  Agastache
  Ajuga
  Ballota
  Blephilia
  Calamintha
  Clinopodium
  Collinsonia
  Cunila
  Dracocephalum
  Galeopsis
  Glechoma
  Hedeoma
  Hyptis
  Hyssopus
  Lamium
  Lavandula
  Leonotis
  Leonurus
  Lepechinia
  Lycopus
  Marrubium
  Melissa
  Mentha
  Moluccella
  Monarda
  Monardella
  Nepeta
  Ocimum
  Origanum
  Perilla
  Phlomis
  Plectranthus
  Prunella
  Pycnanthemum
  Rosmarinus
  Salvia
  Satureja
  Scutellaria
  Stachys
  Teucrium
  Thymus
  Trichostema

Lardizabalaceae
  Akebia

Lauraceae
  Cinnamomum
  Laurus
  Licaria
  Lindera
  Litsea
  Ocotea
  Persea
  Sassafras
  Umbellularia

Leeaceae
  Leea

Leitneriaceae
  Leitneria

Lemnaceae
  Lemna
  Spirodela
  Wolffia
  Wolffiella

Lentibulariaceae
  Pinguicula
  Utricularia

Liliaceae
  Agapanthus
  Aletris
  Allium
  Amaryllis
  Asparagus
  Asphodelus
  Aspidistra
  Brodiaea
  Calochortus
  Camassia
  Chlorogalum
  Chlorophytum
  Clintonia
  Colchicum
  Convallaria
  Crinum
  Cyrtanthus
  Dichelostemma
  Disporum
  Eremurus
  Erythronium
  Eucharis
  Fritillaria
  Galanthus
  Gloriosa
  Haemanthus
  Hemerocallis
  Hippeastrum
  Hosta
  Hyacinthoides
  Hyacinthus
  Hymenocallis
  Hypoxis
  Ixiolirion

  Leucocoryne
  Leucocrinum
  Leucojum
  Lilium
  Liriope
  Lloydia
  Lycoris
  Maianthemum
  Medeola
  Muscari
  Narcissus
  Nerine
  Nothoscordum
  Ornithogalum
  Polygonatum
  Scilla
  Smilacina
  Stenanthium
  Sternbergia
  Streptopus
  Trillium
  Triteleia
  Tulbaghia
  Uvularia
  Veratrum
  Xerophyllum
  Zephyranthes
  Zigadenus

Linaceae
  Linum

Loasaceae
  Mentzelia

Loganiaceae
  Buddleja
  Gelsemium
  Polypremum
  Spigelia

Lycopodiaceae
  Lycopodium

Lythraceae
  Ammannia
  Cuphea
  Decodon
  Didiplis
  Lagerstroemia
  Lythrum
  Rotala

Magnoliaceae
  Liriodendron
  Magnolia
  Michelia

Malpighiaceae
  Byrsonima
  Malpighia
  Stigmaphyllon

Malvaceae
  Abelmoschus
  Abutilon

  Alcea
  Althaea
  Anoda
  Callirhoe
  Gossypium
  Hibiscus
  Kokia
  Lavatera
  Malachra
  Malva
  Malvastrum
  Malvaviscus
  Malvella
  Modiola
  Napaea
  Sida
  Sidalcea
  Sphaeralcea
  Thespesia
  Urena

Marantaceae
  Maranta

Mayacaceae
  Mayaca

Melastomataceae
  Clidemia
  Melastoma
  Rhexia
  Tetrazygia
  Tibouchina

Meliaceae
  Lansium
  Melia
  Swietenia
  Toona

Menispermaceae
  Cocculus
  Menispermum

Menyanthaceae
  Menyanthes
  Nymphoides

Molluginaceae
  Mollugo

Monimiaceae
  Peumus

Moraceae
  Artocarpus
  Broussonetia
  Ficus
  Maclura
  Morus

Musaceae
  Musa

Myoporaceae
  Myoporum

Myricaceae
Comptonia
Myrica

Myrsinaceae
Ardisia
Myrsine

Myrtaceae
Acca
Callistemon
Calothamnus
Calyptranthes
Eucalyptus
Eugenia
Leptospermum
Lophostemon
Melaleuca
Metrosideros
Myrciaria
Myrtus
Pimenta
Psidium
Rhodomyrtus
Syzygium

Najadaceae
Najas

Nelumbonaceae
Nelumbo

Nyctaginaceae
Abronia
Allionia
Boerhavia
Bougainvillea
Mirabilis
Pisonia

Nymphaeaceae
Nuphar
Nymphaea
Victoria

Olacaceae
Ximenia

Oleaceae
Chionanthus
Forestiera
Forsythia
Fraxinus
Jasminum
Ligustrum
Olea
Osmanthus
Syringa

Onagraceae
Boisduvalia
Camissonia
Circaea
Clarkia
Epilobium
Fuchsia
Gaura

Ludwigia
Oenothera

Orchidaceae
Aplectrum
Calypso
Corallorrhiza
Cypripedium
Goodyera
Habenaria
Isotria
Liparis
Orchis
Pogonia
Spiranthes
Vanilla

Orobanchaceae
Epifagus
Orobanche

Oxalidaceae
Averrhoa
Oxalis

Paeoniaceae
Paeonia

Pandanaceae
Pandanus

Papaveraceae
Adlumia
Argemone
Chelidonium
Corydalis
Dendromecon
Dicentra
Eschscholzia
Fumaria
Meconopsis
Papaver
Roemeria
Sanguinaria
Stylophorum

Passifloraceae
Passiflora

Pedaliaceae
Proboscidea
Sesamum

Phytolaccaceae
Phytolacca
Rivina

Pinaceae
Abies
Cedrus
Larix
Picea
Pinus
Pseudolarix
Pseudotsuga
Tsuga

Piperaceae
Peperomia
Piper

Pittosporaceae
Pittosporum

Plantaginaceae
Plantago

Platanaceae
Platanus

Plumbaginaceae
Armeria
Limonium
Plumbago

Poaceae
Aegilops
Agropyron
Agrostis
Aira
Alopecurus
Ammophila
Andropogon
Anthoxanthum
Arctagrostis
Aristida
Arrhenatherum
Arthraxon
Arundinaria
Arundo
Avena
Axonopus
Bambusa
Beckmannia
Blepharoneuron
Bothriochloa
Bouteloua
Brachiaria
Brachypodium
Briza
Bromus
Buchloe
Calamagrostis
Calamovilfa
Catabrosa
Cenchrus
Chasmanthium
Chloris
Chrysopogon
Cinna
Coix
Cortaderia
Ctenium
Cymbopogon
Cynodon
Cynosurus
Dactylis
Dactyloctenium
Danthonia
Deschampsia
Dichanthium
Digitaria

Distichlis
Echinochloa
Eleusine
Elymus
Elytrigia
Eragrostis
Eremochloa
Eriochloa
Festuca
Glyceria
Hackelochloa
Hemarthria
Heteropogon
Hierochloe
Hilaria
Holcus
Hordeum
Hyparrhenia
Imperata
Ischaemum
Koeleria
Lagurus
Lamarckia
Lasiacis
Leersia
Leptochloa
Leymus
Lolium
Luziola
Melica
Melinis
Microstegium
Milium
Miscanthus
Molinia
Muhlenbergia
Munroa
Nassella
Neyraudia
Oplismenus
Oryza
Oryzopsis
Panicum
Parapholis
Paspalum
Pennisetum
Phalaris
Phleum
Phragmites
Phyllostachys
Piptochaetium
Pleuropogon
Poa
Polypogon
Pseudoroegneria
Puccinellia
Rhynchelytrum
Rostraria
Rottboellia
Saccharum
Schedonnardus
Schismus
Schizachne

Schizachyrium
Sclerochloa
Scolochloa
Secale
Setaria
Sorghastrum
Sorghum
Spartina
Sphenopholis
Sporobolus
Stenotaphrum
Stipa
Taeniatherum
Tridens
Triplasis
Tripsacum
Trisetum
Triticum
Uniola
Urochloa
Ventenata
Vetiveria
Vulpia
Zea
Zizania
Zizaniopsis
Zoysia

Podocarpaceae
Podocarpus

Polemoniaceae
Gilia
Leptodactylon
Navarretia
Phlox
Polemonium

Polygalaceae
Polygala

Polygonaceae
Brunnichia
Coccoloba
Emex
Eriogonum
Fagopyrum
Homalocladium
Oxyria
Polygonella
Polygonum
Rheum
Rumex

Pontederiaceae
Eichhornia
Heteranthera
Monochoria
Pontederia

Portulacaceae
Calandrinia
Calyptridium
Claytonia
Lewisia
Montia

Portulaca
Portulacaria
Talinum

Potamogetonaceae
Potamogeton
Ruppia

Primulaceae
Anagallis
Androsace
Cyclamen
Dodecatheon
Glaux
Hottonia
Lysimachia
Primula
Samolus
Trientalis

Proteaceae
Grevillea
Macadamia
Protea
Stenocarpus

Psilotaceae
Psilotum

Pteridophyta
Acrostichum
Adiantum
Aglaomorpha
Asplenium
Athyrium
Blechnum
Botrychium
Ceratopteris
Cheilanthes
Cibotium
Cryptogramma
Cyathea
Cyrtomium
Cystopteris
Davallia
Dennstaedtia
Dicranopteris
Dryopteris
Gymnocarpium
Lygodium
Marsilea
Matteuccia
Microsorium
Nephrolepis
Notholaena
Onoclea
Ophioglossum
Osmunda
Pellaea
Pityrogramma
Platycerium
Polypodium
Polystichum
Pteridium
Pteris

Rumohra
Sphenomeris
Tectaria
Thelypteris
Woodsia
Woodwardia

Punicaceae
Punica

Ranunculaceae
Aconitum
Actaea
Adonis
Anemone
Aquilegia
Caltha
Cimicifuga
Clematis
Consolida
Coptis
Delphinium
Helleborus
Hepatica
Hydrastis
Myosurus
Ranunculus
Thalictrum
Trautvetteria
Trollius
Xanthorhiza

Resedaceae
Reseda

Rhamnaceae
Berchemia
Ceanothus
Colubrina
Condalia
Hovenia
Krugiodendron
Rhamnus
Ziziphus

Rhizophoraceae
Rhizophora

Rosaceae
Adenostoma
Agrimonia
Alchemilla
Amelanchier
Aphanes
Aronia
Aruncus
Cercocarpus
Chaenomeles
Chamaebatia
Chamaebatiaria
Cowania
Crataegus
Cydonia
Dalibarda
Dryas
Duchesnea

Eriobotrya
Exochorda
Filipendula
Fragaria
Geum
Gillenia
Heteromeles
Holodiscus
Kerria
Lyonothamnus
Malus
Mespilus
Oemleria
Peraphyllum
Photinia
Physocarpus
Potentilla
Prunus
Pseudocydonia
Pyracantha
Pyrus
Rhaphiolepis
Rhodotypos
Rosa
Rubus
Sanguisorba
Sorbus
Spiraea
Vauquelinia
Waldsteinia

Rubiaceae
Cephalanthus
Chiococca
Cinchona
Coffea
Coprosma
Diodia
Galium
Gardenia
Genipa
Guettarda
Hamelia
Hedyotis
Ixora
Mitchella
Morinda
Paederia
Pentas
Pinckneya
Psychotria
Randia
Richardia
Rubia
Sherardia
Spermacoce

Rutaceae
Casimiroa
Citrofortunella
Citrus
Clausena
Dictamnus
Fortunella

Limonia
Murraya
Poncirus
Ptelea
Ruta
Severinia
Triphasia
Zanthoxylum

Salicaceae
Populus
Salix

Salviniaceae
Salvinia

Santalaceae
Comandra
Geocaulon

Sapindaceae
Blighia
Cardiospermum
Dodonaea
Exothea
Koelreuteria
Litchi
Melicoccus
Nephelium
Sapindus
Ungnadia

Sapotaceae
Manilkara
Mimusops
Pouteria
Sideroxylon
Synsepalum

Sarraceniaceae
Darlingtonia
Sarracenia

Saururaceae
Anemopsis
Saururus

Saxifragaceae
Astilbe
Boykinia
Heuchera
Lithophragma
Mitella
Parnassia
Penthorum
Saxifraga
Tellima
Tiarella
Tolmiea

Scrophulariaceae
Agalinis
Antirrhinum
Aureolaria
Bacopa
Bartsia
Besseya

Buchnera
Calceolaria
Castilleja
Chaenorhinum
Chelone
Collinsia
Cymbalaria
Dasistoma
Digitalis
Dopatrium
Euphrasia
Gratiola
Hebe
Kickxia
Leucophyllum
Limnophila
Linaria
Lindernia
Mazus
Melampyrum
Micranthemum
Mimulus
Odontites
Orthocarpus
Parentucellia
Paulownia
Pedicularis
Penstemon
Rhinanthus
Russelia
Schwalbea
Scrophularia
Seymeria
Striga
Verbascum
Veronica
Veronicastrum

Selaginellaceae
Selaginella

Simaroubaceae
Ailanthus
Picramnia
Simarouba

Simmondsiaceae
Simmondsia

Smilacaceae
Smilax

Solanaceae
Atropa
Brunfelsia
Capsicum
Cestrum
Chamaesaracha
Cyphomandra
Datura
Fabiana
Hyoscyamus
Lycium
Lycopersicon
Mandragora

Nicandra
Nicotiana
Petunia
Physalis
Salpiglossis
Solanum

Sparganiaceae
Sparganium

Sphenocleaceae
Sphenoclea

Staphyleaceae
Staphylea

Sterculiaceae
Brachychiton
Firmiana
Fremontodendron
Melochia
Theobroma
Waltheria

Strelitziaceae
Ravenala
Strelitzia

Styracaceae
Halesia
Styrax

Symplocaceae
Symplocos

Tamaricaceae
Tamarix

Taxaceae
Taxus
Torreya

Taxodiaceae
Cryptomeria
Cunninghamia
Metasequoia
Sciadopitys
Sequoia
Sequoiadendron
Taxodium

Theaceae
Camellia
Franklinia
Gordonia
Stewartia

Thymelaeaceae
Daphne
Dirca

Tiliaceae
Corchorus
Muntingia
Tilia
Triumfetta

Trapaceae
Trapa

Tropaeolaceae
Tropaeolum

Turneraceae
Turnera

Typhaceae
Typha

Ulmaceae
Celtis
Planera
Trema
Ulmus
Zelkova

Urticaceae
Boehmeria
Parietaria
Pellionia
Pilea
Soleirolia
Urtica

Valerianaceae
Centranthus
Valeriana
Valerianella

Verbenaceae
Aloysia
Avicennia
Callicarpa
Clerodendrum
Duranta
Lantana
Lippia
Phryma
Phyla
Stachytarpheta
Tectona
Verbena
Vitex

Violaceae
Hybanthus
Viola

Viscaceae
Arceuthobium
Phoradendron

Vitaceae
Ampelopsis
Cissus
Parthenocissus
Tetrastigma
Vitis

Xyridaceae
Xyris

Zamiaceae
Zamia

Zannichelliaceae
Zannichellia

Zingiberaceae
Alpinia
Curcuma
Elettaria
Hedychium
Zingiber

Zosteraceae
Zostera

Zygophyllaceae
Guajacum
Larrea
Peganum
Porlieria
Tribulus
Zygophyllum